Geophysical Monograph Series

Including

IUGG Volumes

Maurice Ewing Volumes
Mineral Physics Volumes

GEOPHYSICAL MONOGRAPH SERIES

Geophysical Monograph 81

Solar Wind Sources of Magnetospheric Ultra-Low-Frequency Waves

M. J. Engebretson
K. Takahashi
M. Scholer

Editors

American Geophysical Union

Library of Congress Cataloging-in-Publication Data

Solar wind sources of magnetospheric ultra-low-frequency waves / M. J.
  Engebretson, K. Takahashi, M. Scholer, editors.
    p.   cm. — (Geophysical monograph ; 81)
    Includes bibliographical references.
    ISBN 0-87590-040-2
    1. Magnetosphere. 2. Radio waves. 3. Solar wind.
  I. Engebretson, M. (Mark) II. Takahashi, K. (Kazue) III. Scholer,
  M., 1940-   . IV. Series.
  QC809.M35S645 1994
  538'.766—dc20   94-2008
                                                                 CIP

ISSN: 0065-8448

ISBN 0-87590-040-2

This book is printed on acid-free paper. ∞

Printed in the United States of America.

CONTENTS

CONTENTS

CONTENTS

PREFACE

This volume provides a comprehensive overview of externally generated ULF waves and transients in the Earth's space environment, including practically all the latest advances as of the date of writing. Tutorials are included to give a complete introduction to time-dependent phenomena in the magnetosphere in the ULF frequency range. The material may therefore be used by the interested reader as a starting point for getting involved in this research field, as an overview of the research in progress, and/or as a convenient reference for advanced background knowledge.

One of the main foci of space physics is studying the energy transfer from the solar wind into the Earth's magnetosphere. This volume addresses the wave aspect of that energy transfer: which waves in the magnetosphere are ultimately due to the solar wind-magnetosphere interaction process and what their significance is in the energy transfer process. Although specialists must continue to address individual aspects of the solar wind-geospace system, it is our purpose in this volume to review the interaction in a larger and more fundamental framework, and to present new work stressing the interconnectedness of the various regions of geospace. Among the many subjects discussed are several new efforts to study upstream wave production and subsequent transmission through the bow shock, the magnetosheath, and the magnetopause.

The interaction of the solar wind with the Earth's plasma and magnetic field environment is the source of a wide variety of time-dependent phenomena in the Earth's magnetosphere and ionosphere. Although the importance of these interactions has been recognized for years, recent studies in several areas of space physics have called renewed attention to the importance of transient and sustained wave events at ULF frequencies in transferring momentum and energy from the solar wind into the Earth's space environment. It has become apparent that the bow shock, foreshock, magnetosheath, and magnetopause act as a closely coupled system in mediating or even modifying these interactions.

During the next few years a significant multination fleet of spacecraft is planned for launch to investigate many aspects of solar-terrestrial interactions (the International Solar-Terrestrial Program, ISTP), including the character of solar wind-bow shock-magnetosheath-magnetosphere interactions. It is thus appropriate now to provide a forum for discussion of the physics of upstream, bow shock, and magnetosheath processes and the transport of the resulting fluctuations and waves from various theoretical and observational perspectives, to attempt to reach a consensus on our current knowledge and its limits, and to attempt to clarify the major issues to be addressed by forthcoming missions.

This volume is the result of an AGU Chapman Conference held September 14-18, 1992, in Williamsburg, Virginia. The aim of the meeting was to discuss physical processes leading to magnetospheric ULF waves due to the solar wind-geospace interaction, and to determine the current status, aims, and future directions of research in this area. The conference brought together nearly 70 scientists from 14 nations active in space research.

The conveners of the conference were assisted by a Program Committee consisting of Wolfgang Baumjohann, Max Planck Institute for Extraterrestrial Physics; David Burgess, Queen Mary and Westfield College, London; Don Fairfield, NASA/Goddard Space Flight Center; Stephen Fuselier, Lockheed Palo Alto Research Laboratory; Jack Gosling, Los Alamos National Laboratory; Gene Greenstadt, TRW; W. Jeffrey Hughes, Boston University; Hermann Lühr, Technical University of Braunschweig; Janet Luhmann, Institute for Geophysics and Planetary Physics, University of California, Los Angeles; David Sibeck, The Johns Hopkins University Applied Physics Laboratory; Valeria Troitskaya, LaTrobe University, Australia; Allan Wolfe, City University of New York and AT&T Bell Labs; and Kiyohumi Yumoto, Nagoya University.

We gratefully acknowledge the funds provided by the U.S. National Science Foundation and National Aeronautics and Space Agency in support of this conference.

This conference and book would also not have come to be without the support and cooperation of many scientists. We thank in particular Bruce Tsurutani of the Jet Propulsion Laboratory, Andrew Nagy of the University of Michigan, Tom Potemra of The Johns Hopkins University Applied Physics Laboratory, and especially Chris Russell of University of California, Los Angeles for their encouragement and support. We also wish to thank the following scientists, listed in alphabetical order, who served as referees for the papers submitted to this volume.

B. J. Anderson	R. L. Arnoldy	W. Baumjohann
D. H. Burgess	L. J. Cahill, Jr.	C. R. Clauer
N. U. Crooker	R. E. Denton	R. E. Erlandson
M. P. Freeman	S. Fujita	S. A. Fuselier
K.-H. Glassmeier	J. T. Gosling	E. W. Greenstadt
D. Hubert	W. J. Hughes	D.-H. Lee
N. Lin	W. Lotko	H. Lühr
J. G. Luhmann	R. L. Lysak	C. G. Maclennan
G. Mann	D. L. Matthews	F. W. Menk
K. Morrison	P. T. Newell	T. Ogino
T. Oguti	J. V. Olson	Y. Omura
D. Orr	A. Otto	G. Paschmann
V. L. Patel	W. K. Peterson	A. S. Rodger
T. J. Rosenberg	C. T. Russell	J. C. Samson
N. Sckopke	D. G. Sibeck	H. J. Singer
J. A. Slavin	P. Song	D. J. Southwood
M. A. Temerin	R. Treumann	B. T. Tsurutani
V. A. Troitskaya	H. Wiechen	D. Winterhalter
A. Wolfe	T. Yeoman	K. Yumoto

Mark J. Engebretson
Augsburg College
Minneapolis

Kazue Takahashi
Nagoya University
Japan

Manfred Scholer
Max-Planck-Institut für
Extraterrestrische Physik
Germany

Editors

Magnetospheric ULF Waves: A Tutorial With a Historical Perspective

W. Jeffrey Hughes

Center for Space Physics, Boston University, Boston, MA 02215

Abstract: This paper introduces basic ideas about magnetospheric ULF waves by tracing the history of research in this field. Special emphasis is placed on the concept of a field line resonance, on the early attempts to find solar wind sources, on the coupling of surface waves to field line resonances, and on cavity modes. In the final section we address differences between groundbased and spacecraft observations and why cavity modes have not yet been observed directly.

1. Introduction

The ULF frequency range covers about three decades, from roughly 1 mHz to 1 Hz. Physically this range spans frequencies from the lowest frequencies the magnetospheric cavity can support up to the various ion gyrofrequencies. Waves in this frequency range are endemic within the magnetosphere. Their amplitudes are often large enough that they can be detected quite easily on the ground with a simple magnetometer. For this reason these waves have been known about and classified for more than 130 years, long before the ionosphere, let alone the magnetosphere, was even postulated. Indeed, early observations of ULF waves led to one of the first suggestions that electric currents must flow in the upper atmosphere.

This monograph is aimed at understanding the sources of magnetospheric ULF waves in the solar wind. How can waves generated at or beyond the magnetopause, in the magnetosheath or at or upstream of the bow shock, be carried into or excite waves in the magnetosphere? This paper attempts to set the scene by describing how ideas about ULF waves have evolved, especially over the last 25 years or so. The aim is to provide the relative newcomer with a summary of a long and often complex history so that he can put current research into context.

One obvious characteristic of magnetospheric ULF waves, or micropulsations as they were originally called, is that they are often quite sinusoidal, with very well defined frequencies. In this review I will trace attempts to understand why pulsations are so monochromatic. This question will lead me to focus on understanding the propagation of

these waves rather than their ultimate source of energy, a focus that reflects the emphasis of past research. I will leave it to other reviewers to describe in more detail the various morphological classes of magnetic pulsations (as they are now called) and their probable sources. Nor will I describe theory in detail, but instead refer the reader to the review by Southwood and Hughes [1983].

2. Early History

The first published account of an observation of a ULF wave is contained in the paper entitled "On the great magnetic disturbance which extended from August 2 to September 7, 1859 as recorded by photography at the Kew Observatory" by Balfor Stewart [1861]. Stewart reported quasi-sinusoidal magnetic field oscillations as well as field changes of several hundred nanotesla that occurred on a time scale of a few minutes. These observations led Stewart to suggest that the upper atmosphere must carry electric currents, an idea that led on directly to the concept of the Heaviside layer or ionosphere. We learn two things from the title of this paper (other than Victorian scientists liked long titles). The world was magnetically a lot quieter 130 years ago, allowing magnetic measurements to be made at Kew (which is today well within metropolitan London); and the advent of photography revolutionized geomagnetic research by making possible the continuous recording of magnetic variations. This advance helped foster the beginning of solar-terrestrial research.

For the next century the field advanced observationally, but there was little understanding of where the waves came from. Magnetic recordings became widespread, rapid-run magnetometers were introduced, and different classes of micropulsation were recognized in the data and described in the literature. These various classes were often named after the scientist who first pointed out their features, though

Solar Wind Sources of Magnetospheric Ultra-Low-Frequency Waves
Geophysical Monograph 81

they are usually identifiable as a type of pulsation we recognize today. Pi 2's and giant pulsations were both recognized early [Angenheister, 1920; Rolf, 1931].

3.   STANDING WAVES

The first real understanding of geomagnetic pulsations came in 1954 when Dungey [1954, see also 1963], proposed that the long but regular periods of these oscillations might be the result of standing Alfven waves being excited on geomagnetic field lines. In the previous year, Storey [1953] had explained the dispersion of whistlers. He had shown that the observed dispersion would occur if whistlers were guided from one hemisphere to the other along geomagnetic field lines as plasma waves, which required a finite plasma density along the entire field line, where previously a vacuum had been assumed. Dungey recognized the importance of this result, realizing that this magnetoplasma would also support Alfven or hydromagnetic waves.

In a cold uniform plasma (i.e., where plasma thermal pressure is much less than magnetic pressure), two modes of hydromagnetic wave exist, the fast mode and the intermediate or shear mode. Although the magnetosphere is not a uniform plasma on the scale of hydromagnetic wavelengths, concepts from these basic modes are useful. The fast mode has $\mathbf{j} \cdot \mathbf{B} = 0$, so carries no current along magnetic field lines. However the magnetic perturbation vector, $\mathbf{b}$, does in general, have a component parallel to $\mathbf{B}$ so the fast mode can transmit pressure variations. The fast mode dispersion relation, $\omega^2 = k^2 V_A^2$, where $V_A^2 = B^2/\mu_0\rho$ is the Alfven speed, shows that the mode propagates isotropically. The group velocity, and hence energy propagation, is always parallel to the phase velocity. The shear mode on the other hand can carry a finite field-aligned current, but the magnetic perturbations are always perpendicular to $\mathbf{B}$, as $\mathbf{b} \cdot \mathbf{B} = 0$, so the mode carries no pressure perturbations. The dispersion relation for the shear mode is $\omega^2 = k_{\parallel}^2 V_A^2$, where $k_{\parallel}$ is the component of $\mathbf{k}$ parallel to $\mathbf{B}$. Thus propagation does not depend on $k_{\perp}$, or the variation of the signal across the ambient magnetic field. The group velocity and hence energy propagation of the shear mode is always parallel to $\mathbf{B}$ regardless of the direction of $\mathbf{k}$, and for this reason this mode is also known as the guided mode, for energy is guided along the ambient magnetic field.

Now in the magnetosphere the Alfven speed is typically 1000 km/s, while typical periods of geomagnetic pulsations are 10-600 s. Thus typical wavelengths are 10^4 - 10^6 km or 1-100 R_E, comparable with the size of the magnetosphere itself. So uniform plasma theory is clearly inadequate. What Dungey did was to derive hydromagnetic wave equations in a general axisymmetric field geometry (a dipole being a special case of this).

In cylindrical coordinates (r, ϕ, z)

$$(\omega^2 \mu_0 \rho - \frac{1}{r}(\mathbf{B} \cdot \nabla)r^2(\mathbf{B} \cdot \nabla))(\frac{u_\phi}{r}) = \omega m (\frac{\mathbf{B} \cdot \mathbf{b}}{r}) \qquad (1)$$

$$(\omega^2 \mu_0 \rho - rB^2(\mathbf{B} \cdot \nabla)\frac{1}{r^2 B^2}\mathbf{B} \cdot \nabla)(rE_\phi) =$$

$$i\omega B^2 (\mathbf{B} \times \nabla)_\phi (\frac{\mathbf{B} \cdot \mathbf{b}}{B^2}) \qquad (2)$$

$$i\omega \mathbf{B} \cdot \mathbf{b} = \frac{1}{r}(\mathbf{B} \times \nabla)_\phi(rE_\phi) - imB^2 \frac{u_\phi}{r} \qquad (3)$$

where a variation of the form $\exp(im\phi - i\omega t)$ was assumed. Historically a lot of attention has been directed towards these equations, so it's worth discussing what they mean.

The left hand side (LHS) of (1) and (2) both have the form of a one-dimensional wave equation with the only spatial operator being $\mathbf{B} \cdot \nabla$, the derivative along the direction of $\mathbf{B}$. The equations are coupled by the terms of the right hand side (RHS) which depend on $\mathbf{B} \cdot \mathbf{b}$, the compressional part of the magnetic perturbation. Equation (3) shows how $\mathbf{B} \cdot \mathbf{b}$, and E_ϕ and u_ϕ are related and closes the set. Since the transverse dispersion relation depends only on $k_{\parallel}$, it is tempting to think of the LHS of (1) and (2) as representing pseudo-transverse mode oscillations, and the RHS as representing coupling due to the fast mode which has a compressional component.

Dungey showed that these equations decouple in two limits. If the wave is axisymmetric ($m = 0$) the RHS of (1) vanishes. The LHS of (1) then describes a mode in which the electric field is purely radial and the magnetic and velocity perturbations are azimuthal. Magnetic shells (L shells) decouple and each shell oscillates azimuthally independently of all others. This is the toroidal mode. In this limit equation (2) describes a mode in which the whole magnetospheric cavity pulsates coherently. The other limit occurs as $m \to \infty$. For the RHS of (1) to remain finite, $\mathbf{B} \cdot \mathbf{b} \to 0$ so the RHS of (2) vanishes. Equation (2) then describes a mode in which $\mathbf{E}$ is azimuthal and $\mathbf{u}$ and $\mathbf{b}$ are contained in a meridian plane. Now the oscillations of each meridian plane decouple. This is the poloidal mode.

The two lowest frequency poloidal and toroidal modes are illustrated in Figure 1. Here we have assumed that the ionosphere is a perfect conductor, so $E = u = 0$ there, which is equivalent to saying that field lines are tied in the ionosphere. This is usually a good zero order approximation, especially when the ionosphere is sunlit. All the modes have electric field nodes and magnetic field antinodes at the ends of the field line. However at the equator both fundamentals and all odd harmonics have a magnetic field node and electric field antinode (field tilt is zero and displacement large) whereas the second and all even harmonics have magnetic field antinodes and electric field nodes (large field tilt but zero displacement).

Dungey recognized that such standing waves, which have a discrete set of allowed frequencies on each field line, could explain the quasimonochromatic nature of magnetic pulsations. He even calculated the variation of the fundamental eigenperiod with magnetic latitude, but since knowledge of

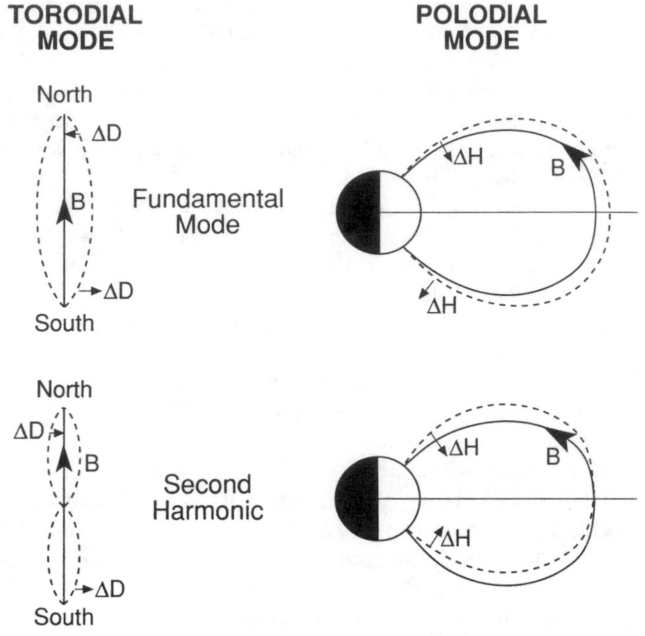

TORODIAL MODE

North

ΔD

B

Fundamental Mode

ΔD

South

North

ΔD

B

Second Harmonic

ΔD

South

POLODIAL MODE

ΔH

B

ΔH

ΔH

B

ΔH

Fig. 1. Cartoons showing the oscillation of a field line in the two lowest frequency toroidal modes (left) and poloidal modes (right). On the left the field line is drawn stretched out from north to south, and we look towards the Earth as the field line is displaced east and west. On the right, the dipolar field line is displaced within its meridian plane so all magnetic perturbations are radial. Note that for the toroidal mode, the magnetic perturbations (ΔD) have the opposite sense north and south in the fundamental and the same sense in the second harmonic. The opposite is true of the ΔH perturbation in the poloidal mode.

magnetospheric plasma density distribution was rudimentary (he used a uniform plasma density corresponding to the outer part of the plasmasphere which had been derived from whistler measurements) his numerical estimates of the eigenperiods are poor.

Soon after Dungey's pioneering paper, space physics received a double impetus, for 1959 was both the dawn of the satellite era and the International Geophysical Year (IGY). It was the latter that gave the biggest initial boost to pulsation research, for magnetic recordings were made around the globe and systematically studied. One outcome of this growth in research was the recognition of the need for consistent terminology. An international committee of four was charged with recommending a pulsation classification [Jacobs et al., 1964]; their solution is shown in Table 1, which is included here as it should be given somewhere early in a book such as this. They divided pulsations into two basic types, continuous, which meant quasisinusoidal signals lasting more than several cycles, and irregular, which meant broad banded or short lived pulsations. They split these two major classes into subclasses based on the wave period or frequency, designating the subclass by a number. Hence the terms Pc 1-5 and Pi 1 and 2. Originally morphological characteristics were associated with each type. But as

Table 1: The Classification of ULF Pulsations

Continuous Pulsations		Irregular Pulsations	
Type	Period Range (sec)	Type	Period Range (sec)
Pc 1	0.2-5	Pi 1	1-40
Pc 2	5-10	Pi 2	40-150
Pc 3	10-45		
Pc 4	45-150		
Pc 5	150-600		

new observations, particularly spacecraft observations, introduced new morphological classes of waves, this classification has come to indicate primarily a frequency band. The main exception is Pi 2, which is used for a very particular type of pulsation associated with substorms and auroral brightenings (and therefore not featured in this monograph).

During the 1960's, Dungey's basic idea of standing waves was verified. First Sugiura [1961] showed that waves are observed simultaneously at both ends of the same field line (see Figure 2). This result shows that waves were guided along field lines but did not prove their standing nature. Next Nagata et al. [1963], who had the advantage of rapid-run magnetograms, showed that pulsations at conjugate points could be matched cycle for cycle. They examined the conjugate phase relationships and found that the H components oscillated in phase whereas the D components oscillated out of phase (see Figure 3). Returning to Figure 1, we see that for the fundamental modes, the poloidal mode ΔH perturbations are in phase whereas the toroidal mode ΔD perturbations are out of phase. The opposite is true for the second harmonics. Thus Nagata et al. could conclude that they were observing odd mode standing waves.

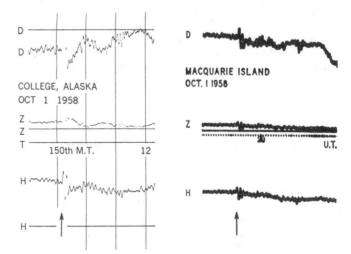

Fig. 2. A damped wave train seen simultaneously on the Earth at the two ends of the same field line, at College, Alaska, in the northern hemisphere and MacQuarie Island in the south. (After Sugiura, 1961).

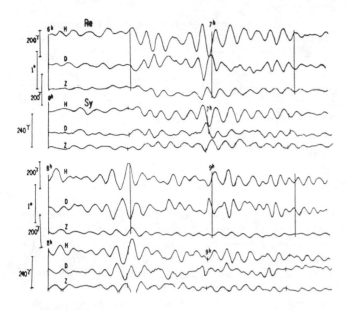

Fig. 3. Giant pulsations observed at the conjugate points Reykjavik, Iceland, and Syowa Base, Antarctica. Note how the H component oscillations are in phase at the two sites while the D component oscillations are out of phase. (After Nagata et al., 1963).

During the 1960's, ULF waves were observed in space by satellites for the first time. Patel [1965], using data from Explorer 12 which provided 15 s samples of the magnetic field, made a number of discoveries. He reported seeing both transverse and compressional waves in space, with periods of 2-3 min, and also made the first ground/satellite correlation, confirming that the waves he saw in space were the same as magnetic pulsations seen on the ground. Cummings et al. [1969] made extensive observations of long period waves with the magnetometer on board the geosynchronous satellite ATS 1. They showed that the waves were long-lived and that they occurred often. Cummings et al. were also the first to numerically integrate equations (1) and (2) obtaining good estimates of field line eigenperiods. Thus by the end of the 1960's, standing waves on geomagnetic field lines were established observationally, confirming Dungey's original idea.

4. SOLAR WIND SOURCES

Soviet scientists pioneered the search for sources of pulsations in the solar wind in the 1960's. They looked for correlations between solar wind parameters and the properties of pulsations observed on the ground, reporting their results to an initially very skeptical western audience. This early history is described in detail elsewhere in this monograph [Troitskaya, 1993; Greenstadt and Russell, 1993], but since these results are central to the theme of this volume, I will briefly review some of this history here. An important consequence of this early work was the realization that different classes of pulsation have different sources. Only cer-

tain types of pulsation correlate with the solar wind. And whereas many types or classes of pulsation are distinguished by their period, not all classes are. This can lead to confusing results, especially when parameters are correlated with the occurrence of pulsations of a given period, or with wave power in a given frequency range, without regard for the type of pulsation observed.

The occurrence of pulsations in the Pc 3 band is modulated globally. Some of this modulation can be explained by the sudden change in the size of the magnetosphere following a negative sudden impulse (Si⁻) [e.g. Troitskaya et al., 1969] but the source of the modulation remained a mystery until Bolshakova and Troitskaya [1968] showed that the occurrence of Pc 3 pulsations depends on the orientation of the interplanetary magnetic field (IMF). This result was confirmed first by further Soviet efforts [Troitskaya et al., 1971; Plyasova-Bakunina, 1972; Plyasova-Bakunina et al., 1978] and later by western scientists [Greenstadt and Olson, 1977; Greenstadt et al., 1979; Wolfe et al., 1980; Russell et al., 1983]. What emerged from this work was that if the IMF direction is within about 50° of the Earth-Sun line, Pc 3 is generally observed on the ground, whereas if the IMF is roughly orthogonal to the Earth-Sun line, Pc 3 is not observed.

Another important result was that the period of Pc 3 observed at Borok (L=3) depends on the magnitude of the IMF according to the relation $\tau(\sec) = 160/B(nT)$ [Troitskaya et al., 1971]. Again, this result was first verified in the Soviet Union [Plyasova-Bakunina, 1972; Gul'elmi et al., 1973; Plyasova-Bakunina et al., 1978] and only later by the initially skeptical western scientists. When these correlations were discovered they were simply observational results with no theoretical explanation, but theories followed soon after [Gul'elmi, 1974; Kovner et al., 1976].

Independently of this Soviet work, Greenstadt et al. [1968, 1970a,b] were investigating waves observed around the bow shock and in the magnetosheath. Their observations led Greenstadt to suggest these waves as a source of pulsations. The whole picture finally came together when Russell and Hoppe [1981] found that the periods of the waves observed upstream of the bow shock are related to the magnitude of the IMF in exactly the same way as the periods of Pc 3 observed at Borok (or indeed anywhere on the ground).

Upstream waves are generated by an ion cyclotron instability driven by ions reflected off the bow shock. Since the ion cyclotron frequency depends only on B, the frequency of the waves correlates with the strength of the IMF. The reflected ions can travel upstream and generate the waves only where the IMF has a significant component normal to the local shock plane. This occurs at the nose of the bow shock when the IMF is roughly parallel to the Earth-Sun line. Only then are these waves advected through the magnetosheath to impinge directly on the magnetopause, to enter the magnetosphere and generate pulsations. This mechanism explains both the dependence of pulsation period on the strength of the IMF and the occurrence on the direction of the IMF.

Two other relationships between the solar wind and pulsations were discovered in the 1960's. First Saito [1964] showed that pulsation amplitudes were correlated with solar wind speeds, a result later confirmed by both Russian and American investigations [Vinogradov and Parkhomov, 1974; Kovner et al, 1976; Greenstadt et al., 1979]. Second Gringauz et al. [1970] found a relationship between solar wind number density and pulsation period. As the solar wind number density increases, they found Pc 3 periods decrease and Pc 4 periods increase. Whereas the first of these results can be fit into the upstream wave picture, the second result remains an enigma.

5. THE DRIVEN FIELD LINE RESONANCE

We return now to Dungey's idea of a field line resonance. This picture could explain why waves observed at a single site are quasi-sinusoidal and have a well defined frequency. But if all field lines can be excited with equal ease, on the ground at the feet of the field lines one should see the wave period or frequency vary with latitude as the observer moves to longer or shorter field lines. Except on rare occasions this is not the case. When waves are observed, their amplitude, but not their period, varies with latitude. Usually the wave amplitude peaks at some latitude, which was thought to coincide with the field line that resonates at the wave frequency. Thus a new question arises: Why are some field lines preferentially excited? The same question can be rephrased as: What determines the frequency at which waves are excited? In this section we deal with the earlier response to this question.

Around 1970 a second revolution in recording technology, as major as the advent of photography a century before, totally changed ULF wave research. This change was digital recording of data in the field. Now wave signals could be analyzed and correlated by computer without the need for laborious digitization of analog records. Furthermore, systematic arrays of magnetometers were deployed that took advantage of this new technology. One of the earliest of these arrays was fielded in Western Canada [Samson et al., 1971]. This array spanned a critical 10° of latitude from about 60° to 70°, which maps from about the plasmapause to near the magnetopause on the dayside. Pc 5's with periods around 200-300 s are common at these latitudes. Samson et al. used this array to determine the typical variation of Pc 5 polarization and amplitude with latitude and local time. Their result is shown in Figure 4.

Samson et al. found that the latitude at which the signal amplitude was largest varied with local time, but remained between about 60° and 65°. At this latitude the waves were linearly polarized, and the sense of polarization of the waves changed across this line from right handed to left handed. The sense of polarization also changed near local noon. Further poleward the data suggested another switch in polarization associated with an amplitude minimum (dashed line).

The polarization switch around local noon had been no-

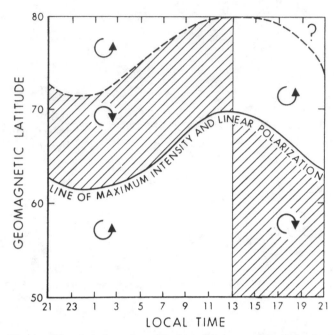

Fig. 4. The variation of amplitude and the sense of polarization of Pc 5's seen at high latitudes with latitude and magnetic local time. (After Samson et al., 1974).

ticed earlier, and explained by surface waves excited on the magnetopause by the magnetosheath flow via a Kelvin-Helmholtz instability. In fact Dungey [1954] had suggested this mechanism as a source of wave energy. As Figure 5 shows, the surface waves would travel westward in the morning and eastward in the afternoon, switching their direction of propagation and sense of polarization around noon. Mapping the polarization shown in Figure 5 down to the ground gives the sense of polarization in the lower part of Figure 4, or that at the very top if the more poleward switch in polarization is real. Samson et al.'s results suggested that the surface waves might excite a field line resonance deep within the magnetosphere on the field line corresponding to the latitude of the wave amplitude maximum.

Southwood [1974] and Chen and Hasegawa [1974] separately developed the theory to describe this mechanism, and in the process explained the polarization changes. They solved equations (1)-(3) in a simpler geometry, keeping the essential physics – a field line eigenfrequency that varied across field lines. Figure 6 illustrates their result. The amplitude of the surface wave on the magnetopause (right hand boundary) decays evanescently away from the boundary. However, at some point inside the magnetosphere the period of the Kelvin-Helmholtz surface wave will match the eigenperiod of the local field line, which changes monotonically with field line length (top panel). At this resonant L shell, the evanescent fast mode wave couples strongly with the shear mode wave and energy is transferred from the surface wave into the field line resonance (bottom panel). In a pure evanescent mode there is no energy transport (the poynting flux is identically zero). But the coupling introduces a small

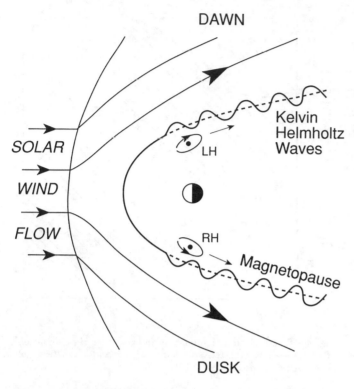

Fig. 5. The polarization of surface waves driven by the Kelvin-Helmholtz instability on the magnetopause changes sense near noon when the magnetopause flow divides to flow around the magnetopause. Waves in the morning travel westward and those in the evening travel eastward.

phase change, so introducing energy transport. The wave energy tunnels through the evanescent wave in much the same way as an evanescent wave function tunnels through a finite potential barrier. This coupling only occurs if there is some dissipation in the system which mathematically moves the singularity off the real axis to a complex frequency. But any real system has dissipation. Later work [e.g., Newton et al., 1978] showed that the primary source of wave damping is joule dissipation in the conducting ionosphere at the foot of the field line. The width in latitude or L-shell of the resonance is determined by the amount of dissipation.

Southwood [1974] showed that the phase of the radial and azimuthal components changed in such a way that the wave polarization changed sense at both the resonant field line, and at the amplitude minimum between the resonant field line and the magnetopause, so explaining Samson et al.'s observations (Figure 4). Figure 6 has deliberately been kept simple. Only one field line eigenperiod is shown whereas higher harmonics also exist. Thus a single frequency surface wave could excite a fundamental harmonic at some L shell, and a higher harmonic at another L shell, creating two amplitude maxima.

This theory seemed to fit the observations very well. The question of why only certain discrete frequencies were excited was answered by these being the frequencies of the

magnetopause surface waves that were feeding energy into the field line resonance. Work over the next ten years or so was aimed at confirming and verifying these ideas and at measuring the latitudinal width of the resonances.

Hughes and Southwood [1976a,b] showed that if the resonance was narrower than about 100 km in the ionosphere, it could not be measured with ground based magnetometers, as short scale variations are shielded from the ground by the atmosphere. So the measurement had to be done either in space or in the ionosphere. A chance close conjunction of two geosynchronous spacecraft allowed Hughes et al. [1978] to obtain an estimate of the width of a resonance by measuring the phase difference in the signal at the two spacecraft. They found that a resonance with a period of about 30 s near geosynchronous orbit had a width of about 0.3 R_E. This estimate compared favorably with another obtained by Walker et al. [1979] using the STARE ionospheric radar. STARE measured the ionospheric electric field of a wave with a period of about 260 s at many points simultaneously. Walker et al. showed that the elec-

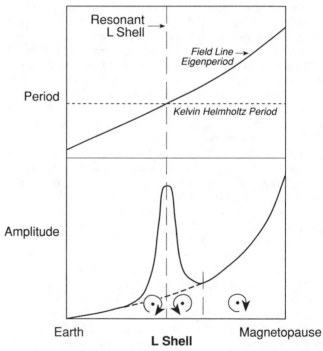

Fig. 6. A schematic representation of a field line resonance driven by a magnetopause surface wave. The top panel shows period as a function of L shell. The surface wave has one period independent of L shell. The field line eigenperiod varies with L shell. At one particular L shell these periods match. The lower panel shows wave amplitude versus L shell. The surface wave amplitude decays exponentially with distance from the magnetopause. Near the resonant field line, where the two periods match, coupling between the modes creates a narrow amplitude maximum. The sense of wave polarization switches both at the amplitude maximum and at the amplitude minimum between the field line resonance and the magnetopause.

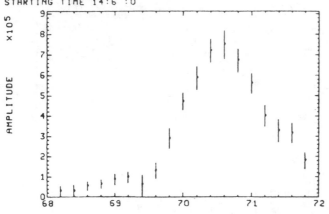

LATITUDE DEPENDENCE: YEAR 1977, DAY 303
LONGITUDE RANGE 16.0 TO 18.0
FREQUENCY 3.906 MILLIHERTZ
STARTING TIME 14:6 :0

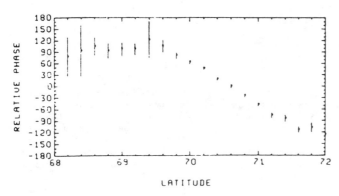

Fig. 7. STARE radar observations of a Pc 5 field line resonance. The top panel shows the amplitude of the electric field oscillations with a period of 250 s as a function of latitude. Note the peak confined between 70° and 71°. The lower panel shows the phase of these oscillations relative to the phase of the largest amplitude oscillation. The phase of the signal changes by almost 180° between 70° and 71° and is nearly constant with latitude both north and south of this. These are both features predicted for field line resonance. (After Walker et al., 1979).

tric field oscillations changed in phase by 180° over about 1° of latitude, and that this latitude corresponded to the half width of the wave amplitude maximum (Figure 7). Both Hughes et al. and Walker et al. found that the wave phase led at smaller L shell, as predicted by the field line resonance theory.

Hughes et al. [1978] also measured the east-west phase variation in the waves and showed that near geosynchronous orbit the waves propagated away from noon on both sides of noon. Olson and Rostoker [1978] confirmed this result using high latitude ground based data. The change in sense of propagation agreed with the Kelvin-Helmholtz picture. However, if the wave azimuthal phase velocity measured by Hughes et al. and Olson and Rostoker was mapped out to the magnetopause, speeds of 500-1000 km/s were obtained. These speeds are well in excess of typical magnetosheath ve-

locities, whereas the phase speed of Kelvin-Helmholtz surface waves should be slower than magnetosheath velocities. This fact remained an enigma for several years.

Finally it was in situ observations that confirmed the field line resonance model. Singer et al. [1982], using data from ISEE 1, proved that these pulsations are standing resonant waves. They showed that the electric and magnetic field oscillations measured by ISEE 1 were 90° out of phase, as expected for a standing wave, and that the observed wave period matched perfectly an eigenperiod of the local field line calculated using the measured plasma parameters. Then Takahashi went back to reexamine some of the older geosynchronous satellite data. Through careful spectral analysis, helped by displaying the results as frequency/time spectrograms, Takahashi and McPherron [1982] and Takahashi et al. [1984] found that each field line resonated at a low amplitude at a series of its own harmonic frequencies. It was not unusual to find 4 or 5 amplitude peaks in a spectrum at frequencies corresponding to the local field line eigenfrequencies. Figure 8 shows simultaneous spectra from three geosynchronous spacecraft. Up to 5 peaks can been seen in each spectrum, corresponding to harmonics ranging from the fundamental to the sixth harmonic. Note that the frequencies of the peaks are slightly different at each spacecraft, as they correspond to the eigenfrequencies of the local field line at each spacecraft location. Subsequently this result was confirmed by eccentrically orbiting spacecraft, first AMPTE/ CCE [Engebretson et al., 1986, 1987] and then CRRES. Their more sensitive magnetometers, better color graphic displays, and eccentric orbits which emphasized the L shell dependence of the eigenfrequencies, all helped to make the point much more clearly. These continuously varying eigenoscillations cover a broad band of frequencies. Hasegawa [1983] in response to Takahashi and McPherron's result, showed that a broad band source of

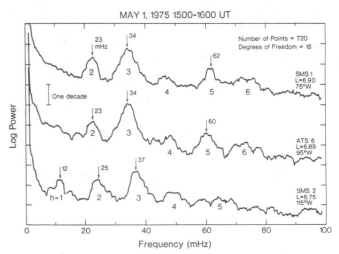

Fig. 8. Spectra of pulsations observed simultaneously by three geosynchronous spacecraft (SMS1, ATS6, and SMS2). Each spacecraft observes a series of harmonically related peaks, but the peaks in each spectrum occur at slightly different frequencies. (After Takahashi et al., 1984).

compressive waves could excite the individual eigenmodes of each field line. Such broad band compressive power is often observed together with the harmonic oscillations.

Thus by the early 1980's the picture looked very secure. Field line resonances had been observed and their width measured. But there remained two inconsistencies with the Kelvin-Helmholtz picture. Although the direction of azimuthal propagation was consistent with this picture, speeds were too high at the magnetopause. In addition, in spite of numerous magnetopause crossings by ISEE 1 and 2, regular oscillations of the magnetopause corresponding to the Kelvin-Helmholtz waves were not seen.

6. CAVITY MODES

Although low level continuously varying resonances are seen in space, on the ground individual monochromatic resonances are the norm (cf. Figure 7). So the same question needed to be raised again. If Kelvin-Helmholtz surface waves were not providing the monochromatic source needed to excite isolated field resonances, what was?

An answer was provided by Kivelson et al. [1984]. In order to explain a monochromatic Pc 5 pulsation seen by ISEE 1, they proposed that the magnetospheric cavity as a whole can ring and has its own set of cavity eigenperiods. This mode is described by equation (2) in the $m = 0$ or small m limit. It is a compressive or fast mode, and if m is small but not equal to zero, these modes will couple to individual field line resonances (described by equation (1)) in just the same way as the evanescent surface wave does. Figure 9 shows this in cartoon form. In the upper panel the periods of the lowest three cavity eigenmodes are shown together with the variation of the fundamental field line resonance eigenperiod with L shell. (Higher harmonic field line eigenperiods are omitted for simplicity.) The period of each of the three cavity eigenmodes matches the field line resonance eigenperiod at some L shell. Here strong coupling transfers energy out of the cavity mode into a field line resonance. Thus three field line resonances are excited, one corresponding to each of the three cavity eigenperiods. In the lower panel the wave amplitude as a function of L shell is shown for each of these three frequencies. The three field line resonances, each at its own frequency and on its own field line, stand out clearly.

To summarize, the general buffeting of the magnetosphere by variations in the solar wind dynamic pressure, or perhaps by sporadic magnetic reconnection, provides a broad band energy source to the magnetosphere. The magnetospheric cavity as a whole rings at its own eigenfrequencies, thus efficiently transporting energy at just those frequencies to field lines deep in the magnetosphere. Those field lines whose eigenfrequencies match one of the cavity eigenfrequencies couple to the cavity mode and resonate strongly, producing the classic field line resonance signatures.

This scenario was explored by a number of authors. Kivelson and Southwood [1985, 1986] developed the idea

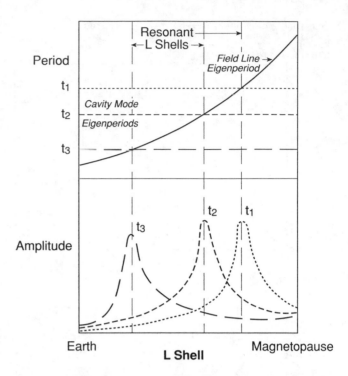

Fig. 9. A schematic representation of field line resonances driven by resonant magnetospheric cavity modes. The top panel shows the periods of three harmonics of cavity resonance, which do not vary with L-shell, and the variation of the fundamental field line eigenperiod with L shell. There are three L shells where the fundamental field line eigenperiod matches one of the cavity mode eigenperiods. In the lower panel the variation of wave amplitude at each of the three cavity eigenperiods with L shell is drawn. Note how a field line resonance is driven each time a field line eigenperiod and cavity eigenperiod match.

analytically. Allan et al. [1986] illustrated the effects in a model magnetosphere. More recently Lee and Lysak [1991] solved the equations numerically in a full dipole geometry and explored the structure of the resonance along the field line. Thus the model is on fairly strong ground theoretically. But the numerical estimates of the cavity mode periods calculated in these papers are too high for them to excite Pc 5 pulsations. Recently Harrold and Samson [1992] proposed that the outer boundary of the cavity is formed not by the magnetopause, but by the bow shock, and modeled this idea numerically. The idea has some merit for Harrold and Samson can explain the very low frequencies (1-3 mHz) of the very lowest frequency field line resonances more easily than if the cavity ends at the magnetosphere. Moreover the bow shock is probably a better reflector of MHD waves than the magnetopause.

Observationally the evidence is rather more circumstantial. Crowley et al. [1987] observed a field line resonance with the EISCAT radar. They were able to estimate the amount of energy leaving the resonance to heat the ionosphere. Yet the wave damped more slowly than expected given this energy loss. They claim that the most plausible explanation is energy coupling into the field line resonance from a cavity mode, which was in turn excited by a sudden impulse (SI). Samson et al. [1991; see also Walker et al., 1992] have shown that there is a definite preference for high latitude field line resonances to be excited at certain frequencies. They claim that these frequencies are the eigenfrequencies of the cavity. These two observations provide circumstantial evidence at best, and we are still waiting for a direct observation by a spacecraft of a cavity mode resonance.

7. Reconciling Space and Ground Based Observations

The cavity mode model is able to explain many ground based observations rather well, but we should probably be worried that cavity modes have not been observed by spacecraft. Anderson et al. [1989] point out several inconsistencies between the AMPTE/CCE data and the cavity mode model that we will try to reconcile.

Anderson et al. [1989] base their arguments primarily on the wave spectrograms routinely produced from the AMPTE/CCE data. This type of display was introduced into spacecraft ULF wave observations by Takahashi and McPherron [1982] but was fully developed by the AMPTE team. Perhaps without our realizing it these spectrograms have changed how we view ULF waves in space, and it is worth comparing the AMPTE/CCE observations with both earlier spacecraft observations and ground based observations. The spectrograms (cf. Anderson et al., 1989, Plate 1) emphasize waves beyond about L=5 with frequencies above about 10 mHz. Nearer perigee the spacecraft moves too fast and the background field is too large to make the spectrogram technique very effective. Takahashi et al. [1990], through very careful data handling, have extended the use of spectrograms inside L=4, but little work has been done in this region. Longer period waves can be observed near the bottom of the spectrograms, but their frequencies are not well resolved. The most prominent waves seen in the spectrograms are toroidal resonances with continuously varying, L dependent periods. Often several higher harmonics are seen, less frequently the fundamental which has a frequency below 10 mHz much the time. Spectrograms are very sensitive to small amplitude signals. Often these resonances have amplitudes of a few tenths of a nanotesla. These waves are in the Pc 3/4 frequency band and their occurrence is controlled by the direction of the IMF [Engebretson et al., 1987].

Anderson et al. [1989] point out that AMPTE/CCE does not observe large amplitude resonances on discrete L shells such as are thought to be produced by the cavity mode, and as are commonly seen on the ground; nor does it see narrow band signals with frequency independent of L in the compressional component, the expected signature of a cavity mode.

However, if Samson et al. [1991] are correct in identifying low order cavity waves with frequencies of a few milli-

hertz, then the cavity mode model is probably only relevant to Pc 5 pulsations. Unlike field line resonances, which are described by a single eigennumber, cavity modes have three eigennumbers, corresponding to the number of wavelengths along a field line, azimuthally around the Earth, and radially between the magnetopause (or bow shock) and the inner reflection point. Thus, if the lowest order mode has a frequency of, say, 2 mHz, then above perhaps 10 mHz the higher order modes will be so numerous and close in frequency that they could probably not be resolved in the data given their inherent width and the frequency resolution of a typical spectrum. Rather they would appear as a continuum. As a continuum, they can not be responsible for exciting discrete frequencies in this higher frequency band (corresponding to Pc 3/4).

Prior to AMPTE/CCE, spacecraft studies had relied on wave form plots to identify ULF waves. This method tends to select only larger amplitude waves. As a result the small amplitude higher harmonic toroidal resonances seen by AMPTE/CCE were missed. Instead, studies emphasized waves in the Pc 5 band and larger amplitude waves in the Pc 4 band. The few ground-satellite correlations [e.g., Allan et al., 1985] do show that the Pc 5's interpreted as field line resonances in ground-based (or radar) data appear as resonant oscillations at geosynchronous orbit also.

Unlike AMPTE/CCE, ground based studies of ULF waves have, on the whole, concentrated on waves whose periods are consistent with being a fundamental or at least a very low order harmonic field line resonance. Multiple frequencies are not often observed at a single station, and when they are, authors tend to attribute them to the effects of resonances on different L shells. The region best sampled by AMPTE/CCE, L=5-9, maps to a small range of latitude on the ground, between 63° and 70°. Here the dominant pulsation type is Pc 5 (see Figures 4 and 7). Usually a single, L independent frequency is observed. On the rare occasions when L dependent frequencies are seen [e.g., Siebert, 1964; Voelker, 1968; Rostoker and Samson, 1972; Poulter and Nielsen, 1982] the waves are excited by an impulsive source, usually an SSC. But at these high latitudes ground based observations are hampered by the convergence of field lines in the high latitude ionosphere. The atmospheric screening mentioned earlier screens from the ground all variations on scales less than about 100 km. Since 100 km in the high latitude ionosphere maps to a very much larger distance in the distant magnetosphere, much spatial detail must be lost. This spatial integration was analyzed quantitatively by Poulter and Allan [1985].

Ground based data are excellent for studying pulsations at lower latitudes, corresponding to L shells largely inaccessible to spacecraft observation since spacecraft move too quickly near perigee. Here Pc 3/4 pulsations are dominant. Often, though not exclusively, waves with frequency independent of L are seen. Waters et al. [1991, see also Menk et al., 1993], making use of interstation phase differences to identify local field line eigenfrequencies, found that the dominant peaks in the wave spectrum rarely occur at the local field line eigenfrequency. This observation is consistent with a single or small group of frequencies, presumably resonant on some L shell, dominating the power spectrum. Only by invoking such a model can we explain the dependence of Pc 3 periods observed at a single station on the strength of the IMF.

8. Conclusions

The search for the monochromatic energy source of magnetospheric ULF waves has led us through an involved intellectual history. As ever, we think we now have the answer. But there remain problems, and only time will tell if our current answer is correct. Work now focuses on the external energy sources that excite the magnetospheric resonances, both cavity and field line resonances. That work is the subject of this monograph. We hope that this tutorial will help put that work into context as you read on.

Acknowledgments. I acknowledge interesting discussions with Valeria Troitskaya and helpful comments from Mark Engebretson and the referees. The writing of this review was supported by the U.S. Air Force under contract F-19628-90-K-0003.

References

Allan, W., E. M. Poulter, K.-H. Glassmeier, and J. Junginger, Spatial and temporal structure of a high-latitude transient ULF pulsations, *Planet. Space Sci., 33*, 159, 1985.

Allan, W., S. P. White, and E. M. Poulter, Impulse-excited hydromagnetic cavity and field-line resonances in the magnetosphere, *Planet. Space Sci., 34*, 371, 1986.

Anderson, B. J., M. J. Engebretson, and L. J. Zanetti, Distortion effects in spacecraft observations of MHD toroidal standing waves: theory and observations, *J. Geophys. Res., 94*, 13425, 1989.

Angenheister, G., Ueber die Fortpflanzungs-Geschwindigkeit erdmagnetischer stoerungen und pulsationen, *Terr. Mag., 25*, 26, 1920.

Bolshakova, O.V., and V.A. Troitskaya, Relation of the IMF direction to the system of stable oscillations, *Dokl. Akad. Nauk. SSSR, 180*, 343, 1968.

Chen, L., and A. Hasegawa, A theory of long-period magnetic pulsation, 1, Steady state excitation of field line resonance, *J. Geophys. Res., 79*, 1024, 1974.

Crowley, G., W. J. Hughes, and T. B. Jones, Observational evidence of cavity modes in the Earth's magnetosphere, *J. Geophys. Res., 92*, 12233, 1987.

Cummings, W. D., R. J. O'Sullivan, and P. J. Coleman, Standing Alfven waves in the magnetosphere, *J. Geophys. Res. 74*, 778, 1969.

Dungey, J. W., Electrodynamics of the outer atmospheres, *Rep. 69*, Ions. Res. Lab. Pa. State Univ., University Park, 1954.

Dungey, J. W., The structure of the exosphere or Adventures in velocity space, in *Geophysics, The Earth's Environment*, C. Dewitt, Ed., p. 537, Gordon & Breach, New York, 1963.

Engebretson, M. J., L. J. Zanetti, T. A. Potemra, and M. H. Acuna, Harmonically structured ULF pulsations observed by the AMPTE CCE magnetic field experiment, *Geophys. Res. Lett., 13*, 905, 1986.

Engebretson, M. J., L. J. Zanetti, T. A. Potemra, W. Baumjohann, H. Luhr, and M. H. Acuna, Simultaneous observation of Pc 3-4 pulsations in the solar wind and in the Earth's magnetosphere, *J. Geophys. Res., 92*, 10,053, 1987.

Greenstadt, E. W., and J. V. Olson, A contribution to ULF

activity in the Pc 3-4 range correlated with IMF radial orientation, *J. Geophys. Res., 82*, 4991, 1977.

Greenstadt, E. W., and C. T. Russell, Stimulation of exogenic, daytime geomagnetic pulsations: A global perspective, in *Solar Wind Sources of Magnetospheric ULF Waves*, AGU Geophys. Monogr. Ser. (this volume). American Geophysical Union, Washington, DC, 1993.

Greenstadt, E. W., I. M. Green, G. T. Inoye, A. J. Hundhausen, S. J. Bame, and I. B. Strong, Correlated magnetic field and plasma observations of the Earth's bow shock, *J. Geophys. Res., 73*, 51, 1968.

Greenstadt, E. W., I. M. Green, D. S. Colburn, J. H. Binsack, and E. F. Lyon, Dual satellite observations of Earth's bow shock: II. Field aligned upstream waves, *Cosmic Electrodyn., 1*, 279, 1970a.

Greenstadt, E. W., I. M. Green, D. S. Colburn, J. H. Binsack, and E. F. Lyon, Dual satellite observations of Earth's bow shock: III. Field determined shock structure, *Cosmic Electrodyn., 1*, 316, 1970b.

Greenstadt, E. W., H. J. Singer, C. T. Russell, and J. V. Olson, IMF orientation, solar wind velocity, and Pc 3-4 signals: A joint distribution, *J. Geophys. Res., 84*, 527, 1979.

Gringauz, K. I., V. A. Troitskaya, E. K. Solomatina, and R. V. Shchepetnov, The relationship of solar wind variables to periods of continuous micropulsations of electromagnetic field of the earth, *Dokl. Akad. Nauk. SSSR, 5*, 1069, 1970.

Gul'elmi, A. V., Diagnostics of the magnetosphere and interplanetary medium by means of pulsations, *Space Sci. Rev., 16*, 331, 1974.

Gul'elmi, A. V., T. A. Plyasova-Bakunina, and R. V. Shchepetnov, Relationship between period of geomagnetic pulsations Pc 3,4 and the parameters of the interplanetary medium at the earth's orbit, *Geomag. Aeron., 13*, 382, 1973.

Harrold, B. G., and J. C. Samson, Standing ULF modes of the magnetosphere: A theory, *Geophys. Res. Lett., 19*, 1811, 1992.

Hasegawa, A., K. H. Tsui, and A. S. Assis, A theory of long period magnetic pulsations, 3, Local field line oscillations, *Geophys. Res. Lett., 10*, 765, 1983.

Hughes, W. J., and D. J. Southwood, The screening of micropulsation signals by the atmosphere and ionosphere, *J. Geophys. Res., 81*, 3224, 1976a.

Hughes, W. J., and D. J. Southwood, An illustration of modification of geomagnetic pulsation structure by the ionosphere, *J. Geophys. Res., 81*, 3241, 1976b.

Hughes, W. J., R. L. McPherron, and J. N. Barfield, Geomagnetic pulsations observed simultaneously on three geostationary satellites, *J. Geophys. Res., 83*, 1109, 1978.

Jacobs, J. A., Y. Kato, S. Matsushita, and V. A. Troitskaya, Classification of geomagnetic micropulsations, *J. Geophys. Res., 69*, 180, 1964.

Kivelson, M. G., and D. J. Southwood, Resonant ULF waves: A new interpretation, *Geophys. Res. Lett., 12*, 48, 1985.

Kivelson, M. G., and D. J. Southwood, Coupling of global magnetospheric MHD eigenmodes to field line resonances, *J. Geophys. Res., 91*, 1986.

Kivelson, M. G., J. Etcheto, and J. G. Trotignon, Global compressional oscillations of the terrestrial magnetosphere: The evidence and a model, *J. Geophys. Res., 89*, 9851, 1984.

Kovner, M. S., V. V. Lebedev, T. A. Plyasova-Bakunina, and V. A. Troitskaya, On the generation of low frequency waves in the solar wind in the front of the bow shock, *Planet. Space Sci., 24*, 261, 1976.

Lee, D.-H. and R. L. Lysak, Impulsive excitation of ULF waves in the three-dimensional dipole model: The initial results, *J. Geophys. Res., 95*, 3479, 1991.

Menk, F. W., B. J. Fraser, C. L. Waters, C. W. S. Ziesolleck, A.

Q. Feng, S. H. Lee, and P. W. McNabb, Ground measurements of low latitude magnetospheric field line resonances, in *Solar Wind Sources of Magnetospheric ULF Waves*, AGU Geophys. Monogr. Ser. (this volume). American Geophysical Union, Washington, DC, 1993.

Nagata, T., S. Kokubun, and T. Ijima, Geomagnetically conjugate relationships of giant pulsations at Syowa Base, Antarctica, and Reykjavik, Iceland, *J. Geophys. Res.*, *68*, 4621, 1963.

Newton, R. S., D. J. Southwood, and W. J. Hughes, Damping of geomagnetic pulsations by the ionosphere, *Planet. Space Sci.*, *26*, 201, 1978.

Olson, J. V., and G. Rostoker, Longitudinal phase variations of Pc 4-5 micropulsations, *J. Geophys. Res.*, *83*, 2481, 1978.

Patel, V. I., Low frequency hydromagnetic waves in the magnetosphere: Explorer XII, *Planet. Space Sci.*, *13*, 485, 1965.

Plyasova-Bakunina, T.A., Effect of the IMF on the characteristic of Pc 2-4 pulsations, *Geomag. Aeron.*, *12*, 675, 1972.

Plyasova-Bakunina, T.A., Yu.V. Golikov, V.A. Troitskaya, and P.C. Hedgecock, Pulsations in the solar wind and on the ground, *Planet. Space Sci.*, *26*, 457, 1978.

Poulter, E. M., and W. Allan, Transient ULF pulsation decay rates observed by ground based magnetometers: The contribution of spatial integration, *Planet Space Sci.*, *33*, 607, 1985.

Poulter, E. M., and E. Nielsen, The hydromagnetic oscillation of individual shells of the geomagnetic field, *J. Geophys. Res.*, *87*, 10,432, 1982.

Rolf, B., Giant pulsations at Abisko, *Terr. Mag.*, *36*, 9, 1931.

Rostoker, G., and J. C. Samson, Pc micropulsations with discrete, latitude-dependent frequencies, *J. Geophys. Res.*, *77*, 6249, 1972.

Russell, C. T., and M. M. Hoppe, The dependence of upstream wave periods on the interplanetary magnetic field strength, *Geophys. Res., Lett.*, *8*, 615, 1981.

Russell, C. T., J. G. Luhmann, T. J. Odera, and W. F. Stuart, The rate of occurrence of dayside Pc 3,4 pulsations: The L-value dependence of the IMF cone angle effect, *Geophys. Res. Lett.*, *10*, 663, 1983.

Saito, T., A new index of geomagnetic pulsations and its relation to solar M-region, Part 1, *Rep. Ionos. Space Res., Japan*, *18*, 260, 1964.

Samson, J. C., R. A. Greenwald, J. M. Ruohoniemi, T. J. Hughes, and D. D. Wallis, Magnetometer data and radar observations of magnetohydrodynamic cavity modes in the Earth's magnetosphere, *Can. J. Phys.*, *69*, 939, 1991.

Samson, J. C., J. A. Jacobs, and G. Rostoker, Latitude dependent characteristics of long-period geomagnetic micropulsations, *J. Geophys. Res.*, *76*, 3675, 1971.

Siebert, M., Geomagnetic pulsations with latitude-dependent periods and their relation to the structure of the magnetosphere, *Planet. Space Sci.*, *12*, 137, 1964.

Singer, H. J., W. J. Hughes, and C. T. Russell, Standing hydromagnetic waves observed by ISEE 1 and 2: Radial extent and harmonic, *J. Geophys. Res.*, *87*, 3519, 1982.

Southwood, D. J., Some features of field line resonances in the magnetosphere, *Planet. Space Sci.*, *22*, 483, 1974.

Southwood, D. J., and W. J. Hughes, Theory of hydromagnetic waves in the magnetosphere, *Space Sci. Rev.*, *35*, 301, 1983.

Stewart, B., On the great magnetic disturbance which extended from August 2 to September 7, 1859 as recorded by photography at the Kew Observatory, *Phil. Trans. Roy. Soc. Lond.*, *11*, 407, 1861.

Storey, L. R. O., An investigation of whistling atmospherics, *Phil. Trans. Roy. Soc.*, *246*, 113, 1953.

Sugiura, M., Evidence of low-frequency hydromagnetic waves in the exosphere, *J. Geophys. Res.*, *66*, 4087, 1961.

Takahashi, K., and R. L. McPherron, Harmonic structure of Pc 3-4 pulsations, *J. Geophys. Res.*, *87*, 1504, 1982.

Takahashi, K., R. L. McPherron, and W. J. Hughes, Multi-spacecraft observations of the harmonic structure of Pc 3-4 magnetic pulsations, *J. Geophys. Res.*, *89*, 6758, 1984.

Takahashi, K., B. J. Anderson, and R. J. Strangeway, AMPTE/CCE observations of Pc 3-4 pulsations at L=2-6, *J. Geophys. Res.*, *95*, 17179, 1990.

Troitskaya, V. A., Discoveries of sources of ULF waves – middle up to polar latitude studies, in *Solar Wind Sources of Magnetospheric ULF Waves*, AGU Geophys. Monogr. Ser. (this volume). American Geophysical Union, Washington, DC, 1993.

Troitskaya, V.A., R. V. Shchepetnov, and A. V. Gul'elmi, Effect of sudden disappearance of geomagnetic Pc 2-4 pulsations, *Geomag. Aeron.*, *9*, 363, 1969.

Troitskaya, V.A., T.A. Plyasova-Bakunina, and A. V. Gul'elmi, Relationship between Pc 2-4 pulsations and the interplanetary field, *Dokl. Aka. Nauk. SSSR*, *197*, 1313, 1971.

Vinogradov, P. A., and V. A. Parkhomov, MHD waves in the solar wind – a possible source of geomagnetic Pc 3 pulsations, *Geomagn. Aeron.*, *15*, 109, 1974.

Voelker, H., Observations of geomagnetic pulsations: Pc 3, 4 and Pi 2 at different latitudes, *Ann. Geophys.*, *24*, 24, 1968.

Walker, A. D. M., R. A. Greenwald, W. F. Stuart, and C. A. Green, STARE auroral radar observations of Pc 5 geomagnetic pulsations, *J. Geophys. Res.*, *84*, 3373, 1979.

Walker, A. D. M., J. M. Ruohoniemi, K. B. Baker, R. A. Greenwald, and J. C. Samson, Spatial and temporal behavior of ULF pulsations observed by the Goose Bay HF radar, *J. Geophys. Res.*, *97*, 12187, 1992.

Waters, C. L., F. W. Menk, and B. J. Fraser, The resonance structure of low latitude Pc 3 geomagnetic pulsations, *Geophys. Res. Lett.*, *18*, 2293, 1991.

Wolfe, A., L. J. Lanzerotti, and C. G. Maclennan, Dependence of hydromagnetic energy spectra on solar wind velocity and interplanetary magnetic field direction, *J. Geophys. Res.*, *85*, 114, 1980.

W. Jeffrey Hughes, Center for Space Physics, Boston University, Boston, MA 02215

Stimulation of Exogenic, Daytime Geomagnetic Pulsations:
A Global Perspective

E. W. Greenstadt

Electromagnetic Technology Dept., TRW, Redondo Beach,
California

C. T. Russell

Dept. of Geophysics and Planetary Physics, Univ. of
Calif., Los Angeles

The long history of observations and concepts in the process of trying to understand the linkage between solar wind parameters and daytime ULF pulsations in the range of f between, roughly, 7 and 70 mHz, has produced statistical studies persuasive of that linkage, but with weak correlations and imprecise, even unspecific, conclusions regarding the locales and mechanisms of physical connection between perturbations outside the magnetopause and oscillations inside the magnetosphere. We believe part of the difficulty lies in the asymmetries and temporal variabilities that affect, individually and together, the foreshock, bow shock, magnetosheath, magnetopause, and magnetosphere through which the progress of ULF stimulation and maintenance must pass. We summarize the asymmetric distributions of hypothesized sources and magnetospheric sites of ULF excitation and the continual, diurnal, and seasonal variation that must affect such excitation, including the asymmetries of foreshock, bow shock, and sheath, and the time-dependent orientation of the magnetosphere to the solar wind. We suggest that the nonuniform and nonstationary factors described, although adding complexity to the overall solar-geomagnetic system, should be taken into account by future investigations with appropriately selected subsets of data. Then the relationship between at least some magnetospheric ULF phenomena and their external sources will be clarified.

Introduction

Oscillations of a few tens of seconds period and striking quasinusoidal regularity are commonly, but not continuously, detected on Earth's surface and in the magnetosphere, magnetosheath, and solar wind outside Earth's bow shock by magnetometers sensitive to changes of ambient magnetic flux greater than one nanotesla. Consequently, ULF waves occurring in the geomagnetic environment are subjects of observation, theory, and speculation as old as the instruments capable of recording them. Magnetospheric waves that occur in local daylight hours have been linked many times to various combinations of solar wind parameters. Although the solar wind can exert control on the occurrence and properties of ULF pulsations in many ways, however, the history of the subject has usually given us widely

Solar Wind Sources of Magnetospheric Ultra-Low-Frequency Waves
Geophysical Monograph 81
Copyright 1994 by the American Geophysical Union.

scattered points in observational correlations and conflicting conclusions in theoretical studies, so that many questions remain. Are all spectrally comparable oscillations unrelated to one another but coincidentally similar? Or are they related to each other by a common generator, and if so, how? Or are they in fact all displays of the same phenomenon travelling through different locations with slightly different local characteristics?

Despite the many uncertainties, some classes of magnetospheric waves have been fairly well explained; the reader will not be tranquilized with these here. The subject of this essay is day time geomagnetic pulsations in what has traditionally been designated the Pc3/Pc4 period ranges 15-45/45-150s, where no one today doubts that an external source of Pc3-4's operates, but where uncertainty and conflict still prevail. A little bit of history in the next section will, first, illustrate the initial difficulties in establishing linkage between waves external and internal to the magnetosphere; second, introduce the possible nonwave generation of ULF in the magnetosphere by pressure variations in, or instigated by, the solar wind, and third, justify the argument presented later for refinements that might help to resolve the remaining uncertainties. Our historical summary of external linkage is brief, intended more to make a point than to be complete. The rich contributions of numerous researchers to either the background or the mechanisms of generation are covered more thoroughly by other reviews and tutorials.

A MICROHISTORY

Soviet scientists, led by the Mother of all Pulsations, Valeria Troitskaya, were among the first to note and demonstrate that values of some parameters of the solar wind, particularly the latitudinal and longitudinal directions *lat*(**B**) and *long*(**B**) of the interplanetary magnetic field (IMF), were correlated with the occurrence of Pc3 and Pc4 magnetic pulsations measured on the Earth's surface [Bol'shakova and Troitskaya, 1968; Troitskaya et al., 1971; Plyasova-Bakunina, 1972]. A theoretical justification for the apparent connection between the solar wind and the magnetosphere was postulated by Kovner et al. [1976], who estimated that the maximal growth rate of Pc3,4-type waves upstream, which they assumed to penetrate to the magnetosphere from the subsolar shock, would occur when $\theta_{XB} \sim 60°$. Kovner et al. also calculated that Pc3 amplitudes (upstream) should rise with increasing solar wind speed V_{SW}, a result compatible with previous observation by Saito [1964] and Vinogradov and Parkhomov [1974] and confirmed by themselves and subsequent investigators (references and figures below).

A possibly direct connection based on observations of quasisinusoidal and quasiperiodic wavetrains arose separately with the appearance of oscillations in the Pc3-4 frequency range in the bow shock, upstream from the bow shock, and downstream from it in the magnetosheath. Since no two of the external and internal oscillations were simultaneous, these data provided only the implication that wavetrains resembling one another might also be occuring in two or more regions at the same time by propagation or excitation from one region to the next. The general pattern of daytime field vibrations, without resolution of waveforms, was statistically similar in occurrence in the the subsolar magnetosheath and on the ground, as seen in Figure 1, where standard deviations from Explorer 34's magnetometer, representing frequencies roughly between .05 and .2 Hz (20 > T > 5 s) [Fairfield and Ness, 1970], showed the same preference for morning hours as surface instruments measurementing Pc3 or Pc4 activity.

The observed behavior of regular Pc-like periodicities in and around the shock defined a relationship between their occurrence and the orientation of the IMF [Greenstadt et al., 1970a,b]. The global implications of that relationship led to a postulated connection between shock associated waves and magnetospheric pulsations [Greenstadt, 1973] that obligingly merged with the relationships found by the Russian pioneers. The key parameter of the IMF's influence on the ULF component of the shock was the angle θ_{Bn} between the field and the local shock normal. This parameter in turn implied a significance for the angle θ_{XB} in determining the pattern of ULF occurrence around the curved bow shock, where θ_{XB} is the "cone angle"

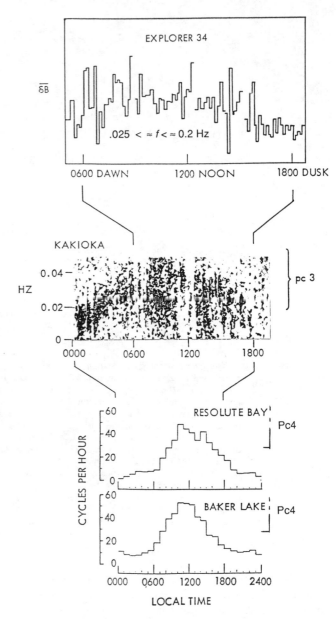

Fig. 1. Distributions of various measures of wave activity across local daylight longitudes recorded by Explorer 34, in the magnetosheath; at Kakioka magnetic observatory; and at Resolute Bay and Baker Lake magnetic observatories. The panels document the tendency of daytime pulsations to peak in amplitude and Pc3,4 frequencies around midmorning to midday.

between the field vector and the solar wind flow. The flow direction is approximated by the solar ecliptic ±X direction independent of sense.

The cone angle concept is that wave phenomenology attendant on the ULF component of quasiparallel shock structure increasingly dominates the subsolar region of the bow shock as θ_{XB} drops below about 51°. Oscillating and turbulent, or quasiparallel ($Q_{\parallel}$), structure characterizes the bow shock where the angle θ_{Bn} between the IMF **B** and the local shock normal **n** is less than ~ 51°; a steplike, quasiperpendicular ($Q_\perp$), structure relatively free of ULF signals characterizes the shock where $\theta_{Bn} > ~ 51°$. Subsolar $Q_\parallel$ structure fills with oscillations the central solar wind flux tube whose plasma later grazes the magnetopause, transferring its waves to the magnetosphere.

The model assumes rotational symmetry, around the X-axis, of the shock and wave-pulsation relationship. Thus the cone angle combines the latitude and longitude of the IMF vector favored by the early Russian work, since cos θ_{XB} = cos($lat\mathbf{B}$) × cos($long(\mathbf{B})$). Sketches illustrating this postulate accompany a later section of this report. The point here is that scatter diagrams developed by different investigators with different data bases, attempting to quantify pulsation activity by amplitude as well as occurrence, supported this model, but with rather poor statistical correlation. This can be readily appreciated, even for the best relationships of occurrence or amplitudes vs. V_{SW}, in Figure 2.

Compressional wave energy in the magnetosheath, derived from wave-particle interactions in the foreshock and shock, can presumably be transferred to the magnetosphere simply by periodic pressure variation at the magnetopause. Such stimulation could also arise from wavetrains delivered from the solar wind itself, but periodic oscillations are not the the only available pressure phenomena. Nonperiodic pressure variations, including pulses and discontinuites, are intrinsic to the solar wind; also, long term (T > 1000s) variations of the IMF cause

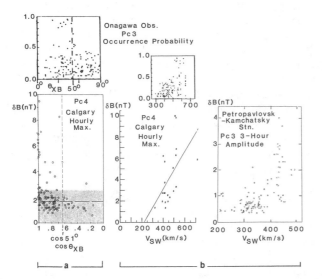

Fig. 2. Scatter diagrams showing weak, but consistent correlations of Pc3,4 occurrence and amplitude at Earth's surface with IMF cone angle, a, and solar wind speed, b. Top of a, hourly occurrence probability of Pc3 at Onagawa vs. cone angle θ_{XB} [Saito et al., 1979]; bottom of a, hourly maximal amplitude of Pc4 at Calgary vs. cos(θ_{XB}) [Greenstadt and Olson, 1977]. Top of b:, hourly occurrence probability of Pc3 at Onagawa vs. solar wind speed V_{SW} [Saito et al., 1979]; bottom left of b, hourly maximal amplitude of Pc4 at Calgary vs. V_{SW} [Greenstadt et al., 1979]; bottom right, amplitude in three-hour intervals of Pc3 for 20 < T < 40 s at Petropavlovsk-Kamchatsky vs. V_{SW} [Kovner et al., 1976].

unsteadiness and relocation of the foreshock. Either may impinge on the magnetopause. There can also be inherent unsteadiness in the interaction between the shocked solar wind and the magnetopause.

All aspects of ULF wave stimulation are subject to prevailing conditions in the global magnetospheric environment. We describe this environment before discussing the specific sources of wave generation and transfer in which they must operate.

THE GLOBAL WAVE-TRANSFER ENVIRONMENT

The lingering uncertainty surrounding generation of daytime ULF activity stems from the inherent complexity of space plasmas and the recognized proliferation of subregions in the Earth's interaction with the solar wind. The subregions in this overview, within their dynamic environment, are the rotating magnetosphere, enclosed by the magnetopause, within the solar wind flowing by in the magnetosheath, modified by having passed through the bow shock/foreshock system, whose wave profiles are controlled by the continuously varying interplanetary magnetic field. We shall work outward from the magnetopause.

Magnetopause

The magnetopause is nonuniform. It is approximately bilaterally (geomagnetically East-West) symmetric with respect to the X-Z plane of the solar wind flow (X) and the magnetic pole (Z_M), but does not in general present a symmetric boundary to the magnetosheath flow within or parallel to that plane. There are even dissimilar boundary zones in the conventional symmetric configuration, as follows:

The symmetric case. Figure 3E, upper right (E for equinoctal), is a simplified, conceptual X-Z cross section of the magnetosphere when it is in a nominally static, symmetric, magnetically untilted orientation behind the bow shock. Notation "1, 2, 3" calls attention to three distinct regions of the boundary that may have distinct responses to the imposition of perturbed magnetosheath plasma on the magnetopause. The three boundary areas are:

1. Equatorial (Axial Stagnation). In this region, the flow of the solar wind close to the subsolar point where the axis of symmetry intersects the magnetopause is presumably at or near zero velocity. Waves present here, particularly pressure waves, or pulses, may stimulate the magnetopause at their prevailing frequencies and propagate directly into the equatorial magnetosphere as compressional waves crossing the earth's field lines.

2. Midlatitude (Anaxial Streamflow). Away from the stagnation region, the solar wind has nonzero speed parallel to the magnetopause. Waves crossing at any point may impart to their magnetospheric successors a tangential **k**-vector component influenced by their convection velocity along the boundary. Propagation might be primarily along outer field lines toward higher latitudes (or, in an equatorial X-Y cross section, toward dawn and dusk terminators).

3. Cusp (Penetration). In the cusp region, tangentially flowing, hot solar wind plasma may penetrate the magnetopause directly, carrying any periodic modulations within it deep into

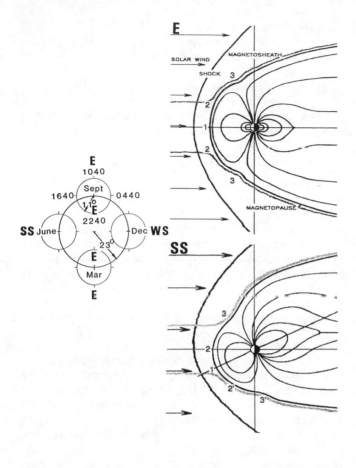

Fig. 3. Two sketches of the magnetosphere in the noon-meridian plane at equinox (E, upper right) and summer solstice (SS, lower right), and a diagram of the annual cycles of the magnetic pole's orientation with respect to the solar wind flow from the sun, (left center). Arrows in the right hand sketches indicate the solar wind direction; shaded curves show approximate flow lines in the magnetosheath that cross the terminator at about 1 R_E from the magnetopause.

the magnetosphere at auroral and cusp latitudes.

Each of the above combinations of local magnetopause configuration and magnetosheath activity acts on its own local version of the magnetopause boundary layer or mantle; each version has its own effects, if any, on the transmission of waves. Additional regions and modes of transfer can be postulated, and, of course, counterparts of these around an equatorial cross section may also be suggested. Not all parts of the magnetosphere's boundary can be expected to receive the same stimulation, or have equal sensitivity to whatever outside perturbations come to them, even within this depiction of what has been the traditional zero-order conception of a static and symmetric magnetosphere.

The asymmetric case. In general the magnetosphere is neither static nor symmetric with respect to the Sun-Earth line. More representative, asymmetric diagrams are becoming common [Walker et al, 1989; Kivelson and Hughes, 1990], and recognition of the magnetopause's time-variable orientations to its

external wave environment may be as important as recognition of its subregions. Just as we expect substorm processes and reconnection to depend on the instantaneous orientation of the magnetosphere to the solar wind, we should not be surprised if geomagnetic pulsations too depend on orientation of the subregions to the magnetosheath flow.

The diagram at the left center of Figure 3 is a schematic projection of the magnetic pole's trajectory on the ecliptic plane through an annual cycle, combining both its seasonal and diurnal dynamics. In this diagram, borrowed from Kivelson and Hughes [1990], the effect of annual and daily tilt on the orientation of the magnetosphere to the solar wind is explicit for some sample times a quarter-year and six hours from one another, and can be inferred for any other times. If we imagine ourselves looking down from the North ecliptic pole with the sun at left of the page, we see that the North magnetic pole can tilt from 34° (23° + 11°) toward the sun at northern summer solstice in June (left point SS, at 1640) to 34° away from the sun at northern winter solstice in December (right point WS, at 0440). Such differing orientations expose the cusp, say, to different regions of the interaction boundary at different times. Thus, entirely different sections of the magnetopause may be affected at one time compared to another, even for identical distributions of magnetosheath disturbance. This is the argument illustrated explicitly in the two large diagrams of the figure, that is, in the pair of meridian cross sections E and SS. Magnetosphere E corresponds to the equinox times E, and magnetospheric sketch SS to noontime of summer solstice of the left-center diagram.

In Figure 3SS, region 2 of the tilted magnetosphere, not region 1 as in 4E, faces the stagnation point of the magnetosheath and region 3 is also reoriented to confront a less deflected, perhaps more penetrating, solar wind than it does in E. Below the axis of symmetry, notations 2′ and 3′ have been substituted for 2 and 3 to emphasize the asymmetry in SS of corresponding north and south magnetic latitudes with respect to the solar wind, unlike their symmetric counterparts in Figure 3E. We can easily imagine a mirror image of Figure 3SS, rotated 180° around the X-axis, showing the approximate magnetospheric orientation at the opposite, winter solstice, (WS), with the south geographic and magnetic poles tilted sharply toward the sun.

Clearly the seasonally corrected magnetosphere presents the different areas (1,2,3) of its envelope to the subsolar flow and its convected magnetic oscillations in quite different postures, and we may reasonably expect waves in the magnetosphere to show different patterns of response, most clearly distinguishable during the seasonal extremes and perhaps more subtly during intermediate times. Seasonal variation of Pc3 activity in the central magnetosphere was derived statistically from data recorded by the geosynchronous satellite ATS 6 [Takahashi et al., 1981], and data bases from such sources may still prove valuable if revisited in the context of Figure 3.

The shaded streamlines through the subsolar area of the shock in Figures 3E & SS, adapted from calculations by Spreiter and Alksne [1969], illustrate a sample projected boundary of a column, or cylinder, of solar wind that flows through the subsolar region and spreads in the magnetosheath to flow within 1 R_E of the flank magnetopause at the terminators, 90° from subsolar point. We may imagine the actual activity in the magnetosphere to depend on combinations of periodic diurnal and seasonal circumstances, such as those of Figure 3 and on the sources of disturbance in the magnetosheath: that is, on the transient global distributions of activity in the layer of plasma flowing over the magnetosphere's surface.

Magnetosheath

The source of those waves in the magnetosphere that derive from external stimulation lies in the magnetosheath. At any given time the magnetosheath contains a nonuniform, global distribution of perturbations whose relationship to the magnetopause determines where and how much, if any, of their energy will be transferred into the magnetosphere. What are the prevailing distributions of disturbance around the magnetopause when the waves are in progress--and when they are not?

We know the answer to this question for certain extreme and specialized situations, namely when the magnetopause is likely to be engulfed in waves convected from upstream, and when it is likely to be surrounded by no convected waves at all [Greenstadt, 1973; Russell et al., 1983]. In the former case, daytime waves are abundant; in the latter, they are absent [Wolfe, 1980; Wolfe et al., 1980]. Figure 4 depicts these two extremes, exhibiting incidentally the contrasting specialized orientations of the interplanetary magnetic field (IMF) that support them. The sketches are derived from the MHD calculations of Spreiter and Alksne. [1969].

In 4*a*, at left, the IMF and the SW flow are parallel (IMF cone angle $\theta_{XB} = 0°$); the field and plasma follow identical flow lines in the magnetosheath, with the lines deflected around

the magnetopause. Only lines near the axis of symmetry, i.e., near the X-axis, come close to the magnetopause, as illustrated in the previous figure.

In 4*b*, at right, the IMF is oriented perpendicular to its orientation in 4*a* ($\theta_{XB} = 90°$), while the SW flow, not depicted explicitly, is the same as in 4*a*. Here, the field lines are draped around the magnetopause, and all of them come near the subsolar magnetopause in their turn, slipping around the magnetopause above or below the plane of the sketch or merging with magnetospheric field lines as the sense of the magnetic vectors might dictate (northward as in the panel, or at times southward). The field lines are drawn as wavy to illustrate the paths along which foreshock and shock perturbations should be expected to reach the daylight magnetopause easily in 4(*a*) but with difficulty in 4*b*. The dashed lines in 4*b* represent the boundaries of the ULF foreshocks for the 90° IMF orientation.

Cases 4*a* and *b* are strongly associated with presence or absence of daytime magnetospheric pulsations, respectively. These extremes are infrequent, however, and statically idealized. On ordinary days the more general magnetosphere is moderately excited, and its waves may be related to the time-varying distributions in the sheath of each parameter of each component of the sheath plasma, such as density of the ions, temperature of the electrons, or strength of the magnetic field, and of the pattern of spectral power of the variations of each of these parameters everywhere in proximity to the magnetopause. Meanwhile, there are additional complications that affect the magnetosheath.

Bow Shock and Cone Angle

Recent study of the divisions between $Q_\perp$ and $Q_\parallel$ structures on the bow shock make it possible to replace or supplement the generic shock-origin sketches of Figure 4, [Greenstadt 1973; Russell et al. 1983; and the widely circulated cover of the Upstream Waves and Particles issue of *J. Geophys. Res. 86*, June 1, 1981] with more refined configurations. One such configuration, for the most probable 45° cone angle of the IMF, will be described here, along with a pair of deviations. More complete sketches of the $Q_\perp/Q_\parallel$ divisions for a selection of cone angles have been published elsewhere [Greenstadt, 1991]; extrapolation of the reasoning below to such configurations can be easily constructed, while the extremes sketched in Figure 4, which correspond, to $\theta_{XB} = 0°$ and 90° remain essentially unaffected.

The most probable or "typical" state of the bow shock is illustrated in Figures 5*a* and *b*. The IMF is assumed to be in the ecliptic plane at the Parker spiral angle, $\theta_{XB} = 45°$, and the transition between $Q_\perp$ and $Q_\parallel$ structures is assumed to occur wherever on the shock the local $\theta_{Bn} = 51°$ [Diodato et al., 1976; Greenstadt, 1991]. Panels *a* and *b* of the figure show the projections on the X-Y and Y-Z planes, respectively, of the boundaries between $Q_\perp$ and $Q_\parallel$ sections of the shock, with the fully $Q_\parallel$ section darkly shaded and the section timeshared between $Q_\parallel$ and $Q_\perp$ signified by darker shading. The timeshared section is meant to represent the area where waves at the shock modify θ_{Bn} locally so that the shock is $Q_\perp$ during some part of each wave period and $Q_\parallel$ during the other part.

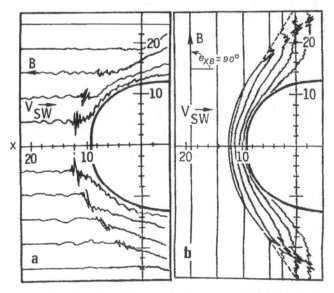

Fig. 4. Two sketches of the extreme patterns of ULF activity in the subsolar magnetosheath, when the IMF is parallel (or antiparallel) to the solar wind flow, at left; and in the flanks when the IMF is across the flow, at right.

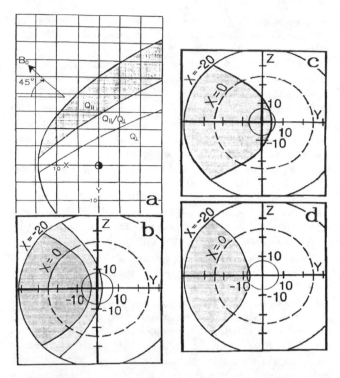

Fig. 5. Projections of boundaries between $Q_\perp$ (clear) and $Q_\parallel$ (shaded) regions of Earth's bow shock for various IMF cone angle (θ_{XB}) orientations; a and b present X-Y and Y-Z meridian projections for $\theta_{XB} = 45°$; c and d present Y-Z projections for $\theta_{XB} = 30°$ and $60°$. In a and b lighter shading indicates a region of alternating $Q_\perp$ and $Q_\parallel$ shock structures according to variable θ_{Bn} between the IMF and the local shock normal n introduced by large amplitude foreshock waves impinging on the bow shock.

We see that a portion of the morning side of the shock is $Q_\parallel$, and that there is a small intrusion of the timeshared structure into the afternoon shock around the subsolar point. The circle around the origin in b represents the boundary of the cylinder whose flow passes within 7 R_E of the subsolar point and then, according to the drawings of Spreiter and Alksne [1969], approximately 1 R_E from the magnetopause terminator. We see that a portion of the plasma in the central flow tube is $Q_\parallel$, another portion borderline $Q_\perp /Q_\parallel$, and yet another entirely $Q_\perp$. Some $Q_\parallel$ activity is carried to within 1 R_E of the dawn terminator and some borderline wave activity bathes the entire magnetopause, but no continuous $Q_\parallel$ wave activity is convected to the dusk terminator.

Figures 5c and d represent the Y-Z projections for less probable, but by no means rare, stream angles 30° and 60°. The sketch in 5c encourages the idea that most of the subsolar flow column, perhaps all of it, if we were to include the comparable $Q_\perp /Q_\parallel$ timeshared region, as in 5a, would be occupied by ULF wave signals in contact with most, if not all, of the daytime magnetopause when the stream angle is 30°. The sketch of 5d, in contrast, supports the expectation that even with an additional $Q_\perp /Q_\parallel$ region, only a section of the subsolar flow column and a corresponding section of the morning magnetopause would be in

contact with outside ULF activity when the stream angle is 60°. These two cases represent the limits, more or less, of stream angles below or above which geomagnetic pulsations might be stimulated all day or at no daylight hours by solar wind-convected ULF activity.

Two more complications merit attention in this synopsis of the ULF environment. First, the wavy horizontal lines in the upper panel of Figure 6 define another subsection of the shock where the shock is presumably $Q_\perp$ because $\theta_{Bn} > 51°$, but where observations have shown foreshock waves generated by reflected ions are convected back to the shock for the 45° orientation of the IMF [Greenstadt, 1991]. The Y-Z projection of this region is indicated by the wavy lines in the lower panel. Second, the three irregular and nonuniform grades of shading in both sketches is intended to portray the nonuniformity of the $Q_\parallel$ structure, with darker shade meaning more parallel geometry and larger amplitude, more compressional oscillations than the lighter shades. In the lower panel, the larger circle marked M and the smaller circle marked F indicate the nominal sizes of the Y-Z cross section of the magnetopause and the subsolar flow tube passing within 1 R_E of M in the upper panel. The nonuniform shading emphasizes superposition of these qualitative complications on the computed $Q_\perp /Q_\parallel$ boundary and on magnetosheath flow lines downstream. Thus the global distribution of shock- or foreshock-related disturbance in the subsolar flow tube of the magnetosheath and consequently around the magnetopause is unmistakably asymmetric.

We must imagine that some waves of small to moderate amplitudes in the leading edge of the $Q_\parallel$ region might reach the magnetopause largely in the morning sector above and below the ecliptic plane, for the illustrated configuration. Their power would be averaged over the expanded cross section of the tube at the magnetopause, so we would not expect an overwhelming response inside the magnetosphere. Decreasing cone angles would place an increasing supply of larger amplitude waves inside the tube on both evening and morning sides, and at high (polar) as well as low (ecliptic) latitudes. Increasing cone angles would remove significant waves from the subsolar tube and therefore from the vicinity of the magnetopause. The response of the magnetosphere would then be presumed to depend on which regions of the magnetopause would be exposed to the perturbations in the flow, as implied in Figure 3. Some regions might be as sensitive to small disturbance when, say, $\theta_{XB} = 50°$ as others to larger disturbance when $\theta_{XB} = 30°$.

Temporal change

Before proceeding to the actual machinery of wave energy-transfer, we emphasize that all the global configurations and processes are in continuous variation caused, in simplest terms, by the rotations of the Sun and Earth, so that the former presents a dynamic plasma source of the rapidly changing solar wind, while the latter offers a diurnal and seasonally changing asymmetry to the portion of solar wind conveyed to it. Thus the actual situation is a time-dependent selection from an infinite number of configurations such as the static cases sketched in the Figures.

Wave transfer

The large and intricate subject of the actual mechanisms by which ULF energy might be physically transferred from the solar wind to the magnetosphere cannot be detailed here, but some of the evidence supporting the main hypotheses can be outlined. Broadly speaking, there are two instigators of magnetospheric oscillation: more or less continuous wavetrains and pressure impulses or irregular transients impinging on the magnetopause. Wavetrains may occur as any combination of transverse and compressional, narrow or broadband, oscillations.

The conceptually simplest hypotheses of wave transfer apply to wave trains. The transverse components of a monochromatic wave vector can be imagined, given the appropriate orientation of **k**, to alternately reinforce and oppose the geomagnetic field lines at the magnetopause, creating small scale, local signals, or even disconnections and reconnections or mini-flux transfer events [FTEs, Russell and Elphic, 1978; see next subsection, below]. These signals might then propagate along the lines and/or deep into the magnetosphere where, upon encountering lines of the same natural frequency, they would initiate or drive a geomagnetic resonance oscillation. Alternatively, or supplementally, the compressional part of a magnetosheath wavetrain might initiate a pressure signal that would penetrate the magnetosphere with the same result.

The weaknesses of this simple concept lie in satellite observations, which have not revealed monochromatic wavetrains in the deep magnetosheath that would correspond to the common occurrence of such waves inside the magnetosphere, and, moreover, have revealed harmonic structures of pulsations inside the magnetosphere that suggest a multichromatic source. The more probable form of external wave stimulation is by the typically broadband signals in the sheath, from which the magnetosphere allows internal propagation and selects resonances when the appropriate field lines or shells are crossed by the penetrating signal.

Nothing in the above concepts enlightens us regarding the actual site(s) of transfer of external to internal wave energy, which could be anywhere from subsolar to flank or polar locations. We do know, both theoretically [Verzariu, 1973; Wolfe and Kaufmann, 1975] and observationally [Greenstadt et al., 1983; Tomomura et al., 1983] that wave power just inside the dayside magnetopause can be orders of magnitude less than the power just outside in the sheath, at least for certain well defined magnetopause models or a few chance locations of measurement. Other models, locations, or roundabout transfer routes are not ruled out, however. Recent reports of investigations by Engebretson et al. [1991] and Lin et al. [1991] have included summaries of such proposed mechanisms.

One model would have compressional oscillations in the foreshock propagate directly through the shock, sheath, and subsolar magnetopause into the lower magnetosphere [Yumoto and Saito, 1983; Yumoto et al., 1984; Russell et al., 1983]. Another suggests that waves enter along cusp/cleft/boundary layer field lines [Troitskaya and Bol'shokova, 1984] and then transfer to the interior dayside magnetosphere via an ionospheric process [Lanzerotti et al., 1972; Engebretson et al., 1990]. Either or both

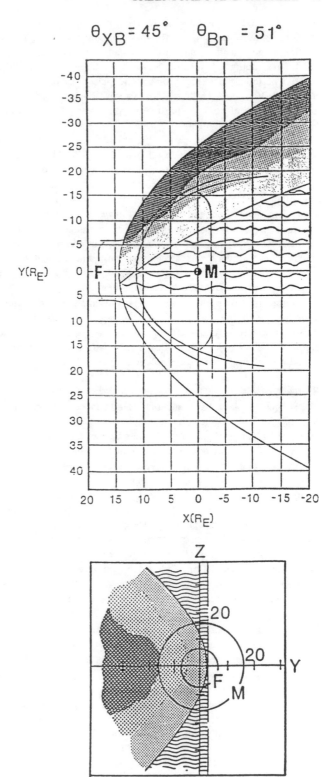

Fig. 6.   X-Y and Y-Z projections of Earth's bow shock for rough subdivisions of $\theta_{XB} = 45°$, with variable shading showing rough subdivisions of $Q_\perp/Q_\parallel$ structure of the shock and, by implication, the magnetosheath, in relation to the solar wind flow tube that encloses the magnetopause.

are supported to some degree by satellite evidence. Figure 7 presents some of the support for these suggestions. All plots in the figure show scatter diagrams consistent with the well known and long standing relationship between the strength of the IMF and the period of daytime geomagnetic pulsations [Gul'elmi et al., 1973; Vero and Hollo, 1978]. Compatible versions of the relationship are offered in the three upper panels [Yumoto et al., 1984] at geosynchronous (GOES) and two low-L surface stations, and in the bottom panel [Engebretson et al., 1986] at a polar station south of the auroral oval. The f-B_{SW} dependence results from particle-wave interactions at local cyclotron frequencies in the solar wind around the bow shock. It's difficult to explain such uniform action within the magnetosphere, where local resonance frequencies should have differed from one another significantly, without invoking direct transfer of some of the original signal to all sites. Of course, the data don't tell us whether the pathway was around the cusp and inward to the equator and the pole via the ionosphere or through the subsolar magnetosphere and outward to the cusp field and beyond.

If either model applies, the seasonal presentation of the equator or cusps to the patterns of disturbance in the sheath must surely play a role in the efficiency of these pathways and the parts of the magnetosphere most readily affected. Seasonality would contribute scatter to the relationships.

Pressure variation

Changes in the solar wind and its embedded IMF could be responsible for pulsations inside the magnetosphere stimulated by irregular, rather than periodic, pressure variations at the magnetopause. Statistical evidence for a dual dependence of Pc3-4 period on solar wind density N_{SW}, decreasing with N_{SW} for Pc3, increasing for Pc4, was developed, for example, by Gringauz et al. [1970]. Since we treat here mainly with sustained, rather than transient pulsations, we skip lightly over impulses, such as interplanetary shocks, as sources. Daytime pulsations in the frequency range we describe have not been shown to correlate strongly with solar wind pressure pulses or discontinuities. Oscillations induced by impulses are essentially transient or damped [Saito and Matsushita, 1967] and, moreover, when sustained, may result from other changes in IMF geometry that happen to accompany the impulse. There are, however, inherent pressure changes at the magnetopause and driven pressure changes from the shock system that, though irregular, are not isolated and might be responsible for sustained effects in the magnetosphere.

Pressure variation associated with unsteadiness of the magnetopause could be due to a fluid instability such as the Kelvin-Helmholtz (K-H) instability or due to a magnetic effect such as time varying reconnection. Since motion of the magnetopause appears to be controlled by the fluctuations in the momentum flux of the solar wind and the southward IMF, as discussed by Song et al. [1988] and illustrated in Figure 8 (their Figure 2), and K-H phenomenology seems to be associated principally with Pc 5 oscillations more at dawn and dusk than midday, we shall not pursue the K-H source here. We consider instead time-varying, "patchy" reconnection in terms of flux transfer

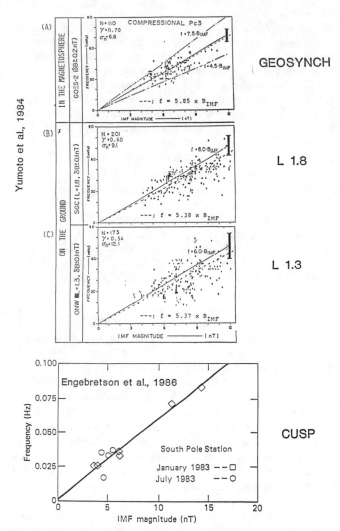

Fig. 7.   Examples of the highly replicable linkage $f \sim 6B$ between geomagnetic pulsation frequency and IMF magnitude. Upper three panels: scatter diagrams for geosynchronous and surface observations at low magnetic latitudes; bottom panel: measurements at a station connected to a polar lobe at higher latitude than the auroral zone.

events, as sketched in Figure 9. While this model is certainly idealized, it shows that time varying reconnections can lead to 3-dimensional structures that in turn apply time-varying pressure to the magnetopause. This time-varying pressure in turn can lead to magnetospheric oscillations that decay in amplitude away from the magnetopause. The few studies that have been done show that waves are generated from Pc 3 to Pc 5 frequencies, that compressional fluctuations penetrate the magnetosphere further than transverse fluctuations, and that low frequencies penetrate further than high frequencies [Wolfe et al., 1989].

Since the properties of the magnetosheath plasma which abut the magnetopause are expected to control FTE generation, and since the magnetosheath plasma in turn is controlled by the shock and solar wind, we expect that FTE properties are controlled by the solar wind, but we have little information on this other than that the FTE occurrence rate is controlled by the

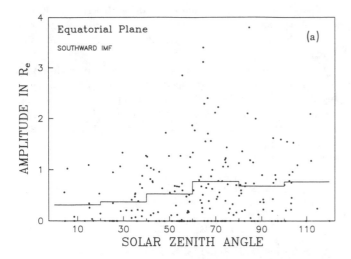

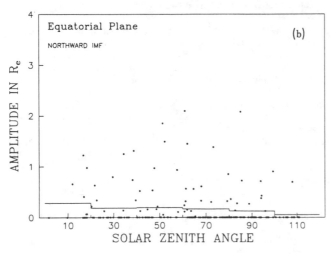

Fig. 8. Amplitude of magnetopause oscillation vs. solar zenith angle near the equatorial plane. The data are within ±30° of the GSM equatorial plane. The solid line is the mean value. *a*) When the IMF is southward. *b*) When the IMF is northward.

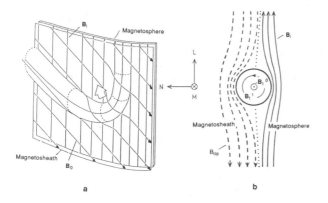

Fig. 9. The distortion of the magnetosheath and magnetospheric magnetic fields by a flux transfer event. The left hand panel shows the view from the magnetosheath. The right hand panel shows a cross-section of the magnetopause [after Russell and Elphic, 1978; Cowley, 1982].

north-south component of the IMF [Berchem and Russell, 1984]. One mode of disguising the FTE effect that we can rule out is that of triggering by fluctuations in the IMF direction. As Figure 10 shows, the IMF can be very steady when an FTE arises. As far as we can tell FTEs arise spontaneously at the magnetopause when the IMF is southward.

Finally, we speculate on an additional pressure effect at the magnetopause, so far unresearched. The solar wind pressure is reduced inside the compressional wave foreshock all the way to the bow shock. An example of lowered pressure in the foreshock is shown in Figure 11 [Le, 1991]. If this effect is transmitted through the shock and sheath to the magnetopause, it could initiate pulsations in the magnetosphere. Recall that the foreshock continually moves around in front of the curved shock in response to changes in orientation of the IMF brought by the solar wind. Transmission of the solar wind/foreshock pressure boundary to the magnetopause should therefore occur according the patterns of the foreshock's footprint on the shock, as illustrated in the examples of Figures 5 and 6. The magnetopause would then be subject to a pressure differential continually mov-

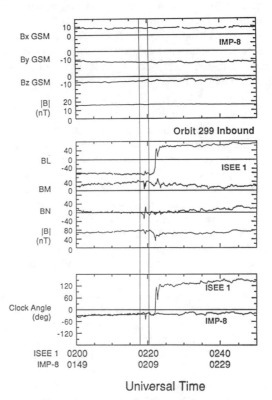

Fig. 10. Interplanetary conditions during FTE occurrence showing that FTEs are not triggered by fluctuations in the IMF. Top panel shows the IMF measured at IMP 8. The middle panel shows the magnetic field measured by ISEE 1 as it moved from the magnetosheath to the magnetosphere. The bottom panel shows the clock angle of the IMF about the solar direction. The field in the solar wind and at ISEE have the same clock angle except at the FTE and after the crossing of the magnetopause. The steadiness of the IMF B vector around the time of the FTE indicates that there was no FTE trigger in the solar wind, but that the FTE arose spontaneously at the magnetopause during southward IMF.

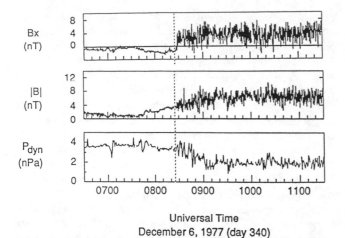

Universal Time
December 6, 1977 (day 340)

Fig. 11.   Three panels from Figure 3.6 of Le [1991] showing a drop in solar wind pressure following entry of the foreshock (dotted line) that accompanied a sudden shift in direction of the IMF, synopsized by B_X in the top panel

ing, sometimes gradually, sometimes suddenly, across it, possibly generating oscillatory responses inside the magnetosphere.

Pressure effects, with the possible exception of FTE activity, like the wave effects summarized earlier, should all be influenced by the transient, global configuration of the Earth's magnetoplasma environment.

DISCUSSION

The central thesis of this report is that the nonuniformity of the shock and foreshock, including the section of the latter outside the locally $Q_\perp$ geometry, is manifested in the magnetosheath and ultimately in the magnetosphere in ways that can be understood only by combining the shock's nonuniformities with the stream flow in the sheath and the nonuniform sensitivities of the magnetopause, taking into account the time variability of all these elements. With this view in mind, steps can be easily enumerated that might advance the investigation of daytime pulsations using the large, compatible satellite and surface data bases currently available.

Inside the magnetosphere, wave statistics need to be divided into subsets according to common combinations of diurnal, seasonal, locational (i.e. by geomagnetic latitude and L-value), and concurrent solar wind parameters. Progressions of event onsets or correlation coefficients might also be instructive organizers of pulsation data.

Outside the magnetosphere, solar wind observations need to be divided into subsets according to common combinations of IMF cone angle and solar wind plasma parameters. The distribution of foreshock and bow shock generated waves and pressure differentials in the magnetosheath, especially in the part of the subsolar flow tube near the magnetopause, must be computed for these subsets and corroborated where possible by observations in the sheath. Promising pairs of satellite observations in two regions at a time, but without benefit of computed distributions, have been described by Engebretson et al. [1991] and Lin

et al. [1991b]; also earlier references therein]. Recently, Lin et al. [1991a], referring to transfer across the magnetopause, concluded that "from five events...it seems that the transfer rate differs from one event to another."

Finally, the two classes of subsets, IMF direction and plasma flow, must be compared systematically. It seems promising at this time that by taking into account as many of the seeming complications of shock, sheath, and magnetospheric nonuniformity as possible, the task of explaining the global roots and routes of stimulation of geomagnetic pulsations may be appreciably simplified and hastened.

Acknowledgement. Preparation of this review was supported at TRW by NASA contracts NASW-4414 and -4749, and at UCLA by the National Science Foundation under grant ATM 91-11913.

REFERENCES

Berchem, J., and C. T. Russell, Flux transfer events on the magnetopause: Spatial distribution and controlling factors, *J. Geophys. Res.*, *89*, 6689, 1984.

Bol'shakova, O. V., and V. A. Troitskaya, Relation of the interplanetary magnetic field direction to the system of stable oscillations, *Dokl. Akad. Nauk SSSR*, *180*, 4, 1968.

Cowley, S. W. H., The causes of convection in the Earth's magnetosphere: A review of developments during the IMS, *Rev. Geophys. Space Phys.*, *20*, 531, 1982.

Diodato, L., E. W. Greenstadt, G. Moreno, and V. Formisano, A statistical study of the upstream wave boundary outside the earth's bow shock, *J. Geophys. Res.*, *81*, 199, 1976.

Engebretson, M. J., C-I. Meng, R. L. Arnoldy, and L. J. Cahill, Jr., Pc 3 pulsations observed near the South polar cusp, *J. Geophys. Res.*, *91*, 8909, 1986.

Engebretson, M. J., L. J. Cahill, Jr., R. L. Arnoldy, B. J. Anderson, T. J. Rosenberg, D. L. Carpenter, U. S. Inan, and R. H. Eather, The role of the ionosphere in coupling upstream ULF wave power into the dayside magnetosphere, *J. Geophys. Res.*, *96*, 1527, 1991a.

Engebretson, M. J., N. Lin, W. Baumjohann, H. Luehr, B. J. Anderson, L. J. Zanetti, T. A. Potemra, R L. McPherron, and M. G. Kivelson, A comparison of ULF fluctuations in the solar wind, magnetosheathg, and dayside magnetosphere 1. magnetosheath morphology, *J. Geophys. Res.*, *96*, 3441, 1991b.

Fairfield, D. H., and N. F. Ness, Magneticfield fluctuations in the Earth's magnetosheath, *J. Geophys. Res.*, *75*, 6050, 1970.

Greenstadt, E. W., Field-determined oscillations in the magnetosheath as possible source of medium-period, daytime micropulsations, in *Proceedings of Conference on Solar Terrestrial Relations, April 1972*, 515, University of Calgary, 1973.

Greenstadt, E. W., Quasi-perpendicular/quasi-parallel divisions of Earth's bow shock, *J. Geophys. Res.*, *96*, 1697, 1991.

Greenstadt, E. W., and J. V. Olson, A contribution to ULF activity in the Pc 3-4 range correlated with IMF radial orientation, *J Geophys. Res.*, *82*, 4991, 1977.

Greenstadt, E. W., H. J. Singer, C. T. Russell, and J. V. Olson, IMF orientation, solar wind velocity, and Pc 3-4 signals: A joint distribution, *J. Geophys. Res.*, *84*, 527, 1979.

Greenstadt, E.W., I.M. Green, D.S. Colburn, J. H. Binsack, and E.F. Lyon, Dual satellite observations of Earth's bow shock: II. Field aligned upstream waves, *Cosmic Electrodyn.*, *1*, 279, 1970a.

Greenstadt, E.W., I.M. Green, D.S. Colburn, J. H. Binsack, and E.F. Lyon, Dual satellite observations of Earth's bow shock: III. Field determined shock structure, *Cosmic Electrodyn.*, 1, 316, 1970b.

Greenstadt, E. W., M. M. Mellott, R. L. McPherron, C. T. Russell, H. J. Singer, and D. J. Knecht, Transfer of pulsation-related wave activity across the magnetopause: Observations of corresponding spectra by

ISEE-1 and ISEE-2, *Geophys. Res. Lett.*, *10*, 659, 1983.

Gringauz, K. I., V. A. Troitskaya, E. K. Solomatina, and R. V. Shchepetnov, The relationship of solar wind variables to periods of continuous micropulsations of electromagnetic field of the earth, *Dokl. Akad. Nauk USSR*, *5*, 1061, 1970.

Gul'elmi, A. V., T. A. Plyasova-Bakunina, and R. V. Shchepetnov, Relation between the period of geomagnetic pulsations Pc 3,4 and the parameters of the interplanetary medium at the earth's orbit, *Geomagn. Aeron.*, *13*, 331, 1973.

Kivelson, M. G., and W. J. Hughes, On the threshold for triggering substorms, *Planet. Space Sci.*, *38*, 211, 1990.

Kovner, M. S., V. V. Lebedev, Y. A. Plyasova-Bakunina, and V. A. Troitskaya, On the generation of low frequency waves in the solar wind in the front of the bow shock, *Planet. Space Sci.*, *24*, 261, 1976.

Lanzerotti, L. J., H. P. Lie, N. A. Tartaglia, Ionospheric Effects on the transmission of ultralow-frequency plasma waves, *Science*, *178*, 499, 1972.

Le, G., Generation of upstream ULF waves in the Earth's foreshock, *Dissertation*, Inst. of Geophys. and Planet. Phys., University of California, Los Angeles, 1991.

Lin, N., M. J. Engebretson, W. Baumjohann, and H. Luehr, Propagation of perturbation energy fluxes in the subsolar magnetosheath: AMPTE IRM observations, *Geophys. Res. Lett.*, *18*, 1667, 1991a.

Lin, N., M. J. Engebretson, R. L. McPherron, M. G. Kivelson, W. Baumjohann, H. Luehr, T. A. Potemra, B. J. Anderson, and L. J. Zanetti, A comparison of ULF fluctuations in the solar wind, magnetosheath, and dayside magnetosphere 2. field and plasma conditions in the magnetosheath, *J. Geophys. Res.*, *96*, 3455, 1991b.

Plyasova-Bakunina, T. A., Effect of the interplanetary magnetic field on the characteristics of Pc 3-4 pulsations, *Geomag. and Aeron.*, *12*, 675, 1972.

Russell, C. T., and R. C. Elphic, Initial ISEE magnetometer results: Magnetopause observations, *Space Sci. Rev.*, *22*, 681, 1978.

Russell, C. T., J. G. Luhmann, T. J. Odera, and W. F. Stuart, The rate of occurrence of dayside PC 3,4 pulsations: The L-value dependence of the IMF cone angle effect, *Geophys. Res. Lett.*, *10*, 663, 1983.

Saito, T., A new index of geomagnetic pulsations and its relation to solar M-region, Part 1, *Rep. Ionos. Space Res.*, *Japan*, *18*, 260, 1964.

Saito, T., and S. Matsushita, Geomagnetic pulsations associated with sudden commencements and sudden impulses, *Planet. Space Sci.*, *15*, 573, 1967.

Saito, T., K. Yumoto, K. Takahashi, T. Tamura, and T. Sakurai, Solar wind control of Pc 3, Magnetospheric Study 1979, in *Proceedings of International Workshop on Selected Topics of Magnetospheric Physics*, p155, Japanese IMS Committee, Tokyo, 1979.

Song, P., R. C. Elphic, and C. T. Russell, ISEE 1 and 2 observations of the oscillating magnetopause, *Geophys. Res. Lett.*, *15*, 744, 1988.

Spreiter, J. R., and A. Y. Alksne, Plasma flow around the magnetosphere, *Rev. Geophys.*, *1*, 11, 1969.

Takahashi, K., R. L. McPherron, E. W. Greenstadt, and C. A. Neeley, Factors controlling the occurrence of Pc 3 magnetic pulsations at synchronous orbit, *J. Geophys. Res.*, *86*, *A7*, 5472, 1981.

Tomomura, K., T. Sakurai, and Y. Kato, Satellite observation of magnetic fluctuations in the magnetosheath and the magnetoaphere, *Proc. Faculty Eng.*, *9*, Tokai Univ., 1983.

Troitskaya, V. A. and O. V. Bol'shakova, The relation of the high-latitude maximum of Pc3 intensity to the dayside cusp, *Geomag. and Aeron.*, *7*, 633, 1984.

Troitskaya, V. A., Y. A. Plyasova-Bakunina, and A. V. Gul'yel'mi, Relationship between Pc 2-4 pulsations and the interplanetary magnetic field, *Dokl. Akadl. Nauk. SSSR*, *197*, 1312, 1971.

Vero, J. and L. Hollo, Connections between interplanetary magnetic field and geomagnetic pulsations, *J. Atmos. Terr. Phys.*, *40*, 857, 1978.

Verzariu, P., Reflection and refraction of hydromagnetic waves at the magnetopause, *Planet. Space Sci.*, *21*, 2213, 1973.

Vinogradov, P. A., and V. A. Parkhomov, MHD waves in the solar wind--a possible source of geomagnetic Pc3 pulsations, *Geomagn. Aeron.*, *15* 109, 1974.

Walker, R. J., T. Ogino, and M. Ashour-Abdalla, Simulating the magnetosphere: The structure of the magnetotail, *Solar System Plasma Physics, AGU Monograph 54*, J. H. Waite, J. L. Burch, and R. L. Moore, eds., 61, 1989.

Wolfe, A., Dependence of mid-latitude hydromagnetic energy spectra on solar wind speed and interplanetary field direction, *J. Geophys. Res.*, *85*,

Wolfe, A., and R. L. Kaufmann, MHD wave transmission and production near the magnetopause, *J. Geophys. Res.*, *80*, 1764, 1975.

Wolfe, A., L. J. Lanzerotti, and C. G. Maclennan, Dependence of hydromagnetic energy spectra on solar wind velocity and interplanetary magnetic field direction, *J. Geophys. Res.*, *85*, 114, 1980.

Wolfe, A., C. Uberoi, C. T. Russell, L. J. Lanzerotti, C. G. Maclennan, and L. V. Medford, Penetration of hydromagnetic wave energy deep into the magnetosphere, *Planet. Space Sci.*, *39*, 1317, 1989.

Yumoto, I., and T. Saito, Relation of compressional HM waves at GOES 2 to low-latitude Pc 3 magnetic pulsations, *J. Geophys. Res.*, *88*, 10041, 1983.

Yumoto, K., T. Saito, B. T. Tsurutani, E. J. Smith, and S-I. Akasofu, Relationship between the IMF magnitude and Pc 3 magnetic pulsations in the magnetosphere, *J. Geophys. Res.*, *89*, 9731, 1984.

E. W. Greenstadt, TRW Inc., R-1/2144, One Space Park, Redondo Beach, CA 90278.

C. T. Russell, IGPP, University of California, Los Angeles, CA 90024-1567.

An Overview of Spacecraft Observations of 10 s to 600 s Period Magnetic Pulsations in the Earth's Magnetosphere

B. J. ANDERSON

The Johns Hopkins University, Applied Physics Laboratory, Johns Hopkins Road, Laurel, Maryland

Extensive studies of magnetic ULF pulsations in the magnetosphere have been made which inform our understanding of pulsation source mechanisms. Upstream waves associated with the quasi-parallel shock have been linked to both compressional Pc 3 and toroidal multi-harmonic field line resonances. Whether these different pulsations are manifestations of different steps in a single energy flow or reflect multiple pathways for transmission of upstream wave energy into the magnetosphere remains to be definitively established. Toroidal Pc 5 are continuously present in space at the dawn flank (dusk is not yet ruled out) and are due to excitation by a low frequency source near the flank(s) of the magnetospheric boundary. The Kelvin-Helmholtz instability, upstream dynamic pressure variations, and the mechanism generating traveling ionospheric vortices (if distinct from dynamic pressure changes), all probably contribute to toroidal Pc 5 excitation. Which mechanism predominates is not yet known. A link between compressional Pc 5 and drift waves is well established and these pulsations draw their energy from ions energized in the magnetotail. Bounce resonance of energetic ions has been shown to occur in poloidal Pc 4 suggesting that a gyrokinetic instability is the excitation mechanism for these pulsations. A significant fraction of pulsations observed in space are incoherent. Their intensity is correlated with geomagnetic activity but specific excitation mechanisms have not been identified.

INTRODUCTION

Magnetic pulsations with 10 s to 600 s period (the Pc 3, 4, and 5 frequency ranges) include a great variety of phenomena with distinct properties. In addition to differences in frequency, the pulsations are distinguished by polarization, harmonic structure, and particle signatures. They also have distinct regions of occurrence. The purpose of this paper is to review the results of spacecraft observations and highlight those results which inform consideration of pulsation energy sources, and specifically their relationship to solar wind sources. A number of excellent reviews of geomagnetic pulsations are available which provide extensive background of the subject [Saito, 1969; McPherron, 1972; Orr, 1973; Hughes, 1983; Odera, 1986; Verö, 1986; Takahashi, 1988].

In general, one can divide ULF pulsations according to whether they result from a local wave-particle instability or from coupling of wave energy propagating through the magnetosphere that is produced either in the solar wind/magnetosheath or at the magnetopause/boundary layer. Ultimately of course all pulsations derive their energy from the solar wind, but the knowledge that energetic ions in the outer magnetosphere ultimately derive their energy from the convection electric field imposed by the solar wind is of little help in identifying the relevant instability responsible for certain pulsations. In the spirit of utility therefore, a distinction will be made here between pulsations which result from wave-particle instabilities excited within the magnetosphere and those which result from coupling to upstream wave energy that propagates through the magnetosphere. In this discussion, upstream is considered to include the solar wind, the magnetosheath, and the magnetopause/boundary layer. By assembling the evidence about each pulsation type, one can assess which are driven by energetic particles and which are more directly related to upstream wave energy. Distinctions among those with an upstream wave energy source can also be made.

For the purposes of this discussion it is useful to organize the pulsations into six categories: compressional Pc 5; poloidal Pc 4; compressional Pc 3; toroidal field line resonance harmonics; toroidal Pc 5; and incoherent noise. In discussing polarization, it is customary to refer to radial, compressional, and azimuthal components, corresponding essentially to an r, ϕ, n coordinate system where $\hat{e}_n$ is parallel to the local field direction, $\hat{e}_\phi$ is eastward, and $\hat{e}_r$ is normal to the field in the meridional plane, $\hat{e}_r = \hat{e}_\phi \times \hat{e}_n$. Compressional Pc 5 pulsations typically have observed periods of about 5 minutes and are polarized in the meridional plane. Poloidal Pc 4 waves are typically observed to be predominantly radially polarized and appear to be the second harmonic poloidal field line resonance. Compressional Pc 3 pulsations are polarized predominantly along the

Solar Wind Sources of Magnetospheric Ultra-Low-Frequency Waves
Geophysical Monograph 81

field and in space are observed most readily close to the magnetic equator. Toroidal harmonic pulsations are azimuthally polarized and generally appear with frequencies corresponding to a harmonic series of field line resonances. Toroidal Pc 5 pulsations appear to be fundamental mode toroidal field line resonances and can occur with or without higher harmonic resonances suggesting that they are driven separately. Finally, incoherent pulsations appear to have power spectra characteristic of noise and appear not to have a strongly preferred polarization structure.

One section is devoted to each of the above pulsation types considering the following questions. What is their spatial occurrence distribution? Under what specific solar wind or magnetospheric conditions are they most likely, or least likely, to be present? What are they, that is, are they a resonance, a propagating wave, or a convecting structure in the plasma? And finally, are they the product of a local plasma instability and if so what instability? After considering each pulsation type, the implications of these results for solar wind sources and outstanding issues are discussed.

Compressional Pc 5

Compressional Pc 5 waves were first reported in space by Sonnerup et al. [1969]. They typically exhibit large, $\delta B/B \approx 0.2$ to 0.5, quasi-sinusoidal oscillations in field magnitude with periods of 5 to 10 min. The compressional perturbation is often accompanied by a comparable radial perturbation while the azimuthal fluctuation amplitude is typically a factor of five smaller. A typical wave-form illustrating the polarization structure is shown in Figure 1, together with the latitudinal and L/MLT occurrence distributions, adapted from Anderson et al. [1990].

The primary occurrence regions of compressional Pc 5 are on the nightside toward the flanks of the magnetosphere at L typically greater than 7 or 8. At geosynchronous orbit the waves occur primarily in the afternoon and early evening [Barfield and McPherron, 1972; Kokubun, 1985]. Observations from elliptically orbiting spacecraft have shown that compressional Pc 5 are the dominant pulsations occurring beyond $L = 8$. Using HEOS-1 data, Hedgecock [1976] first reported the occurrence of compressional Pc 5 at large L (> 10). Takahashi et al. [1990a] reported events clustered at both dawn and dusk. As shown in Figure 1, compressional Pc 5 are rarely observed much inside geosynchronous orbit but are very common beyond $L = 7$, being present nearly 50% of the time within about 5° of the magnetic equator on the nightside toward the flanks of the magnetosphere in both the evening and morning [Anderson et al., 1990]. Zhu and Kivelson [1991] used ISEE data to confirm the dominance of dawn and dusk flank occurrence regions and to show that the compressional pulsation power continues to increase out to the magnetopause.

A close link between ion injections and compressional Pc 5 at dusk is well established. At 6.6 R_E, dusk compressional Pc 5 onsets are correlated with the storm time development of the partial ring current [Barfield et al., 1972; Barfield and McPherron, 1978] and a diamagnetic depression in field magnitude [Higuchi and Kokubun, 1988; Woch et al., 1990]. The correlation of

compressional Pc 5 and substorm onset was demonstrated by Kokubun [1985] and that result is shown in Figure 2. The delay from substorm onset indicates that Pc 5 start in response to the arrival from the east of energetic ions injected during substorms.

Analysis of the particle signatures associated with compressional Pc 5 shows that ions play a significant role in their generation. The ion and magnetic field pressures have been clearly shown to be in anti-phase both for particular events and statistically [e.g. Sonnerup et al., 1969; Kremser et al., 1981; Zhu and Kivelson, 1991]. Figure 3 shows a time series of ion flux and magnetic field magnitude for an event analyzed by Kremser et al. [1981] as well as the histogram distribution of the phase difference of ion and magnetic field pressures observed by Zhu and Kivelson [1991]. It has also been shown that compressional Pc 5 propagate in the westward sense with speeds comparable to energetic (10-30 keV) proton drift speeds [Lin et al., 1988]. In addition, compressional Pc 5 have short azimuthal wavelengths, $m \approx 50 - 100$, where m is the azimuthal harmonic index as in $\exp(im\phi)$ and ϕ is the azimuth angle [Lin and Barfield, 1985; Takahashi et al., 1985]. Together with the propagation speed, the short azimuthal wavelength suggests that the observed pulsation frequency results from propagation of spatial structures past the observation point.

The early results, which have been confirmed by subsequent analyses, led to the conclusion that the waves are most probably drift mirror waves driven by positive plasma pressure anisotropy [Hasegawa, 1969]. Recent theoretical work including the effect of curvature in the magnetic field has been shown to reduce the mirror instability threshold giving better agreement between theory and observation [Woch et al., 1988, 1990; Hasegawa and Chen, 1989; Chen and Hasegawa, 1991].

It is increasingly clear that compressional Pc 5 waves are driven by the pressure anisotropy of the ions and that their occurrence near geosynchronous orbit reflects the results of substorm energization and injection of ion populations and hence, compressional Pc 5 are not directly indicative of upstream solar wind conditions. These pulsations therefore fall in the category of pulsations driven locally by wave particle interactions. None the less, compressional Pc 5 are important for purposes of characterizing pulsations driven directly by upstream wave energy because compressional Pc 5 may overwhelm the signatures of other pulsations.

Poloidal Pc 4

Radially polarized Pc 4 pulsations are distinguished by their transverse and predominantly radial perturbation [Arthur and McPherron, 1981] and are clearly distinguished from compressional Pc 5 by the absence of large compressional components. Generally these events have periods near 100s placing them in the Pc 4 range. Occasionally however individual events can have periods in the Pc 5 range [Takahashi et al., 1990b], so the period designation should not be taken too literally. Figure 4 shows a poloidal Pc 4 together with the L/MLT and latitudinal occurrence statistics as observed by AMPTE/CCE [adapted from Anderson et al., 1990].

Compressional Pc 5

(a)   Example wave-form plot:

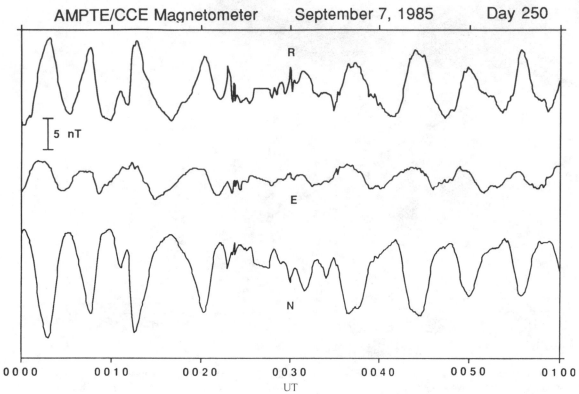

AMPTE/CCE Magnetometer   September 7, 1985   Day 250

5 nT

R

E

N

0000   0010   0020   0030   0040   0050   0100

UT

(b)   L/MLT Occurrence Distribution
Full Scale = 32%

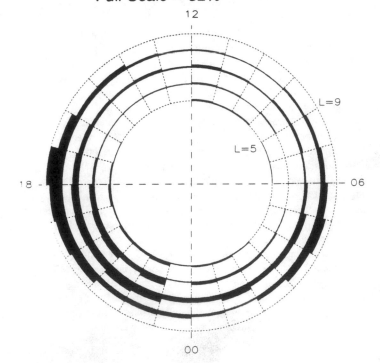

12

L=9

L=5

18   06

00

(c)   MLAT Distribution

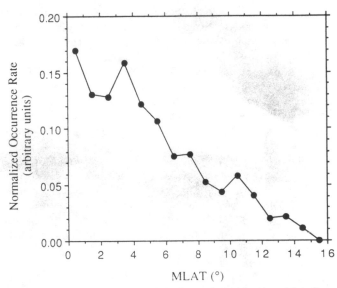

Fig. 1. Summary of compressional Pc 5 occurrence showing: (a) typical wave-form plot; (b) *L*/MLT distribution; (c) magnetic latitude distribution. The *L*/MLT distribution shows the percentage of occurrence of compressional Pc 5 in each spatial bin. Adapted from Anderson et al. [1990].

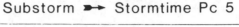

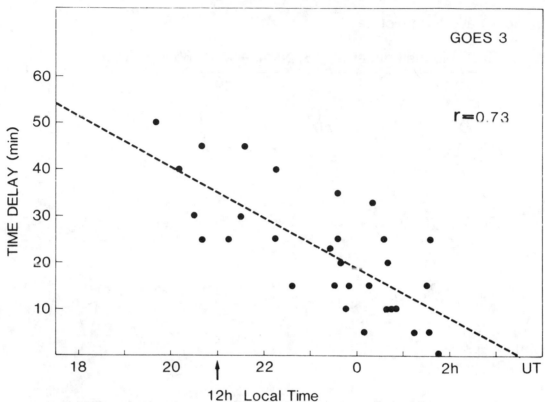

Fig. 2. Time delay between substorm onset and the beginning of compressional Pc 5 at the GOES-3 satellite as a function of UT. Local noon at GOES-3 is indicated by the arrow; the regression line yields zero time delay near local midnight. Adapted from Kokubun [1985].

At geosynchronous orbit these pulsations are found primarily in the afternoon [Arthur and McPherron, 1981; Takahashi and McPherron, 1984; cf. Figure 10 below]. The occurrence pattern deduced from AMTPE/CCE observations reveals that the events occur predominantly within $10°$ of the equator in a band 1 to 2 R_E in radial extent beginning near noon within $L = 6$ and progressing outward through dusk to $L \approx 8$ at midnight. The narrow radial character of poloidal Pc 4 was shown convincingly by Singer et al. [1982] using the pair of ISEE 1 and 2 spacecraft to show that the occurrence region for particular events was localized to regions 0.2 to 1.6 R_E in radial extent. By using two geosynchronous spacecraft together with AMPTE/CCE, Engebretson et al. [1992] have shown that despite the narrow radial width, the longitudinal extension of the occurrence region can be large, up to 8 hours in local time, implying that the occurrence region of individual events mimics the statistical occurrence distribution.

There is some evidence to suggest that poloidal Pc 4 occur preferentially during quiet geomagnetic conditions but following substantial auroral activity. Figure 5 shows the result of a superposed epoch analysis of poloidal Pc 4 with Dst, AE and IMF B_z [Anderson et al., 1991]. In this analysis, pulsation events were divided into two intensity categories: ≤ 3 and ≥ 4

denoting lower and higher amplitude fluctuations, respectively [see Anderson et al., 1991, for the details]. The average AE was found to drop sharply within about 1 hour before poloidal Pc 4 occurrence and was a minimum at the time of the events. In addition, poloidal Pc 4 tended to be preceded by increased auroral activity, especially for the wave events with the largest amplitudes, the ≥ 4 category. The results for average IMF B_z indicate that the events tend to occur for northward IMF and are most often preceded by southward IMF, consistent with the AE behavior. Engebretson et al. [1992] found a similar relationship between AE and their events. The correlation with prior substorm activity suggests that significant populations of freshly injected ions are required to drive the pulsations, but the fact that the events tend to occur during low geomagnetic activity suggests that low convection electric fields also play an important role. Engebretson et al. [1992] also proposed that plasmaspheric refilling may be important as indicated by the broad longitudinal but narrow radial extent of the events.

Poloidal Pc 4 have been found to occur at the frequency of the second harmonic field line resonance. By performing dynamic spectral analysis and comparing the frequency of poloidal Pc 4 with the spectral structure of toroidal multi-harmonic pulsations, Takahashi and McPherron [1984] were able to asso-

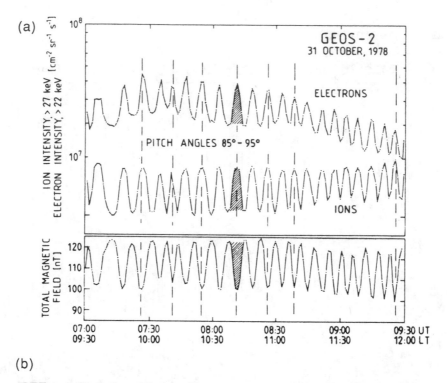

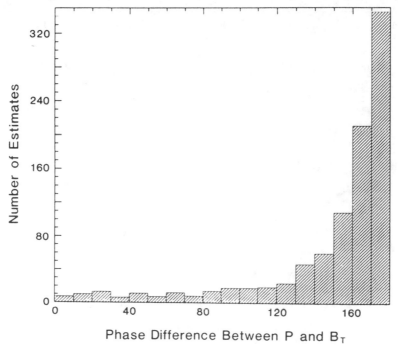

Fig. 3. Anti-phase relationship between ion pressure and magnetic field strength in compressional Pc 5. a) Ion and electron flux plotted with magnetic field magnitude during a compressional Pc 5 event observed on 31 October, 1978, by GEOS-2. Adapted from Kremser et al. [1981]. b) Histogram distribution of the phase difference between variations of the magnetic field and plasma pressures observed by the ISEE 1-2 spacecraft. Adapted from Zhu and Kivelson [1991].

Poloidal Pc 4

(a) Example wave-form plot:

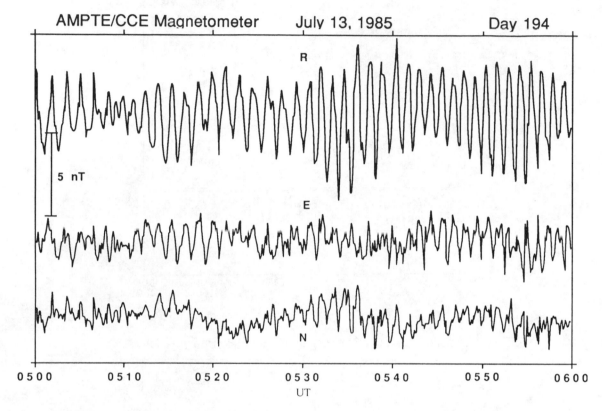

(b) L/MLT Occurrence Distribution
Full Scale = 9%

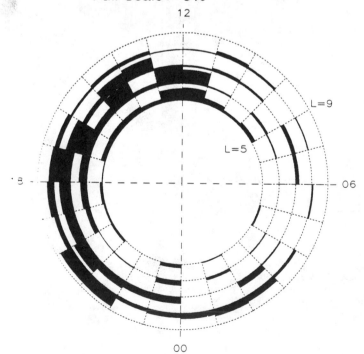

(c) MLAT Distribution

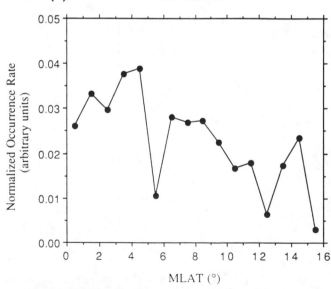

Fig. 4. Summary of poloidal Pc 4 occurrence showing: (a) typical wave-form plot; (b) L/MLT distribution; (c) magnetic latitude distribution. The L/MLT distribution shows the percentage of occurrence of poloidal Pc 4 in each spatial bin. Adapted from Anderson et al. [1990].

Poloidal Pc 4
Superposed Epoch Analysis of average Dst, AE, IMF Bz

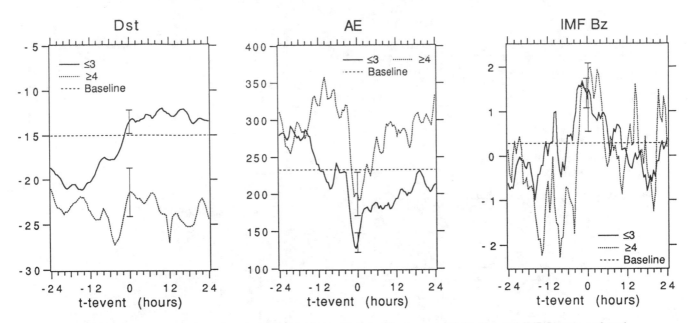

Fig. 5. Results of superposed epoch analysis for poloidal Pc 4 events. The averages of AE, Dst, and IMF B_z were evaluated relative to event occurrence with negative (positive) times corresponding to before (after) wave events. Adapted from Anderson et al. [1991].

ciate the poloidal pulsation with the second harmonic field line resonance frequency (cf. Figure 9 below). Other investigators also found poloidal Pc 4 to be second harmonic resonances [Singer et al., 1982; Hughes and Grard, 1984].

The interaction of poloidal Pc 4 with ions has been studied in considerable detail by Takahashi et al. [1990b] and Figure 6 is adapted from their work. The oscillation of ion fluxes at the same frequency as the wave is clear, as is the energy dependent change in phase between the ion flux oscillations and the magnetic field. The fluxes from the north and south look directions have opposite phase indicating a bounce resonant motion. The fact that the flux modulation phase matches the field oscillations at $\approx$ 150 keV indicates that protons with this energy are in bounce resonance with the waves. Westward propagation is indicated by the phase lag between fluxes in the eastside and westside particles. Takahashi et al. also showed that the wave had $m \approx 100$, consistent with a wave-particle energy source since this azimuthal wavelength satisfies the bounce resonance condition for 150 keV protons.

That poloidal Pc 4 are radially polarized and occur at the second harmonic field line resonance frequency indicate that poloidal Pc 4 are likely to be guided poloidal mode waves which are expected to occur with high azimuthal wave numbers [e.g. Radoski, 1967] and are thought to be driven by bounce resonance with energetic ions [Southwood et al., 1969; Southwood, 1976; Chen and Hasegawa, 1988, 1991]. Although it is clear that upstream conditions ultimately control the factors responsible for poloidal Pc 4 occurrence, they appear to

derive their free energy from the local ion population and are therefore not driven directly by the upstream fluctuations. Since their occurrence rates are low relative to other types of pulsations, they are of secondary concern here.

COMPRESSIONAL PC 3

Compressional Pc 3 are relatively low amplitude (a few nT) fluctuations of field amplitude and have typical periods of 20–30 s. Careful examination of the correspondence between compressional Pc 3 at geosynchronous orbit and low latitude ($L \leq 2$) Pc 3 has been made [Yumoto and Saito, 1983] and Figure 7 shows a simultaneous compressional Pc 3 event observed in space by GOES 3 and at $L = 1.8$ on the ground. Figure 7 also presents the local time distribution of simultaneous ground Pc 3 and space compressional Pc 3 showing a strong dayside occurrence pattern and predominance of similar periods. A comparison of ISEE data with higher latitude stations ($L = 4$ to 6) shows that a close correspondence between compressional Pc 3 in space and on the ground applies to higher latitude ground stations as well [Odera et al., 1991]. Takahashi and Anderson [1992] have found that compressional Pc 3 within geosynchronous orbit occur throughout the dayside close to the magnetic equator but are strongest in an island near the average plasmapause in the late morning.

Low latitude Pc 3 observed on the ground have been known to exhibit a preference to occur for low IMF cone angles, $\cos^{-1}(B_x/B)$, [cf. Verö, 1986; Odera, 1986]. Figure 8, adapted from Yumoto et al. [1984] shows that the period of space com-

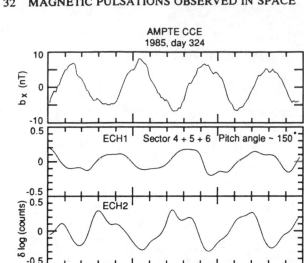

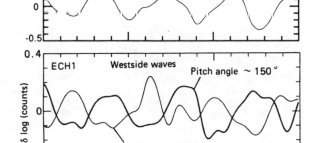

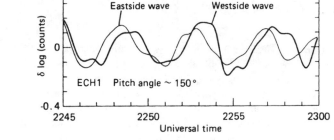

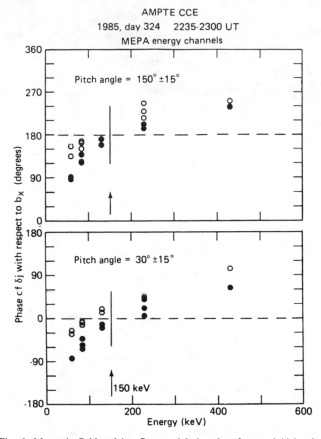

Fig. 6. Magnetic field and ion flux modulation data for a poloidal pulsation event observed on 20 November, 1985 by the AMPTE/CCE satellite. Ion data are shown as deviations from the spin average in log counts. The top four traces show the energy dependent phase relationship between radial field component: 50-68 keV (ECH1); 68-100 keV (ECH2); 100-160 keV (ECH3). The fifth line trace shows the antiphase relationship between ions with northward and southward velocities parallel to the background field. The bottom trace shows the phase difference between ions with gyrocenters east (eastside) and west (westside) of the spacecraft showing the westward propagation of the wave-form. The right hand panels show the phase difference between the ion flux modulation and radial field component as a function of energy (east and west side ions are indicated by open and filled circles, respectively). The energy of strict in-phase (northward particles) or anti-phase (southward particles) indicates bounce resonance at about 150 keV. Adapted from Takahashi et al. [1990b].

pressional Pc 3 and ground Pc 3 depend on the IMF magnitude in the same way. In as much as low latitude Pc 3 and compressional Pc 3 in space are closely correlated, the dependence of low latitude ground Pc 3 on IMF cone angle applies to the compressional Pc 3 as well.

These results imply a close link between the occurrence and the oscillation frequency of compressional Pc 3 and upstream IMF. The correlation with low IMF cone angles suggests that the source population of waves is associated with the quasi-parallel region of the shock since the magnetopause is exposed to the quasi-parallel region of the shock only for sufficiently radial IMF [e.g. Greenstadt and Olson, 1977]. The correlation of compressional Pc 3 frequency with IMF magnitude supports this inference since the numerical correlation is very close to

what one expects for waves generated by cyclotron resonance of protons back-streaming from the quasi-parallel shock into the upstream solar wind.

TOROIDAL HARMONICS

On the dayside magnetosphere, field line resonance periods range from 20 s to 500 s depending on the harmonic of the resonance, the mass density on the field line, and the L shell [Cummings et al., 1969]. Toroidal multi-harmonic pulsations occur commonly on the dayside with up to 6 harmonics simultaneously present over a range of L shells [e.g. Takahashi and McPherron, 1984; Engebretson et al., 1986]. Because the most

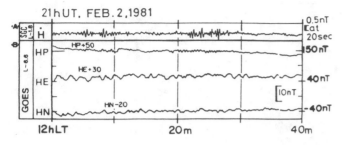

21hUT. FEB.2,1981

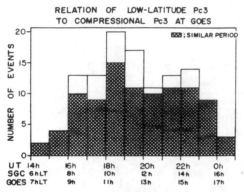

RELATION OF LOW-LATITUDE Pc3
TO COMPRESSIONAL Pc3 AT GOES

Fig. 7. Correlation between compressional Pc 3 at GOES-3 and a low latitude ground station, San Gabriel Canyon. Top panel shows traces from ground and satellite records showing a compressional Pc 3 at 1205-1210 and 1220-1230 UT simultaneous with enhancements in Pc 3 activity in the ground record. Lower panel shows the local time occurrence distribution of simultaneous ground and space events. Shaded area indicates the fraction of events which had overlapping period ranges in space and on the ground. Adapted from Yumoto and Saito [1983].

prominent toroidal harmonics tend to occur in the 20 s to 40 s period range they are called toroidal Pc 3. Figure 9 shows a dynamic spectrogram adapted from Takahashi and McPherron [1984] showing the simultaneous presence of toroidal harmonics throughout the day. At geostationary orbit these toroidal field line resonances are the most commonly observed coherent pulsations in the data records and dominate the dayside [Arthur et al., 1977; Takahashi and McPherron, 1984]. Figure 10 shows the local time distribution of toroidal resonances at geostationary orbit obtained by Takahashi and McPherron. The occurrence distribution out to $L = 9$ for toroidal harmonics has also been determined [Anderson et al., 1990; Takahashi and Anderson, 1992] and the same pattern holds out to $L = 9$, with typical occurrence rates of 50% in the mid-morning.

A correlation of toroidal multi-harmonic pulsations with the solar wind indicates that these pulsations also have a close connection with the upstream waves associated with the quasi-parallel region of the shock. Support for an upstream mechanism was found in the modest correlation between toroidal Pc 3 power and IMF cone angle [Arthur and McPherron, 1977; and Takahashi et al., 1984]. Engebretson et al. [1987, 1989, 1991a] report good correlation between radial IMF and multi-harmonic toroidal dayside resonances. Takahashi et al. [1984] and Engebretson et al. [1991a] also find that the center frequency of the

envelope of Pc 3 power correlates with the magnitude of the IMF. The prompt association of toroidal multi-harmonic and radial IMF was also demonstrated in superposed epoch analysis of IMF and harmonic Pc 3 events [Anderson et al., 1991]. Figure 11 shows the result obtained by these authors indicating a pronounced minimum in cone angle at the time of multi-harmonic events. The minimum is deeper for the more intense events, the ≥3 category.

Both toroidal multi-harmonics and compressional Pc 3 are associated with a quasi-parallel shock geometry. The two types of Pc 3 may represent different steps in the same energy flow, or they may reflect different flow pathways. This issue is discussed below.

TOROIDAL PC 5

It is well established that toroidal Pc 5 pulsations are fundamental mode toroidal field line resonances and consideration of the fundamental mode separately from the multi-harmonics is based on the difference in occurrence distribution [cf. Figure 12, and Anderson et al., 1990]. Statistical studies of ground based data include magnetometer analyses of Gupta [1976 and references therein] and Gupta and Niblett [1979]. Radar studies have also been made [Walker and Greenwald, 1979, 1981] and have shown that the local time distribution of Pc 5 pulsations is peaked at dawn and dusk with comparable occurrence. This is generally reproduced in the magnetometer studies, although in some reports a significant preference of dawn over dusk has been found [e.g. Gupta, 1976]. The relative parity of dawn and dusk in the radar observations was taken as evidence that the pulsations are driven by energy sources at the flanks of the magnetopause, and the Kelvin-Helmholtz instability was invoked as a likely candidate [Kivelson and Pu, 1984; Yumoto, 1984].

Observations in space have been somewhat inconclusive with regard to toroidal Pc 5. Studies of magnetic field data have generally found a strong dawn maximum in event occurrence probability. This was clearly found at geosynchronous orbit [Takahashi and McPherron, 1984; Kokubun, 1985] and is apparent in Figure 10. By contrast, the statistical study of electric field data by Junginger et al. [1984] showed a broad dayside distribution of toroidal (poloidal electric field) Pc 5 power. It is not clear how the magnetic field and electric field observations can be reconciled. Anderson et al. [1990] compiled occurrence statistics of toroidal fundamental mode resonances from $L = 5 - 9$ and their results are shown in Figure 12. The regularity of the azimuthal variations are typical of these waves. The nodal structure of the pulsations is clearly demonstrated by the latitude distribution. The L/MLT distribution shows both a strong preference for wave occurrence at dawn and an occurrence rate increasing monotonically with L. Superposed epoch analysis of toroidal Pc 5 occurrence with IMF and geomagnetic indices showed no discernible temporal relationship and only shows that stronger events occurred during higher average solar wind velocity [Anderson et al., 1991].

It is important that wave activity of other types is strong at dusk and weak at dawn [Anderson et al., 1990; Takahashi and

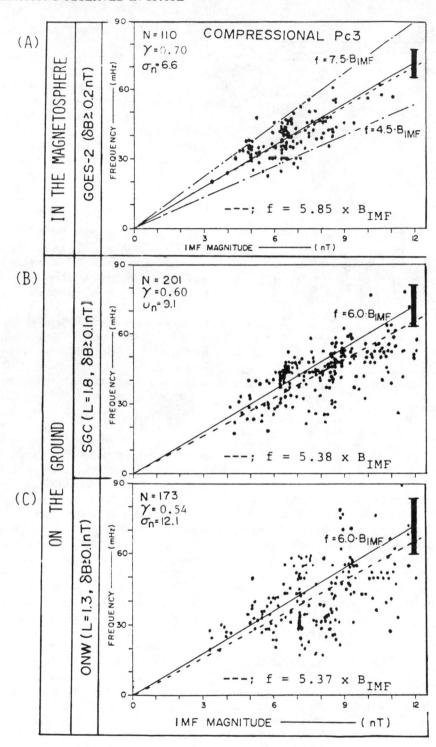

Fig. 8. Correlation of compressional Pc 3 frequency and low latitude Pc 3 frequency with IMF magnitude. The same relationship between frequency and B_{IMF} holds for pulsations observed in both regions. Adapted from Yumoto et al. [1984].

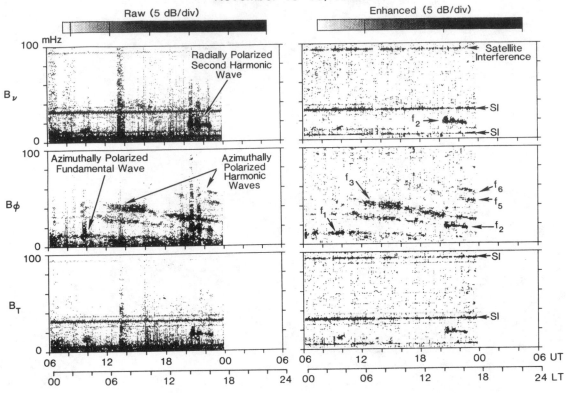

Dynamic Power Spectra
UCLA Magnetometer ATS 6
November 18–19, 1974

Fig. 9. Dynamic spectrogram of ATS-6 magnetometer data showing the prevalence of toroidal harmonics and a poloidal Pc 4 event. The correspondence between the poloidal Pc 4 frequency and the second harmonic field line resonance frequency is clear. Permission from Pergamon press to adapt this figure from Takahashi and McPherron [1984] is gratefully acknowledged.

Anderson, 1992]. It is possible that the toroidal wave signatures are present at dusk but are more difficult to separate from the other waves also occurring at dusk in magnetic field data. Ground radar and magnetometer measurements conversely are most sensitive to toroidal Pc 5. The issue is further complicated by the fact that some ground magnetometer [e.g. Gupta 1976] and radar [Walker and Greenwald, 1981] local time distributions of Pc 5 occurrence do not agree. Hence, the dawn/dusk asymmetry observed in space may be real, but it is not difficult to imagine that observational limitations may conspire to predispose the magnetic field studies in space to observe the toroidal Pc 5 primarily in the relatively quiet region near dawn.

INCOHERENT PULSATIONS

Fluctuations that appear as noise in spectral analysis have received relatively little attention, possibly because they are difficult to understand. Perturbations that exhibit no predominant oscillation frequency were found to be a prominent class of pulsations however, accounting for 25% of all pulsations in the AMPTE/CCE magnetometer data occurring in the $L = 5 - 9$ region [Anderson et al., 1990]. Figure 13, adapted from Ander-

son et al. shows a representative noise pulsation event. Dynamic spectral display reveals the broadband character of the field perturbations indicating that the waves are probably not merely superpositions of the coherent waves discussed above. To measure the correlation of incoherent wave power with geomagnetic activity, averages of Dst, B_z, and AE have been made using the data base of Anderson et al. [1990] for events in five power levels of incoherent waves and the results are shown in Figure 14. Incoherent waves were sub-divided into 5 intensity categories designated as NSE0, NSE1, NSE2, NSE3 and NSE45, where the numbers refer to an index of spectral power which increases logarithmically with wave power. The NSE0 events correspond to the absence of discernible wave power. The events were further subdivided into those occurring on the dayside or on the nightside. Each measure of geomagnetic activity indicates that increased noise wave power correlates with increased geomagnetic activity for both daytime and nighttime events.

SUMMARY AND CONCLUSIONS

Of the above six pulsation types, three have reasonably well

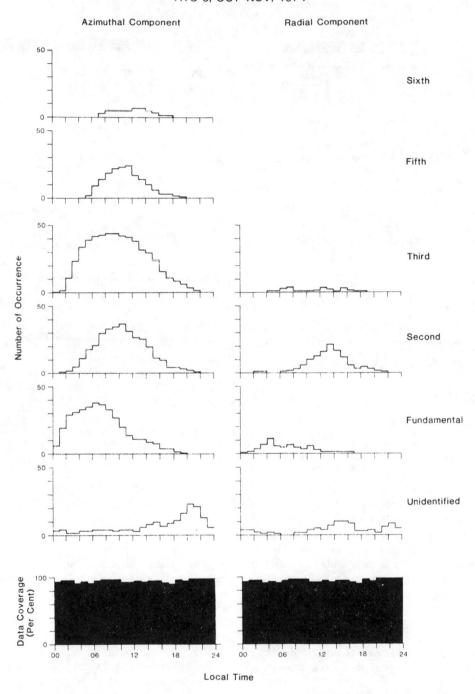

HARMONIC MODE OF PULSATIONS
ATS 6, OCT–NOV, 1974

Fig. 10. Local time distributions of multi-harmonic toroidal and poloidal events observed by ATS-6. The broad dayside occurrence of toroidal harmonics ($n \geq 2$) together with the strong prevalence of toroidal fundamentals in the morning are clear. The afternoon population of poloidal second harmonic events is also clear. Permission from Pergamon press to adapt this figure from Takahashi and McPherron [1984] is gratefully acknowledged.

Toroidal Harmonics
Superposed Epoch Analysis of average IMF Cone Angle

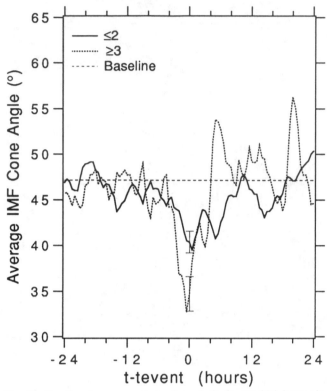

Fig. 11. Results of superposed epoch analysis for toroidal harmonic events. The average IMF cone angle was evaluated relative to event occurrence with negative (positive) times corresponding to before (after) wave events. Adapted from Anderson et al. [1991].

established direct links to energy sources outside or at the magnetopause. Compressional Pc 3 and toroidal multi-harmonics are both associated with quasi-parallel shock geometries. The fact that toroidal Pc 5 occurrence increases monotonically with L and maximizes away from noon indicates that the energy source for toroidal Pc 5 is most likely at or near the magnetospheric boundary and appears to be strongest away from the subsolar region. Compressional Pc 5 and toroidal Pc 4 have been reasonably well identified with local instabilities of energetic protons and are therefore not directly relevant to discussions of solar wind sources of ULF pulsations. Incoherent pulsations are not sufficiently well studied to say whether some of these ULF signals may derive their energy from fluctuations in the upstream magnetosheath or near the magnetopause.

Compressional Pc 3 and Toroidal Multi-harmonic Resonances

The unresolved questions about these pulsations concern how the upstream energy propagates and becomes manifested in the pulsations. Compressional Pc 3 are most likely to be the direct result of compressional energy incident on the magnetopause which launches compressional fast mode waves propagating through the magnetospheric cavity [e.g. Yumoto, 1985]. Since both compressional and toroidal Pc 3 frequencies are correlated with IMF magnitude it seems reasonable to attribute the toroidal resonances to coupling between compressional Pc 3 and field line resonances as proposed theoretically [Chen and Hasegawa, 1974; Southwood, 1974; Hasegawa et al., 1983]. Models of global resonances exhibit coupling between compressional and toroidal modes indicating that the continuum of resonances are readily excited by broadband compressional power [Allan et al., 1987a, b; Lee and Lysak, 1991].

There is some evidence to suggest that Pc 3 frequency power can gain access to the ionosphere in the cusp region [Engebretson et al., 1991a; Olson and Fraser, this volume]. Engebretson et al. [1989] report Pc 3 power and Pc 3 frequency oscillations in auroral luminosity at cusp latitudes coincident with radial IMF and toroidal harmonics in space. Subsequently, Engebretson et al. [1991a] proposed an ionospheric link whereby fluctuating field aligned currents associated with sheath turbulence typical of quasi-parallel shock geometry, drive multi-harmonic pulsations through ionospheric coupling of region 1 and region 2 current systems. In this scheme, the disturbed sheath conditions associated with quasi-parallel shocks are invoked to account for the fluctuating Birkeland currents via fluctuating precipitation and hence ionospheric conductivity.

At this point, it does not appear that the high-latitude entry mechanism can be ruled out as a candidate for driving toroidal multi-harmonic pulsations in the outer magnetosphere. That compressional Pc 3 power is present in the dayside magnetosphere is well documented, as is the coupling of this power to Alfvén resonances. The generation of low latitude Alfvén resonances is sufficient to show the efficacy of this mechanism. On the other hand, Pc 3 modulated cusp/cleft precipitation is also well documented corresponding to modulations in the high latitude Birkeland currents, which couple to latitudes equatorward of the cusp/cleft via Pedersen currents, and hence map to field lines in the dayside outer magnetosphere. Hence, it may be that field lines in the outer magnetosphere should be subject to a Pc 3 frequency perturbation unrelated to the compressional Pc 3. At issue is not whether one or the other mechanism operates, but which predominates in the generation of toroidal multi-harmonic pulsations in the outer magnetosphere.

Untangling this issue may not prove easy. It is unlikely that the properties of the toroidal field line resonances themselves will provide useful guidance since they reflect the properties of the resonance. Further correlations with upstream parameters whether cone angle, IMF magnitude, or other parameters, are also unlikely to yield significant new insight. It is already known that the quasi-parallel shock region is effective in generating Pc 3 frequency power which is manifested throughout the magnetosheath and is implicated in both energy pathways [Engebretson et al., 1987; 1991b; Lin et al., 1991]. A more profitable approach may be to apply multi-point observations of pulsations, using ground stations and spacecraft. Interpretation of these analyses will need to be closely tied to realistic modeling of ULF wave propagation in the magnetosphere. It may

Toroidal Pc 5

(a)   Example wave-form plot:

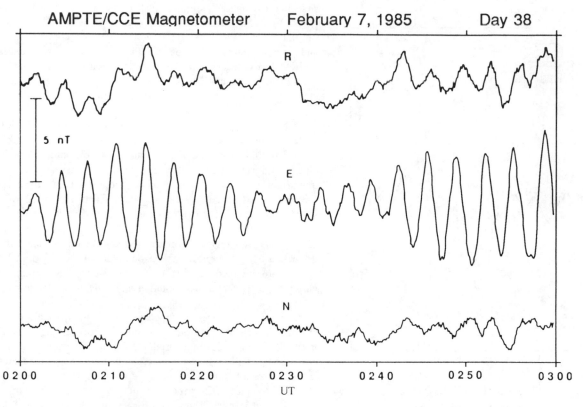

AMPTE/CCE Magnetometer   February 7, 1985   Day 38

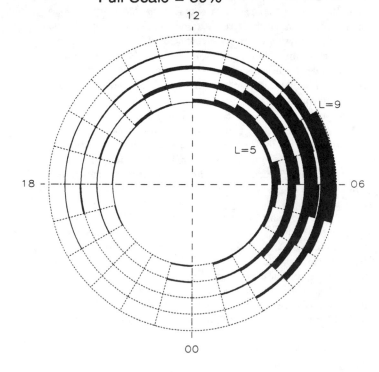

(b) L/MLT Occurrence Distribution
Full Scale = 39%

(c)   MLAT Distribution

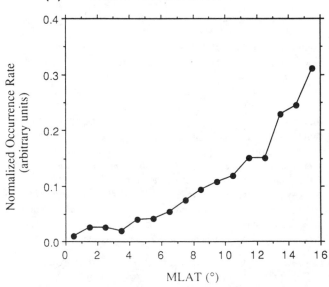

Fig. 12. Summary of toroidal fundamental occurrence showing: (a) typical wave-form plot; (b) *L*/MLT distribution; (c) magnetic latitude distribution. The *L*/MLT distribution shows the percentage of occurrence of toroidal fundamental in each spatial bin. Adapted from Anderson et al. [1990].

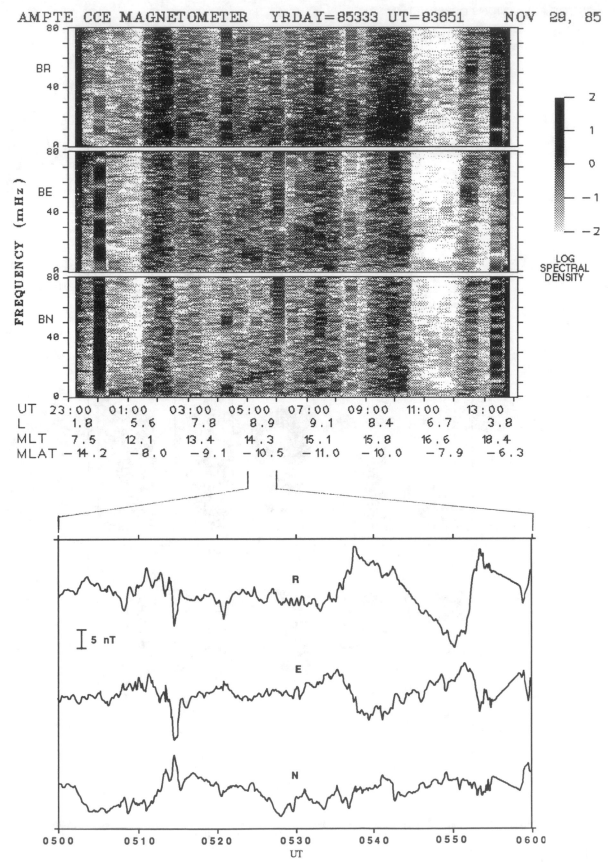

Figure 13. Dynamic spectrogram and wave-form plot of magnetometer data from the AMPTE/CCE satellite showing an interval of incoherent wave activity. Adapted from Anderson et al. [1990].

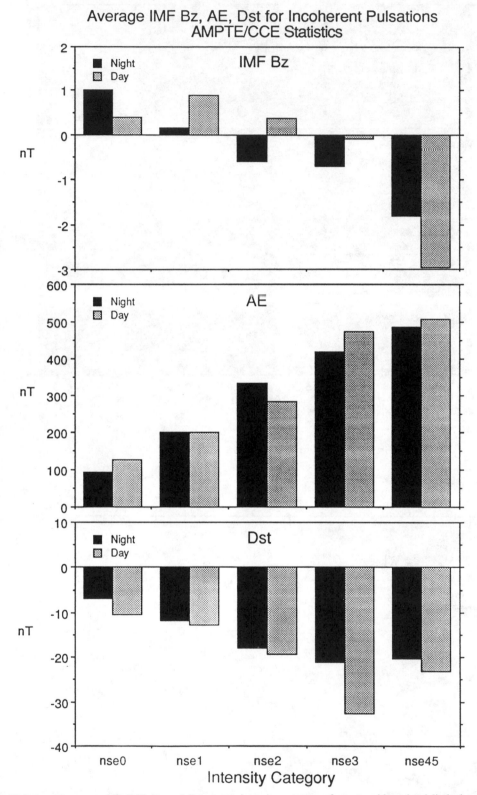

Figure 14. Variation of average AE, IMF B_z, and Dst versus intensity category of events without band limited spectral enhancements. Incoherent waves correspond only to categories NSE2, NSE3 and NSE45 and do not include NSE0 and NSE1. The suffix numbers of these categories refer to an index of spectral power which increases logarithmically with wave power [cf. Anderson et al., 1990 for details].

be that only by refining a quantitative understanding of ULF wave propagation in the magnetosphere and the concomitant excitation of field line resonances will one be able to distinguish which energy pathway is most important.

Toroidal Fundamental Resonances

Toroidal Pc 5 appear to be excited by a low frequency, <15 mHz source and the excitation appears to be strongest at higher L shells indicating a source near the magnetopause boundary. There appears to be no dependence of this excitation on solar wind conditions except that the intensity of the resonances increases as the solar wind velocity (and density) increase. Since increasing power with higher solar wind velocity (and density) is found for all other pulsations as well, even those driven by local ion instabilities [Anderson et al., 1991], this correlation cannot be invoked as definitive evidence for any particular source mechanism. Since toroidal Pc 5 appear to be present nearly all of the time at the dawn flank (the occurrence rate is ≈ 80% for $L > 8$ and MLAT > 10°, [Anderson et al., 1990]), the solar wind velocity dependence may well be a magnetospheric compression effect, that is, a compressed magnetosphere presents a magnetopause source closer to a given observation point in the magnetosphere leading to the appearance that the waves are driven more strongly when in reality the dominant effect may merely be a movement of the source [cf. Takahashi et al., 1991]. Hence, the energy source for toroidal Pc 5 may be related to the dynamics of the magnetospheric boundary rather than the upstream solar wind per se, and it appears that the dynamics are suitable for creating a <15 mHz perturbation source nearly continuously.

The low latitude magnetopause region has been shown to be very rich in ULF wave activity [e.g. Takahashi et al., 1991], and the Kelvin-Helmholtz instability is customarily invoked to explain low frequency perturbations at the flanks [Kivelson and Pu, 1984; Yumoto, 1984]. Observations somewhat down the tail of the magnetosphere at the flanks strongly suggest that some form of wave structure commonly propagates anti-sunward producing undulating motions of the magnetopause boundary [Fairfield, 1976]. Travelling ionospheric convection vortices appear to be a commonplace phenomenon at middle to high latitudes corresponding to changes in the magnetospheric boundary in response to changes in solar wind dynamic pressure and/or IMF orientation [Friis-Christensen et al., 1988]. These ionospheric vortices often occur continuously over many hours and seem to be associated with velocity shears in the low latitude boundary layer [McHenry et al., 1990]. It has also been shown that fundamental mode toroidal pulsations are excited by solar wind pressure variations [Potemra et al., 1989], suggesting that transient variations in the external pressure on the magnetopause and/or IMF orientation may provide an additional source of low frequency energy which could also play a role in toroidal Pc 5 generation. More extensive observations near the magnetopause boundary may be required to make dramatic progress in sorting out which source mechanism operates most effectively in producing the dawn flank toroidal Pc 5.

Incoherent Pulsations

It remains to be seen whether upstream sources play any significant role in these noise-like pulsations. The only piece of evidence in hand at present is a clear correlation with magnetic activity, which may be indicative of any number of factors. If it turns out that local wave-particle instabilities are not responsible for a significant fraction of these pulsations, the correlation of incoherent intensity with magnetic activity at least implies that either the source or transmission of the fluctuations is intimately dependent on the solar wind magnetosphere interaction. That is, the magnetopause may be a significant source of broadband ULF energy when reconnection is occurring, or magnetosheath turbulence may propagate significantly more readily across a rotational than a tangential discontinuity [Kwok and Lee, 1984]. Clearly, much work remains to be done to characterize and understand the sources of incoherent pulsations.

Acknowledgments: Preparation of this paper was supported by NASA under the AMPTE/CCE Missions Operation and Data Analysis Program, by NSF and by ONR. The author gratefully acknowledges extensive discussions with K. Takahashi.

REFERENCES

Allan, W., E. M. Poulter, and S. P. White, Hydromagnetic wave coupling in the magnetosphere—magnetic fields and Poynting fluxes, *Planet. Space Sci., 35,* 1181–1192, 1987a.

Allan, W., E. M. Poulter, and S. P. White, Structure of magnetospheric MHD resonances for moderate "azimuthal" asymmetry, *Planet. Space Sci., 35,* 1193–1198, 1987b.

Anderson, B. J., M. J. Engebretson, S. P. Rounds, L. J. Zanetti, and T. A. Potemra, A statistical study of Pc 3–5 pulsations observed by the AMPTE/CCE magnetic fields experiment 1. Occurrence distributions, *J. Geophys. Res., 95,* 10495–10523, 1990.

Anderson, B. J., T. A. Potemra, L. J. Zanetti, and M. J. Engebretson, Statistical correlations between Pc 3–5 pulsations and solar wind/IMF parameters and geomagnetic indices, in *Physics of Space Plasmas (1990) SPI Conference Proceedings and Reprint Series,* Vol. 10, edited by T. Chang, G. B Crew, and J. R. Jasperse, Scientific Publishers Inc., Cambridge, Massachusetts, pp. 419–429, 1991.

Arthur, C. W., R. L. McPherron, and W. J. Hughes, A statistical study of Pc 3 magnetic pulsations at synchronous orbit, ATS 6, *J. Geophys. Res., 82,* 1149–1157, 1977.

Arthur, C. W., and R. L. McPherron, Interplanetary magnetic field conditions associated with synchronous orbit: observations of Pc 3 magnetic pulsations, *J. Geophys. Res., 82,* 5138–5142, 1977.

Arthur, C. W., and R. L. McPherron, The statistical character of Pc 4 magnetic pulsations at syncrhonous orbit, *J. Geophys. Res., 86,* 1325–1334, 1981.

Barfield, J. N., R. L. McPherron, P. L. Coleman, Jr., and D. J. Southwood, Storm-associated Pc 5 micropulsation events observed at the synchronous equatorial orbit, *J. Geophys. Res., 77,* 143–158, 1972.

Barfield, J. N., and R. L. McPherron, Statistical characteristics of storm-associated Pc 5 micropulsations observed at synchronous equatorial orbit, *J. Geophys. Res., 77,* 4720–4733, 1972.

Barfield, J. N., and R. L. McPherron, Stormtime Pc 5 magnetic pulsations observed at synchronous orbit and their correlation with the partial ring current, *J. Geophys. Res., 83,* 739–743, 1978.

Chen, L., and A. Hasegawa, A theory of long–period magnetic pulsations: 1. steady state excitation of field line resonances, *J. Geophys. Res., 79,* 1024-1032, 1974.

Chen, L., and A. Hasegawa, On magnetospheric hydromagnetic waves excited by energetic ring-current particles, *J. Geophys. Res., 93,*

8763–8767, 1988.

Chen, L., and A. Hasegawa, Kinetic theory of geomagnetic pulsations 1. internal excitations by energetic particles, *J. Geophys. Res.*, 96, 1503–1512, 1991.

Cummings, W. D., R. J. O'Sullivan, and P. J. Coleman, Jr., Standing Alfvén waves in the magnetosphere, *J. Geophys. Res.*, 74, 778–793, 1969.

Engebretson, M. J., L. J. Zanetti, T. A. Potemra, and M. H. Acuna, Harmonically structured ULF pulsations observed by the AMPTE CCE magnetic field experiment, *Geophys. Res. Lett.*, 13, 905–908, 1986.

Engebretson, M. J., L. J. Zanetti, T. A. Potemra, W. Baumjohann, H. Lühr, and M. H. Acuna, Simultaneous observation of Pc 3-4 pulsations in the solar wind and in the Earth's magnetosphere, *J. Geophys. Res.*, 92, 10,053–10,062, 1987.

Engebretson, M. J., B. J. Anderson, L. J. Cahill, Jr., R. L. Arnoldy, P. T. Newell, C.–I. Meng, L. J. Zanetti, and T. A. Potemra, A multipoint case study of high latitude daytime ULF pulsations, *J. Geophys. Res.*, 94, 17,143–17,160, 1989.

Engebretson, M. J., L. J. Cahill, Jr., R. L. Arnoldy, B. J. Anderson, T. J. Rosenberg, D. L. Carpenter, U. S. Inan, and R. H. Eather, The role of the ionosphere in coupling upstream ULF wave power into the dayside magnetosphere, *J. Geophys. Res.*, 96, 1527–1542, 1991a.

Engebretson, M. J., N. Lin, W. Baumjohann, H. Luehr, B. J. Anderson, L. J. Zanetti, T. A. Potemra, R. L. McPherron, and M. G. Kivelson, A comparison of ULF fluctuations in the solar wind, magnetosheath and dayside magnetosphere 1. Magnetosheath morphology, *J. Geophys. Res.*, 96, 3441–3454, 1991b.

Engebretson, M. J., D. L. Murr, K. N. Erickson, R. J. Strangeway, D. M. Klumpar, S. A. Fuselier, L. J. Zanetti, and T. A. Potemra, The spatial extent of radial magnetic pulsation events observed in the dayside near synchronous orbit, *J. Geophys. Res.*, 97, 13,741-13,758, 1992.

Fairfield, D. H., Waves in the vicinity of the magnetopause, in *Magnetospheric Paricles and Fields*, pp. 67–77, B. M. McCormac ed., D. Reidel Pub. Co., Dordrecht-Holland, 1976.

Friis-Christensen, E., M. A. McHenry, C. R. Clauer, and S. Vennerstrøm, Ionospheric traveling convection vortices observed near the polar cleft: a triggered response to sudden changes in the solar wind, *Geophys. Res. Lett.*, 15, 253–256, 1988.

Greenstadt, E. W., and J. V. Olson, A contribution to ULF activity in the Pc 3–4 range correlated with IMF radial orientation, *J. Geophys. Res.*, 82, 4991–4996, 1977.

Gupta, J. C., Some characteristics of large amplitude Pc5 pulsations, *Aust. J. Phys.*, 29, 67, 1976.

Gupta, C.J., and E. R. Niblett, On the quiet–time Pc 5 pulsation events (spacequakes), *Planet. Space Sci.*, 27, 131-143, 1979.

Hasegawa, A., Drift mirror instability in the magnetosphere, *Phys. Fluids*, 12, 2642-2650, 1969.

Hasegawa, A., and L. Chen, Theory of the drift mirror instability, in *Plasma Waves and Instabilities at Comets and in Magnetospheres*, *Geophys. Monogr. Ser.*, vol. 53, edited by B. Tsurutani and H. Oya, p. 173, AGU, Washington, D. C., 1989.

Hasegawa, A., K. H. Tsui, and A. S. Assis, A theory of long period magnetic pulsations, 3. Local field line oscillations, *Geophys. Res. Lett.*, 10, 765-767, 1983.

Hedgecock, P. C., Giant Pc5 pulsations in the outer magnetosphere: a survey of HEOS-1 data, *Planet. Space Sci.*, 24, 921–935, 1976.

Higuchi, T., and S. Kokubun, Waveform and Polarization of Compressional Pc 5 Waves at Geosynchronous Orbit, *J. Geophys. Res.*, 93, 14433–14443, 1988.

Hughes, W.J., Hydromagnetic waves in the magnetosphere, *Rev. Geophys. Space Phys.*, 21, 508–520, 1983.

Hughes, W. J., and R. J. L. Grard, A second harmonic geomagnetic field line resonance at inner edge of the plasma sheet: GEOS 1, ISEE 1, and ISEE 2 observations, *J. Geophys. Res.*, 89, 2755–2764, 1984.

Junginger, H., G. Geiger, G. Haerendel, F. Melzner, E. Amata, and B.

Higel, A statistical study of dayside magnetospheric electric field fluctuations with periods between 150 and 600 s, *J. Geophys. Res.*, 89, 5495-5505, 1984.

Kivelson, M. G., and Z.–Y. Pu, The Kelvin–Helmholtz instability on the magnetopause, *Planet. Space Sci.*, 32, 1335–1341, 1984.

Kokubun, S., Statistical character of Pc5 waves at geostationary orbit, *J. Geomag. Geoelectr.*, 37, 759–779, 1985.

Kremser, G., A. Korth, J. A. Fejer, B. Wilken, A. V. Gurevich, and E. Amata, Observations of quasi-periodic flux variations of energetic ions and electrons associated with Pc 5 geomagnetic pulsations, *J. Geophys. Res.*, 86, 3345–3356, 1981.

Kwok, Y. C., and L. C. Lee, Transmission of magnetohydrodynamic waves through the rotational discontinuity at the Earth's magnetopause, *J. Geophys. Res.*, 89, 10,697–10,708, 1984.

Lee, D.-H., and R. L. Lysak, Impulsive excitation of ULF waves in the three-dimensional dipole model: the initial results, *J. Geophys. Res.*, 96, 3479–3486, 1991.

Lin, C. S., and J. N. Barfield, Azimuthal propagation of storm time Pc 5 waves observed simultaneously by geostationary satellites GOES 2 and GOES 3, *J. Geophys. Res.*, 90, 11075–11077, 1985.

Lin, N., R. L. McPherron, M. G. Kivelson, and D. J. Williams, An unambiguous determination of the propagation of a compressional Pc 5 wave, *J. Geophys. Res.*, 93, 5601–5612, 1988.

Lin, N., M. J. Engebretson, R. L. McPherron, M. G. Kivelson, W. Baumjohann, H. Luehr, T. A. Potemra, B. J. Anderson, and L. J. Zanetti, A comparison of ULF fluctuations in the solar wind, magnetosheath, and dayside magnetosphere 2. field and plasma conditions in the magnetosheath, *J. Geophys. Res.*, 96, 3455–3464, 1991.

McHenry, M. A., C. R. Clauer, E. Friis-Christensen, P. T. Newell, and J. D. Kelly, Ground observations of magnetospheric boundary layer phenomena, *J. Geophys. Res.*, 95, 14,995–15,005, 1990.

McPherron, R.L., C.T. Russell, and P.J. Coleman, Jr., Fluctuating magnetic fields in the magnetosphere: II. ULF waves, *Space Sci. Rev.*, 13, 411-454, 1972.

Odera, T. J., Solar Wind Controlled Pulsations: A Review, *Rev. Geophys.*, 24, 55, 1986.

Odera, T. J., D. Van Swol, C. T. Russell, and C. A. Green, Pc 3,4 magnetic pulsations observed simultaneously in the magnetosphere and at multiple ground stations, *Geophys. Res. Lett.*, 18, 1671–1674, 1991.

Olson, J. V., and B. J. Fraser, Pc 3 pulsations in the cusp, this volume.

Orr, D., Magnetic pulsations within the magnetosphere: A review, *J. Atm. Terr. Phys.*, 35, 1–50, 1973.

Potemra, T. A., H. Lühr, L. J. Zanetti, K. Takahashi, R. E. Erlandson, G. T. Marklund, L. P. Block, L. G. Blomberg, and R. P. Lepping, Multisatellite and ground-based observations of transient ULF waves, *J. Geophys. Res.*, 94, 2543–2544, 1989.

Radoski, H. R, Highly asymmetric MHD resonances: the guided poloidal mode, *J. Geophys. Res.*, 72, 4026-4027, 1967.

Saito, T., Geomagnetic pulsations, *Space Sci. Rev.*, 10, 319–412, 1969.

Singer, H. J., W. J. Hughes, and C. T. Russell, Standing hydromagnetic waves observed by ISEE 1 and 2: radial extent and harmonic, *J. Geophys. Res.*, 87, 3519–3529, 1982.

Sonnerup, B., L. J. Cahill, and L. R. Davis, Resonant vibration of the magnetosphere observed from Explorer 26, *J. Geophys. Res.*, 74, 2276–2288, 1969.

Southwood, D. J., J. W. Dungey, and R. L. Etherington, Bounce resonant interaction between pulsations and trapped particles, *Planet. Space Sci.*, 17, 349–361, 1969.

Southwood, D. J., Some features of field line resonances in the magnetosphere, *Planet. Space Sci.*, 22, 483–491, 1974.

Southwood, D. J., A general approach to low-frequency instability in the ring current plasma, *J. Geophys. Res.*, 81, 3340–3348, 1976.

Takahashi, K., and R. L. McPherron, Standing hydromagnetic waves in the magnetosphere, *Planet. Space Sci.*, 32, 1343–1359, 1984.

Takahashi, K. T., R. L. McPherron, and T. Terasawa, Dependence of the spectrum of Pc 3-4 pulsations on the interplanetary magnetic field, *J. Geophys. Res.*, 89, 2770–2780, 1984.

Takahashi, K., P. R. Higbie, and D. N. Baker, Azimuthal propagation and frequency characteristic of compressional Pc 5 waves observed at geostationary orbit, *J. Geophys. Res.*, *90*, 1473–1485, 1985.

Takahashi, K., Multisatellite studies of ULF waves, *Adv. Space Res.*, *8*, 427–436, 1988.

Takahashi, K., C. Z. Cheng, R. W. McEntire, and L. M. Kistler, Observation and theory of Pc 5 waves with harmonically related transverse and compressional components, *J. Geophys. Res.*, *95*, 977–989, 1990a.

Takahashi, K., R. W. McEntire, A. T. Y. Lui, and T. A. Potemra, Ion flux oscillations associated with a radially polarized transverse Pc 5 magnetic pulsation, *J. Geophys. Res.*, *95*, 3717–3731, 1990b.

Takahashi, K., D. G. Sibeck, P. T. Newell, and H. E. Spence, ULF waves in the low-latitude boundary layer and their relationship to magnetospheric pulsations: a multisatellite observation, *J. Geophys. Res.*, *96*, 9503–9519, 1991.

Takahashi, K., and B. J. Anderson, Distribution of ULF energy (f < 80 mHz) in the inner magnetosphere: a statistical analysis of AMPTE CCE magnetic field data, *J. Geophys. Res.*, *97*, 10751–10773, 1992.

Verö, J., Experimental aspects of low-latitude pulsations—A review, *J. Geophys.*, *60*, 106, 1986.

Walker, A. D. M., and R. A. Greenwald, STARE auroral radar observations of Pc 5 geomagnetic pulsations, *J. Geophys. Res.*, *84*, 3373–3388, 1979.

Walker, A. D. M., and R. A. Greenwald, Statistics of occurrence of hydromagnetic oscillations in the Pc5 range observed by the STARE auroral radar, *Planet. Space Sci.*, *29*, 293-305, 1981.

Woch, J., G. Kremser, A. Korth, O. A. Pokhotelov, V. A. Pilipenko, and J. M. Nezlina, Curvature-driven drift mirror instability in the magnetosphere, *Planet. Space Sci.*, *36*, 383–393, 1988.

Woch, J., G. Kremser, and A. Korth, A comprehensive investigation of compressional ULF waves observed in the ring current, *J. Geophys. Res.*, *95*, 15113–15132, 1990.

Yumoto, K., and T. Saito, Relation of compressional HM waves at GOES 2 to low-latitude Pc 3 magnetic pulsations, *J. Geophys. Res.*, *88*, 10041–10052, 1983.

Yumoto, K., Long–period magnetic pulsations generated in the magnetospheric boundary layers, *Planet. Space Sci.*, *32*, 1205–1218, 1984.

Yumoto, K., Low–frequency upstream waves as a probable source of low–latitude Pc 3–4 pulsations, *Planet. Space Sci.*, *22*, 239–249, 1985.

Yumoto, K., T. Saito, B. T. Tsurutani, E. J. Smith, and S.-I. Akasofu, Relationship between the IMF magnitude and Pc 3 magnetic pulsations in the magnetosphere, *J. Geophys. Res.*, *89*, 9731–9740, 1984.

Zhu, X., and M. G. Kivelson, Compressional ULF waves in the outer magnetosphere 1. statistical study, *J. Geophys. Res.*, *96*, 19451–19467, 1991.

B. J. Anderson, The Johns Hopkins University, Applied Physics Laboratory, Laurel, MD 20723.

Discoveries of Sources of Pc 2-4 Waves—A Review of Research in the Former USSR

V. A. TROITSKAYA

Department of Physics, La Trobe University, Bundoora, Vic., 3083, Australia

Intensive development of national and international studies of ULF waves in the former USSR began in the early 1950's. Interest in these studies was aroused by both scientific and 'relevant' problems. In the 1960's and early 1970's it was discovered that the occurrence of quasi-regular Pc3-4 pulsations depended on the orientation of the interplanetary magnetic field (IMF), their period on the value of IMF and their amplitudes on the velocity of the solar wind (SW). These discoveries were facilitated by the existence of a wide network of ground-based ULF stations in the USSR as well as by a variety of international cooperative studies. The recognition of "pure" (quasi-monochromatic) regimes of Pc3 and Pc4 was an important methodical step, which lead to results which were later confirmed by numerous ground-based (GB) and satellite studies. The discovery of global modulation of Pc3 amplitudes lead to the elaboration of a method of determining IMF irregularities in the SW. It was found that the quasi-monochromatic Pc3 regime is most typical for solar streams from equatorial coronal holes. In the first half of the 1980's an unexpected result was found as regards the distribution of Pc3 around the globe - namely that the maximum of their amplitudes occurs in the vicinity of the projection of dayside cusp.This was the first indication of the possibility of direct penetration of Pc3 waves through the cusp.

INTRODUCTION

Interest in detailed systematic studies of pulsations, their distribution in time and space, the range of their frequencies, etc., grew significantly in the 1950's before the IGY (International Geophysical Year) and specially during the IGY. In the former Soviet Union this interest was aroused by the development of the theory of hydromagnetic waves and of several relevant problems:

1.There was a need for detailed knowledge of spectra of pulsations, their distribution in time and space in different geophysical regions, for elaborating practical recommendations for the magneto-telluric method of prospecting [Troitskaya, 1963]. In this method the ratio of amplitudes of oscillations of different frequencies in magnetic and electric field gives information on the distribution of resistivity of rocks with depth. This method was used at this time by more than a hundred field parties working in different regions of the USSR.

2.In this decade high altitude nuclear explosions were conducted for the first time. They were carried out by United States in the South Atlantic under the code name 'Argus'. The altitude of the explosions was around 500 km. In spite of the fact that the yield of each explosion was relatively small, (1 kilotonne), they could be identified by ULF waves generated by this artificial release of particles. The signals produced by the explosions in the short period range of ULF waves were registered practically over the whole network of stations established in the former USSR [Troitskaya, 1960]. Consequently it became clear that observations of ULF waves can be used for surveillance of high altitude nuclear explosions, and that ULF waves present as well a very sensitive tool for tracing any artificial or natural injections of particles into outer space.

3.New important correlations were found between different types of short period (T ~ 1-15 sec) pulsations and processes in the upper atmosphere and magnetosphere, which allowed formulation for the first time of the problem of diagnosis of outer space by means of ULF waves [Troitskaya and Melnikova, 1959; Troitskaya, 1961; Troitskaya et al., 1962].

4.In the late 40's and beginning 50's there occurred several devastating earthquakes in the middle Asian republics of

Solar Wind Sources of Magnetospheric Ultra-Low-Frequency Waves
Geophysical Monograph 81

Tadjikistan, Turkmenistan and Kazachstan. A significant intensification of geophysical studies took place in these regions with the hope of finding precursors to earthquakes. Consequently, the sensitivity of magnetic and Earth current records in these seismo-active regions was increased by two to three orders of magnitude, and a relatively dense network of stations was installed. No clear effect preceding earthquakes was traced, but a wealth of data showing a variety of types of pulsations was obtained.

In 1953 the concept of two main types of oscillatory regimes (Pc and Pi) was introduced [Troitskaya, 1953] which later was developed by the definition of several subclasses of continuous and irregular pulsations. These observations were of great help in the later preparation of a special album of examples of different types of pulsations. This album was later used as a foundation for the classification of pulsations adopted by a Working Group on rapid variations chaired by Father Romana. This classification was considered by a special group nominated by the International Association of Geomagnetism and Aeronomy (IAGA) at the International Union of Geodesy and Geophysics (IUGG) meeting in Berkeley (USA) in 1963. After excited and sometimes fierce discussions the classification was recommended for all pulsation studies [Jacobs et al., 1964]. The main objection at that time to this classification was that it was based mainly on morphological properties of pulsations and to a considerable extent on intuition, especially as regards dividing the period range of continuous pulsations. This was an important step in co-ordinating pulsations research, enabling comparison of the results of their investigations obtained in different countries. Despite initial criticisms this classification is still used not only by the geomagnetic pulsation community but also by a significantly wider scientific community dealing with ionospheric and magnetospheric and solar wind phenomena. There have been many attempts to improve the classification scheme, but without general agreement on how best to do it.

During the IGY and later in the 60's and 70's, Russian scientists were able to study local and worldwide regularities in pulsations behaviour due to the existence of an extensive network of permanent and temporary stations. Seventeen stations were starting operations at the beginning of IGY, and three of them were located in the Arctic and three in the Antarctic. Later stations were established in India, Cuba, at conjugate points in Sogra (Archangelsk region) and the French base Kerguelen and in conjugate points between Australia and Kamchatka. Temporary simultaneous coordinated measurements were carried out between antipodal points Dallas and Garm (Tadjikistan) and on a latitudinal profile (from Gottingen to Irkutsk) with German scientists. During several years there was close cooperation in pulsation studies in Arctic and Antarctica including studies at the geomagnetic poles (Thule-Vostok) between the Geophysical Institute, Alaska and the Institute of the Physics of the Earth, Moscow. Temporary observations of pulsations in different regions of the USSR were conducted on meridional and latitudinal profiles, on ice flow stations and by field parties engaged in prospecting using the magneto-telluric method. These temporary observations required efforts of different Institutes and Universities. As a result of the development of international cooperation in geophysics we had access to some of the first measurements of solar wind parameters with high resolution in time and later to the measurements of waves and particles in the magnetosphere. This facilitated the progress of research.

SEARCH FOR CONNECTION OF Pc2-4 PROPERTIES WITH THE PARAMETERS OF MAGNETOSPHERE AND SOLAR WIND

Since the beginning of their studies, Pc3-4 pulsations were mainly thought to be the result of resonances either of the field lines or in cavities of magnetosphere. Therefore the investigations of connections of Pc2-4 properties with the parameters of the magnetosphere and of the solar wind have begun with the studies of dependence of their periods on dimensions of the magnetosphere. It was expected that the closer is located the magnetopause to the Earth, the smaller should be the periods of resonance oscillations. The result of this comparison lead to the conclusion that this trend is indeed observed mainly for pulsations with periods less than 30-40 sec [Pc-3] but is not systematically held for continuous pulsations with greater periods [Troitskaya et al., 1966; Troitskaya et al., 1968; Troitskaya et al., 1970]. This observation was one of the first indications that inside the Pc3-4 range of pulsations there are at least two different classes of oscillations having different mechanisms of excitation.

Another direction of research of the influence of the position of the magnetopause on the properties of Pc2-4 regimes was connected with a most curious effect of their "sudden ending" after negative sudden impulses (Si$^-$). Investigations of Pc2-4 regimes after positive and negative sudden impulses (which correspond to contractions and expansions of the magnetosphere) showed that sometimes Pc2-4 periods sharply changed, namely a decrease after Si$^+$ and an increase after Si$^-$. After a big negative Si$^-$, Pc2-4 can suddenly disappear worldwide. This effect of Pc2-4 'sudden ending' was analysed using stations Petropavlovsk Kamchatsky, Soroa (Cuba) and Borok, that is practically around the whole globe [Troitskaya et al., 1969; Troitskaya and Gul'elmi, 1967]. In Figure 1a are given examples of 'sudden endings' of pulsations on records of the station Borok and the corresponding Si$^-$, at the standard magnetic records of the equatorial station Bangi. In Figure 1b the solid lines show the distribution of all cases of analysed negative sudden impulses on R_E and the dashed part of the figure shows cases when after Si$^-$ pulsations disappeared.

The main goal was to clarify the situation for which Pc2-4 disappearance is most probable. It is seen from the figure that the probability of their disappearance is greater for R_E equal ~ 11-13. It follows from this result that 'sudden endings' of Pc2-4 correspond to sudden expansions of the magnetosphere, and should be observed for longer periods of this range, that is mainly for Pc4 pulsations. The analysis of data showed also that the probability of "sudden endings" increases with the amplitude of Si$^-$.

Later checking of the IMF orientation at the moments of "sudden endings" did not show the expected IMF direction perpendicular to the Sun-Earth line, (see next paragraph) when the probability of at least Pc3 occurrence should be minimal. The state of the ionosphere analysed at that time seemed not to have a significant role in this disruption of the Pc2-4 regime.

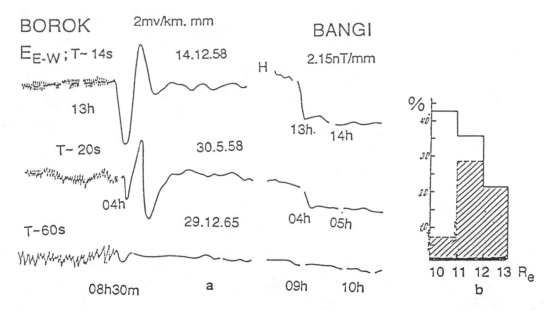

Fig. 1(a)   Examples of sudden disappearance of Pc2-4 oscillations at station Borok coinciding with negative sudden impulses at the equatorial station Bangi.
(b) Dependence in % of disappearances of Pc2-4 on radius of magnetosphere R_E

Before the discovery of the dependence of Pc3-4 periods on the value of the IMF, intensive studies were carried out on their connections with other solar wind parameters. In particular, interesting results were obtained from comparison of Pc3-4 periods with the density of solar wind fluxes measured on Venus 2 (1965), Venus 4 (1967) and Venus 5 and 6 (1969) [Gringauz et al., 1970] and on IMP 1 for the period December 1963 to March 1964. (IMF data were kindly provided by N. Ness.) Pulsation periods were determined for three observatories - Petropavlovsk Kamchatsky, Soroa (Cuba), and Borok. The result of comparison with data of Venus is shown in Figure 2a, where pulsation periods are plotted as a function of ion flux (nV cm^{-2}se^1) of the solar wind. The result obtained was unexpected, and presented another hint of the different nature of Pc3 and Pc4 pulsations. At the same time it gave support to the originally postulated division (~ 40 sec) between Pc3 and Pc4 pulsations. In order to check the phenomenon shown in Figure 2a the same comparison was done using data of IMP-1. The results are given in Figure 2b. Two different dependences of periods of Pc on nV were confirmed - one for T ≤ 40 sec (T diminishing with the growth of nV) and the other for T ≥ 40 sec (T rising with the growth of nV). Figure 2c gives the dependence of Pc periods on changing solar wind speed (V) for fixed value of n and Figure 2d gives the dependence of Pc periods on changing density of the solar wind n for fixed value of solar wind speed. It is seen that the change in Pc periods is influenced mainly by the changes of density n. This existence of a strong dependence on n was at that time also unexpected, because it was assumed that changes in solar wind pressure (nV^2) were governed primarily by variations in V because the dynamic pressure varies as square of V, whereas it is linear in n. However it is known that V in the vicinity of the Earth's orbit changes at most by a factor of four while n can vary by over two orders of

magnitude. This result emphasised the importance of studies of the variations of density of the solar wind on magnetospheric phenomena. The dependencies obtained indicate the possibility of simultaneous generation of two superposed Pc regimes inside the magnetosphere, namely Pc3 and Pc4, which is observed rather frequently. The first indication of dependence of Pc amplitudes on solar wind speed was shown by Saito (1964) and obtained in the former USSR in the 1970's [Vinogradov and Parchomov, 1970]

DISCOVERY OF CONNECTION OF IMF DIRECTION WITH OCCURRENCE OF Pc3-4 AND THEIR FREQUENCY WITH THE VALUE OF THE IMF

Already in the beginning of the 60's, it was realised that in addition to the control of continuous pulsation amplitudes by local time, there is an unquestionable modulation of their excitation in universal time. It was known that in some cases Pc can begin or end simultaneously all around the world without any Si$^-$. Such cases were enigmatic, and attempts to interpret them using correlations with traditional indices characterising the state of magnetic field, disturbances on magnetograms, or other available geophysical data did not produce any satisfactory result. The data of the network of stations encompassing roughly 120° in longitude were used to study the control of universal time of the Pc regime. Later the data obtained in Cuba were used to approach the global coverage. Besides the puzzle of global modulation, there were "mystical" days when continuous pulsations were absent all around the world. Such days occurred relatively seldom (1% - 2% of all cases observed) but this fact did not fit at all the prevailing theories of that time. These theories presumed that Pc pulsation excitation is primarily produced by continuous flow of solar wind at the magnetospheric boundary, with the subsequent generation of resonance oscillations which should occur every day. That lead to the thought that the source of

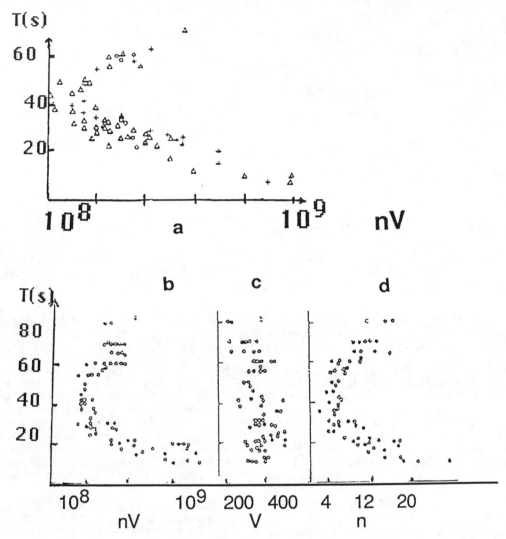

Fig. 2(a) Dependence of Pc2-4 periods on solar fluxes nV (part/cm^{-2}sec^{-1}),obtained using data of Venera 2-4, 5-6 (upper part of the figure).
(b) The same dependence of periods of Pc2-4 on nV using data of IMP-1 from December 1963 to March 1964.
(c) The dependence of Pc2-4 periods on solar wind velocity for fixed value of density n.
(d) The same dependence on n for fixed value of V.

The symbol (Δ) in this figures corresponds to the station Borok, (o) to Soroa, and (+) to Petropavlovsk.

Pc pulsations should depend on some distant parameter, the changes of which would be crucial to Pc excitation. Elucidation of this parameter was obtained from the comparison of pulsation records with one of the first available detailed data sets.(1-14 December 1963) of IMF parameters obtained on IMP-1. In this set of data averaged for every 5.46 min. there were days when the predominant direction of the IMF was perpendicular to the Sun-Earth line. In Fig.3a. are shown copies of records of Earth currents for October 23, 1963 at the station Petropavlovsk-Kamchatsky together with the orientation of IMF in the plane of the ecliptic shown by arrows. The local hours for these cases were around noon, that is the most favourable ones for appearance of Pc3-4. Nevertheless, suddenly the pulsations disappeared following the change in direction of the IMF not only at this station but at all other stations considered. For the same direction of the IMF on records of December 14, 1963 the pulsations were also absent. It was the first striking experimental evidence of the connection of Pc3-4 occurrence with the orientation. of the IMF. Fig. 3b shows the distribution of all directions of IMF for the available set of data (around 1500 cases) organised for observed cases of Pc3, Pc3-4, and absence of Pc3. This picture in addition to direct comparisons gave overwhelming statistical evidence of

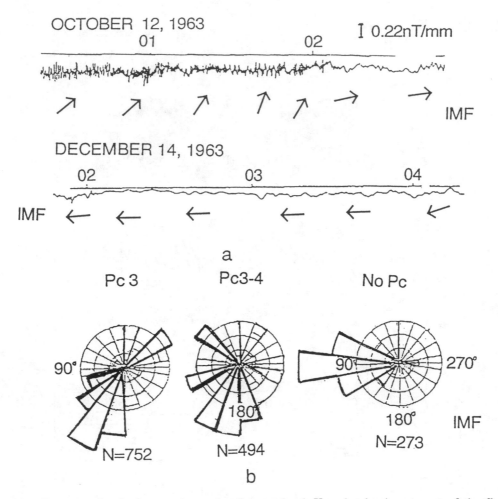

Fig. 3(a) Records of pulsations at the station Petropavlovsk Kamchatsky (upper part of the figure) and the orientation of IMF in the plane of ecliptic shown by arrows. Direction to the sun is upward.
(b) Distribution of direction of IMF for analyzed period organised by three plots: One (752 cases) corresponds to occurrence of Pc3, the second one (494 cases) corresponds mainly to Pc3-4 regime and the third plot shows the situation when Pc3 are absent (273 cases). (One case corresponds to the average value of IMF for each 5.46 minutes.)

the connection [Bolshakova,1966;Bolshakova and Troitskaya 1968],

Comparison of pulsations and directions of IMF obtained on ISEE-1 satellite and occurrence of Pc3-4 pulsations on the ground (Figure 4) on widely separated (in longitude and latitude) stations, showed clearly expressed global modulation both for ground based observations of Pc3-4 and Pc3-4 measured on ISEE-1. The data from ISEE-1 were kindly provided by C.T. Russell. They confirmed without doubt the early, original results [Troitskaya and Bolshakova ,1988].

It was suggested later that the intervals of absence of Pc3-4 on the records can give information on the average size of IMF-irregularities. For the cases when the IMF is directed perpendicularly to the Sun-Earth line the dimensions of irregularities L are equal to the product of solar wind speed and the time of Pc3-4 absence. The obtained average values of L for the dimensions of interplanetary irregularities (~

1.4×10^{11} cm) agreed well with those estimated from cosmic ray anisotropy [Bolshakova et al., 1976] and some theoretical estimates existing at that time.

Investigation of Pc3-4 regimes generated by high speed solar streams (HSS) from different sources on the sun (equatorial and polar coronal holes, disappearance of filaments) showed that there are significant differences in Pc3-4 signatures corresponding to different types of the source of the streams [Bolshakova and Troitskaya 1984]. The striking feature for solar streams from equatorial coronal holes is the presence during several days of a quasi-monochromatic Pc3 regime without significant modulation of their amplitudes. In the light of obtained results it meant that the direction of IMF for these intervals was relatively stable. A quasi-monochromatic Pc3 regime was found not to be typical for HSS from polar coronal holes or disappearance of filaments. But it is a regularly observed feature for the

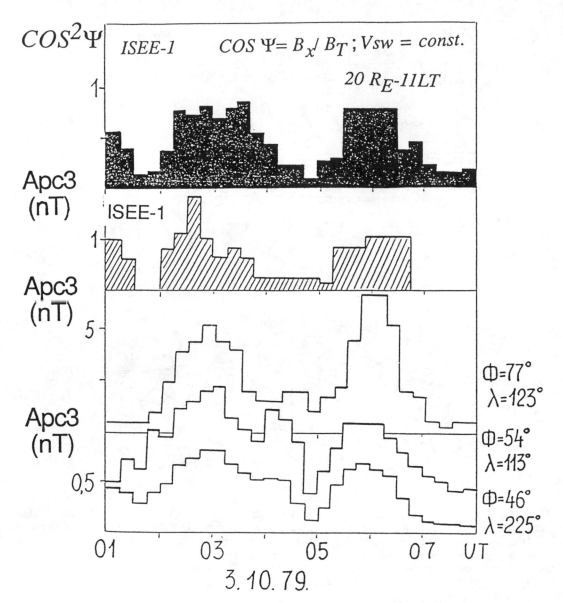

Fig. 4 Example of global modulation of Pc3 regime. Upper part -the records of Pc3 and angle ψ measured on ISSE-1. B_x - component of B towards the sun. B_T - amplitude of B.Vertical axis points in direction of sun. Lower part -Pc 3 amplitude modulations at widely separated ground based stations.

recurrent disturbances, connected with equatorial holes occurring in the declining years of solar activity. Another peculiar feature of the effect of HSS from equatorial coronal holes is the absence in most cases of quasi-monochromatic Pc4 regime. This quasi-monochromatic regime Pc4 however is quite typical at the end of HSS from polar coronal holes and especially for HSS from disappearance of filaments. It is suggested that such differences in generation of Pc3-4 regimes by HSS from different sources may be due to different distributions of density and orientation of the IMF in these HSS [Bolshakova and Troitskaya, 1984].

The dependence of Pc3-4 periods on the magnitude of IMF was discovered in 1971 [Troitskaya et al., 1971] The

investigation of this dependence was based on data of IMF obtained by IMP-3 for September-November 1966, by IMP-4 for August-November 1966 and on ground-based observations of the station Borok. The results of this investigation showed clear inverse dependence of Pc2-4 periods on the magnitude of the IMF. First , an empirical relation between these quantities was established in the form [Troitskaya et al., 1971]

$$T_s = 160/B_{IMF}(nT). \qquad (1)$$

The obvious conclusion from this relation was the suggestion that the source of Pc3-4 pulsations is outside the magnetosphere and that their period is related in some way to the Larmor rotation of protons in the IMF. Verification of this result was carried out by Plyasova-Bakunina [1972], Gul'elmi et al. [1973] and Gul'elmi and Bolshakova [1973] and in additional studies using Explorer 34 observations [Plyasova-Bakunina et al., 1978].

Finally this relationship was usually presented in the form:

$$f(Hz) = cB_{IMF}(nT) \qquad (2)$$

Here c is the constant. In Troitskaya and Bolshakova [1988] are given the values obtained for this constant in numerous ground based studies. They showed a remarkable agreement with its first determination in 1971:

$$f(H_z) = 0.006 \, B_{IMF}(nT) \qquad (3)$$

Direct measurement of wave frequencies and IMF values in the upstream region of the interplanetary medium [Russell and Hoppe, 1981; Hoppe and Russell, 1982] gave the following relation:

$$f(Hz) = 0.0058 \, B_{IMF}(nT). \qquad (4)$$

This result showed that the linear relationship between the magnitude of the IMF and Pc3-4 pulsation frequency first established by ground based studies holds also for the upstream waves in front of the bow shock at the Earth. Moreover the same relationship is approximately maintained during planetary encounters [Hoppe and Russell, 1982]. In Gul'elmi and Troitskaya [1973],Gul'elmi [1974],Kovner et al. [1976], and Troitskaya and Bolshakova [1988] are summarised these and some other early results of the investigations of Pc2-4 pulsations.

DISCOVERY OF MAXIMUM OF Pc3 AMPLITUDES IN THE REGION OF PROJECTION OF THE DAYSIDE CUSP

Investigations of geomagnetic pulsations in Antarctica have been carried out since the IGY. The main attention was given to a variety of previously unknown types of pulsations occurring in the polar cap. Only in 1984, detailed investigations of Pc3 pulsations at several Antarctic stations quite unexpectedly showed that the maximum of their amplitudes is located in the vicinity of the projection of the dayside cusp [Bolshakova et al., 1984]. Depending on the magnetic activity and in accordance with the displacement of the cusp, this maximum of Pc3 shifts correspondingly.

Direct comparison of the position of the cusp and Pc3 amplitudes was carried out for a relatively quiet day (Kp ~ 0-2) - February 15, 1980 and for a disturbed day - February 16, 1980 (Kp ~ 6). The results of this comparison are shown in Fig. 5a. On the quiet day Pc3 amplitudes were ~ 10 nT at the latitude of Mirny ($\phi = 77°$) where presumably the cusp was located. The Pc3 amplitudes at Novolazarevskaya ($\phi \sim 62°$) on this day barely reached 2 nT. On the disturbed day of February 16 the cusp was located at extremely low latitudes ($\phi \sim 62°$) and the amplitudes of Pc3 at Novolazarevskaya amounted to 10-15 nT, whereas at Mirny there were observed only isolated bursts of Pc3 with amplitudes ~ 2-3 nT. The cusp location was identified from data of DMSP - F2 and F4 satellites.

The ratio of average amplitudes of Pc3 at $\phi = 77°$ (A_1), to the average of Pc3 amplitudes at $\phi = 67°$ (A_2) (Station Molodezhnaya) as a function of geomagnetic activity is given on Fig. 5b. For Kp = 0-2, Pc3 amplitude at 77° is higher than at $\phi = 67°$ by 1.5-2 times, the amplitudes are practically comparable for Kp ~4 and the ratio of amplitudes A_1/A_2 becomes less than unity for higher activity. This result was the first indication of the possibility of direct penetration of Pc3 waves through the cusp.

It is interesting and important to note that the ratio of amplitudes of Pc3 in the vicinity of the cusp and in the auroral zone to the amplitudes of Pc3 at geomagnetic pole - Vostok and to the amplitudes at the subauroral station Kerguelen is always greater than unity for any activity levels. This finds a natural explanation in the fact that both of these stations are virtually never located under the projection of the dayside cusp.

It should be noticed that for low magnetic activity, high latitude Pc3 pulsations are a very common phenomenon in the region of the projection of the dayside cusp. Both for the cases of their presence or absence the solar wind speed is relatively low (300-400 km/sec). However their appearance in the region of the cusp for low magnetic activity, in addition to the requirement of the favourable cone angle of solar wind IMF, seems to depend strongly on dynamic pressure, and more precisely on the density n of the solar wind.

The existence of this 'superhigh' latitude maximum of Pc2-4 intensity was independently confirmed in 1985 by observation of pulsations on a network of stations in the Arctic [Plyasova-Bakimina et al., 1986]. In this paper an attempt was made to divide all observed oscillations in the Pc2-4 range into two groups: one for which the value of B derived from the relation $T_s = 160/B_{IMF}(nT)$ coincided with the value of B_{IMF} measured by satellites, and the other for which this relation gave a completely different value from B_{IMF}. The main result of this study was the establishment of a difference in location of the maximum of amplitude for these two groups. The Pc2-4 of presumably solar wind origin have a maximum of their intensity within the region of the equatorial boundary of the dayside cusp. The other group of Pc2-4, thought to be of magnetospheric origin, have their maximum of intensity inside the magnetosphere at $\phi \sim 70°$.

CONCLUDING REMARKS

Almost thirty years have passed since the classification of small oscillations of the Earth's magnetic field in the range of frequencies from several Hz to ~ 1 mHz was introduced. At first these oscillations were called geomagnetic micropulsations, then pulsations, and finally ULF waves. This change was required as a consequence of the discovery in the same frequency range of oscillations in the fluxes of ions and electrons in the magnetosphere, in the aurora, in the electron content of the ionosphere, in the interplanetary magnetic field, etc., which often were correlated with geomagnetic pulsations. As a result, the same classification which was elaborated for geomagnetic pulsations was used for a broad variety of phenomena classified under the general term - ULF waves.

Solar wind sources of the ULF waves can be either solar streams with different properties acting on the

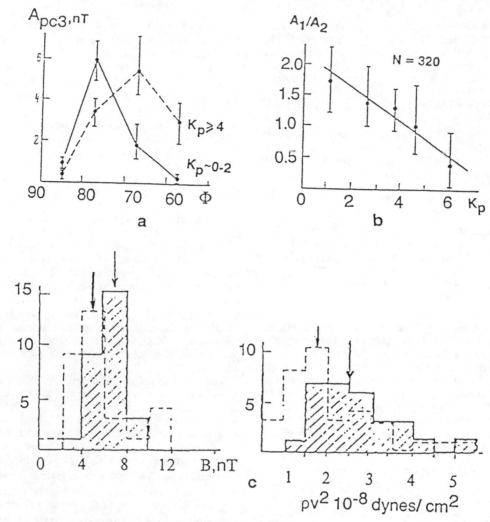

Fig. 5(a) Displacement of Pc3 amplitude maximum with magnetic activity - Antarctic stations.
(b) Dependence on magnetic activity of statistically determined ratio of Pc3 average amplitudes at station with $\phi = 77°$ to the Pc3 amplitudes at station with $\phi = 67°$.
(c) Dependence of occurrence of Pc3 in cusp region on the value of IMF-B and solar wind dynamic pressure. Most typical values of the dynamic pressure and the value of B of the solar wind for Pc3 occurrence at the cusp region are shown by dashed portions of the plot.

magnetosphere with the subsequent generation of ULF waves in the magnetosphere, or ULF waves generated "in situ" which penetrate into the magnetosphere. The only ULF waves generated in the solar wind and observed at ground based stations belong to the Pc2-4 type of continuous pulsations, which were described in this paper. It is interesting to note that the discovery of waves in the upstream region of the bowshock [Greenstadt et al., 1968; Fairfield, 1969] was done practically simultaneously with and independently of the discovery of the connection of properties of continuous pulsations observed at the earth's surface with orientation of interplanetary magnetic field [Bolshakova,1966; Bolshakova and Troitskaya, 1968]. In 1971, using ground-based data, the relationship: $T_s = 160/B_{IMF}(nT)$ or, $f(Hz) = 0 .006B_{IMF}(nT)$ was found

[Troitskaya et al., 1971].In 1981, this relationship was remarkably confirmed by in situ measurements in front of the Earth's bow shock as $f(Hz) = 0.0058B_{IMF}(nT)$ [Russell and Hoppe, 1981].

The theory of generation of these pulsations was also developed independently from ground based and satellite data. Later, comparison of records of ISEE-1 and ground based data [Bolshakova et al., 1987] showed that the value of the IMF can be determined with the greatest reliability from ground based data in the interval of local time between 09 - 13 hrs.

Many of the early results of ULF studies contain information on some regularities of Pc2-4 behaviour, which are still unexplained. For instance, many studies now are dedicated to investigation of ULF waves generated after a positive impulse (Si+) or an SSC. But the global disappearance of the Pc regime after a negative impulse (SI⁻) (Figure 1), which is possibly connected with stabilisation of the magnetopause, is still a mystery. Of considerable interest is the dual dependence of Pc2-4 period on density of the solar wind (Figure 2). This shows the possibility of simultaneous generation of continuous pulsations in the two frequency ranges, where possibly strict conditions exist for "forbidden" and "allowed" combinations of Pc2-3 and Pc4.

Studies of Antarctic data lead to the conclusion that direct penetration of Pc3 oscillations via the cusp explains one of the mechanisms of transfer of waves from the solar wind to the earth's surface.

In conclusion, the variety of papers on ULF observations, and theory presented at this Conference, demonstrate that the field is very active and attracts more and more interest from much wider areas of space physics than could have been envisaged thirty years ago.

REFERENCES

Bolshakova, O.V., Stable geomagnetic micropulsations and solar corpuscular streams *Geomag. Aeron,* 6 879-851, 1966.

Bolshakova, O.V., and V.A. Troitskaya, Relation of the IMF direction to the system of stable oscillations, *Dokl. Akad. Nauk. USSR, 180,* 343-346, 1968.

Bolshakova, O.V., V.A. Troitskaya, and L.Mirochnitchenko, Characteristic of irregularities of IMF using data on Pc pulsations and solar cosmic rays, *Bulletin Cosmic Rays,* Nauka, Moscow, *10,* 78-86, 1976.

Bolshakova, O.V., and V.A. Troitskaya, Diagnostic of high speed streams and coronal holes by means of geomagnetic pulsations, *Geomag. Aeron., 20,* 87-94, 1980.

Bolshakova, O.V., and V.A. Troitskaya, The relation of high-latitude maximum of Pc-3 intensity to the dayside cusp, *Geomag. Aeron., 7,* 633, 1984.

Bolshakova, O.V., V.A. Troitskaya, and T.B. Rusakova , Interplanetary magnetic field control of Pc3 pulsations in the solar wind and on the Earth, *Geomag. Aeron., 27,* 306-307, 1987.

Fairfield, D.H., Bow associated waves observed in the far upstream interplanetary medium., *J. Geophys. Res., 74,* 3451, 1969.

Greenstadt,E.W.,I.M.Green,G.T.Inoue, A.J.Hundhausen, S.J. Bame, and I.B.Strong, Correlated magnetic field and\ plasma observations of the Earth's bow shock. *J. Geophys.Res., 73,* 51-60, 1968.

Gringauz, K.I., V.A. Troitskaya, E.K. Solomatina, and R.V. Shchepetnov, The relationship of solar wind variables to periods of continuous micropulsations of electromagnetic field of the earth, *Dokl. Akad. Nauk USSR, 5,* 1069-1061, 1970.

Gul'elmi, A.V., T.A. Plyasova-Bakunina, and R.V. Shchepetnov, Relationship between the period of geomagnetic pulsations Pc3,4 and the parameters of the interplanetary medium at the earth's orbit, *Geomag. Aeron., 13,* 382, 1973.

Gul'elmi, A.V., and O.V. Bolshakova, Diagnostics of the IMF from ground based data on Pc2-4 micropulsations, *Geomag. Aeron., 18,* 535-539, 1973.

Gul'elmi, A.V., and V.A. Troitskaya, Geomagnetic pulsations and the diagnostics of the magnetosphere, *Monography,* Nauka, Moscow, 208 pp, 1973.

Gul'elmi, A.V., Diagnostics of the magnetosphere and interplanetary medium by means of pulsations, *Space Sci. Rev., 16,* 331, 1974.

Hoppe, M., and C.T. Russell, Particle acceleration at planetary bow shock waves, *Nature, 235,* N5844, 41-42, 1982.

Jacobs, J.A., T. Kato, S. Matsushita, and V.A. Troitskaya, Classification of geomagnetic pulsations, *J. Geophys. Res., 69,* 180-181, 1964.

Kovner,M.S.,Lebedev,V.V.,Plyasova-Bakunina,T.A., and V.A.Troitskaya,On the generation of low frequency waves in the solar wind in the front of the bow-shock,Planet.Space Sci.,24,261, 1976.

Plyasova-Bakunina, T.A., Effect of IMF on the characteristics of Pc2-4 pulsations, *Geomag. Aeron., 12,* 675-676, 1972.

Plyasova-Bakunina, T.A., Yu.V. Golikov, V.A. Troitskaya, and P.C. Hedgecock, Pulsations in the solar wind and on the ground, *Planet. Space Sci., 26,* 457,1978.

Plyasova-Bakunina, T.A., V.A. Troitskaya, J.W. Munch, Super high latitude maximum of Pc2-4 intensity, *ActaGeodaet. Geophys. et Montanist. Hung., 21,* 143-153, 1986.

Russell, C.T., and M.M. Hoppe, The dependence of upstream wave periods on the interplanetary magnetic field strength, *Geophys. Res. Lett., 8,* 615-617, 1981.

Saito,T., A new index of geomagnetic pulsation and its relation to solar M-regions, *Rept Ionosph. Space Res. Japan,18,* 260-274, 1964.

Troitskaya, V.A., Two oscillatory regimes of the electromagnetic field of the Earth and their diurnal variation in universal time, *Dokl. Akad. Nauk USSR, 91,* N2, 181-183, 1953.

Troitskaya, V.A., and M.V. Melnikova, On IPDP (10-1 sec) in the electromagnetic field of the Earth and their connection with phenomena in high atmosphere. Dokl.Akad.Nauk.USSR,91,181-183,1959.

Troitskaya, V.A., Effects in Earth currents produced by high altitude nuclear explosions, *Izvestia Akad. Nauk. USSR,* Geophys. Series, 9, 65-70,1960.

Troitskaya,V.A.,Pulsations of the electromagnetic field with periods 1-15 sec. and their connection with phenomena in the high atmosphere, *J. Geophys. Res.,66,* 5-19, 1961.

Troitskaya, V.A., L.S. Alperovitch, and N.V. Djordjio, Connection of short period oscillations of electromagnetic field of the Earth and aurora, *Izvestia Akad. Nauk. SSSR,* Geophys. Series, 2, 267-178, 1962.

Troitskaya, V.A., Dependence of intensity and frequency of occurrence of short period oscillations of the electromagnetic field of the Earth on Solar activity, *J. Applied Geophys.* (Moscow), *37,* 60-63, 1963.

Troitskaya, V.A., O.V. Bolshakova, and E.T. Matveeva, Rapid variations of the electromagnetic field as an indication of the state of the radiation belts and geomagnetosphere., *Geomag. Aeron., 6,* 292-298, 1966.

Troitskaya, V.A., and A.V. Gul'elmi, Diagnostics of the magnetosphere by means of geomagnetic pulsations, Space Sci.Revs.7, 689, 1967.

Troitskaya, V.A., R.V. Shchepetnov, and A.V. Gul'elmi, Dimensions of subsolar magnetosphere from geomagnetic micropulsation ground observations, *Dokl. Akad. Nauk. USSR, 189*, 1069-1066, 1968.

Troitskaya, V.A., R.V. Shchepetnov, and A.V. Gul'elmi, Effect of sudden disappearance of geomagnetic Pc2-4 pulsations.*Geomag. Aeron., 9*, 363-366, 1969.

Troitskaya, V.A., and T.A. Plyasova-Bakunina, Relationship between the period of Pc2-4 pulsations and the position of the boundary of the magnetosphere. *Geomag. Aeron., 10*, 902-904, 1970.

Troitskaya, V.A., T.A. Plyasova-Bakunina, and A.V. Gul'elmi, Relationship between Pc2-4 pulsations and the interplanetary field, *Dokl. Akad. Nauk USSR*, *197*, 1312, 1971.

Troitskaya, V.A., and O.V. Bolshakova, Hydromagnetic diagnostics of the nonhomogeneous structure of the solar wind, *Acta Geodact. et Montanist. Hung., 19(3-4)*, 273-283, 1984.

Troitskaya, V.A., and O.V. Bolshakova, Diagnostics of the magnetosphere using multipoint measurments of ULF wave. *Adv. Space Res.,8*, 413-429,1988 , *Research, 8-10*, 713-726, 1988.

Vinogradov, P.A., and V.A. Parkhomov, *Issledovaniya po geomagnetismu aeronomii i fisike Solntsa*, (Irkutsk), *6*, 71, 1970.

V.A.Troitskaya, Physics Department, La Trobe University, Bundoora, Victoria, 3083, Australia.

Upstream Waves and Field Line Resonances — Pulsation Research at the Nagycenk Observatory During Three Solar Cycles

JÓZSEF VERŐ AND BERTALAN ZIEGER

Geodetic and Geophysical Research Institute
of the Hungarian Academy of Sciences, Sopron, Hungary

The Geophysical Observatory Nagycenk has been established for the International Geophysical Year in 1957 in Western Hungary. Pulsation recording has been continuous since that date, thus it covers three complete solar cycles. This long series of data has enabled us to investigate changes of the pulsation activity with solar activity and solar wind parameters, respectively. It is interesting to note that not only the pulsation activity is connected to solar activity, but the latter influenced actual problems of research, too. Thus e.g. during high solar activity high electron concentration in the ionosphere/plasmasphere (f_oF2 more than 11 MHz) results in some damping of the pulsations. This situation occurs at solar maximum, during local winter.

Less disturbed conditions during solar minimum create favourable conditions for the study of connections between solar wind and interplanetary magnetic field parameters and pulsation activity. These connections may be destroyed during high geomagnetic activity intervals which occur mainly during solar maximum.

Pulsation studies carried out on arrays proved the existence, sometimes co-existence of slightly transformed upstream waves and of shell resonances. A few data from the last solar cycle indicated a change in the occurrence of the two types, without a clear connection with solar activity.

HISTORICAL REVIEW

The Geophysical Laboratory of the Hungarian Academy of Sciences decided in 1956 to build up a new geophysical observatory for the International Geophysical Year. Due to the activity of, and to connections with, the French Schlumberger Company, geophysicists had experience in earth current (or 'telluric') measurements, even an instrument was produced for a future expedition to China. Two instruments of the same type were installed in the observatory, at a very carefully selected site near the village Nagycenk:

$$\varphi = 47°38', \quad \lambda = 16°43'$$
$$\Phi = 47.2°, \quad \Lambda = 98.3°$$
$$L \sim 1.9$$

(φ, λ latitude and east longitude, Φ, Λ magnetic latitude and longitude).

Solar Wind Sources of Magnetospheric Ultra-Low-Frequency Waves
Geophysical Monograph 81

In this village Count István Széchenyi, a reformer in the 19th century who had given an immense sum for the foundation of the Hungarian Academy, had a family home. Continuous recording started on August 2, 1957, just at the beginning of the IGY. Soon the great geomagnetic storms in September 1957 meant a great surprise with all supposed limits of the activity surpassed.

At that time pulsation research had just restarted, due to activities in France, England and in the Soviet Union. The French Schlumberger Company carried out measurements at different places of the world, and Kunetz [1952] confirmed the simultaneous occurrence of certain pulsations at very great distances. Dungey [1954] was the first to produce a promising theory for the pulsations, using hydromagnetic waves in the magnetosphere — in spite of the fact that neither the term 'magnetosphere', nor its structure were known. Troitskaya [1956, 1957] in Central Asia started with earthquake prediction; she set up stations for recording pulsations, and deduced a most intriguing idea about UT-dependence of the pulsation activity. Some years later the first measurements on arrays, e.g. by Jacobs and Sinno [1960, 1961], introduced the idea of the latitude dependence of pulsation periods, supporting Dungey's theory on the hydromagnetic origin of the pulsations. But even the morphology was hardly known, and the names Pc, Pi, came just into general use [Jacobs et al., 1964].

This was the situation at the beginning of the activity of the Nagycenk observatory. The first years, perhaps even decades went by with efforts to establish a classification for the pulsations. The merits of the earliest method are illustrated by Figs. 1 and 2, where the daily distribution of pulsations with different periods and an average 'spectrum' interpreted as the sum of two distributions of the pulsation periods are shown [Verő, 1958]. And at this point the connection between history of the pulsation research and actual solar — geomagnetic activity started. Namely, during the late fifties the solar activity was very high. As we know today, in such cases pulsation amplitudes are strongly damped in the F2 region of the ionosphere or in the plasmasphere above it, in winter of the corresponding hemisphere, and perhaps also in the December halfyear at both hemispheres. It was curious to note that in contradiction to most previous results, pulsation activity had a very significant yearly wave, with maximum in summer and minimum in winter. Simultaneously with this, the first long series of plasma densities at the equatorial plane deduced from whistler observations were published. A comparison showed that during high solar activity, the correlation between the two series (monthly means) was very high, while with decreasing solar activity (and decreasing ionospheric-plasmaspheric plasma densities) nothing remained from this correlation (Fig. 3), [Verő, 1965].

Simultaneously with pulsation research, geomagnetic induction studies were also started in Hungary. At first a so-called 'absolute earth current method' was used which supposed homogeneous subsurface structure at a carefully selected site used as a basis. It was obvious to use the Nagycenk observatory as such a basis, but the homogeneity of the subsurface proved soon to be an illusion. Nevertheless, measurements made at about 100 sites in Hungary produced valuable material for the study of the pulsations, too, among others ideas about changes of pulsation periods with latitude. [Ádám and Verő, 1964; Ádám et al., 1972].

In the late sixties, interest decreased worldwide in pulsations. Connections between magnetospheric parameters and pulsations were looked for in vain, groups in Western and Eastern countries worked rather separately, with few connections among the groups and it seemed that some 'random' source plus hydromagnetic waves along field lines give sufficient explanation to all observed facts. This was mainly valid for Pc 3-4. Saito [1961] has already found the basic questions of these pulsations: one of them was whether Pc 3 and Pc 4 differ from each other in their origin, the other being the daily variation of pulsation amplitudes.

The data of the Nagycenk observatory helped us to try to partially answer these questions. First of all, we have introduced a classification of the pulsations [Holló and Verő, 1967; Holló et al., 1972; Verő, 1972; Verő and Holló, 1972; Tátrallyay and Verő 1973b] which included a subjective de-

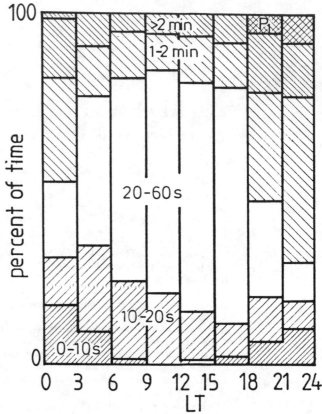

Fig. 1. Daily distribution of the pulsation activity vs. period in 1957–1958. The predominance of Pc 3 (as periods 20–60 s) during daytime, and Pi 1 and Pi 2 (as periods 0–10 s, 1–2 min > 2 min, and P1) during night are evident [Verő, 1958].

scription of their regularity. 'O' denoted regular sinusoids, oscillations, then followed 'Q', 'W' and irregular 'T'. Regular waveforms were most often found around periods of 20 to 25 s (see Fig. 2, too) (we used 12 period ranges in this classification (Table 1)). Periods longer than 30 s were found to be seldom regular. It was, however, clear that this limit did not coincide with the limit between Pc 3 and Pc 4 set at 45 s. Not only the regularity changed at 30 s, but also e.g. the connection with the geomagnetic activity: below it, the activity increased with increasing Kp, above it the connection was opposite. A further point was that the non-Markovian character of daily pulsation indices was found, i.e. some kind of memory is to exist, most likely in the plasmasphere's dimensions, for past geomagnetic activity (Fig. 4), [Verő, 1974].

Our first experiences with long distance correlations were obtained around 1955, between Sopron (Hungary) and Beijing (China), [Ádám, 1958], then comparisons partly with observatory records (e.g. from Tamanrasset, M'Bour, Niemegk, Budkov), partly with records obtained from geophysical expeditions (e.g. from Mongolia) followed [Ádám et al., 1966; Cz. Miletits and Verő, 1971]. These have shown a global temporal coincidence of certain pulsation impulses

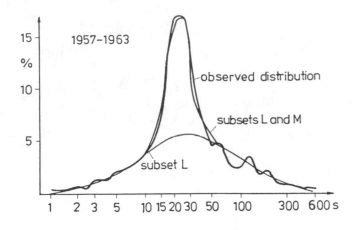

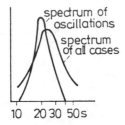

Fig. 2. An early experiment to distinguish two kinds of signals in the (Pc 3–4) pulsation activity. The lower, thin line refers to slightly transformed upstream waves according to the present knowledge, the area between the two curves pulsations originating from to field line resonances. The latter approximates the spectrum of regular (0) events (insert). 'Spectra' mean here subjective estimations from analog records [Verő, 1980; from unpublished thesis, 1964].

TABLE 1. Period Ranges Used in the Pulsation Catalogue Nagycenk

P1	1– 5 s	P5	20–25 s	P9	60– 90 s
P2	5–10 s	P6	25–30 s	P10	90–120 s
P3	10–15 s	P7	30–40 s	P11	120–300 s
P4	15–20 s	P8	40–60 s	P12	300–600 s

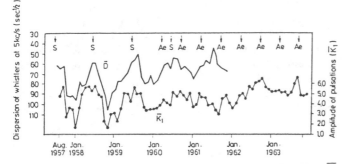

Fig. 3. Correlation between equatorial particle densities $\overline{D}$ (monthly average whistler dispersion) and pulsation activity, i.e. monthly average pulsation indices $\overline{K}_1$ (from NCK) at and after the solar maximum in 1957 [Verő, 1965]. Whistler dispersion is plotted with inverted axis (increasing downward).

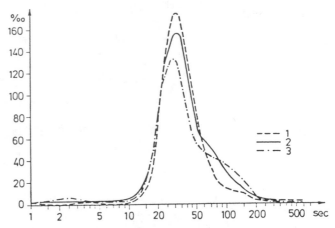

Fig. 4. 'Memory' for past geomagnetic activity in the magnetosphere. The three spectra are for days of low geomagnetic activity, when the average geomagnetic activity of three previous days was (1) high, (2) medium, (3) low, for the years 1966 to 1972 [Verő, 1974].

(Figs. 5, 6), thus confirming Troitskaya's assumption on the UT-dependence, but at the same time showing the much stronger influence of local time. Moreover, indications were obtained that the more regular the pulsations are, the more pronounced the latitude dependence of the periods is. For example, regular Nagycenk events with 20 to 25 s period coincided with 10 to 13 s activity at the low-latitude station Tamanrasset. In the case of the observatories Niemegk and Nagycenk, the value 10 percent change of the period for a degree of geomagnetic latitude was established.

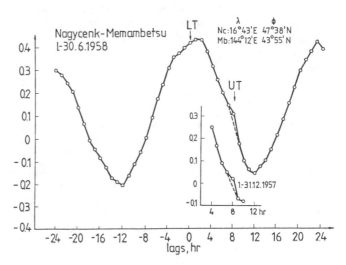

Fig. 5. Cross-correlation functions between hourly pulsation amplitudes at NCK and Memambetsu, Japan. The function for June 1958 is shown completely, for December 1957, only the most significant part. The difference in LT between the two stations is about 8h. Zero lag means the same local time at both stations. [Cz. Miletits and Verő, 1971].

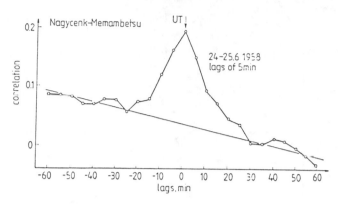

Fig. 6. Cross-correlation function between NCK and Memambetsu for 5 min average pulsation amplitudes. With the higher time resolution, the UT peak gets sharper. Zero lag means the same local time at both stations [Cz. Miletits and Verő, 1971]. Lower correlation at zero lag is due to the shorter interval of averaging.

The idea that the whole range of periods of the pulsations 'shifts' with the latitude (or L-value) found no acceptance at that time. In spite of Dungey's theoretical results, the limit 45 s between Pc 3 and Pc 4 was supposed to be valid everywhere. This limit was by no means a merely morphological one, it was supposed to be a physical limit, too. It was sometimes forgotten that when establishing this classification, mostly observatories at L-values 2.5 to 3 were used which are, as we know today, just in the region of the strongest field line resonances, exactly with periods of 35 to 45 s. At a lower latitude station, like at Nagycenk at $L \sim 2$, this limit shifts to about 25 s.

With the advent of the availability of interplanetary and magnetospheric data from in situ measurements a new era began in the pulsation research, too. As soon as the first Soviet sputnik was launched in 1957, and some data became available, comparisons were carried out between these data and pulsations. The existence of such correlations was supported by changes in the pulsation activity with solar activity, by connections (even if they were complicated ones) with the geomagnetic activity, and not to forget, with particle densities deduced from whistler observations. As soon as solar wind velocity data became available, the connection between pulsation activity and solar wind velocity became evident. The solar wind velocity could be estimated from pulsation data, from amplitudes as well as from periods, even more precisely than solely from geomagnetic activity indices (Fig. 7), [Verő, 1975a]. Variables to be used in such an estimation are activities at different periods, as different period ranges gave different responses to changes in the solar wind velocity. It was somewhat surprising, and it remains so even today that periods corresponding to field line resonances indicate more exactly the solar wind velocity than less transformed upstream waves from the interplanetary medium do [Verő and Holló, 1978].

It is worth mentioning why the connection between pulsations and ionospheric-magnetospheric plasma densities had

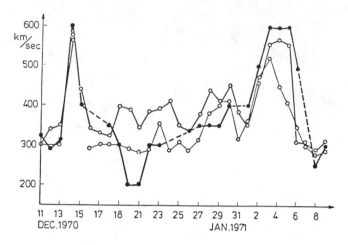

Fig. 7. Daily average solar wind velocity derived from in situ measurements (thick line), from Kp-indices (medium line) and from NCK pulsation data (thin line). Mean square deviation between the first two is ± 76 km/s, between measured data and those computed from pulsation indices ± 66 km/s [Verő, 1975a].

not been studied further at that time. The solar maximum following that during IGY, namely that in 1968–69 was less strong and accordingly neither the geomagnetic activity, nor that of the pulsations responded as clearly as during the previous solar maximum. The yearly wave of the pulsation activity due to damping in local winter was less evident, and other connections with the interplanetary medium met more interest, as partly they referred to the origin of the pulsations, partly they could be more easily studied, without long data series.

The effect of geomagnetic sudden impulses (SI-s) was first detected by Veldkamp [1960]. Later for the explanation of the changes of the pulsation activity during SI-s, Troitskaya et al. [1968] proposed a model where the change of the size of the magnetosphere played the main role. According to our observations [Tátrallyay and Verő, 1973a], the most important change in connection with SI-s occurred some time after the impulse (Fig. 8). The activity before SI-s was mostly longer period Pc 4 which were switched off during the impulse, and some minutes after the impulse, shorter period activity, typically in the Pc 3 range, appeared and continued for some time, for about one hour or so. Such events could be traced at higher latitudes, in the vicinity of the auroral zone, where accompanying Pc 1 activity was also detected (Fig. 9).

At that time, in the seventies, the connection between the parameters direction and scalar magnitude of the interplanetary magnetic field, and pulsation amplitudes and periods, respectively, became known as well [Gul'elmi, 1974]. As Interplanetary Data Books made great amounts of solar wind and interplanetary magnetic field (IMF) data available, we could also start to investigate this problem, and based on the complete material of two years, we could confirm, in some points even supplement the previous results. As this investigation included local time and geomagnetic activity,

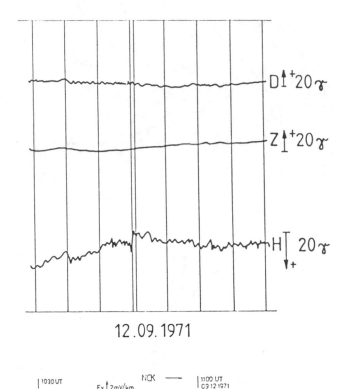

Fig. 8. An impulse in the geomagnetic field (NCK record, top) and on quick-run electric record (bottom, E_y) [Tátrallyay and Verő, 1973a]. Longer period Pc 4 activity stops at the impulse, some minutes later, shorter period Pc 3 activity starts.

too, we could look for different forms of the connections between pulsations and the interplanetary medium in different conditions [Verő and Holló, 1978; Verő, 1980].

The approximate connection, T = 160/|B|, was found to be valid as an average in most situations, however, the accuracy of the approximation changed. Its validity included |B|-values between a few nT-s and values up to or even slightly over ten nT (Fig. 10). At even higher values of |B|, geomagnetic activity is generally high, too, and in such conditions, pulsation spectra are broad, often without significant peaks or with many peaks, and short periods prevail (Fig. 11). At very low values of |B|, a secondary peak was noticed at 20 to 25 s (Fig. 12), coinciding with the local field line resonant period. Besides, average 'spectra' (deduced from subjectively estimated values of periods and amplitudes, then averaged over certain sets of data) were not very narrow indicating that deviations from the rule T = 160/|B| do occur.

Concerning the influence of the direction of the IMF, our data set (Fig. 13) hinted at a maximum of pulsation amplitudes at cone angles of about 30° deviating from the Sun-Earth line. The same result was obtained by Gul'elmi et

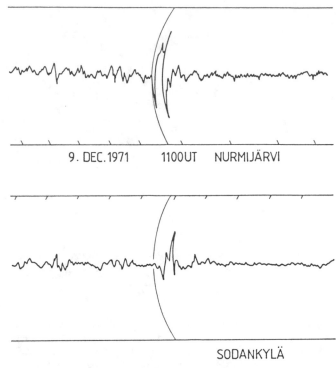

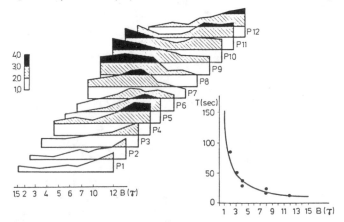

Fig. 9. The geomagnetic impulse presented in Fig. 8 at the stations Nurmijärvi $L \sim 3.5$ and Sodankylä $L \sim 5.1$ [Verő, 1975b]. At NUR, below the auroral zone, the situation is similar to NCK, at SOD, Pc 3 is lacking after the impulse, and some Pc 1 activity is present (not shown). (The time marks are for 5 min intervals).

Fig. 10. Connections of the activity of the 12 period bands of Table 1 with the scalar magnitude of the interplanetary magnetic field (IMF). The periods of the bands with maximum activity for each range of |B| values (activity scale 1.0–5.0, the insert top left gives the location of the band with maximum activity at different IMF scalar magnitudes determined for the complete year 1972) fit well to the formula $T = 160/|B|$ [Verő and Holló, 1978].

al., [1973] and a theoretical paper by Kovner et al. [1976] postulated just these results: maximum activity at 30° and some delay in the excitation of the pulsations after impulses which cause a change in the conditions of the upstream re-

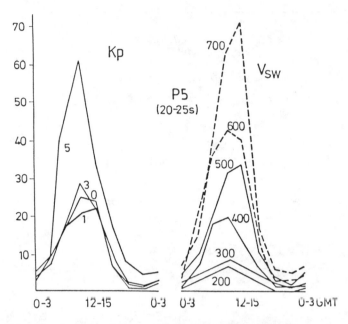

Fig. 11. Activity of the band P 5 (20-25 s) vs. local time, geomagnetic activity, and solar wind velocity (V_{SW}). Each curve is identified by the lowest value of the parameter for the actual curve. Geomagnetic activity hardly influences the activity of this range till Kp=6, at even greater activity, amplitudes grow suddenly. The increase of the amplitude with V_{SW} is also evident [Verő, 1980]. Data of the year 1973.

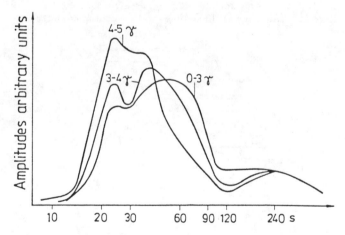

Fig. 12. Subjective pulsation spectra at very low IMF scalar magnitudes (1972–1973). In addition to the peak at T=160/|B|, an additional peak appears at the NCK local field line resonance, i.e. at 20–25 s [Verő, 1980].

gion. Some minutes would be necessary to build up again the upstream source corresponding to the new conditions as found by Tátrallyay and Verő [1973a]. Since the maximum at a cone angle of 30° did not occur again in any other investigation, this idea was dropped; if it was correct or not, remains a question. A new, more exact investigation would need very long data series to be used.

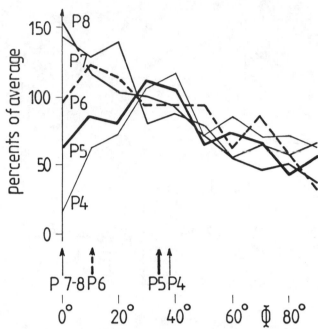

Fig. 13. Average hourly pulsation relative amplitudes in different bands vs. cone angle for all data and for 6 to 18h LT. At the field line resonance (20–25 s, Pc 5), maxima are at cone angles of 30°. Same data as for Fig. 10. [Verő and Holló, 1978].

The third parameter of primary importance for the pulsation activity is solar wind velocity. We have already mentioned the long-term connection between pulsation amplitudes and solar wind velocity. The connection is valid for shorter intervals, e.g. for hourly averages, too. The connection, however, did change with the period of the pulsations. We found that a connection of the form $A = c \cdot V_{SW}^n$ is a proper approximation for a wide range of periods, where A is the amplitude in the corresponding period range, c a constant, and n an exponent which changes with the period. The greatest values of the exponent n, about 3, were found for periods at about 15 to 25 s (Fig. 14), just at the period of the field line resonances, thus solar wind would influence most effectively the field line resonances.

An indication of the source of this close connection came from a comparison between NCK and satellite data. One year of the magnetometer data of the satellite ATS 6 were used to derive daily indices of the pulsation activity [Holló and Verő, 1987]. ATS activities were found to depend on the solar wind velocity. A certain daily ATS 6 value yielded, however, only a lower limit for the pulsation activity to be expected at the surface station (Fig. 15). Any higher activity level could be observed. And the surplus activity at the surface was again correlated with the solar wind velocity, i.e. it could be supposed that this surplus activity was due to field line resonances in the plasmasphere, between the satellite and the surface, and this activity originating from the field line resonance is closely correlated with the solar wind speed, as noticed previously (Fig. 16).

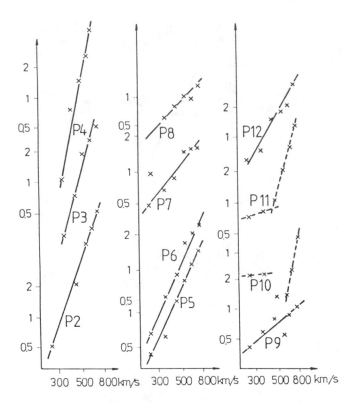

Fig. 14. Average amplitudes in the period ranges of Table 1 vs. V_{SW}, double logarithmic scale. Best fitting linear approximations are also plotted [Verő, 1980].

Several authors, e.g. Plyasova-Bakounina have noted that the connection $T=160/|B|$ is valid only if the period of the pulsations is the same at stations lying rather far from each other in meridional and/or longitudinal directions. That means that pulsations of a world-wide, or at least hemispheric extension correspond to this rule, whereas 'local' pulsations do not obey it. Comparing this with the fact that regular pulsations have periods which change with the L-value, we arrive at the supposition that irregular pulsations would correspond to the rule $T=160/|B|$, and regular ones not. However, it was found that regular pulsations have also periods which change with IMF. This contradiction can be resolved by supposing that the source of the pulsations (upstream waves) has a broad-band spectrum, and when propagating through the magnetosphere, different parts of this spectrum correspond to local field line resonances and excite them [Varga, 1980]. Thus a certain range of the field lines are excited by the incoming wave, and the envelope of these local spectra would correspond to the original broadband spectrum. This supposition has been confirmed in fact by Varga's [1980] calculations. In other cases either the upstream source spectrum does not contain sufficient energy in certain period ranges to excite the field line resonances, or these resonances cannot be excited at all (e.g. if plasmaspheric particle concentrations are high).

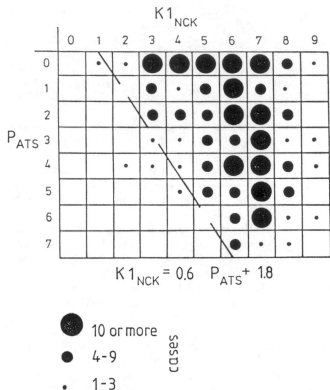

$$K1_{NCK} = 0.6 \; P_{ATS} + 1.8$$

Fig. 15. Daily NCK pulsation indices ($K1_{NCK}$, quasi-exponential scale for average amplitudes) vs. pulsation indices of ATS 6 (P_{ATS} based on the number of hours with pulsations). The dotted line is the lower limit of the NCK index for a given ATS 6 index [Verő and Holló, 1983].

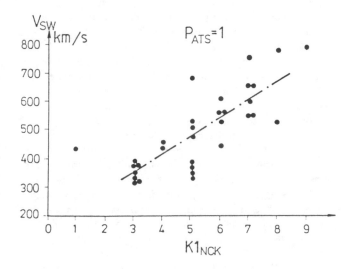

Fig. 16. Dependence of the median of NCK pulsation indices on V_{SW} for an ATS 6 pulsation index of 1. As the NCK index increases with V_{SW} even in this case, the surplus activity at the surface — supposedly due to field line resonances — is influenced by V_{SW}, too [Holló and Verő, 1987].

The selection of pulsations with periods corresponding to the rule $T=160/|B|$ and of those not corresponding to the rule led us to another problem. This is essentially the same as the separation of Pc 3 and Pc 4 pulsations, those with regular and irregular waveforms. In 1977, an array of stations was used for a direct investigation of the latitude dependence of pulsation periods [Cz. Miletits, 1980]. The southernmost station of the array was L'Aquila in Italy, and the northern end was in Finland at auroral latitudes. Several hundred pulsation events were studied from analogue records. Periods were measured with the greatest possible accuracy for intervals containing a dozen cycles. Plots of period vs. latitude (L-value) revealed several facts. The greatest part of the pulsations, especially in the range around 20 to 25 s proved to have latitude dependent periods. The average rate of the change was roughly 10 percent change in the period for a degree of geomagnetic latitude (at latitudes 45 to 55°), (Fig. 17). The change of the period with latitude was, however, mostly not smooth: in certain parts of the array the period changed rather quickly, then in a certain section of the array it remained constant. These sections with constant period were in different parts of the array not bound to a special section, and most frequently they were found at the stations lying nearest to each other, at a distance of about 1.5°. A plausible explanation for these sections with constant period is the existence of resonating shells of field lines, not independent field lines.

The earlier idea that the more regular the pulsations are, the more likely it is that they have latitude dependent periods, could also be verified in this investigation (Fig. 18). Thus, the regularity seemed to be a parameter which can be used at a single station to decide if a pulsation event belongs to the upstream wave-type, or to the shell resonance type. It was supposed that upstream waves correspond more or less to the traditional Pc 4, and shell resonances to Pc 3, at least at latitudes below $L \sim 2.5$, but in this sense the limit between the two types should be on the one hand changing with geomagnetic latitude (L-shell), representing longer periods at higher latitudes, and on the other hand the limit may change from event to event, or even short period events may belong to the Pc 4-type.

The solar maximum around 1980 proved to be again of a very high intensity. In accordance with this high activity, the summer maximum and the winter minimum reappeared. Verő and Menk [1986] have carried out an investigation using two geomagnetic stations at opposite hemispheres (NCK and BEVeridge in Australia). The somewhat surprising result was that the yearly wave of the pulsation activity has both a global component (with maximum in July) and a hemispheric one (with maximum in summer). More correctly, in December and in winter, pulsations are sometimes damped. The condition for the damping can be given in form of a limit in critical frequency f_oF2 at about 11 MHz at both hemispheres (Figs. 19-21). If this limit is surpassed, then pulsation amplitudes decrease at the corresponding hemisphere. The curious point of the decrease is

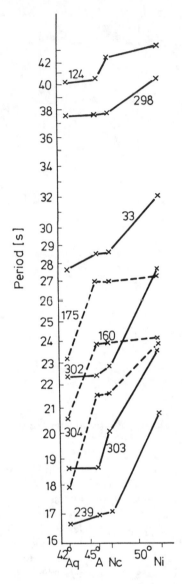

Fig. 17. Change of the period vs. latitude (L-value) for a few pulsation events from the data of the 1977 array. In most cases the increase of the period with latitude is not smooth the period remains constant for a section of the array (Numbers indicate events of a few minutes duration in the original sample set and earlier abbreviations are used for the stations, Aq for L'Aquila, A for a temperary station, Nc for Nagycenk, and Ni for Niemegk) [Cz. Miletits, 1980].

that pulsation amplitudes decrease in a similar ratio during day and night conditions, only the (midday) maximum of f_oF2 counts. Even on days with damping, f_oF2 might be rather low as well during the late afternoon, nevertheless, the damping remains during the whole day. Remembering that in equatorial particle concentrations, deduced from whistler measurements, a similar limit was found [Verő, 1965], the double (global and hemispheric) effect seemed to be confirmed.

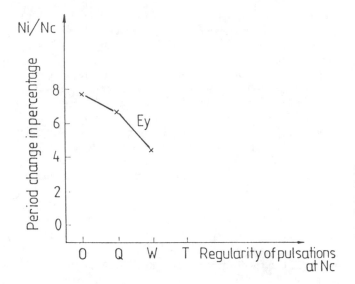

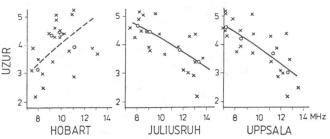

Fig. 20. Monthly average pulsation activity at Uzur (near Irkutsk, Sibiria) vs. monthly median f_0F2 maximum at Juliusruh, Uppsala (Sweden) and Hobart. Uzur pulsation amplitudes depend rather strongly on Juliusruh and Uppsala f_0F2 in spite of the great longitudinal distance, showing the hemispheric character of the effect [Verő and Menk, 1986].

Fig. 18. Change of the period per degree of latitude for Pc 3 plotted against the 'regularity' of these pulsations [Cz. Miletits, 1980].

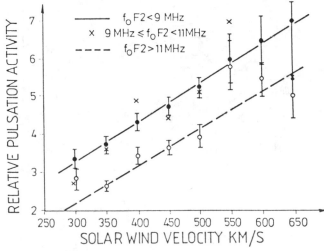

Fig. 21. Average pulsation activity at NCK for low, medium and high values of f_0F2 at Juliusruh [Verő and Menk, 1986].

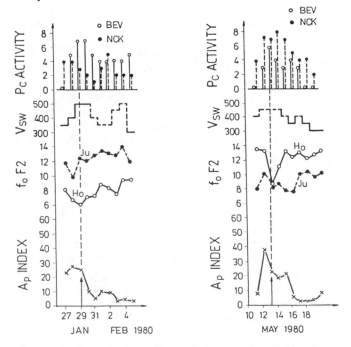

Fig. 19. Daily pulsation indices of the observatories NCK and BEV (eridge), Australia, V_{SW}, f_0F2 at the ionospheric observatories Juliusruh (for NCK) and Hobart (for BEV), and geomagnetic activity, for one northern and one southern hemispheric event (when f_0F2 surpasses 11 MHz) [Verő and Menk, 1986].

Using experience gained in the processing of geomagnetic induction measurements (magnetotellurics), high resolution dynamic spectra were obtained from electromagnetic records. This technique was applied for the first time in a comparison between NCK and KVB (Kvistaberg), a station in Sweden on the same meridian [Cz. Miletits et al., 1988].

This sample contained again a large amount of L-dependent period events, and it was also found that certain structures, e.g. smooth shifts of the period occurred only in case of one type of pulsations, i.e. upstream waves, and never in case of shell resonances, in addition, they are simultaneous at least at the two stations studied (Fig. 22).

Information collected with different measurements in arrays and networks led to several questions, among others to the occurrence of upstream waves and shell resonance-events at different latitudes. A new network initially consisted of seven stations in Eastern Europe, which was supplemented with several stations in Northern Europe, and later with data of GOES 5 and AMPTE/CCE, too [Cz. Miletits et al., 1990]. These measurements were made in 1984, with the southernmost station in Bulgaria, and including auroral stations, such as Tromsø and Kevo, too. From the records of about one month, 24 events were selected, each with a length

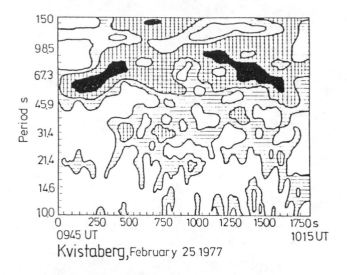

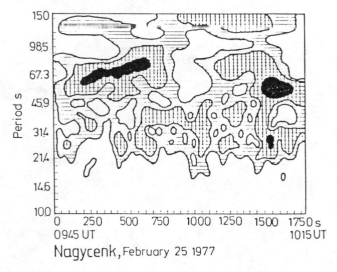

Fig. 22. Simultaneous high resolution dynamic spectra of a pulsation event at NCK and Kvistaberg (Sweden), both on the same meridian. Note the period changes around 50 to 70 s at both stations simultaneously. Contours differ in amplitude by a factor of 2 [Cz. Miletits et al., 1988].

of normally 30 to 180 minutes. The first surprise with this array was the low occurrence of events with L-dependent periods; they were clearly in minority against constant period events, opposite to the 1977 sample. A similar change in the occurrence of latitude dependent periods was observed by Hattingh and Sutcliffe [1987], too. The change in the ratio of the two types is not related to the solar cycle, as far as it can be judged from the relatively low number of data points obtained hitherto.

As a second observation, even in the case of the few events when the period changed with the L-value, often two signals could be distinguished in the period vs. L-value or latitude plots, one with constant period and the other with L-dependent periods. The simultaneous occurrence of the two types was a new fact, it could not be observed with the previous technique, when only one period could be identified at each station. Thus it is possible that upstream waves and shell resonances are simultaneously active.

The selection of pulsation events for the 1984 array at higher latitude stations in the auroral zone, especially at Tromsø [Verő et al., 1990], led to the fact that regular pulsation events at these latitudes belong in each case to the type with L-independent periods. In geomagnetically quiet times, some regular pulsation events were found in these records, which proved to belong to this type without exception. The same observation was made by other authors, too, who supposed that upstream waves have a higher latitude maximum than magnetospheric pulsations (shell resonances). At the poleward edge of the auroral zone, i.e. at the southern boundary of the polar cap, a regular change was found in the period of the pulsation events, which corresponded to theoretical predictions [Warner and Orr, 1979; Singer et al., 1981].

At the other extremity of latitudes, near the geomagnetic equator, pulsation activity was found to be rather different from that at mid-latitudes. Both at the geomagnetic equator (Baclieu in Vietnam) and just outside of the equatorial electrojet (Dhargapur in India) [Verő et al., 1991] shorter period pulsations were absent, only longer period Pc 4 could be observed. There was no correlation of this activity with mid-latitude pulsations, more exactly, there was a slight similarity with higher latitude stations (NGK at $L \sim 2.6$) where the period of the shell resonance corresponded to the observed periods of the equatorial pulsations, while at lower latitudes (around $L \sim 2$) the correlation was completely absent. The situation was very similar at both equatorial stations, thus some leakage of energy cannot be excluded from shell resonances around $L \sim 2.5$ to the equatorial region.

Recently an exhaustive study has been published on the long-term variations of pulsation activity at the Nagycenk Observatory, which confirmed the previous results [Zieger, 1991]. Significant 11-year, annual and semiannual variations of the pulsation activity was reported and compared with solar wind velocity, geomagnetic activity, and F2 region electron density. The long-term variation of the total pulsation activity (K1 index) exactly follows the so-called solar wind cycle, which is the 11-year cyclic variation of the solar wind velocity with a maximum 1-3 years before solar activity minimum (Fig. 23). The annual variation of K1 reflects the combination of two effects: one is the annual variation of the solar wind, and the other is some damping mechanism related to F2 region electron density. The annual variation of the total pulsation activity is the highest in the years of solar activity maximum, and it has a maximum in June showing strong anticorrelation with the annual variation of f_oF2. However, in the years of solar activity minimum when there exist significant annual variations in the solar wind velocity, the annual variation of K1 correlates with the solar wind velocity (Fig. 24). The semiannual variation of the total pulsation activity, however, seems to be

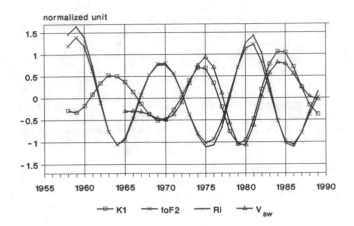

Fig. 23. Solar cycle variation of the total pulsation activity at NCK (K1), comparing with F2 region electron density (f_oF2), solar activity (R_i: relative sunspot number), and solar wind velocity (V_{SW}) [Zieger, 1991].

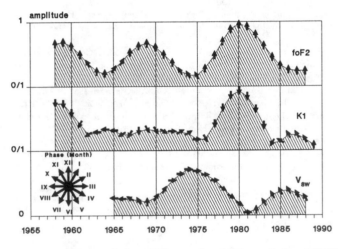

Fig. 24. Annual variation of the total pulsation activity at NCK (K1), comparing with F2 region electron desnity (f_oF2), and solar wind velocity (V_{SW}). The phase of the annual variation is indicated by arrows [modified, Zieger, 1991].

controlled mainly by the geomagnetic activity. The analysis of occurrence and amplitude data for pulsations in 12 period bands revealed that the total pulsation activity refers mainly to Pc 3 pulsations, and that the pulsations with a period of 20-25 s corresponding to the eigenperiod of the local field-line at Nagycenk, have the strongest correlation with solar wind velocity.

At present, the most significant investigations are: searching for different periodicities in the pulsation data series of the NCK observatory covering three solar cycles now, a German-Italian-Hungarian cooperation to identify factors governing upstream wave penetration and shell resonances, and studies of the transmission of pulsations from satellite altitudes to the surface.

We think that in spite of the progress in the understanding of the excitation and propagation of geomagnetic pulsa-

tions, many problems remained unexplained. Thus further research is needed to explain e.g. the occurrence of pulsations with latitude dependent periods and of those whose periods do not depend on the latitude, or to find the causes for the winter damping during high solar activity.

Acknowledgements. The authors are grateful to the two anonymous referees who significantly helped to improve the quality of the present paper. The research presented in this paper is supported by the Hungarian state grant OTKA No. 1171.

REFERENCES

Ádám, A., Über ein modifiziertes tellurisches Schurfgerät und dessen Anwendung zu tellurischen Untersuchungen grossen Ausmasses. *Freiberger Forschungshefte*, C45, 52, 1958

Ádám, A. and J. Verő, Ergebnisse der regionalen tellurischen Messungen in Ungarn. *Acta Technica*, 47, 63-77, 1964

Ádám, A., J. Verő and Á Wallner, Die räumliche Verteilung der Pulsationen des elektromagnetischen Feldes der Erde. *Acta Geod. Geoph. Mont. Hung.*, 1, 379-417, 1966

Ádám, A., J. Cz. Miletits and J. Verő, Das Mikropulsationsfeld in Osteuropa (Ergebnisse der KAPG-Synchronmessungen vom Jahre 1969). *Acta Geod. Geoph. Mont. Hung.*, 7, 289-304, 1972

Cz. Miletits, J., Microstructure of the latitude dependence of Pc-type pulsation periods. *J. Atmos. Terr. Phys.*, 42, 563-567, 1980

Cz. Miletits, J. and J. Verő, Correlated Pc-type micropulsations observed at separate places. *J. Atmos. Terr. Phys.*, 33, 967-970, 1971

Cz. Miletits, J., J. Verő and W.F. Stuart, Dynamic spectra of pulsation events at L 1.9 and L 3.3. *J. Atmos. Terr. Phys.*, 50, 649-656, 1988

Cz. Miletits, J., J. Verő, J. Szendrői, P. Ivanova, A. Best and M. Kivinen, Pulsation periods at mid-latitudes. A seven-station study. *Planet. Space Sci.*, 38, 85-95, 1990

Dungey, J.W., Electrodynamics of the outer atmosphere. Ionosphere Research. Lab., Sci. Rept. No. 69, 1954

Gul'elmi, A.V., Diagnostics of the magnetosphere and interplanetary medium by means of pulsation. *Space Sci. Res.*, 16, 331, 1974

Gul'elmi, A.V., T.A. Plyasova-Bakounina and R.V Shepetnov, O svjazi perioda geomagnitnikh pulsatsii Pc3, 4 s parametrami mezhplanetnoj sredi na orbite zemli *Geomagn. Aeron.*, 13, 382, 1973

Hattingh, S.K.F. and P.R. Sutcliffe, Pc 3 pulsation eigenperiod determination at low latitudes. *J. Geophys. Res.*, 92, 12433, 1987

Holló, L. and J. Verő, Zur Charakterisierung der Pulsationstätigkeit des Erdmagnetischen Feldes für einzelne Tage und deren Anwendung. *Acta Geod. Geoph. Mont. Hung.*, 2, 335-349, 1967

Holló, L. and J. Verő, Comparison of one year of data on geomagnetic pulsations at geostationary orbit and at a low latitude ground station *J. Atmosph. Terr. Phys.*, 49, 1147-1153, 1987

Holló, L., M. Tátrallyay, and J. Verő, Experimental results with the characterization of geomagnetic micropulsations I. *Acta Geod. Geoph. Mont. Hung.*, 7, 155-166, 1972

Jacobs, J.A. and K. Sinno, World-wide characteristics of geomagnetic micropulsations. *Nature*, 188, 285, 1960

Jacobs, J.A. and K. Sinno, The morphology of geomagnetic micropulsations pc. Symposium on Rapid Magnetic Variations, Utrecht, 82, 1959, 1961

Jacobs, J.A., Y. Kato, S. Matsushita and V.A. Troitskaya, Classification of geomagnetic micropulsations. *Geophys. J.*, 8, 341, 1964

Kovner, M.S., V.V Lebedev, T.A. Plyasova-Bakounina and V.A. Troitskaya, On the generation of low-frequency waves in the solar wind in the front of the bow shock. *Planet. Space Sci.*, 24, 261, 1976

Kunetz, G. and H. Richard, Comparison des variations rapides du champ tellurique entre stations située à grande distance. Atti del settimo convegno nazionale del metano e del petrolio, I. (1952), Taormina, 511

Saito, T., Oscillation of geomagnetic field with the progress of pt-type pulsation. *Sci. Rept. Tohoku Univ.*, Ser. 5., 13, 53, 1961

Singer, H.J., D.J. Southwood, R.F. Walker and M.G. Kivelson, Alfven wave resonances in a realistic magnetospheric magnetic field geometry. *J. Geophys. Res.*, 86, 4589, 1981

Tátrallyay, M. and J. Verő, Changes of geomagnetic micropulsations following sudden impulses. *J. Atm. Terr. Phys.*, 35, 1507-1515, 1973a

Tátrallyay, M. and J. Verő, Experimental results with the characterisation of geomagnetic micropulsations. IV. *Acta Geod. Geoph. Mont. Hung.*, 8, 217-225, 1973b

Troitskaya, V.A., Korotkoperiodicheskie vozmushenija elektromagnitnogo polja Zemli. *Tr. geof. inst.*, 32, 26, 1956

Troitskaya, V.A., Zakonomernosti vozbuzhdenija po mirovonnu vremeni dvukh osnovnikh tipov korotkoperiodicheskikh kolebannij elektromagnitnogo polja Zemli i ikh svjaz s korpuskuljarnimi potokami. Tr. konferentsii po fizike solnechnikh korpuskuljarnikh potokov 246, 1957

Troitskaya, V.A., R.V. Shepetnov and A.V. Gul'elmi, Opredelenie razmerov podsolnechnoj magnitosferi po dannim nazemnikh nabljudenij geomagnitnikh mikropulsatsii. *Dokladi*, AN SZSZSZR, 182, 1063, 1968

Varga, M., A numerical study of the excitation of Pc 2-4 type pulsations. *J. Atm. Terr. Phys.*, 42, 4, 365-369, 1980

Veldkamp, J., A giant geomagnetic pulsation. *J. Atm. Terr. Phys.*, 17, 320, 1960

Verő, J., Über einige Ergebnisse des Erdstrom-Observatoriums bei Nagycenk in Ungarn. *Zschr. f. Geoph.*, 24, 214, 1958

Verő, J., Seasonal distribution of geomagnetic pulsations. *J. Geophys. Res.*, 70, 2254, 1965

Verő, J., Experimental results with the cheracterization of geomagnetic micropulsations III. *Acta Geod. Geoph. Mont. Hung.*, 7, 177-190, 1972

Verő, J., Pulsation activity as a non-Markovian process. *Acta Geod. Geoph. Mont. Hung.*, 9, 291-297, 1974

Verő, J., Determination of the solar wind velocity from pulsation indices. *J. Atm. Terr. Phys.*, 37, 561-564, 1975a

Verő, J., Geomagnetic pulsations around impulses (Role of the plasmapause). *Acta Geod. Geoph. Mont. Hung.*, 10, 247-253, 1975b

Verő, J., Geomagnetic pulsations and parameters of the interplanetary medium. *J. Atm. Terr. Phys.*. 42, 371-380, 1980

Verő, J. and L. Holló, Experimental results with the characterization of geomagnetic micropulsations II. *Acta Geod. Geoph. Mont. Hung.*, 7, 167-176, 1972

Verő, J. and L. Holló, Connections between interplanetary magnetic field and geomagnetic pulsations. *J. Atm. Terr. Phys.*, 40, 857-867, 1978

Verő, J. and L. Holló, A comparison of Pc magnetic pulsations on the ground and at synchronous orbit. *Acta Geod. Geoph. Mont. Hung.*, 10, 400-410, 1980

Verő, J. and F.W. Menk, Damping of geomagnetic Pc3-4 pulsations at high F2-layer electron concentrations. *J. Atm. Terr. Phys.*, 48, 231-245, 1986

Verő, J., L. Holló, A. Egeland and A. Brekke, Connections between high- and middle-latitude pulsations *J. Atm. Terr. Phys.*, 52, 789-796, 1990

Verő, J., L. Holló and B.P. Singh, Geomagnetic pulsations at low- and mid-latitudes. *Acta Geod. Geoph. Mont.*, 26 253-263, 1991

Warner, M. and D. Orr, Time of flight calculations for high latitude geomagnetic pulsations. *Planet. Space Sci.*, 27, 679, 1979

Zieger, B., Long-Term variations in pulsation activity and their relationship to solar wind velocity, geomagnetic activity, and F2 region electron density. *J. Geophys. Res.*, 96, 21115-21123, 1991

J. Verő and B. Zieger, Geodetic and Geophysical Research Institute of the Hungarian Academy of Sciences, POB 5, H-9401 Sopron, Hungary.

Long Term Variations in the Solar Wind
of Importance to ULF Phenomena

J. G. Luhmann, S. M. Petrinec and C. T. Russell

Institute of Geophysics and Planetary Physics, University of California,
Los Angeles, California

Although solar wind control of ULF pulsation activity can be difficult to detect on a case-by-case basis, statistical studies sometimes show correlations of power, period or occurrence rate with interplanetary quantities such as solar wind velocity or magnetic field magnitude or orientation. If these correlations indeed reflect a mechanism that is controlled primarily by the solar wind, then the pulsations should also exhibit long term (e.g., solar cycle) trends similar to those of the causative parameter. In this paper we show long term trends of some potentially key solar wind properties as observed in the IMP-8 plasma and field data from the last ~ 1.5 solar cycles (1973-1988). We point out that most of the published multi-year statistical studies of pulsations are over a decade old. They need to be updated and extended before they can be used to gain further insights regarding both the sources of pulsations and the use of pulsations as diagnostics of solar wind and magnetospheric states.

Introduction

The frequent act of dividing magnetospheric ULF pulsations into "internal" and "external" categories (e.g., B. J. Anderson, this volume) implies that there are classes of these phenomena that result solely from magnetospheric processes. However, magnetospheric processes generally exist as a result of the distortion of the Earth's magnetic field by the solar wind. Indeed, it is worth asking whether any of the known ULF waves would occur in a dipole-dominated magnetosphere in the absence of the solar wind. The usual rule is to distinguish "external" from "internal" sources according to whether the wave energy enters directly from the upstream region as opposed to being produced within the magnetopause. However, if one considers proposed "internal" drivers: charged particle precipitation (Pi 1 and some Pc 4); impulsive processes in the magnetotail (Pi2); and instabilities of the particle distributions in the ring current (IPDP and some Pc 5), it seems clear that the solar wind interaction plays a fundamental, albeit sometimes indirect, role in generating and/or controlling the properties of all of these types of

pulsations. Here we adopt the viewpoint that because the solar wind is responsible for determining the basic configuration of the magnetosphere, its size (through dynamic pressure variations), its internal energy state (through interplanetary magnetic field (IMF) orientation) and the stability of its boundary (through velocity - shear induced instabilities, IMF cone angle effects, and dynamic pressure pulses), all ULF activity is potentially relatable to solar wind properties at least on a long time scale. Probably the only internal property that is separately controlled to a major extent is the density and composition of the magnetospheric plasma, for which solar EUV radiation is a key parameter through its effects on the upper atmosphere and ionosphere.

Like the EUV intensity, the major solar wind factors influencing ULF pulsation activity vary due to the changes in the sun that produce the solar cycle. One thus expects pulsation activity or properties to show temporal trends reflecting the long-term behavior of the associated solar wind parameters. However, most studies of solar cycle variations in pulsation activity [e.g., examples given by Saito, 1969] only make comparisons with the sunspot number which, though an adequate representation of the time variation in the EUV radiation intensity, does not necessarily reflect the trends in solar wind velocity, dynamic pressure, or magnetic field parameters. Moreover, these early studies were carried out before all of the modern classification schemes existed. The present paper is devoted to the presentation of results on the long term behavior of solar wind plasma and field parameters as seen over the last ~ 1.5 solar cycles on the IMP-8 spacecraft. When combined with future long-term surveys, these observed variations can be used to reinforce or reconsider current thinking on the factors controlling magnetospheric ULF pulsations.

DESCRIPTION OF THE DATA

For the purpose of the present study, we used the 5-minute resolution archived IMP-8 plasma and magnetic field data for the period December 1973-December 1987 available from the NSSDC. As shown by Figure 1, this period includes the 1979-80 solar maximum (cycle 21) and the 1974-75 and 1986-87 solar minima (from cycles 20 and 21, respectively). A description of the IMP-8 magnetometer and plasma instruments can be found in reports by King [1974; 1982]. The plasma data used are moments (ion density and velocity) obtained with the LANL instrument. The 3-axis Goddard Space Flight Center magnetometer operated in the range ± 36 nT per sensor, which was adequate for typical IMF variations. The magnetic vector data are here presented in standard GSE coordinates. Magnetosphere and magnetosheath data were removed by using the plasma data as a mask since the plasma data were included in the archive only for periods when the spacecraft was upstream of the bow shock. To determine long term trends, these 5-minute resolution data were used to find first daily averages, and then annual medians of field and plasma parameters. Occurrence histograms of the 5-minute data were also constructed to examine the details of the distributions of the parameter values.

ANALYSIS

Primary solar wind parameters in pulsation theories include the interplanetary magnetic field magnitude (B_T), which has been found to affect the periods of Pc3s and certain classes of Pc4s, the cone angle of the field (the angle between the Earth-sun line and the field) which controls the occurrence and/or power of Pc3 and Pc4 oscillations, and the north-south component (B_z) which controls the importance of the magnetosphere's interconnection with the solar wind and the degree of energy transfer between them. Figure 2 shows the time series of the annual medians of these key field parameters. The absolute value of B_z is used since the sources of B_z in the solar wind (coronal disturbances and stream interactions) should not produce preferred polarities upon averaging over sector structure. Also shown on these plots are the 25% and 75% quartiles, and 10% and 90% deciles of the distributions of points going into each annual median (the 50% level). These give an idea of both the scatter of points around the median and the extremes of the values in each annual data interval. Figure 3 shows similar plots for three important plasma parameters: the velocity (V),

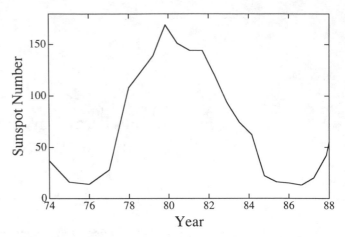

Fig. 1. Annual sunspot numbers during the period of the present study.

which should affect the growth of shear-driven instabilities at the flanks of the magnetopause [e.g., Southwood et al., 1982], the density (n), which together with the velocity determines the incident dynamic pressure, and the dynamic pressure (nmV^2, where m is here set equal to the proton mass), which both controls the size of the magnetosphere's resonant "cavity" that is available to support pulsations, and can itself act as a trigger of pulsations if it increases suddenly [e.g., Sibeck et al., 1990; Kivelson et al., 1990]. Of course, the annual medians do not give information about the history of rapid changes.

One of the most striking features about Figures 2 and 3 is that only the medians of $|B_z|$ (Figure 2b) appear to vary in phase with the sunspot number (although there seems to be a double peak near sunspot maximum that was also observed in the previous cycle $|B_z|$ history by Siscoe et al. [1978]), while the plasma parameters are generally out of phase with the sunspot number. This behavior can be understood as follows. The solar cycle variation of the field magnitude (Figure 2a), which has been studied by others [e.g., see Slavin et al., 1986; Smith, 1990] apparently results from changes in the solar source surface field magnitude at its ecliptic plane intersection. That field generally increases with distance from the magnetic neutral sheet, which becomes increasingly tilted from the ecliptic as the cycle progresses [e.g., Hoeksema, 1992], but it is otherwise a fairly unpredictable quantity given the uniqueness of the field configuration for each cycle. Indeed, Slavin et al. [1986] has shown that the behavior of B_T is quite different for the two cycles (20 and 21) for which we have spacecraft measurements. The different B_z component history, on the other hand, is a reflection of the greater numbers of B_z-enhancing coronal mass ejections (CMEs) that occur near sunspot maximum [e.g., see Gosling et al., 1992]. As has previously been pointed out by Gosling et al. [1992] enhanced B_z levels in the solar wind are statistically related to bidirectional electron events, the most sensitive indicators of

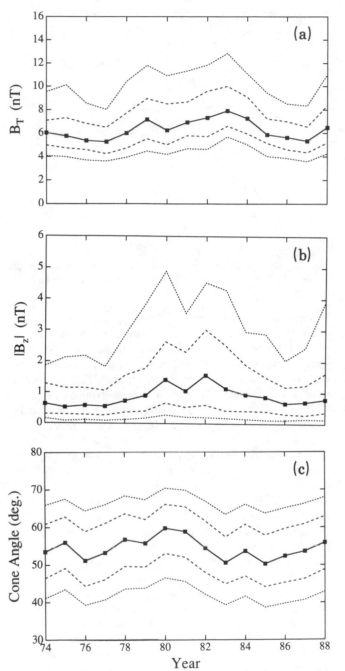

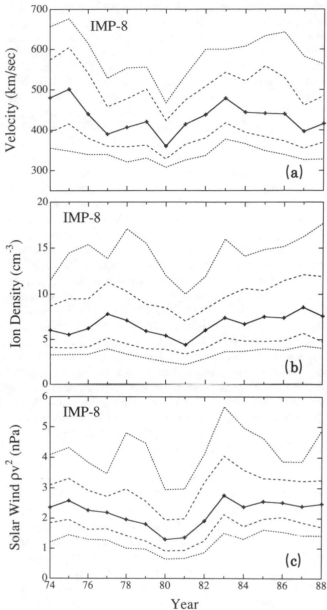

Fig. 2. (a) Variation of annual medians of the IMF magnitude, B_T, for the period of solar activity covered in Figure 1. The dashed lines show the 25% and 75% quartiles of the annual distributions of the 5-minute IMP-8 field data. The dotted lines show the 10% and 90% deciles. Departures of the trends in the quartiles and deciles from those in the medians signify an IMF distribution shape that is changing with time. (b) Same as (a) but for annual medians of the absolute value of the out-of-ecliptic component of the IMF, $|B_z|$. (c) Same as (a) but for annual medians of the IMF cone angle (= $\cos^{-1}(B_{radial}/B_T)$).

Fig. 3. (a) Variation in the annual medians of the solar wind velocity during the period of Figure 1. The format is the same as in Figure 2. (b) Same as (a) but for annual medians of the solar wind density. (c) Same as (a) but for the solar wind dynamic pressure nmV^2 where nm is the approximate mass density obtained from multiplying the measured number density by the proton mass.

CMEs. The last field parameters shown, the cone angle medians, are more related to the behavior of the solar wind velocity than to the solar magnetic field. The solar wind velocity is known to have a strong heliomagnetic latitude dependence [e.g., Newkirk and Fisk, 1985]. The changing inclination and configuration of the magnetic neutral sheet with the solar cycle is expected to bring the high-latitude, high speed wind to the ecliptic plane most often in the periods between sunspot maxima and minima. At solar minimum, the neutral sheet is fairly flat and most nearly parallel to the ecliptic plane, while at solar maximum the high-speed wind sources, the coronal holes, tend to be smaller and weaker [see Gazis et al., 1992]. Thus the lowest median velocities are expected both at solar maximum and at solar minimum, as seen in Figure 3a. The density exhibits the complementary property of being largest near the neutral sheet [e.g., Gazis et al., 1992] and so the density trends (Figure 3b) are roughly opposite except near solar maximum when this order breaks down. As a result, the ecliptic plane dynamic pressure (Figure 3c) is observed to be high just past solar maximum due to the aforementioned velocity enhancement, and again around solar minimum due to the density enhancement. Thus the general "anticorrelation" of the plasma parameters with sunspot number, as well as the slightly different phases of the velocity, density and dynamic pressure histories, can be understood. The cone angle medians in Figure 2c are seen to vary in opposition to the velocity medians because the cone angle is largely determined by the "Parker spiral" angle $\alpha = \tan^{-1}(r\Omega/V)$ (where r is heliocentric distance and Ω is the solar rotation rate). Since large velocities produce small Parker spiral angles, the median cone angles are expected to be smallest for the periods when the high speed streams enhance the median velocities. This expectation is borne out in Figure 4 which shows the cone angle distributions in two velocity ranges for a two year period when classic high and low speed streams were observed.

It is sometimes the case that the extremes of a parameter are more important for the pulsation generation process than the median values. The sometimes different histories of the extrema can be seen, for example, in the 90% decile of nmV^2 (Figure 3c), which follows the general trend of the medians but shows pronounced peaks in the rising and declining phases of the solar cycle. Such features are better understood by examining the complete distributions of the 5-minute data. To improve the statistics of the distributions around the extrema, we here consider 2-year (rather than 1-year) intervals representing the first solar minimum in the period studied (1973-75), the rising phase of solar cycle 21 (1976-78), the solar maximum period (1979-81), the declining phase (1982-84), and the second solar minimum (1985-87). These histograms for the field data (used for Figure 2) and for the plasma data (used for Figure 3) are shown in Figures 5 and 6, respectively. Except in the cone angle histograms, Log_{10} scales are used for the occurrence in order to emphasize the extrema.

One can generally see the behavior of the medians in the positions of the peaks in the B_T and B_z histograms (Figure 5a, b). The "tails" of these distributions, on the other hand, sometimes behave independently. In particular, the high value tail of B_T occurrence (Figure 5a) is clearly most pronounced

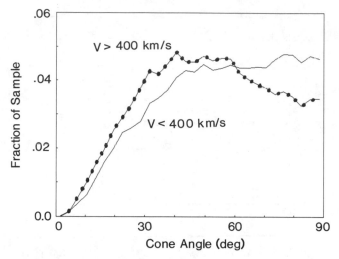

Fig. 4. Cone angle distributions from the 5-minute data from the years 1976-78 for velocities greater than and less than 400 km/s.

in the solar maximum period (1979-81), although median B_T peaks later (1982-84, also see Figure 2a). As one would expect if the high field tail is caused primarily by CMEs (as opposed to stream interaction regions), it is weakest in the earliest and latest (solar minimum) periods. Thus, while the solar source surface field variation controls median B_T, the CMEs determine the high B_T occurrence rate. The median B_z values can be seen in the half-widths of the B_z distributions (Figure 5b). Although the high B_z extrema of these distributions are generally asymmetric with respect to $B_z = 0$, their importance generally parallels that of the half-widths. This behavior, which is seen in Figure 2b from the fact that the 90% deciles of $|B_z|$ track the medians, is expected since all B_z components ideally have their origin in similar (disturbance) sources with the exception of small values introduced by the dipole tilt. The plot in Figure 7, from the same time interval used to construct Figure 4, further suggests that higher $|B_z|$ values are also associated with the highest velocities. The cone angle histograms (Figure 5c) show the effect of the velocity changes over the solar cycle in that the distribution at solar maximum, when median V is lowest, is skewed toward larger angles compared to the solar minimum distributions. The CMEs that occur primarily at solar maximum also act to increase the cone angles due to both the piling-up of the interplanetary field in the CME sheath regions and the structure of the fields within the CMEs.

The velocity histograms (Figure 6a) are particularly enlightening because there is a high velocity "hump" on the solar maximum (1979-81) distribution that can be attributed to CMEs which does not show up in the deciles in Figure 3a. Also notable is the observation that the main body of the distribution from the solar maximum period falls off much more rapidly with increasing velocity than the other distributions. The especially strong contribution of high speed

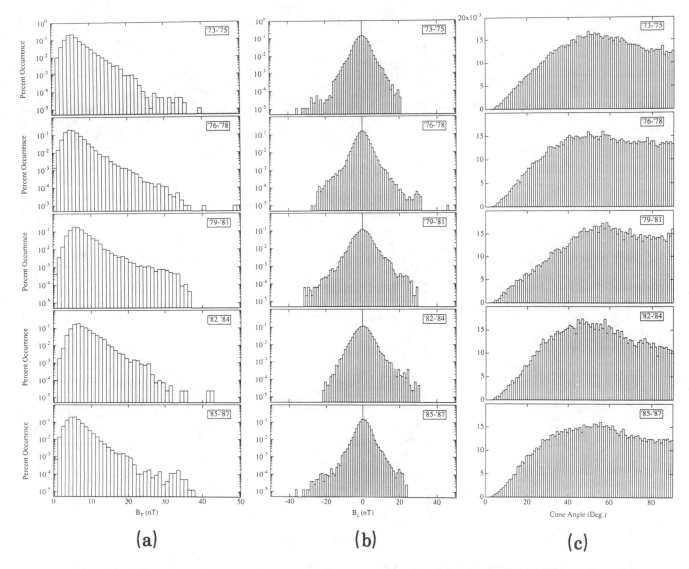

Fig. 5. (a) Biannual distributions of IMF magnitude (5-minute averages) obtained from the IMP-8 archive for the period covered in Figure 1. The high-field tail of the distribution is more prominent during the solar maximum years 1979-81. (b) Same as (a) but for the B_Z component of the IMF. Here the sign is retained to show the degree of departure from symmetry with respect to $B_Z=0$. (c) Same as (a) but for the IMF cone angle calculated from the 5-minute averages of B_{radial} and B_T.

streams to the 1973-75 solar minimum period can be seen from its relatively flat profile to velocities ~ 600 km. The density histograms (Figure 6b) suggest that occasional extreme density values, perhaps from stream interactions, occurred most frequently in the rising (1976-78) and declining (1982-84) phases when the neutral sheet was most inclined and coronal holes were well-developed. The rate of fall-off of the distributions, which is related to the medians, appears fastest for the solar maximum distribution in spite of a probable contribution at higher densities from coronal mass ejections. The symmetry in behavior with respect to the rising and declining phase distributions that is displayed by the densities

is not found in the dynamic pressure distributions in Figure 6c because the distributions of both the velocity and the density are effectively combined here. The rates of fall-off of the distributions with increasing nmV^2 again mirror the behavior of the medians (Figure 3c), but in this case a "hump" in the decline in occurrence between ~ 10 and 15 nPa in 1976-78 and 1982-84 affects the 90% deciles more than do the extended tails of the distributions.

DISCUSSION

As mentioned in the introduction, Saito [1969] summarizes a number of examples of studies of solar cycle trends in

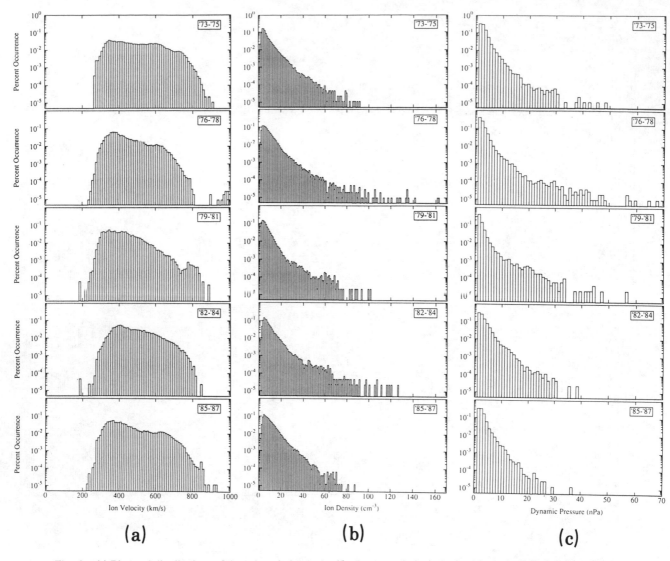

(a)   **(b)**   **(c)**

Fig. 6.   (a) Biannual distributions of the solar wind velocity (5-minute resolution) obtained from the IMP-8 archive for the period covered in Figure 1. A contribution from CMEs is apparent in the solar maximum years 1979-81, while high speed streams flatten the distributions during the 1973-75 and 1985-87 solar minimum periods. (b) Same as (a) but for the solar wind density. (c) Same as (a) but for the solar wind dynamic pressure calculated from the data that were used in (a) and (b).

pulsation characteristics. Although he shows results mainly for the 1955-66 period of solar cycle 19 we here reproduce some of his plots under the assumption that there are some basic similarities in long term solar wind behavior from cycle to cycle. These plots, shown in Figures 8-10, suggest that: pulsation pearl (PP or Pc1) activity is highest during the declining phase of the solar cycle as seen in the sunspot number (Figure 8); Pc 3 periods are highest around sunspot minimum, while Pc 4 periods are lowest then (Figure 9); Pi 2 frequency of occurrence is approximately anticorrelated with sunspot number (Figure 10).

The proposed mechanism for the generation of Pc 1 oscillations, and by inference PP events, invokes a cyclotron

instability of energetic proton distributions [Cornwall, 1965] that depends on the density and energy of the resonant protons. It is not clear whether the solar wind plays a role in this mechanism other than to modulate the occurrence of the substorms during which anisotropic protons are injected into the ring current from the plasma sheet. The peak occurrence of PP activity during the declining phase of the sunspot cycle seen in Figure 8 suggests that higher solar wind velocities either enable or accompany plasma sheet proton injections (see Figure 3a). This interpretation is consistent with the observation by Crooker et al. [1979] and others that V can be used to predict the geomagnetic activity index "aa".

Considering another example, if the periods of the Pc 3s are

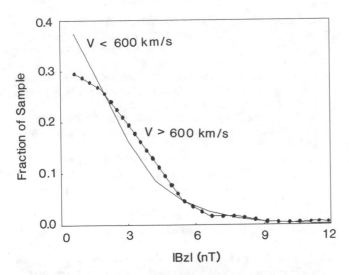

Fig. 7. Similar to Figure 4 but showing magnetic field z-component magnitude distributions. In this case the higher velocity of 600 km/s was used to separate the data since the 400 km/s value gave barely distinguishable results.

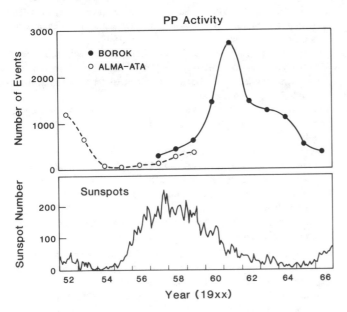

Fig. 8. Figure adapted from Saito [1969], originally from Troitskaya [1967] showing an apparent solar cycle variation in PP event occurrence.

indeed related to B_T, as many studies have shown [Troitskaya et al., 1977; Yumoto et al., 1984], then the implication of Figure 9 is that the solar cycle variation of B_T during cycle 19 was different than that in the period analyzed here, with maximum fields occurring during the end of the declining phase and into solar minimum. (Indeed, it may be that records of Pc 3 frequencies can be used as a proxy for B_T behavior before the age of solar wind monitors such as IMP-8.) The well known relationship of both Pc 3s and Pc 4s to

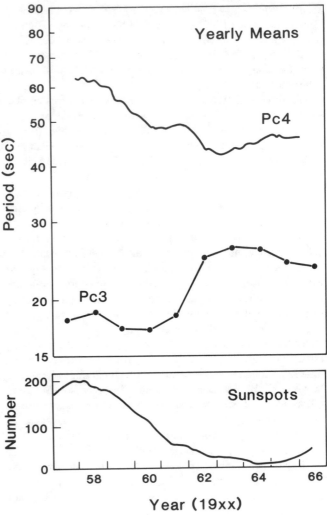

Fig. 9. Figure adapted from Saito [1969] showing results of long-term studies of the periods of Pc3 and Pc4 oscillations (yearly means).

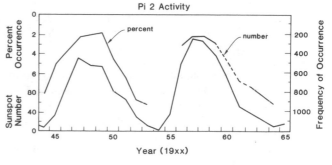

Fig. 10. Figure adapted from Saito [1969] showing results from a study of the occurrence frequency of Pi2 (annual sums).

small interplanetary field cone angles [Troitskaya et al., 1971; Greenstadt et al., 1979] suggests that both types of pulsations may have their source in upstream waves, the frequency of which depend on B_T [e.g., Le and Russell, 1990], but then the opposite solar cycle behavior of the period of Pc 4s shown in Figure 9 is difficult to explain.

The anticorrelation of Pi 2 occurrence with sunspot number suggested in Figure 10 could be consistent with the proposed source of impulsive processes in the tail neutral sheet if those particular processes occur primarily in the undisturbed magnetosphere [e.g., Singer et al., 1988]. The close phasing with sunspot number suggests the possibility of EUV control, perhaps through its effect on ionospheric conductivity. However, it is not apparent why the higher ionospheric conductivities at solar maximum should discourage tail neutral sheet disruptions. There is also some dichotomy in Pi2 behavior since, as pointed out in Saito's [1969] review, their occurrence tends to be positively correlated with geomagnetic activity in the form of bay disturbances on short time scales (e.g., hourly, daily) though negatively correlated with activity when analyzed over long time intervals (e.g., years). Of course, a fundamental problem with all of these efforts at interpretation may be that in the early studies of solar cycle effects during which the data in Figures 8-10 were compiled, little attention may have been paid to spatial and/or temporal biases (e.g., high versus low latitude; day versus night), or to differences between wave properties (e.g., compressional versus transverse). Clearly, new retrospective analyses using updated pulsation classification schemes are necessary if we are to invoke long-term solar wind behavior to help test the validity of generation hypotheses. Such retrospectives could lead to new appreciation of the distinction between, or more likely the coupling of, "external" and "internal" excitation mechanisms.

Acknowledgements. This work was supported, in part, by NSF grant ATM90-16900 to C. T. Russell.

REFERENCES

Cornwall, J. M., Cyclotron instabilities and electromagnetic emission in the ultra low frequency and very low frequency ranges, *J. Geophys. Res.*, *70*, 61, 1965.

Crooker, N. U., J. Feynman and J. T. Gosling, On the high correlation between long-term averages of solar wind speed and geomagnetic activity, *J. Geophys. Res.*, *82*, 1933, 1977.

Gazis, R. P., A. Barnes, J. D. Mihalov and A. J. Lazarus, The structure of the inner heliosphere from Pioneer Venus and IMP observations, *Solar Wind Seven*, edited by E. Marsch and R. Schwenn, pp.183-186, Pergamon Press, New York, 1992.

Gosling, J. T., D. J. McComas, J. L. Phillips and S. J. Bame, Geomagnetic activity associated with Earth passage of interplanetary shock disturbances and coronal mass ejections, *J. Geophys. Res.*, *96*, 7831, 1991.

Gosling, J. T., D. J. McComas, J. L. Phillips and S. J. Bame, Counterstreaming solar wind halo electron events: Solar cycle variations, *J. Geophys. Res.*, *97*, 6531, 1992.

Greenstadt, E. W., H. J. Singer, C. T. Russell, J. V. Olson, IMF orientation, solar wind velocity and Pc3-4 signals: A Joint distribution, *J. Geophys. Res.*, *84*, 527, 1979.

Hoeksema, T., Large scale structure of the heliospheric magnetic field, in *Solar Wind Seven*, edited by E. Marsch and R. Schwenn, pp.191-196, Pergamon Press, New York, 1992.

King, J. H., Interplanetary medium data book, NSSDC/WDC-A-R&S, 77-04, National Space Sci. Data Center, Greenbelt, MD, 1974.

King, J. H., Availability of IMP-7 and IMP-8 data for the IMS period, p.10-15 in *The IMS Source Book*, edited by C. T. Russell and D. J. Southwood, American Geophysical Union, Washington DC, 1982.

Le, G. and C. T. Russell, Observations of the magnetic fluctuation enhancement in Earth's foreshock region, *Geophys. Res. Lett.*, *17*, 905, 1990.

Newkirk, G. Jr. and L. A. Fisk, Variation of cosmic rays and solar wind properties with respect to the heliospheric current sheet 1. Five GeV protons and solar wind speed, *J. Geophys. Res.*, *90*, 3391, 1985.

Saito, T., Geomagnetic pulsations, *Space Science Reviews*, *10*, 319, 1969.

Sibeck, D. G., W. Baumjohann and R. E. Lopez, Solar wind dynamic pressure variations and transient magnetospheric signatures, *Geophys. Res. Lett.*, *16*, 13, 1989.

Singer, H. J., E. W. Hones Jr. and T. J. Rosenberg, Multipoint measurements from substorm onset to recovery: The relation between magnetic pulsations and plasma sheet thickening, *Adv. Space Res.*, *8*, 443, 1988.

Siscoe, G. L., N. U. Crooker and L. Christopher, A solar cycle variation of the interplanetary magnetic field, *Solar Phys.*, *56*, 449, 1978.

Slavin, J. A., G. Jungman and E. J. Smith, The interplanetary magnetic field during solar cycle 21: ISEE-3/ICE observations, *Geophys. Res. Lett.*, *13*, 513, 1986.

Smith, E. J., Interplanetary magnetic field over two solar cycles and out to 20 AU, *Adv. Space Res.*, *9*, (4)159, 1989.

Southwood, D. J., The hydromagnetic stability of the magnetosphere boundary, *Planet. Space Sci.*, *16*, 587, 1968.

Southwood, D. J. and M. G. Kivelson, The magnetohydrodynamic response of magnetospheric cavity to changes in solar wind pressure, *J. Geophys. Res.*, *95*, 2301, 1990.

Takahashi, K., ULF waves in the magnetosphere, *Rev. of Geophys.*, *Supp.*, U.S. National Report to IUGG 1987-1990, pp.1066-1074, 1991.

Troitskaya, V. A., T. A. Plyasova-Bakunina and A. V. Gulyelmi, Relationship between Pc2-4 pulsations and the interplanetary magnetic field, *Dokl. Akad. Nauk. SSSR*, *197*, 1312, 1971.

Yumoto, K., T. Saito, B. T. Tsurutani, E. J. Smith, S-I. Akasofu, Relationship between the IMF magnitude and Pc3 magnetic pulsations in the magnetosphere, *J. Geophys. Res.*, *89*, 9731, 1984.

J. G. Luhmann, S. M. Petrinec, and C. T. Russell, Institute of Geophysics and Planetary Physics, University of California, Los Angeles, CA 90024-1567.

Planetary Upstream Waves

C. T. RUSSELL

Institute of Geophysics and Planetary Physics, University of California, Los Angeles

Waves are observed upstream of the bow shock of all the magnetized planets, the unmagnetized planets and comets. Some of these waves are generated at the bow shock and propagate upstream. Other waves are generated by electrons and ions accelerated at the bow shock and reflected back into the solar wind or leaked from the magnetosheath back upstream. These backstreaming particles generate waves through various instabilities and these waves are then convected with the solar wind flow toward the shock. Still other waves originate as newly created ions scatter and thermalize both in the extended coronas surrounding comets and in the exosphere of the unmagnetized planets.

The variation of solar wind properties with heliocentric distance causes stronger bow shocks in the outer solar system than in the inner solar system. Thus the flux of backstreaming particles is expected to be higher in this region. The size of planetary magnetospheres and their foreshocks also varies considerably through the solar system. As a result of both of these factors the foreshocks of the planets present us with a spectrum of conditions. Further, the very large change in field strength from the inner to outer heliosphere accompanied by a proportional change in the upstream wave period clearly demonstrates that the dependence of wave frequency on the strength of the interplanetary magnetic field observed on the surface of the Earth in the Pc3 band is seen in planetary foreshocks throughout the solar system.

A particularly fruitful place to study upstream waves is the Venus foreshock whose configuration is quite stable and for which there are over a decade of observations. Because the amplitude of waves in the Venus foreshock generally remains at a small fraction of the background field, they can be used to test linear theories. These studies show that the two-fluid, Hall-MHD dispersion relation does not properly describe ULF waves in a moderate or high beta plasma. Instead, the full kinetic Vlasov dispersion relation must be used to treat these waves.

INTRODUCTION

All the planets visited thus far by our planetary spacecraft have bow shocks associated with the deflection of the solar wind around either a magnetospheric cavity or a highly conducting ionosphere. This shock wave develops in front of each planet because the information needed to deflect the solar wind plasma around the planetary obstacle travels at a velocity that is less than that of the solar wind flow. In general three waves are needed to deflect the flow: a slow magnetosonic wave that increases the density and slows the flow while it bends both the field and the flow and decreases the field magnitude; an intermediate wave that bends the flow and the magnetic field but does not compress either; and a fast magnetosonic wave which compresses and deflects both field and flow. The slow wave will stand in the flow close to the magnetopause where the flow is the slowest.

The intermediate wave will stand in the flow upstream of the slow wave where the flow is faster and the fast wave will stand further still upstream, at a distance that is about 30% greater than the size of the obstacle to the flow. The presence and properties of this last wave in the form of the bow shock are well established [see for example Russell, 1985] but the presence of the former two waves in planetary magnetosheaths has only recently been appreciated [Song et al., 1990].

These waves may affect strongly the properties of the plasma and the other waves in the plasma that pass through them. In particular they may convert wave modes into other wave modes and they themselves may be secondary wave generators. Thus the relationship between waves seen in the solar wind, and those in the magnetosheath which may be convected to the magnetosphere, depends strongly on the nature of processes at the bow shock and in the magnetosheath and therefore on the Mach number of the bow shock. The strength of the coupling between waves upstream of the bow shock and the magnetosphere is more than just a question of the geometry imposed by the direction of the interplanetary magnetic field

Solar Wind Sources of Magnetospheric Ultra-Low-Frequency Waves
Geophysical Monograph 81

(IMF).

The bow shock is the location where the fastest of the three MHD waves stands in the incoming flow. Nevertheless, as discovered initially in front of the Earth by Greenstadt et al. [1970], the influence of the Earth on the incoming flow can extend farther into the flow along field lines connected to the bow shock. This upstream region, or foreshock, in some senses is part of an extended bow shock in which heating, slowing and deflection of the flow takes place, but at times this region is a small perturbation on the incoming flow of no major consequence to the first order properties of the solar wind. Which of these two situations prevails depends on the Mach number of the flow. When the Mach number is very large, the effect of the foreshock on the solar wind is significant, but when the Mach number is low, the effect is small.

The range of the variation of the individual parameters of the solar wind is very large, but they vary in such a way that the Mach number of the fast mode bow shocks associated with the flow of the solar wind relative to the planets varies in a fairly narrow range at any particular heliocentric distance. Thus, to study foreshocks and shocks over a significant range of Mach numbers one needs to examine the behavior over a range of heliocentric distances. Figure 1 shows how Mach number and beta should vary with heliocentric distance for steady solar wind conditions [Russell et al., 1982] [See also Greenstadt and Fredricks, 1979]. At Mach numbers of 3-4, as are typically observed at Mercury and Venus, the foreshock perturbations are weak, while at Jupiter and beyond with Mach numbers of 8 and greater, the foreshock is generally a very strong perturbation on the solar wind.

Another factor that affects the nature of the foreshock is the size of the interaction region. Figure 2 shows the relative sizes of the magnetospheres of the magnetized planets. The Earth's magnetosphere stands off the solar wind about 65,000 km above the surface, making a magnetosphere close to 20 times as large as that of Mercury. Jupiter in turn has a magnetosphere that is 100 times larger than that of the Earth, so large in fact that the Sun could easily fit inside, as is illustrated. Since the region of interaction in the foreshock is roughly the region in which the IMF is connected to the bow shock, the foreshock interaction region of Jupiter and Saturn can make significant changes in the solar wind properties, while at Mercury there should be very little effect. The two unmagnetized planets, Venus and Mars, present small obstacles to the solar wind flow, similar to Mercury, and hence also only mildly perturb the solar wind.

A final factor that affects the foreshock is the stability of the foreshock geometry. This mainly affects our ability to study the spatial variations within the foreshock. The sizes of the magnetospheres of the magnetized planets are set by a balance between the momentum flux of the solar wind and the magnetic pressure exerted against the solar wind flow by the intrinsic magnetic field of the planet. This size can vary significantly and is sometimes hard to predict. The sizes of the solar wind obstacles of the unmagnetized planets are much less variable, usually changing but slowly over the solar cycle. Since one can determine where in the foreshock a particular observation was

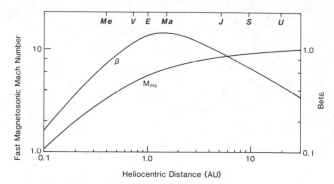

Fig. 1. The variation of the fast magnetosonic Mach number and beta, the ratio of thermal to magnetic pressure, as a function of heliocentric distance in an idealized solar wind. [Russell et al., 1982].

obtained it is advantageous to study the foreshocks of the unmagnetized planets, all else being equal.

In this review we will examine the processes occurring in planetary foreshocks, discussing in turn waves generated by the shock, by backstreaming electrons, by backstreaming ions and by ions picked up by the solar wind from the neutral atmosphere. We will also examine how wave modes are identified in the foreshock.

THE GEOMETRY OF THE FORESHOCK

The geometry of the foreshock is determined principally by the size of the bow shock and the direction of the interplanetary magnetic field [see for example, Greenstadt et al., 1987]. Figure 3 shows the foreshock geometry in the plane containing the solar wind flow, the IMF and the center of the planet. However, one should not forget that the foreshock is a three dimensional region. Slices through the 3-dimensional foreshock above or below this plane are thought to be very similar in their properties and we will assume in this review that they are. However, there is one difference between these planes that may prove to be important. The shock normal lies in the plane shown in Figure 3 but it does not lie in corresponding planes above or below this one, moving further out of the plane with increasing distance from the B-V plane containing the center of the planet. If the physics of the foreshock is controlled only by the angle of the field in the B-V plane, then this difference is not important but, if the full angle between the IMF and the shock normal is important, then parallel slices of the foreshock may not be equivalent.

The key magnetic field line in the foreshock geometry is the IMF field line that is tangent to the bowshock. At the tangent point the shock normal is at 90° to the IMF and the bow shock is exactly perpendicular. At this point particles accelerated by the bow shock or leaking from the solar wind have their first access to the solar wind. Empirically we find that electrons are accelerated here but not ions. Any accelerated ions seem to drift with the solar wind plasma into the magnetosheath, but a fraction of the solar wind electrons reach energies along the magnetic field that allow them to escape this fate. The acceleration process appears to be due to what has been called fast Fermi

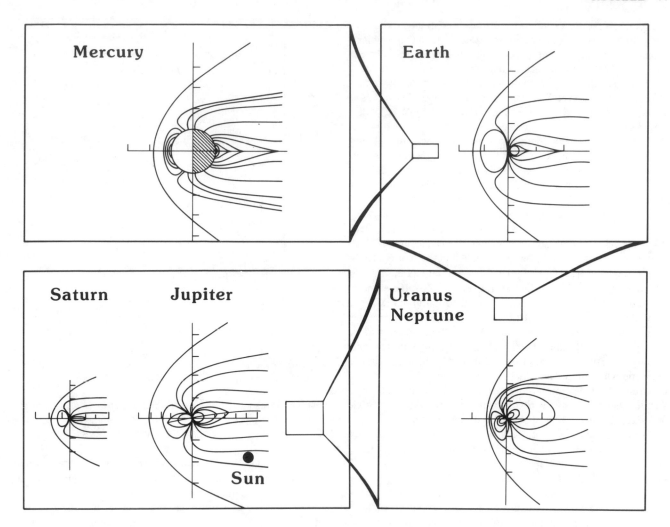

Fig. 2. A comparison of the sizes of the magnetospheres of the magnetized planets. The magnetosphere of Jupiter is so large that the sun could fit inside it as illustrated.

Foreshock Geometry

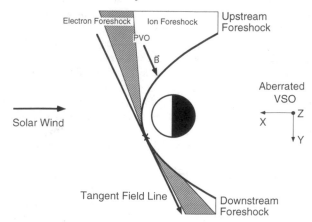

Fig. 3. Schematic diagram of a planetary foreshock in the plane containing the upstream magnetic field, the solar wind velocity vector and the center of the planet. [Crawford et al., 1993].

acceleration in which electrons sense a moving magnetic mirror as the field line is convected through the perpendicular shock [Leroy and Mangeney, 1984]. A small fraction of the solar wind electrons have pitch angles, measured in the frame of the magnetic mirror, that lead to reflection through conservation of the first adiabatic invariant. These electrons are turned around when they reach the shock ramp and return upstream. Since electrons have thermal velocities much greater than their convected velocities, this process happens on both sides of the tangent point, in the upstream and downstream foreshocks as labelled in Figure 3. However, this process may not be entirely symmetric about the tangent point.

The accelerated electrons have finite velocity and they are affected by the electric field of the solar wind which causes the electrons to drift downstream perpendicular to the solar wind flow. Thus the electron foreshock is swept back slightly from the tangent field line as illustrated in the figure.

Ions seem not to be appreciably affected by fast Fermi acceleration because their ion gyro radii are larger than the shock thickness. Nevertheless not all ions pass through the bow shock. Some do not have sufficient energy to pass through the combined magnetic and electric field barrier of the shock and are reflected. The so called non-coplanar component of the magnetic field within the shock ramp is thought to play an important role in this reflection [Goodrich and Scudder, 1984; Jones and Ellison, 1987; Thomsen et al., 1987; Gosling et al., 1988]. At the perpendicular shock these reflected ions are turned around by the interplanetary magnetic field and convected back through the shock by the interplanetary electric field. However, as one moves away from the perpendicular shock ions can begin to gradient drift along the shock and become energized as they cross the electric potential of the solar wind. Eventually the geometry changes sufficiently as one moves away from the tangent point that ions receive sufficient velocity parallel to the magnetic field that they begin to move upstream along the magnetic field line. Thus some distance away from the tangent point at approximately where the angle between the bow shock normal and the IMF is 50° [Gosling et al., 1978; Le and Russell, 1992] sufficient ions flow upstream to cause a noticeable perturbation in the IMF.

Another source of upstream ions is leakage from the magnetosheath [Tidman and Krall, 1971]. When electrons and ions pass through the bow shock they become heated and although initially quite anisotropic in their distribution about the magnetic field, they eventually become thermalized. If the shock is strong, the temperature downstream can be sufficiently high that the thermal velocity of the particles can take them back upstream across the bow shock and into the solar wind.

As mentioned above, the Mach number of the solar wind flow relative to the planets increases with heliocentric distance. However, there is another important change in the geometry of the foreshock which affects observations. The geometry shown in Figure 3 is appropriate for the average IMF direction seen at Venus. As one moves out in the solar system the magnetic field becomes more perpendicular to the flow. Figure 4 shows the geometry appropriate for Saturn together with the Voyager 1 and 2 trajectories through this region. The tangent line is now almost orthogonal to the solar wind flow and the ion foreshock is well behind the nose of the shock. The reason for this change is that the radial component of the IMF decreases as the inverse square of the heliocentric distance while the azimuthal component varies as the inverse first power. Thus for the outer planets the foreshock can usually be observed only well downstream of the planets, and the distinction between upstream and downstream foreshocks become moot. Furthermore, since upstream waves are seldom formed in front of the dayside magnetopause, upstream waves convected against the dayside magnetopause are not expected to be an important source for ULF waves in the outer planet magnetospheres.

Finally we note that since there have been no orbiters of the outer planets as yet our information on the outer planets foreshocks is meager, principally coming from the Voyager 1 and 2 spacecraft with some observations from Pioneer 10 and 11 and Ulysses. Also we note that much of our information about foreshocks comes from the upstream foreshock rather than the downstream foreshock because of the much greater size of the

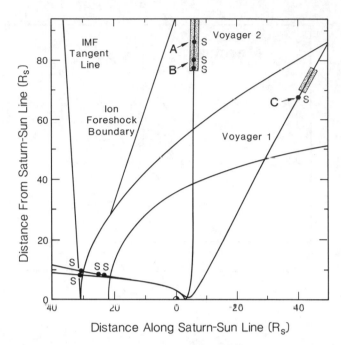

Fig. 4. The foreshock geometry at Saturn where the interplanetary magnetic field is nearly orthogonal to the solar wind flow. Shaded areas show where upstream waves were analyzed by Orlowski et al. [1993b].

upstream foreshock in the inner solar system.

THE SOURCES OF UPSTREAM WAVES

The ions and electrons which stream along field lines into the solar wind are a source of free energy for the generation of waves. Many different processes can be responsible for this generation and these different processes produce waves with different properties. The topic of the identification of wave modes is left to the end of the review. For now it suffices to point out what some of these instabilities can do to the plasma. For example, if the intensity of the back streaming ions along the field is sufficiently strong relative to the magnetic pressure the firehose instability can be excited in which the magnetic field line oscillates back and forth about its mean position. If there is an anisotropy in the temperature for the ions parallel and perpendicular to the field (in the solar wind frame), then the ion cyclotron instability can be excited. This instability is sometimes called the ion-ion instability. The electrons streaming back into the solar wind create a bump on the tail of the solar wind electron distribution. Such electron distributions can cause electrostatic oscillations near the electron plasma frequency.

In addition to particles streaming back into the solar wind from the shock, there are two other sources of waves. The first is associated with newly created ions which suddenly experience the electric field of the solar wind upon becoming ions and drift in the solar wind electric field so that they move relative to the solar wind ions. This process is most important at comets and produces instabilities much like those seen in the ion beams backstreaming from the shock. The second source is waves produced by the shock itself. Upstream whistler waves (at close to 1 Hz is the spacecraft frame at Earth) for example are thought to be caused by the instability of electrons accelerated in the potential drop across the shock [Tokar et al., 1984]. While the shock itself cannot move further into the solar wind because the solar wind is quite supersonic, whistler mode waves can move at velocities much higher than those of the solar wind, and travel upstream against the solar wind flow with minimal damping. In the sections to follow we will first describe the shock as a source of upstream waves.

WAVES GENERATED AT THE SHOCK

The bow shock is the source of many waves (see for example Farris et al., this volume). Some of these waves are generated propagating downstream toward the planet. These are carried away from the shock. Some of these waves are generated propagating upstream. If the component of their velocity pointing upstream along the shock normal is less than the velocity of the shocked solar wind along the normal, these waves too will be swept downstream. Some waves are generated with phase velocities that match the flow of the solar wind across the shock normal. These waves form nearly monochromatic standing wave patterns at the bow shock but are generally not seen far from the shock, at least not far from the quasi-perpendicular shock. Still other waves propagate more nearly along the magnetic field where damping is weak and can reach large distances from the shock if the field line is properly oriented. These electromagnetic waves must propagate in the so-called whistler mode because it is the only mode that has a sufficiently high velocity to allow this propagation. It is important to note that a wave has two velocities: a phase velocity at which phase fronts travel and a group velocity at which the energy of the waves travels. For whistler mode waves the group velocity can reach twice the value of the phase velocity. This additionally assists the upstream propagation of the whistler mode.

Upstream whistlers were first discovered by Russell et al. [1971] in front of the Earth's bow shock and later studied in detail by Fairfield [1974] who concentrated on the observations at the shock itself. While Fairfield [1974] concluded that these waves were upstream whistlers generated at the shock, later work led to some controversy about this interpretation. Hoppe et al. [1982] dismissed the shock source interpretation and sought a source in the backstreaming ions populations but could not find one. Sentman et al [1983] believed that they did identify a particle source in the distribution function of backstreaming electrons, but they did not determine whether the resonance feature they saw was caused by the waves or was responsible for the wave. Wong and Goldstein [1987; 1988] have postulated yet a third particle source, reflected ions. Thus, despite the initial work of Fairfield [1974] much doubt lingered about the source of these waves.

Figure 5 shows examples of these upstream whistler waves at three planets, Mercury, Venus, and Earth [Orlowski et al., 1990]. The bottom panels show expansions of the top panels to a common 30 second scale. It is obvious that the same phenomenon is being observed at the three planets although the frequency is somewhat higher at Mercury than at Earth and Venus. This is consistent with the higher background magnetic field strength at Mercury. Similar waves have also been observed at Saturn at a much lower frequency, 0.1 Hz, again consistent with the lower magnetic field strength in the outer solar system [Orlowski et al., 1993b].

These waves can be both right handed and left handed but it is possible to show that both polarizations are due to whistler mode waves. The left handed waves occur when the polarization of the waves is reversed by Doppler shifting of waves propagating nearly along the solar wind velocity vector. This has been demonstrated both at Earth [Fairfield, 1974] and Venus [Orlowski and Russell, 1991].

Another indication that the waves are created at the shock is that the amplitude of the waves decreases with distance from the shock with the waves propagating at the largest angle to the magnetic field decreasing in amplitude the fastest. This is to be expected if the waves are undergoing Landau damping by the solar wind electrons. The waves are also confined to propagate principally in the plane defined by the local shock normal and the IMF, a situation highly unlikely unless the waves are generated at the shock [Orlowski et al., 1993a]. We note that these latter studies at Venus were made possible by the stability of the foreshock region which enables one to be confident in the geometric constructions used in the statistical studies.

WAVES GENERATED BY ELECTRONS

While the ULF waves discussed in the previous section do not appear to be generated by backstreaming electrons, there are many VLF emissions associated with these electrons. Figure 6 shows a sketch of the geometry of the terrestrial foreshock illustrating where VLF emissions are found. The left hand inset shows the spectrum of Langmuir oscillations seen around the electron plasma frequency. These waves occur in conjunction with the electron bump-on tail distributions near the tangent field line. The right hand inset shows the spectrum of what have been called down-shifted emissions. These waves have a broader spectrum which peaks well below the electron plasma frequency.

To first order these waves seem to be controlled by the observer's location relative to the tangent field line. We can quantify this by measuring the distance behind the tangent field line along the solar wind flow. This distance has been called the foreshock depth. Figure 7 shows the amplitude of VLF waves in the Venus foreshock as a function of foreshock depth. The lines shows the amplitude of the electric field exceeded 10%, 20%, 30%, etc of the time. The solid line indicates the amplitude near the electron plasma frequency. It peaks on the tangent field line. The dashed line indicates the amplitude of downshifted emissions. They peak in amplitude deep in the ion foreshock [Crawford et al., 1993]. We note that these waves are quite ubiquitous in solar system foreshocks. In addition to Venus they have been observed at Earth [Anderson et al., 1981], Mars [Trotignon et al., 1992], Jupiter [Moses et at., 1984], Saturn [Scarf et al., 1983], Uranus [Gurnett et al., 1986], and Neptune

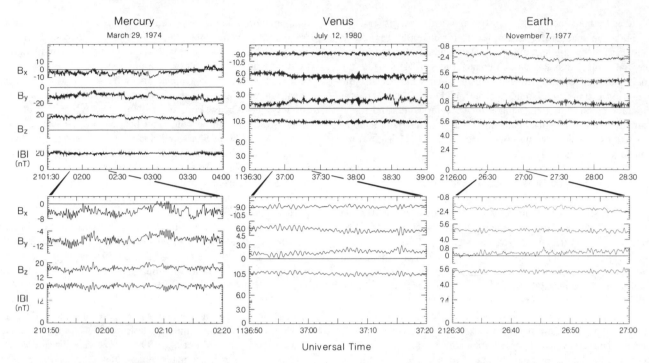

Fig. 5. Upstream whistler mode waves seen in the foreshocks of Mercury, Venus, and Earth. Each panel shows the three components of the magnetic field in solar ecliptic coordinates together with the total magnetic field strength. The bottom panels each show 30 seconds of data [Orlowski et al., 1990].

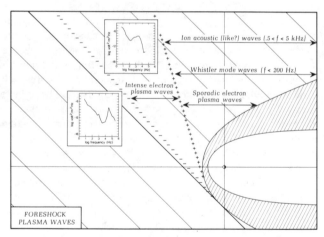

Fig. 6. Schematic diagram of the Earth's foreshock region showing where various VLF emissions are observed. The insets show spectra of Langmuir waves near the tangent field line (left) and downshifted waves (right). The Langmuir waves maximize near the electron plasma frequency and the downshifted waves maximize below the electron plasma frequency. [Russell and Hoppe, 1983].

[Gurnett et al., 1989].

WAVES GENERATED BY BACKSTREAMING IONS

As discussed above ions backstream into the solar wind for many reasons. They can gradient drift along the shock front and cross electric equipotentials. They can be reflected by the electric and magnetic fields of the shock. They can leak back upstream from the thermalized post shock population. Moreover, when these particles going back upstream generate fluctuations in the strength of the magnetic field, these field compressions are carried toward the shock and with the shock provide converging magnetic mirrors which further accelerate the particles. Many wave-particle instabilities are generated in the ion foreshock. These instabilities are discussed detail by Le and Russell [this volume] for the Earth. We will discuss only a few points about these waves here.

Figure 8 compares waves seen at Venus at VLF frequencies with their counterparts in the terrestrial foreshock [Russell and Hoppe, 1983]. The same phenomena appear to be seen at both planets except for the so-called 3-second waves [Le at el., 1992]. While the same phenomena appear, the waves appear to be significantly smaller at Venus than at the Earth. Orlowski and Russell [1993] report that the waves at Venus reach only about 15% of the background field strength compared with about 50% for Earth. This difference could be related to the smaller growth region for the waves and the shorter time of connection of the field line to the shock.

The frequency of the ULF waves which occur with periods near 30 seconds at Earth seems to be controlled by the strength of the IMF in a manner similar to the upstream whistler waves discussed above. While a similar relation had been known for waves seen on the surface of the Earth for some time [Troitskaya et al., 1971], it was not until much later that this relationship was

Intensity vs Foreshock Depth

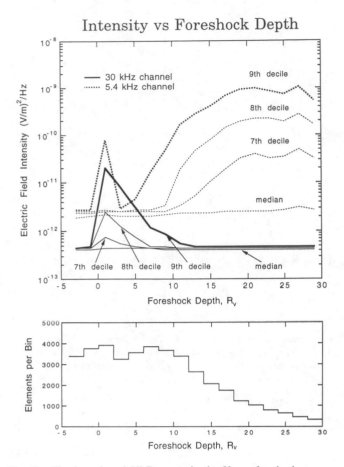

Fig. 7. The intensity of VLF waves in the Venus foreshock as a function of the depth of the penetration along the solar wind flow. Langmuir waves in the 30 kHz channel maximize at the tangent field line. The downshifted waves at 5.4 kHz maximize well behind the tangent field line in the ion foreshock. Shown in the top panel are the amplitudes exceeded by 10%, 20%, 30% etc. of the signals. The bottom panel shows the number of samples used in the creation of the top panel. [Crawford et al., 1993].

demonstrated for waves in the foreshock of the Earth [Russell and Hoppe, 1981] and of other planets [Hoppe and Russell, 1982]. Figure 9 shows an example of upstream waves at Uranus [Russell et al., 1990]. Similar waves have been observed recently at Saturn [Bavassano-Cattaneo et al., 1991] and at Neptune [Zhang et al., 1991]. These waves are in many respects like those observed in the Earth's foreshock except that the periods are close to 1000 seconds instead of 30 seconds. Figure 10 shows the frequency of the peak of the spectrum of these waves as seen from Mercury to Uranus. The frequency dependence found by Troitskaya et al., [1971] on the IMF field strength seems to be valid throughout the solar system. The reason for this consistency appears to be that the acceleration of the backstreaming protons is similar at each of the planets. As shown by Hoppe and Russell [1982], if the protons are simply reflected at the shock without loss of energy, the Doppler-shifted frequency of the waves they produce should be 0.006 B where the field is expressed in nT and the frequency in Hz.

WAVES GENERATED BY PICKUP IONS

When neutral atoms find themselves in the solar wind plasma, they may become ionized through photoionization by the solar EUV flux, through charge exchange with the protons in the solar wind or through impact, ionization by the solar wind electrons. Sources of neutral atoms in the solar wind include both the expanding comas of comets, interstellar neutrals, and the exospheres of both the magnetized and unmagnetized planets. The newly created ion is immediately affected by the electric field of the solar wind and executes a cycloidal motion perpendicular to the IMF. This motion is equivalent to a drift with the component of the solar wind velocity perpendicular to the magnetic field and a thermal velocity around the magnetic field equal to that of its drift. This population of particles is unstable to the generation of waves. Figure 11 shows a schematic of how these picked up ions affect the solar wind flow past a comet [Neugebauer, 1990]. The panel on the upper left shows the distribution function of the ions. The horizontal direction is the direction of the magnetic field. The vertical direction is perpendicular to the IMF. The lines shows contours of phase space density. The contours resembling the end of a sawn log give the distribution function of the solar wind. The dark blob shows the newly created ions picked up by the solar wind. They have a component of motion parallel to the magnetic field that is different than that of the solar wind ions. This leads to instability and the generation of waves. The two lines show paths along which the particles would diffuse if they resonated with waves moving to the right or the left.

As shown in the schematic the addition of mass to the solar wind results in a slowing of the solar wind causing the magnetic field to be draped about the obstacle and forming a magnetotail. The plasma also heats and, if the rate of mass addition is sufficiently high, a bow shock will form. The waves created in this process also scatter and accelerate particles.

A wide variety of waves are generated by these picked up ions, most of them resembling waves seen in the terrestrial foreshock [Smith et al., 1986; Tsurutani and Smith, 1986; Russell et al., 1987]. Figure 12 shows examples of these waves. The mirror mode waves illustrated in the bottom panel are seen closest to the nucleus in the cometary coma or magnetosphere proper. At Earth similar waves are seen in the magnetosheath.

The neutral atmospheres of the unmagnetized planets also extend, albeit tenuously, into the solar wind. Searches for ions and ion pick-up associated waves at Venus have proven to be negative, although such a search did find ions accelerated by the bow shock (Moore et al., 1989). The strong gravitational pull of Venus keeps the majority of the Venus atmosphere close the planet and very little exosphere is ionized and picked up by the solar wind.

Mars is only about 15% of the mass of Venus and has a weaker gravitational pull on its atmosphere. Thus one might expect the exosphere to extend to a greater height and the effect on the solar wind to be greater at Mars. Moreover, Mars has

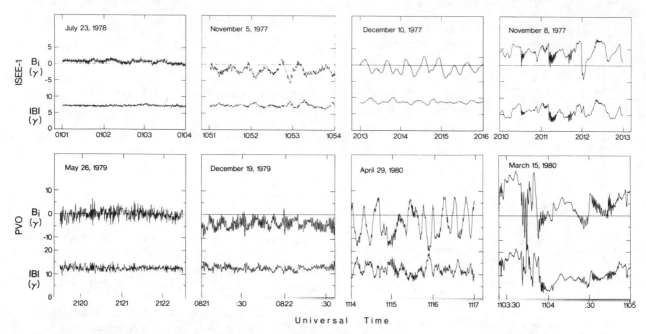

Fig. 8. Examples of the various classes of ULF waves seen in the Venus and the terrestrial bow shocks. The left hand panels show the upstream whistler discussed above. The next panel shows one Hertz waves together with the usual ion foreshock waves. The next panel shows principally ion foreshock waves. The last panel shows what have been called discrete wave packets and shocklets [Russell and Hoppe, 1983].

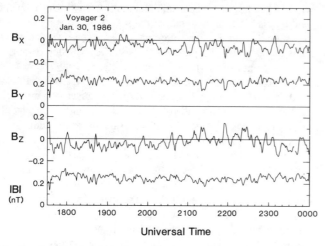

Fig. 9. An example of ion foreshock waves at Uranus. Shown are the three components of the magnetic field and the total field [Russell et al., 1990].

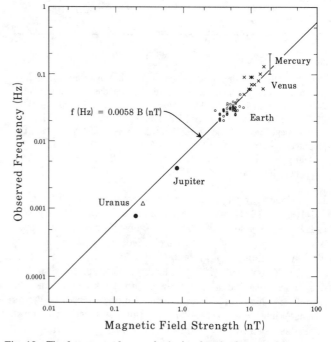

Fig. 10. The frequency of waves in the ion foreshock seen at Mercury, Venus, Earth, Uranus versus the magnetic field strength [Russell et al., 1990].

two moons Deimos and Phobos which spend much of each orbit in the solar wind. Initially it was believed that these moons might be outgassing or might be providing a dust torus that interacted with the solar wind (Riedler et al., 1989). However, later studies showed that the disturbances seen near the orbits of these moons were probably just due to backstreaming ions [Russell et al., 1990]. We note that generally at Mars the amplitude of upstream waves seems to be reduced over the amplitude seen at Venus. This could be due to the smaller growth region in front of the Martian bow shock.

Mars does, however, exhibit a very weak ion pick up signature as shown in Figure 13. In the region close to the nose of the bow shock, where the neutral atmospheric density and the ion pick up rate should be the greatest, a weak signal at the proton

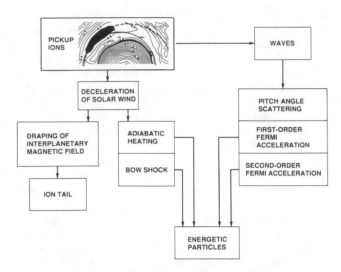

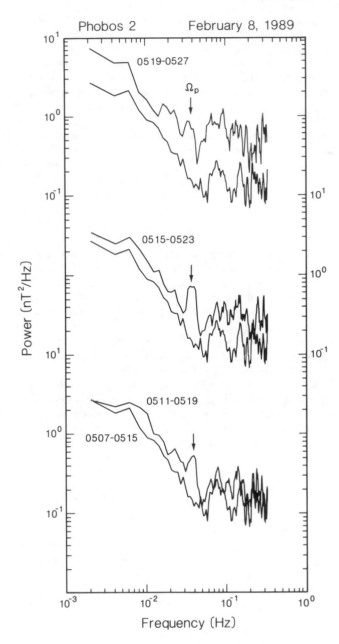

Fig. 11. Schematic diagram showing the effect of picked up ions on the solar wind. The top panel shows contours of constant phase space density of newly born ions in the coma of comet Halley, together with the solar wind. The solid dark area shows the picked up ions. The contours that represent the solar wind ions resemble cross section of a log. The dashed and dotted lines show the paths followed by diffusing ions resonating with waves moving to the right and the left. The direction of the magnetic field is horizontal in this display. [Neugebauer, 1990].

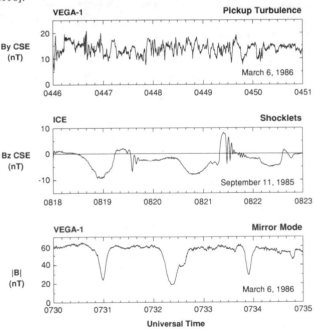

Fig. 12. Examples of ULF waves seen in the vicinity of comets. The top panel shows turbulence seen in the foreshock of comet Halley by Vega 1. The middle panel shows shocklets and discrete wave pockets seen in the foreshock of comet Giacobini-Zinner by the ISEE-3 spacecraft also called ICE. The bottom panel shows mirror mode waves seen deeper in the coma of comet Halley by the Vega 1 spacecraft. Waves similar to those in the middle panel are frequently seen in the Earth foreshock. Waves similar to those in the bottom panel are frequently seen in the terrestrial magnetosheath.

Fig. 13. Power spectra of fluctuations in the solar wind near Mars showing waves at the proton gyrofrequency, possibly associated with newly created planetary protons. The bottom trace is obtained further from the planet and is shown for reference.

gyro frequency is seen [Russell et al., 1990]. Evidence for the picked up ions themselves has also been found [Barabash et al., 1991] but in both instances the signals are weak.

THE IDENTIFICATION OF WAVE MODES

The objective of our studies of the waves upstream of planetary bow shocks is to understand the physical processes taking place. In the discussion thus far we have examined principally the existence of these waves and their first order properties but we have made little attempt to quantitatively examine the waves. In fact this is extremely difficult for a variety of reasons. First most of our measurements are made on a single platform, the ISEE 1 and 2 pair of co-orbiting spacecraft being the major exception. The solar wind moves close to or above the speed of most of the waves in the foreshock. Thus waves are observed with a different speed and possibly a different sense of polarization than they possess in the plasma (solar wind) frame. Secondly the waves are often strong enough that linear theories of small amplitude waves are insufficient to treat the waves. Third, the models that treat the propagation of these waves often neglect the resonances of the wave with the motion of the particles. This may be permissible at low frequencies well away from the resonances and in plasmas that are cold but it may be quite misleading in warm plasmas. Finally, in the presence of beams of backstreaming particles, the properties of the waves may be different than in the absence of the beams so that the beams and their properties must be taken into account.

The tools that we have at our disposal are the dispersion relations of various wave modes developed by theorists under a variety of approximations. Cold plasma theory is perhaps epitomized by the treatment of Stix [1962]. This theory is appropriate for a plasma with no thermal energy, and therefore has no damping or wave growth. It is most often used for waves in the Earth's magnetosphere and to treat the whistler mode. This theory has been used in the study of planetary foreshocks to determine the frequency of the upstream whistlers in the plasma frame [Orlowski and Russell, 1991]. Since plasma measurements are seldom available at frequencies high enough to be cross-correlated with the wave as seen in the magnetometer data, one has to rely on the properties of the wave as deduced from the magnetometer itself. The property of the upstream whistlers that was used, was the elliptical cross section of the magnetic perturbation of the field in the plane perpendicular to the direction of propagation. Cold plasma theory works quite well in this situation possibly because most of these waves propagate nearly parallel to the magnetic field and there is very little damping for parallel propagating waves.

It is often assumed that at low frequencies well below the ion cyclotron frequency one can safely treat waves with ideal magnetohydrodynamic (MHD) theory which treats the plasma as a magnetized fluid. In this theory there are three modes: fast, intermediate and slow. The intermediate mode is incompressible, bending the field and flow but not changing the plasma density or field strength. The fast mode is associated with a density and field compression that is in phase and the slow mode with one that is out of phase. The ideal MHD treatment assumes isotropy of the plasma and ignores all plasma resonances. The MHD theory clearly does not work for the low frequency upstream waves since the theory predicts that the waves have linear polarization, while the majority of upstream waves have a distinct elliptical polarization.

An improvement on the MHD model for low beta plasmas was developed by Stringer [1963], and is called the two-fluid model or the Hall-MHD model. This model was also later developed

by Formisano and Kennel [1969] who applied it to high beta plasma. This model also ignores the resonances of the plasma with the wave but might be expected to work in a low beta plasma where the plasma pressure is dominated by the magnetic field and not the thermal motions of the particles. In order to test the Hall-MHD model we can use both an improved theory and data in the foreshock. If the improved theory and the data agree and disagree with the approximate theory, than we can conclude that the approximation is not adequate. When we do this using the foreshock region, as we will see below, the Hall-MHD dispersion relation is found to have serious limitations.

Our test uses the fully electromagnetic dispersion relation derived from linear Vlasov theory [Gary, 1991 and references therein]. We also use measurements of the waves in the Venus foreshock which are small amplitude, $\delta B/Bo \leq 15\%$, and which appear to have steady amplitude to which the linear theory seems applicable. Since we do not have plasma observations even close to the frequency at which the waves oscillate we use strictly magnetic measurements as our test. Theory suggests two such parameters are useful for distinguishing wave modes: the magnetic field polarization which is the ratio of the field perturbations in the two directions perpendicular to the wave vector; and the normalized magnetic compression ratio which measures the ratio of the field perturbations along the magnetic field to that perpendicular to the field. Figure 14 shows the results of such a calculation from the two theories and from Venus foreshock measurements [Orlowski et al., 1993c]. The full Vlasov kinetic treatment works well and the two fluid Hall-MHD model does not. Thus previous interpretations of the nature of upstream waves based on the Hall-MHD approach should be viewed with some caution [e.g. Orlowski et al., 1993c]. We are presently re-examining these observations [Orlowski et al., 1993d].

SUMMARY AND CONCLUSIONS

The solar system provides us with a wealth of data on planetary foreshocks. The information from the different planets is quite complementary because of the variation of solar wind properties with heliocentric distance and because of the varying sizes of the foreshock regions which allow much greater wave growth at some planets rather than others. The unmagnetized planets, Venus and Mars have foreshocks similar to those of the magnetized planets and their fixed sizes are a help in undertaking statistical studies. Moreover, the small size of the Venus foreshock supports only weak wave growth and allows us to study upstream waves of small amplitude which can be more easily compared with linear theory.

A number of different waves have been discovered in the foreshock region. One quite prevalent wave generated by the bow shock itself has been found at Mercury, Venus, Earth, and Saturn. This wave propagates in the whistler mode, well above the ion gyro frequency at velocities that allow it to travel upstream against the solar wind flow. The waves which travel nearly along the field seem to be little damped but those at large angles seem to damp rapidly.

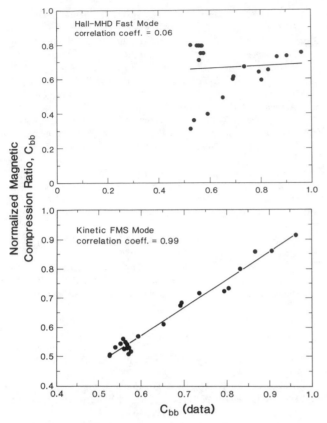

Fig. 14. The compressibility of ULF waves in the Venus bow shock as predicted by two theories Hall-MHD (top) and kinetic Vlasov (bottom) versus the observed compressibility. The fast magnetosonic mode as given by the Vlasov theory is in almost complete agreement with observations [Orlowski et al., 1993c].

At lower frequencies, below the proton gyrofrequency, another type of wave also seems to be pervasive in planetary foreshocks. This wave, which has a period close to 30 seconds at 1AU, seems to be associated with similar waves seen inside the Earth's magnetosphere because the same frequency - IMF strength relationship is found for the magnetospheric waves as for the planetary foreshock waves. This wave appears to be a fast magnetosonic wave driven by backstreaming ions. The waves grow in the foreshock region and are carried back toward the shock as they grow.

Ion cyclotron waves due to the instability of picked up ions are strong at comets but only at Mars are such waves discernible near a planet. However, even at Mars the waves are very weak.

Finally we have learned that even for a moderate beta plasma the Hall-MHD theory is extremely limited. Thus to study waves in planetary foreshocks we must use the full kinetic Vlasov treatment.

Acknowledgements. This work was supported by the National Aeronautics and Space Administration under research grant NAGW-2886.

REFERENCES

Anderson, R. R., G. K. Parks, T.E. Eastman, D. A. Gurnett, and L. A. Frank, Plasma waves associated with energetic particles streaming into the solar wind from the Earth's bow shock, *J. Geophys. Res., 86,* 4493, 1981.

Barabash, S., E. Dubinin, N. Pissarenko, R. Lundin, and C. T. Russell, Picked up protons near Mars: Phobos observations, *Geophys. Res. Lett. 18,* 1805-1808, 1991.

Bavassano-Cattaneo, M. B., P. Cattaneo, G. Moreno, and R. P. Lepping, Upstream waves in Saturn's foreshock, *Geophys. Res. Lett., 18,* 797-800, 1991.

Crawford, G. K., R. J. Strangway, and C. T. Russell, VLF emissions in the Venus foreshock: Comparison with terrestrial observations, *Proceedings 4th COSPAR Colloquium,* submitted, 1993.

Fairfield, D. H., Whistler waves observed upstream from collisionless shocks, *J. Geophys. Res. 79,* 1368 - 1378, 1974.

Formisano, V., and C. F. Kennel, Small amplitude waves in high β plasma, *J. Plasma Phys., 3,* 55 - 74, 1969.

Goodrich, C. C., and J. D. Scudder, The adiabatic energy change of plasma electrons and the frame dependence of the cross-shock potential at collisionless magnetosonic shock waves, *J. Geophys. Res., 89,* 6654-6662, 1984.

Gosling, J. T., J. R. Asbridge, S. J. Bame, G. Paschmann, and N. Skopke, Observation of two distinct populations of bow shock ions in the upstream solar wind, *Geophys. Res.Lett., 5,* 957-960, 1978.

Gosling, J. T., D. Winske, and M. F. Thomsen, Noncoplar magnetic fields at collilsionless shocks: A test of a new approach, *J. Geophys. Res., 93,* 2735-2740, 1988.

Greenstadt, E. W., and R. W. Fredricks, Shock systems in collisionless space plasmas, in *Solar System Plasma Physics, III,* edited by L. J. Lanzerotti, C. F. Kennel, and E. N. Parker, pp. 5-43, North-Holland, New York, 1979.

Greenstadt, E. W., I. M. Green, D. S. Colburn, J. H. Binsack, and E. F. Lyon, Dual satellite observations of Earth's bow shock: II Field aligned upstream waves, *Cosmic Electrodyn., 1,* 279, 1970.

Greenstadt, E. W., L. W. Baum, K. F. Jordan, and C. T. Russell, The compressional ULF foreshock boundary of Venus: Observations by the PVO magnetometer, *J. Geophys. Res., 92,* 3380-3384, 1987.

Gurnett, D. A.,W. S. Kurth, F. L. Scarf, and R. L. Poynter, First plasma wave observations at Uranus, *Science, 233,* 106, 1986.

Gurnett, D. A., W. S. Kurth, R. L. Poynter, L. J. Granroth, I. H. Cairns, W. M. Macek, S. L. Moses, F. V. Coroniti, C. F. Kennel, and D. D. Barbosa, First plasma wave observations at Neptune, *Science, 246,* 1494, 1989.

Hoppe, M.M., and C. T. Russell, Particle acceleration at planetary bow shock waves, *Nature, 295,* 41-42, 1982.

Hoppe, M.M., C. T. Russell, T. E. Eastman, and L. A. Frank, Characteristics of the ULF waves associated with upstream ion beams, *J. Geophys. Res., 87,* 643-650, 1982.

Jones, F. C., and D. C. Ellison, Non coplanar magnetic fields at collisionless shocks: A test of new approach, *J. Geophys. Res., 93,* 11205-11207, 1987.

Le, G., and C. T. Russell, A study of ULF wave foreshock morphology I. ULF foreshock boundary, *Planet. Space Sci., 40,* 1203-1213, 1992.

Le, G., C. T. Russell, M. F. Thomsen, and J. T. Gosling, Observation of a new class of upstream waves with periods near 3 seconds, *J. Geophys. Res., 97,* 2917-2925, 1992.

Leroy, M. M., and A. Mangeney, A theory of energization of solar wind electrons by the Earth's bow shock, *Annales Geophys., 2,* 449-456, 1984.

Moore, K. R., D. J. McComas, C. T. Russell, and J. D. Mihalov, Suprathermal ions observed upstream of the Venus bow shock, *J. Geophys. Res., 94,* 3743-3748, 1989.

Moses, S. L., F. V. Coroniti, C. F. Kennel, and F. L. Scarf, Strong electron heat flux modes in Jupiter's foreshock, *Geophys. Res. Lett., 11,* 869, 1984.

Neugebauer, M., Spacecraft observations of the interaction of active comets with the solar wind, *Rev. Geophys., 28,* 231-252, 1990.

Orlowski, D. S., G. K. Crawford, and C. T. Russell, Upstream waves at Mercury, Venus, and Earth: Comparison of the properties of one Hertz waves, *Geophys. Res. Lett., 17,* 2293-2296, 1990.

Orlowski, D. S., and C. T. Russell, ULF waves upstream of the Venus bow shock: Properties of the one Hertz waves, *J. Geophys. Res., 96,* 11271, 1991.

Orlowski, D. S., and C. T. Russell, Properties of ULF waves with frequencies below the proton gyrofrequency in the Venus foreshock, *J. Geophys. Res.,* submitted, 1993.

Orlowski, D.S., C. T. Russell, D. Krauss-Varban, and N. Omidi, On the source of upstream whistlers in the Venus foreshock, *Proceedings of the 4th COSPAR Colloquium,* submitted, 1993a.

Orlowski, D. S., C. T. Russell, and R. P. Lepping, Wave phenomena in the upstream region of Saturn, *J. Geophys. Res.* in press, 1993b.

Orlowski, D. S., C. T. Russell, D. Krauss-Varban, and N. Omidi, ULF waves in Venus upstream region: Critical test for two-fluid model, in *Proceedings 4th COSPAR Colloq.* submitted, 1993c.

Orlowski, D. S., C. T. Russell, D. Krauss-Varban, and N. Omidi, Growing Alfvenic modes in the upstream region of Saturn, *J. Geophys. Res.,* submitted, 1993d.

Reidler, W. et al. Magnetic fields near Mars: First results, *Nature, 341,* 604-607, 1989.

Russell, C. T., Planetary bow shocks, in *Collisionless Shocks in the Heliosphere: Current Research,* edited by B. T. Tsurutani, and R. G. Stone p.109-130, American Geophysical Union, 1985.

Russell, C. T., D. D. Childers, and P. J. Coleman, OGO-5 observations of upstream waves in the interplanetary medium: Discrete wave packets, *J. Geophys. Res., 76,* 845-861, 1971.

Russell, C. T., and M. M. Hoppe, The dependence of upstream wave periods on the interplanetary magnetic field strength, *Geophys. Res. Lett. 8,* 615-617, 1981.

Russell, C. T., M. M. Hoppe, and W. A. Livesey, Overshoots in planetary bow shocks, *Nature, 296,* 45-58, 1982.

Russell, C. T., and M. M. Hoppe, Upstream waves and particles, *Space Sci. Rev., 34,* 155-172, 1983.

Russell, C. T., W. Riedler, K. Schwingenschuh, and Ye. Yeroshenko, Mirror mode instability in the magnetosphere of comet Halley, *Geophys. Res. Lett., 14,* 644-647, 1987.

Russell, C. T., R. P. Lepping, and C.W. Smith, Upstream waves at Uranus, *J. Geophys. Res., 95,* 2273-2279, 1990.

Russell, C. T., J. G. Luhmann, K. Schwingenschuh, W. Riedler, and Ye. Yeroshenko, Upstream waves at Mars: Phobos observations, *Geophys. Res. Lett., 17,* 897-900, 1990.

Scarf, F. L., D. A. Gurnett, W. S. Kurth, and R. L., Poynter, Voyager plasma wave measurements at Saturn, *J. Geophys. Res., 88,* 8971, 1983.

Sentman, D. D., M. F. Thomsen, S. P. Gary, W.C. Feldman, and M. M. Hoppe, The oblique whistler instability in the Earth's foreshock, *J. Geophys. Res., 88,* 2048-2056, 1983.

Smith, E. J., B. T. Tsurutani, J. A. Slavin, D.E. Jones, G. L. Siscoe, and D. A. Mendis, International Cometary Explorer encounter with Giacobini-Zinner: Magnetic field observations, *Science, 232,* 383-385, 1986.

Song, P., C. T. Russell, J. T. Gosling, M. F. Thomsen, and R. C. Elphic, Observations of the density profile in the magnetosheath near the stagnation streamline, *Geophys. Res. Lett., 17,* 2035-2038, 1990.

Stix, T. H., *The Theory of Plasma Waves,* McGraw-Hill, New York, 1962.

Stringer, T. E., Low frequency waves in an unbounded plasma, *Plasma Phys. (J. Nucl. Energy), 5,* 89-117, 1963.

Thomsen, M. F., J. T. Gosling, S. J. Bame, K. B. Quest, D. Winske, W. A. Livesey, and C. T. Russell, On the non-coplanarity of the magnetic field within a fast collisionless shock, *J. Geophys. Res., 92,* 2305-2314, 1987.

Tidman, D.A., and N. A. Krall, *Shock waves in collisionless plasmas,* 175 pp., Wiley and Sons, New York, 1971.

Tokar, R. L., D. A. Gurnett, and W. C. Feldman, Whistler mode turbulence generated by electron beams in Earth's bow shock, *J. Geophys. Res., 89,* 105-114, 1984.

Troitskaya, V. A., T. A. Plyasova-Bakounina, and A. V. Gulyelmi, Relationship between Pc 2-4 pulsations and the interplanetary magnetic field, *Dokl. Akad. Nauk SSSR, 197,* 1312, 1971.

Trotignon, J. G., A. Skalsky, R. Grard, C. M. C. Nairn, and S. Klimov, Electron density in the Martian foreshock as a by-product of the electron plasma oscillation observations, *J. Geophys. Res., 97,* 10831, 1992.

Tsurutani, B. T., and E. J. Smith, Hydromagnetic waves and instabilities associated with cometary ion pick up: ICE observations, *Geophys. Res. Lett., 13,* 263, 1986.

Wong, H. K., and M. L. Goldstein, Proton beam generation of whistler waves in the Earth's foreshock, *J. Geophys. Res., 92,* 12, 419-12424, 1987.

Wong, H. K., and M. L. Goldstein, Proton beam generation of oblique whistler mode waves, *J. Geophys. Res., 93,* 4110-4114, 1988.

Zhang, M., J. W. Belcher, J. D. Richardson, V. M. Vasyliunas, R. P. Lepping, N. F. Ness, and C. W. Smith, Low frequency waves in the solar wind near Neptune, *Geophys. Res. Lett., 18,* 1071-1074, 1991.

C. T. Russell, Institute of Geophysics and Planetary Physics, University of California, Los Angeles, CA 90024-1567.

The Morphology of ULF Waves in the Earth's Foreshock

G. LE AND C. T. RUSSELL

Institute of Geophysics and Planetary Physics, University of California at Los Angeles

The Earth's foreshock is the region upstream from the bow shock where the interplanetary magnetic field intersects the bow shock. It is characterized by backstreaming electrons and ions, as well as associated electrostatic and electromagnetic waves over a wide frequency range. One class of upstream electromagnetic waves, large-amplitude compressional ULF waves in the ion foreshock region, has long been postulated as a major source of magnetospheric ULF waves. In this paper, we discuss recent observations of properties of ULF waves. First the general morphology of the foreshock at different IMF orientations and the different types of ULF waves are reviewed. Then observations of critical wave properties of importance to the Earth's magnetosphere, such as frequency, bandwidth, coherence length and spatial evolution in the foreshock, are discussed.

1. INTRODUCTION

The generation of upstream waves, especially the ultra-low-frequency (ULF) waves, in planetary foreshocks is both an active and an important topic of space plasma physics investigations. These waves are ubiquitous, having been observed in front of the bow shock of all of the planets and in front of comets Halley and Giacobini-Zinner. They modify both the solar wind and the turbulence spectrum convected to the Earth's magnetosphere, altering the nature of the shock jump, the pressure fluctuations on the magnetopause and ultimately the spectrum of waves in the magnetosphere. Moreover, these foreshock waves lead to the acceleration of some particles to very high energies by means of wave-particle interactions and provide us with a testbed in which to study processes thought to be responsible for the acceleration of cosmic ray particles. Finally, the instabilities which drive these waves and the processes associated with the waves provide an excellent test of the basic theories of waves and instabilities in plasma physics.

Upstream waves have been found in front of the bow shocks of all of the planets visited to date. Terrestrial upstream waves have been studied extensively by many authors and much of the early work has been well documented in the 1 June, 1981 special issue of *J. Geophys. Res.*. Upstream waves have also been investigated at Mercury by Fairfield and Behannon [1976] and Orlowski et al. [1990]; at Venus by Hoppe and Russell [1982] and Orlowski and Russell [1991]; at Mars by Russell et al. [1990a]; at Jupiter by Smith et al. [1983] and Goldstein et al. [1985]; at Uranus by Russell et al. [1990b], Smith et al. [1989] and Zhang et

al. [1991a]; and at Neptune by Zhang et al. [1991b]. Based on the early work, as reviewed by Russell and Hoppe [1983], we have developed a picture of the morphology of the upstream wave region. In that picture, wave and particle phenomena upstream of the shock arise from kinetic effects occurring in the shock transition region (see Sonnerup [1969]; Paschmann et al. [1980]; Armstrong et al. [1985]; Scholer [1985]; Gosling and Robson [1985] and references therein). These effects include: reflection, shock drift acceleration, and heating of post-shock plasma. The effectiveness of the above mechanisms as well as the characteristics of the backstreaming populations (at the shock) are strongly dependent on the direction of the shock normal relative to the plasma flow and the interplanetary magnetic field (IMF) orientation.

For a stationary IMF we are able to define regions (in the shock frame) with characteristic backstreaming populations. Figure 1 shows a schematic of the foreshock geometry. First we can divide the unshocked plasma into two regions, one magnetically unconnected where no IMF lines connect to or cross the shock and one so-called magnetic foreshock region where all IMF lines are connected to the bow shock. The IMF lines tangent to the bow shock form the boundary between those two regions. The superposition of the upstream motion along the magnetic field line and the convection associated with the interplanetary electric field, results in the formation of electron and ion foreshocks within the region of unshocked solar wind connected to the bow shock. Because of the finite velocity of these backstreaming particles and the convection associated with the interplanetary electric field, the electron and proton foreshocks do not fill the entire magnetically connected region, nor are they fully coincident. Moreover, the properties of the waves and plasma are position dependent within the foreshock.

The electron foreshock boundary is defined by the fastest electrons accelerated along the tangent magnetic field lines. Since

Solar Wind Sources of Magnetospheric Ultra-Low-Frequency Waves
Geophysical Monograph 81

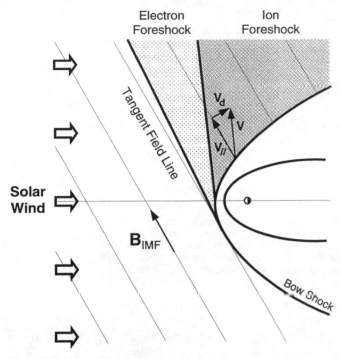

Fig.1. Schematic of foreshock geometry.

the convection velocity is small compared to the typical bulk velocity of escaping electrons (which can be many times of their thermal velocity) the electron foreshock boundary is only slightly displaced relative to the IMF tangent lines. A variety of associated waves has been found in the electron and ion foreshocks [Hoppe et al., 1981]. In the electron foreshock in addition to electrostatic (Langmuir) mode [Gurnett, 1985] an electromagnetic mode at ULF frequencies has also been identified: the so-called one-Hertz whistlers [Russell et al., 1971; Fairfield, 1974; Hoppe et al., 1981, 1982]. One example of one-Hz waves is shown in Figure 2(a).

The ion foreshock boundary in contrast appears to be defined by position-dependent acceleration processes that do not allow protons to move upstream (in the shock frame) until some distance downstream of the electron foreshock [Thomsen, 1985]. It is located much deeper in the magnetic foreshock, beginning close to the region called the quasi-parallel part of the bow shock. Studies of the ISEE 1 and 2 data have revealed that the ion foreshock in turn can be divided into three regions characterized by beam-like, intermediate and/or gyrating, and diffusive ion distributions depending on the local θ_{Bn}, the angle between the IMF and the bow shock normal [Gosling et al., 1978; Paschmann et al., 1979, 1981]. ULF waves in the ion foreshock have been observed at frequencies below the proton gyrofrequency [Greenstadt et al., 1968; Russell et al., 1971; Hoppe et al., 1981; Le et al., 1992]. These observations revealed several types of ULF waves in the ion foreshock: sinusoidal waves, shocklets and discrete wave packets, and three-second waves. Figure 2(b) shows an example of low-frequency waves which are nearly sinusoidal. They typically have amplitudes of a few nT, are primarily transverse, and exhibit predominantly left-handed polarization in the spacecraft frame.

ULF waves in the ion foreshock are most frequently observed in large-amplitude, highly compressional forms. They steepen into small shocklike waveform and have been called shocklets. Discrete wave packets, which have higher frequency, are often associated with the steepening edges of the shocklets, as shown in Figure 2(c). Another type of foreshock waves has frequency near three seconds and is always right-handed and nearly circularly polarized in the spacecraft frame. One example of three-second waves is shown in Figure 2(d).

One type of upstream ULF waves, the large-amplitude compressional waves, is particularly important to the Earth's magnetosphere. These waves must propagate obliquely with respect to the magnetic field since they are highly compressional. Since they were first observed by Greenstadt et al. [1968] and Fairfield [1969], several generation mechanisms related to electromagnetic ion/ion instabilities from linear theory have been proposed to explain the source of these waves [Gary, 1991]. The backstreaming ions have a density of $\sim$ 1% of the solar wind density. Their bulk velocities range from the order of the Alfven velocity ($\sim$ 50 km/sec at 1 AU) for diffuse ions to the order of solar wind speed for intermediate, gyrating and beam-like ions [cf. Thomsen, 1985 for a review]. Under these conditions, the most obvious instability is the anomalous Doppler-shifted ion/ion resonant instability, in which the right-handed polarized waves (fast magnetosonic mode) resonate with the backstreaming ion beam [Barnes, 1970; Gary, 1978; Sentman et al., 1981]. This instability generates right-handed waves in the upstream region which propagate in the same direction as the beams (upstream) in the solar wind frame. Since they are convected downstream by the solar wind flow, they will be Doppler-shifted to left-handed waves in the spacecraft frame. When the backstreaming ions are sufficiently fast ($v_b > 10\ v_A$) and dense ($n_b \sim$ 10% of solar wind density), besides the above resonant instability, the non-resonant firehose-like instability will excite ULF waves in the right-handed magnetosonic branch which propagate in the direction opposite to the ion beam, i.e., downstream toward the bow shock. These waves are also right-handed in the spacecraft reference frame. Gary et al. [1984] have shown that the non-resonant instability has larger growth rate than the resonant instability if v_b/v_A and n_b/n_{sw} are sufficiently large. In another extreme case when the backstreaming ions are extremely hot with thermal speeds greater than their streaming velocity ($v_{th} > v_b$), the left-handed Alfven/ion resonant mode is also unstable. The growth rate for the two resonant instabilities can be comparable [Sentman et al., 1981; Gary, 1985]. This left-handed resonant instability generates waves propagating upstream from the bow shock and will be Doppler-shifted to right-handed waves in the spacecraft frame as they are convected downstream by the solar wind. Thus, according to linear theory, the upstream ULF waves will be more often left-handed in the spacecraft frame under typical conditions and also can be right-handed in the spacecraft frame under extreme conditions of fast, dense beams or hot beams. However, when the waves are right-handed in the spacecraft frame, they can be intrinsically either right-handed magnetosonic mode or left-handed Alfven mode in the solar wind frame. Observations have shown that the left-handed ULF waves (in the spacecraft frame)

Upstream ULF Waves - ISEE Observations

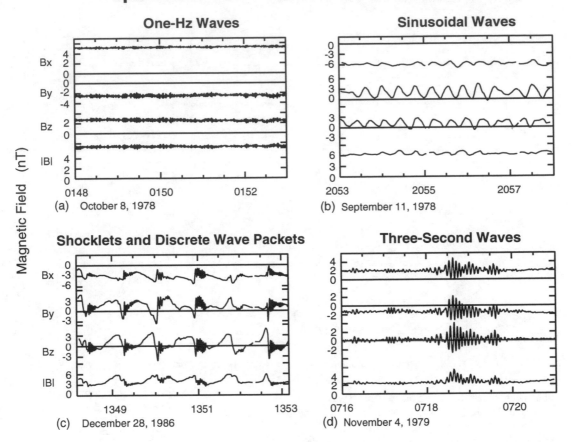

Fig.2. Observations of different types of ULF waves in the Earth's foreshock region. (a) One-Hz waves; (b) Sinusoidal waves; (c) Shocklets and discrete wave packets; (d) Three-second waves.

are indeed the dominant mode in the foreshock region [Hoppe et al., 1981; Hoppe and Russell, 1983]. They have frequencies of $\sim 0.1\ \Omega_{ci}$ (Ω_{ci} is the local ion cyclotron frequency) and wavelengths of the order of 1 R_E. For the right-handed waves (in the spacecraft frame), their intrinsic modes have not been identified.

One important fact of obliquely propagating magnetosonic waves is that there is a first-order density perturbation associated with the magnetic perturbation. In the MHD limit, the density and magnetic perturbation for a magnetosonic waves propagating at an angle θ_{Bk} are related by [Siscoe, 1983]:

$$\frac{\delta n}{n} = \frac{\delta B}{B} \times \frac{v_{ph}^2}{(v_{ph}^2 - c_s^2 cos\theta_{Bk})}$$

where δn is the perturbation in plasma number density, δB is the perturbation in field strength, v_{ph} is the wave phase velocity and c_s is the sound speed in the plasma. From the equation, we note that the $\delta n/n$, the relative perturbation in density, is always greater or equal to $\delta B/B$, the relative field perturbation, since $v_{ph}^2 \geq v_{ph}^2 - c_s^2 cos\theta_{Bk}$ for fast magnetosonic waves. For upstream waves which are strongly compressional, the density fluctuations associated with the waves are very significant as they

cause large fluctuations in solar wind dynamic pressure.

Observations show that the ULF waves generated in the region upstream of the bow shock are convected downstream by the solar wind flow because their group velocity is much smaller than the solar wind flow speed [Hoppe et al., 1981; Hoppe and Russell, 1983]. The downstream convection of compressional ULF waves can have a profound impact on the Earth's magnetosphere. The upstream fluctuations associated with these waves are carried downstream along solar wind streamlines into the magnetosheath. The streamlines in the magnetosheath which approach most closely to the magnetopause pass through the subsolar region of the bow shock. During intervals when the IMF cone angle is small ($\leq 45°$), these waves fill the subsolar upstream region, as illustrated in Figure 3 (adapted from Russell et al. [1983]). Under the condition of small IMF cone angle, the solar wind carries the upstream ULF fluctuations to the magnetopause boundary. The magnetopause responds to these pressure fluctuations and transfers wave energy into the dayside magnetosphere. The waves on the magnetopause can excite field line resonances in the magnetosphere, as described by Southwood [1974] and Chen and Hasegawa [1974]. Although the process by which the energy enters into the magnetosphere is still not well

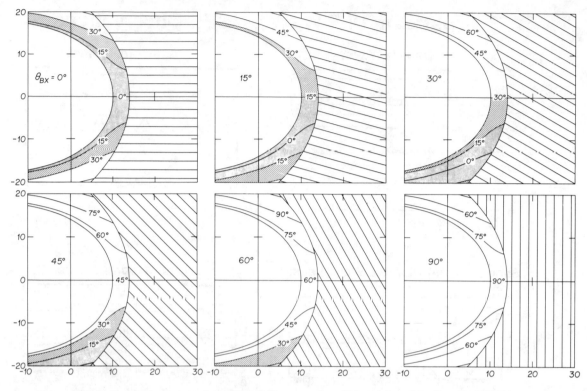

Fig.3. Foreshock geometry for various IMF cone angles. (Adapted from Russell et al. [1983])

understood, it has been generally agreed that upstream waves are a major source for dayside Pc 3 and 4 magnetic pulsations in the Earth's magnetosphere (cf. Odera, [1986] for a review).

There is much observational evidence to support this idea. Many studies of the magnetic pulsations in the dayside magnetosphere have indicated that their occurrence is controlled by the IMF cone angle [eg. Troitskaya et al., 1971; Russell et al., 1983; Yumoto et al., 1985; Engebretson et al., 1986, 1987; Yumoto, 1988]. Luhmann et al. [1986] found that magnetosheath turbulence is sensitive to the IMF cone angle and is enhanced in the subsolar region during periods of small cone angles. By using multiple spacecraft observations, Engebretson et al. [1991] found on a case-by-case basis that small cone angles are well correlated with large turbulence in the magnetosheath and the simultaneous excitation of Pc 3 and 4 pulsations in the dayside outer magnetosphere. Other evidence is that the frequencies of both upstream and magnetospheric ULF waves have a similar dependence on the IMF strength, as shown in Figure 4 (adapted from Russell and Hoppe [1981]). In Figure 4, the solid circles and solid lines are from observations of upstream ULF waves. Dashed, dash-dot, and dotted lines are from observations of ground-based Pc 3–4 waves as given by Gul'yel'mi et al. [1973], Gul'yel'mi [1974], and Gul'yel'mi and Bol'shakova [1973], respectively.

In this paper, we discuss recent advances in the understanding of the foreshock ULF morphology and wave properties, as well as remaining problems. We present our studies of critical upstream ULF wave properties of importance to the magnetospheric ULF waves.

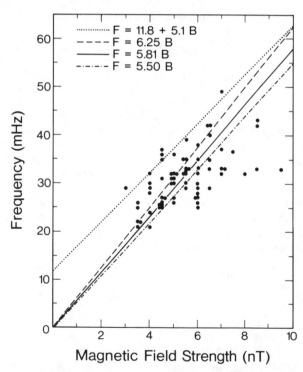

Fig.4. Relationship between the wave frequency and the IMF strength for upstream ULF waves (solid circles and solid line) and for ground-based Pc 3–4 waves (dashed, dash-dotted and dotted lines). (Adapted from Russell and Hoppe [1981])

2. GEOMETRY OF ULF FORESHOCK

The schematic of the foreshock in Figure 1 shows that its geometry depends mainly on the IMF cone angle. The first requirement in studying the general morphology of the foreshock is to quantitatively determine the upstream boundary of the ULF foreshock for different IMF cone angles. This boundary represents the motion of backstreaming ions in the foreshock region since the ULF waves are the consequences of instabilities generated by these backstreaming ions. If the ions originating in the bow shock have velocity components upstream away from the bow shock, the ions will leave the bow shock with the equation of motion to the first order as:

$$dv/dt = (\Omega_i/B) \cdot (-V_{sw} \times B + v \times B)$$

where B is the IMF, Ω_i is the ion gyrofrequency, and $-V_{sw} \times B$ is the solar wind convection electric field. The above equation shows that the motion of backstreaming is confined in the plane containing the solar wind flow and the IMF, called V-B plane. The ions' net guiding center velocity in the V-B plane is the vector sum of parallel velocity along the IMF upstream and the downstream $E \times B$ drift:

$$v = v_{//} + v_d$$

as shown in Figure 1. The ion foreshock boundary in the V-B plane is parallel to the guiding center velocity [Greenstadt, 1976]. But the starting point of the foreshock boundary on the bow shock is controlled largely by θ_{Bn}, the angle between the IMF and the local bow shock normal, because θ_{Bn} controls the ion reflection process. In the case of small θ_{Bn}, or quasi-parallel shock, the guiding center velocity makes a large angle to the bow shock surface, and thus, particles can leave the bow shock very easily. On the other hand, if θ_{Bn} is very large, or quasi-perpendicular shock, the gyromotion of the ions around the magnetic field lines may bring the ions back to the bow shock before they finish one gyration around the magnetic field line if their pitch angles are appropriate. Gosling et al. [1982] have demonstrated that the reflected ions can escape upstream only when $\theta_{Bn} < 45°$ in the case of specular reflection.

The pioneering work of locating the ULF foreshock boundary can be found in Greenstadt et al. [1970] and Greenstadt [1972], in which this boundary was determined based on detailed case studies. Later this boundary was determined statistically by Greenstadt and Baum [1986], in which they used the ISEE 1 magnetometer data to find actual crossings of the ULF foreshock boundary. Their study clearly showed the IMF cone angle control of the ULF foreshock boundary. They displayed the locations of ISEE 1 at ULF foreshock boundary crossings in the V-B plane for moderate cone angles of 40°–50° and for small cone angles of 20°–30°. They found that the patterns of the scatter plots of the crossings defined a boundary for each of the two subsets, but the slopes of the two boundaries are different. The ULF wave foreshock boundary determined in this study also inferred the backstreaming ion velocity of $1.6V_{sw}$.

In our recent study of determining the ULF foreshock boundary, we identified many foreshock boundary crossings in the upstream region when the IMF cone angle was nearly constant for extended time periods and thus the foreshock boundary was steady in space [Le and Russell, 1992a]. The work consisted of two steps, first to determine θ_{Bn} of the ULF foreshock on the bow shock, and second to determine the slope of the ULF foreshock boundary. In the first step, we examined ISEE bow shock crossings at various positions to determine the source point on the bow shock which separated disturbed (with ULF waves) and undisturbed (without ULF waves) upstream magnetic field. The statistical study of the bow shock crossings showed that the ULF foreshock started at $\sim 50° \theta_{Bn}$. In the second step, we found that the ULF foreshock boundary was less sensitive to small changes of the IMF direction at larger cone angle. The ULF foreshock boundary was well defined in the V-B plane for cone angles > 40°. Figure 5 shows the ISEE positions in the V-B plane for five ULF foreshock boundary crossings identified at 50°±5° IMF cone angles. The bow shock was scaled by the solar wind dynamic pressure and Mach number, and then, normalized to the same size for each crossing. The spacecraft positions form a clear boundary in the V-B plane. From this boundary, we infer that the backstreaming ions had a velocity of $1.3V_{sw}$ along the IMF and a net guiding center velocity of $1.5V_{sw}$ in the Earth's frame. On average, the ULF foreshock boundary corresponds to the trajectory of backstreaming ions with a streaming velocity of $\sim 1.4V_{sw}$ along the IMF in the Earth's frame and a source point at $\sim 50° \theta_{Bn}$ when the IMF cone angle is moderate ($\theta_{BV} > 40°$). When the IMF cone angle is small ($20° < \theta_{BV} < 30°$), the ULF foreshock boundary is not well defined.

3. ULF WAVES FOR RADIAL IMF

The foreshock geometry for nearly radial IMF is different from that at moderate and large cone angles. The solar wind convection electric field is very small and the backstreaming ions move along the magnetic field in a nearly flow-aligned IMF condition. It is the most favorable condition for the generation of upstream waves since the particles can go upstream more easily from the bow shock. In this case, most of the day side upstream region is inside the ion foreshock region, although it is still not clear if there is a distinct boundary which separates the ULF foreshock from the undisturbed solar wind. We emphasize that the large-amplitude ULF waves for nearly radial IMF can be convected close to the Earth's magnetosphere under this IMF configuration. Observations also show that it is the most favorable geometry for the occurrence of magnetospheric ULF waves [Russell et al., 1983].

ULF waves observed for nearly radial IMF are typically in the form of steepened shocklets which sometimes have discrete wave packets at the steepening edges. Figure 6 shows examples of ULF waves for nearly radial IMF where each panel consists of 10 minutes of high resolution data within an interval of cone angle <10° which lasted at least one hour. These waves are similar in form to those observed well downstream from the foreshock

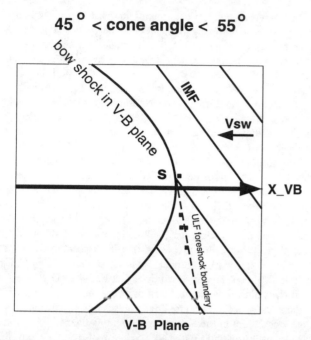

45° < cone angle < 55°

V-B Plane

Fig.5. The spacecraft positions at the ULF foreshock boundary crossings for 50°±5° cone angles are plotted in the **V-B** plane. The bow shock is scaled by the solar wind dynamic pressure and the magnetosonic Mach number, and then, normalized to the same size for each crossings. (Adapted from Le and Russell [1992a])

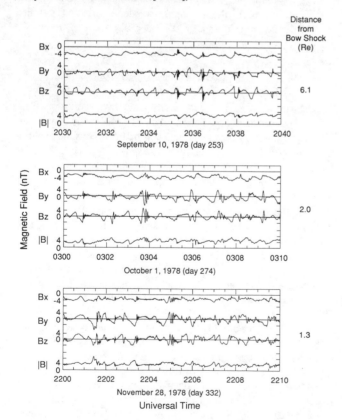

Fig.6. Examples of ULF waves for nearly radial IMF. (Adapted from Le [1991])

boundary at moderate and large cone angle. As shown in Figure 7, the ULF wave region extends upstream with a scale length of ~ 23 R_E for this geometry. The top panel of Figure 7 shows the normalized ULF wave spectral amplitude as a function of distance from the bow shock along the IMF, which is roughly the same as distance from bow shock along Sun-Earth line for nearly radial IMF. The bottom panel of Figure 7 shows the spatial coverage of these data in the plane which contains the spacecraft and the Sun-Earth line (there is no meaningful **V-B** plane for nearly radial IMF). Although there is a tendency of decreasing wave amplitude with increasing distance from the bow shock, the decrease of the amplitude is very slow with a scale of ~ 23 R_E. Ipavich et al. [1981] found that the upstream 30 keV proton intensity varied exponentially with the radial distance from the bow shock and had a scale length of 7 ± 2 R_E. The IMF direction during the time of peak intensity was within $\sim 15°$ of the radial direction for 90% of their events. The two scale lengths are qualitatively consistent to the extent that the upstream particles may have different scale lengths at different energies and the correlation between the beam density and wave amplitude is not exactly linear.

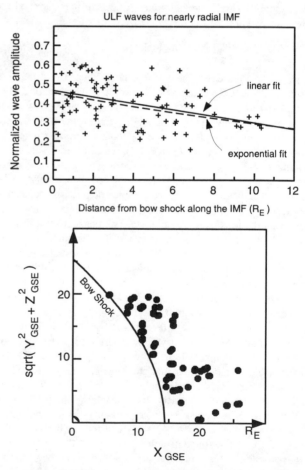

Fig.7. The upper panel is the normalized ULF wave spectral amplitude as a function of the distance from the bow shock for nearly radial IMF. The solid line is a linear fit to the data and the dotted line is an exponential fit. The lower panel is the data coverage in the plane containing the spacecraft and the Earth-Sun line. (Adapted from Le [1991])

Right-handed polarized waves (in the spacecraft frame) are more frequently observed at nearly radial IMF than at moderate and large cone angles. Under conditions typical of the ion foreshock, the most unstable mode is the right-handed polarized magnetosonic wave due to the resonant instability [Barnes, 1970; Gary, 1978]. These waves will be Doppler-shifted to left-handed polarizations in the spacecraft frame. Left-handed ULF waves (in the spacecraft frame) are indeed the dominant mode observed in the upstream region [Hoppe et al., 1981; Hoppe and Russell, 1983]. However both left-handed and right-handed modes are observed with equal probability for nearly radial IMF and their ellipticity seems to be correlated with the wave amplitude. Figure 8 shows the wave ellipticity versus the normalized wave amplitude. Negative ellipticity indicates left-handed polarization in the spacecraft frame and positive ellipticity corresponds to right-handed polarization in the spacecraft frame. There is a positive correlation between the ellipticity and the wave amplitude with a correlation coefficient of 0.57. A similar correlation between ellipticity and wave amplitude has been reported by Russell et al. [1987] for moderate IMF cone angles. The right-handed waves are stronger than the left-handed waves (in the spacecraft frame).

The correlation between polarization and amplitude suggests that different mechanisms generate waves with different amplitudes. As we discussed in the introduction, according to linear theory, the left-handed polarized waves in the spacecraft frame are favored under typical backstreaming ion condition ($n_b \sim n_{sw}$, $v_b \lesssim 10\, v_A$) via the ion/ion right-handed resonant instability. When the backstreaming ions are very dense and fast ($n \sim 10\, n_{sw}$, $v > 10\, v_A$) or very hot ($v_{th} > v_b$), right-handed polarized waves are observed in the spacecraft frame, that are generated either by the ion/ion nonresonant instability or the ion/ion left-handed resonant instability. The facts 1) that right-handed waves in the spacecraft frame are more often observed and 2) that these waves are stronger for nearly radial IMF suggest that this upstream configuration is a favorable condition for faster and denser ion beams (nonresonant instability) or hotter ion beams (resonant instability). Energetic particle observations have shown that the upstream ions are of diffuse type and have small bulk velocity under radial IMF [Ipavich et al., 1981]. It seems that the nonresonant instability can be dismissed. However, this is just speculation since little has been done to identify the intrinsic mode for the waves which have right-handed polarization in the spacecraft frame.

4. PROPERTIES OF UPSTREAM WAVES RELEVANT TO ULF WAVES IN THE MAGNETOSPHERE

Compressional ULF waves are an intrinsic feature of quasi-parallel shocks. Their existence in front of the bow shock modifies both the solar wind and the turbulence spectrum convected downstream to the magnetopause. In this section, we review recent observations of the wave properties, especially those of importance to the Earth's magnetosphere, including the magnitude of the pressure fluctuations associated with the waves, the coherence length, and the bandwidth.

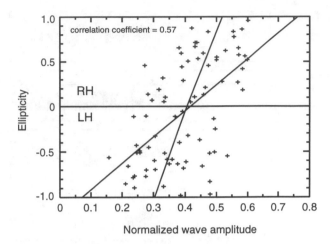

Fig.8. The wave ellipticity versus the normalized amplitude for nearly radial IMF. The correlation coefficient is 0.57, which is significant at almost 100% level. (Adapted from Le [1991])

Pressure Fluctuations Associated with the Waves

MHD theory predicts that ULF waves cause significant density fluctuations, and thus dynamic pressure fluctuations as well in the unshocked solar wind. Although the fluctuating magnetic field of these waves is the most readily measured aspect of the waves, it is the associated pressure fluctuations which can cause the magnetopause to oscillate in response to the waves. Observations show that density and dynamic pressure fluctuations have been significantly enhanced in the ULF foreshock [Paschmann et al., 1979; Bame et al., 1980; Le, 1991]. The density and dynamic pressure fluctuations associated with the ULF waves are ~ 20% of the average background value. In comparison, fluctuations in the undisturbed solar wind at these frequencies are only ~ 5% of the background average.

Figure 9 shows an example of how the ULF foreshock can modify the solar wind itself. It contains 5 hours data from the ISEE 1 magnetometer and Cross-Fan Solar Wind Experiment, including magnetic field, solar wind ion density, bulk velocity and dynamic pressure in the upstream region. As already noted, the foreshock geometry is very sensitive to the IMF direction. When the IMF changes its direction, the spacecraft may suddenly find itself located inside the foreshock region. In Figure 9, the IMF changes its direction as well as magnitude near 0825UT. Following the onset of the ULF waves, enhanced fluctuations in solar wind ion density, ion bulk velocity and dynamic pressure (ρv^2) are present that are clearly associated with the ULF waves. The bulk of the solar wind is also slowed down (by ~ 38 km/sec in this example) in the foreshock region. Such decelerations are common within the foreshock and are caused by the interaction of the solar wind with backstreaming ions which slow down the incoming solar wind by transferring the momentum flux [Bame et al., 1980; Bonifazi et al., 1980; see also the review by Thomsen, 1985]. Thus varying IMF direction can modify the geometry of the foreshock and alter the distribution of pressure on the

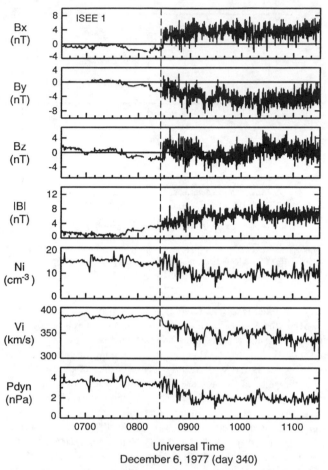

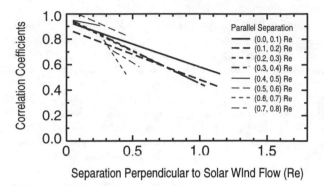

Fig.10. The cross-correlation coefficients as a function of spacecraft separation perpendicular to the solar wind flow for each 0.1 R_E separation parallel to the solar wind flow. (Adapted from Le and Russell [1990a])

Fig.9. The time series of magnetic field components, magnetic field strength, ion density, bulk velocity, and dynamic pressure from 0630 to 1130 UT on December 6, 1977. The IMF changes its direction near 0825 UT. There are enhanced fluctuations in ion density and dynamic pressure associated with the ULF waves. (Adapted rom Le [1991])

despite different separations along the solar wind flow. The coherence length is on the order of an Earth radius transverse to the solar wind flow, a scale similar to the wavelength. This result is consistent with that estimated from the bandwidth of the power spectra. The limited coherence length in the direction transverse to the flow is mainly due to the solar wind convection effect. In the direction along the solar wind flow, the coherence length is at least several Earth radii. Thus, ULF waves are large-scale coherent structures, and should induce similar scale-size coherent oscillations in the bow shock and magnetopause, and in turn in the magnetosphere itself.

magnetopause. In this way IMF directional fluctuations can cause compressions of the magnetosphere and may explain some pressure pulses seen there.

Coherence Length of ULF Waves

ULF waves are carried downstream towards the bow shock and magnetopause along solar wind streamlines and modulate the structures of both the bow shock and the magnetopause if they have sufficient amplitude and scale size. The coherence length of the ULF waves has been investigated using simultaneous observations from the dual ISEE 1 and 2 spacecraft [Le and Russell, 1990a]. In that study, we examined the correlation between these simultaneous observations for different separations of the two spacecraft. Figure 10 shows the cross-correlation coefficients as a function of separation distance perpendicular to the solar wind flow where each line corresponds to a different separation parallel to the solar wind flow. The cross-correlation coefficients decrease as the separation perpendicular to the flow increases. However the cross-correlation coefficients are similar

Bandwidth of ULF Waves

The ULF foreshock modifies the turbulence spectrum convected downstream to the magnetopause. Thus, in order to predict the magnetospheric effects of these waves it is of interest to determine the frequency range of enhanced power of magnetic fluctuations in the foreshock region. We have compared the ULF wave power spectrum with the background solar wind spectrum and found that the enhanced wave power has limited bandwidth [Le and Russell, 1990b]. In that study, simultaneous observations from two largely separated spacecraft that are located on either side of the ULF foreshock were examined. The data indicate that the foreshock ULF wave power spectrum has a clear low-frequency cutoff, below which the power spectra are similar in the undisturbed solar wind and in the foreshock. This low-frequency cutoff occurs above ∼ 5 mHz. Figure 11 shows one example in which ISEE 1 is inside the foreshock and IMP-8 is in the undisturbed solar wind. From this figure it is apparent that ISEE 1 observes enhanced power at frequencies higher than 7 mHz. There is no significant power enhancement or damping below 7 mHz. This result does not support the suggestion that upstream shock-related pressure oscillations drive magnetopause surface waves and magnetospheric oscillations with periods of ∼ 200–600 seconds [Sibeck et al., 1989].

In short the solar wind flow is significantly perturbed in the foreshock region. The perturbation is unsteady due to frequent variations of IMF orientations, perhaps explaining some pressure pulses in the magnetosphere. Wave processes in the foreshock are however band limited and thus can be directly responsible for

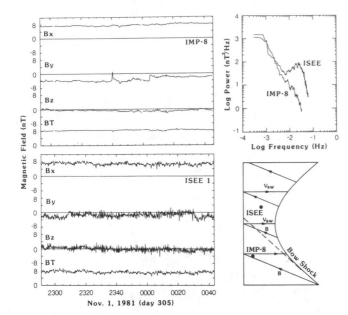

Fig. 11. The magnetic field time series, power spectra, and foreshock geometry from simultaneous observations of ISEE 1 and IMP-8 on November 1, 1981. (Adapted from Le and Russell [1990b])

only a fraction of the waves in the magnetosphere.

5. Spatial Variation of ULF Waves

It is well known that the properties of upstream ULF waves are position dependent within the foreshock and the waveform varies from nearly sinusoidal to highly steepened as the θ_{Bn} increases [Hoppe et al., 1981; Russell et al., 1987]. The region populated by the intermediate ions is the forward boundary of the ULF wave foreshock. The waves there are fairly monochromatic and typically last many cycles. They are left-handed polarized in the spacecraft frame. Further downstream from the ULF foreshock boundary in the region populated by diffuse ions, the ULF waves often appear in the form of the steepened shocklets. They exhibit both left-handed and right-handed polarization in the spacecraft frame, and the left-handed and right-handed waves are similar in form, frequency, and wavelength.

Russell et al. [1987] used simultaneous observations from two well separated spacecraft to show that the properties of the ULF waves depend on the their location in the foreshock region. The properties include waveform, amplitude, polarization and power spectrum. Early observations of backstreaming ions also revealed that the properties of backstreaming ions varied with respect to point of origin along the bow shock [Paschmann et al., 1981; Ipavich et al., 1981; Thomsen, 1985].

Since solar wind conditions, especially the IMF direction, are highly variable, early observations of the ULF waves were made under various solar wind conditions and did not reveal the structure of the region behind the foreshock boundary, or spatial variation of ULF waves within this region. When the IMF cone angle changes, the foreshock geometry changes accordingly. Thus it is essential to study the evolution of ULF waves under similar conditions of IMF and bow shock strength, which is another controlling factor. During a long interval of nearly constant solar wind conditions, the motion of backstreaming ions can reach a nearly steady state, as do the waves generated by them. In this case, the changes of wave properties are primarily related to the geometry in the foreshock, i.e., the different depth from the foreshock boundary and distance from the bow shock.

We have examined cases in which the IMF cone angle was nearly constant over an extended time period and ISEE 1 and 2 spacecraft were traveling a large distance within the foreshock [Le and Russell, 1992b]. Figure 12 shows one of the cases examined. The top panel shows the foreshock geometry and ISEE trajectory in the **V-B** plane. The lower panels show the magnetic field data during this interval. Large-amplitude ULF waves with periods near 40 seconds were present throughout this interval and these ULF waves exhibited predominantly a steepened form (shocklets). The amplitudes of the waves increase gradually as the spacecraft travels to large depth from the foreshock boundary and closer to the bow shock. Following the increase of wave amplitude, the waves vary from elliptically polarized to more linearly polarized. The wave power spectra (Figure 13) shows that the peak of power spectrum became broader towards both lower and higher frequencies while the peak frequency stayed nearly the same as the bow shock is approached. In addition the spectra had a limited bandwidth and a clear low frequency cutoff throughout the interval.

From the bottom panels of Figure 12, it is evident that the discrete wave packets found at the steepening edges of the lower-frequency waves, or shocklets, become more intense and develop more cycles associated with the increasing intensity of the shocklets. Figure 14 shows the wave duration in cycles and the peak-to-peak amplitude of the discrete wave packets (the largest cycle) as a function of distance from the bow shock along a solar wind streamline. The scale length given by the exponential fit is 1.9 R_E for the wave duration and 2.8 R_E for the wave amplitude. Based on the facts that these wave packets were convected downstream with the solar wind and that their phase velocity was much less than the solar wind velocity, the estimated growth rate for the discrete wave packets was - 0.035 Ω_i for the particular geometry and solar wind conditions of this interval.

6. Remaining Problems

Despite intensive studies of the upstream waves over the past two decades, there are still many remaining problems which need further investigation. First of all, the present theoretical models for wave generation are not successful in predicting all the properties of observed ULF waves. From the linear theories the maximum growth rates for all three instabilities always occur for parallel propagation (**k**//**B**) over all parameter space, although the growth rate for oblique propagation waves can be significant [Montgomery et al., 1976; Gary, 1991]. However, the large-amplitude ULF waves are observed as highly compressional and steepening. Thus they propagate obliquely to the magnetic field. Hada et al. [1987] proposed a refraction mechanism to account for the observed obliquely propagating waves. In their mechanism

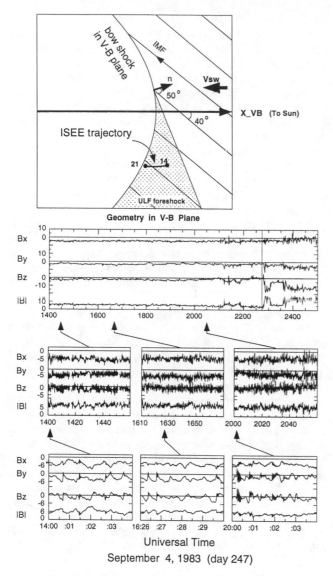

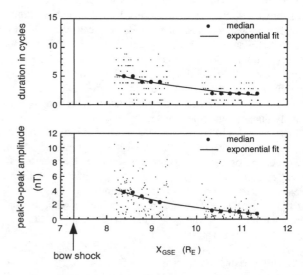

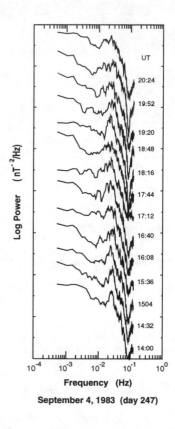

Fig.13. The power spectra of 32–minute magnetic field data with starting times indicated in the figure on September 4, 1983. Each spectrum is the total power summed over three components, and is plotted by shifting one decade upward from the previous one. (Adapted from Le and Russell [1992b])

Fig.12. The upper panel shows the ISEE trajectory in the **V-B** plane from 1400 to 2100 UT on September 4, 1983 when the IMF cone angle was nearly constant over an extended time period. The lower panels are magnetic field observations. (Adapted from Le and Russell [1992b])

Fig.14. The spatial variation of the duration and the peak-to-peak amplitude of discrete wave packets on September 4, 1983. (Adapted from Le and Russell [1992b])

the parallel propagating right-handed and left-handed waves are excited by the backstreaming ions in the upstream region. These waves are refracted to oblique propagation as they are convected downstream by the solar wind due to the non-uniform index of refraction caused by the spatial variation of the backstreaming ions. As a result, the waves become compressional and steepen into the shocklets. In this mechanism, geometry is very important and shocklets should not be observed in the nearly radial IMF geometry because the effect of refraction is minimum for parallel propagating waves. However, we have shown evidence that both shocklets and discrete wave packets are observed when the IMF is nearly radial (Figure 6). The linear theories also predict that ULF waves exhibit both left-handed and right-handed polarization in the spacecraft frame. Observations have shown that polarization and wave amplitude are correlated. For the right-handed waves in the spacecraft frame, their intrinsic mode can be either right-handed or left-handed. Thus the identification of the intrinsic mode and conditions for which they occur will help us to understand their generation mechanisms.

Solving these problems also relies largely on a better understanding of upstream particles. Combined data sets from many instruments (both field and plasma) will provide detailed information on the physics of the underlying processes. This is still a weak point in the foreshock study despite of the fact that existing ISEE data sets with high time resolution are available for over a decade. Previous work used either primarily magnetic field data or primarily plasma data to study the morphology of the foreshock region and the spatial variation of waves properties. For example, previous attempts to determine the ULF foreshock boundary were mainly based on field observations and properties of the backstreaming ions were inferred from these observations. However, we do not know the behavior of the backstreaming ions near the foreshock boundary identified from field data. Do the backstreaming ions move upstream at a velocity inferred from magnetic field observations? What is the difference, if any, between the boundaries of backstreaming ions and ULF waves and how does the backstreaming ion distribution function change across the ULF wave boundary? We note that the growth of waves in the unstable backstreaming ions is not instantaneous and in fact waves grow in a parcel of plasma as it convects towards the bow shock. The onset of waves may not simply reflect the variation in backstreaming ion properties. We studied the spatial variation of ULF waves under steady solar wind conditions, but we do not know the spatial variation of the backstreaming ions under these conditions. We do not know how the change of wave properties depends on the change of streaming velocity, density, and distribution function of the backstreaming ions.

The importance of upstream phenomena is not limited to the Earth's upstream region. The comparative study of a variety of foreshocks is very important for understanding the underlying physical processes. Waves have been observed upstream from bow shocks of Mercury, Venus, Mars, Jupiter, Saturn and Uranus. Their similarities and differences are not currently very well understood. First, the solar wind Mach number increases with increasing heliocentric distance. The significant increases in the strength of planetary shocks with increasing distance from the sun may induce considerable changes in the relative efficiencies of the various processes such as leakage and reflection that generate the backstreaming ions. Another important factor that may influence the wave and particle signatures observed in the upstream region is the varying time of connection of the field lines to the planetary shock and the varying radius of curvature of the bow shock relative to characteristic scale lengths, (eg., the proton gyroradius and ion inertial length). For the above reasons it is essential to study a variety of foreshocks and to determine the dependence of the microphysics of the foreshock phenomena on varying boundary conditions.

Acknowledgments. This work was supported by the National Aeronautics and Space Administration under research grant NAGW-2067.

REFERENCES

Armstrong, T. P., M. E. Pesses, and R. B. Decker, Shock drift acceleration, in *Collisionless Shocks in the Heliosphere: Review of Current Research*, eds. B. T. Tsurutani and R. G. Stone, AGU Geophysical Monograph 35, 271–286, 1985.

Bame, S. J., J. R. Asbridge, W. C. Feldman, J. T. Gosling, G. Paschmann, and N. Sckopke, Deceleration of the solar wind upstream from the earth's bow shock and the origin of diffuse upstream ions, *J. Geophys. Res.*, 85, 2981, 1980.

Barnes, A., A theory of generation of bow–shock–associated hydromagnetic waves in the upstream interplanetary medium, *Cosmic Electrodyn.*, 1, 90, 1970

Bonifazi, C., A. Egidi, G. Moreno, and S. Orsini, Backstreaming ions outside the earth's bow shock and their interaction with the solar wind, *J. Geophys. Res.*, 85, 3461, 1980.

Chen, L., and A. A. Hasegawa, Theory of long–period magnetic pulsations 1, Steady state of excitation of field line resonance, *J. Geophys. Res.*, 79, 1024, 1974

Engebretson, M. J., L. J. Zanetti, T. A. Potemra, and M. A. Acuna, Harmonically structured ULF pulsations observed by the AMPTE/CCE magnetic field experiment, *Geophys. Res. Lett.*, 13, 905, 1986.

Engebretson, M. J., L. J. Zanetti, T. A. Potemra, W. Baumjohann, H. Luehr, and M. H. Acuna, Simultaneous observation of Pc 3–4 pulsations in the solar wind and in the Earth's magnetosphere, *J. Geophys. Res.*, 92, 10,053, 1987.

Engebretson, M. J., N. Lin, W. Baumjohann, H. Luehr, B. J. Anderson, L. J. Zanetti, T. A. Potemra, R. L. McPherron, and M. G. Kivelson, A comparison of ULF fluctuations in the solar wind, magnetosheath, and dayside magnetosphere 1. Magnetosheath morphology, *J. Geophys. Res.*, 96, 3441, 1991.

Fairfield, D. H., and K. W. Behannon, Bow shock and magnetosheath at Mercury, *J. Geophys. Res.*, 81, 3891, 1976.

Fairfield, D. H., Bow shock associated waves observed in the far upstream interplanetary medium, *J. Geophys. Res.*, 74, 3541, 1969.

Fairfield, D. H., Whistler waves observed upstream of collisionless shocks, *J. Geophys. Res.*, 79, 1368, 1974.

Gary, S. P., Electromagnetic ion beam instability and energy loss of fast alpha particles, *Nucl. Fusion*, 18, 327, 1978.

Gary, S. P., The electromagnetic ion beam instabilities: Hot beams at interplanetary shocks, *Astrophys. J.*, 88, 65, 1985.

Gary, S. P., Electromagnetic ion/ion instabilities and their consequences in space plasma: A review, *Space Sci. Rev.*, 56, 373, 1991.

Gary, S. P., C. W. Smith, M. A. Lee, M. L. Goldstein, and D. W. Forslund, Electromagnetic ion beam instabilities, *Phys. Fluids*, 27, 1852, 1984. (Correction, *Phys. Fluids*, 28, 438, 1985.)

Goldstein, M. L., H. K. Wong, A. F. Vinas, and C. W. Smith, Large amplitude MHD waves upstream of Jovian bow shock: Reinterpretation, *J. Geophys. Res.*, 90, 302, 1985.

Gosling, J. T., J. R. Asbridge, S. J. Bame, G. Paschmann, and N. Sckopke, Observations of two distinct populations of bow shock ions in the upstream solar wind, *Geophys. Res. Lett.*, 5, 957, 1978.

Gosling, J. T., M. F. Thomsen, S. J. Bame, W. C. Feldman, G. Paschmann, and N. Sckopke, Evidence of specularly reflected ions upstream from the quasi-parallel bow shock, *Geophys. Res. Lett.*, 9, 1333, 1982.

Gosling, J. T., and A. E. Robson, Ion reflection, gyration and dissipation at supercritical shocks, in *Collisionless Shocks in the Heliosphere: Review of Current Research*, eds. B. T. Tsurutani and R. G. Stone, AGU Geophysical Monograph 35, 207, 1985.

Greenstadt, E. W., A binary index for assessing local bow shock obliquity, *J. Geophys. Res.*, 77, 5467, 1972.

Greenstadt, E. W., Energies of backstreaming protons in the foreshock, *Geophys. Res. Lett.*, 3, 553, 1976.

Greenstadt, E. W., and L. W. Baum, Earth's compressional foreshock boundary revisited: Observations by ISEE 1 magnetometer, *J. Geophys. Res.*, 91, 9001, 1986.

Greenstadt et al., Correlated magnetic field and plasma observations of the Earth's bow shock, *J. Geophys. Res.*, 73, 51, 1968.

Greenstadt, E. W., I. M. Green, D. S. Colburn, J. H. Binsack, and E. F. Lyon, Dual satellite observations of the Earth's bow shock, II, The thick pulsation shock, *Cosmic Electrodyn., 1,* 279, 1970.

Gul'yel'mi, A. V., Diagnostics of the magnetosphere and interplanetary medium by means of pulsations, *Space Sci. Rev., 16,* 331, 1974.

Gul'yel'mi, A. V., and O. V. Bol'shakova, Diagnostics of the interplanetary magnetic field from ground-based data on Pc 2–4 micropulsations, *Geomag. Aeron., 13,* 535, 1973.

Gul'yel'mi, A. V., T. A. Plyasova-Bakounina and R. V. Shchepetnov, Relation between the period of geomagnetic pulsations Pc3, 4 and the parameters if the interplanetary medium at the Earth's orbit, *Geomag. Aeron., 13,* 382, 1973.

Gurnett, D. A., Plasma waves and instabilities, in *Collisionless Shocks in the Heliosphere: Review of Current Research,* eds. B. T. Tsurutani and R. G. Stone, AGU Geophysical Monograph 35, 207, 1985.

Hada, T., C. F. Kennel, and T. Terasawa, Excitation of compressional waves and the formation of shocklets in the Earth's foreshock, *J. Geophys. Res., 92,* 4423, 1987.

Hoppe, M. M., and C. T. Russell, Particle acceleration at planetary bow shock waves, *Nature, 295,* 41, 1982.

Hoppe, M. M., and C. T. Russell, Plasma rest frame frequencies and polarizations of the low-frequency upstream waves: ISEE 1 and 2 observations, *J. Geophys. Res., 88,* 2021, 1983.

Hoppe, M. M., C. T. Russell, L. A. Frank, T. E. Eastman, and E. W. Greenstadt, Upstream hydromagnetic waves and their association with backstreaming ion population: ISEE 1 and 2 observations, *J. Geophys. Res., 86,* 4471, 1981.

Hoppe, M. M., C. T. Russell, L. A. Frank, and T. E. Eastman, Characteristics of ULF waves associated with upstream ion beams, *J. Geophys. Res., 87,* 643, 1982.

Ipavich, F. M., A. B. Galvin, G. Gloeckler, M. Scholer, and D. Hovestadt, A statistical survey of ions observed upstream of Earth's bow shock: Energy spectra, composition and spatial variation, *J. Geophys. Res., 86,* 4337, 1981.

Le, G., Generation of upstream waves in the Earth's foreshock, Ph.D. Dissertation, University of California, Los Angeles, 1991.

Le, G., and C. T. Russell, A study of the coherence length of ULF waves in the Earth's foreshock region, *J. Geophys. Res., 95,* 10,703, 1990a.

Le, G., and C. T. Russell, Observations of the magnetic fluctuation enhancement in the Earth's foreshock region, *Geophys. Res. Lett., 17,* 905, 1990b.

Le, G., and C. T. Russell, A study of ULF wave foreshock morphology, 1. ULF foreshock boundary, *Planet. Space Sci., 40,* 1203, 1992a.

Le, G., and C. T. Russell, A study of ULF wave foreshock morphology, 2. Spatial variation of ULF waves, *Planet. Space Sci., 40,* 1215, 1992b.

Le, G., C. T. Russell, M. F. Thomsen, and J. T. Gosling, Observations of a new class of upstream waves with periods near 3 seconds, *J. Geophys. Res., 97,* 2917–2925, 1992.

Luhmann, J. G., C. T. Russell, and R. C. Elphic, Spatial distributions of magnetic field fluctuations in the dayside magnetosheath, *J. Geophys. Res., 91,* 1711, 1986.

Montgomery, M. D., S. P. Gary, W. C. Feldman, and D. W. Forslund, Electromagnetic instabilities driven by unequal beams in the solar wind, *J. Geophys. Res., 81,* 2743, 1976.

Odera, T. J., Solar wind controlled pulsations: A review, *Rev. Geophys., 24,* 55, 1986.

Orlowski, D. S., and C. T. Russell, ULF waves upstream of the Venus bow shock: Properties of the one Hertz waves, *J. Geophys. Res., 96,* 11,271, 1991.

Orlowski, D. S., G. K. Crawford, and C. T. Russell, Upstream waves at Mercury, Venus and Earth: Comparison of the properties of one Hertz waves, *Geophys. Res. Lett., 17,* 2293, 1990.

Paschmann, G., N. Sckopke, S. J. Bame, J. R. Asbridge, J. T. Gosling, C. T. Russell, and E. W. Greenstadt, Association of low frequency waves with suprathermal ions in the upstream solar wind, *Geophys. Res. Lett., 6,* 209, 1979.

Paschmann, G., N. Sckopke, J. R. Asbridge, S. J. Bame, and J. T. Gosling, Energization of solar wind ions by reflection from the Earth bow shock, *J. Geophys. Res., 85,* 4598, 1980.

Paschmann, G., N. Sckopke, I. Papamastorakis, J. R. Asbridge, S. J. Bame, and J. T. Gosling, Characteristics of reflected and diffuse ions upstream from the Earth bow shock, *J. Geophys. Res., 86,* 4355, 1981.

Russell, C. T., and M. M. Hoppe, The dependence of upstream wave periods on the interplanetary magnetic field strength, *Geophys. Res. Lett., 8,* 615, 1981.

Russell, C. T., and M. M. Hoppe, Upstream waves and particles, *Space Sci. Rev., 34,* 155, 1983.

Russell, C. T., D. D. Childers, and P. J. Coleman, Jr., OGO 5 observations of upstream waves in interplanetary medium: discrete wave packets, *J. Geophys. Res., 76,* 845, 1971.

Russell, C. T., J. G. Luhmann, T. J. Odera, and W. F. Stuart, The rate of occurrence of dayside Pc 3, 4 pulsations: The L-value dependence of the IMF cone angle effect, *Geophys. Res. Lett., 10,* 663, 1983.

Russell, C. T., J. G. Luhmann, R. C. Elphic, D. J. Southwood, M. F. Smith and A. D. Johnstone, Upstream waves simultaneously observed by ISEE and UKS, *J. Geophys. Res., 92,* 7354, 1987.

Russell, C. T., J. G. Luhmann, K. Schwingenschuh, W. Riedler, and Y. Yeroshenko, Upstream waves at Mars: Phobos observation, *Geophys. Res. Lett., 17,* 897, 1990a.

Russell, C. T., R. P. Lepping, and C. W. Smith, Upstream waves at Uranus, *J. Geophys. Res., 95,* 2273, 1990b.

Scholer, M., Diffusive acceleration, in *Collisionless Shocks in the Heliosphere: Review of Current Research,* eds. B. T. Tsurutani and R. G. Stone, AGU Geophysical Monograph 35, 287, 1985.

Sentman, D. D., J. P. Edmiston, and L. Frank, Instabilities of low frequency, parallel propagating electromagnetic waves in the Earth's foreshock region, *J. Geophys. Res., 86,* 7487, 1981.

Sibeck, D. G., W. Baumjohann, R. C. Elphic, D. H. Fairfield, J. F. Fennell, W. B. Gail, L. J. Lanzerotti, R. E. Lopez, H. Luehr, A. T. Y. Lui, C. G. Maclennan, R. W. McEntire, T. A. Potemra, T. J. Rosenberg, and K. Takahashi, The magnetospheric response to 8–minute period strong-amplitude upstream pressure variations, *J. Geophys. Res., 94,* 2505, 1989.

Siscoe, G. L., Solar system magnetohydrodynamics, in *Solar–Terrestrial Physics,* edited by R. L. Carovillano and J. M. Forbes, pp. 64, 1983.

Smith, C. W., M. L. Goldstein, and W. H. Matthaeus, Turbulence analysis of Jovian upstream wave phenomenon, *J. Geophys. Res., 88,* 5581, 1983.

Smith, C. W., M. L. Goldstein, and H. K. Wong, Whistler wave bursts upstream of Uranian bow shock, *J. Geophys. Res., 94,* 17,035, 1989.

Sonnerup, B. U. O., Acceleration of particles accelerated in a shock, *J. Geophys. Res., 74,* 1301, 1969.

Southwood, D. J., Some features of field line resonance in the magnetosphere, *Planet. Space Sci., 22,* 483, 1974.

Thomsen, M. F., Upstream suprathermal ions, in *Collisionless Shocks in the Heliosphere: Review of Current Research,* eds. B. T. Tsurutani and R. G. Stone, AGU Geophysical Monograph 35, 141, Washington, D. C., 1985.

Troitskaya, V. A., T. A. Plyasova-Bakounina, and A. V. Gul'elmi, Relationship between Pc 2–4 pulsations and the interplanetary magnetic field, *Dokl. Akad. Nauk. SSSR, 197,* 1312, 1971.

Yomoto, K., External and internal sources of low–frequency MHD waves in the magnetosphere – A review, *J. Geomagn. Geoelec., 40,* 291, 1988.

Yomoto, K., T. Saito, S.–I. Akasofu, B. T. Tsurutani, and E. J. Smith, Propagation mechanism of dayside Pc 3–4 pulsations observed at synchronous orbit and multiple ground–based stations, *J. Geophys. Res., 90,* 6439, 1985.

Zhang, M., J. W. Belcher, J. D. Richardson and C. W. Smith, Alfven waves and associated energetic ions downstream from Uranus, *J. Geophys. Res., 96,* 1647, 1991a.

Zhang, M., et al., Low frequency waves in the solar wind near Neptune, *Geophys. Res. Lett., 18,* 1071, 1991b.

G. Le and C. T. Russell, Institute of Geophysics and Planetary Physics, University of California, Los Angeles, CA 90024–1567.

Wave Activity Associated With the Low Beta Collisionless Shock

M. H. FARRIS AND C. T. RUSSELL

Institute of Geophysics and Planetary Physics, Department of Earth and Space Sciences, University of California, Los Angeles

M. F. THOMSEN

Los Alamos National Laboratory, Los Alamos, New Mexico

ISEE-1 and ISEE-2 magnetic field data are used to analyze wave activity associated with the low beta ($\beta \leq 0.33$) shock. Upstream precursor whistler mode waves are found to phase stand in the solar wind, propagating upstream along the shock normal for low Mach number quasi-perpendicular and quasi-parallel shocks. Smaller amplitude whistler mode waves having frequencies near 1 Hz in the spacecraft frame are observed propagating upstream at angles within 45° of the magnetic field direction for subcritical, quasi-perpendicular shocks. Downstream of quasi-perpendicular shocks, ion cyclotron waves begin to grow more rapidly as the ratio of criticality (M/M_C) increases. These waves, which appear to result from the excitation of the Alfvén ion cyclotron instability, can act to pitch angle scatter ions in order to remove a temperature anisotropy created by the reflection of ions in front of the shock ramp. In front of supercritical, quasi-parallel shocks, field aligned fast magnetosonic waves are observed to propagate upstream in the plasma rest frame. However, the phase fronts of these waves are swept back into the shock by the solar wind and interfere with the structure of the shock. The wave frequencies, ellipticities, and propagation directions are conserved through the shock, and the downstream phase velocities correspond to the Alfvén mode. The upstream fast mode waves appear to be mode converted to Alfvén mode waves through the shock. Mach number dependence of the behavior of these waves and the possible effect these waves may have on the magnetosphere is stressed.

INTRODUCTION

It has been clearly shown that the solar wind controls the behavior of some of the Pc 3 and Pc 4 magnetic pulsations in the Earth's magnetosphere [see reviews by Gul'el'mi, 1974; Odera, 1986]. However, it is important to note that the solar wind must pass through the collisionless bow shock before it encounters the outward reaches of the magnetosphere, the magnetopause. The Earth's bow shock stands in front of the Earth's magnetosphere to slow, heat, and divert the solar wind flow around the impenetrable magnetospheric obstacle. The shock changes the properties of the solar wind in accordance with the Rankine-Hugoniot conservation relations [e.g. Tidman and Krall, 1971] to accomplish this task. The mechanisms by which the bow shock changes the properties of the plasma vary as the upstream parameters of the solar wind vary. Therefore, it is important to take into consideration how the shock affects the plasma under varying solar wind conditions to better understand solar wind control of magnetic pulsations in the magnetosphere.

Several parameters of the solar wind plasma govern the behavior of the collisionless bow shock which forms in front of the Earth. The three parameters which are generally considered to be most important are the upstream magnetosonic Mach number, M_{MS} (the ratio of the solar wind velocity to the speed at which magnetosonic waves can propagate), the upstream plasma beta, β (the ratio of the solar wind thermal pressure to the magnetic pressure), and θ_{BN} (the angle between the magnetic field and shock normal directions). When the magnetic pressure of the solar wind exceeds the thermal pressure by a factor of three or more (the solar wind plasma beta is low), the plasma can be accurately described by magnetohydrodynamic theory. In this regime, the shock is laminar (or quasi-laminar) [Formisano, 1977], and the processes occurring at the shock do not appear to be turbulent in nature. For this reason, the low beta regime is a good area to analyze the wave behavior and examine the processes which lead to dissipation at the shock.

In this analysis, we perform cross-correlation analysis using data from the ISEE spacecraft mission to determine the behavior of waves associated with the low beta ($\beta \leq 0.33$) shock. ISEE-1 and ISEE-2 UCLA high resolution magnetometer data [Russell, 1978] are used in conjunction with solar wind data accumulated from the ISEE-1 Solar Wind Experiment (SWE) [Bame et al.,

Solar Wind Sources of Magnetospheric Ultra-Low-Frequency Waves
Geophysical Monograph 81

1978] and ion distribution data from the ISEE-2 Fast Plasma Experiment (FPE) [Bame et al., 1978] to examine the wave activity and any solar wind parameter dependence under low beta conditions.

ANALYSIS AND OBSERVATIONS

ISEE-1 magnetometer data from the period of late October 1977 to the end of December 1980 have been examined to identify all terrestrial bow shock crossings made by the ISEE spacecraft pair, which consists of over 1100 individual crossings for this period [Farris et al., 1993]. Accompanying solar wind measurements from the ISEE-1 SWE [Bame et al., 1978] have been used to determine the solar wind parameters relevant for each crossing. With the magnetic field measurements from the ISEE-1 magnetometer, the density, velocity, and ion temperature measurements from the ISEE-1 SWE, and time-delayed ISEE-3 electron temperature measurements (when available), the upstream solar wind beta, magnetosonic Mach number, and θ_{BN} can be determined. In the cases where no electron temperature measurements are available (and all other measurements of the plasma can be made), an estimate of 150,000° K for the electron temperature has been used. Another parameter which is calculated for each of the shocks is the ratio between the upstream Mach number and the first critical Mach number [Edmiston and Kennel, 1984], or ratio of criticality (M/M_C). The first critical Mach number is, by definition, the upstream Mach number at which the downstream flow speed along the shock normal is equal to the downstream sound speed, for a specific β and θ_{BN}. When the ratio of criticality is greater than one, it is believed that additional dissipation mechanisms must be invoked for the shock to properly meet the heating requirements set forth by the Rankine-Hugoniot relations [e.g., Woods, 1969; Coroniti, 1970].

Magnetic field measurements in this report will be displayed in L-M-N coordinates, in which the upstream and downstream magnetic fields are contained within the L-N plane and the shock normal direction (N) points upstream. Cases used in this analysis make up a small subset of the database of ISEE shock crossings. Each individual shock crossing used has an upstream magnetic pressure greater than the upstream thermal pressure by at least a factor of three ($\beta \leq 0.33$), where these conditions have around a 10% occurrence rate at 1 AU [Farris et al., 1993]. Also, we require that the separation between the two spacecraft allow for accurate determinations of the shock velocities and wave phase velocities. Separations cannot be too small, yielding large errors in measurements, or too large, possibly exceeding characteristic coherence lengths for specific phenomena. These discrimination criteria substantially reduce the number of individual events for study from around 1100 to 26. However, the 26 events used in this analysis range in plasma beta from 0.04 to 0.33, Mach number from 1.5 to 3.8, and θ_{BN} from 10° to 81°, giving a wide range of events with differing parameters and ranges of criticality for low beta, quasi-perpendicular and quasi-parallel shocks.

Precursor Whistler Mode Waves

Figure 1 displays the magnetic field measurements for a typical subcritical, quasi-perpendicular shock ($M/M_C < 1$) in shock normal coordinates. We can easily see a large amplitude, low frequency wave upstream of the shock ramp. Formisano et al. [1971] and Greenstadt et al. [1975] first reported these waves in front of the bow shock and described them as whistler waves. In Figure 1, we see this wave is 90° out of phase in the L and M components and has no component in the N direction. This coordinate system makes it easy to see that this elliptically polarized wave propagates along the shock normal direction. The phase of this precursor with respect to the shock ramp is also of interest. We notice that here the ramp of the shock arises from the minimum of the L component of the field.

Detailed wave analysis sheds light onto the behavior of this wave. In analyzing the handedness of the waves, the ellipticities change from right-handed to left-handed, depending on whether the spacecraft was traveling into or out of the magnetosheath. Also, in computing the cross-correlation between ISEE-1 and -2, the time separation between phase encounters is equal to the time separation between shock crossings for the two spacecraft, to within a few tenths of a second. After removing Doppler shifts in the frequencies of these waves, we find that these waves propagate upstream along the shock normal direction, are right-hand elliptically polarized, and have frequencies of about 5-20 f_{ci} in the plasma rest frame. These results are similar to those of Mellott and Greenstadt [1984], who examined the precursor waves in front of subcritical shocks and found that the waves phase stand in the solar wind.

In the analysis performed by Mellott and Greenstadt [1984], two comparisons were made. One compared the observed

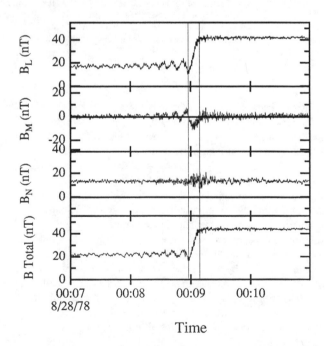

Fig. 1 Magnetic field measurements of a low beta, subcritical, quasi-perpendicular shock in shock normal coordinates. Except within the shock ramp, the magnetic field is contained within the L-N plane, and the shock normal points upstream along the positive N direction. The two solid lines indicate the shock thickness, as defined in this study.

wavelength of the precursor wave with the predicted wavelength of whistler mode waves, while another compared the shock thickness to the precursor wavelength. The plasma rest frame frequency of the precursor wave is used in the cold plasma dispersion relation [Stix, 1962] to determine the phase velocity and, subsequently, the wavelength of whistler mode waves propagating in a plasma, given the parameters of the plasma. A comparison of the observed wavelength with the wavelength determined by the cold plasma dispersion relation is shown in Figure 2a. The solid line indicates the one-to-one correlation line. As we can see, the observed wavelength of the precursor wave matches well with the wavelength predicted by the cold plasma dispersion relation.

In comparing the precursor wavelength with the shock thickness, we test the assumption that the shock is merely the largest amplitude cycle of the precursor wave. If this is the case, the thickness of the shock, from minimum to maximum, should correspond to one-half of the precursor wavelength. For this reason, we define the shock thickness as beginning at the point of lowest magnetic field strength immediately preceding the shock ramp and ending at the point of highest magnetic field strength immediately following the shock ramp. We use lowpass filtered data to determine these thickness boundaries, thus eliminating higher frequency fluctuations. The comparison of the precursor wavelength with the thickness of the shock is shown in Figure 2b. The solid line indicates a shock thickness equal to one half of the precursor wavelength. We note that the shock thickness is, in some cases, appreciably greater than the precursor wavelength and in all cases greater than the test length. The comparison made by Mellott and Greenstadt [1984] showed that the shock thickness was equal to the wavelength of the precursor wave. The discrepancy between our results and those of Mellott and Greenstadt [1984] can be explained by the use in this study of additional shocks having higher Mach numbers. The phase-standing precursors were observed in this study for subcritical and marginally critical shocks, while the Mellott and Greenstadt [1984] study focused only on subcritical shocks. Subcritical shocks from recent simulations performed by Wilkinson [1991] and Krauss-Varban and Omidi [1991] have thicknesses equal to one-half the precursor wavelength. Figure 2a shows that the observed value for the wavelength matches well with the predicted wavelength; however, Figure 2b shows that the shock is actually thicker than the test length of one-half the precursor wavelength. It is possible that other factors, such as the nonlinearity of the wave or the slowing of the flow through the ramp, may help determine the thickness of the shock.

It is very rare for a spacecraft to encounter the shock when it is in a subcritical state for $\theta_{BN} < 45°$. The limited occurrence of low Mach number solar wind velocities, the geometrical effect of encountering less quasi-parallel shocks than quasi-perpendicular shocks, and the fact that in this regime the critical Mach number is quite low ($M_C \sim 1.5$) [Edmiston and Kennel, 1984] make the probability of encountering a subcritical, quasi-parallel shock extremely low. However, it is observed that when the magnetic field lies in a quasi-parallel orientation to the shock normal, the precursor whistler wave is still evident in front of the shock when

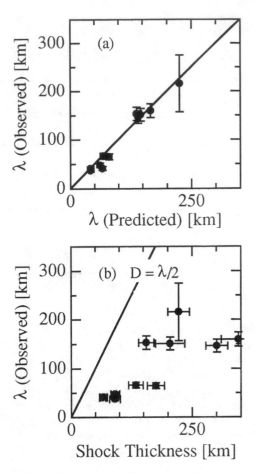

Fig. 2 (a) Comparison of observed wavelength of precursor waves to wavelength predicted from cold plasma dispersion relation for whistler mode waves. The solid line represents the one-to-one correlation line of observed wavelength to predicted wavelength. (b) Comparison of precursor wavelength to shock thickness. The solid line represents the one-to-one correlation line of shock thickness to one-half the precursor wavelength.

the ratio of criticality is less than one. Figure 3 displays the magnetic field measurements for a subcritical, quasi-parallel shock ($\theta_{BN} \sim 17°$). We note that the precursor whistler wave is evident in front of the shock and has a similar phase relation to the shock as seen in the quasi-perpendicular case. The wave, however, is not damped nearly as much and extends very far upstream, since the wave is propagating generally along the magnetic field direction. This shock is a "switch-on" shock [Kantrowitz and Petschek, 1966], where the tangential component of the magnetic field is greatly increased through the shock, or "switches on". We can see in Figure 3 that the L component of the magnetic field increases by a substantial amount compared to its average value upstream and is, in fact, larger than the normal component of the field downstream.

'One Hz' Whistler Mode Waves

A second type of upstream wave is noticeable in subcritical shocks. In Figure 1, higher frequency waves ($f \sim 1$ Hz) can be seen

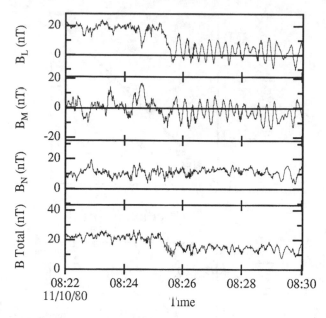

Fig. 3 Magnetic field measurements of a low beta, subcritical, quasi-parallel shock in shock normal coordinates. The format to this figure is similar to Figure 1.

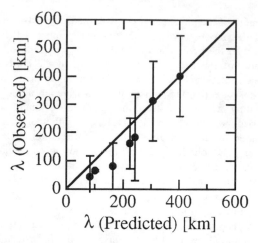

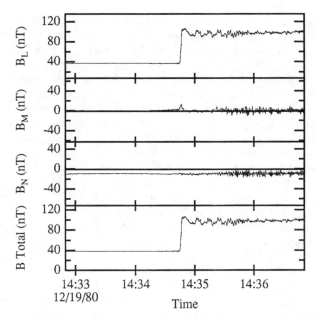

Fig. 4 Comparison of observed wavelength of upstream '1 Hz' waves to wavelength predicted from cold plasma dispersion relation for whistler mode waves. The solid line represents the one-to-one correlation line of observed wavelength to predicted wavelength.

Fig. 5 Magnetic field measurements of a low beta, marginally critical, quasi-perpendicular shock in shock normal coordinates. The format to this figure is similar to Figure 1.

in all three components of the field. These 1 Hz waves were first observed in front of the Earth's bow shock by Russell et al. [1971]. A comparison between the observed wavelength and the wavelength predicted by cold plasma theory is shown is Figure 4. The 1 Hz waves seen in this study are propagating upstream with angles between the propagation direction and the magnetic field less than 45°, have plasma rest frame frequencies higher than those of the precursor whistlers, and are right-hand elliptically polarized in the plasma rest frame. This agrees with previous analyses [e.g., Fairfield, 1974; Hoppe et al., 1982] which showed that these waves are whistler mode waves. A study by Orlowski and Russell [1991] shows that the group velocities of these 1 Hz waves, as derived from the cold plasma dispersion relation for whistler mode waves, are greater than the solar wind velocity; thus the wave energy is able to propagate upstream. However, the phase velocity of the waves studied by Orlowski and Russell [1991] was slower than the solar wind speed; thus the phase fronts of those 1 Hz waves were convected back toward the shock, and the polarization was reversed in the spacecraft frame compared to the plasma rest frame. In the present study, the phase and group velocities of the 1 Hz waves are found to exceed the solar wind velocity, and both the phase fronts and the Poynting flux move upstream. This is not necessarily inconsistent with the study by Orlowski and Russell [1991]: For these lower Mach number shocks, the solar wind velocity is much lower relative to the wave velocities than in the study by Orlowski and Russell [1991]. Thus it is plausible for the phase velocity of whistlers to move upstream from their bow shock source.

Ion Cyclotron Waves

Figure 5 displays magnetic field measurements in the shock normal coordinate system for a marginally critical shock

($M/M_C \sim 1$). At about 1435:30 UT, there is a clear onset of downstream waves. These waves have frequencies that are a significant fraction of the ion cyclotron frequency (about 0.2–0.8 f_{ci} in the plasma rest frame) and are left-hand elliptically polarized, propagating within about 10° of the magnetic field direction. In the shock coplanarity coordinate system, it can be seen that the waves propagate mostly along the magnetic field direction: Downstream of the shock, the magnetic field points mostly in the L direction, and the power of the downstream waves is largely in the M and N components of the field. Note that these ion cyclotron waves do not begin immediately following the shock; there is a significant separation between the shock ramp and the wave activity.

Figure 6 displays magnetic field measurements in the shock normal coordinate system for a supercritical shock ($M/M_C>1$). It can be seen in Figure 6 that the ion cyclotron waves noted above at the marginally critical shock now begin immediately following the shock transition region; the waves grow to an appreciable amplitude much faster than in the marginally critical case. It appears that these ion cyclotron waves which are seen downstream of the marginally critical and supercritical shocks can be explained by the excitation of the Alfvén ion cyclotron (AIC) instability [Davidson and Ogden, 1975; Gary et al., 1976]. Figure 7 displays the normalized dispersion relation for these waves from Davidson and Ogden [1975], as well as the normalized frequencies and wavelengths (wavenumbers) observed. This instability is driven by a $T_{i\perp}>T_{i\parallel}$ temperature anisotropy [Gary et al., 1976; see also Gary, 1992], which can be created by the reflection of solar wind ions in front of the bow shock. Figure 8 shows the magnetic field strength (on a distance scale) of a low beta, supercritical shock with corresponding ion phase space distributions. The first panel shows the cold solar wind beam flowing antisunward upstream of the bow shock. The second panel, within the shock foot, shows the solar wind beam accompanied by specularly reflected ions, which are moving upstream from the shock. As the plasma moves past the shock, displayed by the third panel, we note that there remains a significant anisotropy in the ion population ($T_{i\perp}>T_{i\parallel}$), and this enables the AIC instability to be excited. Ion cyclotron waves are seen downstream of the shock and pitch angle scatter the ions to create the more isotropized core and shoulder distribution seen in the fourth panel. It can be seen from the transition between the second and third ion distribution panels that the shock does not heat the core solar wind ions sufficiently to produce the downstream state seen in panel 4 [see also Thomsen et al., 1985]. Therefore, if the ion temperature anisotropy is to be removed, some additional dissipation mechanism is required.

Increasing the ratio of criticality seemed to bring the onset of the waves closer to the shock ramp, which would indicate that the growth rate of the ion cyclotron waves is faster as the ratio of criticality increases. However, the AIC instability is predicted to have a growth rate for these waves which is dependent upon the beta of the plasma (the higher the beta, the faster the waves grow) [Davidson and Ogden, 1975]. Figure 9 displays the relationship between the downstream beta of the plasma with the upstream ratio of criticality for shocks in this study. For a given upstream Mach number, beta, and θ_{BN}, the downstream beta can be determined using the Rankine-Hugoniot equations. Also, for those same upstream parameters, the ratio of criticality is calculated. We can see that as the ratio of criticality increases, the downstream beta increases. The growth rate of the waves increasing as the ratio of criticality increases in consistent with the growth rate increasing as the downstream beta increases.

Finally, the last panel in Figure 8 shows a residual ion temperature anisotropy which exists into the magnetosheath. Other observations of supercritical shocks by Sckopke et al. [1990] and Farris et al. [1992] indicate that the ions are quickly

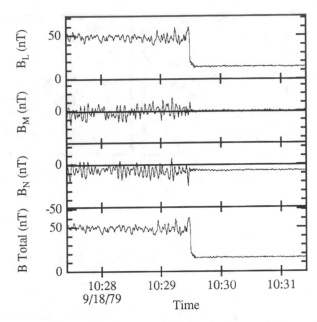

Fig. 6 Magnetic field measurements of a low beta, supercritical, quasi-perpendicular shock in shock normal coordinates. The format to this figure is similar to Figure 1.

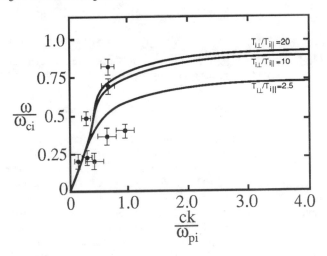

Fig. 7 Normalized dispersion relation for Alfvén ion cyclotron instability from Davidson and Ogden [1975]. Different lines indicate dispersion relations for differing ion temperature anisotropies and solid dots indicate normalized frequencies and wavelengths (wavenumbers) observed.

isotropized and any temperature anisotropy is limited to the foot and shock ramp. The theory of the ion cyclotron instability indicates that the threshold anisotropy that can be held decreases as the plasma beta increases [Gary et al., 1976]; thus, lower beta plasmas can hold higher temperature anisotropies. Our observations appear to be consistent with this result, since it is likely that the downstream betas for the other observed supercritical shocks are higher than those analyzed in this study; this is very true in the latter instance since the Farris et al. [1992] study involved high beta shocks.

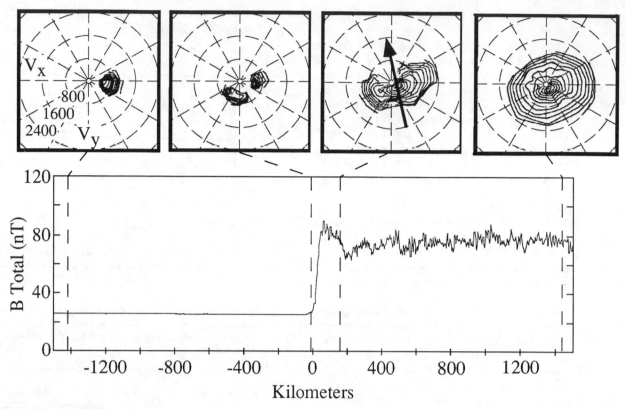

Fig. 8   Magnetic field strength and ion phase distributions for a low beta, supercritical, quasi-perpendicular shock. The dotted lines indicate where the phase space distribution corresponds to the magnetic field trace.

ULF Fast Magnetosonic Waves and Mode Conversion Downstream

When the shock becomes supercritical and quasi-parallel, the structure of the shock is not as clear. Figure 10 displays magnetic field measurements for a supercritical, quasi-parallel shock (The shock crossing is at 2340 UT). We see that the upstream region of the shock (to the left of 2340 UT) has a substantial amount of wave activity. Figure 11 compares the observed wavelength with that of fast magnetosonic waves using cold plasma dispersion. These waves propagate upstream with a right-hand elliptical polarization and have frequencies of about 0.1 f_{ci} in the plasma rest frame. The phase velocities of these waves are slower than the oncoming solar wind velocity; thus, the phase fronts of these waves are swept back into the shock and are seen to have a left-hand polarization in the shock frame. We can see in Figure 10 that most of the wave power is in the L and M components of the field. The waves are swept back into the shock and amplified, preserving many of its characteristics. The handedness of the waves and the propagation vector remain relatively unchanged, as well as the frequency of the wave. These waves have phase velocities in the plasma rest frame which match closely to the velocity of Alfvén mode waves. This phenomenon can be explained by mode conversion of fast mode waves through the shock to Alfvén mode waves downstream. Figure 12 displays a schematic of mode conversion from Krauss-Varban and Omidi

[1991] showing the process. We can see that an upstream fast mode wave conserves its rest frame frequency through the shock, and, since the fast mode is not accessible downstream of the shock at this frequency, the wave energy is converted to the Alfvén mode.

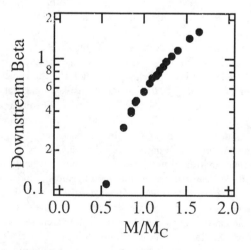

Fig. 9   Rankine-Hugoniot predicted downstream plasma beta versus ratio of criticality.

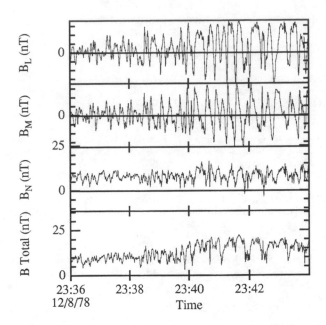

Fig. 10 Magnetic field measurements of a low beta, supercritical, quasi-parallel shock in shock normal coordinates. The format to this figure is similar to Figure 1.

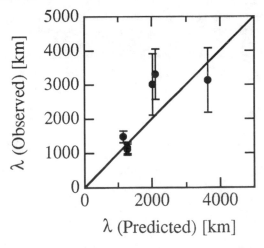

Fig. 11 Comparison of observed wavelength of upstream ULF waves to wavelength predicted from cold plasma dispersion relation for fast magnetosonic mode waves.

DISCUSSION AND CONCLUSIONS

We should note that the type of wave activity associated with the bow shock is highly dependent upon the upstream parameters of the solar wind plasma. When the shock is subcritical, we see that the upstream precursor waves are clearly evident in front of the shock. When the Mach number of the solar wind is low, or, more accurately, when the ratio of criticality is below unity, these whistler mode waves dominate the upstream region. It is also important to note that there is no significant wave activity downstream of the shock when the ratio of criticality is below unity. Therefore, any waves in the magnetosheath that would

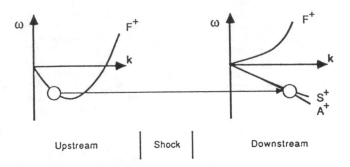

Fig. 12 Schematic from Krauss-Varban and Omidi [1991] showing mode conversion of fast mode waves upstream to Alfvén mode waves downstream.

interact with the magnetosphere under these solar wind conditions are most likely not generated by processes at the bow shock.

However, when the ratio of criticality increases to values above one, we see that the shock does affect downstream wave activity. For quasi-perpendicular shocks, ion cyclotron waves are generated to isotropize the downstream ion population, and these waves are convected toward the magnetosphere. For quasi-parallel orientations of the magnetic field, upstream ULF fast magnetosonic waves are convected back to the shock and subsequently amplified in the downstream region. The wave energy appears to be converted to the Alfvén mode, since the fast mode is not accessible downstream at the frequency with which these waves are propagating. Supercritical upstream conditions change the behavior of the shock and associated wave activity. Waves which are generated by the processes acting at the shock can be convected downstream and can couple with the magnetosphere; however, we have seen that "shock generated" downstream waves tend to only become important for marginally and supercritical upstream conditions of the solar wind plasma. Therefore, it is very important to take into consideration what might be occurring at the shock when attempting to infer how the magnetosphere couples with changes in solar wind conditions.

Acknowledgments. This work at UCLA was supported by the National Aeronautics and Space Administration under grant NAGW-3477 and by a grant from the Institute of Geophysics and Planetary Physics, Los Alamos National Laboratory.

REFERENCES

Bame, S. J., J. R. Asbridge, H. E. Felthauser, J. P. Glore, G. Paschmann, P. Hemmerich, K. Lehmann, and H. Rosenbauer, ISEE-1 and ISEE-2 Fast Plasma Experiment and the ISEE-1 Solar Wind Experiment, *IEEE Trans. Geosci. Electron.*, GE-16(3), 216, 1978.

Coroniti, F. V., Dissipation discontinuities in hydromagnetic shock waves, *J. Plasma Phys.*, 4, 265, 1970.

Davidson, R. C., and J. M. Ogden, Electromagnetic ion cyclotron instability driven by ion energy anisotropy in high-beta plasmas, *Phys. Fluids*, 18, 1045, 1975.

Edmiston, J. P., and C. F. Kennel, A parametric study of the first critical Mach number for a fast MHD shock, *J. Plasma Phys.*, *32*, 429, 1984.

Fairfield, D. H., Whistler waves observed upstream from collisionless shocks, *J. Geophys. Res.*, *79*, 1368, 1974.

Farris, M. H., C. T. Russell, and M. F. Thomsen, Magnetic structure of the low beta, quasi-perpendicular shock, *J. Geophys. Res*, in press, 1993.

Farris, M. H., C. T. Russell, M. F. Thomsen, and J. T. Gosling, ISEE 1 and 2 observations of the high beta shock, *J. Geophys. Res.*, *97*, 19121, 1992.

Formisano, V., The physics of the earth's collisionless shock wave, *J. Phys.*, *38* , C6-65, 1977.

Formisano, V., P. C. Hedgecock, G. Moreno, J. Sear, and D. Bollea, Observations of the earth's bow shock for low Mach numbers, *Planet. Space Sci.*, *19*, 1519, 1971.

Gary, S. P., The mirror and ion cyclotron anisotropy instabilities, *J. Geophys. Res.*, *97*, 8519, 1992.

Gary, S. P., M. D. Montgomery, W. C. Feldman, and D. W. Forslund, Proton temperature anisotropy instabilities in the solar wind, *J. Geophys. Res.*, *81*, 1241, 1976.

Greenstadt, E. W., C. T. Russell, F. L. Scarf, V. Formisano, and M. Neugebauer, Structure of the quasi-perpendicular laminar bow shock, *J. Geophys. Res.*, *80*, 502, 1975.

Gul'el'mi, A. V., Diagnostics of the magnetosphere and interplanetary medium by means of pulsations, *Space Sci. Rev.*, *16*, 333, 1974.

Hoppe, M. M., C. T. Russell, T. E. Eastman, and L. A. Frank, Characteristics of the ULF waves associated with upstream ion beams, *J. Geophys. Res.*, *87*, 643, 1982.

Kantrowitz, A., and H. E. Petschek, *Plasma Physics in Theory and Application*, McGraw-Hill, New York, p. 181, 1966.

Krauss-Varban, D., and N. Omidi, Structure of medium Mach number quasi-parallel shocks: upstream and downstream waves, *J. Geophys. Res.*, *96*, 17715, 1991.

Mellott, M. M., and E. W. Greenstadt, The structure of oblique subcritical bow shocks: ISEE 1 and 2 observations, *J. Geophys. Res.*, *89*, 2151, 1984.

Odera, T. J., Solar wind controlled pulsations: A review, *Rev. Geophys.*, *24*, 55, 1986.

Orlowski, D. S., and C. T. Russell, ULF waves upstream of the Venus bow shock: Properties of one-Hertz waves, *J. Geophys. Res.*, *96*, 11271, 1991.

Russell, C. T., The ISEE-1 and 2 fluxgate magnetometers, *IEEE Trans. Geosci. Electron.*, *GE-16(23)*, 239, 1978.

Russell, C. T., D. D. Childers, and P. J. Coleman, OGO 5 observations of upstream waves in the interplanetary medium: Discrete wave packets, *J. Geophys. Res.*, *76*, 845, 1971.

Sckopke, N., G. Paschmann, A. L. Brinca, C. W. Carlson, and H. Lühr, Ion thermalization in quasi-perpendicular shocks involving reflected ions, *J. Geophys. Res.*, *95*, 6337, 1990.

Stix, T. H., *The Theory of Plasma Waves*, p. 11, McGraw-Hill, New York, 1962.

Thomsen, M. F., J. T. Gosling, S. J. Bame, and M. M. Mellott, Ion and electron heating at collisionless shocks near the critical Mach number, *J. Geophys. Res.*, *90*, 137, 1985.

Tidman, D. A., and N. A. Krall, *Shock Waves in Collisionless Plasmas*, chap. 1, Interscience, New York, 1971.

Wilkinson, W. P., Ion kinetic processes and thermalization at quasi-perpendicular low Mach number shocks, *J. Geophys. Res.*, *96*, 17675, 1991.

Woods, L. C., On the structure of collisionless magnetoplasma shock waves at supercritical Alfvén-Mach numbers, *J. Plasma Phys.*, *3*, 435, 1969.

M. H. Farris and C. T. Russell, Institute of Geophysics and Planetary Physics, University of California, Los Angeles, Los Angeles, CA 90024-1567.

M. F. Thomsen, Los Alamos National Laboratory, Los Alamos, NM 87545.

Suprathermal Ions Upstream and Downstream From the Earth's Bow Shock

STEPHEN A. FUSELIER

Lockheed Palo Alto Research Laboratory, Palo Alto, California

Suprathermal ions distributions upstream and downstream from the Earth's bow shock are an integral part of the plasma physics in these regions. These ion distributions have energies (thermal + kinetic) from 1 to several keV/e and temperatures at least an order of magnitude greater than the "thermal" solar wind. The suprathermal ion distributions are the free energy source for essentially all the low frequency magnetic field fluctuations observed upstream from the shock. In addition, suprathermal ions specularly reflected off the bow shock are the prime source for ion dissipation within the shock. These reflected distributions return to the downstream region where they determine the downstream temperature and temperature anisotropy and therefore are the prime free energy source for low frequency fluctuations generated in the downstream region. Suprathermal ion distributions are reviewed here with particular emphasis on the topics of interest to magnetospheric ULF wave research.

INTRODUCTION

Early spacecraft observations upstream from the Earth's bow shock revealed a dynamic region containing large amplitude, low frequency waves [e.g., Greenstadt et al., 1968; Fairfield, 1969] and suprathermal ions backstreaming from the Earth's bow shock [Asbridge et al., 1968]. From these initial observations, it was apparent that the low frequency waves and ions were related, probably through electromagnetic ion instabilities [Barns et al., 1970]. By the early 1970's, a significant fraction of the qualitative phenomena associated with the region containing the backstreaming ions from the Earth's bow shock and the low frequency waves (called the foreshock region [Greenstadt et al., 1976]) was identified. However, the realization that the evolution of the suprathermal backstreaming ions in the wave magnetic fields could be observed in the foreshock region [Gosling et al., 1978], the role suprathermal ions play in the structure of the bow shock, and the variety of source mechanisms for the low frequency waves upstream and downstream from the shock had to wait the improved particle detectors flown on the International Sun Earth Explorer (ISEE) spacecraft. The ISEE mission lead directly to a renewed focus on the foreshock region, resulting in the discovery of several different types of ion suprathermal distributions [e.g., Gosling et al., 1978; Paschmann et al., 1981] including the so-called gyrophase-bunched ion distributions [Eastman et al., 1981; Gurgiolo et al., 1981] critical to the shock structure [e.g., Leroy et al., 1982; Sckopke et al., 1983]. The combination of the improved particle

Solar Wind Sources of Magnetospheric Ultra-Low-Frequency Waves
Geophysical Monograph 81
Copyright 1994 by the American Geophysical Union.

detectors on the ISEE spacecraft and considerable advancement in computer simulations resulted in a reasonably good understanding of collisionless shocks in space and the suprathermal ion and electrons they generate, first in the quasi-perpendicular regime [Leroy et al., 1982; Sckopke et al., 1983] and recently in the quasi-parallel regime [Quest, 1988; Burgess; 1989; Gosling et al., 1989a; Thomsen et al., 1990; Onsager et al., 1990]. Finally, analysis of ion composition data from the ISEE mission and the more recent Active Magnetospheric Particle Tracer Explorers (AMPTE) mission has extended the suprathermal ion measurements in the foreshock region to species (such as solar wind He^{2+}) other than the dominant solar wind protons. The analysis has revealed similarities between these species and protons but has also resulted in some surprises regarding the evolution of ion distributions in the foreshock region [Ipavich et al., 1984, 1988; Fuselier et al., 1990; Fuselier and Thomsen, 1992].

In this paper, suprathermal ion observations upstream and downstream from the Earth's bow shock will be reviewed. There are already several excellent reviews that cover much of this topic (in particular, the more recent reviews by Gosling and Robson [1985] and Thomsen [1985]). While these reviews were written almost ten years ago, the framework and theoretical development contained in them remain current. Instead of restating much of what is discussed in these reviews, this paper will focus on more recent developments, especially those associated with the quasi-parallel bow shock. The primary emphasis will be on topics of interest to magnetospheric ULF wave research, namely the differences between the suprathermal ions in quasi-parallel and quasi-perpendicular shock regions and how these differences affect the plasma that ultimately convects to the Earth's magnetopause. Waves associated with the suprathermal ion distributions are reviewed separately [Le and Russell, this volume] although some of the low frequency waves associated with the suprathermal ion distributions will be discussed here.

In the next section the foreshock region is introduced, the quasi-parallel and quasi-perpendicular bow shock regions are defined, and cone angle effects are discussed (where the cone angle is defined as the angle between the magnetic field and the Earth-sun line). Following the introductory section, the next two sections contain discussions of the suprathermal ions associated with the quasi-perpendicular shock and the quasi-parallel shock, respectively. The last section contains a discussion of possible future studies that would be applicable to magnetospheric ULF wave research. Unless explicitly stated, the use of the term "ion" in this paper applies to protons, the dominant species in the solar wind.

SUPRATHERMAL IONS IN THE FORESHOCK AND MAGNETOSHEATH

The foreshock region contains a variety of suprathermal ion distributions. The spatial location of these distributions and the location of the quasi-parallel and quasi-perpendicular bow shock regions are shown schematically in Figure 1 for the Parker spiral interplanetary magnetic field (IMF) orientation. From early observations it was clear that the regions upstream and downstream from the quasi-parallel and quasi-perpendicular bow shocks differed considerably. Here, the shock is considered quasi-perpendicular (-parallel) shock when ϑ_{Bn} greater than (less than) $45°$, where ϑ_{Bn} is the angle between the average upstream magnetic field

(averaged over an appropriate time interval to remove the fluctuating magnetic field) and the shock normal. In Figure 1, magnetic field lines are shown upstream from the shock, while in the magnetosheath downstream from the shock, streamlines are labeled with their value of ϑ_{Bn} at the shock [see also Russell et al., 1983]. The region containing ions backstreaming from the Earth's bow shock is called the foreshock region. It is bounded by the bow shock and an upstream edge called the ion foreshock boundary. The ion foreshock boundary is not parallel to the magnetic field because ions backstreaming from the shock have a velocity along the magnetic field of only about two times the solar wind speed in the Earth's rest frame [Bonifazi and Moreno, 1981] and therefore a substantial component of their total velocity is convection with the solar wind. Two sets of terms are introduced here as a shorthand notation for the discussion of the foreshock and magnetosheath. The regions of the foreshock that are upstream from the quasi-perpendicular (-parallel) bow shock are referred to as the quasi-perpendicular (-parallel) foreshock. Since many of the effects (in particular wave turbulence levels) that characterize these upstream regions are reflected in the magnetosheath downstream from the shock, it is appropriate to divide the magnetosheath into quasi-parallel and quasi-perpendicular regions as well. The regions downstream from the shock that are on streamlines that connect to the quasi-perpendicular (-parallel) bow shock are referred to as the quasi-perpendicular (-parallel) magnetosheath.

In the quasi-perpendicular foreshock, field-aligned beams (FAB in Figure 1) are observed. Just upstream from the shock and also

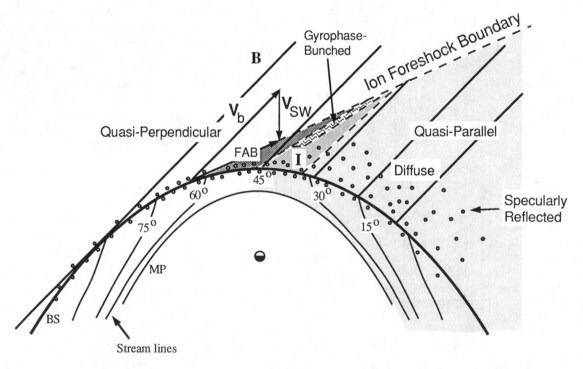

Fig. 1. Schematic of the foreshock and magnetosheath for a cone angle of $45°$. The magnetic field is shown upstream from the bow shock (BS) and streamlines labeled with their appropriate ϑ_{Bn} are shown downstream from the shock. The ion foreshock is bounded by the Earth's bow shock and the ion foreshock boundary. The foreshock boundary is not parallel to the magnetic field because ion beam speeds along the magnetic field (V_b) are comparable to the convection speed of the solar wind (V_{sw}). In the foreshock region, field-aligned beams (FAB), gyrophase-bunched, intermediate (I), diffuse, and specularly reflected distributions are observed.

in the quasi-perpendicular magnetosheath, specularly reflected ion distributions are observed. The quasi-parallel foreshock contains a variety of distributions including intermediate, (I in Figure 1), gyrophase-bunched, diffuse, and specularly reflected ion distributions. In the quasi-parallel magnetosheath, specularly reflected and diffuse ion distributions are observed.

Many ion distributions in the foreshock region in Figure 1 evolve from other distributions. In most instances this evolution is relatively smooth so that the distinction between some types of distributions, such as specularly reflected and diffuse distributions in the quasi-parallel magnetosheath far from the shock, is sometimes a matter of interpretation. However, simultaneous observations of disparate distributions or distributions with different origins, such as field-aligned beams and diffuse distributions or field-aligned beams and specularly reflected distributions in the quasi-perpendicular foreshock, have not been reported.

For Parker spiral IMF conditions in Figure 1 (i.e., for an IMF cone angle of 45°), the subsolar point is at $\vartheta_{Bn}=45°$ or at the division between the quasi-parallel and quasi-perpendicular bow shocks. The quasi-parallel bow shock extends over the entire dawnside while the quasi-perpendicular bow shock extends over the entire duskside. Since the thermal and suprathermal solar wind ions that pass closest to the magnetopause cross the shock in the subsolar region, the magnetopause is downstream from the nominal division between the quasi-parallel and quasi-perpendicular bow shock for the Parker spiral IMF. As will be seen in the next two sections, there are profound differences between the suprathermal ion distributions in these two regions.

For cone angles other than 45°, the subsolar magnetopause will be downstream from either the quasi-parallel bow shock or the quasi-perpendicular bow shock. This cone angle effect has been discussed in detail by Luhmann et al., [1984, 1986] although they chose a value of $\vartheta_{Bn}=30°$ as the division between the quasi-parallel and quasi-perpendicular geometry. Here, the more traditional definition of $\vartheta_{Bn}=45°$ is used and the cone angle effect on the location of the quasi-parallel and quasi-perpendicular regions is illustrated in Figure 2. The panels in this figure are similar to Figure 1. The top panel in Figure 2 shows the quasi-parallel and quasi-perpendicular foreshock and magnetosheath for a 15° cone angle while the bottom panel shows the same for a 60° cone angle. For a cone angle of 15°, the quasi-parallel foreshock and magnetosheath (shaded regions in Figure 2) cover nearly the entire dayside and the subsolar magnetopause is clearly downstream from the quasi-parallel bow shock. For this geometry, the quasi-perpendicular foreshock containing field-aligned beams is located on the flanks of the bow shock, a fact that has implications for field-aligned beam evolution discussed in the next two sections. By extrapolating the trend in Figure 1 and the top panel in Figure 2, it is clear that for a cone angle of 0°, nearly the entire dayside bowshock is quasi-parallel. However, it is important to realize that the bow shock is still quasi-perpendicular on the flanks for this IMF orientation [Fuselier, 1989; Gosling et al., 1989b].

For a cone angle of 60° (Figure 2, bottom panel), the quasi-perpendicular foreshock and magnetosheath cover nearly the entire dayside and the subsolar region is clearly downstream from the quasi-perpendicular bow shock. Field-aligned beams originating

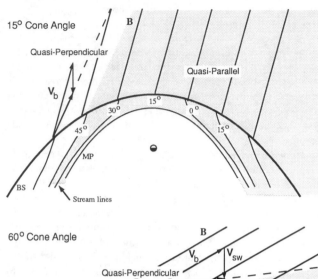

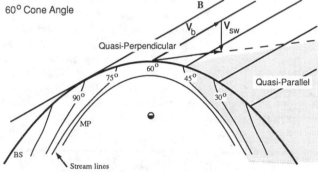

Fig. 2. (Top panel) Schematic for a cone angle of 15°. (Bottom panel) same for a cone angle of 60°. For 15° cone angles, the subsolar magnetopause (MP) is downstream from the quasi-parallel bow shock. Field-aligned beams originating in the quasi-perpendicular region on the flanks do not have access to most of the quasi-parallel region near the nose of the bow shock since their convection speeds in the -Y direction are small. For 60° cone angles, the magnetopause is downstream from the quasi-perpendicular bow shock.

from the quasi-perpendicular bow shock region have little motion in the $+X_{GSE}$ direction. Thus the foreshock boundary (dashed line in the bottom panel of Figure 2) is nearly in the $-Y_{GSE}$ direction and the quasi-parallel foreshock and magnetosheath are confined to the dawn flank of the bow shock.

To summarize the cone angle effects, Figures 1 and 2 show that solar wind ions that pass closest to the magnetopause cross the quasi-parallel bow shock in the subsolar region for cone angles <45° and cross the quasi-perpendicular bow shock in the subsolar region for cone angles >45°. In general, the ϑ_{Bn} of the shock in the subsolar region is equal to the cone angle of the IMF. Since the occurrence distribution of IMF cone angles peaks at 45° at the Earth's orbit, the subsolar bowshock and magnetosheath are typically on the dividing line between the quasi-perpendicular and quasi-parallel geometry as illustrated in Figure 1.

THE QUASI-PERPENDICULAR SHOCK

Field-aligned beams

Field-aligned beams were the first type of suprathermal ion distribution observed in the upstream region [Asbridge et al., 1968].

An example of a field-aligned beam is shown in Figure 3 (after Fuselier and Thomsen [1992]). The contours in this figure show the phase space density of ions (assumed to be protons) in the ecliptic plane with two contours per decade of phase space density. The V_x direction (toward the sun) is to the left and the V_y direction (duskward) is down. The center of the contour plot is zero velocity in the spacecraft frame. The tightly spaced contours in the $-V_x$ direction are the solar wind ion distribution which is not well resolved by the ISEE-2 Fast Plasma Experiment that made the measurements. The magnetic field direction is identified by the arrow labeled **B** in each contour plot and this magnetic field line is connected to the Earth's bow shock in the **-B** direction. The contours in the upper part of the plot along the magnetic field are field-aligned ion beams propagating away from the shock with a speed of approximately two to three times the solar wind speed in the spacecraft frame. Field-aligned beams in the upstream region (e.g., Figure 3) have temperatures on the order of 2-3 times the solar wind proton temperature, densities are on the order of 1% of the solar wind proton density [Bonifazi and Moreno, 1981a,b] and typical beam energies (mainly kinetic energy) are on the order of 1 keV/e.

Field-aligned beams such as the one in Figure 3 were first thought to originate through reflection of a small portion of the incident solar wind proton distribution at the quasi-perpendicular bow shock [Sonnerup, 1969]. Probably for this reason, they were first called "reflected ions" [Gosling et al., 1978]. However, there is now evidence that field-aligned proton beams may be produced by either a reflection mechanism at the shock or by leakage of a small portion of the hot proton population from the magnetosheath into the upstream region. Figure 4 show some of this evidence [from Schwartz and Burgess, 1984]. Plotted versus ϑ_{Bn} are the observed beam speeds normalized by the incident velocity parallel to the magnetic field in the deHoffman-Teller frame of reference. This is the frame moving perpendicular to the shock normal in which the magnetic field is parallel to the incident velocity and the electric field on either side of the shock vanishes. It is a convenient frame for discussing field-aligned beams because, as evidenced in Figure 4, the ϑ_{Bn} dependence of the beam speed is different for reflection off the shock and leakage from the magnetosheath. Figure 4 shows that ion beams observed typically in the quasi-perpendicular foreshock are consistent with leakage and reflection.

Thus far, only proton beams, such as the one in Figure 3, have been discussed. Recent analysis of field-aligned beams have shown that they are indeed composed almost entirely of protons. In particular, they contain very little solar wind He^{2+} when compared to the $\sim 4\%$ contribution this ion makes to the solar wind density [Ipavich et al., 1988; Fuselier and Thomsen, 1992]. Figure 5 compares the percentages of He^{2+} and protons in field-aligned beams [from Fuselier and Thomsen, 1992]. Plotted are the He^{2+} versus the proton density for 15 field-aligned beams normalized

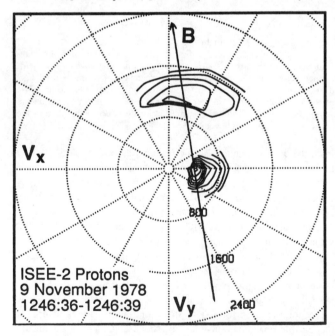

Fig. 3. A field-aligned beam observed upstream from the Earth's bow shock. In this and subsequent contour plots, contours of constant phase space density (assuming all ions are protons) are plotted with one contour for each half decade of phase space density. The two-dimensional plot was produced by integrating over $\pm 55°$ from the ecliptic plane. The center of the panel is zero velocity in the spacecraft frame. Positive V_x and V_y point sunward and duskward, respectively. **B** in the figure is the magnetic field direction and the direction to the shock is along **-B**. The tightly spaced contours in the $-V_x$ direction are the solar wind. The more loosely space contours in the $-V_y$ direction are the field-aligned beam.

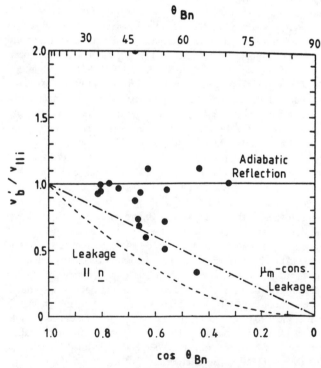

Fig. 4. Observed beam speeds in the deHoffman-Teller frame versus $\cos(\vartheta_{Bn})$ for several beam events. The three theoretical curves show the expected behavior with $\cos(\vartheta_{Bn})$ for one reflection and two leakage origins of the field-aligned beams. Events are consistent with "adiabatic" reflection and leakage from the hot downstream magnetosheath.

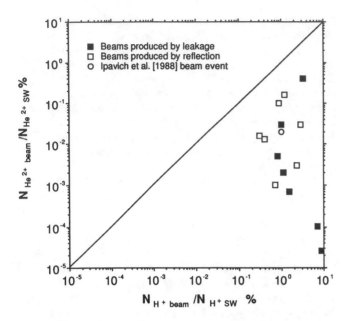

Fig. 5. He^{2+} and H$^+$ beam densities (normalized by their respective solar wind densities) for 15 field-aligned beam events. Field-aligned beams contain an average of about 1% of the solar wind proton densities but very little He^{2+} (at least a factor of 10 below the proton beam density percentages), independent of how the H$^+$ beams were formed.

by their respective densities in the solar wind. Proton beam concentrations average about 1% of the proton density in the solar wind but He^{2+} beam concentrations average two orders of magnitude below that. The low He^{2+} concentrations are independent of the origin of the field-aligned beams, either through the reflection or leakage mechanisms discussed above. The observations in Figure 5 have implications for the source of other ion distributions in the quasi-parallel foreshock because, as will be seen in the next section, some of these other ion distributions have much higher He^{2+} concentrations.

Field-aligned beams are a source of free energy in the foreshock. These beams are unstable to the generation of low frequency electromagnetic waves primarily through the right hand resonant ion beam instability [Gary et al., 1981]. These waves eventually saturate and scatter the more tenuous beam component of the foreshock plasma. The time scale for this beam disruption is long compared to the time it takes the beam to propagate away from the shock, thus field-aligned beams propagate a significant distance (up to several R$_E$) from the shock and convect with the solar wind into the quasi-parallel foreshock as they are scattered, as illustrated in Figure 1. The evolution of these beams into other suprathermal ion distributions is discussed in the next section on the quasi-parallel bow shock.

Specularly reflected ions

Field-aligned beams are not the only type of ion distribution produced by the quasi-perpendicular shock. Also, field-aligned beams propagate away from the shock and into the quasi-parallel foreshock and do not enter the quasi-perpendicular magnetosheath

(or re-enter, in the case of the beams leaked from the magnetosheath) . Another type of suprathermal ion distribution that is observed near and even downstream from the quasi-perpendicular bow shock is the specularly reflected ion distribution [Paschmann et al., 1982]. An example of a specularly reflected ion distribution near the quasi-perpendicular bow shock is shown in Figure 6 [from Fuselier et al., 1986a]. The format is similar to that of Figure 3 with the tightly spaced contours representing the solar wind distribution and the more open contours representing the suprathermal ion distribution. In contrast to the field-aligned beams in Figure 3, the specularly reflected suprathermal ion distribution in Figure 6 is clearly non-gyro-tropic.

The term "specularly reflected" identifies the process that created the suprathermal distribution. In specular reflection, the component of the incident solar wind ion velocity normal to the shock is reversed but the component tangential to the shock and the total velocity remains the same. The resulting ion distribution can have a large component of its motion perpendicular to the upstream magnetic field and is non-gyrotropic since the vector direction of a reflected ion is independent of the upstream magnetic field direction. For specular reflection, the guiding center motion and the gyro-motion depend only on the incident flow speed and the shock geometry [Gosling et al., 1982]. The guiding center motion, V$_g$ is particularly important because, as shown in Equation 1 (from Gosling et al., [1982] Equation 7), the net guiding center motion is back toward the shock only if $\cos^2(\vartheta_{Bn})$ is less than 1/2 or $\vartheta_{Bn} > 45°$

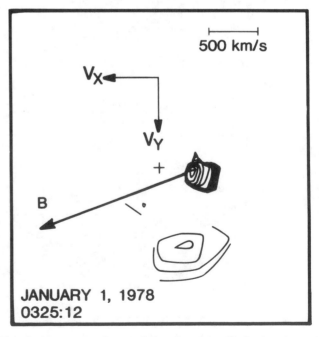

Fig. 6. An example of a specularly reflected ion distribution observed near the quasi-perpendicular bow shock. The format is the same as in Figure 3 except that the cross in the center represents zero velocity in the spacecraft frame. The suprathermal distribution is clearly non-gyrotropic, distinct from the field-aligned beam in Figure 3.

$$V_{gc} \cdot \mathbf{n} = V_i cos(\vartheta_{Vn})(1 - 2cos^2(\vartheta_{Bn})) \qquad 1$$

In this equation, ϑ_{Vn} is the angle between the incident velocity of the ion and the shock normal and is always less than zero for solar wind ions incident on the shock. Thus, in the quasi-perpendicular geometry, specularly reflected ion distributions such as the one in Figure 6 are confined to within one gyro-radius upstream from the shock. When the finite size of the gyroradius is taken into account, specularly reflected ions with guiding center motion directed upstream still return to the shock when $39° < \vartheta_{Bn} < 45°$ [Schwartz et al., 1983]. However, these ions return with less energy normal to the shock surface than their initial normal energy and therefore probably reflect off the shock again. The ϑ_{Bn} dependence of the guiding center motion of specularly reflected ions as defined in Equation 1 is one of the fundamental differences between quasi-perpendicular and quasi-parallel shocks.

For quasi-perpendicular shocks, the specularly reflected ions gyrate into the upstream region, gain energy by moving parallel to the solar wind electric field, and return to the shock with considerably more energy than when they left it. These higher energy ions cross the shock and gyrate into the downstream region. In the quasi-perpendicular magnetosheath, they form a shoulder on the ion distribution at energies of approximately a few keV/e (i.e., several hundred km/s) as illustrated in Figure 7. Specularly reflected ion densities range from a few percent of the solar wind ion density for low Mach number shocks ($M_f \sim 2$, where M_f is the fast mode speed) to more than 20% of the solar wind ion density

for high Mach number shocks ($M_f > 6$) [Paschmann and Sckopke, 1983; Paschmann et al., 1982; Sckopke et al., 1983; Gosling and Robson, 1985]. Because of their high densities and high effective thermal temperatures, specularly reflected ions are the dominant contributors to ion heating at supercritical shocks like the Earth's bow shock. Not only do they determine the downstream temperature, but they also determine the downstream temperature anisotropy and therefore are the free energy source for low frequency waves such as mirror mode and ion cyclotron waves observed in the quasi perpendicular magnetosheath [see e.g., Anderson and Fuselier, 1993; Gary et al., 1993].

There is indirect evidence that ions other than protons also specularly reflect off the quasi-perpendicular bow shock and enter the downstream region. A shoulder has been observed on the He^{2+} distribution in the quasi-perpendicular magnetosheath similar to the shoulder on the H^+ distributions produced by specularly reflected H^+ [Peterson et al., 1979; Gloeckler et al., 1986; Fuselier et al., 1988; 1991]. The shoulder on the He^{2+} distribution helps determine the downstream He^{2+} temperature and anisotropy, two quantities that are also important for the generation of low frequency ion cyclotron waves in the quasi-perpendicular magnetosheath.[e.g., Gary et al., 1993; Denton et al., 1993]

In summary, there are two basic types of suprathermal ion distributions upstream and downstream from the Earth's quasi-perpendicular shock. Field-aligned beams consisting almost entirely of protons are produced by reflection of a portion ($\sim$1%) of the solar wind ion distribution incident on the shock or by leakage of a portion of the hot magnetosheath plasma back into the up-

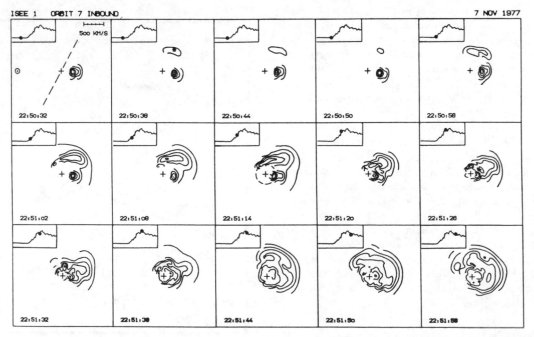

Fig. 7. A series of snapshots of ion distributions across a quasi-perpendicular bow shock (ϑ_{Bn}=85°, from Paschmann et al. [1982]). The format is similar to that in Figures 3 and 6. The dashed line in the first panel shows the approximate orientation of the shock and the small circle shows the approximate orientation of the magnetic field. The plasma density profile is shown as insets. Solid dots in these insets indicate the relative position of the spacecraft. The top row of panels are just upstream from the shock and (after the first panel) specularly reflected ions are observed primarily in the $-V_y$ direction. The second row of panels shows the change of these specularly reflected distributions through the shock ramp while the third row of panels shows the scattering and isotropization of the specularly reflected ions that have penetrated into the downstream region.

stream. These beams are a free energy source for low frequency waves as they propagate into the upstream region and convect with the solar wind into the quasi-parallel foreshock. The evolution of the beams is discussed in the next section.

Within one gyroradius of the shock, a fairly large fraction (up to ~20%) of the solar wind ions specularly reflect off the quasi-perpendicular shock. Because these ions have guiding center motion directed downstream, they perform approximately one half gyro-orbit in the upstream, gain energy in the solar wind electric field, and return to the shock. The higher energy allows these ions to cross the shock and enter the downstream region where they are the primary source of ion heating and ion free energy in the quasi-perpendicular magnetosheath.

THE QUASI-PARALLEL SHOCK

Gyrophase-bunched and intermediate ion distributions

The scattering of field-aligned beams and the generation of low frequency waves nominally begins in the quasi-perpendicular foreshock but the evolution of field-aligned beams and the waves they generate really occurs in the quasi-parallel foreshock for cone angles ~45°. Initially, a field-aligned beam generates a small amplitude monochromatic transverse wave which in turn phase traps the beam. The phase trapping process produces a gyrophase-bunched distribution first predicted from computer simulations of the beam disruption process [Hoshino and Terasawa, 1985] and later confirmed by spacecraft observations [Thomsen et al., 1985, Fuselier et al., 1986c]. An example of a gyrophase-bunched ion distribution produced by phase trapping of an initially field-aligned beam is shown in Figure 8 [from Fuselier et al., 1986a]. The format is the same as Figure 3. The open contours in the $+V_x, +V_y$ direction are the phase bunched distribution, which gyrates around the average magnetic field direction with a period equal to that of the low frequency wave that traps the ions. By comparing Figure 8 and 7, it is clear that the gyrophase-bunched distribution and the non-gyrotropic specularly reflected distribution are very similar. However, two features of these distributions allow their distinction.

First, since gyrophase-bunched distributions are produced by disruption of field-aligned beams, they have densities on the order of 1% of the solar wind ion density and rarely have densities above a few percent of the solar wind density. In contrast, specularly reflected distributions can have densities up to 20% of the solar wind density depending on the Mach number of the upstream ion flow and the location within the "foot" of the shock [Paschmann et al., 1982]. Second, because the wave growth and phase trapping takes time, gyrophase-bunched distributions and the monochromatic low frequency waves that trap these distributions are observed somewhat downstream from the ion foreshock boundary and relatively far (several Earth radii, R_E) from the shock (see Figure 1 and Fuselier et al., [1986a]). In contrast, non-gyrotropic specularly reflected ion distributions are observed within one gyroradius of the quasi-perpendicular shock (and, as will be shown below, within one gyroradius of the quasi-parallel shock as well). The dependence on distance from the shock is probably the most important distinction between gyrophase-

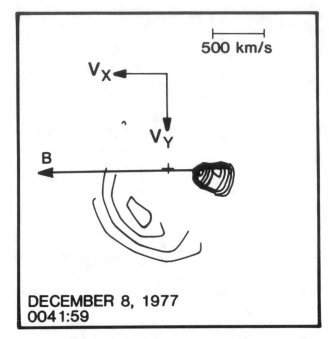

Fig. 8. An example of a gyrophase-bunched ion distribution produced by phase trapping of a field-aligned beam. Phase trapping occurs because the initial field-aligned beam generates a highly monochromatic wave which phase traps the field-aligned beam.

bunched distributions produced by beam disruption far from the shock (Figure 8) and non-gyrotropic ion distributions produced by specular reflection at the shock (Figures 6 and 7).

The gyrophase-bunched distributions produced by disruption of field-aligned beams are still a free energy source in the plasma and contribute to continued wave growth. Eventually, these waves scatter the gyrophase-bunched distributions into presumably intermediate ion distributions [Fuselier et al., 1986a]. An example of an intermediate ion distribution is shown in Figure 9 [from Fuselier et al., 1986a]. Comparing the intermediate ion distribution and the field-aligned beams in Figure 3, it is evident that the difference between the two types of distributions is simply that the intermediate ion distribution extends over a much larger angular range than the field-aligned beam. Thus, the net effect of the early stages of the disruption of field-aligned beams is pitch angle scattering of beam ions with little change in their ~1 keV/e energy in the solar wind frame. This pitch angle scattering is accomplished through the growth of low frequency electromagnetic waves [Gary et al., 1981] observed in conjunction with gyrophase-bunched and intermediate ion distributions and is similar to the pitch angle scattering of new born cometary ion distributions upstream from a cometary bow shock [see e.g., Neugebauer et al., 1989].

Diffuse ion distributions

From the early observations, it was believed that diffuse ion distributions represented the final evolution of field-aligned beams initially produced at the quasi-perpendicular shock [e.g., Gosling et al., 1978; Paschmann et al., 1979; Bame et al., 1980]. Several features of these distributions seem to support this belief. Field-

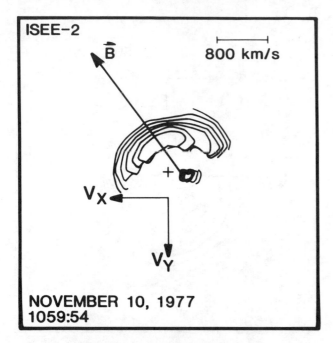

Fig. 9. An example of an intermediate ion distribution. This distribution evolved from a field-aligned beam similar to the one shown in Figure 3. The evolution when through an intermediate state shown in Figure 8 but the net effect of this evolution was pitch angle scattering of the initial field-aligned beam.

aligned beams, intermediate, and diffuse ion distributions all have typical densities of ~1% of the solar wind proton density, are never observed simultaneously, and typically exhibit smooth transitions from one type to another. Diffuse ion distributions such as the example in Figure 10 [from Gosling et al., 1989b, Figures 5] appear to represent fully scattered intermediate distributions. The format of Figure 10 is similar to that of Figure 3, except in the lower panel a cut in the distribution along the sun-Earth direction (the X axis) is shown. Diffuse ion distributions like the one in Figure 10 extend to much higher energies (~150 keV/e) than the intermediate distributions and field-aligned beams. Also, most of the energy of the distribution is thermal energy and typical energies of these distributions are several keV/e. The higher energies of the diffuse distributions, compared to the ~1 keV/e mainly kinetic energy of the field-aligned beams and intermediate distributions, suggest energy as well as angular diffusion of the initial field-aligned beam. Finally, the fluctuating magnetic field amplitudes increase from nearly zero in association with field-aligned beams to their highest levels in association with diffuse ion distributions and all distributions have free energy required to generate these low frequency waves [e.g., Gary et al., 1981]. The large amplitude fluctuations in the quasi-parallel foreshock ($\delta B/B \sim 1$) are responsible for the acceleration of solar wind ions to high energies to form the energetic ion tail of the diffuse ion distributions [e.g., Lee, 1982].

While it might be true that the gyrophase-bunched and intermediate ion distributions evolve from field-aligned beams, it is clear the beams cannot be the only source of diffuse ion distributions in the quasi-parallel foreshock. There are two pieces of evidence

that support this conjecture. First, Figures 1 and 2 show that as the cone angle decreases, the convection velocity of field-aligned beams in the $-Y_{GSE}$ direction also decreases. This implies that for small cone angles such as in the top panel of Figure 2, field-aligned beams will not propagate far into the quasi-parallel foreshock. For zero cone angle, field-aligned beams originating at the quasi-perpendicular shock will not propagate into the quasi-parallel foreshock at all since their $-Y_{GSE}$ convection velocity is zero [Ellison and Möbius, 1987]. Yet suprathermal ion distri-

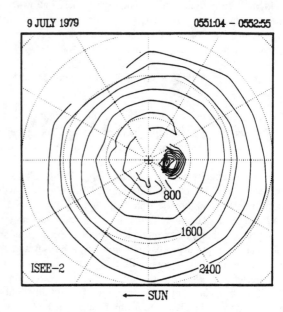

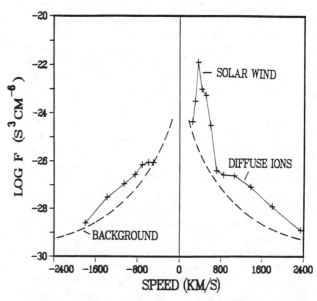

Fig. 10. An example of a diffuse ion distribution observed when the magnetic field cone angle was nearly zero. The format of the top panel is similar to that of Figure 3 and the bottom panel shows a cut through the diffuse distribution in the plasma flow direction. The diffuse distribution represents a nearly fully evolved ion distribution (i.e., one that has undergone severe pitch angle and energy diffusion).

butions are still observed in these quasi-parallel regions even for very small cone angles. Indeed, the diffuse ion distribution in Figure 10 was observed when the cone angle was nearly zero and therefore probably originated from the shock as something other than a field-aligned beam.

Second, composition measurements suggest that field-aligned beams may not be a significant source for diffuse ion distributions even for larger cone angles. As pointed out in the previous section, field-aligned beams are composed almost entirely of protons with very little solar wind He^{2+} present (Figure 5). In contrast, diffuse ion distributions like the one in Figure 10 contain a significant fraction of solar wind He^{2+}, especially for higher Mach numbers [Ipavich et al., 1984]. When compared on an equal velocity basis, the composition of diffuse ion distributions is similar to that of the solar wind, that is they contain approximately 4% solar wind He^{2+}. This is one of the most important pieces of evidence that the diffuse ion distributions in the energy range from a few keV/e to ~150 keV/e are predominately from the solar wind and not predominately from a source inside the Earth's magnetosphere as suggested by some authors [e.g., Sarris and Krimigis, 1988]. (However, ions with energies above ~150 keV/e predominantly from the Earth's magnetosphere are observed at times in the quasi-parallel foreshock [Gosling et al., 1979; Scholer et al., 1981].) The approximately solar wind concentration of the diffuse ion distribution in the energy range from a few keV/e to ~150 keV/e is also important evidence that field-aligned beams, which contain very little solar wind He^{2+} (Figure 4) may not be an important source of diffuse ion distributions.

Specularly reflected ions (discussed below) are another possible solar wind source for diffuse ion distributions. Whatever the source or sources, it is clear that the diffuse ion distributions have undergone considerable pitch angle and energy scattering in the turbulent regions upstream and downstream from the quasi-parallel bow shock and therefore do represent the final evolution of suprathermal ion distributions in the foreshock region.

Specularly reflected ions

As discussed in the previous section, the guiding center motion of ions specularly reflected off the Earth's bow shock is determined by the incident solar wind velocity and the shock geometry (see Equation 1). For a steady quasi-parallel geometry, $\vartheta_{Bn} \lesssim 45°$, the guiding center motion of specularly reflected ions is directed upstream and these ions will not return to the shock. Specularly reflected ion distributions that fit this description have been observed at large distances (i.e., more than several ion gyro-radii) from the shock [Gosling et al., 1982]. An example of such a distribution is shown in Figure 11 [from Fuselier et al., 1986a]. Unlike the specularly reflected ion distributions near the quasi-perpendicular shock (Figures 6 and 7), the specularly reflected distributions seen further than a few ion gyro-radii from the quasi-parallel bow shock have typical densities on the order of a few percent of the solar wind density and are gyrotropic. Large densities, such as the ~20% densities of specularly reflected ions at the quasi-perpendicular shock, have not been reported for the gyrotropic specularly reflected ions in the quasi-parallel foreshock.

Gyrotropic distributions such as the one in Figure 11 probably

did not originate at the shock as a gyrotropic distribution. Instead, it is believed that they are initially non-gyrotropic but rapidly become gyrotropic once they have left the shock. The initial thermal velocity of the non-gyrotropic specularly reflected distribution could lead to phase mixing within a few gyro-radii of the shock [Gurgiolo et al., 1983; Burgess and Schwartz, 1984] or low frequency electromagnetic waves already present in the quasi-parallel foreshock and/or generated by the specularly reflected ion distribution can rapidly cause the non-gyrotropic distribution to become gyrotropic [Winske et al., 1984; Thomsen et al., 1985]. The high degree of gyrotropy distinguishes these specularly reflected distributions from the gyrophase-bunched ion distributions observed as an initial step in the disruption of field-aligned beams (see Figure 8). Gyrophase-bunched distributions remain non-gyrotropic because they are phase-trapped in a self-generated highly monochromatic low frequency wave. Distributions specularly reflected from the quasi-parallel bow shock have no corresponding self-generated monochromatic low frequency wave to phase trap them. Instead, they enter an upstream region characterized by large amplitude, low frequency waves and form gyrotropic distributions such as the example in Figure 11 within a few ion gyroradii of the shock.

Gyrotropic distributions like the one in Figure 11 are a free energy source for further wave growth in the upstream region. Their density and ability to propagate upstream into the quasi-parallel foreshock make them prime candidates for a source of diffuse ions in that region. Composition measurements appear to support this contention. Unlike field-aligned beams upstream

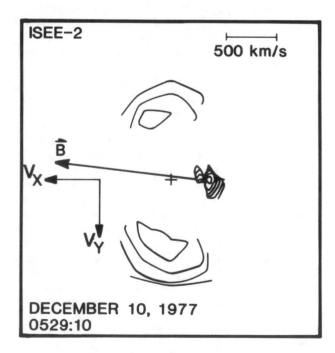

Fig. 11. An example of a gyrotropic distribution produced through specular reflection at the quasi-parallel bow shock and phase mixing in the upstream region. The format is similar to that of Figure 3. The suprathermal distribution is nearly gyrotropic but appears as two separate distributions because of the 2-dimensional nature of the measurements.

from the quasi- perpendicular bow shock, which contain little or no solar wind He^{2+}, suprathermal ion distributions near the quasi-parallel bow shock appear to contain He^{2+} at approximately solar wind concentrations [Fuselier et al., 1990]. Further research is needed to determine if specularly reflected distributions are the source of diffuse ions in the quasi-parallel foreshock.

One of the most puzzling problems of shock physics a decade ago was how downstream ion thermalization occurred at the quasi-parallel bow shock. At that time, it was known that specularly reflected ions provided the bulk of the downstream ion thermalization at the quasi-perpendicular shock. However, the guiding centers of these ions are directed upstream for a steady quasi-parallel bow shock and therefore they could not provide the necessary downstream thermalization for shocks with $\vartheta_{Bn} < 45°$. The observations of gyrotropic distributions consistent with specular reflection but several gyro-radii from the quasi-parallel bow shock seemed to support this interpretation [Gosling et al., 1982; Gosling and Robson, 1985]. The key to the resolution of this problem was the realization that large amplitude waves upstream from the quasi parallel bow shock cause variations in the local ϑ_{Bn} of the shock [Gosling and Robson, 1985] and that the quasi-parallel bow shock structure was not steady. Simulations of the quasi-parallel bow shock structure revealed that the shock is in a state of constant reformation. Observations supported this picture and showed that the local ϑ_{Bn} of the shock varied considerably as large amplitude magnetic field fluctuations in the quasi-parallel foreshock convected into the shock structure [Greenstadt and Mellott, 1985]. Specularly reflected ion distributions similar to those observed at the quasi-perpendicular bow shock were observed near the quasi-parallel bow shock when the local ϑ_{Bn} was greater than 45° [Gosling et al., 1989a; Onsager et al., 1990]. As a result of this recent work and some recent computer simulations [e.g., Burgess, 1989; Scholer and Terasawa, 1990; Thomas et al., 1990; Winske et al., 1990] a consensus picture of the reforming quasi-parallel bow shock was developed.

In this picture, the shock undergoes a cyclic reformation process characterized by strong temporal and spatial variations in the magnetic field and plasma parameters. This cyclic reformation is linked to the fraction of ions that are specularly reflected off the shock. These specularly reflected ions contribute to heating downstream from the quasi-parallel bow shock possibly because they return to the shock when the instantaneous ϑ_{Bn} is greater than 45° [Greenstadt and Mellott, 1985; Fuselier et al., 1986b]. Ions that specularly reflect off the shock when ϑ_{Bn} is less than 45° may propagate upstream and represent the gyrotropic distributions observed several gyro-radii from the quasi-parallel bow shock.

As a result of this cyclic reformation, the downstream region is characterized by two states, a cooler, denser state where the proton distributions appear to be similar to those observed downstream from the quasi-perpendicular shock (at least to a few keV/e) and a hotter, less dense state where the proton distribution appears to be very maxwellian like out to suprathermal energies. An example of these alternating two states downstream from a quasi-parallel bow shock is shown in Figure 12 [from Thomsen et al., 1990, Figure 2]. The proton distributions at the top of the fig-

ure show the hotter, less dense state and the cooler, denser state, respectively. Below these distributions are the plasma moments in the quasi-parallel magnetosheath from 1640 to 1652 UT. The outbound quasi-parallel bow shock was crossed at ~1654 UT. With the exception of the hot diffuse proton distribution at velocities greater than about 800 km/s, the cooler, denser state in the upper right-hand panel of Figure 12 is very similar to the distribution downstream from the quasi-perpendicular shock in Figure 7. The shoulder on the distribution in Figure 12, represented by the more widely space contours in the 200 to 600 km/s is range, is caused by the specularly reflected ions that have returned to the downstream region. The distribution observed during the hotter, denser state (upper left hand panel) is maxwellian-like out to ~800 km/s. Above this velocity, the diffuse ions are also isotropic and maxwellian-like but with a higher temperature than the core distribution.

There is some evidence that the hotter and cooler states evolve as the plasma convects downstream so that the distinction between the two states becomes less clear [Thomsen et al., 1990]. However, further study is required to determine if these variations in the plasma parameters still exist when the plasma in the subsolar region arrives at the magnetopause.

Summary

There are a wide variety of suprathermal ion distributions in the quasi-parallel foreshock and magnetosheath. Gyrophase-bunched distributions far from the shock and intermediate distributions (Figures 8 and 9) are produced by disruption of field- aligned beams that originated in the quasi-perpendicular foreshock. Diffuse ion distributions (Figure 10) have several possible sources including the final state of the disruption of field- aligned beams and the final state of disruption of gyrotropic distributions (Figure 11). Composition measurements appear to support the latter source as the primary source for diffuse ion distributions.

The quasi-parallel bow shock itself appears to undergo cyclic reformation as a result of a time varying amount of ions reflected off the shock. Ion distributions reflected off the shock when the instantaneous ϑ_{Bn} is less than 45° appear to propagate upstream and become more gyrotropic (Figure 11), while during times when the instantaneous ϑ_{Bn} is greater than 45°, the reflected distributions may return to the shock and be observed in the downstream region as a shoulder on the shocked solar wind proton distribution. In addition to the specularly reflected ion distributions, diffuse ion distributions convected downstream by the solar wind are also observed (Figure 12). Cyclic reformation of the quasi-parallel bow shock results in a two state downstream region with an alternating higher density, lower temperature state and lower temperature, higher density state. There is some evidence that the temperature and density fluctuations become less pronounced as the ion distributions convects downstream from the shock.

BOW SHOCK AND MAGNETOSPHERIC RELATIONS

The basic features of the suprathermal ion distributions observed upstream and downstream from the Earth's bow shock have been discussed above. The understanding of these regions has been developed using several very large data bases as well as

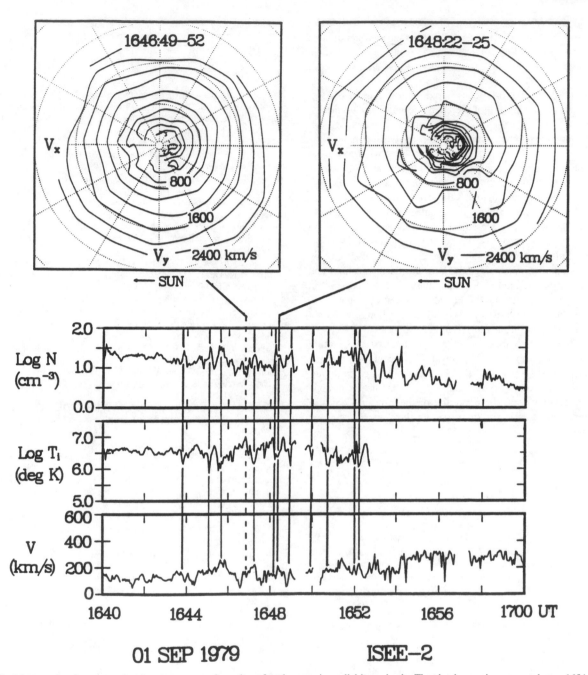

Fig. 12. An example of two-state heating downstream from the reforming quasi-parallel bow shock. The shock crossing occurred at ~1654 UT and the magnetosheath region ins to the left of this time. The two distributions shown in the upper panel were taken in the hotter, less dense and the cooler, denser regions downstream from the shock. With the exception of the diffuse ions at velocities greater than ~800 km/s, the distribution in the cooler, denser region is similar to the distributions in Figure 7 observed downstream from the shock. The distribution in the hotter, denser region is nearly isotropic and maxwellian-like out to velocities on the order of 800 km/s. The diffuse ions convected from the upstream region are observed at velocities greater than 800 km/s.

computer simulations and analytic theory. There is overwhelming evidence that particle distributions as well as wave phenomena are ordered by ϑ_{Bn} of the shock. Unfortunately for magnetospheric physics considerations, specifically ultra-low frequency (ULF) pulsation studies, this evidence has lead to an emphasis on

ϑ_{Bn} rather than the IMF cone angle. As is evident in Figures 1 and 2, the cone angle and ϑ_{Bn} are identical only at the subsolar point. Thus, large statistical studies of bow shock phenomena and its relation to ϑ_{Bn} discussed above are not necessarily directly applicable to magnetospheric pulsation research. To better understand the relationship between bow shock phenomena and magnetospheric ULF pulsations requires further consideration of the subsolar region and an emphasis on the cone angle rather than ϑ_{Bn}. Studies of this region can be done with existing spacecraft data bases.

As an example of the difference between cone angle and ϑ_{Bn}, consider the high energy tail of the energetic ion distribution produced by first order fermi acceleration in the turbulent regions upstream and downstream from the Earth's quasi-parallel bow shock. The maximum energy gained by a solar wind ion is related to the time it spends in the acceleration region. This time is related to the time the magnetic field is connected to the Earth's quasi-parallel bow shock region as the field convects with the solar wind. For large cone angles, this connection time is relatively short and solar wind ions are accelerated to modest energies. In the limiting case of zero cone angle, the connection time is essentially infinite, solar wind ion acceleration is limited only by the extent of the turbulent region along the Earth-sun line, and these ions are accelerated to the maximum energies possible in the Earth's foreshock (about 150 keV/e). Thus, at the same ϑ_{Bn}, different particle spectra (and presumably two different turbulent wave fields) are possible for different cone angles.

To be applicable to magnetospheric pulsation studies, future work should concentrate on the subsolar region and its relation to the Earth's magnetopause. Such studies should include understanding how previous work related to ϑ_{Bn} is applicable to cone angle considerations and the evolution of the two state density-temperature structures in the quasi-parallel bow shock for small cone angles.

Acknowledgments. Research at Lockheed was performed under the NASA Guest Investigator program through contract NAS5-31217 and Lockheed Independent Research.

REFERENCES

Anderson, B. J., and S. A. Fuselier, Magnetic pulsations from 0.1 to 4.0 Hz and associated plasma properties in the Earth's subsolar magnetosheath and plasma depletion layer, *J. Geophys. Res., 98,* 1461, 1993.

Asbridge, J. R., S. J. Bame, and I B. Strong, Outward flow of protons from the Earth's bow shock, *J. Geophys. Res., 73,* 5777, 1968.

Bame, S. J., J. R. Asbridge, W. C. Feldman, J. T. Gosling, G. Paschmann, and N. Sckopke, Deceleration of the solar wind upstream from the Earth's bow shock and the origin of diffuse upstream ions, *J. Geophys. Res., 85,* 2981, 1980.

Barnes, A., Theory of generation of bow-shock-associated hydromagnetic waves in the upstream interplanetary medium, *Cosmic Electrodyn., 1,* 90, 1970.

Bonifazi, C., and G. Moreno, Reflected and diffuse ions backstreaming from the Earth's bow shock, 1, Basic Properties, *J. Geophys. Res., 86,* 4397, 1981a.

Bonifazi, C., and G. Moreno, Reflected and diffuse ions backstreaming from the Earth's bow shock, 2, Origin, *J. Geophys. Res., 86,* 4405, 1981b.

Burgess, D., Cyclic behavior at quasi-parallel collisionless shocks, *Geophys. Res. Lett., 16,* 345, 1989.

Burgess, D., and S. J. Schwartz, The dynamics and upstream distributions of ions reflected at the Earth's bow shock, *J. Geophys. Res., 89,* 7407, 1984.

Denton, R. E., M. K. Hudson, S. A. Fuselier, and B. J. Anderson, Electromagnetic ion cyclotron waves in the plasma depletion layer, *J. Geophys. Res.,* submitted, 1992.

Eastman, T. E., R. R. Anderson, L. A. Frank, and G. K. Parks, Upstream particles observed in the Earth's foreshock region, *J. Geophys. Res., 86,* 4379, 1981.

Ellison, D. C., and E. Möbius, Diffusive shock acceleration: comparison of a unified shock model to bow shock observations, *Astrophys. J., 318,* 474, 1987.

Fairfield, D. H., Bow shock associated waves observed in the far upstream interplanetary medium, *J. Geophys. Res., 74,* 3541, 1969.

Fuselier, S. A., Comment on: "Upstream energetic ions under radial IMF: A critical test of the Fermi model", *Geophys. Res. Lett., 16,* 109, 1989.

Fuselier, S. A., and M. F. Thomsen, He^{2+} in field aligned beams: ISEE results, *Geophys. Res. Lett., 19,* 437, 1992.

Fuselier, S. A., M. F. Thomsen, J. T. Gosling, S. J. Bame, and C. T. Russell, Gyrating and intermediate ion distributions upstream from the Earth's bow shock, *J. Geophys. Res., 91,* 91, 1986a.

Fuselier, S. A., J. T. Gosling, and M. F. Thomsen, The motion of ions specularly reflected off a quasi-parallel shock in the presence of large-amplitude, monochromatic MHD waves, *J. Geophys. Res., 91,* 4163, 1986b.

Fuselier, S. A., M. F. Thomsen, S. P. Gary, S. J. Bame, C. T. Russell, and G. K. Parks, The phase relationship between gyrophase-bunched ions and MHD-like waves, *Geophys. Res. Lett., 13,* 60, 1986c.

Fuselier, S. A., E. G. Shelley, and D. M. Klumpar, AMPTE/CCE observations of shell-like He^{2+} and O^{6+} distributions in the magnetosheath, *Geophys. Res. Lett., 15,* 1333-1336, 1988.

Fuselier, S. A., O. W. Lennartsson, M. F. Thomsen, and C. T. Russell, Specularly reflected He^{2+} at high mach number quasi-parallel shocks, *J. Geophys. Res., 95,* 4319, 1990.

Fuselier, S. A., O. W. Lennartsson, M. F. Thomsen, and C. T. Russell, He^{2+} heating at a quasi-parallel shock, *J. Geophys. Res., 96,* 9805-9810, 1991.

Gary, S. P., J. T. Gosling, and D. W. Forslund, The electromagnetic ion beam instability upstream of the earth's bow shock, *J. Geophys. Res., 86,* 6691, 1981.

Gary, S. P., S. A. Fuselier, and B. J. Anderson, Ion anisotropy instabilities in the magnetosheath, *J. Geophys. Res., 98,* 1481, 1993.

Gloeckler, G. F. M. Ipavich, D. C. Hamilton, B. Wilken, W. Stüdemann, G. Kremsher, and D. Hovestadt, Solar wind carbon, nitrogen, and oxygen abundances measured in the Earth's magnetosheath with AMPTE, *Geophys. Res. Lett., 13,* 793-796, 1986.

Gosling, J. T., and A. E. Robson, Ion reflection, gyration, and dissipation at supercritical shocks, in *Collisionless Shocks in the Heliosphere: Reviews of Current Research, Geophys. Monogr. Ser., vol 35,* edited by B. T. Tsurutani and R. G. Stone, pp. 141-152, AGU, Washington, D. C., 1985.

Gosling, J. T., J. R. Asbridge, S. J. Bame, G. Paschmann, and N. Sckopke, Observations of two distinct populations of bow shock ions in the upstream solar wind, *Geophys. Res. Lett., 5,* 957, 1978.

Gosling, J. T., J. R. Asbridge, S. J. Bame, and W. C. Feldman, Ion acceleration at the Earth's bow shock: A review of observations in the upstream region, in *Particle Acceleration Mechanisms in Astrophysics,* edited by J. Arons, C. Max, and C. McKee, PP. 81-99, American Institute of Physics, New York, 1979.

Gosling, J. T., M. F. Thomsen, S. J. Bame, W. C. Feldman, G. Paschmann, and N. Sckopke, Evidence for specularly reflected ions upstream from the quasi-parallel bow shock, *Geophys. Res. Lett., 9,* 1333, 1982.

Gosling, J. T., M. F. Thomsen, S. J. Bame, and C. T. Russell, Ion reflection and downstream thermalization at the quasi-parallel bow shock, *J. Geophys. Res., 94,* 10,027, 1989a.

Gosling, J. T., M. F. Thomsen, S. J. Bame, and C. T. Russell, On the source of diffuse, suprathermal ions observed in the vicinity of the Earth's bow shock, *J. Geophys. Res., 94,* 3555, 1989b.

Greenstadt, E. W., Phenomenology of the Earth's bow shock system. A summary description of experimental results, in *Magnetospheric Particles and Fields,* edited by B. M. McCormac, pp. 13-28, D. Reidel Publishing Company, Dordrecht-Holland, 1976.

Greenstadt, E. W., and M. M. Mellott, Variable field-to-normal shock-foreshock boundary observed by ISEE-1 and -2, *Geophys. Res. Lett.,*

12, 129, 1985.

Greenstadt, E. W., I. M. Green, G. T. Inouye, A. J. Hundhausen, S. J. Bame, and I. B. Strong, Correlated magnetic field and plasma observations of the Earth's bow shock, *J. Geophys. Res., 73*, 51, 1968.

Gurgiolo, C., G. K. Parks, B. H. Mauk, C. S. Lin, K. A. Anderson, R. P. Lin and H. Réme, Non-$\mathbf{E \times B}$ ordered ion beams upstream of the Earth's bow shock, *J. Geophys. Res., 86*, 4415, 1981.

Gurgiolo, C., G. K. Parks, and B. H. Mauk, Upstream gyrophase bunched ions: A mechanism for creation at the bow shock and the growth of velocity space structure through gyrophase mixing, *J. Geophys. Res., 88*, 9093, 1983.

Hoshino, M., and T. Terasawa, Numerical study of the upstream wave excitation mechanism 1. Nonlinear phase bunching of beam ions, *J. Geophys. Res., 90*, 57, 1985.

Ipavich, F. M., J. T. Gosling, and M. Scholer, Correlation between the He/H ratios in upstream particle events and in the solar wind, *J. Geophys. Res., 89*, 1501, 1984.

Ipavich, F. M., G. Gloeckler, D. C. Hamilton, L. M. Kistler, and J. T. Gosling, *Geophys. Res. Lett., 15*, 1153, 1988.

Le, G., and C. T. Russell, The morphology of ULF waves in the Earth's forshock, [This volume], 1993.

Lee, M. A., Coupled hydromagnetic wave excitation and ion acceleration upstream of the Earth's bow shock, *J. Geophys. Res., 87*, 5063, 1982.

Leroy, M., M. D. Winske, C. C. Goodrich, C. S. Wu, and K Papadopoulos, The structure of perpendicular bow shocks, *J. Geophys. Res., 87*, 5081, 1982.

Luhmann, J. G., R. J. Walker, C. T. Russell, J. R. Spreiter, S. S. Stahara, and D. H. Williams, Mapping the magnetosheath field between the magnetopause and the bow shock: Implications for magnetospheric particle leakage, *J. Geophys. Res., 89* 6829, 1984.

Luhmann, J. G., C. T. Russell, and R. C. Elphic, Spatial distributions of magnetic field fluctuations in the dayside magnetosheath, *J. Geophys. Res., 91*, 1711, 1986.

Neugebauer, M., A. J. Lazarus, H. Balsiger, S. A. Fuselier, F. M. Neubauer, and H. Rosenbauer, The velocity distributions of cometary protons picked up by the solar wind, *J. Geophys. Res., 94*, 5227, 1989.

Onsager, T. G., M. F. Thomsen, J. T. Gosling, S. J. Bame, and C. T. Russell, Survey of coherent ion reflection at the quasi-parallel bow shock, *J. Geophys. Res., 95*, 2261, 1990.

Paschmann, G., and N. Sckopke, Ion reflection and heating at the Earth's bow shock, in *Topics in Plasma-, Astro-, and Space Physics,* edited by G. Haerendel and B. Battrick, P. 139, Max-Planck-Institut für Physik und Astrophysik, Garching, Germany, 1983.

Paschmann, G., N. Sckopke, S. J. Bame, J. R. Asbridge, J. T. Gosling, C. T. Russell, and E. W. Greenstadt, Association of low-frequency waves with suprathermal ions in the upstream solar wind, *Geophys. Res. Lett., 6*, 209, 1979.

Paschmann, G., N. Sckopke, I. Papamastorakis, J. R. Asbridge, S. J. Bame, and J. T. Gosling, Characteristics of reflected and diffuse ions upstream from the Earth's bow shock, *J. Geophys. Res., 86*, 4355, 1981.

Paschmann, G., N. Sckopke, S. J. Bame, and J. T. Gosling, Observations of gyrating ions in the foot of the nearly perpendicular bow shock, *Geophys. Res. Lett., 9*, 881, 1982.

Peterson, W. K., E. G. Shelley, R. D. Sharp, R. G. Johnson, J. Geiss, and H. Rosenbauer, H^+ and He^{++} in the dawnside magnetosheath, *Geophys. Res. Lett., 6*, 667-670, 1979.

Quest, K. B., Theory and simulation of collisionless parallel shocks, *J. Geophys. Res., 93*, 9649, 1988.

Russell, C. T., J. G. Luhmann, T. J. Odera, and W. F. Stuart, The rate of occurrence of dayside PC 3,4 pulsations: The L-value dependence of the IMF cone angle effect, *Geophys. Res. Lett., 10*, 663, 1983.

Sarris, E. T., and S. M. Krimigis, Upstream energetic ions under radial IMF: A critical test of the Fermi model, *Geophys. Res. Lett., 15*, 233, 1988.

Schwartz, S. J., and D. Burgess, On the theoretical/observational comparison of field-aligned ion beams in the Earth's foreshock, *J. Geophys. Res., 89*, 2381, 1984.

Schwartz, S. J., M. F. Thomsen, and J. T. Gosling, Ions upstream of the Earth's bow shock: A theoretical comparison of alternative source populations, *J. Geophys. Res., 88*, 2039, 1983.

Scholer, M., and T. Terasawa, Ion reflection and dissipation at quasi-parallel collisionless shocks, *Geophys. Res. Lett., 17*, 119, 1990.

Scholer, M., F. M. Ipavich, G. Gloeckler, D. Hovestadt, and B. Klecker, Leakage of magnetospheric ions into the magnetosheath along reconnected field lines at the dayside magnetopause, *J. Geophys. Res., 86*, 1299-1304, 1981.

Sckopke, N., G. Paschmann, S. J. Bame, J. T. Gosling, and C. T. Russell, Evolution of ion distributions across the nearly perpendicular bow shock: Specularly and non-specularly reflected ions, *J. Geophys. Res., 88*, 6121, 1983.

Sonnerup, B. U. Ö., Acceleration of particles reflected at a shock front, *J. Geophys. Res., 74*, 1301, 1969.

Thomas, V., D. Winske, and N. Omidi, Re-forming supercritical quasi-parallel shocks 1. One- and two-dimensional simulations, *J. Geophys. Res., 95*, 18809, 1990.

Thomsen, M. F., Upstream suprathermal ions, in *Collisionless Shocks in the Heliosphere: Reviews of Current Research, Geophys. Monogr. Ser., vol 35,* edited by B. T. Tsurutani and R. G. Stone, pp. 253-270, AGU, Washington, D. C., 1985.

Thomsen, M. F., J. T. Gosling, S. J. Bame, and C. T. Russell, Gyrating ions and large-amplitude momochromatic MHD waves upstream of the Earth's bow shock, *J. Geophys. Res., 90*, 267, 1985.

Thomsen, M. F., J. T. Gosling, S. J. Bame, T. G. Onsager, and C. T. Russell, Two-state ion heating at quasi-parallel shocks, *J. Geophys. Res., 95*, 6363, 1990.

Winske, D., C. S. Wu, Y. Y. Li, and G. C. Zhou, Collective capture of released lithium ions in the solar wind, *J. Geophys. Res., 89*, 7327, 1984.

Winske, D., N. Omidi, K. B. Quest, and V. A. Thomas, Re-forming supercritical quasi-parallel shocks 2. Mechanism for wave generation and front re-formation, *J. Geophys. Res., 95*, 18,821, 1990.

Stephen A. Fuselier, Dept 91-20 Bldg 255, Lockheed Palo Alto Research Laboratory, 3251 Hanover St., Palo Alto, CA 94304.

Bow Shock and Magnetosheath Simulations:
Wave Transport and Kinetic Properties

D. Krauss-Varban

Department of Electrical and Computer Engineering, University of California at San Diego

In quasi-parallel geometry, the interaction of the solar wind with planetary magnetospheres gives rise to upstream perturbations that are carried back into the magnetosheath and (at Earth) have long been held responsible for magnetospheric pulsations. The kinetic aspects of the upstream wave generation, transmission, and conversion are reviewed and elucidated using results of hybrid (kinetic ions, fluid electrons) simulations and linear kinetic theory. First, quasi-parallel shock re-formation and its ensuing downstream, compressional perturbations are examined. The temporal and spatial scales set significant constraints on this process as a means of generating the magnetosheath fluctuations in question. On the other hand, for $\theta_{Bn} \sim 30°$ and medium Mach numbers ($M_A \sim 2.3$ to ~ 3), one-dimensional hybrid simulations have shown that upstream fast magnetosonic (F/MS) waves are directly converted into downstream Alfvén/ion-cyclotron (A/IC) waves in the Pc 3–4 band. Here, θ_{Bn} is the acute angle between the upstream magnetic field and the shock normal. It is demonstrated that this process is viable also at higher Mach numbers ($M_A = 4.5$), at smaller angles ($\theta_{Bn} \sim 10°$), and in two-dimensional simulations. Understanding the relevant wave processes requires a good characterization of the low-frequency, warm plasma mode properties in a collisionless plasma. The mode structure is reviewed, and some important differences between kinetic and 2-fluid mode properties are shown. How Alfvénic magnetosheath wave energy eventually transmits into the magnetosphere is not currently understood. However, three examples are given which demonstrate that kinetic aspects need to be taken into account in the wave interaction with the (open) magnetopause. When modelled as a rotational discontinuity (RD), the interaction with downstream propagating A/IC waves leads to reflected and converted F/MS waves that are not expected from MHD theory. The wave conversion also depends on the handedness of the RD. Furthermore, the wave interaction can trigger the break-up of RDs with rotation angle larger than 180°.

1. INTRODUCTION

Much progress has been made in the past decade in the understanding of collisionless fast mode shocks [Leroy et al., 1982; Paschmann et al., 1982; Sckopke et al., 1983; Goodrich, 1985; Greenstadt, 1985; Kennel et al., 1985; Mellott, 1985; Quest, 1985, 1988; Burgess, 1989; Gosling et al., 1989; Thomsen et al., 1990a,b; Mandt, 1992]. At sufficient strength (supercritical shocks), the thermalization of directed flow energy invariably involves the reflection of part of the incoming ion distribution. This reflected population is responsible for the generation of low-frequency waves via plasma instabilities. Due to the ion kinematics in the respective field topology, such waves are restricted to the downstream at quasi-perpendicular shocks ($\theta_{Bn} > 45°$), but are generated upstream or in an extended shock transition

region at quasi-parallel shocks ($\theta_{Bn} \lesssim 45°$, where θ_{Bn} is the acute angle between the upstream magnetic field and the shock normal). As a consequence of the highly super-Alfvénic and super-fast solar wind speed, the Earth's bow shock as well as other planetary bow shocks are nearly always of supercritical nature. Thus, in addition to intrinsic solar wind turbulence and large scale perturbations, the principal interaction of planetary magnetospheres with the solar wind is by no means of a steady, quiet nature, but instead leads to upstream and downstream perturbations that are ultimately transmitted into the magnetosphere. It is the kinetic aspects of this interaction that we are interested in.

The idea that bow-shock associated waves cause magnetosheath fluctuations and may be responsible for magnetospheric pulsations has been a topic of intense investigation in the past two decades [Barnes, 1970; Fairfield, 1976; Greenstadt et al., 1983; Takahashi et al., 1984; Watanabe and Terasawa, 1984; Yumoto et al., 1985; Luhmann et al., 1986; Fairfield et al., 1990; Engebretson et al., 1991; Lin et al., 1991]. Observations have long indicated a strong corre-

Solar Wind Sources of Magnetospheric Ultra-Low-Frequency Waves
Geophysical Monograph 81

lation of magnetospheric wave activity with the IMF cone angle [Bol'shakova and Troitskaya, 1968; Troitskaya et al., 1971; Greenstadt and Olson, 1976, 1977; Wolfe, 1980; Russell et al., 1983a; Wolfe et al., 1985]. Together, the findings stress the importance of the quasi-parallel and nearly parallel shock geometry. Compressional waves generated at or downstream of the quasi-perpendicular shock are predominantly mirror waves [Moustaizis et al., 1986; Hubert et al., 1989; Lacombe et al., 1990, 1992]. Because of their convection with the magnetosheath plasma, in the shock or magnetopause rest frame they have typically a higher frequency than the energetic low frequency waves of magnetospheric interest (here: Pc 3 to part of Pc 4 frequency range, periods 20 to $\sim 100\,s$). Consequently, waves downstream of quasi-perpendicular shocks are here only mentioned in passing.

It is the goal of the present paper to point out kinetic aspects of the low frequency waves, concerning their upstream generation and propagation, their interaction with the bow shock and magnetopause, and regarding their identification. Prime methods of investigation are hybrid simulations (kinetic ions, fluid electrons) and the use of linear kinetic (Vlasov) theory. Before the advent of computer simulations, the interaction of low frequency waves with the bow shock and magnetosphere was investigated by studying the reflection and transmission of MHD waves at fast mode shocks [McKenzie and Westphal, 1970] and other discontinuities [Fejer, 1963; McKenzie, 1970; Verzariu, 1973; Wolfe and Kaufmann, 1975; Lee, 1982; Kwok and Lee, 1984]. For several reasons, theoretical analyses that make use of hybrid simulations are quite different from such MHD calculations. First, the simulations study the upstream and downstream plasma as well as the shock (or the magnetopause) as a complete, coupled system, which leads to consistent wave generation and propagation [Omidi, 1991; Krauss-Varban and Omidi, 1991; Omidi et al., 1993]. Second, in agreement with observations, the upstream waves usually reach large amplitudes and can no longer be treated with a linearized theory. Third, the scale sizes of the shock (or magnetopause) transition region and wavelength are often of the same order of magnitude, precluding simplifying analyses. Fourth, in the usually medium to high beta plasma, a description using MHD or two-fluid theory breaks down and gives wrong results concerning the frequency of the modes as well as other mode properties, such as their compressibility and polarization.

Here I present results of both one and two-dimensional simulations. A recent description of the hybrid simulation code has been given by Winske and Omidi [1993]. Typical simulation parameters for one-dimensional simulations are a box size of 600 c/ω_{pi}, a cell size of 0.25 c/ω_{pi}, and 100,000 to 200,000 macro particles. Here, c/ω_{pi} is the upstream ion inertial length, and times are measured in terms of the (upstream) inverse ion cyclotron frequency, Ω_{ci}^{-1}. Details of the simulation geometry and set-up are given by Krauss-Varban and Omidi [1991, 1993]. Linear theory calculations were performed with the newly developed code "DIS", which is described by Krauss-Varban et al. [1993].

The paper is structured as follows. Low frequency waves upstream and downstream of quasi-parallel shock simulations are discussed in section 2. Section 3 clarifies the kinetic mode properties and nomenclature in a warm plasma and points out some relevant differences to two-fluid theory. Aspects of the wave characteristics that can only be elucidated in two-dimensional simulations are examined in section 4. Section 5 briefly discusses simulations that address kinetic effects in the interaction of magnetosheath waves with the magnetopause. A summary is given in section 6.

2. THE QUASI-PARALLEL SHOCK

2.1. *Upstream Waves and Shock Re-formation*

The first simulations of quasi-parallel shocks were conducted by Kan and Swift [1983], Leroy and Winske [1983], Quest et al. [1983], and Mandt and Kan [1985, 1988]. Interest in these shocks revived when both simulations [Burgess, 1989] and observations [Gosling et al., 1989; Thomsen et al., 1990a,b] showed an interesting unsteady, cyclical behavior. It became soon clear that both the steepening of far upstream waves as well as the coupling of specular reflected and heated downstream ions with the incoming solar wind were involved in a rather complicated shock re-formation processes [Scholer and Terasawa, 1990; Thomas et al., 1990, Winske et al., 1990].

The above findings of shock re-formation are valid for typical Mach numbers of about $M_A \sim 4$ to 10. Simulations of very low Mach number shocks ($M_A \lesssim 1.5$), on the other hand, agree with the classical picture of laminar, steady shock transitions [Tidman and Northrop, 1968], in which the shock is the last cycle of an upstream phase standing whistler wave, which limits further steepening. Omidi et al. [1990] performed simulations at just slightly higher Mach numbers, and showed that after sufficient time the upstream region is dominated by longer wavelength fast-magnetosonic waves. These waves are generated by backstreaming ions, and appear to be the same waves that contribute to shock re-formation at higher Mach number. However, these shocks ($M_A \sim 2$) are still steady and possess a quiet downstream.

From the above, three questions arise in the context of this paper. First, how does the transition take place from steady shocks to re-forming shocks with turbulent downstream states? Second, what is the role and consequence of the upstream waves? Third, can shock re-formation in itself be responsible for periodic fluctuations in the downstream (magnetosheath)?

To answer the first question, we performed simulations at increasing Mach number, from $M_A = 1.5$ to 3.2 [Krauss-Varban and Omidi, 1991]. We found the following scenario at successively stronger shock transitions: Backstreaming ions, which are already present at low Mach numbers ($M_A \lesssim 2.0$), excite fast magnetosonic waves via the resonant electromagnetic ion/ion instability. In the reference frame of the shock, the group velocity of the dominant waves changes. While the group velocity is at first oriented upstream, at increasing Mach number it becomes Doppler shifted to group standing and finally has downstream orientation, until there is sufficient downstream directed energy flux associated with the waves to disrupt the shock. It is important to emphasize that it is not the flux of backstreaming ions nor simply the amplitude of upstream waves, but the Doppler shift associated with the shock Alfvén Mach number that is decisive in determining when re-formation takes place [Krauss-Varban and Omidi, 1991] (see also the comparison to observations by Greenstadt et al., 1991]).

Figure 1 illustrates the characteristics of the generated waves, as calculated from linear Vlasov theory (see e.g., Gary [1991] for a recent review on ion-beam instabilities).

The top panels show the dispersion and growth of the waves, the bottom shows the growth rate plotted against the group velocity; both are in the shock rest frame. Figure 1a is for $M_A = 2$, Figure 1b for $M_A = 3$. In both cases I have assumed a heated population of backstreaming ions (subscript b) with a beam velocity in the shock frame equal to their thermal velocity, and a temperature of $T_b/T_o = 4$ with respect to the incoming solar wind (subscript o). For comparison, results for two beam densities n_b/n_o are given, as indicated. The angle of propagation with respect to the magnetic field is $\theta = 30°$, but the results are (apart from the magnitude of the growth rate) not very sensitive to this

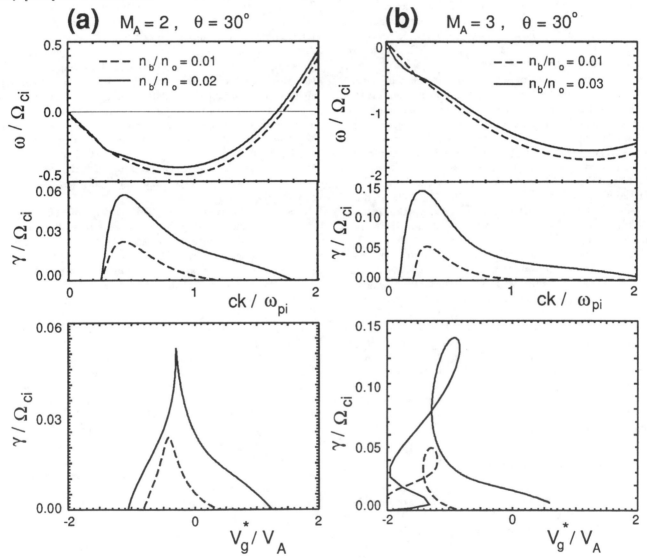

Fig. 1. Dispersion $\omega(k)$, growth rate γ and group velocity $v_g{}^* = \partial\omega/\partial k - M_A v_A$ of upstream generated waves for Alfvén Mach number (a) $M_A = 2$ and (b) $M_A = 3$, respectively, from linear kinetic theory. The frequency ω is normalized with the upstream ion cyclotron frequency, Ω_{ci}, and the wavenumber is normalized with the upstream ion inertial length, c/ω_{pi}. The background plasma (subscript o) has $\beta_i = \beta_e = 0.5$. A warm beam of backstreaming ions (subscript b) has $T_b/T_o = 4$ and varying n_b/n_o, as indicated, for $\theta = 30°$. Note that for $M_A = 2$, the generated waves are predominantly group standing or propagating upstream, whereas for $M_A = 3$ the group velocity has a substantial component directed downstream, in agreement with simulations and observations.

angle. The plotted group velocity is $v_g^* = \partial\omega/\partial k - M_A v_A$, normalized with the Alfvén velocity v_A, which is the Doppler shifted projection of the group velocity in the direction of the wave vector $\mathbf{k}$. This quantity is relevant in one-dimensional simulations. For small θ, it does not differ much from the projection of the Doppler shifted group velocity onto the shock normal, $\mathbf{v}_g^* \cdot \mathbf{n}$.

Figure 1a shows a broad region of growth of approximately groupstanding waves. The beam flattens the dispersion such that at increasing beam density, a larger fraction of waves around maximum growth becomes groupstanding or attains upstream directed group velocity. This effect is even more pronounced close to the shock, at yet higher beam density. Only at sufficient Mach number does the Doppler shift become large enough such that all wave energy around maximum growth gets convected downstream. Figure 1b shows that at $M_A = 3$, even for a beam density of $n_b/n_o = 0.03$ (and corresponding flattening of the dispersion) the waves have a substantial downstream directed group velocity. Thus, when the effect of the beam on the dispersion is taken into account, linear theory suggests that a Mach number somewhat larger than 2 is needed for the wave energy to get convected downstream and disrupt the shock. Indeed, a detailed analysis of simulations has shown mostly groupstanding waves and a steady shock at $M_A \sim 2$, and the onset of shock unsteadiness at $M_A \gtrsim 2.3$ [Krauss-Varban and Omidi, 1991]. This is also in good agreement with observations by Greenstadt et al. [1991] at the flank of the Earth's bow shock.

The upstream generated and downstream convected waves have a twofold role concerning magnetosheath turbulence. They can initiate shock re-formation, which in turn causes compressional downstream perturbations. This is described in the next subsection 2.2. Such perturbations have their main fluctuating magnetic field in the plane of the ambient magnetic field $\mathbf{B}_o$ and wave vector $\mathbf{k}$. In the subsequent subsection 2.3 it is demonstrated that the upstream waves also mode convert into Alfvén waves, which carry predominantly transverse perturbations. I use the adjectives "compressional" and "Alfvénic" in the simulation·results to distinguish between the two field components in and out of the $\mathbf{B}_o$–$\mathbf{k}$ plane, respectively.

2.2. Temporal and Spatial Scales Associated With Shock Re-formation

At medium Mach numbers, it has been shown that there are no characteristic time scales associated with shock re-formation [Krauss-Varban and Omidi, 1991; Krauss-Varban, 1993b]. Higher Mach number simulations have demonstrated that the generation of energetic, backstreaming ions, the excitation of upstream waves, and shock re-formation cannot be viewed as independent. Are they coupled such as to cause periodic perturbations well within the magnetosheath? For example, at high Mach numbers backstreaming ion fluxes depend on the instantaneous gradient of the re-forming shock [Thomas et al., 1990; Krauss-

Varban, 1993b]. Conceivably, this may result in a feedback between the waves and the ions, and eventually lead to periodic shock re-formation. For now, it appears that different processes may be more important. At high Mach number shocks, the larger convection speed, amplification at the shock interface [Winske et al., 1990], and (at small θ_{Bn}) the shorter growth times associated with specular reflected, non-gyrotropic ions [McKean et al, 1993] versus far upstream, diffuse ions [Kucharek and Scholer, 1991] all lead to a much narrower shock transition region. They cause a predominance of shorter time scales, which superficially gives the impression of periodicity [Krauss-Varban, 1993b]. This is demonstrated in Figure 2, which shows (as a rough measure of shock re-formation) the Fourier transform of the fluctuations in the maximum density gradient of the one-dimensional shock transition. The figure shows that the density gradient is only affected once the group velocity of the upstream generated waves points downstream, causing an unsteady shock ($M_A \gtrsim 2.3$). The dashed line indicates a spectral index of one, implying that there are no dominant modes over a wide range of frequencies for medium Mach numbers. At the highest Mach number, very low frequency fluctuations are weaker, in support of the time-scale arguments given above. In conjunction with the spectral decline at higher frequencies, seemingly quasi-periodic shock re-formation results.

As described by Thomas et al. [1990] and Thomsen et al. [1990a,b], shock re-formation causes alternating zones of largely thermalized plasma and areas that contain relatively cold solar wind plasma. These are convected downstream and carry a strong magnetic field and density signature with them. The immediate question one may ask is, whether these compressional structures may be directly responsible for periodic pulsations in the magnetosphere. Unfortunately, there are several problems with such an interpretation. In the first place, I have shown above that the time scale for shock re-formation becomes shorter at higher (i.e., more common) Mach numbers. Frequencies that correspond to the long period/high power part of the Pc 3–4 range become suppressed, and instead periods close to a few upstream Ω_{ci}^{-1} dominate (see also Thomas et al. [1990]). Second, the simulations show that the convected structures do not survive very long; they typically only extend $\sim 100\,c/\omega_{pi}$ or approximately one Earth radius downstream. An example of this is given in Figure 3. Figure 3a shows a stackplot of the z (compressional) component of the magnetic field in a one-dimensional simulation, at $M_A = 4.5$, over time. The heavy, dashed line is located about $\sim 100\,c/\omega_{pi}$ downstream from the shock. The downstream compressional structures that are approximately stationary in the simulation (downstream) frame thermalize in front of this line. Figure 3b shows the density profiles from the same simulation in a similar format. It is visible how the upstream waves steepen and lead to compressional structures. They are convected downstream, but only survive for 30 to 50 Ω_{ci}^{-1} with significant amplitude. These results also hold in two-dimensional simulations and for smaller θ_{Bn}, see section 4. Such behavior is theoretically expected, because the perturbations are not on a normal mode of the plasma. Instead, the structures' borders resemble contact discontinuities, which disintegrate in a warm, kinetic plasma [Barnes, 1979]. Such an evolution is consistent with detailed observations of downstream ion distributions by Gosling et al. [1989] and Thomsen et al. [1990a].

There is a further reason why shock re-formation is not the best candidate for the production of periodic pulsa-

Power Spectra

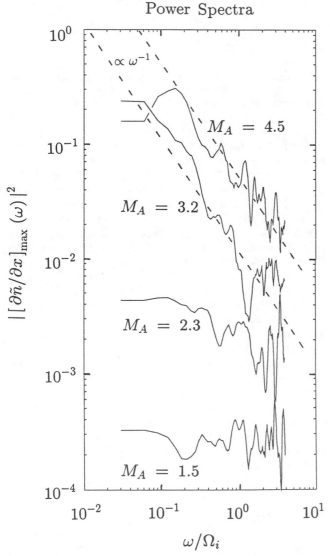

Fig. 2. Power spectra of the fluctuations in the density gradient at the shock with M_A as parameter, taken from Krauss-Varban [1993b]. At $M_A = 1.5$ the shock is steady, with a flat power spectrum characteristic of low-amplitude, white noise. When the group velocity of the upstream waves points downstream ($M_A \gtrsim 2.3$), the shock becomes unsteady, but without characteristic time scale ($M_A = 3.2$). At the harder driven high Mach numbers ($M_A = 4.5$) low frequency re-formations are suppressed. Throughout the paper, when not otherwise noted, the upstream parameters are: $\beta_i = \beta_e = 0.5$, $\omega_{pi}/\Omega_{ci} = 4000$, and $\theta_{Bn} = 30°$.

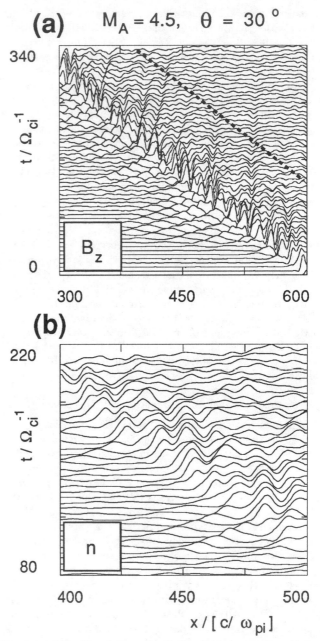

Fig. 3. (a) Profile of the compressional (z) component of the magnetic field for $M_A = 4.5$ over time. Note how the compressional perturbations of the re-formation cycle are convected downstream, but thermalize within a distance of $\sim 100\ c/\omega_{pi}$ downstream from the shock (dashed line). The shock is moving from right to left, and the downstream plasma is stationary (simulation frame). (b) Stackplot of the density, central portion of (a). Again, the compressional structures thermalize in a short time (30 to 50 Ω_{ci}^{-1}).

tions. Recent two-dimensional simulations have shown various phases of shock re-formation happening simultaneously along the transverse direction, only short distances apart (a few $10\ c/\omega_{pi}$, [Thomas et al., 1990; Krauss-Varban and Omidi, 1993]). On the other hand, upstream waves have a transverse coherence length very large compared to kinetic length scales [Krauss-Varban and Omidi, 1993], of the order of an Earth radius [Le and Russell, 1990]. Thus, they may be more effective in eventually causing sizable and coherent perturbations in the magnetosphere.

From this one may conclude that shock re-formation in itself is not likely the cause of the magnetosheath fluctuations we are interested in here. I should mention one caveat, however. At very small θ_{Bn} (nearly parallel shock), the differences between wave transmission, local wave generation at the shock, and shock re-formation become less clear. For example, Winske et al. [1990] have shown that the interface between the incoming solar wind and the downstream plasma can directly amplify compressional fluctuations. In their high Mach number simulations, these perturbations survive longer downstream at smaller θ_{Bn}. Putting aside the other arguments given above, such compressional perturbations could convectively transport fluctuation energy to the magnetopause, e.g., in a fashion as discussed in Lin et al. [1991]. Yet, Engebretson et al. [1991] have demonstrated that it is not the compressional, but the transverse (Alfvénic) wave power in the magnetosheath that is correlated with magnetospheric pulsations. Clearly, further simulation studies are required to help understand the detailed wave properties at high Mach number, nearly parallel shocks.

2.3. Upstream Waves and Magnetosheath Turbulence

When the Mach number of the shock simulations is increased, at the same time when the shock becomes unsteady, there is also a notable increase in downstream wave activity. Using one-dimensional simulations, we have previously shown that the upstream fast mode waves are converted into downstream Alfvén waves [Krauss-Varban and Omidi, 1991]. Similar to the upstream fast mode waves, these waves have an upstream directed wavevector, but are convected downstream. When the frequency is conserved in the transition to downstream, the fast mode (propagating in the same direction and with the same helicity) is no longer accessible. A sketch of this process is presented in Figure 4. While the slow magnetosonic mode dispersion may be quite similar to that of the Alfvén mode downstream, the slow magnetosonic mode is heavily damped under normal conditions and thus should play no role in the downstream turbulence. (A detailed discussion of the kinetic mode properties and nomenclature is deferred to section 3.)

A question that arises is whether this picture, that was derived for medium Mach numbers, also holds at higher M_A, which are more common at the Earth's (and other planetary) bow shocks. In that case, the upstream waves steepen to highly non-linear structures (compare Figure 3). This fact as well as other processes involved in shock re-formation could affect the formation of downstream waves. Evidently, large-scale shock-generated upstream pressure variations may also find a more direct way of entering the magnetosphere [Fairfield et al., 1990].

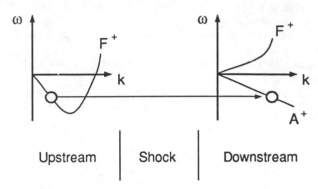

Conversion from Fast Magnetosonic to Alfvén Mode

Fig. 4. Sketch of the conversion of upstream fast magnetosonic waves to downstream Alfvén waves at passage through the shock. Downstream, the fast magnetosonic mode is no longer accessible when the frequency is conserved in the shock frame.

Figure 5 shows the power spectra of the upstream (compressional field component, B_z) and downstream waves (Alfvénic component, B_y) for a $M_A = 4.5$ shock, $\theta_{Bn} = 30°$. The upstream Fourier transform was executed over an area of 220 $\Omega_{ci}^{-1} \times$ 220 c/ω_{pi}, at a distance $\sim$ 50 c/ω_{pi} upstream from the shock. The downstream evaluation area was 150 $\Omega_{ci}^{-1} \times$ 120 c/ω_{pi}, and $\sim$ 30 c/ω_{pi} downstream from the shock. As at low and medium Mach numbers [Krauss-Varban and Omidi, 1991], the wave power fits the dispersion curves calculated from linear kinetic theory (not shown). The upstream power spectrum corresponds to the fast magnetosonic mode (propagating upstream, but convected downstream; F^+). The downstream spectrum shows both propagation branches of the Alfvén mode. However, the vast majority ($\gtrsim$ 90%) of wave energy is in the branch that is propagating upstream, but convected downstream (A^+).

Thus, the results confirm that the mode conversion process is still operative at high M_A. While this conversion from upstream waves leads to wave activity in the Pc 3–4 frequency range in the magnetosheath, it is currently not understood how these (in the sheath) Alfvénic perturbations could propagate through the magnetopause.

3. TWO-FLUID THEORY VERSUS KINETIC THEORY

The β of the magnetosheath plasma is usually of the order of unity or larger. Under such conditions, the relatively well-known properties of low-frequency modes as calculated from 2-fluid theory [Stringer, 1963; Formisano and Kennel, 1969] are no longer correct. Clearly, a proper characterization of the mode structure and properties is extremely important in understanding the upstream and magnetosheath wave processes. A key problem is the fact that the nomenclature in 2-fluid theory is based on the phase velocity of the waves (slow mode: S, intermediate mode: I, fast mode: F). In the correct theory of plasma waves in a collisionless plasma (lin-

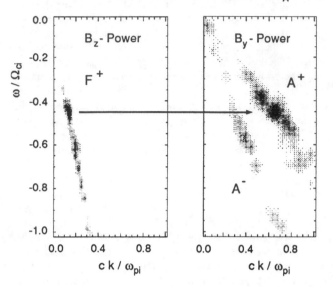

Upstream to Downstream Mode Conversion (M_A = 4.5)

Fig. 5. Upstream compressional (B_z) and downstream Alfvénic (B_y) power spectra from one-dimensional hybrid simulations at $M_A = 4.5$, in the shock frame. Upstream spectrum corresponds to the fast magnetosonic mode (propagating upstream, but convected downstream; F^+), downstream spectrum shows both propagation directions of the Alfvén mode (A^+, A^-). The gray scale is linear and slightly enhanced for the downstream plot to make the weaker, downstream propagating branch of the Alfvén mode (A^-) visible. The spectra show that even at high M_A the upstream F^+ wave energy is mode converted into downstream A^+ waves.

ear kinetic, or Vlasov, theory) modes cross each other in parameter space, and a unique and useful nomenclature should be based on the physical properties of the modes, instead. These properties may or may not correspond to expectations from fluid theory. For reasons that will become clear shortly we call the three propagating low-frequency waves the Fast/Magnetosonic (F/MS), Slow/Sound (S/SO), and Alfvén/Ion cyclotron (A/IC) modes.

3.1. Downstream Waves

In our one-dimensional simulations [Krauss-Varban and Omidi, 1991], we found that the helicity, compressibility, and magnetic field–density correlation of the S/SO and A/IC mode may be quite similar for typical downstream parameters. Other quantities remain more useful in distinguishing between the modes. For example, the S/SO mode remains magnetosonic in the sense that its magnetic field perturbation is predominantly in the $\mathbf{B}_o$-k plane (compressional component). Conversely, the A/IC mode's magnetic field perturbation points mostly out of this plane (Alfvénic component). This makes it easy to verify that the downstream waves are indeed A/IC waves in the simulations. This mode property can also be used in observations. For example, the magnetic compression ratio C_{BB} [Krauss-Varban et al., 1993], which is based on the direction of the magnetic field

perturbation, has been recently successfully used to identify upstream waves at Venus [Orlowski et al., 1993].

3.2. Mode Properties and Nomenclature

At sufficiently long wavelengths, kinetic theory gives a separate Alfvén/Ion-Cyclotron (A/IC) wave decoupled from the other solutions [Barnes, 1979], with phase speed $\omega/k \approx v_A \cos\theta$ for $\omega \ll \Omega_{ci}$. Let me present one example of the differences to 2-fluid theory that simultaneously illustrates a typical problem encountered in mode identification. A comprehensive review and comparison of the mode properties has recently been given by Krauss-Varban et al. [1993].

A mode property that would in principle be of interest to distinguish between modes encountered in the magnetosheath is the phase between the (parallel) magnetic field and density perturbations, $C_\parallel$. Generally, $C_\parallel$ for a species j may be defined as

$$C_\parallel{}^{(j)} = \frac{n_j B_\parallel{}^* / (n_{jo} B_o)}{B_\parallel B_\parallel{}^* / B_o{}^2}.$$

Here the subscript o refers to the undisturbed ambient value, and the asterisk designates the complex conjugate. In linear Vlasov theory, the perturbation quantities are calculated from the dielectric tensor elements. In 2-fluid theory (neglecting electron inertia), the parallel compressibility can be expressed as [Lacombe et al., 1990; Krauss-Varban et al., 1993]

$$C_\parallel = \frac{v_A{}^2}{v_{ph}{}^2 - c_s{}^2}.$$

Thus, the magnetic field and density are either exactly correlated or anticorrelated, and $C_\parallel$ changes sign depending on the ratio between v_{ph} and c_s, the latter being the sound speed. The intermediate mode correlation changes sign at the same time when its helicity (and other polarizations) change sign, at a critical angle $\cos\theta_c = \sqrt{\frac{1}{2}\gamma\beta}$ (in the case of equal specific heat ratio γ and temperature for the ions and electrons). Figure 6 shows that the behavior is different for the A/IC mode. A value of $-90° < \text{Phase}(C_\parallel) < 90°$ implies correlation, with the strongest correlation at $0°$, while a value of $\text{Phase}(C_\parallel)$ outside this range corresponds to anticorrelation, strongest close to $\pm 180°$. The figure was calculated for a wavenumber $ck/\omega_{pi} = 0.1$, but remains qualitatively the same for a wide range of wavenumbers. The A/IC mode is only weakly correlated in the high β–small θ quadrant. Also at large angles and β, the phase is not $180°$, but approximately $140°$. In other words, while the phase of the correlation is well defined in 2-fluid theory, n and $B_\parallel$ are essentially uncorrelated for the high β A/IC mode, making this quantity quite useless with regard to mode identification. Close to parallel propagation, this holds also for the F/MS mode.

How about the dispersion of the other two modes? It turns out that for a given wavenumber, the magnetosonic mode is a double-valued solution with a singular point in θ, β parameter space [Krauss-Varban et al., 1993]. At low β,

2-Fluid Theory: Intermediate Mode

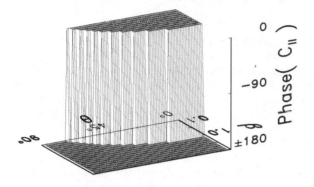

Vlasov-Theory: Alfvén/Ion-Cyclotron Mode

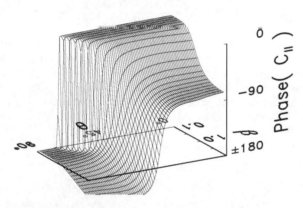

Fig. 6. Phase of the parallel compressibility $C_\parallel$ for the intermediate mode of 2-fluid theory, and for the A/IC mode of linear Vlasov theory, as a function of propagation angle $0° < \theta < 90°$ and beta $0.1 < \beta < 10.0$ (logarithmic scale). The view is from the large θ, β direction.

one of the solutions corresponds to the S mode, the other one to the F mode of fluid theory. However, when β is not restricted, kinetic theory requires selection of a branch cut to separate the modes. We have found that a branch cut starting at the singular point $\theta \sim 30°$, $\beta \sim 3$ and leading to larger β gives a practical and consistent separation of this double-valued magnetosonic solution [Krauss-Varban et al., 1993]. Selection of this branch cut results in a moderately damped fast/magnetosonic (F/MS) and a heavily damped slow/sound (S/SO) wave.

Figure 7a shows the phase velocity for the magnetosonic solution(s) of kinetic theory as a function of the angle of propagation θ and the plasma β, again at $ck/\omega_{pi} = 0.1$. The results show that the magnetosonic mode is in fact a double-valued solution that consists of two interconnected, semi-infinite planes. In this situation, selection of an (at first arbitrary) branch cut is required to obtain two separate solutions. For physical reasons, we have chosen the branch cut as evident in the figure. At the branch cut the two results

fit together, creating an "infinite staircase" structure with periodicity 4π. When β is increased at small angles, the "slow" magnetosonic mode increases its phase velocity and goes smoothly into the sound wave (hence: S/SO). What was the "fast" magnetosonic mode remains a right-handed electromagnetic mode at about constant v_{ph}. At large angles, the "slow" magnetosonic mode phase velocity increases with β above that of the A/IC mode but remains below the F/MS solution. Returning to smaller angles at high β, the F/MS solution connects across the branch cut with the S/SO wave, whereas the S/SO solution merges with the right-hand, nearly-circular polarized electromagnetic part of the F/MS solution.

Figure 7b shows that the damping decrement γ/ω_r of the modes is consistent with our selection of the branch cut. We call the only physical (i.e., not heavily damped) magnetosonic solution the F/MS mode, and the heavily damped "slow" and sound wave the S/SO (quasi-) mode. The F/MS mode can be consistently viewed as one entity also based on other important mode properties [Krauss-Varban et al., 1993]. In the context of upstream wave generation and steepening it is of special interest that this is the mode that

Magnetosonic Mode

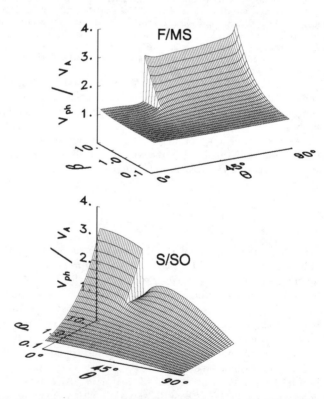

Fig. 7a. Double-valued solution of the magnetosonic mode phase velocity in similar format as Figure 6. The mode structure requires selection of a branch cut to separate the solutions. The cut is chosen for physical reasons (see text) as displayed here and separates the F/MS mode (top) from the S/SO mode (bottom).

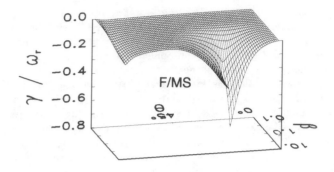

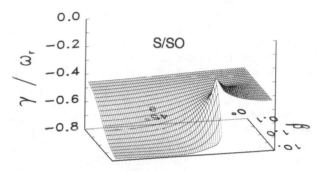

Fig. 7*b*. Damping decrement γ/ω_r for the double-valued magnetosonic solution in similar format as phase velocity in Figure 7*a*. Selection of the branch cut separates the modestly damped F/MS mode from the heavily damped S/SO mode.

continues into the whistler wave at large wavenumbers, and it is the magnetosonic solution that is unstable to ion beams. I would like to emphasize that this nomenclature is different from what is traditionally employed: previously, with the exception of the A/IC mode [Gary, 1986] the kinetic modes have been named according to their phase velocities regardless of their properties [Barnes. 1966].

4. TWO-DIMENSIONAL SIMULATIONS

One-dimensional hybrid simulations leave a variety of questions open, because the waves are restricted to propagate along the shock normal (simulation direction). However, recent two-dimensional simulations of quasi-parallel shocks have confirmed the essence of one-dimensional results [Krauss-Varban and Omidi, 1993; Scholer and Fujimoto, 1993]. A clear difference is that, as may be expected from linear theory, the upstream F/MS waves that are generated by field-aligned backstreaming ions are approximately propagating along the magnetic field direction, and not along the shock normal. This is in agreement with observations at medium Mach number interplanetary shocks [Russell et al., 1983b].

Figure 8 shows the compressional magnetic field component of the shock transition region for a $M_A = 3.0$, nearly parallel shock ($\theta_{Bn} = 10°$) at the end of the simulation. In the two-dimensional simulations presented here, the shock surface defines the y-z plane, whereas the upstream mag-

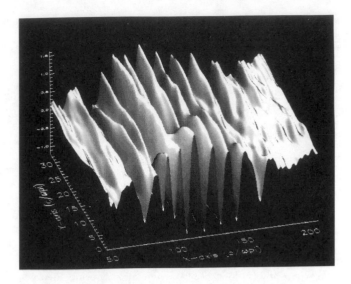

Fig. 8. Surface plot of the compressional magnetic field component B_y in the shock transition region of a two-dimensional simulation ($\theta_{Bn} = 10°$, $M_A = 3.0$). The region shown is $34 \times 160\ c/\omega_{pi}$ wide. Normalized with the upstream field, the ordinate is $-1.2 \leq B_y/B_0 \leq 1.6$. Note that the subscripts y, z for the compressional and Alfvénic components are reversed in the two-dimensional set-up.

netic field is contained in the x-y plane. Thus, different from the one-dimensional simulations, the dominantly compressional magnetic field component is B_y, the Alfvénic component is B_z. The shock normal and upstream flow direction are along the x coordinate. The simulation box in x-y is $300 \times 34\ (c/\omega_{pi})^2$. There are 2 cells per ion inertial length along x, and one cell per c/ω_{pi} along y. The simulation is followed for for 7500 time steps corresponding to $150\ \Omega_{ci}^{-1}$. There are approximately $1.5 \cdot 10^6$ macro particles in the box. More details of the set-up and the results are given in Krauss-Varban and Omidi [1993].

Figure 8 demonstrates that the shock transition is still dominated by large-amplitude F/MS waves. It can be seen that the downstream compressional fluctuations (B_y is shown) are fairly small. In fact, the Alfvénic power is about four times as large downstream. Figure 9*a* shows the Alfvénic downstream spectrum (B_z component) in the shock frame. Here, the units are the local downstream cyclotron and plasma frequency, and the solutions from linear Vlasov theory are also shown (A^+: upstream propagating A/IC mode, A^-: downstream propagating A/IC mode). The pairs of theoretical solutions correspond to waves propagating parallel to **B** and along the normal **n**, respectively. While the wave power is clearly on the A/IC branch that has upstream directed **k** (but is convected downstream), the propagation angle cannot be easily distinguished in this representation. Figure 9*b* shows the power as a function of k_x and k_y. Evidently, most of the wave power is in the normal direction, thus confirming the one-dimensional results. In addition, some waves are propagating field aligned and also perpendicular to the field. Similar results have been

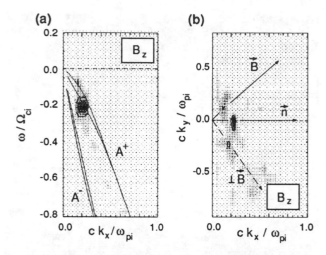

Fig. 9. Alfvénic (B_z) downstream fluctuation spectrum in (a) ω, k_x-space and (b) k-space of the simulation shown in Figure 8, in the shock frame. Here, c/ω_{pi} and Ω_{ci}^{-1} are local downstream units. Solid lines: solutions from linear Vlasov theory (A^+: upstream propagating A/IC mode, A^-: downstream propagating A/IC mode). The pairs of lines correspond to waves propagating parallel to **B** and along the normal **n**, respectively. The majority of wave power is on the A^+ mode, with propagation angle discernible in (b). Arrows in (b) indicate the direction of **B**, **n**, and the direction perpendicular to **B**. The majority of wave power is propagating along **n**, confirming the one-dimensional simulation results.

obtained for a $\theta_{Bn} = 30°$ shock [Krauss-Varban and Omidi, 1993].

5. INTERACTION OF WAVES WITH THE MAGNETOPAUSE

From a kinetic viewpoint, the interaction of low-frequency magnetosheath waves with the magnetopause has not been studied much. Some of the answers of how the wave energy enters the magnetosphere may lie in the gradients and curvature of the flow around the magnetopause, which certainly requires at least two-dimensional simulations for a proper treatment. Another problem of one-dimensional simulations is again the fact that it can only address wave propagation normal to any simulated discontinuity. For example, if the closed magnetopause is modelled as a tangential discontinuity, only perpendicular propagating waves can be studied. An example of such a simulation is the work by Mandt and Lee [1991], who studied the penetration of a fast mode shock through a tangential discontinuity. The anisotropy $T_\perp > T_\parallel$ behind the shock generates A/IC waves that may be the cause of magnetospheric fluctuations in the Pc 1 frequency range.

Omidi et al. [1993] have recently investigated the interaction between solar wind-type plasma and magnetospheric plasma in one-dimensional hybrid simulations. Such calculations lead to a self-consistent formation of a shock, magnetosheath, and magnetopause boundary. In addition to the usual mirror waves downstream of the shock, they ob-

tained interesting, nearly phase-standing waves at the magnetopause that showed many similarities to observations by Song et al. [1990]. While the magnetopause in the simulation had a finite normal component and thus was open, the geometry was nearly perpendicular.

To understand the kinetic structure of an open magnetopause, several hybrid simulation studies of rotational discontinuities (RD's) have been carried out over the past years [Swift and Lee, 1983; Lee et al., 1989; Richter and Scholer, 1991; Goodrich and Cargill, 1991; Omidi, 1991; Krauss-Varban, 1993a]. Such set-ups can also be used to investigate the interaction of obliquely propagating waves with the (open) magnetopause. Kinetic effects may be anticipated to be important, since the magnetopause transition is observed to take place over ion kinetic length scales. Here, I give three examples that illustrate such kinetic effects. The simulations were carried out in one dimension with 180,000 macroparticles, a time step of 0.04 Ω_{ci}^{-1}, cell size of 0.5 c/ω_{pi}, for $\beta_e = 0.25$ and $\beta_i = 1.0$. The initial width of the RDs is 2 c/ω_{pi}, and θ_{Bn} is 60°.

Figure 10a shows stack plots of the Alfvénic (B_y) and compressional (B_z) magnetic field components over the whole simulation box, as a function of time. The field rotation is $\alpha = 180°$ and in the ion sense (left-handed hodogram, see e.g., Krauss-Varban [1993a]). The simulation frame is propagating with $v_A \cos\theta_{Bn}$ upstream and is approximately the frame of the RD. An Alfvén wave pulse with wavelength of 30 c/ω_{pi} (~ 1000 to 1500 km) and $\Delta B_y/B \sim 20\%$ has been initialized at the upstream (left-hand) side. Over time, it separates into an upstream propagating A/IC wave (A^+, standing in the flow) and a downstream propagating part (A^-). The latter penetrates the RD without affecting its width or structure. As can be seen from Figure 10a, part of the wave is converted into upstream and downstream propagating F/MS waves (F^+, F^-). Other small-amplitude perturbations that are visible are caused by a non-perfect initialization. Note that a simple MHD analysis predicts that there should be no conversion to fast mode waves, and no reflected waves in this coplanar geometry [Lee, 1982].

Figure 10b shows the compressional field component for an electron-sense (right-handed) rotation of $\alpha = -180°$, with otherwise same parameters, and in the same format. In addition to slightly larger perturbations from the initialization in this case, the main difference is the lack of creation of F/MS waves at passage of the A/IC wave through the RD. (The two weak F/MS signatures that may be distinguished downstream originate from the RD initialization and from a F/MS pulse at the beginning of the simulation. Note that no F/MS waves are emanating from the disturbed region of the RD, closer to the center of the simulation, where Figure 10a shows the F/MS signature.)

In other words, the handedness of the RD has an important consequence concerning the generation of downstream compressional waves during transmission of Alfvénic waves. An even more dramatic kinetic effect is presented in Figure 11, which shows an ion-sense rotation with the same parameters as before, except that the field rotation is $\alpha = 270°$.

Ion Sense RD, α = 180°

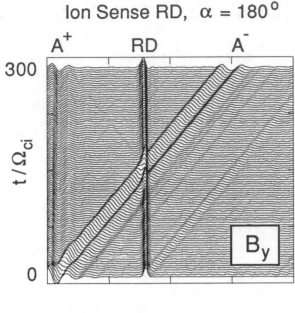

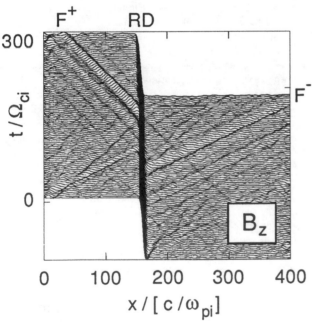

fact that there is little observational evidence of such RDs [Berchem and Russell, 1982; Neugebauer and Buti, 1990]. Ion-sense rotations $\alpha > 180°$ show internal ion-heating in the simulations, which eventually leads to a delayed, explosive disruption of the current layer. For the parameters shown in Figure 11, the RD should be metastable for times $> 300\ \Omega_{ci}^{-1}$ [Krauss-Varban, 1993a]. Here, however, the simulation shows that the incoming A/IC wave disrupts the RD, leading to an ensemble of mostly downstream propagating A/IC and F/MS waves, in addition to those generated at the original transmission point. A relatively quick and orderly transformation into a thin, right-handed RD with $\alpha = -90°$ takes place. This transformation is quite different from the violent disruptions seen without incoming waves, in which part of the plasma is strongly heated and a wide RD results. In Figure 11, there is also a convected, diamagnetic structure (C), whose properties are not fully understood at this point. The important point here is that Alfvénic turbulence in the magnetosheath (or in the solar wind, in case of RDs there) may be quite effective in restricting ion-sense rotations to angles $\alpha \leq 180°$. Electron-sense rotations have been shown to follow a different route to rotation angles $|\alpha| \leq 180°$ [Krauss-Varban, 1993a]. The transmitting A/IC waves do not influence that evolution (not shown).

To conclude, examples have been shown which demonstrate important kinetic effects in the transmission of waves through the current layer that makes up part of the magnetopause transition.

Electron Sense RD, α = -180°

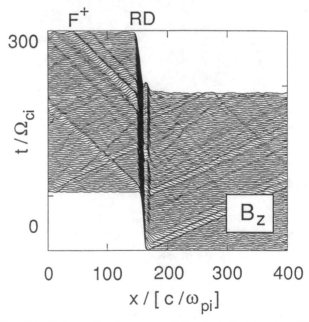

Fig. 10a. Stackplots of a one-dimensional simulation of an ion-sense rotational discontinuity (RD) with rotation angle $\alpha = 180°$ and $\theta_{Bn} = 60°$. Alfvénic component B_y and compressional component B_z. An A/IC wave pulse with $\lambda \sim 30\ c/\omega_{pi}$ is initialized at the left-hand side and separates into an upstream propagating wave (A^+, standing in the flow) and a downstream propagating wave (A^-). The compressional component B_z shows a reflected (F^+) and converted F/MS wave (F^-).

Note that in this case the respective roles of the Alfvénic and compressional components are reversed, downstream. RDs with rotation angles larger than 180° have previously been shown to be unstable [Swift and Lee, 1983; Richter and Scholer, 1991; Krauss-Varban, 1993a], which may explain the

Fig. 10b. Compressional component B_z of an electron-sense RD ($\alpha = 180°$) in the same format as Figure 10a. Disregarding some compressional noise, this sense of rotation does not show the downstream converted F/MS wave (F^-) the ion-sense RD has.

Ion Sense RD, α = 270°

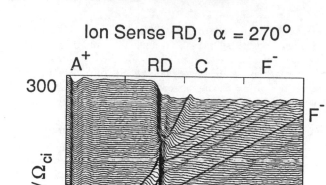

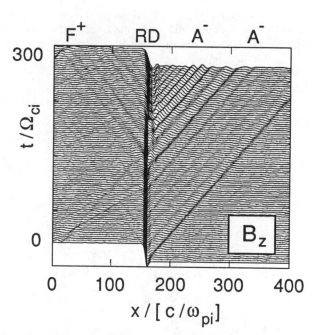

Fig. 11. Stackplots of an ion-sense RD with rotation angle α = 270° in the same format as Figure 10a. The A/IC wave triggers transformation into an α = −90° electron-sense RD. Note that the role of the compressional and Alfvénic components are reversed, downstream. In addition to transmitted and converted A^- and F^- waves, there are additional downstream-propagating perturbations due to the RD break-up.

6. SUMMARY

Perturbations originating upstream of the quasi-parallel bow shock have long been held responsible for magnetospheric pulsations. In this paper it was emphasized that the processes that are involved in upstream wave generation, wave transmission and conversion, and their effect on the magnetopause, require a kinetic treatment. Several distinct topics were touched which are, however, all connected by the common theme regarding their kinetic nature and their importance in the transfer of perturbations from the upstream region into the magnetosphere.

Reviewing and discussing recent results (which are partially published elsewhere in more detail) of hybrid simulations and linear kinetic theory, the following points were made:

1. The temporal and spatial scales do not favour quasi-parallel shock re-formation *per se* as a means of generating magnetosheath fluctuations of sufficient amplitude and coherence in the Pc 3–4 frequency range. Nevertheless, more work involving high M_A, nearly parallel shock simulations is required to understand the respective roles of upstream and local wave generation, shock re-formation, and the partition between compressional and Alfvénic perturbations.

2. For $\theta_{Bn} \sim 30°$ and medium Mach numbers ($M_A \sim$ 2.3 to $\sim$ 3), one-dimensional hybrid simulations had self-consistently shown that upstream generated F/MS waves are converted into downstream A/IC waves, at frequencies in the Pc 3–4 band. This process is also viable at higher Mach numbers ($M_A = 4.5$), at smaller angles ($\theta_{Bn} \sim 10°$), and in two-dimensional simulations.

3. The mode structure and properties of the relevant low-frequency waves in a warm, collisionless plasma were reviewed. It was emphasized that a proper characterization of the upstream and magnetosheath modes is extremely important regarding their generation, identification, and effects.

4. Three examples demonstrated important kinetic aspects of the interaction of low-frequency waves with the open magnetopause. The interaction of downstream propagating A/IC waves with an RD leads to reflected and converted F/MS waves that are not expected from MHD theory. In fact, the wave conversion depends on the handedness of the RD. Also, the wave interaction can trigger the break-up of RDs with rotation angle larger than 180°.

To conclude, kinetic aspects are very important in the processes enabling the transfer of upstream generated perturbations to the magnetosphere. It is hoped that the review and discussion will stimulate further work on this topic in the future.

Acknowledgments. Much of the work presented in this paper derives from collaboration with N. Omidi and K. B. Quest, whom the author would like to thank. Useful conversations with H. Karimabadi, D. Orlowski, C. T. Russell, and M. Engebretson are also acknowledged. This work was supported by the Space Physics Theory Program of the National Aeronautics and Space Administration, research grant NAG 5–1492. The computational facilities were provided by the NSF San Diego Supercomputer Center.

REFERENCES

Barnes, A., Collisionless damping of hydromagnetic waves, *Phys. Fluids, 9,* 1483–1495, 1966.

Barnes, A., Theory of generation of bow-shock-associated hydromagnetic waves in the upstream interplanetary medium, *Cosmic Electrodynamics, 1,* 90–114, 1970.

Barnes, A., Hydromagnetic waves and turbulence in the solar wind, in *Solar System Plasma Physics*, vol. 1, edited by E. N. Parker, C. F. Kennel, and L. J. Lanzerotti, pp. 249–319, North-Holland, New York, 1979.

Berchem, J., and C. T. Russell, Magnetic field rotation through the magnetopause: ISEE 1 and 2 observations, *J. Geophys. Res.*, *87*, 8139–8148, 1982.

Bol'shakova, V. O., and V. A. Troitskaya, Relation of the interplanetary magnetic field direction to the system of stable oscillations (in Russian), *Dokl. Akad. Nauk. SSSR*, *180*, 4–6, 1968.

Burgess, D., Cyclical behavior at quasi-parallel collisionless shocks, *Geophys. Res. Lett.*, *16*, 345–349, 1989.

Engebretson, M. J., N. Lin, W. Baumjohann, H. Lühr, B. J. Anderson, L. J. Zanetti, T. A. Potemra, R. L. McPherron, and M. G. Kivelson, A comparison of ULF fluctuations in the solar wind, magnetosheath, and dayside magnetosphere, 1., Magnetosheath morphology, *J. Geophys. Res.*, *96*, 3441–3454, 1991.

Fairfield, D. H., Magnetic fields in the magnetosheath, *Rev. Geophys. Space Phys.*, *14*, 117–134, 1976.

Fairfield, D. H., W. Baumjohann, G. Paschmann, H. Lühr, and D. G. Sibeck, Upstream pressure variations associated with the bow shock and their effects on the magnetosphere, *J. Geophys. Res.*, *95*, 3773–3786, 1990.

Fejer, J. A., Hydromagnetic reflection and refraction at a fluid velocity discontinuity, *Phys. Fluids*, *6*, 508–512, 1963.

Formisano, V., and C. F. Kennel, Small amplitude waves in high β plasma, *J. Plasma Phys.*, *3*, 55–74, 1969.

Gary, S. P., Low-frequency waves in a high-beta collisionless plasma: Polarization, compressibility and helicity, *J. Plasma Phys.*, *35*, 431–447, 1986.

Gary, S. P., Electromagnetic ion/ion instabilities and their consequences in space plasmas: A review, *Space Sci. Rev.*, *56*, 373–415, 1991.

Goodrich, C. C., Numerical simulation of quasi-perpendicular collisionless shocks, in *Collisionless Shocks in the Heliosphere: Reviews of Current Research*, Geophys. Monogr. Ser., vol. 35, edited by B. T. Tsurutani and R. G. Stone, pp. 153–168, AGU, Washington, D. C., 1985.

Goodrich, C. C., and P. J. Cargill, An investigation of the structure of rotational discontinuities, *Geophys. Res. Lett.*, *18*, 65–68, 1991.

Gosling, J. T., M. F. Thomsen, S. J. Bame, and C. T. Russell, Ion reflection and downstream thermalization at the quasi-parallel bow shock, *J. Geophys. Res.*, *94*, 10,027–10,037, 1989.

Greenstadt, E. W., Oblique, parallel, and quasi-parallel morphology of collisionless shocks, in *Collisionless Shocks in the Heliosphere: Reviews of Current Research*, Geophys. Monogr. Ser., vol. 35, edited by B. T. Tsurutani and R. G. Stone, pp. 169–184, AGU, Washington, D. C., 1985.

Greenstadt, E. W., and J. V. Olson, Pc 3,4 activity and interplanetary field orientation, *J. Geophys. Res.*, *81*, 5911–5920, 1976.

Greenstadt, E. W., and J. V. Olson, A contribution to ULF activity in the Pc 3–4 range correlated with IMF radial orientation, *J. Geophys. Res.*, *82*, 4991–4496, 1977.

Greenstadt, E. W., M. M. Mellott, R. L. McPherron, C. T. Russell, H. J. Singer, and D. J. Knecht, Transfer of pulsation-related wave activity across the magnetopause: Observations of corresponding spectra by ISEE-1 and ISEE-2, *Geophys. Res. Lett.*, *10*, 659–662, 1983.

Greenstadt, E. W., F. V. Coroniti, S. L. Moses, B. T. Tsurutani, N. Omidi, K. B. Quest, and D. Krauss-Varban, Weak, quasiparallel profiles of Earth's bow shock: A comparison between numerical simulations and ISEE 3 observations on the far flank, *Geophys. Res. Lett.*, *18*, 2301–2304, 1991.

Hubert, D., C. C. Harvey, and C. T. Russell, Observations of magnetohybrodynamic modes in the Earth's magnetosheath at 0600 LT, *J. Geophys. Res.*, *94*, 17,305–17,309, 1989.

Kan, J. R., and D. W. Swift, Structure of the quasi-parallel bow shock: Results of numerical simulations, *J. Geophys. Res.*, *88*, 6919–6925, 1983.

Kennel, C. F., J. P. Edmiston, and T. Hada, A quarter century of collisionless shock research, in *Collisionless Shocks in the Heliosphere: A Tutorial Review*, Geophys. Monogr. Ser., vol. 34, edited by R. G. Stone and B. T. Tsurutani, pp. 1–36, AGU, Washington, D. C., 1985.

Krauss-Varban, D., Structure and length scales of rotational discontinuities, *J. Geophys. Res.*, *98*, 3907–3917, 1993a.

Krauss-Varban, D., Unsteady and reforming quasi-parallel collisionless shocks, in *Research Trends in Nonlinear Space Plasma Physics*, edited by R. Z. Sagdeev, pp. 289–308, AIP, New York, 1993b.

Krauss-Varban, D., and N. Omidi, Structure of medium Mach number quasi-parallel shocks: Upstream and downstream waves, *J. Geophys. Res.*, *96*, 17,715–17,731, 1991.

Krauss-Varban, D., and N. Omidi, Two-dimensional simulations of quasi-parallel shocks, *Geophys. Res. Lett.*, *20*, 1007–1010, 1993.

Krauss-Varban, D., N. Omidi, and K. B. Quest, Mode properties of low frequency waves: Kinetic theory versus Hall-MHD, *J. Geophys. Res.*, *98*, in press, 1993.

Kucharek, H., and M. Scholer, Origin of diffuse superthermal ions at quasi-parallel supercritical collisionless shocks, *J. Geophys. Res.*, *96*, 21,195–21,205, 1991.

Kwok, Y. C., and L. C. Lee, Transmission of magnetohydrodynamic waves through the rotational discontinuity at the Earth's magnetopause, *J. Geophys. Res.*, *89*, 10,697–10,708, 1984.

Lacombe, C., E. Kinzelin, C. C. Harvey, D. Hubert, A. Mangeney, J. Elaoufir, D. Burgess, and C. T. Russell, Nature of the turbulence observed by ISEE 1–2 during a quasi-perpendicular crossing of the Earth's bow shock, *Ann. Geophys.*, *8*, 489–502, 1990.

Lacombe, C., F. G. E. Pantellini, D. Hubert, C. C. Harvey, A. Mangeney, G. Belmont, and C. T. Russell, Mirror and Alfvénic waves observed by ISEE 1–2 during crossings of the Earth's bow shock, *Ann. Geophys.*, *10*, 1–13, 1992.

Le, G., and C. T. Russell, A study of the coherence length of ULF waves in the earth's foreshock, *J. Geophys. Res.*, *95*, 10,703–10,706, 1990.

Lee, L. C., Transmission of Alfvén waves through the rotational discontinuity at magnetopause, *Planet. Space Sci.*, *30*, 1127–1132, 1982.

Lee, L. C., L. Huang, and J. K. Chao, On the stability of rotational discontinuities and intermediate shocks, *J. Geophys. Res.*, *94*, 8813–8825, 1989.

Leroy, M. M., and D. Winske, Backstreaming ions from oblique Earth bow shocks, *Ann. Geophys.*, *1*, 527–536, 1983.

Leroy, M. M., D. Winske, C. C. Goodrich, C. S. Wu, and K. Papadopoulos, The structure of perpendicular bow shocks, *J. Geophys. Res.*, *87*, 5081–5094, 1982.

Lin, N., M. J. Engebretson, R. L. McPherron, M. G. Kivelson, W. Baumjohann, H. Lühr, T. A. Potemra, B. J. Anderson, and L. J. Zanetti, A comparison of ULF fluctuations in the solar wind, magnetosheath, and dayside magnetosphere, 2., Field and plasma conditions in the magnetosheath, *J. Geophys. Res.*, *96*, 3455–3464, 1991.

Luhmann, J. G., C. T. Russell, and R. C. Elphic, Spatial distributions of magnetic field fluctuations in the dayside magnetosheath, *J. Geophys. Res.*, *91*, 1711–1715, 1986.

Mandt, M. E., A review of recent developments in collisionless shock research, in *Physics of Space Plasmas*, edited by T. Chang, G. B. Crew, and J. R. Jaspers, pp. 297–309, Scientific Publishers, Cambridge, MA, 1992.

Mandt, M. E., and J. R. Kan, Effects of electron pressure in quasi-parallel collisionless shocks, *J. Geophys. Res., 90,* 115, 1985.

Mandt, M. E., and J. R. Kan, Ion equation of state in quasi-parallel shocks: A simulation result, *Geophys. Res. Lett., 15,* 1157–1160, 1988.

Mandt, M. E., and L. C. Lee, Generation of Pc 1 waves by the ion temperature anisotropy associated with fast shocks caused by sudden impulses, *J. Geophys. Res., 96,* 17897–17901, 1991.

McKean, M. E., D. Winske, M. F. Thomsen, and T. G. Onsager, Near-specular reflection of ions at quasi-parallel shocks, *J. Geophys. Res., 98,* 3859–3873, 1993.

McKenzie, J. F., Hydromagnetic wave interaction with the magnetopause and the bow shock, *Planet. Space Sci., 18,* 1–23, 1970.

McKenzie, J. F., and K. O. Westphal, Interaction of hydromagnetic waves with hydromagnetic shocks, *Phys. Fluids, 13,* 630–640, 1970.

Mellott, M. M., Subcritical collisionless shock waves, in *Collisionless Shocks in the Heliosphere: Reviews of Current Research, Geophys. Monogr. Ser.,* vol. 35, edited by B. T. Tsurutani and R. G. Stone, pp. 131–140, AGU, Washington, D. C., 1985.

Moustaizis, S., D. Hubert, A. Mangeney, C. C. Harvey, C. Perche, and C. T. Russell, Magnetohydrodynamic turbulence in the Earth magnetosheath, *Ann. Geophys., 4,* 355–362, 1986.

Neugebauer, M., and B. Buti, A search for evidence of the evolution of rotational discontinuities in the solar wind from nonlinear Alfvén waves, *J. Geophys. Res., 95,* 13–20, 1990.

Omidi, N., Rotational discontinuities in anisotropic plasmas, *Geophys. Res. Lett., 19,* 1335–1338, 1991.

Omidi, N., K. B. Quest, and D. Winske, Low Mach number parallel and quasi-parallel shocks, *J. Geophys. Res., 95,* 20,717–20,730, 1990.

Omidi, N., A. O'Farrell, and D. Krauss-Varban, Sources of magnetosheath waves and turbulence, *Adv. Space Res.,* in press, 1993.

Orlowski, D. S., C. T. Russell, D. Krauss-Varban, and N. Omidi, Critical test for Hall-MHD model: Application to low frequency upstream waves at Venus, *J. Geophys. Res., 98,* in press, 1993.

Paschmann, G., N. Sckopke, S. J. Bame, and J. T. Gosling, Observations of gyrating ions in the foot of the nearly perpendicular bow shock, *Geophys. Res. Lett., 9,* 881–884, 1982.

Quest, K. B., Simulation of quasi-parallel collisionless shocks, in *Collisionless Shocks in the Heliosphere: Reviews of Current Research, Geophys. Monogr. Ser.,* vol. 35, edited by B. T. Tsurutani and R. G. Stone, pp. 185–194, AGU, Washington, D. C., 1985.

Quest, K. B., Theory and simulation of collisionless parallel shocks, *J. Geophys. Res., 93,* 9649–9680, 1988.

Quest, K. B., D. W. Forslund, J. U. Brackbill, and K. Lee, Collisionless dissipation in quasi-parallel shocks, *Geophys. Res. Lett., 10,* 471–474, 1983.

Richter, P., and M. Scholer, On the structure of rotational discontinuities with large phase angles, *Adv. Space Res., 11,* (9)111–(9)115, 1991.

Russell, C. T., J. G. Juhmann, T. J. Odera, and W. F. Stuart, The rate of occurrence of dayside Pc 3,4 pulsations: The L-value dependence of the IMF cone angle effect, *Geophys. Res. Lett., 8,* 663–666, 1983a.

Russell, C. T., E. J. Smith, B. T. Tsurutani, J. T. Gosling, and S. J. Bame, Multiple spacecraft observations of interplanetary shocks: Characteristics of the upstream ULF turbulence, in *Solar Wind Five,* edited by M. Neugebauer, pp. NASA Conf. Publ., CP-2280, 385–400, 1983b.

Scholer, M., and M. Fujimoto, Low mach number quasi-parallel shocks: Upstream waves, *J. Geophys. Res., 98,* in press, 1993.

Scholer, M., and T. Terasawa, Ion reflection and dissipation at quasi-parallel collisionless shocks, *Geophys. Res. Lett., 17,* 119–122, 1990.

Sckopke, N., G. Paschmann, S. J. Bame, J. T. Gosling, and C. T. Russell, Evolution of ion distributions across the nearly perpendicular bow shock: Specularly and nonspecularly reflected gyrating ions, *J. Geophys. Res., 88,* 6121–6136, 1983.

Song, P., R. C. Elphic, C. T. Russell, J. T. Gosling, and C. A. Cattell, Structure and properties of the subsolar magnetopause for northward IMF: ISEE observation, *J. Geophys. Res., 95,* 6375–6387, 1990.

Stringer, T. E., Low-frequency waves in an unbounded plasma, *J. Nucl. Energy, Part C, 5,* 89–107, 1963.

Swift, D. W., and L. C. Lee, Rotational discontinuities and the structure of the magnetopause, *J. Geophys. Res., 88,* 111–124, 1983.

Takahashi, K., R. L. McPherron, and T. Terasawa, Dependence of the spectrum of Pc 3–4 pulsations on the interplanetary magnetic field, *J. Geophys. Res., 89,* 2770–2780, 1984.

Thomas, V. A., D. Winske, and N. Omidi, Re-forming supercritical quasi-parallel shocks, 1, One- and two-dimensional simulations, *J. Geophys. Res., 95,* 18,809–18,819, 1990.

Thomsen, M. F., J. T. Gosling, S. J. Bame, T. G. Onsager, and C. T. Russell, Two-state ion heating at quasi-parallel shocks, *J. Geophys. Res., 95,* 6363–6374, 1990a.

Thomsen, M. F., J. T. Gosling, S. J. Bame, and C. T. Russell, Magnetic pulsations at the quasi-parallel shock, *J. Geophys. Res., 95,* 957–966, 1990b.

Tidman, D. A., and T. G. Northrop, Emission of plasma waves by the Earth's bow shock, *J. Geophys. Res., 73,* 1543, 1968.

Troitskaya, V. A., T. A. Plyasova-Bakunina, and A. V. Gul'elmi, Relationship between Pc 2–4 pulsations and the interplanetary magnetic field (in Russian), *Dokl. Akad. Nauk. SSSR, 197,* 1312, 1971.

Verzariu, P., Reflection and refraction of hydromagnetic waves at the magnetopause, *Planet. Space Sci., 21,* 2213–2225, 1973.

Watanabe, Y., and T. Terasawa, On the excitation mechanism of the low-frequency upstream waves, *J. Geophys. Res., 89,* 6623–6630, 1984.

Winske, D., and N. Omidi, Hybrid codes: Methods and applications, in *Computer Space Plasma Physics: Simulation Techniques and Software,* edited by H. Matsumoto and Y. Omura, in press, Terra Scientific Publishing Company, Tokyo, Japan, 1993.

Winske, D., N. Omidi, K. B. Quest, and V. A. Thomas, Reforming supercritical quasi-parallel shocks, 2, Mechanism for wave generation and front re-formation, *J. Geophys. Res., 95,* 18,821–18,832, 1990.

Wolfe, A., Dependence of mid-latitude hydromagnetic energy spectra on solar wind speed and interplanetary magnetic field direction, *J. Geophys. Res., 85,* 5977–5982, 1980.

Wolfe, A., and R. L. Kaufmann, MHD wave transmission and production near the magnetopause, *J. Geophys. Res., 80,* 1764–1775, 1975.

Wolfe, A., A. Meloni, L. J. Lanzerotti, C. G. Maclennan, J. Bamber, and D. Venkatesan, Dependence of hydromagnetic energy spectra near L = 2 and L = 3 on upstream solar wind parameters, *J. Geophys. Res., 90,* 5117–5131, 1985.

Yumoto, K., T. Saito, S.-I. Akasofu, B. T. Tsurutani, and E. J. Smith, Propagation mechanism of daytime Pc 3–4 pulsations observed at synchronous orbit and multiple ground-based stations, *J. Geophys. Res., 90,* 6439–6450, 1985.

Dietmar Krauss-Varban, ECE Department, EBU1; Room 2803, University of California-San Diego, 9500 Gilman Drive, La Jolla, CA 92093-0407.

Pc3 Pulsations: From the Source in the Upstream Region to Alfven Resonances in the Magnetosphere. Theory and Observations

A. S. Potapov and V. A. Mazur

Institute of Solar-Terrestrial Physics, Irkutsk, Russia

A review is given of questions relating to the generation, propagation and structure of geomagnetic Pc3 pulsations. The pulsation spectrum and the dependence of the frequency and amplitude on the magnitude and direction of the IMF are formed ahead of the bow shock front in the process of excitation of upstream waves. The effectiveness of penetration of the waves into the magnetosphere depends on solar wind velocity. The interaction of the penetrating field with shear Alfven waves results in the formation of Pc3 resonance structure. Interaction features are described. Finally, ground-based observation techniques for Pc3 field structure are discussed.

Introduction

It should be recognized that the last decade has not brought such significant theoretical and experimental advances in understanding the nature of dayside low frequency geomagnetic pulsations as did the previous decade [see Southwood and Hughes, 1983]. The concept developed in 1974 by Chen and Hasegawa [1974] and Southwood [1974] was accepted by 1980 almost entirely by most investigators. In the past decade they concentrated mainly on comparisons of their satellite and ground observations with theoretical predictions and the development of various modifications of the basic theory. Nevertheless, a more complete understanding of the Pc3 origin, propagation and spatial structure was obtained. The experiments became greatly sophisticated and problem-oriented. As the theory is concerned, efforts were concentrated on investigating field line resonance in terms of more complicated and adequate magnetospheric models.

This review deals with a rather limited range of questions relating to theoretical and experimental considerations of the excitation, propagation and field structure of Pc3. Primary attention is paid to Russian authors, not well known among Western scientists. We start by recalling the picture of wave generation in front of the Earth's bow shock due to a reflected proton beam (so-called upstream waves in the foreshock region). Attention is focused on the formation mech-

anisms for dependences of pulsation characteristics on solar wind parameters and the IMF observed on the ground. The next two sections are devoted to the contribution of Russian authors to the modification and development of driven field line resonance theory. Finally, we review some new methods of observing the Pc3 field structure. In the Conclusions, some possible lines of future research are discussed.

Upstream Waves

The idea of the extramagnetospheric origin of geomagnetic Pc3 pulsations was first suggested in some papers by Guglielmi as early as the beginning of the 1970s [Guglielmi, 1972, 1974; Troitskaya et al., 1971]. This idea was based on experimental data on a close relationship between Pc3 characteristics and solar wind parameters. The following connections have been established: increase in amplitude with increasing solar wind velocity, increase in amplitude with decreasing angle between the IMF-vector and the Sun-Earth direction, and increase of the frequency with increasing scalar magnitude of the IMF. All these connections were first observed by Bol'shakova and Troitskaya [1968] and later confirmed by a large number of authors [Troitskaya et al., 1971; Guglielmi et al., 1973; Guglielmi and Bol'shakova, 1973; Guglielmi, 1974; Vinogradov and Parkhomov, 1975; Verö and Hollo, 1978; Potapov and Polyushkina, 1979; Potapov et al., 1979; Greenstadt et.al., 1980; Polyushkina and Potapov, 1983; Verö et. al., 1985].

In terms of Guglielmi's hypothesis, the observed connections were interpreted under the assumption that upstream waves excited ahead of the bow shock front due to a beam instability of reflected protons, subsequently penetrate the

Solar Wind Sources of Magnetospheric Ultra-Low-Frequency Waves
Geophysical Monograph 81

magnetosphere and are observed on the ground as geomagnetic Pc3 pulsations. Initially, the hypothesis was met by the scientific community with much scepticism [D'Angelo, 1975], then it was conceived with some doubt about it [Russell and Fleming, 1976]. A decade later it had a strong development, but the pioneer authors were forgotten (see a review by Yumoto [1986]).

Generation Mechanism

A proton flux reflected from a bow shock was detected for the first time by Asbridge et al. [1968]. What is important is that protons are a permanent property of the near-terrestrial plasma; they are observed almost in all cases when interplanetary magnetic field lines pierce through the bow shock front. Fairfield [1969], for the same region of space, reported the detection of transverse hydromagnetic waves that are quasi-circularly left-handedly polarized in the magnetospheric frame of reference and have periods in the 20–100 s range. He demonstrated that these waves are brought about by the interaction of reflected protons with a solar wind stream. The excitation mechanism for these waves was treated first by Barnes [1970]. Potapov [1974] analyzed this mechanism in more detail with applications to Pc3.

A coordinate system moving with the solar wind velocity was used in this study. The plasma ahead of the bow shock front was modeled in the form of a stationary background (electrons (e) and protons (p) of the solar wind flow) and a longitudinal beam along magnetic field lines (reflected protons). A bi-Maxwellian distribution of each of the plasma components was assumed. Two cases, field-aligned propagation and propagation at a small angle to an external magnetic field, were analysed. The dispersion equation in the case of a strictly field-aligned propagation has the form:

$$k_z{}^2 A^2 (1 - \beta_{\|p}\Delta_p - \beta_{\|e}\Delta_e) - \omega^2 - \alpha\Omega_p{}^2$$

$$\cdot \left\{ \left(\Delta_p' - \frac{\omega - k_z u}{\omega - k_z u + \Omega_p}\right) \left[Z\left(\frac{\omega - k_z u + \Omega_p}{k_z W_p'}\right) - 1 \right] \right.$$

$$\left. + \frac{\omega - k_z u}{\Omega_p(\omega - k_z u + \Omega_p)} \right\} = 0,$$

(1)

where k_z is wave vector component along the external magnetic field **B**, $A = B/(4\pi\rho)^{1/2}$ is the Alfven velocity, $\beta_\|$ is the ratio of longitudinal gas to magnetic pressure, $\Delta = 1 - W_\perp/W_\|$ is the temperature anisotropy, W is the thermal velocity, ω is the frequency, $\alpha = N'/N$ is the number density ratio of the reflected proton beam to solar wind protons, Ω is the gyrofrequency, u is the bulk velocity of reflected protons along the magnetic field, $Z(x) = xe^{-x^2}\left(2\int_0^x e^{x^2} dt - i\sqrt{\pi}\right)$, and primes refer to the beam parameters. From the structure of equation (1) it is evident that the wave-particle interaction is strongest when $\omega - k_z u + \Omega_p = 0$, which corresponds to cyclotron resonance of the beam's particles with the right-handedly polarized wave.

Analytic estimates show that, depending on the relationship between the parameters $\alpha = N'/N$ and $\mu = u/W_{\|p}'$ (the ratio of the beam's velocity to the thermal velocity of reflected protons), there are four main domains of solution of the dispersion equation, and two of them correspond to a hydrodynamic cyclotron instability (cold beam), and the other two correspond to a kinetic instability. The diagram in Figure 1 shows schematically these domains. The strongest is the instability in domain III where the growth rate varies as a one-third power of the relative beam density. However, under typical conditions of the foreshock the development of a kinetic instability (domains I and II) is more probable, when the growth rate is proportional to the first power or the square root of the number density of the beam.

It was possible to reveal some details of the excitation mechanism for upstream waves through numerical calculations in terms of two models: 'a weak beam' and 'a strong beam' [Guglielmi et al., 1976]. The combinations of parameters assumed in these models cover almost entirely the range of observed values of number density, thermal spread and other parameters of the beam. The main conclusions are the following. The growth rate has a rather sharp maximum when $k \sim \Omega_p/u$, and the position of the maximum depends little on N' and $W_{\|p}'$ over a wide range of variation of these parameters. On the other hand, the growth rate maximum displaces toward longer wavelengths with increasing anisotropy of reflected protons, with a simultaneous increase in growth rate. The presence of α-particles (i.e. He^{++}-ions) in the solar wind leads to the fact that an additional maximum when $k \leq 0.5\Omega_p/u$ starts to form on the frequency profile of growth rate. This is illustrated in Figure 2. Solid lines and the left-hand scale correspond to 'a strong beam' model, broken lines and the right-hand scale correspond to 'a weak beam' model. Besides, there is a decrease of the main maximum; however, its position remains

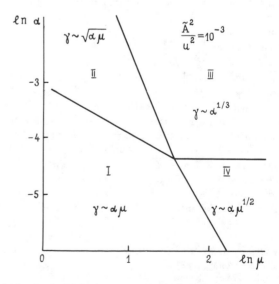

Fig. 1. Regions of solution of equation (1) (see text).

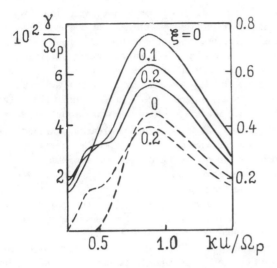

Fig. 2. Deformation of the growth rate frequency profile due to increase of α-particle number density $\xi = N_\alpha'/N$ in the beam.

the same. The influence of the α-particles can become appreciable only at large values of their number density, for example, after chromospheric flares when helium-enriched plasma clouds approach the Earth.

To apply the results obtained to the problem of extramagnetospheric generation of Pc3, it is necessary to take into account the Doppler shift of frequency when passing from the solar wind coordinate system to a system fixed with respect to the magnetosphere: $\omega' = \omega + \mathbf{k}V_{SW}$, $\mathbf{V}_{SW}$ is the solar wind velocity. Since $\omega \approx kA$, and $A \ll V_{SW}$ hence $|\mathbf{k}V_{SW}| \gg \omega$, then $\omega' \simeq \mathbf{k}V_{SW}$. In this case the waves now are left-handedly, rather than right-handedly polarized.

Potapov [1974] considered the formation mechanism for the Pc3 spectrum ahead of the bow shock. First of all it was shown that the growth rate $\gamma(k_z, \theta)$ as the function of wave vector longitudinal component k_z and angle θ between $\mathbf{k}$ and $\mathbf{B}$ has a maximum when $k_z = \Omega_p/u$, and $\theta = 0$. Then an analysis was made of the evolution of a wave packet with arbitrary spectral density at the initial moment of time. A feature inherent in this problem is the fact that the frequency spectrum width depends substantially on the angular spectrum of the waves in the upstream region. Actually, frequency bandwidth is about

$$\Delta\omega' \approx \mathbf{V}_{SW}\Delta\mathbf{k} = \Delta k_z V_{SW}\cos\theta_{XB} + \Delta k_\perp V_{SW}\sin\theta_{XB},$$

θ_{XB} is IMF cone angle (the angle between the IMF vector and the Sun-Earth line). It was shown that the frequency spectrum $\Delta\omega'$ becomes narrower both as a result of decreasing of $\Delta k_\parallel$ and due to a decrease of $\Delta k_\perp$, i.e. as a result of angular spectrum narrowing. Estimations show that as the packet drifts toward the bow shock, the frequency spectrum narrows as $\Delta\omega' \sim l^{1/2}$, where $l \sim V_{SW}t$ is the path length of enhancement along which the packet is transported over to the shock front by the solar wind.

The Formation of Relationships of Pc3 With Solar Wind and the IMF Parameters

In resonance ($k_z = \Omega_p/(V_{SW\parallel} + v_\parallel)$) the oscillation frequency is

$$\omega \approx \frac{V_{SW\parallel}}{V_{SW\parallel} + v_\parallel}\Omega_p, \tag{2}$$

where $V_{SW\parallel}$ and $v_\parallel$ are, respectively, the solar wind and beam velocities (in a coordinate system fixed to the Earth) projected onto the direction of the interplanetary magnetic field. For the typical parameters $\Omega_p \approx 0.5$ s^{-1}, $v \approx 1.6V_{SW}$ the oscillation period ($T \approx 30$ s) lies in the middle of the Pc3 range; in addition, the oscillation frequency is proportional to the modulus of the interplanetary magnetic field strength B: $f = gB$, where $g \approx 6$ if the frequency is measured in milli-Hertz and the field is measured in nano-Tesla [Guglielmi, 1974]. Such is the way in which the dependence of the Pc3 frequency on the IMF strength is formed.

A dependence of the pulsation amplitude on the value of IMF cone angle θ_{XB} arises as a result of a change in the path length of enhancement of the wave packets with a change of θ_{XB}. Indeed, in a coordinate system fixed to the Earth the path length of enhancement is $l \leq R_m \cot\theta_{XB}$, where R_m is the typical size of the magnetosphere. Therefore, the pulsation amplitude must increase with decreasing angle, until the length l exceeds a certain value of l^* at which effects of nonlinear limitation of the amplitude become substantial. In accordance with results of direct measurements [Bonifazi et al., 1983] l^* is about 10 R_E.

The theory of upstream waves does not yield any connections between the pulsation amplitudes and the velocity of the counterstreaming solar wind particles, but there are several mechanisms which could explain such a dependence. Firstly, hydromagnetic noise, which is the seed turbulence for the excitation of beam-cyclotron instability, has its highest amplitudes inside the high-velocity solar wind streams. Secondly, in the case of the high velocities of the stream flowing around the magnetosphere and in case of high Mach numbers, the bow shock front lies nearer to the magnetopause and thus the path from the shock-front to the magnetopause to be traversed by the waves in the strongly attenuating medium of the magnetosheath is shorter. Moreover, wave attenuation in the magnetosheath decreases with increasing solar wind velocity [Barkhatov, 1977]. Finally, higher solar wind velocities mean a more transparent magnetosheath and magnetopause.

Upstream Waves Before Other Shock Fronts

It is appropriate to examine in greater detail the dependence $f(B)$ which is crucial from the point of view of the spectral formation of the upstream waves. This dependence was analyzed by many research groups, and all of them obtained rather close values of g. If all values obtained from both ground-based Pc3 observations and satellite observations of the upstream waves in the Earth's foreshock are averaged, then we get the estimate of $g = 5.8 \pm 0.3$ [Guglielmi, 1984]. More interestingly, this same value to within errors of measurements was also found to be valid for upstream waves observed ahead of foreshocks of other planets (Mercury, Venus and Jupiter) [Hoppe and Russell, 1982], although the conditions for solar wind flow around them vary greatly from planet to planet. This indicates that the generation mechanism of the upstream waves by the cyclotron instability of reflected protons does not depend on the flow-around conditions and is a stable physical regularity typical of all bow shocks. Guglielmi and Ivanov [1984] suggested in this connection that ground-based observations should be used to seek wave precursors of flare streams, because waves could be generated in front of the interplanetary shocks associated with them [Kennel et al., 1982]. Waves in this case will propagate downstream the solar wind leading the flare stream. According to those authors' estimates, the oscillation period must be about 10 s, and the mean lead time is about 8 hours.

Transmission of the Upstream Waves From the Foreshock Into the Magnetosphere

In order to get into the magnetosphere, waves need to traverse three structural regions: bow shock, magnetosheath and magnetopause. The qualitative picture is the following. Waves incident on the shock front can increase considerably in amplitude by deriving energy from the supersonic solar wind stream. Six wave modes are produced in the magnetosheath in this case: two Alfven waves, three magnetosonic waves and one entropy wave. Only some of them can reach the magnetopause as they propagate in the highly turbulent medium of the magnetosheath. On the magnetopause there also occurs the transformation of one wave mode into several different modes, including surface waves, shear Alfven waves and slow magnetosonic waves. In this case, under typical conditions the wave energy is dissipated. However, the transmitted 1–2% of the energy does suffice to make a significant contribution to the overall energy balance of the magnetosphere.

These issues were all addressed in some detail in a review by Yumoto [1986]. We may add here only two publications [Barkhatov, 1977; Barkhatov et al., 1977] which consider the propagation of the waves through the bow shock and the magnetosheath. The best conditions for the propagation correspond to right-handedly polarized waves with wave vectors directed along or opposite to the stream. The transparency of the magnetosheath increases with increasing solar wind velocity.

FIELD STRUCTURE IN THE MAGNETOSPHERE

The Penetration of a Disturbance Deep Into the Magnetosphere

In the frequency range $\omega < \Omega_p$ of our interest, the magnetized plasma includes three pulsation branches: shear Alfven wave, slow magnetosonic or slow compressional wave, and fast magnetosonic or fast compressional wave. Shear Alfven waves are almost incapable of propagating across the magnetic field; therefore, they cannot penetrate deep into the magnetosphere. In a low-pressure plasma, $\beta \ll 1$, the slow magnetosonic wave is a modification to the usual sound wave, the electromagnetic field in it is almost not disturbed, and the propagation velocity is low (of about the sound velocity $c_s \sim \sqrt{T/m_i}$). For that reason, the slow magnetosonic wave is also of no interest as the possible agent that transports the disturbance into the magnetosphere. There remains one wave mode: the fast magnetosonic wave.

It should be emphasized that the disturbance that propagates in the magnetosphere across the geomagnetic field has the structure of a fast magnetosonic wave, whatsoever the mechanism for generating this disturbance. Thus, both upstream waves, and the surface wave excited on the magnetopause by the Kelvin-Helmholtz instability, and oscillations caused by solar wind disturbances, produce inside the magnetosphere a disturbance that represents fast compressional waves.

At the same time, typical frequencies and wavelengths of disturbances depend greatly on the excitation mechanism. But this affects dramatically the propagation character of fast compressional waves in the magnetosphere. As is known, in a homogeneous plasma the dispersion equation for these waves has the form

$$\omega^2 = k^2 A^2, \qquad (3)$$

where k is a wave number. In an inhomogeneous plasma, for oscillations whose wavelength is much less than the irregularity scale, this relationship remains valid, but it should be understood in the sense of the WKB approximation, that is, k^2 should be regarded as the square of a quasi-classical wave vector. Qualitatively, on the order of magnitude, the relationship (3) remains valid even for oscillations with a wavelength of the order of the irregularity scale.

The WKB approximation has the simplest form in the plasma plane layer model. In this model the magnetic field is directed along the z axis, and the plasma is considered inhomogeneous along the x axis, i.e., the Alfven velocity is a function of coordinate x: $A = A(x)$. This model is often used to describe the magnetosphere — whenever one wishes to investigate the influence of a radial inhomogeneity. In this case the x axis is identified with the radial coordinate and is considered to be directed anti-earthward, the y axis is identified with the azimuthal coordinate, and the z axis is identified with the coordinate along the geomagnetic field. Then from (3) for a disturbance of the form $\exp(i \int k_x dx + i k_y y + i k_z z - i\omega t)$ we get the following expression for a quasi-classical wave vector k_x :

$$k_x{}^2(x) = \frac{\omega^2}{A^2(x)} - k_y{}^2 - k_z{}^2. \qquad (4)$$

A fast compressional wave propagates in the region where $k_x{}^2 > 0$. This is called the transparency region. In the opacity region, where $k_x^2 < 0$, the wave cannot propagate, but this does not mean that the wave field in this region is absent. The oscillation amplitude decays exponentially with depth in the opacity region, but a typical length of decrease can be rather large and even comparable with typical magnetospheric scales. The transparency and opacity regions in the coordinate x are separated by the reflection point, at which $k_x^2 = 0$.

Inside the magnetosphere the Alfven velocity, generally speaking, grows from the magnetopause to the Earth, although some departures from this rule are possible on the plasmapause, say. In the model concerned this means that the transparency region, if existent, is located in the outer part of the magnetosphere, and the opacity region lies in the inner part. At the same time, as follows from (4), the position of the reflection point and hence of the transparency and opacity regions depends substantially on the relationship between the frequency ω and the components of wave vectors k_y and k_z. With decreasing ω and increasing k_y and k_z, there is an increase in the distance of the reflection point from the Earth.

On a given magnetic shell $k_y \sim m/L$ and $k_z \sim N/L$, where m is the azimuthal wave number, N is the longitudinal wave number, and L is the field line length. The typical values of longitudinal wave number N lie in the range 1–5 since it is such values that occur for standing Alfven waves excited by the fast mode. The range of observed values of m is much wider: from $m \sim 1$ to $m \sim 10$. For the Pc3–4 pulsations of our interest, the typical value of $\omega \sim 10^{-1}$ s^{-1}. In the outer magnetosphere, where $A \sim 3{\times}10^2$ km/s and $L \sim 10^5$ km, we have $\omega/A \sim 3{\times}10^{-4}$ km^{-1}, $k_y \sim m$ 10^{-5} km^{-1}, and $k_z \sim N\ 10^{-5}$ km^{-1}. Consequently, this region is transparent for waves with $m < 30$. In the middle of the magnetosphere $A \sim 10^3$ km/s, and $L \sim 3{\times}10^4$ km; consequently, $\omega/A \sim 10^{-4}$ km^{-1}, $k_y \sim (m/3)\ 10^{-4}$ km^{-1}, and $k_z \sim (N/3)\ 10^{-4}$ km^{-1}. This region is transparent when $m < 3$ and is opaque when $m > 3$. The reflection point for waves with $m \sim 3$ lies somewhere in this region. Finally, in the inner magnetosphere we have $A \sim 3{\times}10^3$ km/s and $L \sim 10^4$ km, and hence $\omega/A \sim 3{\times}10^{-5}$ km^{-1}, $k_y \sim m \cdot 10^{-4}$ km^{-1}, and $k_z \sim N\ 10^{-4}$ km^{-1}. From these relationships one can see that the inner magnetosphere is an opacity region for the fast mode at all values of m.

This analysis can also be approached somewhat differently. We are interested in the fast mode as the source of shear Alfven waves excited as a result of field line resonance. This resonance takes place on the same magnetic shell on which the Alfven wave frequency $k_z A(x)$ coincides with the frequency of the driving force. Considering that k_z for Alfven and fast modes are equal (in the plane layer model under consideration, precisely this must be the case), we obtain from (4) that at the resonance point $k_x^2 = -k_y^2$. Thus, the fast mode reaches the resonance surface already in the opacity region. The typical scale of wave amplitude decay in the opacity region is of the order of $k_x^{-1} = k_y^{-1} \sim L/m$. From this, and also from the estimates presented in the pre-

ceding paragraph, it follows that when $m \gg 1$ the fast wave cannot penetrate deep into the magnetosphere. It seems as if maximum admissible values of m are about 5–7. For such m, the outer part of the magnetosphere (near the magnetopause) most probably is transparent to the fast mode, while the middle and inner parts are opaque.

Field line resonance

The theory of field line resonance, suggested by Chen and Hasegawa [1974] and Southwood [1974] was initially developed within the framework of a plane layer model (see a review by Southwood and Hughes [1983] for surveying such publications). According to this theory, a fast compressional wave with frequency ω generates a shear Alfven wave near the resonance surface $x = \bar{x}$ determined by the resonance condition $\omega = k_z A(\bar{x})$. The situation can be outlined more accurately in the following manner. Strictly speaking, there is neither a fast compressional wave nor a shear Alfven wave in an inhomogeneous plasma. However, far from the resonance surface the properties of the wave (for example, its polarization) almost coincide with those of fast mode of a homogeneous plasma. Near this surface, however, the field enhances abruptly, and its properties coincide with those of a shear Alfven wave. One of the varieties of the linear transformation of waves occurs here [Stix and Swanson, 1980].

In the approximation of ideal MHD the electric and magnetic field of the oscillation have, on the resonance surface, the singularities:

$$E_x, B_y \sim \frac{1}{x - \bar{x}}. \tag{5}$$

If the dissipation is taken into account, then the singularity is eliminated

$$E_x, B_y \sim \frac{1}{x - \bar{x} + i\varepsilon},$$

where the quantity ε is determined by the dissipation mechanism. Under real magnetospheric conditions the main role is played by the Ohmic dissipation in the ionosphere. There exists also another factor that regulates the singularity (5), namely the effect of finite Larmor ion radius and electron inertia. This effect leads to the transverse dispersion of the shear Alfven wave which in this case is called the kinetic wave [Hasegawa, 1976; see also Goertz, 1984]. The field structure near the resonance surface in this case has a complicated oscillatory character, and the wavelength of these oscillations is determined by the dispersion parameter (by either the Larmor radius of ions, $\rho_\perp$, or the skin length, c/ω_{pe}, depending on which of these parameters is larger). In the general case the field structure near the resonance surface is determined by the combined action of the dissipation and dispersion.

During the past decade there have been several attempts to get beyond the scope of the plasma plane layer model [Krylov et al., 1981; Krylov and Lifshitz, 1984; Southwood and Kivelson, 1986]. A constructive and rather complete theory of field line resonance for an axisymmetrical model of the magnetosphere was developed by Leonovich and Mazur [1989a,b], and we shall give here a brief account of their results. It should be noted that the axisymmetrical model makes it possible to investigate the influence of such factors as the longitudinal and transverse inhomogeneity of plasma and the geomagnetic field, including the curvature of the field lines of the latter. Besides, the theory presented here includes the dissipation and dispersion of shear Alfven waves.

The orthogonal curvilinear coordinate system shown in Figure 3 is used to describe the magnetosphere. The geomagnetic field B_0, plasma density ρ and, hence, Alfven velocity A are assumed to be functions of the coordinates x^1 and x^3. A disturbance in the wave is represented by covariant components of a disturbed magnetic field vector; in this case both the components themselves and their Fourier harmonics

$$B_i(x^1, x^2, x^3, t) = \int_{-\infty}^{\infty} \tilde{B}_i(x^1, x^2, x^3, \omega) e^{-i\omega t} d\omega$$

are used. At first, we shall consider a description of the Fourier harmonic, that is, a monochromatic wave.

The system of equations of magnetic hydrodynamics can be reduced to two equations for the components $\tilde{B}_2$ and $\tilde{B}_3$. The component $\tilde{B}_3$ is associated with a fast compressional wave because in shear Alfven waves the longitudinal component of a disturbed magnetic field is practically absent. And vice versa, a sharp increase of $\tilde{B}_2$ near the resonance

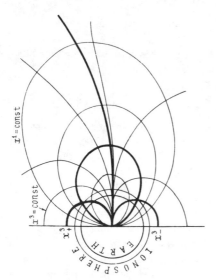

Fig. 3. A curvilinear, orthogonal system of coordinates (x^1, x^3) in the meridional plane (x^2 = const.). Surfaces x^1 = const. coincide with magnetic shells, the coordinate x^2 specifies the line of force on a given shell and x^3 specifies a point on a given line of force. Emphasis is deliberately placed on the possible north-south asymmetry of the magnetosphere. The figure singles out: the equatorial line that is a separatrix for coordinate curves x^3 = const., one of the magnetic shells and coordinate lines x^3 = const., corresponding to the intersection of this shell with the ionospheres in the Northern (x_+^3) and Southern (x_-^3) hemispheres.

surface should be treated as the field of a shear Alfven wave.

Leonovich and Mazur [1989a] showed that the problem of determining the field $\tilde{B}_3$ can be stated internally closely by separating its solution from the problem of determining the field $\tilde{B}_2$. In this case the influence of a shear Alfven wave upon the fast mode reduces to a jump of the derivative $\partial \tilde{B}_3 / \partial x^1$ at the intersection of the resonance surface, and matching conditions for these derivatives can be represented in terms of the field $\tilde{B}_3$ itself. The problem formulated in this manner for the field $\tilde{B}_3$, as has already been noted above, can be solved numerically only. Qualitative properties of the solution have been described in the preceding Section. Subsequently, the field $\tilde{B}_3$ will be considered given.

For the shear Alfven wave field $\tilde{B}_2$, the following solution

$$\tilde{B}_2(x^1, x^2, x^3, t) = \sum_N \tilde{\mu}_N(x^1, x^2, \omega) \tilde{Q}_N(x^1, \omega) H_N(x^1, x^3).$$

was obtained. Let us describe the terms involved.

The functions $H_N(x^1, x^3)$, where the index N runs over a natural series of values of $N = 1, 2, \ldots$, are longitudinal harmonics, describing a standing wave along a geomagnetic field line. They are eigen-solutions of the following one-dimensional problem for eigenvalues:

$$\frac{g_2}{\sqrt{g}} \frac{\partial}{\partial x^3} A^2 \frac{g_1}{\sqrt{g}} \frac{\partial H}{\partial x^3} + \omega^2 H = 0, \quad \frac{\partial H}{\partial x^3}\bigg|_{x^3 = x_\pm^3} = 0.$$

The coordinate x plays in this problem the role of a parameter, on which eigenvalues and eigenfunctions depend:

$$\omega = \Omega_N(x^1), \quad H = H_N(x^1, x^3).$$

The eigenfrequencies defined in this way will play a crucial role in what follows.

For large values of N, the WKB approximation in the coordinate x^1 is applicable. In this case

$$\Omega_N(x^1) = 2\pi N / t_A(x^1),$$

where

$$t_A(x^1) = \oint \frac{\sqrt{g_3} dx^3}{A}$$

is the transit time with Alfven velocity along the field line "there and back".

If the dissipation in the ionosphere is taken into account, then eigenfrequencies become complex:

$$\omega = \Omega_N(x^1) - i\gamma_N(x^1).$$

The expression for the damping decrement γ_N is given in the cited paper. It depends weakly on the number N; at large N this dependence vanishes altogether. Schematic plots of the functions $\Omega(x^1) = 2\pi / t_A(x^1)$, and $\gamma(x^1)$ are given in Figure 4.

The function $Q_N(x^1, \omega)$ satisfies the equation

$$\sigma_N(x^1) \frac{\partial^2 Q_N}{\partial x^{1^2}} + \left[\frac{(\omega + i\gamma_N(x^1))^2}{\Omega_N^2(x^1)} - 1 \right] Q_N = 1. \quad (6)$$

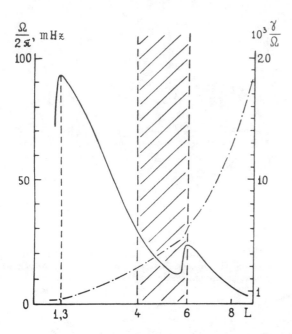

Fig. 4. Plots of the functions $\Omega(x^1)$ (solid line) and $\gamma(x^1)$ (broken line). As the coordinate x^1, the McIlwain parameter L is chosen. The transition region is shaded; in this layer the Alfven wave energy is absorbed by magnetospheric plasma electrons.

Here

$$\sigma_N(x^1) = \oint \Lambda^2 \frac{g_3}{\sqrt{g}} H_N^2 dx^3$$

where Λ^2 is the square of the dispersion length. In limiting cases this parameter has straightforward expressions

$$\Lambda^2 = \begin{cases} -s^2 \equiv -c^2/\omega_{pe}^2, & \text{when} & \beta_e \ll m_e/m_i, \\ \rho_s^2 \equiv T_e/m_i W_i^2, & \text{when} & \beta_e \gg m_e/m_i, \end{cases}$$

where $\beta_e = 8\pi T_e/B_0^2$. When $\beta_e \sim m_e/m_i$, the expression for Λ^2 has a much more complicated form, in particular it is complex, with Im $\Lambda^2 < O$. We put $\sigma_N = \rho_N^2(x^1) \exp[-i\alpha_N(x^1)]$. The quantity ρ_N has the dimension of length, and the phase α_N lies in the range $(0, \pi)$. A schematic plot of the dependences $\rho_N(x^1)$ and $\alpha_N(x^1)$ is given in Figure 5.

Finally, the function $\tilde{\mu}_N$ is specified by the relationship

$$\tilde{\mu}_N(x^1, x^2, \omega) = \oint e_N(x^1, x^3) \frac{\partial \tilde{B}_3(x^1, x^2, x^3, \omega)}{\partial x^2} dx^3, \quad (7)$$

$$e_N(x^1, x^3) = \frac{1}{\Omega_N^2} \frac{g_1}{\sqrt{g}} A^2 \frac{\partial H_N}{\partial x^3}.$$

The quantity $\tilde{\mu}_N$ defines the amplitude of the shear Alfven wave in terms of the amplitude of magnetosound $\tilde{B}_3$. The expression (7) demonstrates, perhaps, the largest difference of the solution in the magnetospheric model under consideration from the plane layer model, in which the depen-

dences of the Alfven wave and magnetosound on the coordinate along the geomagnetic field are the same: $\sim \exp(ik_z z)$. In our axisymmetrical model these dependences are quite different: for the Alfven wave it is defined by the term $H_N(x^1, x^3)$, and for the fast magnetosonic wave $\tilde{B}_3$ it is defined by its equation. The quantity $\tilde{\mu}_N$ is determined by the integral along the field line. Since at large N the function $e_N(x^1, x^3)$ oscillates rapidly along the field line, $\tilde{\mu}_N$ for such N virtually goes to zero. This means that standing Alfven waves with large N ($N > 5$, say) are not excited by field line resonance.

Let us now describe the properties of the function $Q_N(x^1, \omega)$ that is the solution of equation (6) and defines the main dependence of the Alfven wave field on the coordinate x^1. The dispersion parameter Λ^2, and hence the value of $\sigma_N(x^1)$, are very small. In the limit $\sigma_N \to 0$ we have

$$Q_N(x^1, \omega) = \frac{\Omega_N^2(x^1)}{(\omega + i\gamma_N)^2 - \Omega_N^2(x^1)}. \quad (8)$$

If, in addition, $\gamma_N \to 0$ also holds, then the function Q_N has the known singularity on the resonance surface defined by the equation $\Omega_N(x^1) = \omega$. Both a finite value of γ_N and a finite value of σ_N regularize this singularity; in these cases, however, the solution remains concentrated in a narrow vicinity near the resonance surface.

An analysis of the equation (8) permits us to establish the following features. If $\gamma_N \gg \tau_N^{-1}$, where $\tau_N = (l_N/\rho_N)^{2/3} \bar{\Omega}_N^{-1}$ is the so called dispersion time, then the dissipation effect predominates over the dispersion effect. In this case the typical scale of the function $Q_N(x^1, \omega)$ in the variable x^1 is $\Delta x^1 \sim (\gamma_N/\bar{\Omega}_N)l_N$ and in frequency, respec-

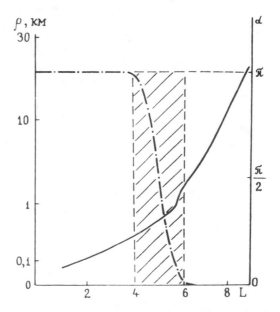

Fig. 5. Plots of the functions $\rho(x^1)$ (solid line) and $\alpha(x^1)$ (broken line).

tively, is $\Delta\omega_N = \gamma_N$. In the opposite case, $\gamma_N \ll \tau_N^{-1}$, the scale in the coordinate x^1 is $\Delta x^1 \sim l_N^{1/3} \rho_N^{2/3}$ and in frequency, accordingly, is $\Delta\omega_N = \tau_N^{-1}$. In the last case $Q_N = (l_N/\rho_N)^{2/3}\phi((x^1 - \bar{x}^1)/\Delta x^1)$, where $\bar{x}^1$ is the co-ordinate of the resonance surface, and a schematic plot of the function $\phi(z)$ is given in Figure 6. Under real magnetospheric conditions the parameter $\gamma_N \tau_N$ varies in the range from 10 in the inner part of the magnetosphere to 0.1 in its outer part, that is, both cases occur.

The solution described for a monochromatic wave is inapplicable directly to the problem of excitation of standing Alfven waves by the extramagnetospheric source of our interest because the reflected proton flux instability excites an upstream wave over a rather wide range of frequencies; moreover, this source should be considered stochastic. Accordingly, a fast magnetosonic wave in the magnetosphere also has a stochastic character. This means that each Fourier-harmonic $\tilde{B}_3(x^1, x^2, x^3, \omega)$ is a random function. It should, however, be clearly kept in mind that in spatial variables $\tilde{B}_3$ is not a random function. It satisfies quite a definite equation, boundary conditions and matching conditions on resonance surfaces. Therefore, it can be written as

$$\tilde{B}_3(x^1, x^2, x^3, \omega) = \eta(\omega)\psi(x^1, x^2, x^3, \omega)$$

where $\psi(x^1, x^2, x^3, \omega)$ is a certain standard solution normalized in a certain manner, and $\eta(\omega)$ is a random complex-valued function of frequency. Assuming the random process under consideration to be a stationary one we have for the correlator

$$\langle \eta(\omega)\eta(\omega') \rangle = \beta(\omega)\delta(\omega - \omega').$$

The function $\beta(\omega)$ can be considered to be the spectral density of fast magnetosonic waves in the magnetosphere. It has a maximum at the frequency $\omega = \bar{\omega} \sim 0.1$ sec and a typical width in ω of the same order of magnitude.

Leonovich and Mazur [1989b] obtained the expressions for spectral density of standing Alfven waves in the magnetosphere and for the mean square of their amplitude. The spectral density $P(x^1, x^2, x^3, \omega)$ is defined by the relationship

$$\langle B_2{}^*(x^1, x^2, x^3, \omega)B_2(x^1, x^2, x^3, \omega') \rangle$$
$$= P(x^1, x^2, x^3, \omega)\delta(\omega - \omega').$$

We have

$$P(x^1, x^2, x^3, \omega) = \sum_N \beta(\Omega_N(x^1))|\tilde{\beta}_N(x^1, x^2, \Omega_N(x^1))|^2$$
$$\cdot |Q_N(x^2, \omega)|^2 H_N{}^2(x^1, x^3)$$

Here

$$\tilde{\beta}_N(x^1, x^2, \omega) = \oint e_N(x^1, x^3)\frac{\partial\psi(x^1, x^2, x^3, \omega)}{\partial x^2}dx^3$$

is an analog of the function $\tilde{\mu}_N$. The dependence of P on x^3 is specified by the term $H_N{}^2$. The dependence on the

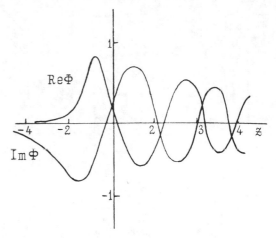

Fig. 6. Plots of the real and imaginary parts of the function $\phi(z)$ describing the spatial structure of a standing Alfven wave perpendicular to the magnetic shells near a resonance shell.

coordinate x^1 is defined by the terms $\beta(\Omega_N(x^1))$ and $|\tilde{\beta}_N|^2$. The former is nonzero if only the eigenfrequency Ω_N lies in the spectral range of the source. As far as the term $|\tilde{\beta}_N|^2$ is concerned, the important feature is its decrease in the earthward direction because of the fast mode penetrating deep into the opacity region. As for the frequency, the dependence on it is described by the functions $|Q_N|^2$ which give sharp peaks near the eigenfrequencies $\omega = \Omega_N(x^1)$.

For the mean square of the Alfven wave amplitude, the expression

$$\langle B_2{}^2(x^1, x^2, x^3, t) \rangle = \sum_N \frac{\Omega_N{}^2(x^1)}{\gamma_N(x^1)}\beta(\Omega_N(x^1))$$
$$\cdot |\tilde{\beta}_N(x^1, x^2, \Omega_N(x^1))|^2 H_N{}^2(x^1, x^3)$$

is obtained. The analysis made in the cited paper has shown that this expression can be used to explain the observed dependence of the pulsation amplitude on the coordinate x^1. The main features of this analysis are evident in Figure 7.

Methods of Experimental Investigation of the Pc3 Wave Field Structure

It should be noted that the picture of geomagnetic pulsation field observed in the magnetosphere and on the ground reflects the entire complicated process of their generation in the foreshock region, propagation through the bow shock, magnetosheath and magnetopause and, finally, interaction inside the magnetosphere with field line resonances.

This is manifested most conspicuously in the spectral composition of pulsations recorded on the ground and satellites. Thus, at each individual station a resonance structure characteristic for that station manifests itself, on the one hand, and the relationship $f \approx 6B$ between the frequency of observed pulsations and the IMF strength is satisfied, on

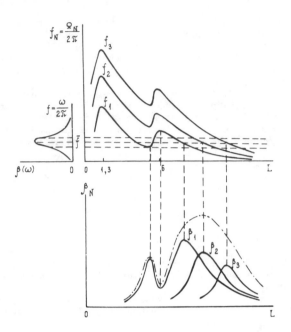

Fig. 7. The upper plot shows the dependences of frequencies of the three first harmonics of the eigenoscillations of the magnetosphere f_N on the parameter L and a spatial function of the source of compressional waves $\beta(\omega)$ (left). The lower plot shows amplitude profiles of appropriate harmonics β_N for a given spectral function $\beta(\omega)$ and the total amplitude profile (broken line).

the other. Moreover the picture on the ground can differ from that occurring in the magnetosphere. This question was discussed in detail by Verö et al. [1985].

The Method of Dynamical Parameters

A theory of field line resonances developed by Chen and Hasegawa [1974] and the interpretation of its results [Hughes and Southwood, 1976a,b] gave strong impetus to research on resonance structure of the geomagnetic pulsation field. Primarily, this refers to the investigations comparing spectral features observed at latitudinally spaced stations, including magnetospheric ones. Other methods based on studying local characteristics of the pulsation field were also developed. The varieties of such an approach are the gradient method and its modification, the gradient-time method of observing resonance structure [Baransky et al., 1985; Best et al., 1986]. They are based on comparing pulsation spectra recorded simultaneously at two closely located (50–80 km) sites lying on the same meridian. Here we shall examine a different method taking into account not only resonance characteristics of the magnetosphere but also the pulsation source properties.

Recently the method of inverse problems of statistical theory of oscillations has received recognition in geophysics. It is based on correlational analysis of fluctuations of the oscillation amplitude and phase. The principal merit of the

method is that using as input data only recordings of the realization of oscillations, it is possible to determine a number of interesting parameters that characterize the equation governing these pulsations. For geomagnetic Pc3 pulsations, unlike the other pulsation types, such a model equation can be considered known. Indeed, an investigation of statistical properties of the amplitude and phase fluctuations [Guglielmi et al., 1983] suggests that oscillations of this type can be described by the equation of a linear oscillator under the action of a periodic external force in the presence of a δ-correlated noise source. This formal model corresponds to the above idea of the origin of Pc3 pulsations. Fast magnetosonic waves excited in front of the bow shock serve as the driving force for field line resonances.

If $x(t)$ is the initial realization of the oscillations (for example, the H-component of the Pc3 magnetic field), then it might be assumed [Polyakov and Potapov, 1989] that $x(t)$ is the solution of a stochastic system of differential equations

$$\frac{dx}{dt} = y,$$

$$\frac{dx}{dt} = -\nu y - \omega_0{}^2 x + A\sin(\omega t + \phi) + \mu F(t), \qquad (9)$$

where ν and ω_0 are, respectively, the attenuation and the eigenfrequency of the oscillator; A, ϕ, and ω are, respectively, the amplitude, phase and frequency of a regular source; $F(t)$ is δ-correlated random noise, and μ is its intensity. Later the solution is evaluated in the limit $\mu \ll 1$.

The mean solution x_0, y_0 of the system (9) coincides with an undisturbed one ($\mu = 0$). After normalization it describes, on the phase plane (x, y), a circle with its center at the origin of coordinates and a unit radius. Phase trajectories of the system (9) will lie in a small vicinity of the mean phase trajectory. If, instead of (x, y), the coordinates (n, γ) are introduced, which are, respectively, normal and tangential deviations from the mean phase trajectory, and instead of time t, the tangential coordinate $\theta = t + \gamma$ is used, then, by discarding terms nonlinear in n and γ, we obtain, instead of (9),

$$\frac{dN}{d\theta} = -N(\theta)n + K(\theta) + F_n(\theta)$$

$$\frac{d\gamma}{d\theta} = Q(\theta)n + P(\theta) + F_\gamma(\theta),$$

where F_n and F_γ are, respectively, the normal and tangential components of force F, and N, Q, K, and P are, respectively, their dynamical coefficients, and their average (over the phase trajectory) values are directly related to physical characteristics of the system:

$$\bar{N} = \bar{P} = \nu/2\omega, \qquad \bar{Q} = -\bar{K} = (\omega_0{}^2/\omega^2 - 1)/2.$$

Polyakov and Potapov [1989] described and investigated a method of determining dynamical coefficients from results of processing of oscillation oscillograms. It permits the determination, using ground-based data, of the resonance frequency and attenuation of the Alfven resonator at the observing point. With several observing points available, it is possible to probe the spatial structure of the field, and through a continuous registration of pulsations it is possible to follow variations in this structure. The method was tested against pulsation observations at three midlatitude sites located along the meridian [Polyakov et al., 1992].

Conclusions

The concept of generation and propagation of geomagnetic Pc3 pulsations finds applications in methods of diagnosing solar wind parameters and the IMF [Guglielmi and Bol'shakova, 1973; Potapov and Polyushkina, 1979] and the state of the magnetosphere [Guglielmi, 1989; Potapov and Polyushkina, 1992] under development on its basis. At the same time, work on this concept can in no way be considered completed. Let us outline the way in which theory and experiment can be further developed.

Among the outstanding theoretical problems, the problem of field line resonances in a three-dimensionally-inhomogeneous magnetosphere is particularly important and challenging. Some simple questions lack clarity here: What is the resonance surface (it, of course, does not coincide with the magnetic shell)? How to determine a toroidal oscillation, etc.? Of great interest is a theoretical investigation of nonlinear processes near the resonance surface where the oscillation amplitude increases sharply. It is quite possible that under certain conditions a maximum value of amplitude will be determined by the nonlinearity rather than the dissipation or dispersion.

The problem of comparing the conclusions of the theory of field line resonances (including the above theory) with data from ground-based and satellite observations should be added to current challenges. One is led to believe that by analyzing data on the six components of the electric and magnetic field of a disturbance, including their phase relationships, it would be possible to gain much insight into fine spatial structure of shear Alfven waves in the magnetosphere even from observations on a single satellite or a single ground-based station. Of course, simultaneous observations by several satellites still are of great interest, however.

Acknowledgments. The authors would like to thank Mr V. G. Mikhalkovsky for his assistance in preparing the English version of the manuscript.

References

Asbridge, J. R., S. J. Bame, and I. B. Strong, Outward flow of protons from the Earth's bow shock, *J. Geophys. Res., 73*, 5777–5782, 1968.

Baransky, L. N., Yu. E. Borovkov, M. B. Gokhberg, and S. M. Krylov, Gradient method of measurement of magnetic field lines resonant frequences (in Russian), *Izv. AN SSSR. Fizika Zemli*, No.8, 74–91, 1985.

Barkhatov, N. A., On an interpretation of Pc3,4 activity dependence on solar wind velocity (in Russian), *Geomagn. Aeron.,* 17, 767–770, 1977.

Barkhatov, N. A., P. A. Bespalov, and M. S. Kovner, On transmission of low-frequency wave packet through the bow shock (in Russian), *Geomagn. Aeron., 17*, 16–20, 1977.

Barnes, A., Theory of generation of bow-shock-associated hydromagnetic waves in the upstream interplanetary medium, *Cosmic Electrodyn., 1*, 90–97, 1970.

Best, A., S. M. Krylov, Yu. P. Kurchashov, Ya. S. Nikomarov, and V. A. Pilipenko, Gradient-time analysis of Pc3 pulsations (in Russian), *Geomagn. Aeron., 26*, 980–984, 1986.

Bol'shakova, O. V., and V. A. Troitskaya, Relationship between direction of the interplanetary magnetic field and behaviour of continuous pulsations (in Russian), *Dokl. Akad. Nauk SSSR, 180*, 343–346, 1968.

Bonifazi, C., G. Moreno, C. T. Russell, A. T. Lazarus, and J. D. Sullivan, Solar wind deceleration and MHD turbulence in the Earth's foreshock region: ISEE 1 and 2 and IMP 8 observations, *J. Geophys. Res., 88*, 2029–2037, 1983.

Chen, L., and A. Hasegawa, A theory of long-period, 1, Steady state excitation of field-line resonance, *J. Geophys. Res., 79*, 1024–1032, 1974.

D'Angelo, N., Have Pc2–4 micropulsations extramagnetospheric origin, or not? (in Russian), *Geomagn. Aeron., 15*, 1062–1065, 1975.

Fairfield, D. H., Bow shock associated waves observed in the far upstream interplanetary medium, *J. Geophys. Res., 74*, 3541–3547, 1969.

Goertz, C. K., Kinetic Alfven waves on auroral field lines, *Planet. Space. Sci., 32*, 1387–1392, 1984.

Greenstadt, E. W., R. L. McPherron, and K. Takahashi, Solar wind control of daytime, midperiod geomagnetic pulsations, *J. Geomagn. Geoeletr., 32*, suppl.II, SII89, 1980.

Guglielmi, A. V., The magnetosphere and interplanetary medium diagnosis using data of geomagnetic pulsations observation (in Russian), paper presented at All-Union Conference on Magnetosphere Physics, Borok, June 1972.

Guglielmi, A. V., Diagnostics of the magnetosphere and interplanetary medium by means of pulsations, *Space Sci. Rev., 16*, 331–345, 1974.

Guglielmi, A. V., Geomagnetic pulsations of the extramagnetospheric origin (in Russian), in *Itogi Nauki i Tekhniki. Geomagn. i Vysok. Sloi Atmos.*, 7, pp.114–151, VINITI, Moscow, 1984.

Guglielmi, A. V., Hydromagnetic diagnostics and geoelectric sounding (in Russian), *Uspehi Fizicheskih Nauk, 158*, 605–637, 1989.

Guglielmi, A. V., and O. V. Bol'shakova, Diagnostics of the IMF by means of ground data on Pc3–4 micropulsations (in Russian), *Geomagn. Aeron., 13*, 535–537, 1973.

Guglielmi, A. V., and K. G. Ivanov, Hydromagnetic precursors of flare-associated interplanetary plasma flow (in Russian), *Geomagn. Aeron., 24*, 489–491, 1984.

Guglielmi, A. V., T. A. Plyasova-Bakunina, and R. V. Schepetnov, On relationship between period of Pc3–4 geomagnetic pulsations and the interplanetary medium parameters on the Earth's orbit (in Russian), *Geomagn. Aeron., 13*, 382–384, 1973.

Guglielmi, A. V., A. S. Potapov, and A. D'Costa, On the theory of the type Pc3 geomagnetic pulsations excitation (in Russian), in *Issled. Geomagn. Aeron. Phys. Solntsa*, Vyp.39, pp.27–32, Nauka, Moscow, 1976.

Guglielmi, A. V., B. I. Klaine, A. S. Potapov, and A. R. Polyakov, Amplitude-and-phase fluctuations of geomagnetic Pc3 pulsations (in Russian), *Issled. Geomagn. Aeron. Phys. Solntsa*, Vyp.66, pp.38–41, Nauka, Moscow, 1983.

Hasegawa, A., Particle acceleration by MHD surface wave and formation of aurora, *J. Geophys. Res., 81*, 5083–5090, 1976.

Hoppe, M. M., and C.T. Russell, Particle acceleration at planetary bow shock waves, *Nature, 295*, 41–42, 1982.

Hughes, W. J., and D. J. Southwood, The screening of micropulsation signals by the atmosphere and ionosphere, *J. Geophys. Res., 81*, 3234–3240, 1976a.

Hughes, W. J., and D. J. Southwood, An illustration of modification of geomagnetic pulsation structure by the ionosphere, *J. Geophys. Res., 81*, 3241–3247, 1976b.

Kennel, C. F., F. L. Scarf, F. V. Coroniti, E. J. Smith, and D. A. Gurnett, Nonlocal plasma turbulence associated with interplanetary shocks, *J. Geophys. Res., 87*, 17–34, 1982.

Krylov, A. L., and A. E. Lifshitz, Quasi-Alfven oscillations of magnetic surfaces, *Planet. Space Sci., 32*, 481–492, 1984.

Krylov, A. L., A. E. Lifshitz, and E. N. Fedorov, On resonance properties of the magnetosphere (in Russian), *Izv.AN SSSR. Fizika Zemli*, No.6, 49–51, 1981.

Leonovich, A. S., and V. A. Mazur, Resonance excitation of standing Alfven waves in an axisymmetric magnetosphere (monochromatic oscilations), *Planet. Space Sci., 37*, 1095–1108, 1989a.

Leonovich, A. S., and V. A. Mazur, Resonance excitation of standing Alfven waves in an axisymmetric magnetosphere (nonstationary oscilations), *Planet. Space Sci., 37*, 1109–1116, 1989b.

Polyakov, A. R., and A. S. Potapov, Dynamic parameters of Pc3 geomagnetic pulsations (in Russian), *Geomagn. Aeron., 29*, 921–925, 1989.

Polyakov, A. R., A. S. Potapov, and B. Tsegmid, An experimental evaluation of resonant frequency and Q-quality of low-latitude Alfven resonators (in Russian), *Geomagn. Aeron., 32*, 156–159, 1992.

Polyushkina, T. N., and A. S. Potapov, The relation of the amplitude of daytime geomagnetic pulsations to solar wind velocity over the solar activity cycle (in Russian), in *Issled. Geomagn. Aeron. Phys. Solntsa*, Vyp.66, pp.54–61, Nauka, Moscow, 1983.

Potapov, A. S., Excitation of Pc3 geomagnetic pulsations in front of the Earth's bow shock by a reflected proton beam (in Russian), in *Issled. Geomagn. Aeron. Phys. Solntsa*, Vyp.34, pp.3–12, Nauka, Irkutsk, 1974.

Potapov, A. S., and T. N. Polyushkina, Diurnal U_C-index of solar wind velocity and analysis of B-index on Irkutsk data (in Russian), in *Issled. Geomagn. Aeron. Phys. Solntsa*, Vyp.46, pp.152–157, Nauka, Moscow, 1979.

Potapov, A.S., and T. N. Polyushkina, A phenomenological study of the D_{st} storm variation, *Planet. Space Sci., 40*, 731–739, 1992.

Potapov, A.S., T. N. Polyushkina, and A. V. Buzevich, Novel data on the relationship of daytime stable geomagnetic pulsations with the parameters of solar wind (in Russian), in *Issled. Geomagn. Aeron. Phys. Solntsa*, Vyp.49, pp.84–88, Nauka, Moscow, 1979.

Russell, C.T., and B. K. Fleming, Magnetic pulsations as a probe of the interplanetary magnetic field. A test of the Borok B index, *J. Geophys. Res., 81*, 5882–5886, 1976.

Southwood, D. J., Some features of field line resonances in the magnetosphere, *Planet. Space Sci., 22*, 483–491, 1974.

Southwood, D. J., and W. J. Hughes, Theory of hydromagnetic waves in the magnetosphere, *Space Sci. Rev., 35*, 301–366, 1983.

Southwood, D. J., and M. G. Kivelson, The effect of parallel inhomogeneity on magnetospheric hydromagnetic wave coupling, *J. Geophys. Res., 91*, 6871–6876, 1986.

Stix, G., and D. Swanson, Alfven resonance (in Russian), in *Osnovy Fiziki Plasmy* edited by A. A. Galeev and R. M. Sudan, p. 333, Energoatomizdat, Moscow, 1980.

Troitskaya, V. A., T. A. Plyasova-Bakunina, and A. V. Guglielmi, Relationship between Pc2–4 pulsations and the interplanetary magnetic field (in Russian), *Dokl. Akad. Nauk SSSR, 197*, 1312–1314, 1971.

Verö, J., and L. Hollo, Connections between interplanetary magnetic field and geomagnetic pulsations, *J. Atmos. Terr. Phys., 40*, 857–865, 1978.

Verö, J., L. Hollo, A. S. Potapov, and T. N. Polyushkina, Analysis of the connections between solar wind parameters and dayside geomagnetic pulsations based on data from the observatories Nagycenk (Hungary) and Uzur (USSR), *J. Atmos. Terr. Phys., 47*, 557–565, 1985.

Vinogradov, P. A., and V. A. Parkhomov, MHD-waves in the solar wind, as a possible source of geomagnetic pusations (in Russian), *Geomagn. Aeron., 15*, 134–137, 1975.

Yumoto, K., Generation and propagation mechanisms of low-latitude magnetic pulsations — A review, *J. Geophys., 60*, 79–105, 1986.

V. Mazur and A. Potapov, Institute of Solar-Terrestrial Physics, P.O.Box 4026, Irkutsk, 664033, Russia

E-mail: root@sitmis.irkutsk.su

Telex: 133163 TAIGA SU

Theory and Observation of Magnetosheath Waves

A. N. FAZAKERLEY[1] AND D. J. SOUTHWOOD

Space and Atmospheric Physics Group, Imperial College, London

Solar wind influences on the magnetosphere are conveyed via the magnetosheath, so a better understanding of the magnetosheath forms an important part of the study of how such influences operate. We review studies of magnetosonic and Alfven waves in the magnetosheath, but the greater part of our paper discusses another phenomenon thought likely to give rise to observations of magnetic disturbances in the ULF range in the magnetosheath i.e. mirror waves. Mirror wave signatures are thought to be the result of the spacecraft travelling through regions of magnetosheath plasma which have assumed an inhomogenous structure due to the action of the mirror instability. We briefly review the linear theories of mirror instability in both the MHD fluid and the more rigorous kinetic forms, and point out the previously overlooked, but crucial role of particles with small speeds along the field. We refer to such particles as resonant particles, since they are a special subset of the distribution distinguished by their velocity, although in some respects the behaviour of these particles differs in character from the behaviour of resonant particles in more familiar resonant instabilities. The improved description of mirror instability is accompanied by a new expression for the growth rate (although the instability condition is not altered). In addition we outline some ideas about the non-linear development of the instability. We also report some recent observational work in which we use an unusually long interval (15 minutes) of well correlated dual spacecraft mirror wave observations to attempt to analyse the three dimensional form of mirror structures.

INTRODUCTION

It seems inevitable that any model which seeks to show that the solar wind has a role in the stimulation of ULF wave activity in the magnetosphere must consider how the solar wind is modified as it enters the magnetosheath, the region bounded by the bow shock wave and the magnetopause. On crossing the bow shock the solar wind plasma is rapidly compressed, thermalised and decelerated, modified to such a degree as to justify a separate classification as magnetosheath plasma. The transformation of the supermagnetosonic solar wind plasma flow into a submagnetosonic magnetosheath flow can be modelled quite effectively by describing the plasma as a magnetohydrodynamic (MHD) fluid (since typical gyroradii are small compared to the size of the magnetospheric obstacle to the solar wind flow). One aspect of the magnetosheath which is not represented in such models is the persistent presence of magnetic fluctuations at levels significantly greater than in the solar wind or magnetosphere (e.g. see Fairfield, [1976] and references therein). The fluctuations have their largest amplitudes in the MHD frequency range (i.e. at frequencies below the proton gyrofrequency) in which MHD wave theory predicts three magnetic field perturbing wave modes (fast and slow magnetosonic, and Alfvenic). The undisputed presence of a fast magnetosonic shock (the bow shock) in the solar wind upstream of the magnetosphere implies the propagation of fast magnetosonic waves through the magnetosheath. Song et al. [1990] have reported data which they interpret as evidence of a slow mode shock in the magnetosheath, which if correct would also imply that slow mode magnetosheath wave activity must be present, at least in the region between the magnetopause and the slow mode shock. The three MHD wave modes can in principle be distinguished from one another in observations using only magnetic field and plasma density data from a single spacecraft, since a different relationship between the plasma density and field perturbations is expected in each case (although an identification of the slow mode using these data alone cannot be regarded as convincing, as we discuss below). An analysis of data acquired by two spacecraft located a suitably small distance apart helps to rule out other possible explanations for observed wave-like signatures and can strengthen tentative identifications of MHD wave modes. Such dual spacecraft studies have been rare but include an examination of foreshock waves [Hoppe et al., 1981] and work by Gleaves and Southwood [1991a] in which both Alfven waves and slow magnetosonic waves are apparently identified in the magnetosheath. The reports of observations of slow mode waves and shocks are of particular interest since they appear to contradict theoretical expectations [e.g. Barnes, 1966;

[1] Now at Mullard Space Science Laboratory, University College London, Holmbury St. Mary, Dorking, Surrey, RH5 6NT

Solar Wind Sources of Magnetospheric Ultra-Low-Frequency Waves
Geophysical Monograph 81

Hasegawa, 1969] that slow mode waves experience heavy damping, precluding the observation of such waves.

The analysis of dual spacecraft data sets can also reveal wave propagation directions, in principle providing the means to trace waves back to their source regions [Gleaves and Southwood, 1991b] and possibly also forward to their destinations. Taking a relevant example, one probable source of magnetosheath fluctuations is upstream wave activity. Upstream waves are generated in the solar wind upstream of the quasi-parallel bow shock and are then carried into the magnetosheath by the solar wind flow. Some upstream waves apparently reach the magnetopause and ultimately stimulate ULF wave activity in the magnetosphere (e.g. see recent work by Engebretson et al. [1991], Lin et al. [1991] and Song et al. [1993] as well as references therein). Studies of transport of upstream wave energy within the magnetosheath have tended to assume that the fluctuations are transported along flow stream lines [e.g. Russell et al., 1983; Luhmann et al., 1986] but may need revising in order to account for the across-flow transport of wave energy implied by the work of Gleaves and Southwood [1991b] which seems to imply that upstream waves can disturb the magnetopause over a larger area than has previously been suggested.

Observations of magnetic fluctuations can also occur if a spacecraft travels through a plasma in which a static non-uniform magnetic field structure exists. The fluctuations will be quasi-periodic if there is a fairly regular distribution of stronger and weaker field regions and may lie in the ULF band if the spacecraft velocity relative to the structure is appropriate (in terms of the size, shape and distribution of the regions of differing magnetic field strength). One way in which such structure could arise is through the magnetic mirror instability. An initially homogenous plasma which becomes vulnerable to the mirror instability can undergo a redistribution of material along the field lines that creates a stable structure in which regions of lower field strength and higher plasma density are separated from one another by regions of higher field strength and lower plasma density. It seems probable that mirror structures may be a fairly common feature of the magnetosheath, since the conditions for instability (below) are often met downstream of the bow shock. Indeed, magnetic and plasma fluctuations categorised as mirror waves have been reported in the terrestrial magnetosheath [e.g. Tsurutani et al., 1982; Tsurutani et al., 1984; Moustaizis et al., 1986; Hubert et al., 1989a,b; Lacombe et al., 1990, 1992] and in the magnetosheaths of other planets [e.g. Tsurutani et al., 1982; Balogh et al., 1992]. It has also been suggested that mirror waves have been observed in the solar wind in association with comets [e.g. Russell et al., 1991] and interplanetary shocks [Tsurutani et al., 1992]. The study of Tsurutani et al. [1982] demonstrates the advantages of an analysis based on a dual spacecraft data set (which they were able to use to eliminate some other possible interpretations of the data) and takes care to demonstrate the existence of the temperature anisotropy without which the plasma could not be mirror unstable. It is difficult to make an unambiguous identification of the mirror mode since slow mode waves propagating at large acute angles to the field

have similar characteristics, as we discuss later. Nevertheless, it is important to be able to distinguish between mirror waves and slow mode waves. Slow mode waves are propagating disturbances which in principle can convey energy through the plasma, whereas mirror waves correspond to a static structure (provided the instability has saturated) in the magnetosheath plasma rest frame. Thus only the slow mode wave seems capable of providing wave energy which could potentially stimulate ULF wave activity in the magnetosphere (although the issue of whether slow mode waves can propagate in the magnetosheath or not is controversial, as noted above). On the other hand, mirror structures may be important if the propagation of magnetosonic or Alfven waves in a mirror structure dominated plasma differs from their propagation in a less inhomogeneous plasma. If so, the presence or absence of magnetosheath mirror structure could affect the transport of wave energy from the solar wind or bow shock to the magnetosphere.

In the following we will discuss work which argues that a kinetic model provides a more satisfactory treatment of mirror instability than a fluid model and we will describe a recent interpretation of the kinetic model which emphasises the physics of the instability. We also discuss the question of what can be learned about the three dimensional form of mirror structures using single and dual spacecraft observations. We report what we believe to be an unusually long interval of well correlated mirror wave signatures and discuss the three dimensional form of the plasma/magnetic structures which might give rise to these signatures. We also consider an example of less well correlated mirror waves for comparison. We argue that although most analyses treat mirror waves in two dimensions (parallel and transverse) and do not consider the possibility that the transverse structure may exhibit differing scale lengths in different directions, real mirror structures are not cylindrically symmetric about the mean magnetic field direction.

MIRROR INSTABILITY THEORY

A mirror geometry magnetic field consists of field lines which pass through two regions of relatively strong magnetic field, between which lies a region of relatively weaker magnetic field. The behaviour of plasma in a static magnetic field of this spatial structure is well known. Charged particles with pitch angles larger than a critical pitch angle (defined in terms of the magnetic field strengths in the weak and strong field regions) are trapped in the weak field region. We are interested in a slightly different problem, namely the question of how a uniformly distributed plasma responds when a very slight perturbation patterned on the mirror geometry magnetic field is applied to the uniform magnetic field in which the plasma is immersed. Under certain circumstances the response of the plasma leads to a growth of the perturbation i.e. mirror instability. In the following we argue that we need to consider the response of the plasma to both the spatial structure of the perturbed magnetic field and the variation of the field with time to fully describe the instability.

Linear Theory - MHD Theory

The MHD fluid treatment of the mirror instability depends on the fact that at very low frequencies, a perturbation in the magnetic field is accompanied by a change ($\delta p_\perp$) in the perpendicular plasma pressure $p_\perp$ given by

$$\delta p_\perp = 2 p_\perp \left(1 - \frac{T_\perp}{T_\parallel}\right) \frac{\delta B}{B} \tag{1}$$

for a bi-Maxwellian distribution where $T_\perp$ and $T_{\parallel}$ are respectively the perpendicular and parallel temperatures, and δB is the perturbation to the magnetic field B [Hasegawa, 1969, Equation 60]. The same temperature anisotropy term relates the parallel pressure perturbation and also the plasma density perturbation to the magnetic field perturbation [e.g. Hasegawa, 1969, Equations 25 and 61] showing that the underlying cause of the pressure variation is the plasma density variation. According to Equation 1 a plasma for which $T_\perp > T_{\parallel}$ responds to a magnetic field perturbation with perturbations of the density and of the parallel and perpendicular pressure which are in the opposite sense to the field perturbation. Such a response occurs because the plasma tends to move along the field away from regions of stronger magnetic field and towards regions of weaker magnetic field, behaviour which is usually described in terms of the magnetic mirror force.

In the fluid model, plasma motions perpendicular to the field will occur if a gradient in total pressure, P_t, exists (where P_t is the sum of the perpendicular plasma pressure, P_p, plus magnetic pressure, P_b). The plasma is unstable if the total pressure perturbation δP_t is in the opposite sense to the magnetic field perturbation δP_b (in other words, δP_p opposes and outweighs δP_b). For example, if P_t is reduced in areas of enhanced field strength, the resulting perpendicular motion of plasma (and the accompanying frozen in magnetic flux) into the region of reduced P_t causes an increase in field strength. This growth of field strength in a region of raised field strength is positive feedback and corresponds to growth of the perturbation. Similarly, regions of weakened field become weaker still, again corresponding to growth of the perturbation. The alternative possibility is that P_t rises with a rise in field strength (i.e. δP_b opposes and outweighs δP_p). Then the theoretical consequence is the occurrence of plasma motions which tend to reduce the field perturbation (i.e. negative feedback) but cause the displaced plasma to overshoot its unperturbed position. Thus in the stable case the theory predicts plasma oscillations of constant amplitude, rather than growth or decay of the perturbation amplitude.

So we see that in an unstable plasma P_t should vary in antiphase with P_b. Equation 2 expresses this condition in terms of the level of temperature anisotropy needed for the plasma to become unstable

$$\frac{T_\perp}{T_{\parallel}} > 1 + \frac{1}{\beta_\perp} \tag{2}$$

where P_p and P_b are represented through $\beta_\perp$ the perpendicular plasma beta [Tajiri, 1967]. An explicit derivation of Equation 2

in the fluid approximation (using Equation 1 and hence valid for the low frequency limit) is given by Southwood and Kivelson [1992].

We have used Equation 1 to describe the response of the anisotropic plasma to the magnetic field perturbation and note above that the magnetic mirror force provides a useful way of picturing the cause of the plasma motion along the field. Strictly speaking, the mirror force is not so much a force as a useful way of describing the behaviour of the guiding centre of a charged particle (and hence is helpful when thinking about the plasma in terms of a fluid). The motion of the notional guiding centre (only a valid concept in adiabatic theory) is actually determined by the motion of the particle, as the particle responds to the action of the $v \times B$ Lorentz force in the mirror geometry magnetic field. The total energy, W, of the particle does not change provided that the Lorentz force is the only force acting. The perpendicular energy, $W_\perp$, changes in order to preserve the first adiabatic invariant μ as the particle moves along the parallel gradient in magnetic field strength and a corresponding change in parallel energy, $W_{\parallel}$, must also occur in order to conserve W. Hence the apparent force acting to accelerate plasma along the field away from regions of stronger magnetic field and towards regions of weaker magnetic field, which we think of as the mirror force. Note that the conservation of μ implies that magnetic field disturbances should have low frequencies and large length scales compared to gyration frequencies and radii.

Some treatments of the mirror instability in an MHD fluid complete the set of fluid equations [e.g. Thompson, 1964] using the double adiabatic CGL equations, which effectively treat the perpendicular and parallel pressure as independent quantities. This approach amounts to a neglect of the adiabatic particle behaviour which underlies the coupling of $W_\perp$ and $W_{\parallel}$ and hence the parallel plasma motions which are the key to the linear instability, and leads to an incorrect expression for the instability condition which differs from Equation 2 by a factor of 6.

Equation 2 shows that a high beta plasma becomes unstable at a smaller level of anisotropy than a low beta plasma. High beta plasmas are commonly found downstream of shock waves, where suitable temperature anisotropies are also usually present (particularly when downstream of a quasi-perpendicular shock). Thus it is not especially surprising that mirror structures are often observed in the magnetosheath (although there is some interest in determining why other instabilities which also require the $T_\perp > T_{\parallel}$ anisotropy, such as the ion-cyclotron instability, do not destroy the anisotropy before the mirror structures can grow, e.g. Gary [1992]) .

The instability theory in the fluid approximation yields the correct instability condition, but is not considered to provide a complete and therefore correct description of the instability. For that we need to approach the problem using kinetic theory which allows a more accurate description.

Linear Theory - Kinetic Theory

A kinetic treatment of low frequency waves in collisionless plasmas by Tajiri [1967] revealed a fourth magnetic field perturbing wave mode (in addition to the familiar fast and slow magnetosonic modes and the Alfven mode known from MHD fluid theory). The fourth wave mode obeys the mirror instability condition (i.e. Equation 2) with the unstable solution corresponding to mirror instability as in the fluid case. The stable solution corresponds to a damped, non-oscillatory motion (i.e. the perturbation dies away) rather than the oscillatory motion predicted by fluid theory. It is important to note that this wave mode, which is considered to describe the mirror instability, is distinct from the slow mode (which is described by a separate solution in kinetic theory as noted above). Belmont et al. [1992] reinforce the distinction between slow and mirror mode waves in a novel theoretical analysis which combines aspects of the fluid and kinetic approaches.

The kinetic theory treatments [Tajiri, 1967; Hasegawa, 1969] show that the fluid theory is unsatisfactory, but do not provide a very accessible description of the physics involved. In the following we will present a picture of the kinetic instability in which we argue that the plasma particle population in a mirror unstable plasma can be divided into two distinct groups which have different responses to a perturbation of the magnetic field. The differing behaviour of these two groups of particles cannot both be described in the fluid approximation - hence the difference between the fluid and kinetic theories. A more mathematical presentation of the following material can be found in Southwood and Kivelson [1992].

At the heart of the argument is the idea that the particles having small parallel speeds ($v_{//}$) behave differently to other particles. The procedure for a linear stability analysis of this problem is to imagine that an initially uniform magnetic field is perturbed slightly, becoming weaker or stronger in different areas in a manner consistent with the anticipated mirror structure. If the plasma distribution under consideration is indeed unstable, the analysis will reveal that the perturbation to the field has a tendency to grow. For an instability of growth time τ and characteristic parallel scale length $\lambda_{//}$ it turns out that particles with $v_{//} < \lambda_{//}/\tau$ do not respond in the same way as other particles to the hypothetical field perturbation. All other particles behave in exactly the manner anticipated on the basis of fluid theory. In other words, their response to the magnetic field perturbation is to alter $W_\perp$ while conserving W, thereby also altering $W_{//}$ with the general effect of altering their distribution along the field. However, an additional effect modifies the behaviour of those particles with $v_{//} < \lambda_{//}/\tau$ which by definition must respond predominantly to the temporal evolution of the field (since they move very little along the field with respect to the scale length of the magnetic field perturbation, during the growth time). The consequence is that their total energy can be changed by the changing magnetic field, raised or lowered depending on whether the field in their locality is growing stronger or weaker. Southwood and Kivelson label the particles which satisfy the condition $v_{//} < \lambda_{//}/\tau$ as resonant particles, since they share the same (negligible) speed along the background field direction as the non-propagating magnetic field perturbation and as a result exhibit distinctive behaviour. The pitch angles of resonant particles tend to lie close to 90° due to their small parallel speeds. The bulk of the plasma population is non-resonant and behaves as outlined in the fluid description given above. Thus the perpendicular pressure exerted by the bulk of the plasma P_{pf} (the non-resonant particles) declines in response to a slight magnetic compression, causing a perturbation in $\delta P_{tf} = \delta P_{pf} + \delta P_b$ equivalent to the total pressure perturbation found in the fluid model. The total energy W of these non-resonant particles is unaffected by the magnetic field perturbation (although energy is redistributed between $W_\perp$ and $W_{//}$). However, the true total pressure P_t remains balanced (here the kinetic picture differs from the fluid picture) due to the behaviour of the resonant particles. Wherever the magnetic field strengthens, conservation of μ increases the total energy W of the resonant particles (i.e. betatron acceleration) and thereby raises the perpendicular pressure due to the resonant particles P_{pr}, so that $\delta P_t = \delta P_{tf} + \delta P_{pr}$ is zero (P_t is similarly conserved where the field weakens). The energy gained by the resonant particles in regions of strengthening magnetic field is balanced by the energy lost by the resonant particles in regions of weakening magnetic field. Thus the particle population as a whole neither gains nor loses energy in the linear case.

Southwood and Kivelson [1992] have provided a formal derivation of the argument summarised here and derived an expression for the linear instability growth rate γ (in the limit of small growth rate γ or large τ) given in terms of the parallel thermal velocity $v_{T//}$, the parallel wavevector $k_{//}$, the temperature anisotropy $A = T_\perp/T_{//}$, and the perpendicular plasma beta $\beta_\perp$ as

$$-\frac{\gamma}{k_{//} v_{T//}} \sqrt{\frac{\pi}{2}} = \frac{1 + \beta_\perp (1-A)}{\beta_\perp A} \tag{3}$$

valid for the special case of a bi-Maxwellian distribution. As ought to be the case, this result is identical to the result given by Hasegawa [1969] (his Equation 23, in the small perpendicular, large parallel wavelength limit) who also considered the special case of a bi-Maxwellian distribution function. The treatment of Southwood and Kivelson also shows that an expression for the growth rate can be written for any plasma velocity distribution function F

$$-\frac{\gamma}{k_{//}} = \frac{B^2}{\mu_0} \frac{1 + \beta_\perp (1-A)}{2\pi T_\perp F_{res}} \tag{4}$$

where F_{res} represents that part of the distribution function for which $v_{//} = 0$ (i.e. the resonant particles in the linear theory, in the limit of small growth rate γ).

The growth rate is found to be inversely proportional to the number of resonant particles. This unusual relationship (most resonant instabilities have γ directly proportional to the number of resonant particles) is due to the pressure balancing role of the resonant particles discussed above. For example, imagine two different distributions in which the non-resonant particles both cause a given δP_{tf} but the number of resonant particles is different. In both cases the resonant particles must provide the same δP_{pr} to prevent a perturbation of the total pressure. However, δP_{pr} depends on the number of resonant particles and on the amount of energy they have gained. Thus the distribution with fewer resonant particles must experience a greater change in resonant particle energy to provide the same resonant pressure perturbation, which can only be achieved by having a larger growth rate. Hence the growth rate is inversely proportional to the number of resonant particles.

Non-linear Development

An instability will usually begin to modify a plasma as soon as the plasma is even marginally unstable and the changes in the plasma often make further phenomena possible. These additional phenomena could not occur to the plasma in its initial state and are termed non-linear. We will discuss two possible mechanisms for heating particles as the mirror structure develops.

As the mirror instability begins to take effect in an unstable plasma the growth rate will start to rise above the very low level we considered in discussing the linear instability analysis (above). At first sight the rise in γ (fall in τ) appears to imply that more particles will have parallel speeds which satisfy the resonant condition $v_{//} < \lambda_{///}/\tau$ so that they cannot escape a given region in a growth time. Thus some particles which are considered non-resonant in the linear analysis may become resonant as the instability develops. The behaviour of particles with parallel velocities in this category needs careful consideration. If they are travelling in the $-gradB$ direction associated with the growing magnetic field perturbation (i.e. in the direction of the mirror force) their small parallel speeds will grow. However, if they are travelling in the opposite direction (i.e. towards a region of strengthening field) they will lose parallel speed until at some point they become categorisable as resonant particles. These particles may ultimately undergo a reflection, but while travelling very slowly parallel to the field they will experience betatron acceleration. Since the particles are initially non-resonant in the virtually unperturbed field (as treated in the linear case) they are unlikely to lose sufficient parallel speed to become resonant unless they move into regions where the magnetic field is growing stronger than the unperturbed level. Thus, when they do become resonant, they will be heated rather than cooled. Unlike the linear case where the distribution of resonant particles among regions of strengthening and weakening field is uniform, and so no net heating occurs, this mechanism implies a net heating of the particle population, i.e. the number of resonant particles which absorb energy appears to outweigh the number which release energy. Such behaviour seems consistent with the expectation that the resonant particles (the only ones able to change their energy) must take up the free energy associated with the initial temperature anisotropy, as it is released through the action of the instability. This makes an interesting contrast with the more usual form of resonant instability in which a wave develops at the expense of the energy of resonant particles.

As the field perturbations grow, a small but increasing fraction of the initially non-resonant particle distribution will be reflected from regions of strengthening magnetic field and hence become trapped, bouncing back and forth along the field between two such regions of strengthened field. Thus the plasma evolves toward a state in which most of the particles are concentrated in weak field regions and stronger field regions are relatively devoid of particles. Those members of the trapped particle distribution which have larger pitch angles will tend to have shorter bounce paths than those with smaller pitch angles. We suggest that particles which are mirroring in the region of strengthening magnetic field (i.e. trapped particles with smaller pitch angles) will find that their mirror points are gradually moved toward the weaker field region, while the reverse applies to the mirror points of particles trapped in the region of weakening field (i.e. trapped particles with larger pitch angles). Thus the former population may be cooled and the latter population heated by Fermi acceleration as they bounce between magnetic mirrors which are respectively receding and approaching one another.

The eventual saturation of the instability, when growth comes to an end, is a non-linear effect. Hasegawa [1969] suggests that growth must end when the perpendicular length scale is reduced to a value of the order of an ion gyroradius (and particle behaviour can no longer be described by adiabatic theory). This is roughly equivalent to the condition that the growth time is of the same order as the bounce time for a typical trapped particle.

OBSERVATIONS

Mirror structures and slow mode magnetosonic waves are both expected to be characterised by predominantly compressional magnetic field perturbations in which the magnetic field strength varies in antiphase with fluctuations in plasma density. As slow mode waves are true propagating waves, we expect oscillatory plasma motion (with respect to the plasma rest frame) to occur and that characteristic relationships should exist between these plasma oscillations, the magnetic field perturbations and the background magnetic field. Some analyses of this sort have been attempted [e.g. Gleaves and Southwood, 1991a]. A simpler test may be to compare the values of μ (first adiabatic invariant) in the strong and weak field regions. Tsurutani et al. [1982] describe observations of mirror waves which are not consistent with conservation of μ throughout the plasma. Instead, strong field regions appear to be adiabatically decoupled from weak field regions. It is not clear that such behaviour can be expected in a plasma through which a slow mode wave is propagating, whereas the development of a mirror structure may lead to adiabatic decoupling of strong and weak field regions due to heating of the trapped plasma (as discussed above). After all, many mirror wave observations are likely to be of plasma in which the mirror instability is saturated and non-linear effects have had a chance to act while stabilising the tendency for growth. Furthermore, some authors believe that strong damping processes cause slow mode waves to decay before they can propagate very far, in which case an observation of compressional waves lasting many cycles is more likely to be related to mirror structures than slow mode waves. In the following we discuss two sets of data which we suggest are observations of mirror waves. We will outline some constraints on the possible three-dimensional form of mirror structures, and then interpret our data in terms of our model mirror structure.

Hypothetical Mirror Structure

The hypothetical structure we describe here is based on the theoretical picture of mirror instability and is constrained by our knowledge of how a mirror structure looks when observed by a single spacecraft. The redistribution of plasma along the magnetic field that occurs when a mirror structure develops is associated with the simultaneous appearance of regions of weaker and stronger magnetic field, along the field direction. Plasma becomes trapped between regions of strong field. We assign a (parallel) scale length l_1 to describe the separation of the field strength maxima (or minima) parallel to the mean magnetic field direction (unit vector b). The parallel gradients in magnetic field strength must be accompanied by perpendicular gradients, in order to satisfy $div.B = 0$, so that there must also be transverse components of B (i.e. perpendicular to b). Theoretical work by Barnes [1966] and Price et al. [1986] suggests that the fastest growing mirror waves have a wave vector inclined at about 70° to b so that the perturbation field vector lies at 20° to b (authors reporting observations of mirror structures with wave vectors oriented at about this angle to b include Tsurutani et al., [1982] and Hubert et al., [1989a]). Observations typically indicate that the compressional perturbation is less than half the average field strength, thus the tilt of the perturbed field vectors (unperturbed plus perturbation) is usually quite slight. The width of a region of trapped plasma (and weak field) measured in the plane containing the field and the field perturbation may be characterised by a second (perpendicular) scale length l_2 which will usually be significantly smaller than l_1. The simulations of Price et al. [1986] offer an idea of how a two dimensional slice through a mirror structure in the plane containing l_1 and l_2 might appear. Theoretical mirror instability studies discuss gyrotropic plasmas and thus implicitly assume that the structures are symmetrical about b, whereas simulations are confined to two dimensions. However we note that single spacecraft observations have indicated that transverse magnetic fluctuations in the magnetosheath tend to lie preferentially on surfaces lying parallel to the local shock or magnetopause surface [Fairfield and Ness, 1970]. These observations suggest that the velocity shears and curved streamlines which characterise the flow of magnetosheath plasma around the magnetosphere and lead to the draping of the magnetosheath magnetic field about the magnetosphere, cause an ordering of the orientation of transverse magnetic fluctuations in the magnetosheath. It therefore seems likely that shears in the flowing plasma may lead to increasing distortion of mirror structures as they are carried through the magnetosheath. Thus the cross-section of a trapped plasma region might be initially circular, but become distorted into a more flattened, elliptical form, in which case we need a second transverse dimension l_3 (orthogonal to l_1 and l_2) to characterise the structure. As noted above, Hasegawa suggests that growth should cease when the perpendicular wavelength becomes of order an ion gyroradius, thus the transverse scale lengths l_2 and l_3 should be at least one or two gyroradii at the time when the instability saturates (care must be taken to define the local magnetic field before defining the gyrodiameter).

Inferences from Observations

In the magnetosheath a spacecraft is immersed in a flowing magnetised plasma. The plasma flow can be resolved into components $v_{//}$ along b and $v_\perp$ normal to b. The natural choice for a coordinate system in the following discussion is the orthogonal set (a,b,c) where b is defined above, a is the direction along $v_\perp$ and c, which completes the set, lies across the flow direction in the transverse plane. In the plasma rest frame the spacecraft has a velocity $(-v_\perp,-v_{//})$ in the (a,b) plane.

We will find it helpful to describe the model mirror structure outlined above in terms of the (a,b,c) coordinate system. Figure 1 shows a flattened ellipsoid, which is intended only as a crude representation of the volume occupied by plasma trapped in a mirror structure, for use in illustrating some of the parameters we define below. The ellipsoid probably corresponds least well to a real mirror structure at $(0,0,0)$ and $(0,l_1,0)$ which represent the two field maxima which cap the trapped plasma region. The model structure has an elliptical cross-section (in the (a,c) plane). The ellipse axes need not be aligned with the a and c directions. Their lengths l_2 and l_3 vary with distance along b according to the local field strength, so that the largest values (say l_2' and l_3') are found in the transverse plane which passes through $(0,l_1/2,0)$ where the field is weakest in our symmetrical model. The axis lengths become smaller nearer the ends of the structure as the field strengthens and field lines converge. The transverse scale length in the plane containing b and the perturbation field vector will probably be significantly smaller than l_1. A spacecraft will detect a trough in magnetic field strength (a weak field region) when it passes through a region of trapped plasma, but in order to estimate the dimensions of the trapped plasma region from such data we need to understand how the spacecraft path relates to the model structure.

If the spacecraft motion in the plasma rest frame is sufficiently slow along the field compared to its motion across the field, the variations in field it observes will be largely due to transverse structure. Then, given $v_\perp$, the duration of a trough can be translated into a distance l_T, the width of the weak field region measured along a in the plasma rest frame. The neglect of parallel structure will be a valid assumption provided that $v_\perp/v_{//} \gg l_T/l_1$, which may often be true if the parallel scale length dominates the perpendicular scale lengths as theories suggest. If the spacecraft passes through $(0,0)$ in the (a,c) plane, the core of the weak field region, l_T will take a value between the long axis length, say l_2, and the short axis length l_3 (depending on the ellipse orientation). If the spacecraft track passes to one side of the (a,c) plane origin the upper and lower limits on the range of possible values for l_T both fall. Thus the inferred value of l_T for a flight through a given structure depends on the path through the structure, and could take a maximum value of l_2' (when $b =$

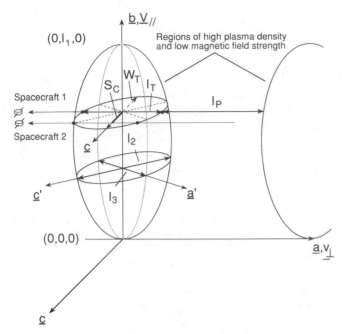

Fig. 1. An illustration of the coordinate systems and parameters used in the text to describe mirror structures and parameters inferred from observations.

$l_1/2$) but may be smaller. For a given structure, we would expect the measured field minimum to be less deep for crossings which miss the origin in the (a,c) plane or crossings for which the b coordinate is not near $l_1/2$, compared to the result for a passage through $(0,l_1/2,0)$. We cannot extend this argument to the interpretation of a sequence of magnetic minima of differing depths in real data, as we cannot assume that all mirror structures observed in a given sequence will be essentially the same size. The duration of a peak of field strength can be translated into a distance l_P. Short lived peaks might be interpreted as passage of the spacecraft through one of the field maxima which caps a region of trapped plasma, but longer lasting ones may indicate an extended region in which mirror structures associated with trapped plasma are absent. If a clear plane of greatest transverse disturbance exists in the data it is reasonable to define directions of maximum and minimum disturbance and to use l_T to estimate values l_2 and l_3 in terms of those directions. Little more can be inferred using data from a single spacecraft data.

The data from a suitably located second spacecraft enables us to make estimates of the dimensions of mirror structures in the c direction (provided that $s_{//} \ll l_1$, where $s_{//}$ is the parallel component of the spacecraft separation, so that we can argue that any differences between the two data sets are due to differences in transverse structure). It is hard to demonstrate that an isolated magnetic field trough seen in the data from one spacecraft is due to the same magnetic structure as a trough seen in data from another spacecraft. However, if the sequence of troughs seen at one spacecraft is repeated at the other, it seems likely that both spacecraft are seeing the same set of mirror structures, and we may then infer correspondence

between individual troughs. The projection of the structure onto the (b,c) plane has a width w_T measured along c. If both spacecraft pass through a given structure, w_T must be greater than the c component of the spacecraft separation s_c. If a long sequence of well correlated troughs is seen, w_T is probably much larger than s_c, otherwise we would expect to occasionally detect a structure at only one spacecraft. When dual spacecraft observations show little or no correlation between the signatures recorded by the two spacecraft, it suggests that it is common for a structure to be seen by one spacecraft but not by the other. Consequently we then infer that w_T must be smaller than s_c, but that the spacing between trapped plasma regions (measured along c) w_P must typically be as large as s_c. If w_P were smaller we might often see a trough at both spacecraft simultaneously, although the troughs would be related to different structures and hence probably of different durations. A well correlated dual spacecraft data set allows us to draw further conclusions as we discuss below with reference to Data Set 2.

We have concentrated here on what can be discovered when we have data from two spacecraft with relatively low parallel separation and when parallel structure can be neglected. If the spacecraft had larger parallel separation we could investigate the variation of transverse structure as a function of distance along b, but only at the expense of knowledge regarding the c dimension. The four spacecraft Cluster mission has the potential to allow us to carry out studies of mirror structures in all three dimensions simultaneously and thus reduce our dependence on assumptions about scale sizes. A useful configuration might be to place the spacecraft at points $(0,0,0)$, $(l_a,0,0)$, $(0,l_b,0)$ and $(0,0,l_c)$ defined in terms of (a,b,c) with l_a, l_b and l_c chosen to suit anticipated values of l_1, l_T and w_T. Ideally the flow velocity would be predominantly transverse to the field. Due to fluctuations in the direction of the background magnetosheath field there would be an element of luck in attaining such a configuration, but if the planning flexibility existed, careful orientation of the spacecraft based on typical field and flow directions could enhance the chances of success.

In the following we will analyse two dual spacecraft data sets in terms of the arguments outlined above, by assuming that the spacecraft data reveal transverse rather than parallel structure. We consider the plausibility of the assumption in each case.

Data Set 1

The first data set is taken from 0216-0232 on 8 November (Day 312) 1977 and has been described previously by Tsurutani et al. [1982] who discussed a variety of possible explanations of the data before selecting the mirror wave interpretation. The observations were made at about (10.4, -3.0, 4.8) R_E GSE, in the magnetosheath near the dayside magnetopause. The mean magnetic field B_m was about 28.3 nT in the direction (0.200, 0.956, -0.103) GSE and s was about 339 (-0.643, 0.631, -0.434) km. The transverse (i.e. perpendicular to B_m) component of the spacecraft separation distance, $s_\perp$, is about 288 km and lies virtually perpendicular to the transverse component of the plasma flow (at about 88°) which had a speed of 73 km/s (see Figure 2). The separation of the spacecraft along the transverse flow direction is thus only about 9 km. The parallel separation of the spacecraft is 179 km and the parallel flow speed only 33 km/s. The gyroradius of an ion λ_i depends on the ion temperature and the magnetic field strength, both of which vary systematically in mirror structures. We estimate the range of values of λ_i to be about 52 km $> \lambda_i >$ 18 km with larger values (λ_{iw}) corresponding to the weaker field regions where plasma is concentrated and smaller values (λ_{is}) to the regions of stronger field which separate the plasma rich regions.

We assume that the data can be interpreted in terms of the mirror structure model discussed above and that the magnetic signatures represent the detection of transverse structure only (note that $v_\perp/v_{//}$ is about 2.2 so we are making the plausible assumption that $l_1 \gg 0.45\ l_T$). As indicated in Figure 3, there is little correlation between the observations of ISEE 1 and 2, though occasionally troughs coincide in a manner consistent with a slight lead by ISEE 2 much as suggested by Tsurutani et al.. The generally poor correlation implies that w_T is usually smaller than s_c and w_P is usually larger than s_c (as we have argued above). Due to the geometry of the spacecraft orientation on this day s_c is virtually equal to $s_\perp$ (= 288 km) which is about 5 λ_{iw} (appropriate if both spacecraft are within the same trough, which is not necessarily the case even if both see a weak field) or 16 λ_{is} (appropriate if no troughs lie between the spacecraft, although in principle one might pass between them undetected). We can be more precise in discussing dimensions measured along a. The magnetic field

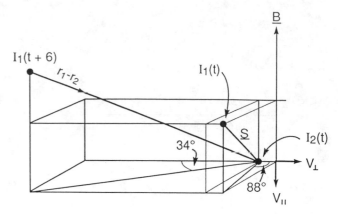

Fig. 2. The relationship between spacecraft position, magnetic field and plasma flow together with the vector r_1-r_2 for Day 312. The magnetic field and plasma flow vectors lie in the plane of the paper.

data (e.g. Figure 3) can be described as a series of troughs, broader at the top than the bottom and separated by regions of strong field which occur as narrow peaks or broad plateaus. Most troughs are deep and of similar depths, but a minority are shallower. Deep troughs have field minima lasting 5 to 10 s (occasionally 20 s) but may also be described as lasting between 10 and 30 s (typically 15 s) in terms of the time the field spends below the typical high value. In terms of distances, l_T for a typical trough lies in the range 730 to 2190 km or 14 to 42 λ_{iw},

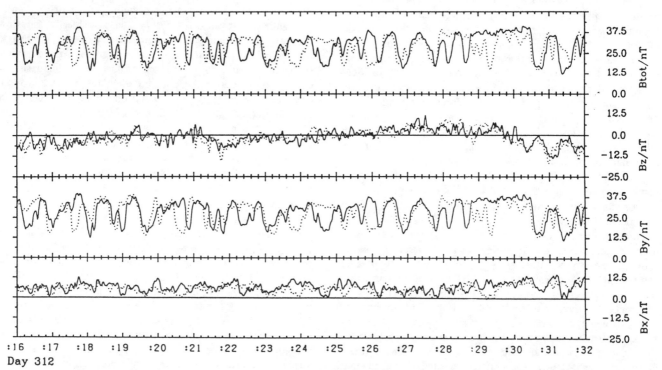

Fig. 3. Magnetic field in GSE coordinates during a 16 minute interval on Day 312. The ISEE 2 trace (dotted) has been shifted by 6 s to remove the observed lead. Note the poor correlation between the ISEE 1 and 2 traces.

with the weakest field found in a region measuring 365 to 1460 km or 7 to 28 λ_{iw} along a. The field typically remains at a high value for 10 to 50 s corresponding to 730 to 3650 km or 40 to 200 λ_{is} (but can last for as short as 5 s or as long as 110 s, 20 to 450 λ_{is}). The shallower troughs tend to last no longer than 10 s (of order 14 λ_{iw} though a slightly larger value may be appropriate to reflect the smaller gyroradii in a shallower trough). Note that if $l_T \ll l_1$ as we have assumed then since $s_{//} \ll l_T$, our second constraint (that $s_{//} \ll l_1$) is easily satisfied.

In summary, our interpretation of the data suggests the spacecraft encountered bodies of plasma that were very narrow (<5 λ_{iw}) in the direction perpendicular to the flow, but stretched out along the plasma flow vector (e.g. 15-40 λ_{iw}) and widely spaced out along the plasma flow direction (40-200 λ_{is}). As we note below, this evidence is insufficient to allow us to describe the cross-sectional shape of the bodies of plasma. Measured along $v_\perp$ the trapped plasma regions seem to consist of a weak field core surrounded by a transition region of about half the width, in terms of gyroradii. The variability in the structure sizes and separations suggest a fairly disordered arrangement of structures in the plasma rest frame.

In the light of these results, it seems likely that the poor correlation is a result of the spacecraft separation vector s lying almost perpendicular to a, which was a rather unsuitable orientation for detecting structures of the dimensions and orientations that we describe above. If s had been aligned with the flow, a good correlation would be expected between the signatures (according to our interpretation of the data) as the plasma flow carried the structures over each spacecraft in turn. In order to test this idea, we searched for examples of well correlated data during those months when the ISEE trajectory seemed likely to lead to a reasonable alignment between s and the flow. Our second data set is an example of well correlated data from our search, but interestingly our analysis suggests that the form of the mirror structures differed from the form we have inferred using the Day 312 data.

Data Set 2

On 3 October (Day 276) 1978 the ISEE 1 and 2 spacecraft were located at about (14.5, 4.7, 2.3) R_E GSE travelling inbound. The mean magnetic field B_m was approximately 14.1 nT in the direction (-0.176, 0.978, -0.046) during the interval 1821-1834. The spacecraft separation vector s, from ISEE 2 to ISEE 1, was about 2232 (0.950,-0.089, 0.299) km. Resolving the spacecraft separation into parallel and transverse (with respect to the field) components, ISEE 2 lies about 600 km down field of ISEE 1 and about 2150 km away across the field. The plasma flow vector can be resolved into components of 78 km/s up field and 108 km/s across the field. The angle between the transverse component of the spacecraft separation vector and the transverse component of the flow vector is 133°, rather than the sought after 0° or 180°.

The spacecraft crossed the bow shock 20-30 minutes before encountering what we believe to be mirror structures. Plasma data indicates that the plasma temperature anisotropy satisfies the mirror instability condition (Equation 2) and that the

magnetic field pressure varies in antiphase with the plasma pressure (and density) while conserving the total pressure. The large number of cycles in the magnetic fluctuations (see Figure 4) and non-conservation of μ also support the view that these are mirror waves (as discussed earlier). We hope to present plasma data in a future paper.

Both spacecraft recorded mirror signatures during the interval 1819-1835. The phase delay between the two signatures can readily be determined after cross-correlating the data sets and is about 40 s, with ISEE 1 leading. These mirror signatures are strikingly similar, as is apparent in Figure 4 where the 40 second time lag has been removed to facilitate a comparison. Thus at any given time t during the mirror interval, the magnetic field strength at the location (r_1) of ISEE 1 is very similar to the field strength at the location of ISEE 2 at time t + 40 s (r_2). If s had been aligned with the flow velocity v, we would expect well correlated, time-delayed signatures as the spacecraft flew along the same path through the plasma rest frame one after the other such that $r_1 = r_2$. If s were not parallel to v, but $s_\perp$ were nevertheless aligned with $v_\perp$, we would still expect well correlated, time-delayed signatures provided parallel structure could be ignored. In that case structure varies with position in the (a,c) plane, and as the spacecraft paths projected onto the (a,c) plane would still be the same the spacecraft should see the same sequence of structures (r_1 and r_2 would have the same (a,c) coordinates).

Surprisingly however, although the data is well correlated, r_1 and r_2 do not coincide on Day 276. Instead, we find that the relative separation vector $r_s = (r_1 - r_2)$ corresponds to a parallel separation (i.e. along b) of 2520 km and a transverse separation of 3260 km in the plasma rest frame. The transverse component of r_s makes an angle of about 29° with $v_\perp$ as shown in Figure 5. Put another way, the spacecraft travel through the plasma rest frame along parallel tracks (in the ($-v_\perp, -v_{//}$) direction) separated in the (a,c) plane by 1570 km (measured along c) and observe the same signatures at points in the plasma rest frame separated by 2855 km along a and 2520 km along b. If parallel structure makes no contribution to the observations, we suggest that the mirror structures are organised in such a way that a given signature might be observed anywhere in the plane defined by r_s and b, at least on length scales of order $r_{s\perp}$ and $r_{s//}$ (the components of r_s measured perpendicular and parallel to b). We will refer to a coordinate system (a',b,c') in the following, in which c' lies along $r_{s\perp}$ and a' completes the set. Thus these new coordinates represent a rotation of the (a,b,c) coordinates about b, and we hypothesise that observers anywhere in the (b,c') plane and in the vicinity of the spacecraft will see the same field. We define parameters such as l_T' in (a',b,c') coordinates by analogy with their counterparts, such as l_T, in (a,b,c) coordinates.

It does not seem unreasonable to assume that parallel structure can be ignored. On Day 276, $v_\perp/v_{//}$ is about 1.4 so that our assumption becomes that $l_1 \gg 0.71\, l_T$ which is not an extreme requirement. If we suppose for a moment that most of the observed magnetic field fluctuations are due to parallel structure then we must explain why the same pattern of peaks

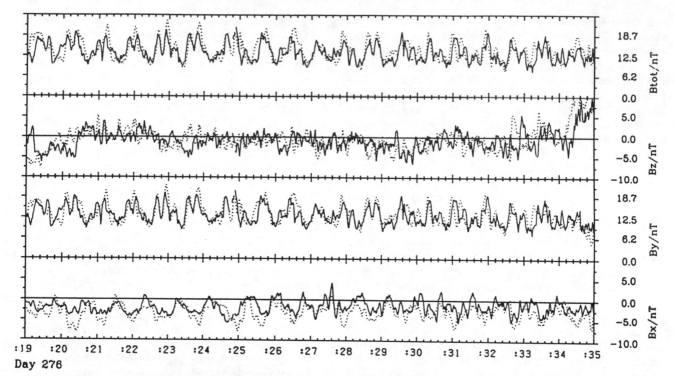

Fig. 4. Magnetic field in GSE coordinates during a 16 minute interval on Day 276. The ISEE 2 trace (dotted) has been shifted by 40 s to remove the observed lag. Note the good correlation between the ISEE 1 and 2 traces.

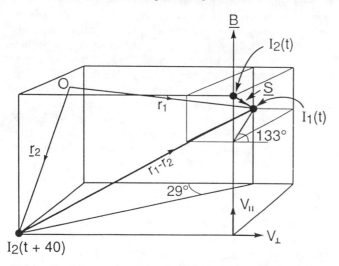

Fig. 5. These figures show the relationship between spacecraft position, magnetic field and plasma flow together with the vector r_1-r_2 for Day 276. An arbitrary origin O is included. The magnetic field and plasma flow vectors lie in the plane of the paper.

and troughs that exists along the field lines passing through r_1 is also present on the field lines passing through r_2, but is displaced along b by 2520 km, a large distance compared to the scale size of the structure (e.g. in the parallel structure interpretation, a 5 s long magnetic peak corresponds to the spacecraft travelling only 390 km along b). The difficulty in explaining how such a phenomenon could arise supports the

view that transverse structure is largely responsible for the observed signature.

We estimate that the range of values of gyroradii throughout the mirror region is 300 km > λ_i > 120 km with larger values corresponding to weaker magnetic fields. The separation of the spacecraft along c' is 3530 km, i.e. larger than 11 λ_{iw}. As virtually all magnetic troughs seen by one spacecraft were seen by the other, we suggest that w_T' is rather larger than 11 λ_{iw} otherwise we would occasionally expect a body of trapped plasma to miss one of the spacecraft. An inspection of the high resolution magnetic field data reveals quite a complicated signature. Peaks of field strength (18 to 23 nT) are typically shortlived (usually no more than 5 s in duration) and separated by intervals of varying length. Shorter intervals (typically 10 to 15 s) between peaks generally correspond to shallower troughs (13 to 15 nT). Longer intervals between peaks (typically 20 to 30 s) tend to be accompanied by deeper troughs (9 to 12 nT). In some cases a shallow trough appears to lie adjacent to a deep trough without a well defined peak separating them. This pattern is a reasonable description of the first 10 minutes or so, but after about 18:30 more of the troughs are both deep (<13 nT) and of shorter duration (<15 sec). Proton gyroradii in the deeper (<13 nT) magnetic troughs are usually in the range λ_{iw} = 200 to 300 km. We estimate that l_T for these troughs is roughly 2200 to 3200 km in the plasma rest frame (8 to 16 λ_{iw}). The shallower troughs, with durations 10-15 s have l_T = 1000 to 1600 km (5 to 10 λ_i, with λ_i in the range 150-200 km). Regions of peak field strength (<20 nT) lasting roughly 5 s correspond to

500 km or perhaps 3 or 4 λ_{is}. Due to the arrangement of deep and shallower troughs, peaks can be separated by as much as 50 s, 20-30 gyroradii. The brief, deep troughs seen after 18:30 seem to correspond to only about 2 λ_{iw}. Since the $(\underline{r}_s, \underline{b})$ planes are tilted with respect to the spacecraft paths in the plasma rest frame, the thickness of the dense plasma regions measured along the normal to the planes must be smaller than the values given above, by a factor of sin(29°) or about 0.5.

Thus we estimate that $l_{T'}$ is 4 to 8 λ_{iw} for a deep, wide trough, which may correspond to the spacecraft crossing the region of weakest field of a structure. A shallower trough may be only 3 to 5 λ_{iw} across. As discussed above, variations in trough field strength could be due both to encounters with structures of different size and core field strength and to encounters with similar structures in which the spacecraft pass at different distances from the core region. We estimate (above) that $w_{T'}$ is no less than 11 λ_{iw} and probably somewhat larger. The dimension of the peak field region, measured along $\underline{a}'$ is only 1 or 2 gyroradii, consistent with the suggestion of *Hasegawa* (see above). However, the brief, deep troughs appear to be surprisingly narrow with $l_{T'}$ of order only a gyroradius wide (although we noted above that measured values of l_T and hence of $l_{T'}$ may underestimate the true transverse dimensions of a structure).

We therefore argue that the most likely explanation of our data is that the spacecraft encountered a series of field aligned bodies of trapped plasma, thinner in one transverse direction than the other and stretched out along the field (perhaps resembling a thick ribbon or possibly sheet depending on their dimension along the plane) separated from one another by regions of lower plasma density. The ribbons, or sheets, are organised in layers tilted at an angle of about 30° to the transverse component of the flow (see Figure 6).

Data Set 1 Revisited

If the intermittent evidence of a correlation (with a 6 s ISEE 2 lead [e.g. *Tsurutani et al.*, 1982]) is meaningful, an analysis in terms of $\underline{r}_s$ suggests that the plasma bodies may be organised in planes tilted at about 34° to the transverse plasma flow vector (see Figure 2) much as on Day 276. However, unlike the Day 276 situation, we must conclude that very few such bodies are broad enough to engulf both spacecraft and so generate correlated signals. In this tilted plane interpretation, our estimates of trough dimensions are modified to account for the tilt. The projection of the spacecraft separation onto the tilted plane $(s_{c'})$ is 514 km, about 10 λ_{iw}. Measuring along $\underline{a}'$ we find the trough width across the plane $l_{T'}$ is about 8 to 24 λ_{iw} (or 4 to 16 λ_{iw} for the interval of minimum field in the trough) and the troughs are separated by strong field regions $(l_{P'})$ of about 20 to 110 λ_{is}. In the our earlier interpretation of this data we set constraints on the size of a mirror structure measured along $\underline{a}$ and $\underline{c}$, but we could not specify a shape for the structure. In this alternative interpretation of the Day 312 data, that difficulty is removed by the introduction of the idea that the structures are organised in planes tilted with respect to the flow (see Figure 7) allowing us to apply the flattened ellipsoid model with $l_{T'}$ and

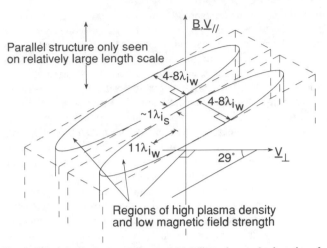

Fig. 6. Sketch indicating our estimates of the dimensions and orientation of the mirror structures observed on Day 276. The regions of trapped plasma are indicated.

$w_{T'}$ as the transverse ellipse axes. Comparing the results with the Day 276 picture, we find that our Day 312 structures are longer and much more spread out along $\underline{a}'$, but much narrower along $\underline{c}'$, in terms of gyroradii. If Day 276 can be compared to closely spaced sheets of trapped plasma, Day 312 is more similar to widely spaced sheets with large holes in them. It is tempting to suggest that there is a progression from well defined structures near the bow shock (e.g. Day 276) to a more fragmented state nearer the magnetopause (e.g. Day 312) perhaps as a result of the shears in the magnetosheath flow field, but we clearly need to study many more intervals before drawing any firm conclusions.

SUMMARY

We have briefly reviewed studies of wave propagation in the magnetosheath, a field which seems likely to be of some

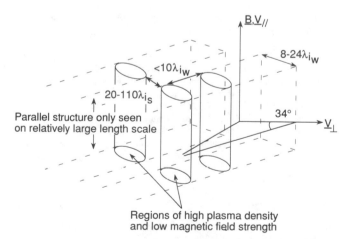

Fig. 7. Sketch indicating our estimates of the dimensions and orientation of the mirror structures observed on Day 312, assuming that the structures are organised to some extent, rather than arranged purely at random.

importance in understanding how the solar wind stimulates magnetospheric ULF wave activity. The magnetosheath is still a relatively poorly understood region. Mirror waves are quite frequently observed in the magnetosheath but are also still rather mysterious in some respects. We have discussed the mirror instability at some length, describing recent work in which a new resonant interpretation of the mirror instability in the linear case has been proposed and mentioning some ideas about the non-linear development of the instability. We have used the dual spacecraft ISEE data set to study the little understood three dimensional form of magnetosheath mirror structures. The analysis has been simplified by reducing the problem to two dimensions (by assuming that we can ignore the parallel structure). We suggest that the mirror data presented here might be explained in terms of trapped plasma confined in bodies with distinctly different widths and breadths, and of large but undetermined length parallel to the field. Our first example, from the vicinity of the magnetopause, is typical of many intervals in which there is at best a rather poor correlation. In our second example, recorded near the bow shock, the plasma is apparently organised in well defined planar structures (on the length scales of the observations) tilted with respect to the flow. Our analysis shows that dual spacecraft data cannot be used to give an unequivocal description of the three dimensional form of mirror structures, but illustrates the potential of the four point simultaneous measurement capability of the forthcoming Cluster mission to resolve such questions. A clearer picture of the growth and form of mirror structures is emerging, which should make it easier to distinguish mirror and slow mode signatures in data, and help us to understand wave propagation in the magnetosheath.

Acknowledgements. We thank C.T. Russell (U.C.L.A.) and M.F. Thomsen (L.A.N.L.) for generously providing magnetic field and plasma data. AF thanks the UK SERC for financial support.

REFERENCES

Balogh, A., M.K. Dougherty, R.J. Forsyth, D.J. Southwood, E.J. Smith, B.T. Tsuritani, N. Murphy and M.E. Burton, Magnetic field observations during the Ulysses flyby of Jupiter, *Science, 257*, 1515-1518, 1992.

Barnes, A., Collisionless damping of hydromagnetic waves, *Phys. Fluids, 9*, 1483-1495, 1966.

Belmont, G., D. Hubert, C. Lacombe, and F. Pantellini, Mirror mode and other compressive modes, *Proceedings of the 26th ESLAB Symposium,* Killarney, ESA SP-346, 263-267, 1992.

Engebretson, M.J., N. Lin, W. Baumjohann, H. Luehr, B.J. Anderson, L.J. Zanetti, T.A. Potemra, R.L. McPherron, and M.G. Kivelson, A comparison of ULF fluctuations in the solar wind, magnetosheath, and dayside magnetosphere, 1, magnetosheath morphology, *J. Geophys. Res., 96*, 3441-3454, 1991.

Fairfield, D.H., and N.F. Ness, Magnetic field fluctuations in the Earth's magnetosheath, *J. Geophys. Res., 75*, 6050-6060, 1970.

Fairfield, D.H., Magnetic fields of the magnetosheath, *Rev. Geophys., 14*, 117-134, 1976.

Gary, S.P., The mirror and ion cyclotron anisotropy instabilities, *J. Geophys. Res., 97*, 8519-8529, 1992.

Gleaves, D.G., and D.J. Southwood, Magnetohydrodynamic fluctuations in the Earth's magnetosheath at 1500 LT: ISEE 1 and ISEE 2, *J. Geophys. Res., 96*, 129-142, 1991a.

Gleaves, D.G., and D.J. Southwood, MHD wave propagation in the magnetosheath: recent results, *J. Geomag. Geoelectr., 43*, 631-644, 1991b.

Hasegawa. A., Drift mirror instability in the magnetosphere, *Phys. Fluids, 12*, 2642-2650, 1969.

Hoppe, M.M., C.T. Russell, L.A.Frank, T.E. Eastman, and E.W. Greentsadt, Upstream hydromagnetic waves and their association with backstreaming ion populations: ISEE 1 and 2 observations, *J. Geophys. Res., 86*, 4471-4492, 1981.

Hubert, D., C. Perche, C.C. Harvey, C. Lacombe and C.T. Russell, Observation of mirror waves downstream of a quasi-perpendicular shock, *Geophys. Res. Lett., 16*, 159-162, 1989a.

Hubert, D., C.C. Harvey and C.T. Russell, Observations of magnetohydrodynamic modes in the Earth's magnetosheath at 0600 LT, *J. Geophys. Res., 94*, 17305-17309, 1989b.

Lacombe, C., E. Kinzelin, C.C. Harvey, D. Hubert, A. Mangeney, J. Elaoufir, D. Burgess, and C.T. Russell, Nature of the turbulence observed by ISEE 1-2 during a quasi-perpendicular crossing of the Earth's bow shock, *Ann. Geophysicae, 8*, 489-502, 1990.

Lacombe, C., F.G.E. Pantellini, D. Hubert, C.C. Harvey, A. Mangeney, G. Belmont, and C.T. Russell, Mirror and Alfvenic waves observed by ISEE 1-2 during crossings of the Earth's bow shock, *Ann. Geophysicae, 10*, 772-784, 1992.

Lin, N., M.J. Engebretson, R.L. McPherron, M.G. Kivelson, W. Baumjohann, H. Luehr, T.A. Potemra, B.J. Anderson, and L.J. Zanetti, A comparison of ULF fluctuations in the solar wind, magnetosheath, and dayside magnetosphere, 2, field and plasma conditions in the magnetosheath, *J. Geophys. Res., 96*, 3455-3464, 1991.

Luhmann, J.G., C.T. Russell, and R.C. Elphic, Spatial distributions of magnetic field fluctuations in the dayside magnetosheath, *J. Geophys. Res., 91*, 1711-1715, 1986.

Moustaizis, S., D. Hubert, A. Mangeney, C.C. Harvey, C. Perche, and C.T. Russell, Magnetohydrodynamic turbulence in the Earth's magnetosheath, *Ann. Geophysicae, 4 A*, 355-361, 1986.

Price, C.P., D.W. Swift, and L.-C. Lee, Numerical simulation of nonoscillatory mirror waves at the Earth's magnetosheath, *J. Geophys. Res., 91*, 101-112, 1986.

Russell, C.T., J.G. Luhmann, T.J. Odera, and W.F. Stuart, The rate of occurence of dayside Pc 3,4 pulsations: the L-value dependence of the IMF cone angle effect, *Geophys. Res Lett., 10*, 663-666, 1983.

Russell, C.T., G. Le, K. Schwingenschuh, W. Riedler and Ye. Yeroshenko, Mirror mode waves at Comet Halley, in: *Cometary Plasma Processes,* Geophysical Monograph 61, AGU, 161-169,1991.

Southwood, D.J., and M.G. Kivelson, Mirror instability 1: the physical mechanism of linear instability, *J. Geophys. Res.,* in press, 1992.

Song, P., C.T. Russell and M.F. Thomsen, Slow mode transition in the frontside magnetosheath, *J. Geophys. Res., 97*, 8295-8305, 1992.

Song, P., C.T. Russell, R.J. Strangeway, J.R. Wygant, C.A. Cattell, R.J. Fitzenreiter and R.R. Anderson, PC 3-4 energy coupling for northward interplanetary magnetic field, *J. Geophys. Res., 98*, 187-196, 1993.

Tajiri, M., Propagation of hydromagnetic waves in collisionless plasma. II. Kinetic approach, *J. Phys. Soc. Japan., 22*, 1482-1494, 1967.

Thompson, W.B., *An Introduction to Plasma Physics,* 214-218, 274 pp., Pergamon Press, 1964.

Tsurutani, B.T., E.J. Smith, R.R. Anderson, K.W. Ogilvie, J.D. Scudder, D.N. Baker and S.J. Bame, Lion roars and non-oscillatory drift mirror waves in the magnetosheath, *J. Geophys. Res., 87*, 6060-6072, 1982.

Tsurutani, B.T., I.G. Richardson, R.P. Lepping, R.D. Zwickl, D.E. Jones, E.J. Smith and S.J. Bame, Drift mirror mode waves in the distant (X ≈ 200 R$_E$) magnetosheath, *Geophys. Res. Lett., 11*, 1102-1105, 1984.

Tsurutani, B.T., D.J. Southwood, E.J. Smith, and A. Balogh, Nonlinear magnetosonic waves and mirror mode structures in the March 1991 Ulysses interplanetary event, *Geophys. Res. Lett., 19*, 1267-1270, 1992.

A.N. Fazakerley, Mullard Space Science Laboratory, University College London, Holmbury St. Mary, Dorking, Surrey, RH5 6NT, UK.

D.J. Southwood, Space and Atmospheric Physics Group, Imperial College, London SW7 2BZ, UK.

Observations of Waves at the Dayside Magnetopause

P. Song

High Altitude Observatory, National Center for Atmospheric Research[1], Boulder, CO

A brief overview of the wave activity over a broad spectrum of frequencies near the magnetopause is given based on ISEE observations. Waves of a period greater than 30 min cause the magnetopause to move in and out, the breathing mode, since these waves have wavelengths larger than the scale of the magnetosphere. Waves in the Pc 4-5 frequency range cause surface waves on the magnetopause. These waves either have a solar wind origin or are the result of the solar wind-magnetosphere interaction. Although many mechanisms have been proposed to generate these waves, observations show that the variations in solar wind dynamic pressure can explain most of the magnetopause oscillations near the subsolar region. The various processes related to magnetic reconnection are additional sources of oscillation when the IMF is southward, and increase in amplitude away from the subsolar region. The waves in the Pc 3-4 frequency range are difficult to study because of technical reasons. New methods are being developed and have been tested for a simple case: near the subsolar region for northward IMF. About 10~20% of the wave energy seen in the magnetosheath can be transmitted into the magnetopause. The transmitted waves propagate predominantly along the magnetic field indicating a mode conversion at the magnetopause. Wave enhancements in the Pc 1 range occur at the electric current region of the magnetopause. For southward IMF, the waves appear to have a greater amplitude and to be more turbulent. In the case of northward IMF, the waves are weaker and linearly polarized. The waves in the VLF range are also discussed briefly. They can be used as diagnostics of the particle distributions, or the status of the field lines.

INTRODUCTION

There is a wide spectrum of waves present at the dayside magnetopause, as summarized in Table 1. The ion gyrofrequency is about 1 Hz near the dayside magnetopause and the electron gyrofrequency about 1 kHz. Wave enhancements are usually found around these two natural frequencies. Between these two frequencies, there are either broad-band or narrow-band wave enhancements depending on the regions at the magnetopause and the upstream conditions. The waves in the magnetosheath have periods typically from 20 sec to 2 min. This period range also applies to magnetospheric compressional eigen-modes. Waves of a longer period may excite the magnetospheric Alfven modes. Observationally the waves of a frequency higher than a few Hertz can be measured by plasma wave experiments. These experiments usually measure both electric and magnetic fields of the wave with reasonably high time and frequency resolutions.

They may also provide the polarization of the perturbed fields. Usually, the waves near the ion gyrofrequency can be studied using the traditional Fourier analysis of the magnetometer data because a magnetopause crossing lasts about a half minute or slightly longer and a magnetometer usually has a time resolution of about 4 Hz or higher. The displacement of the magnetopause for waves of a period longer than 2 min is usually larger than the thickness of the magnetopause current layer which is about 1000 km thick and moves at 10~20 km/s. Therefore these low frequency waves can be studied by looking at multiple crossings of the magnetopause current layer from the magnetometer data. The waves in the Pc 3-4 frequency range are the most difficult to study because the wave displacements are smaller than the thickness of the current layer, hence these waves cannot be identified from multiple magnetopause crossings, and, because there are only a few wave cycles at the magnetopause which are not appropriate for using the traditional Fourier analysis. In this paper, we present an overview of the wave activity at the dayside magnetopause. Technical details can be found in the references. We use Pc to refer to the frequency range of a wave although it may be irrelevant to the Pc (continuous pulsation) wave observed on the ground in places other than the dayside cusp region and the wave may not be in a continuous form. We start with the low frequency waves.

---

[1]The National Center for Atmospheric Research is sponsored by the National Science Foundation.

Solar Wind Sources of Magnetospheric Ultra-Low-Frequency Waves
Geophysical Monograph 81

TABLE 1. Waves near the Magnetopause

Period (Frequency)	Wave	Process	Observation
2~30 min	Pc 4-5	Msph Alfven-time	Surface wave
20 sec~2 min	Pc 3-4	Msph Compr Mode Msheath Wave	Difficult
~1 sec	Pc 1	Ion Gyromotion	Fourier Ana.
~10^{1-2} Hz	VLF	Ele-ion Interaction Ele Gyromotion	Plasma Wave
~kHz	VLF	Ele Gyromotion	Plasma Wave

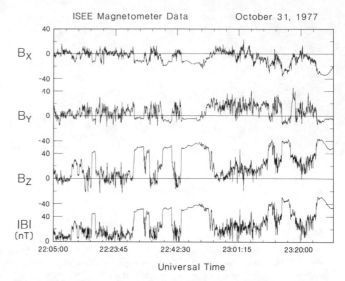

Fig. 1. Magnetic measurements, in GSE, of an ISEE magnetopause inbond-pass with a large number (15) of magnetopause crossings. The first crossing is at 22:11 UT. The last crossing is at 23:25 UT. The location of this pass is (10.2, -1.3, 4.6) R_e GSE. In the GSE coordintes, the x is sunward, y is eastward and z is northward.

Pc 4-5 Waves

The Pc 4-5 waves at the magnetopause correspond to surface waves [Holzer et al., 1966; Russell and Elphic, 1978; Southwood, 1979]. An upper cutoff period can be chosen as the convection time of the magnetosheath plasma flowing over the dayside magnetosphere [Song et al., 1988]. A longer period corresponds to the motion of the magnetopause from an equilibrium position to another.

Observations

Figure 1 shows an ISEE 1 magnetopause pass with multiple crossings of the current layer. As indicated in Figure 2, we can define the amplitude of a magnetopause oscillation as the distance between the first and last magnetopause crossings during a pass projected on the normal of the magnetopause, or

$$A = | \mathbf{n} \cdot (\mathbf{r}_2 - \mathbf{r}_1) |$$

where **n** is the normal of the magnetopause, and $\mathbf{r}_1$ and $\mathbf{r}_2$ are the locations of the first and last magnetopause crossings for a magnetopause pass. From the 10-year ISEE operation, we have collected more than one thousand magnetopause passes, for details see Song et al. [1988]. Figure 3 shows the coverage of these crossings. Only the center location is plotted for each pass and most of the passes are near the equatorial plane. We have divided the data set according to the orientation of the interplanetary magnetic field (IMF) into three categories, i.e., southward, northward and horizontal IMFs. Figure 4 compares the amplitude of the magnetopause oscillation for strongly southward IMF with strongly northward IMF. A zero amplitude indicates a single crossing. Although the magnetopause is moving during a single crossing, it does not oscillate. It is obvious that most of the northward IMF crossings are single but most of the southward IMF crossings are multiple. On average, the amplitude of the oscillations is greater for southward IMF than for northward IMF. Furthermore, the amplitude of the oscillations increases from the subsolar region to the flank for southward IMF but does not for northward IMF. The amplitudes for the two IMF orientations are similar near the subsolar region. The amplitude is slightly larger in the dawn side than in the dusk side.

Implications

There are two major sources of free energy, the solar wind flow and the IMF, which can dump energy into the magnetosphere and therefore cause the magnetopause to move. There are five possible mechanisms which convert the free energy into magnetopause wave energy. They are the Kelvin-Helmholtz (K-H) instability [Dungey, 1955; Southwood, 1979; Pu and Kivelson, 1983], reconnection [Aubry et al., 1970; Russell and Elphic, 1978; Lee and Fu, 1984; Southwood et al., 1988; Scholer, 1988], the upstream waves [Greenstadt et al., 1968; Fairfield, 1976; Greenstadt and Olson , 1977; Russell et al., 1983; Luhmann et al., 1986; Fairfield et al., 1990; Engebretson et al., 1991; Lin et al., 1991], solar wind pressure pulses [Song et al., 1988; Friis-

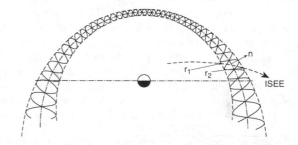

Fig. 2. Definition of the amplitude of the magnetopause oscillation. The dot-and-dash curve indicates the average position of the magnetopause. The two dashed curves indicate the amplitude of the oscillation. For an outbond-pass, a satellite observes the first magnetopause crossing at $\mathbf{r}_1$ and then observes multiple crossings. The last crossing is at $\mathbf{r}_2$ **n** is the normal of the average magnetopause.

MAGNETOPAUSE CROSSINGS

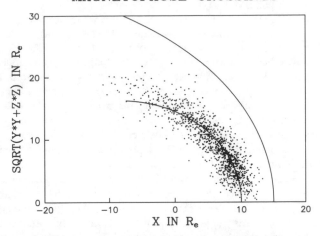

Fig. 3. Location of the magnetopause crossings in GSM coordinates from 10 years of ISEE 1 and 2 orbits. Only center location is plotted for the magnetopause with multiple crossings. The two solid lines are model magnetopause and bow shock respectively.

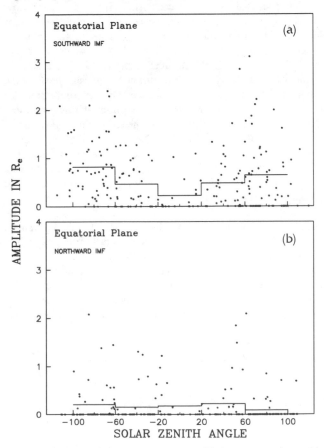

Fig. 4. Amplitude of the magnetopause oscillation versus solar zenith angle near the equatorial plane. The local noon is at 0°, morning terminator is at -90° and the evening terminator is at 90° a) when the IMF is southward, b) when the IMF is northward. The solid lines are the mean value of each bin.

Christensen et al., 1988; Elphic, 1989; Potemra et al., 1989; Farrugia et al., 1989; Southwood and Kivelson, 1990; Sibeck, 1990] and interplanetary corotating discontinuities [Gold, 1955; Singer, 1957; Wilson and Sugiura, 1961; Kaufmann and Konradi, 1969; Wilken et al., 1982; Baumjohann et al., 1983; Sibeck, 1990]. Often there is confusion in the distinction between the corotating discontinuities and the pressure pulses. Although they both are associated with changes in the solar wind properties, they are different physical processes operating at different scales in the solar wind, and hence predict different patterns in the magnetopause motion, as shown in Figure 5. The pressure pulse model was based on a small perturbation in the steady-state flow of a gasdynamic model [e.g., Spreiter 1966]. Because the position of the subsolar magnetopause is determined by the pressure balance between the solar wind dynamic pressure and magnetospheric magnetic pressure, changes in the solar wind dynamic pressure will change the position of the subsolar magnetopause. As the irregularity associated with a solar wind pressure change, a so-called pressure pulse, propagates along the magnetopause, the remaining part of the magnetopause oscillates. A basic assumption in this model is that the pressure variation is small compared with the pressure so that the perturbations in a steady-state streamline are small. Since, in the steady-state, the stagnation streamline forms the magnetopause, a pressure pulse first dents the subsolar magnetopause and then radiates away from this source region. Constrained by the scale of the magnetosphere, the solar wind irregularities of a time scale less than 20 ~ 30 min create surface waves on the magnetopause.

Distinct from pressure pulses, an interplanetary corotating discontinuity is associated with a change in the solar wind thermodynamic state and hence a change in the dynamic pressure. The change associated with a discontinuity is usually so large that the steady-state streamlines are destroyed and the flow downstream of the bow shock no longer follows the previous steady-state streamlines. Thus, the source region of the magnetopause surface waves is where the discontinuity first hits the magnetopause. Since most of the corotating discontinuity fronts align with Parker's spiral, the source region of the magnetopause oscillation should be in the afternoon side.

In these two models, the interaction of the solar wind irregularities or the discontinuities with the bow shock is ignored. Most recently, Wu et al. [1993] pointed out the importance of this interaction. When a tangential discontinuity collides with the fast mode bow shock, a pair of fast mode shocks are created in addition to the tangential discontinuity. One of these fast mode shocks moves forward and the other backward relative to the tangential discontinuity. As the tangential discontinuity propagates toward the magnetopause, the backward propagating fast shock forms the new bow shock and the forward fast shock carries most of the pressure change, originally associated with the tangential discontinuity, moving in front of the tangential discontinuity. Thus, the magnetopause is first hit by the fast mode shock rather than by the irregularity or the interplanetary discontinuity. This raises the possibility that even for the pressure pulse, the first interaction with the magnetopause is in the afternoon side. However, the dynamic pressure normal to the tangent of the bow

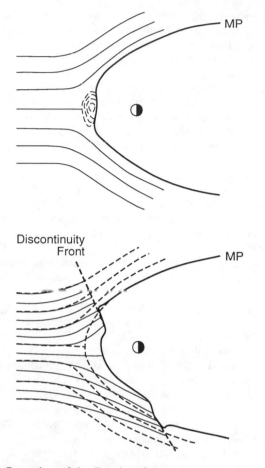

Fig. 5. Comparison of the distortion of the magnetopause (thick solid line) caused by a solar wind pressure pulse, upper panel, and by an interplanetary discontinuity, lower panel. The dashed lines are the steady state streamlines. The thin solid lines are equal flux lines.

shock, $\rho v_x^2 \cos^2\theta$, is about half of the solar wind dynamic pressure, ρv_x^2, if assuming θ, the angle between the bow shock normal and the Sun-Earth line, is about 45° at the tangent. When the dynamic pressure change is small across an interplanetary tangential discontinuity, the waves near the subsolar streamlines carry more pressure variation than the fast mode wave from the

tangent point. Therefore, the first impact could still be at the subsolar point.

Most likely, the magnetopause oscillations generated by a corotating discontinuity is associated with the magnetospheric global response to the pressure change rather than the high pressure bulge shown in Figure 5b. That bulge was proposed as a fast mode wave. However, because the magnetospheric fast mode velocity is much greater than the magnetosheath flow velocity, the pressure imbalance in the magnetosphere will be dispersed quickly and will not be accumulated in front of the discontinuity. As discussed by Wu et al. [1993], the interaction of a discontinuity with the magnetopause occurs first in the form of a fast mode wave front, i.e., then discontnuity front in Figure 5b should be replaced by a fast mode front. It is not possible to have an additional fast mode front in front of this front as proposed in this model. Thus in a better model of the corotating discontinuity, the bulge in front of the discontinuity should not appear and the discontinuity is replaced by a fast mode wave front. The discontinuity itself carries little pressure change and hence causes little displacement of the magnetopause.

Table 2 summarizes the prediction of the magnetopause oscillation caused by each mechanism. Since the sheath flow velocity increases from the subsolar point, and most observations indicate that reconnection takes place near the subsolar region for southward IMF [Russell and Elphic, 1978; Paschmann et al., 1979; Sonnerup et al., 1984; Gosling et al., 1990], the Kelvin-Helmholtz instability and reconnection models predict the oscillations to be symmetric about the subsolar point. Because these two mechanisms are associated with instabilities and more free energy is added to the wave as the instabilities grow and as the plasma flows away from the subsolar point, the amplitude of the wave should increase from the symmetric point. The upstream waves are generated upstream and then convect downstream of the quasi-parallel shocks. These waves may cause pressure variations in the sheath and then oscillate the magnetopause. The source region of the magnetopause oscillations should be where the IMF is parallel to the bow shock normal. These waves have access to the magnetopause only when the cone angle, the angle between the IMF and Sun-Earth line, is small. Since according to Parker's spiral, the quasi-parallel shocks are more likely to occur in the morning side than in the afternoon side, this mechanism predicts greater oscillations on the morning side. Since the pressure pulses, upstream waves and

TABLE 2. Prediction of Five Mechanisms

Source	Prediction	
	Symmetric About	Growing from the Symmetric Point?
K-H	Subsolar Point	Yes
Reconnection	Subsolar Point	Yes
SW Pressure Pulse	Subsolar Point	No
Upstream Wave	$B_{IMF} \parallel N_{BOWSHOCK}$	No
Corotating Discontinuity	Postnoon	(?)

corotating discontinuities are not instabilities, no additional free energy is added in the processes. As the wave radiates from the source region, its amplitude should decrease because the area increases, although one may argue that the amplitude should increase for corotating discontinuities [Sibeck, 1990]. An increase in the amplitude requires additional mechanisms to convert the free energy into wave energy which have not been clearly proposed. Comparing Table 2 with Figure 4, the observation is in favor of reconnection, pressure pulses and upstream waves.

To be more quantitative about the importance of the solar wind pressure pulses, we look at the solar wind spectrum. Figure 6 shows the solar wind dynamic pressure variations for different time periods from three years IMP8 data. The average solar wind dynamic pressure is about 2.6×10^{-8} dyne/cm^2. The pressure variations near $10 \sim 20$ min, $1.7 \sim 2.3 \times 10^{-9}$ dyne/cm^2, are less than 10% of the average pressure. Thus, the assumption for the pressure pulse model applies. We can assume that the magnetospheric field is dipolar and then calculate the displacement of the subsolar magnetopause which is also shown in Figure 6 according to the pressure variation. For a 10 to 20 min period, the subsolar magnetopause displacement is about 0.3 Re. Therefore most of the magnetopause oscillations seen near the subsolar region, Figure 4, can be provided by the solar wind pressure variations. If we subtract the amplitude for northward IMF from that for southward IMF, we have an amplitude of a half Re near the terminators. This is about the scale of a Flux Transfer Event [Saunders et al., 1984]. The difference between the amplitudes in the dusk side and the dawn side, 0.1 Re, may be used to estimate the importance of the upstream waves. However, this is close to the uncertainty of the averages. A further study of the dependence of the amplitude on the IMF cone angle will be useful to resolve the importance of the upstream waves. This observation indicates that the K-H instability is unimportant in causing the dayside magnetopause surface wave in the periods longer than 2 min. Although the oscillation for northward IMF can be explained quantitatively in terms of the solar wind pressure pulses, Belmont and Chanteur [1989] and Miura [1992] argue that the effects of the K-H instability can be seen if the oscillations of 1-10 min are separated from those of from 10 min to 1 hour in this data set because, in their argument, the K-H instability causes shorter period oscillations and the pressure pulses cause longer period oscillations. There has been no indication that the magnetopause oscillates at two distinct frequency ranges. The period of the oscillations may be determined by the properties of the magnetosphere instead of the sources of the waves. Figure 1 is a good example of this point. As shown by Song et al. [1989] the oscillations in this pass were triggered by a pressure pulse. The period for these oscillations is about 10 min or about 7 min if we count the possible missing crossings near 22:24 UT and 23:01 UT, which may be due to a smaller displacement for the cycles. Multiple crossings of a short period but long in duration, i.e., a large amplitude, as expected by Belmont and Chanteur and Miura, have not been reported so far. We believe if such events do exist, they must have been studied because the available data sets have been studied exhaustively.

In summary, the K-H instability is not important to cause

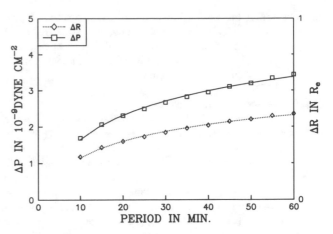

Fig. 6. Squares are the variations in the solar wind dynamic pressure averaged over each period on the x axis measured from IMP 8. The solid line is a fit to a power function. The diamonds are the corresponding displacement of the subsolar magnetopause assuming a dipole magnetospheric field. The dotted line is a fit, e.g., on average the solar wind pressure changes 0.2×10^{-9} dyne/cm^2 in a period of 15 min which causes the subsolar point moves 0.29 R$_e$.

dayside magnetopause oscillation longer than 2 min, although it may be important in the far tail and in phenomena of other types, for example causing the motion of the inner edge of the boundary layer [Sckopke et al., 1981; Ogilvie and Fitzenreiter, 1989; Takahashi et al., 1991], and may generate much shorter period, smaller amplitude motions. We should note that if the K-H instability exists only at the inner edge of the boundary layer, it becomes secondary in the energy transfer at the magnetopause. Other mechanisms are needed to transport the energy in the solar wind across the magnetopause current layer.

Pc 3 - 4 Waves

Presence of large amplitude waves in the Pc 3-4 frequency range is one of the major characteristics of the dayside magnetosheath [Kaufmann et al., 1970; Kaufmann and Horng, 1971; Wolfe and Kaufmann, 1975; Tsurutani et al., 1982]. Although there are many careful studies of the waves simultaneously in the solar wind, magnetosheath and magnetosphere [e.g., Fairfield et al., 1990; Engebretson et al., 1991; Lin et al., 1991a; Takahashi et al., 1991], as we have pointed out in the introduction, the wave behaviors at the magnetopause in the Pc 3-4 frequency range are very difficult to study because of technical reasons. As an attempt, Williams et al. [1979] used the finite Larmor radius effect to remote-sensing the small amplitude motion of the magnetopause. They found that in addition to the Pc 4-5 frequency motion of the magnetopause, there are boundary waves with a 90 sec to 150 sec period and an amplitude of a few hundred kilometers. Paschmann et al. [1990] also found that the magnetopause oscillates with a period of 60 sec to 120 sec when they determined the deHoffmann-Teller frame of the magnetopause. Song et al. [1993a] recently developed and tested some new techniques in order to examine the behavior of the Pc 3-4 wave at the magnetopause. In the following, we briefly review these techniques. Note that since these techniques have been tested for only the simplest case, i.e., subsolar region and strongly northward IMF, more testing is needed to use them for more complicated situations.

Observation and Technique Development

Figure 7 shows a subsolar magnetopause crossing when the IMF was strongly northward. Large amplitude waves are observed in the magnetosheath with a period of about 1 min. These waves can still be seen in the sheath transition layer with a much smaller amplitude and they become invisible in the magnetosphere. Without active reconnection, the magnetopause acts as a tangential discontinuity, and the most effective way to transport energy is to oscillate the boundary through total pressure variations. The total pressure has to be measured with two different instruments, the magnetometer and plasma detectors. This raises a question of whether the two instruments have been intercalibrated properly. Furthermore, even though the instruments may have been adequately intercalibrated before launch, the instruments may age differently after the launch. To investigate this problem we plot the magnetic pressure measured from the magnetometer against the plasma pressure, including thermal and dynamic pressures, measured from the Fast Plasma Experiment in the magnetosheath and the sheath transition layer. There is a good linear relation between the two, but with a slope not equal to -1 as required by the pressure balance. Thus we introduce a linear intercalibration factor, which makes the slope to be -1, to calibrate the plasma density assuming the temperature is constant in the magnetosheath. We compare the calibrated density with the density measured with other instruments, plasma frequency measurements, Vector Electron Spectrometer, and radio transmission experiment. The agreements are highly precise, see details by Song et al. [1993a]. Figure 8 shows the Fourier spectra for the pressures after the intercalibration. Although the thermal pressure and magnetic pressure are out of phase, there is still wave power in the total pressure. The wave power in the total pressure is about 15 ~ 20% of the wave power in the magnetic pressure. Therefore although the wave power in the magnetic field is large in the magnetosheath, it is offset by the thermal pressure with only about 15 ~ 20% of it available for transmission. As expected, the displacement caused by this total pressure variation is smaller than the thickness of the current layer.

Transmission of the wave energy at a boundary can also be studied using the Poynting flux measurements. The Poynting flux measurements have not been used very often for ULF wave studies. As a test, Lin et al. [1991b] used the magnetic field and velocity measurements to calculate the Poynting flux to study the waves in the magnetosheath. The difficulty with the Poynting flux measurements is to isolate the Poynting flux associated with the wave being studied from the residue due to other sources. For example, the slow variations in the background fields may be coupled into the wave Poynting flux through running average. To solve this problem, we use band-pass, centered at the peak in Figure 8, filtered electric and magnetic fields to calculate the

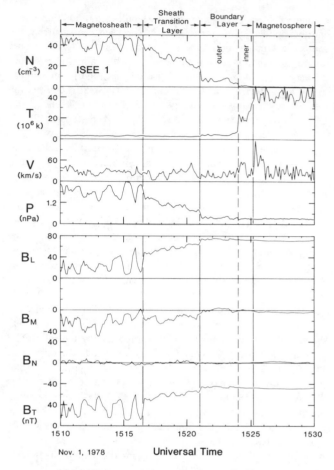

Fig. 7. A magnetopause crossing by ISEE 1 on Nov. 1, 1978. The location of this crossing is at 12:01 LT, 0.0 MLAT and 10.86 R_e. Plasma data with a time resolution of 6 seconds are from the FPE. N, T, V, and P are the ion density, cm⁻³, temperature, 10^6 K, ion flow velocity, km/s, and ion pressure, nPa. The magnetic field data are from the fluxgate magnetometer and are presented in the boundary normal coordinates. The tangential discontinuity technique was used to determine the normal direction of the magnetopause. For this crossing, N is sunward, M is dawnward, and L is northward. Regions are separated by vertical lines. The sheath transition layer is the region the magnetosheath plasma density changes. The outer boundary layer and the inner boundary layer are characterized by the sudden drop in the density and the sudden increase in the temperature.

Poynting flux. This filtered flux is the wave energy flux seen in the plasma frame. The normal component of this flux is the component which can be transmitted across a tangential discontinuity. Figure 9 shows the filtered Poynting flux. The normal component is bi-directional indicating that there are incident and reflected waves. The net flux is inward and about 15 ~ 20% of the total wave flux involved; here we assume the incident wave and reflected wave are not coherent. Within the sheath transition layer, the Poynting flux turns to being more field-aligned and decays rapidly away from the magnetopause

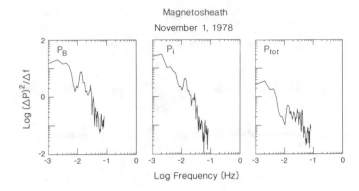

Magnetosheath
November 1, 1978

Fig. 8. Fourier spectra of the magnetic, ion thermal and total pressures for the crossing in Figure 7 in the magnetosheath. A peak about 0.015 Hz is apparent in the magnetic and thermal pressures.

current layer. Little energy is transmitted into the magnetosphere.

The two methods described above provide similar energy transmission coefficient. Furthermore, we can use the Poynting theorem to show that these two methods are quantitatively consistent; one is from the energy flux point of view and the other is from work-power point of view [Song et al., 1993a]. To generalize these methods, one may need to check their

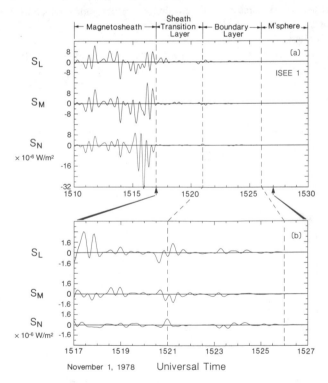

November 1, 1978   Universal Time

Fig. 9. The Poynting flux calculated by the cross product of the filtered electric field and magnetic field with a resolution of 3 sec in boundary normal coordinates for the crossing in Figure 7. Note expanded vertical scale in (b) compared with that in (a). The amplitude of the variations of the Poynting flux is 10.1 x 10⁻⁶ W/m², which provides an estimate of the

applicability. For example, near a reconnection zone, the Poynting flux associated with the flow may also be transmitted into the magnetopause.

Discussion

In the above, we have treated waves regardless of their sources and wave modes. Identification of the wave sources and the wave modes in the magnetosheath is extremely important in solar wind-magnetosphere energy coupling. The quasi-parallel shocks have been proven to be one of the major sources for the magnetosheath waves. These magnetosheath waves are rather broad-band [Engebretson et al., 1991; Lin et al., 1991a]. The case we showed above has a relatively narrow frequency range, as seen in Figure 8, and is associated with a quasi-perpendicular shock. Mirror waves and slow waves are the other two candidates. Although, observationally, these two modes have some similarity, i.e., the density and the field strength change out of phase for both modes, they are distinct. The mirror mode is a non-propagating mode but the slow mode is a propagating mode if the wavevector is not exactly perpendicular to the background field. To resolve this difference in a flow, in general two spacecraft and 3-D velocity measurements are required with a well determined direction of the wavevector. The determination of the direction of the wavevector is not trivial for a nearly pure compressional wave because the medium and minimum eigenvalues are close and hence the wavevector may be in any direction on the plane containing the medium and minimum eigenvectors. Tsurutani et al. [1982] showed some waves in the magnetosheath but not very close to the magnetopause to be mirror modes with adequate measurements and a small uncertainty. This was the only case known. However, since then, a mirror mode has been taken for granted in most observational studies referring to magnetosheath mirror waves. In the context of the solar wind-magnetosphere wave coupling, we note that the mirror mode does not modify the sheath flow and does not carry the Poynting flux in the flow frame. It requires wave of other propagating modes to convert the dynamic energy of the mirror wave seen in the Earth's frame into a form which is available for transmission into the magnetosphere. The function of the propagating modes is to slow the incoming sheath flow. Song et al. [1992] showed that there exists a quasi-standing slow mode wave near the magnetopause and the wave properties near the magnetopause are quite different from those not too close to the magnetopause. The former is propagating, i.e., the slow modes, but the latter is convective, i.e., the mirror modes. For the case shown in this section, it can be shown that the Poynting flux, 10.1 x 10⁻⁶ W/m², is larger compared with the convecting energy flux, 7.4 x 10⁻⁶ W/m², i.e., the propagation nature of the wave is dominant, where the convecting energy flux is defined as the magnetic wave energy times the convection velocity. Therefore the waves are slow modes rather than mirror modes. Recently, a theoretical scheme has been developed to identify the MHD wave modes systematically both spatially and spectrally [Song et al., 1993c]. It has shown that the mirror modes exist in the middle magnetosheath for quasi-perpendicular shocks and these mirror modes become slow modes in the inner sheath.

Pc 1 Waves

Observations

Figure 10 compares the power spectrum in the magnetosheath with those in the sheath transition layer for the crossing discussed in the last section. The magnetosheath spectrum has a higher power in the low frequencies. The sheath transition layer has wave enhancements in the frequencies higher than 0.2 Hz. The dawn-dusk component contains the most power, i.e., the wave is not circularly polarized. Figure 11 shows the high-pass filtered magnetic field data. Most of the wave enhancement is within the sheath transition layer, or the magnetopause current layer, indicating that the source of the wave lies within the sheath transition layer. A small data set survey [Song et al., 1993b] shows that the waves are much stronger for southward IMF than for northward IMF. The waves for southward IMF are also more turbulent, i.e., the wave vector is not well determined, and the wave is weakly coherent and has a broader spectrum. However, most importantly, the polarization is not the same for different IMF orientations. For southward IMF, the waves are left-handed but for northward IMF, they are more likely to be linearly polarized. There is no clear evidence that these waves are associated with either the quasi-perpendicular or quasi-parallel shocks.

Discussion

Obviously, since the wave enhancement near the ion gyrofrequency is concentrated in a relatively narrow region, the sheath transition layer, they must be important to the processes in this region. For southward IMF, the sheath transition layer hosts the field reversal. Thus these waves may play a key role in reconnection processes, namely providing a means to convert the magnetic energy into plasma energy. The processes associated

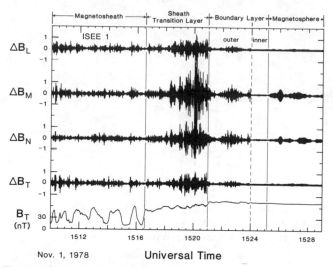

Fig. 11. High-pass filtered magnetic field data on Nov. 1, 1978. The lower cutoff frequency is 0.4 Hz and the Nyquist frequency is 2 Hz. The maximum variation direction is the M, dawn-dusk, direction. Waves are enhanced in the sheath transition layer.

with these waves have not been included in theoretical reconnection models. In this sense, most of the reconnection models need to be advanced further. The processes in the "diffusion region" and the mechanisms that provide the large resistivity required for reconnection are rather poorly understood at the present time. These waves may cast a new light on the problem.

The waves for northward IMF have received more attention. Song [1991] and Anderson et al. [1991, 1993] have shown that in the region with wave enhancements a higher ion temperature anisotropy is observed. Although the ion cyclotron mode is the

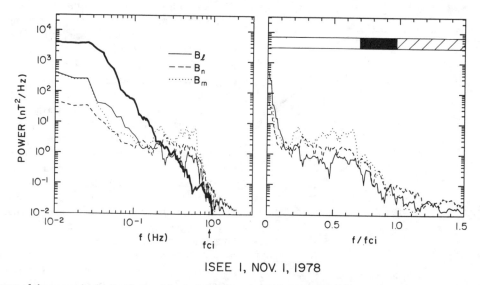

ISEE 1, NOV. 1, 1978

Fig. 10. Power spectra of the waves in the sheath transition layer on Nov. 1, 1978, Figure 7. The thick solid line on the left panel is the spectrum taken from the magnetosheath as reference. Waves in the sheath transition layer are enhanced in the Pc 1 frequency range and linearly polarized in the M direction. The polarization of the wave is indicated by the bars on the right panel: the open bar indicates linear polarization, solid bar left-hand and shaded bar right-hand.

most plausible candidate to generate waves in this frequency range for a high anisotropy, the most puzzling thing is that the observed wave is linearly polarized and not left-handed. This latter polarization is required by the most unstable mode of the ion cyclotron waves. In order to include the mode coupling with other modes and other effects, Gary et al. [1993] referred to these waves as ion cyclotron-like waves. Here we have to consider the contribution from the other free energy source: the gradients, which were observed first [Song et al., 1990a]. Since maximum variation is along the M direction, the wave vector of these waves is most likely to be in the L-N plane, which contains the field and the density gradient. First, consider the role of the anisotropy. For a homogeneous plasma, the growth rate of ion temperature anisotropy instability is above the MHD thresholds in the sheath transition layer if the anisotropy is about 2 and the plasma beta is about 0.5 [Gary et al., 1976]. These plasma conditions are observed in the sheath transition layer. This wave mode is left-handed and grows fastest for waves propagating along the field. Thus, this mode may contribute to waves in the sheath transition layer. Second, consider the role of the density gradient. The instabilities associated with plasma inhomogeneity are the drift wave instabilities. For a plasma with $\beta \sim 1$, the drift wave instabilities have features of Alfven mode and are called drift Alfven instability [Hasegawa, 1975; Mikhailovsky, 1983]. The drift modes are of the electron polarization, i.e., they are right-handed waves. Thus, the observed linearly polarized waves for northward IMF may be a coupling of the two types of the waves, one is left-handed and the other is right-handed. The propagation direction of these waves may not be purely along the field and may have a component along the density gradient. One clue in support of this argument is that the wavelength is comparable to the thickness of the sheath transition layer as discussed by Song et al. [1990a].

Anderson et al. [1991] reported that the power spectrum for these waves from AMPTE/CCE contains two branches separated by a slot at the He^{++} gyromotion. This slot has not been found in the ISEE data and puzzled the author. Recently, Anderson et al. [1993] have shown that the slot exists only in rare cases when the plasma beta is extremely low. The more often observed transverse spectrum has a left hand polarization in the frequencies above 0.5 proton gyrofrequency, f_{cp}, linear polarization below 0.5 f_{cp}, and no slot at 0.5 f_{cp}. This is similar to the spectrum shown in Figure 10. Therefore the observations from ISEE and CCE are generally in good agreement and the spectral slot may be due to the orbital effects that CCE observes the magnetopause under the situations which is not common for ISEE.

Many theoretical investigations have been undertaken to understand these waves and the relationship with the wave in the Pc 3-4 range. Gary et al. [1993] and Denton et al. [1993] proposed that the waves above and below 0.5 f_{cp} are driven by the proton and He^{++} cyclotron anisotropy instabilities independently. Because there are only a few He^{++} particles to provide the free energy for the wave, the lower band wave has less power than that in the upper band in the linear calculations. In contrast, observations show that most wave energy is in the lower band. In nonlinear particle simulations [Gary and Winske, 1993], the wave generated by protons can cascade to the lower band through nonlinear processes. Thus the lower band is driven by both protons and He^{++} and hence contains more power. He^{++} may also play a role in the generation of the magnetosheath mirror waves [Price et al., 1986; Gary et al., 1993]. Without the presence of He^{++}, the proton cyclotron anisotropy instability has a greater growth rate than the mirror mode instability [Gary, 1992] over a large parameter regime. Thus, the mirror wave should be replaced by the proton cyclotron wave in the magnetosheath if there were no He^{++}. When He^{++} is significantly present, the mirror mode instability has greater growth rate than the ion cyclotron anisotropy instability in high plasma beta and low anisotropy regime [Gary et al., 1993]. This explains why the mirror-like waves occur in the magnetosheath where the plasma beta is high and the anisotropy is low, but the ion cyclotron-like waves occur in the sheath transition layer where the beta becomes low and the anisotropy becomes high. Here we should remember that the mirror waves have become slow waves in the region close to the magnetopause.

VLF WAVES

VLF waves have been paid relatively less attention to in the study of the solar wind-magnetosphere coupling. The main reason is that they carry little energy. However, they are important as an information carrier, because they are very sensitive to the state of the plasma, especially the electron distribution. In this section, we will not discuss in detail the properties of these waves but discuss some examples to complete the picture of the waves at the magnetopause. These phenomena were reviewed by Anderson et al. [1982] and by Labelle and Treumann [1988].

Northward IMF

Figure 12 shows the VLF wave measurements for the crossing shown in Figure 7. Significant wave enhancements are indicated by the dark areas on the plot. In the electric field data, there is an intensification centered at about 100 kHz in the magnetosheath and then decreasing in frequency to about 5 kHz in the magnetosphere. The lower boundary of this emission indicates the plasma frequency. It can be compared with the density measurements in Figure 7, see comparison in Song et al. [1993a]. In the magnetosheath, there are electromagnetic waves between 0.01 kHz to 0.1 kHz, best seen in the magnetic field data. These waves are called the lion roars [Smith et al., 1969; Smith and Tsurutani, 1976; Tsurutani et al., 1982] and coincide with the minima of the magnetic field associated with the Pc 3 waves in Figure 7. They are generated by the resonant electrons trapped in these magnetic cavities. These waves provide a unique opportunity to study the scale of the magnetic cavities [Song et al., 1991]. In the sheath transition layer, weak electron gyrofrequency waves can be seen in the electric field at about 1.5 kHz.

There is a distinct electrostatic wave at about 3/2 f_{ce} [Kennel et al., 1970; Kurth et al., 1979; Gurnett et al., 1979] in the boundary layer and magnetosphere. This wave can be generated by a mixture of a cold population with a hot population [e.g., Kennel

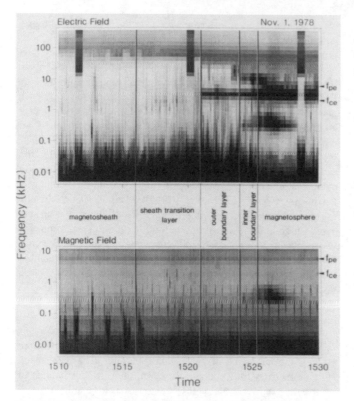

Fig. 12. Plasma waves measured by the Iowa plasma wave instrument for the crossing in Figure 7. The grey scale is Log (Frequency x Intensity). f_{pe} and f_{ce} on the right are the plasma frequency and electron gyrofrequency in the magnetosphere. The values are different in other regions.

and Abdalla, 1982]. Thus, it can be used as a diagnostic for field lines which connect with the ionosphere. An electromagnetic wave occurs in the inner boundary layer and magnetosphere at a frequency which is a fraction of the electron gyrofrequency. This wave is called chorus [Tsurutani and Smith, 1977; Gurnett et al., 1979; Anderson et al., 1982]. Note an interesting feature of this wave is that it has only a finite bandwidth, or has a lower cutoff. According to the particle cyclotron resonance theory, waves of lower frequency resonate with particles of greater parallel energy. The absence of the lower frequency waves indicates that the plasma lacks particles of large parallel energy. This should be caused by particles' loss in the ionosphere. Therefore the chorus indicates a pancake type distribution in high energies. Since it takes some time, about 1 min, to set up this type of distribution, the wave thus indicates that the field line has connected to the earth for a while. In comparison, since the outer boundary layer has the 3/2 f_{ce} electrostatic wave but without the chorus, it must be on newly closed field lines. The above mentioned features in the particle distribution have been reported by Song et al. [1993d].

Southward IMF

Figure 13 shows the plasma wave measurements for a strongly southward IMF magnetopause crossing near the subsolar region. The magnetic field changes about 180° from the magnetosheath

to the magnetosphere. The field and plasma data for this crossing can be found in Song et al. [1993b]. The major change takes place at the two edges of the sheath transition layer. Evidence for particle heating is observed within the sheath transition layer [Song et al., 1990b]. There is no Pc 3 compressional wave in the magnetosheath for this crossing. The magnetosheath is relatively quiet in the VLF band and the lion roars are not present. There is a strong broad-band electromagnetic wave enhancement in the sheath transition layer. This broad-band emission is commonly observed to be associated with a large field shear. These data show that the broadband electromagnetic emission extends from a few Hz to more than 100 Hz. This emission appears to be the high frequency extension of the Pc 1 frequency enhancements observed in the sheath transition layer discussed above. Here we recall that the Pc 1 wave in this region is broadband with a large amplitude for southward IMF. The presence of enhanced wave power above the proton gyrofrequencies suggests that the waves strongly interact with particles of a broad energy range and couple the ions with the electrons. Therefore it should be very important for the microscopic processes of reconnection. In the boundary layer and magnetosphere, only the chorus occurs but not the 3/2 f_{ce} wave. The correlation between the chorus and the pancake distribution at high energies is verified [Song et al., 1990b]; but the absence of the 3/2 f_{ce} is still a mystery. Similar features are also found in other southward IMF crossings.

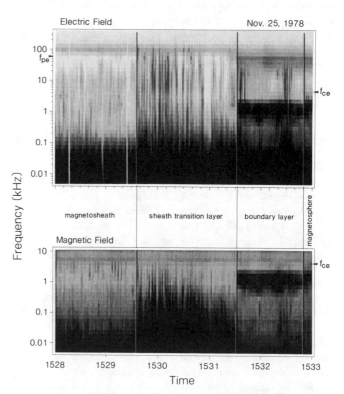

Fig. 13. Plasma waves measured by the Iowa plasma wave instrument for a southward IMF crossing. Plasma data of the crossing can be found in Figure 1 of Song et al. [1993b].

SUMMARY

A variety of wave activities occur at the magnetopause. Each of these waves carries particular information, plays a particular role in the solar wind-magnetosphere coupling, and has to be studied using different methods.

(1) Pc 4-5 waves can be studied simply by looking at the multiple crossings of the magnetopause current layer. The variations in the solar wind dynamic pressure can account for most of the magnetopause oscillations in the subsolar region. The processes related to reconnection appear to add more oscillations for southward IMF with amplitude growing away from the solar region. A slightly higher level of oscillation in the morning side than in the dusk side may indicate the significance of the upstream waves. The Kelvin-Helmholtz instability and the interplanetary corotating discontinuities appear to play a minor role in causing the magnetopause oscillation in the dayside, in periods longer than 2 min.

(2) Pc 3-4 waves at the magnetopause are difficult to study because of technical reasons. New methods are being developed and have been tested for simple cases. The results have been shown that about 10 to 20% of the magnetosheath wave energy flux can be transmitted into the subsolar magnetopause when the IMF is strongly northward. Most of the transmitted flux becomes field-aligned, and hence should be transmitted to the high latitude ionosphere, rather than to the low latitudes through compressional modes.

(3) Pc 1 waves are concentrated within the magnetopause current layer. They are not the same for different IMF orientations in the following sense. For southward IMF, the waves have a large amplitude and appear to be ion cyclotron turbulence. These waves should provide a means to convert the magnetic energy into particle energies in reconnection processes. For northward IMF, the waves have a relatively small amplitude and are linearly polarized in the direction normal to both the background field and its gradient. The mechanisms which generate these waves should be related to that maintaining the thickness of the magnetopause current layer.

(4) VLF waves carry information of, and are sensitive to, the plasma condition. They have several distinct wave modes which depend on IMF orientations. While they can be used as a diagnostic of particle distributions with much higher time resolution than direct particle measurements, little work has been done.

Acknowledgments. The author wishes to thank C. T. Russell for his guidance and stimulating discussions through the past few years during the research reviewed in this paper. The constructive discussions with S. P. Gary about the slow/mirror mode wave and ion gyrofrequency wave are greatly appreciated. He is also grateful to B. C. Low for carefully reading the draft and many useful suggestions, and to Liz Boyd for her assistance in preparing the manuscript.

REFERENCES

Anderson, B. J., S. A. Fuselier, and D. Murr, Electromagnetic ion cyclotron waves observed in the plasma depletion layer, *Geophys. Res. Letts., 18*, 1955, 1991.

Anderson, B. J., S. A. Fuselier, S. P. Gary, R. E. Denton, Magnetic spectral signatures from 0.1 to 4.0 Hz in the Earth's magnetosheath and plasma depletion layer, *J. Geophys. Res.,* submitted, 1993.

Anderson, R. R., C. C. Haravey, M. M. Hoppe, B. T. Tsurutani, T. E.Eastman, and J. Etcheto, Plasma waves near the magnetopause, *J. Geophys. Res., 87*, 2087, 1982

Aubry, M. P., C. T. Russell, and M. G. Kivelson, Inward motion of the magnetopause before a substorm, *J. Geophys. Res., 75*, 7018, 1970.

Baumjohann, W., O. H. Bauer, G. Haerendel, H. Junginger, and E. Amata, Magnetospheric plasma drifts during a sudden impulse, *J. Geophys. Res., 88*, 9287, 1983.

Belmont, G., and G. Chanteur, Advances in magnetopause Kelvin-Helmholtz instability studies, *Phys. Scr., 40*, 124, 1989.

Chen, L., and A. Hasegawa, A theory of long period magnetic pulsation, 1, Steady state excitation of field line resonance, *J. Geophys. Res., 79*, 1024, 1974.

Crooker, N. U., and G. L. Siscoe, A mechanism for pressure anisotropy and mirror instability in the dayside magnetosheath, *J. Geophys. Res., 82*, 185, 1977.

Denton, R. E., M. K. Hudson, S. A. Fuselier, and R. J. Anderson, Electromagnetic ion cyclotron waves in the magnetosheath plasma depletion layer, *J. Geophys. Res.,* in press, 1993.

Dungey, J. W., Electrodynamics of the outer atmosphere, in *Proceedings of the Ionosphere Conference,* The Physical Society of London, p. 225, 1955.

Elphic, R. C., Multipoint observations of the magnetopause: Results from ISEE and IMPTE, *Adv. Space Res., 8*, 223, 1989.

Engebretson, M. J., N. Lin, W. Baumjohann, H. Luehr, B. J. Anderson, L. J. Zanetti, T. A. Potemra, R. L. McPherron, and M. G. Kivelson, A comparison of ULF fluctuations in the solar wind, magnetosheath, and dayside magnetosphere, 1, Magnetosheath morphology, *J. Geophys. Res., 96*, 3441, 1991a.

Engebretson, M. J., L. J. Cahill, Jr., R. L. Arnoldy, B. J. Anderson, T. J. Rosenberg, D. L. Carpenter, U. S. Inan, and R. H. Eather, The role of the ionosphere in coupling upstream ULF wave power into the dayside magnetosphere, *J. Geophys. Res., 96*, 1527, 1991b.

Fairfield, D. H., Magnetic fields of the magnetosheath, *Rev. Geophys. Space Phys., 14*, 117, 1976.

Fairfield, D. H., W. Baumjohann, G. Paschmann, H. Luhr, and D. G. Sibeck, Upstream pressure variations associated with the bow shock and their effects on the magnetosphere, *J. Geophys. Res., 95*, 3773, 1990.

Farrugia, C. J., M. P. Freeman, S. W. H. Cowley, D. J. Southwood, M. Lockwood, and A. Etemadi, Pressure-driven magnetopause motions and attendant response on the ground, *Planet. Space Sci., 37*, 589, 1989.

Friis-Christensen, E., M. A. McHenry, C. R. Clauer, and S. Vennerstrom, Ionospheric traveling convection vortices observed near the polar cleft: A triggered response to sudden changes in the solar wind, *Geophys. Res. Letts., 15*, 253, 1988.

Gary, S. P., The mirror and ion cyclotron anisotropy instabilities, *J. Geophys. Res., 97*, 8519, 1992.

Gary, S. P., S. A. Fuselier, and B. J. Anderson, Ion anisotropy instabilities in the magnetosheath, *J. Geophys. Res., 98*, 1481, 1993.

Gary, S. P. and D. Winske, Simulations of ion cyclotron anisotropy instabilities in the terrestrial magnetosheath, *J. Geophys. Res., 98*, 9171, 1993.

Gary, S. P., M. D. Montgomery, W. C. Feldman, D. W. Forslund, Proton temperature anisotropy instabilities in the solar wind, *J. Geophys. Res., 81*, 1241, 1976.

Gold, T., *Dynamics of Cosmic Clouds,* North-Holland Publishing Co., Amsterdam, p. 103, 1955.

Gosling, J. T., M. F. Thomsen, S. J. Bame, T. G. Onsager, C. T. Russell, The electron edge of the low latitude boundary layer during accelerated flow events, *Geophys. Res. Letts., 17*, 1833, 1990.

Greenstadt, E. W., Field-determined oscillations in the magnetosheath as a possible source of medium period daytime micropulsations, paper presented at the Solar Terrestrial Relations Conference, Univ. of Calgary, Calgary, Alberta, Canada, Aug. 28 to Sept. 1, 1972.

Greenstadt, E. W., I. M. Green, G. T. Inouye, A. J. Hundhausen, S. J. Bame, and I. B. Strong, Correlated magnetic field and plasma observations of the Earth' bow shock, *J. Geophys. Res., 73* 51, 1968.

Greenstadt, E. W., and J. V. Olson, A contribution toi ULF activity in the Pc 3-4 range correlated with IMF radial orientation, *J. Geophys. Res., 82,* 4991, 1977.

Gurnett, D. A., R. R. Anderson, B. T. Tsurutani, E. J. Smith, G. Paschmann, G. Haerendel, S. J. Bame, and C. T. Russell, Plasma waves turbulence at the magnetopause: Observations from ISEE 1 and 2, *J. Geophys. Res., 84,* 7043, 1979.

Hasegawa, A., *Plasma Instabilities and Nonlinear Effects,* 241 pp, New York, 1975.

Holzer, R. E., M. G. Mcleod, and E. J. Smith, Preliminary results from the Ogo 1 search coil magnetometer: Boundary positions and magnetic noise spectra, *J. Geophys. Res.,* 71, 1481, 1966.

Kaufmann, R. L., A. Konradi, Explorer 12 magnetopause observations: Large-scale nonuniform motion, *J. Geophys. Res.,* 74, 3609, 1969.

Kaufmann, R. L., and J. T. Horng, Physical structure of hydromagnetic disturbances in the inner magnetosheath, *J. Geophys. Res.,* 76, 8189, 1971.

Kaufmann, R. L., J. T. Horng and A. Wolfe, Large amplitude hydromagnetic waves in the inner magnetosphere, *J. Geophys. Res.,* 75, 4666, 1970.

Kennel, C. F., M. Ashour-Abdalla, Electrostatic waves and the strong diffusion of magnetosphere electrons, in *Magnetosphere Plasma Physics,* ed. A. Nishida, p. 245, D. Reidel Publishing Co., London, 1982.

Kennel, C. F., F. L. Searf, R. W. Fredricks, J. H. McGehee, and F. V. Coroniti, VLF electric field observations in the magnetosphere, *J. Geophys. Res.,* 75, 6136, 1970.

Kurth, W. S., M. Ashour-Abdalla, L. A. Frank, C. F. Kennel, D. A. Gurnett, D. D. Sentman, and B. G. Burek, A comparison of intense electrostatic waves near f_{UHR} with linear instability theory, *Geophys. Res. Letts.,* 6, 487, 1979.

LaBelle, J., and R. A. Treumann, Plasma waves at the dayside magnetopause, *Space Sci. Rev.,* 47, 175, 1988.

Lee, L. C., and Z. F. Fu, A theory of a magnetic flux transfer at the Earth's magnetopause, *Geophys. Res. Letts.,* 12, 105, 1985.

Lin, N., M. J. Engebretson, R. L. McPherron, M. G. Kivelson, W. Baumjohann, H. Luehr, T. A. Potemra, B. J. Anderson, and L. J. Zanetti, A comparison of ULF fluctuations in the solar wind, magnetosheath, and dayside magnetosphere, 2, Field and plasma conditions in the magnetosheath, *J. Geophys. Res.,* 96, 3455, 1991a.

Lin, N., M. J. Engebretson, W. Baumjohann, and H. Luehr, Propagation of perturbation energy fluxes in the subsolar magnetosheath: AMPTE IRM observations, *Geophys. Res. Letts.,* 18, 1667, 1991b.

Luhmann, J. G., C. T. Russell, R. C. Elphic, Spatial distributions of magnetic field fluctuations in the dayside magnetosheath, *J. Geophys. Res., 91,* 1711, 1986.

Mikhailovsky, A. B., Instabilities in inhomogeneous plasma, in *Handbook of Plasma,* eds. Rosenbluth and Sageev, pp. 588, North-Holland, 1983.

Miura, A., Kelvin-Helmholtz instability at the magnetoshperic boundary: Dependence on the magnetosheath sonic Mach number, *J. Geophys. Res., 97,* 10655, 1992.

Ogilvie, K. W., and R. J. Fitzenreiter, The Kelvin-Helmholtz instability at the magnetopause and inner boundary layer surface, . *Geophys. Res., 94,* 15,113, 1989.

Paschmann, G., B. U. O. Sonnerup, I. Papamastorakis, N. Sckopke, G. Haerendel, S. J. Bame, J. R. Asbridge, J. T. Gosling, C. T. Russell, and R. C. Elphic, Plasma acceleration at the Earth's magnetopause: Evidence for reconnection, *Nature, 282,* 243, 1979.

Paschmann, G., B. Sonnerup, I. Papamastorakis, W. Baumjohann, N. Sckopke, and H. Luhr, The magnetopause and boundary layer for small magnetic shear: Convection electric fields and reconnection, *Geophys. Res. Letts.,* 17, 1829, 1990.

Potemra, T. A., L. J. Zanetti, K. Takahashi, R. E. Erlandson, H. Luehr, G. Marklund, L. P. Block, L. G. Blomberg and R. P. Lepping, Multi-satellite and ground-based observations of transient ULF waves, *J. Geophys. Res.,* 94, 2543, 1989.

Price, C. P., D. W. Swift, and L. C. Lee, Numerical simulation of nonoscillatory mirror waves at the Earth's magnetosheath, *J. Geophys. Res., 91,* 101, 1986.

Pu, Z., and M. G. Kivelson, Kelvin-Helmholtz instability at the magnetopause, *J. Geophys. Res., 88,* 1983.

Russell, C. T., and R. C. Elphic, Initial ISEE magnetometer results: Magnetopause observations, *Space Sci. Rev.,* 22, 681, 1978.

Russell, C. T., J. G. Luhmann, T. J. Odera, and W. F. Stuart, The rate of occurrence of dayside Pc 3,4 pulsations: The L-value dependence of the IMF cone angle effect, *Geophys. Res. Lett.,* 10, 663, 1983.

Saunders, M. A., C. T. Russell, and N. Sckopke, Flux transfer events: Scale size and interior structure, *Geophys. Res. Letts.,* 11, 131, 1984.

Scholer, M., Magnetic flux transfer at the magnetopause based on single x line bursty reconnection, *Geophys. Res. Letts.,* 15, 291, 1988.

Sckopke, N., G. Paschmann, G. Haerendel, B. U. O. Sonnerup, S. J. Bame, T. G. Forbes, E. W. Hones, Jr., and C. T. Russell, Structure of the low-latitude boundary layer, *J. Geophys. Res.,* 86, 2099, 1981.

Sibeck, D. G., A model for the transient magnetospheric response to sudden solar wind dynamic pressure variations, *J. Geophys. Res.,* 95, 3755, 1990.

Singer, S. F., A new model of magnetic storms and aurorae, *Trans. Amer. Geophys. Union,* 38, 175, 1957.

Smith, E. J., R. E. Holzer, and C. T. Russell, Magnetic emissions in the magnetosheath at frequencies near 100 Hz, *J. Geophys. Res.,* 74, 3027, 1969.

Smith, E. J., and B. T. Tsurutani, Magnetosheath lion roars, *J. Geophys. Res., 81,* 2261, 1976.

Song, P., The subsolar magnetopause and surrounding plasma layers, Ph.D. thesis, University of California, Los Angeles, 1991.

Song, P., R. C. Elphic, and C. T. Russell, ISEE 1 and 2 observations of the oscillating magnetopause, *Geophys. Res. Lett.,* 15, 744, 1988.

Song, P., R. C. Elphic, and C. T. Russell, Multi-spacecraft observations of magnetopause surface waves: ISEE 1 and 2 determinations of amplitude, wavelength and period, *Adv. Space Res.,* 8, (9), 245, 1989.

Song, P., R. C. Elphic, C. T. Russell, J. T. Gosling, and C. A. Cattell, Structure and properties of the subsolar magnetopause for northward IMF: ISEE observations, *J. Geophys. Res.,* 95, 6375, 1990a.

Song, P., C. T. Russell, N. Lin, R. J. Strangeway, J. T. Gosling, M. Thomsen, T. A. Fritz, D. G. Mitchell, and R. R. Anderson, Wave and particle properties of the subsolar magnetopause, *Phys. of Space Plasmas* (1989), p. 463, the SPI Conference Proceedings and Reprint Series, eds. T. Chang, G. Crew and J. Jasperse, Scientific Publishers, Inc., Cambridge, MA, USA, 1990b.

Song, P., C. T. Russell, R. J. Strangeway, and R. R. Anderson, Interrelationship between the magnetosheath ULF waves and VLF waves, *Phys. of Space Plasmas* (1991), p. 459, the SPI Conference Proceedings and Reprint Series, eds. T. Chang, G. Crew, and J. Jasperse, Scientific Publishers, Inc., Cambridge, MA, USA, 1991.

Song, P., C. T. Russell, and M. F. Thomsen, Waves in the inner magnetosheath: A case study, *Geophys. Res. Letts.,* 19, 2191, 1992.

Song. P., C. T. Russell, R. J. Strangeway, J. R. Wygant, C. A. Cattell, R. J. Fitzenreiter, and R. R. Anderson, Wave properties near the subsolar magnetopause: Pc 3-4 energy coupling for northward interplanetary magnetic field, *J. Geophys. Res.,* 98, 187, 1993a.

Song, P., C. T. Russell, and C. Y. Huang, Wave properties near the subsolar magnetopause: Pc 1 waves in sheath transition layer, *J. Geophys. Res.,* 98, 5907, 1993b.

Song, P., C. T. Russell, and S. P. Gary, Identification of low-frequency fluctuations in the terrestrial magnetosheath, *J. Geophys. Res.,* submitted, 1993c.

Song, P., C. T. Russell, R. J. Fitzenreiter, J. T. Gosling, M. F. Thomsen, D. G. Mitchell, S. A. Fuselier, G. K. Parks, R. R. Anderson, D. Hubert, Structure and properties of the subsolar magnetopause for northward IMF: Multiple-instrument particle observations, *J. Geophys. Res.,* in press, 1993d.

Sonnerup, B. U. O., Magnetic field reconnection at the magnetopause: An overview, in *Magnetic Reconnection in Space and Laboratory Plasmas,* ed. E. G. Hones, pp. 92, AGU, Washington, DC, 1984.

Southwood, D. J., Some features of field line resonances in the magnetosphere, *Planet. Space Sci.,* 22, 483, 1974.

Southwood, D. J., Magnetopause Kelvin-Helmholtz instability, in *Proceedings of Magnetospheric Boundary Layers Conference,* p. 357,

ESA Scientific and Technical Publications Branch, Noordwijk, The Netherlands, 1979.

Southwood, D. J., C. J. Farrugia, and M. A. Saunders, What are flux transfer events?, *Planet. Space Sci., 36,* 503, 1988.

Southwood, D. J., and M. G. Kivelson, The magnetohydrodynamic response of the magnetospheric cavity to changes in solar wind pressure, *J. Geophys. Res., 95,* 2301, 1990.

Takahashi, K. D., G. Sibeck, P. T. Newell, and H. E. Spence, ULF waves in the low-latitude boundary layer and relationship to magnetospheric pulsations: A multi-satellites observations, *J. Geophys. Res., 96,* 9503, 1991.

Tsurutani, B. T., and E. J. Smith, Postmidnight chorus: A substorm phenomenon, *J. Geophys. Res., 79,* 118, 1974.

Tsurutani, B. T., E. J. Smith, R. R. Anderson. K. W. Ogilvie, J. D. Scudder, D. N. Baker, and S. J. Bame, Lion roars and nonoscillatory drift mirror waves in the magnetosheath, *J. Geophys. Res., 87,* 6060, 1982.

Wilken, B., C. K. Goertz, D. N. Baker, P. R. Higbie, and T. A. Fritz, The SSC on July 29, 1977, and its propagation within the magnetosphere, *J. Geophys. Res., 87,* 5901, 1982.

Williams, D. J., T. A. Fritz, B. Wilken, and E. Keppler, An energetic particle perspective of the magnetopause, *J. Geophys. Res., 84,* 6385, 1979.

Wilson, C. R., and M. Sugiura, Hydromagnetic interpretation of sudden commencements of magnetic storms, *J. Geophys. Res., 66,* 4097, 1961.

Wolfe, A., and R. L. Kaufmann, MHD wave transmission and production near the magnetopause, *J. Geophys. Res., 80,* 1764, 1975.

Wu, B. H., L. C. Lee, M. E. Mandt, and J. K. Chao, Magnetospheric response to solar wind dynamic pressure variations: Interactions of interplanetary tangential discontinuities with the bow shock, *J. Geophys. Res.,* in press, 1993.

P. Song, High Altitude Observatory, National Center for Atmospheric Research, P. O. Box 3000, Boulder, CO 80307-3000

Transient and Quasi-Periodic (5–15 Min) Events in the Outer Magnetosphere

DAVID G. SIBECK

The Johns Hopkins University Applied Physics Laboratory, Laurel, Maryland

This article reviews and summarizes the expected characteristics of transient (~1-2 min) and quasi-periodic (T ~ 5-15 min) events in the outer magnetosphere driven by sudden changes in solar wind dynamic pressure; impulsive penetration; the Kelvin–Helmholtz instability; or patchy, sporadic, merging on the magnetopause. The review emphasizes tests that might be used in case and statistical studies to distinguish between the various competing mechanisms. A well-known case study which has received several contradictory interpretations is reexamined. The conflicting results suggest a need for further theoretical refinements and more extensive observational studies. Until such time as those tasks are completed, it does not seem possible to rule out any mechanism or combination of mechanisms.

INTRODUCTION

Transient (~1–2 min) variations in the outer magnetospheric magnetic field, plasma, and energetic particle distributions are common and often recur stochastically on time scales of ~5–15 min. The transient events can be attributed to transient and quasi-periodic variations in the solar wind dynamic pressure buffeting the magnetopause, impulsive penetration by solar wind filaments, the Kelvin–Helmholtz instability, or patchy sporadic merging of magnetosheath and magnetospheric magnetic field lines.

Current research emphasizes the latter mechanism, for two reasons. First, statistical studies indicate that transient events are observed at the magnetopause during periods of southward interplanetary magnetic field (IMF), and a southward IMF should favor magnetic merging at the dayside magnetopause. Second, transient events at the dayside magnetopause often display many of the signatures expected for magnetic merging: a magnetic field component normal to the nominal magnetopause, a mixture of magnetosheath and magnetospheric plasmas, high-speed plasma flows, and streaming energetic particles. Thus most transient events are interpreted in terms of patchy sporadic merging [e.g., Russell and Elphic, 1978]. Nevertheless, some recent papers advocate other mechanisms, including solar wind dynamic pressure pulses [e.g., Sibeck et al., 1989], impulsive penetration [Roth, 1992], and the Kelvin–Helmholtz instability [Ogilvie and Fitzenreiter, 1989].

Although continual monitoring of the magnetopause is not possible, outer magnetospheric magnetic field lines map to the low-altitude cusp. Following a suggestion by Cowley [1984], considerable effort has been expended to describe and identify the various transient ionospheric signatures observed by ground photometers,

radars, riometers, and magnetometers located under the cusp [e.g., Goertz et al., 1985; Sandholt et al., 1985]. The ground observations may ultimately prove decisive, for they are available on a nearly continual basis. Furthermore, the ground signatures of flux transfer events (FTEs) and pressure pulses differ strikingly: flows at the center of an FTE should be in the direction of the event motion, whereas those at the center of a pressure pulse point in the direction perpendicular to event motion. Combined ground radar and photometer observations of quasi-periodic flows and events moving in the direction predicted on the basis of the IMF orientation conclusively demonstrate the existence of patchy, sporadic merging [e.g., Lockwood et al., 1990]. Other studies provide evidence for transient events driven by variations in the solar wind dynamic pressure [Friis-Christensen et al., 1988]. McHenry et al. [1990] suggested an interpretation in terms of the Kelvin–Helmholtz instability, whereas Heikkila et al. [1989] maintained that one transient ground event was caused by impulsive penetration. Until statistical surveys are performed, the relative importance of the various categories cannot be determined.

If transient events represent evidence for magnetic merging and are the dominant mode of solar wind magnetosphere interaction [e.g., Cowley, 1982], we should make great efforts to determine when and where they occur, as well as their cause. Even if the events do not represent such evidence, they provide an important source for quasi-periodic oscillations in the outer magnetosphere and the high-latitude ionosphere [Chen and Hasegawa, 1974]. This article considers only the characteristics of transient events observed at the magnetopause and omits further discussion of the events in the ionosphere. First, predictions are surveyed for event characteristics and occurrence patterns from each of the various models proposed to date. Then a series of events are reexamined that have been variously interpreted in terms of patchy, sporadic merging, wavy low-latitude boundary layer (LLBL) motion, and pressure-pulse-driven magnetopause motion. Statistical studies offer a means of distinguishing

Solar Wind Sources of Magnetospheric Ultra-Low-Frequency Waves
Geophysical Monograph 81
Copyright 1994 by the American Geophysical Union.

between the various proposed modes of transient solar wind-magnetosphere interactions, but much further work will be required to settle the issues. The article concludes with a discussion of those tests that need to be applied in the nearest future.

SOLAR WIND PRESSURE PULSES

The first step is to consider the relative importances of interplanetary shocks, tangential discontinuities, and bow shock pressure-pulses in driving transient magnetopause motion. Interplanetary shocks and foreshock pressure-pulses typically provide factor of 2 to 3 variations in the solar wind dynamic pressure [e.g., Gosling et al., 1967; Fairfield et al., 1990], whereas tangential discontinuities correspond to much smaller solar wind pressure fluctuations [e.g., Solodyna et al., 1977]. Foreshock pressure pulses (each lasting 1–2 min) can recur each 5–10 min, tangential discontinuities arrive at a rate of about once per hour, and interplanetary shocks occur once every several days.

Based on their frequent occurrence and large amplitudes, the foreshock pressure-pulses must be considered the dominant source of pressure variations that batter the magnetosphere. Several case studies indicate that only spacecraft within the foreshock observe pressure pulses and the associated magnetic field fluctuations [Sibeck et al., 1989; Fairfield et al., 1990; Sibeck, 1992]. Thus high-time-resolution observations by a satellite in the subsolar foreshock are essential to determine whether or not pressure pulses produce individual transient events within the magnetosphere or ionosphere.

For the typical Parker spiral IMF orientation, the foreshock lies upstream of the pre-noon bow shock, and we expect pressure pulses to be more common before, rather than after, local noon. It is assumed that the solar wind sweeps foreshock pressure-pulses downstream, through the bow shock, and into the magnetosheath. If pressure pulses are more common upstream of the dawn than dusk bow shock, their effects should also be more common in the pre-noon magnetosheath. Although there is as yet no survey of pressure pulses in the magnetosheath, the dawnside magnetosheath does exhibit a somewhat higher level of magnetic field fluctuations than the duskside [e.g., Fairfield and Ness, 1970].

Upon crossing the bow shock, the pressure pulses may be converted into fast-mode waves, in which case information about the density enhancements propagates across the draped magnetosheath magnetic field lines and preferentially strikes the pre-noon magnetopause. Alternatively, regions of enhanced plasma pressure may be tied to magnetic field lines confined to regions of depressed magnetic field strength. In this case only pressure pulses on the earth-sun line, the streamline that defines the magnetopause surface, would strike the magnetosphere, and there would be no dawn-dusk asymmetry in the rate of pressure pulses striking the magnetosphere.

Figure 1 shows a pressure pulse which struck the magnetopause in the vicinity of the subsolar point and is now moving antisunward with the magnetosheath flow. Since the magnetopause lies along the locus of points where magnetosheath and magnetospheric pressures balance, transient increases in magnetosheath pressure drive transient magnetopause motion. The magnetopause first moves inward and then outward. Because the time required for the solar wind to sweep from local noon to the dawn/dusk meridians is on the order of 10 min and many pressure pulses last only 1–2 min, the magneto-

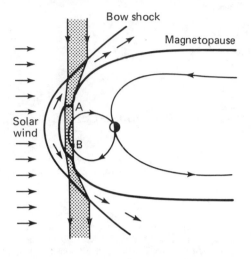

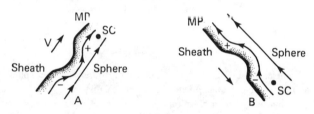

Fig. 1. A region of enhanced solar wind dynamic pressure briefly compresses the magnetopause, thereby producing an inward/outward bipolar oscillation of the magnetic field component normal to the nominal magnetopause south of the equator and an outward/inward bipolar signature north of the equator. Adopted from Sibeck et al. [1989].

pause generally does not reach global pressure balance during a series of pressure pulses. Instead the pulses generate a series of antisunward moving waves on the magnetopause.

As a trough in the magnetopause position passes, the component of the magnetospheric magnetic field perpendicular to the axis of the trough is enhanced, and the total magnetic field strength increases [Sibeck, 1990]. The magnetospheric magnetic field under the trough rotates toward a direction perpendicular to the trough axis. Furthermore, the trough indents the magnetopause and bends the magnetospheric magnetic field orientation in such a manner as to produce a bipolar outward/inward signature normal to the nominal magnetopause during the passage of a northward moving trough and a bipolar inward/outward signature during the passage of a southward moving trough (see Figure 1).

Figure 2 illustrates the pattern of magnetospheric magnetic perturbations associated with these troughs. A series of pressure pulses striking the magnetopause produces a set of concentric antisunward propagating troughs in the magnetopause position [Elphic, 1988]. The magnetospheric magnetic field orientation tilts dawnward under the troughs in the northern dawn and southern dusk quadrants, but duskward under the troughs in the northern dusk and southern dawn quadrants. No tilts are expected in the vicinity of local noon because of symmetry.

The LLBL, a region of mixed magnetosheath and magnetospheric plasmas, lies just inside the equatorial magnetopause, per-

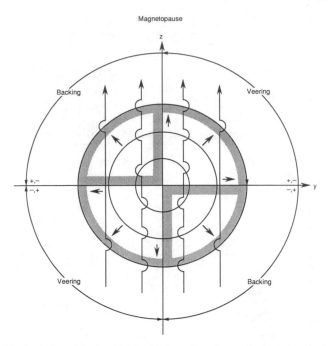

Magnetopause

Fig. 2. A view of the dayside magnetopause from the sun showing patterns of bipolar signatures normal to the magnetopause and azimuthal deflections inside the magnetosphere. The concentric circles represent troughs in the magnetopause position propagating outward from the subsolar point. The magnetospheric magnetic field drapes under each trough. During the passage of a trough, the magnetospheric magnetic field component perpendicular to the trough axis increases. Consequently, the magnetospheric magnetic field veers in the northern dusk and southern dawn quadrants, but backs in the southern dusk and northern dawn quadrants. Each trough produces a bipolar outward/inward magnetic field signature normal to the nominal magnetopause north of the equator, but a bipolar inward/outward signature south of the equator. Taken from Sibeck [1990].

haps on a mixture of open and closed magnetic field lines. As troughs in the magnetopause pass, spacecraft in the magnetosphere may enter the compressed LLBL, or even cross into the magnetosheath (see Figure 1). As a result, they may briefly observe magnetosheath-like plasmas. If the LLBL lies on open magnetic field lines, the spacecraft will briefly observe flows consistent with magnetic merging. Such a transient observation would not necessarily imply that merging was patchy and sporadic.

IMPULSIVE PENETRATION

Lemaire [1977] proposed that the enhanced density regions associated with solar wind pressure pulses not only deform, but also penetrate, the magnetopause. Schindler [1979] noted that under conditions of ideal magnetohydrodynamics (MHD), impulsive penetration was possible only for aligned magnetospheric and magnetosheath magnetic fields. Impulsive penetration, recently discussed by Roth [1992], has the possible advantage over the pressure pulse model in that it might explain the origin of the LLBL. However, Owen and Cowley [1991] have shown that impulsive penetration does not occur within the framework of ideal MHD, and Heikkila [1992] agrees. Discussion continues [Owen and Cowley, 1992]. Detailed predictions of the signatures expected for impulsive penetration are not available. Using a resistive MHD code, Ma et al.

[1991] numerically simulated impulsive penetration and showed that the process is more likely during periods of northward IMF. Magnetosheath filaments penetrate with ease only when their magnetic field orientation is precisely parallel to that in the magnetosphere. When the orientation is antiparallel, a dense plasma builds up at the magnetopause, obstructing progress. When the magnetosheath magnetic field is neither strictly parallel nor antiparallel, a large velocity normal to the magnetopause is required to produce penetration. Otherwise, as shown in Figure 3, the filament disintegrates as it impacts the magnetopause. The latter scenario is quite likely, since solar wind features in the magnetosheath are greatly decelerated and diverted to flow around the magnetosphere and do not reach the magnetopause with large velocities normal to that boundary.

KELVIN–HELMHOLTZ INSTABILITY

Dungey [1955] suggested that the flow of the solar wind past the magnetospheric magnetic field provides the conditions necessary for the onset of the Kelvin–Helmholtz instability. Southwood [1968] studied the linear instability for the general case in which field strength, direction, and plasma parameters change across the magnetopause. The instability criterion is satisfied in the incompressible limit when

$$\mu_0 \rho_1 \rho_2 (\mathbf{V} \cdot \mathbf{k_t})^2 > (\rho_1 + \rho_2)((\mathbf{B}_1 \cdot \mathbf{k_t})^2 + (\mathbf{B}_2 \cdot \mathbf{k_t})^2) \qquad (1)$$

where subscripts 1 and 2 refer to the values of the density (ρ) and magnetic field (**B**) in the magnetosheath and magnetosphere, the magnetosheath velocity is given by **V**, the tangential wave number vector is given by $\mathbf{k_t}$, and μ_0 is the permittivity of free space. Equation 1 shows that the instability conditions are most easily satisfied when (a) waves move in the direction of the magnetosheath flow, (b) waves move perpendicular to both the magnetosheath and magnetospheric magnetic field orientations, and (c) both the magnetosheath and magnetospheric plasma densities are large (a large magnetosheath density alone does not suffice). Thus, the instability is most likely on the magnetospheric flanks, particularly when the magnetosheath magnetic field lies parallel to or opposes the magnetospheric magnetic field. The antisunward moving boundary waves cause outer magnetospheric magnetic field lines to rotate clockwise before local noon and counterclockwise after local noon when viewed from above the northern hemisphere. Figure 4 illustrates the growth of the instability with downstream distance and the reversal in magnetic field and plasma flow polarization which is expected near local noon.

The growth rate increases steadily with decreasing wavelength when an infinitesimally thin magnetopause transition is considered [Lerche, 1966]. However, the inclusion of a boundary layer of finite thickness suppresses the growth of the instability at short wavelengths [Lee et al., 1981]. Under these conditions, one expects peak growth rates at a wavelength $\lambda \sim 2\pi d/0.6$ and a period $T \sim 2\pi d/0.6V$, where d is the thickness of the LLBL. Taking d as ~0.1 to 0.5 R_E [Eastman and Hones, 1979] and V as ~200 km s^{-1} (appropriate to the magnetospheric flanks), one obtains periods of 0.5 to 2.5 min in the magnetospheric rest frame. The growth rate for the same conditions is about 1 min. Since the time for the sheath to move from the subsolar region to the flanks is about 10 min, we may expect observable wave amplitudes at the flanks and further tailward. Inferences that the LLBL is thinner during periods of southward IMF (e.g., Sibeck

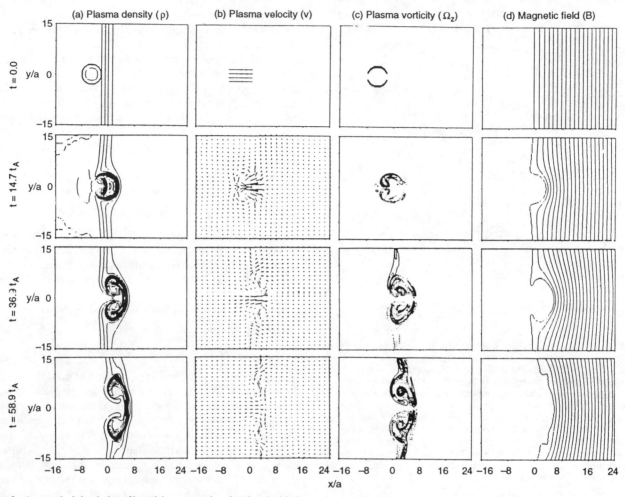

Fig. 3. A numerical simulation of impulsive penetration showing the (a) plasma density, (b) plasma velocity, (c) vorticity, and (d) magnetic vector potential as functions of time. The kinetic energy of the plasma filament is insufficient to cause penetration, spontaneous reconnection does not occur, and the filament disintegrates as it impacts the magnetopause. Taken from Ma et al. [1991].

[1990]) imply that short-period magnetopause motion with high growth rates is also more likely during periods of southward IMF. Lee et al. [1981] reported that the inner edge of the LLBL is almost always unstable. Although the velocity there is much lower than that in the magnetosheath, it is often nearly perpendicular to the magnetospheric magnetic field. Thus the magnetopause itself might remain in a nearly constant position, while the inner edge of the LLBL oscillates.

Numerical models permit study of the nonlinear growth of the instability. Miura [1984; 1987] used realistic parameters, e.g., Alfvénic Mach Number $M_A = 2.5$ and sonic Mach Number $M_S = 1.0$. For conditions appropriate to the equatorial dayside magnetopause, i.e., flow parallel to the magnetosheath magnetic field orientation, Figure 5 shows the development of the instability at two times: the early ($T = 40$) and saturation ($T = 80$) stages. During the early stage, the boundary undulates slightly. During the saturation stage, a pair of eddies appears near $y/a \sim 20$, and the magnetic field is slightly compressed and bent. However, a strong twisting of the magnetic

field lines is precluded by the large tension within them. If we substitute values for the velocity shear scale lengths ranging from 0.1 R_E on the dayside to 0.5 R_E on the flanks and magnetotail magnetopause, the solution indicates a magnetopause gently fluctuating with an amplitude of 0.4 to 1.8 R_E and periods of 80 to 400 s, corresponding to wavelengths of 2.5 to 12.5 R_E moving antisunward at 200 km s^{-1}. The corresponding growth rates range from 130 to 380 s. Again, one may expect waves generated by the instability to reach observable amplitudes on the dawn and dusk flanks. Larger and more intense disturbances occur when the flow is transverse to both the magnetospheric and magnetosheath magnetic fields. Wei et al. [1990] studied the nonlinear growth of the instability for absolutely unstable conditions in which the magnetic field is perpendicular to the flow velocity. Substituting reasonable parameters into their results ($V = 200$ km s^{-1}, half-thickness of the velocity shear scale length, $a = 0.4\ R_E$), one estimates the growth rate as $t = 25aV^{-1} = 320$ s and the largest amplitude magnetopause motion as about 0.5 R_E.

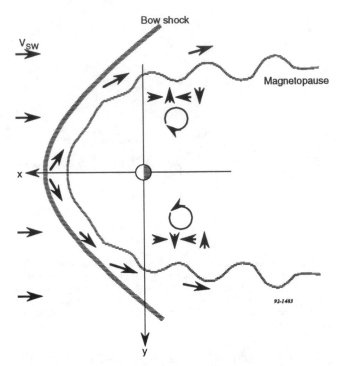

Fig. 4. The Kelvin–Helmholtz instability is expected at the equatorial magnetotail magnetopause. The instability produces waves which move antisunward with the magnetosheath velocity. Viewed from above the ecliptic plane, the magnetospheric magnetic field and plasma oscillations are polarized clockwise on the dawnside and counterclockwise on the duskside.

Flux Transfer Events

It is generally agreed that magnetic merging at the dayside magnetopause occurs along a tilted line passing through the subsolar point when the IMF is strong and has a southward component [e.g., Sonnerup, 1974]. The tilt of the line depends upon the IMF orientation, such that it passes from southern dawn to northern dusk during periods of duskward IMF and from northern dawn to southern dusk during periods of dawnward IMF. Cowley [1982] has suggested that the length of the merging line varies from 6 to 25 R_E. Reconnection may be entirely absent from high β plasmas [Sonnerup, 1974].

Dayside magnetic merging may be steady or sporadic, (and) patchy or extensive. Russell and Elphic [1978] suggested that transient merging at a single location might result in a flux tube of interconnected magnetosheath and magnetospheric magnetic field lines, which they labelled a flux transfer event. Subsequently, Rijnbeek et al. [1982] suggested that FTEs form in pairs. Figure 6 shows a pair of FTEs leaving the merging site. In contrast to the antisunward magnetosheath flow, the direction of motion of an FTE depends upon the balance of pressure gradient and magnetic curvature forces and may be sunward [e.g., Cowley and Owen, 1989]. Magnetosheath and magnetospheric plasmas are free to mix along the open magnetic field lines of the FTE [Russell and Elphic, 1978].

As the FTE moves along the magnetopause, it displaces magnetospheric magnetic field lines inward and magnetosheath magnetic field lines outward. That effect results in simultaneous magnetic field strength increases and bipolar magnetic field signatures normal to the

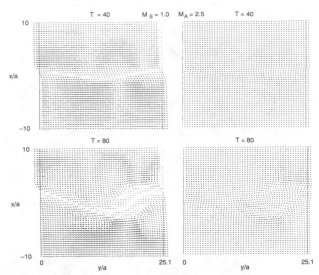

Fig. 5. Flow velocity (left) and magnetic field (right) in the early (T = 40) and saturation (T = 80) stages of the Kelvin–Helmholtz instability when the magnetic field is parallel to the flow. The magnetopause initially lies at rest along x/a = 0. The figure shows a large, twin, eddylike circulation pattern imbeded near y/a ~ 20 during the saturation phase. Taken from Miura [1984].

nominal magnetopause in both the draped magnetosheath and magnetospheric magnetic field lines. The draped magnetic fields surrounding the rope are deflected toward the direction perpendicular to the flux-rope axis [Farrugia et al., 1987]. Since merging and flux-rope formation are expected along an axis whose tilt depends on the IMF orientation, the magnetospheric magnetic field deflections produced by a passing FTE also depend upon the IMF orientation. As shown in Figure 7, when the IMF has a positive B_y component, the FTEs tilt from southern dawn to northern dusk (like the merging line), and the deflection within the magnetosphere (dashed lines) is dawnward ($-B_y$). Conversely, when the IMF has a negative B_y component, FTEs run from northern dawn to southern dusk (like the merging line) and the magnetospheric magnetic field deflection is duskward ($+B_y$).

The magnetic field structure inside FTEs has been a topic of great interest. Cowley [1982] suggested that a current flows along the flux-tube axis, and that this current produces a bipolar signature within a flux rope. However, Scholer [1988a] considered the expected results of a burst of (enhanced) merging at a single merging line and showed that it does not result in the formation of flux ropes but rather a bubble-like or bulge structure of bent and interconnected magnetosheath and magnetospheric magnetic field lines with a weak core magnetic field strength. Scholer [1988a] modelled a case with equal but opposite magnetosheath and magnetospheric magnetic fields and equal magnetosheath and magnetospheric densities, to show that the dimension of the bubble normal to the magnetopause grows with time and can exceed 0.3 R_E. The dimension in the direction of motion is 1 R_E.

Scholer [1988b] argued that merged lines could collect unmerged magnetosheath magnetic field lines, compress them, and produce a strong core field in the direction of the magnetosheath magnetic field. Scholer [1989] showed that the bulge produces a bipolar signature normal to the nominal magnetopause as it moves away from the

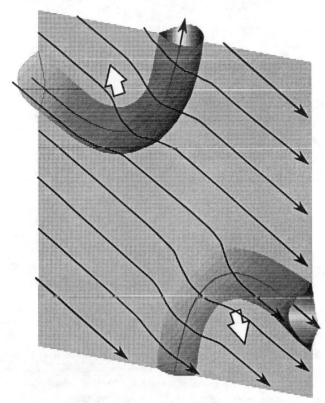

Fig. 6. A sketch of the northern and southern ends of open FTE flux tubes shown shortly after reconnection at the dayside magnetopause. The magnetic tension of the open tubes results in their contraction along the magnetopause relative to the overlying magnetosheath plasma and field in the directions indicated by the broad arrows. Consequently, the magnetospheric portions of the open tubes tilt towards magnetosheath orientations. The magnetosheath magnetic field component in the direction normal to the magnetopause is deflected outward, then inward, during passage of the northern tube, but inward, then outward, during passage of the southern tube. Adopted from Rijnbeek et al. [1982].

reconnection site with a velocity approximately 70% that of the magnetosheath Alfvén velocity. As shown in Figure 8, the bulge lies almost exclusively on the magnetosheath side of magnetospheric current layer when ratios of magnetospheric to magnetosheath magnetic field strengths exceed 1.5. Thus entries into, and remote observations of, magnetosheath FTEs should be much more common than for magnetospheric FTEs. Magnetic field signatures are much larger on the trailing edge of such events, resulting in a nearly monopolar signature [e.g., Ding et al., 1991]. Finally, there is a large jet in the direction of FTE motion on the magnetospheric side of the FTE.

Production of a true flux rope requires at least two merging lines [Lee and Fu, 1985]. Magnetic islands, or flux ropes in three dimensions, form between tilted pairs of merging lines. The most unstable tearing mode produces flux ropes with dimensions of 0.5 R_E along the magnetopause and 0.17 R_E along the direction normal to the magnetopause [Lee and Fu, 1985], although flux ropes may rapidly coalesce and grow in size. Fu and Lee [1985] and Ding et al. [1986] presented two-dimensional simulations in the X–Z plane and showed that FTEs with dimensions 1.5 R_E along the magnetopause and

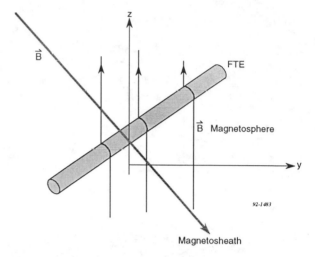

Fig. 7. A view of the subsolar dayside magnetopause from the sun. The magnetosheath magnetic field has a southward and duskward orientation ($+B_y$, $-B_z$). Merging results in the formation of an FTE on the magnetopause with an axis which runs from southern dawn to northern dusk. Magnetospheric magnetic field lines under the FTE (dashed lines) are deflected dawnward ($-B_y$) as the FTE passes.

0.5 R_E normal to the magnetopause recur each 5 to 15 min. In the presence of an IMF B_y component, merging forms twisted flux ropes. Currents flow parallel to the strong axial core flux rope magnetic field when the IMF B_y has a positive component and antiparallel to the magnetic field when the IMF B_y has a negative component.

Shi et al. [1988] reported the results of a numerical simulation in which FTEs had dimensions of 1–1.5 R_E normal to the magnetopause and 2–3 R_E in the direction of motion along the magnetopause. Each event could be observed for a period of 2 to 3 min. The dimensions of the events grow in time as they move away from the point where they were formed. Shi et al. [1988] argued that because the magnetic Reynolds number at the dayside magnetopause is large, magnetic merging should occur in the form of multiple X lines, rather than at a single X line. Furthermore, they suggested that merging and FTE formation are subsolar during periods of low solar wind Alfvénic Mach number, but may shift poleward during periods of high Mach number. Although none of the above numerical simulations predicted the longitudinal extent of FTEs, the simulation done by Ogino et al. [1989] indicates that they extend all the way across the dayside magnetopause, from dawn to dusk.

A CASE STUDY OF TRANSIENT EVENTS IN THE OUTER MAGNETOSPHERE

As shown in Figure 9, ISEE-1 observed a series of transient (~1–2 min) plasma density enhancements (pulses) as it traversed the mid-latitude dawnside magnetopause outbound on November 6, 1977. The spacecraft began the interval in the magnetosphere proper (Region 1 in Figure 9) and ended it in the magnetosheath (Region 4). In between lies a magnetospheric boundary layer (Region 2) and the series of plasma density pulses (Region 3). Sckopke et al. [1981] interpreted the pulses as antisunward moving blobs of LLBL plasma produced by the Kelvin–Helmholtz instability acting at the inner edge of the LLBL. Ogilvie and Fitzenreiter [1989] confirmed that

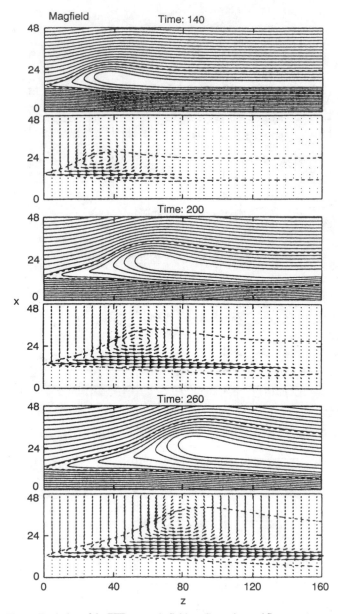

Fig. 8. Evolution of the FTE magnetic field configuration and flow pattern as a function of time. The undisturbed magnetospheric magnetic field strength is twice as strong as that in the magnetosheath. As a result, the FTE lies almost entirely on the magnetosheath side of the magnetopause current layer. Adopted from Scholer [1989].

conditions on that day satisfied the Kelvin–Helmholtz instability criteria.

In contrast, Cowley [1982] and Saunders [1983] suggested that the blobs represented a series of FTEs moving antisunward past the spacecraft. Saunders [1983] noted the following blob characteristics: a magnetic field tilted antisunward ($\alpha'_B > 0$), energetic (E > 25 KeV) ions streaming antiparallel to the magnetic field, and a flux of energetic (E > 20 KeV) electrons depressed greatly from magnetospheric levels. Velocities within the events exceeded those within the nearby magnetosheath. These features were interpreted as indicating

the passage of a series of twisted magnetic flux-ropes consisting of newly merged open magnetic field lines connected to the northern hemisphere.

Sibeck et al. [1990] provided a third explanation for the blobs. They noted that the blobs occurred during a period of northward and/ or ecliptic, rather than southward, IMF and argued that they represented a series of brief crossings into the magnetosheath caused by solar wind pressure pulses.

The latter interpretation requires the magnetosheath magnetic field to have pointed very strongly northward during the events. Sckopke [1991] noted that the observed IMF lay nearly in the ecliptic plane and invoked gasdynamic models of the magnetosheath magnetic field pattern to suggest that strongly northward magnetosheath magnetic field orientation would have been unlikely. However, recent work suggests that magnetohydrodynamic and gasdynamic magnetosheath flow patterns differ significantly [Chen et al., 1993] and we must also expect patterns of magnetic field line draping to differ. Magnetic pressure and curvature forces, neglected in gasdynamic models, accelerate magnetosheath magnetic field lines and plasma in the antisunward direction and allow them to slip more easily over the magnetopause. To slip over the magnetopause, magnetic field lines must be deflected out of the plane of the IMF. Therefore an ecliptic IMF may gain a strong poleward component outside the northern dawn magnetopause. Until such time as a comprehensive survey of magnetosheath magnetic field line draping patterns is performed, we maintain that the magnetic field observations within the density pulses on November 6, 1977 may be consistent with brief crossings into the magnetosphere.

Sckopke [1991] also argued that the flux of energetic (75 < E < 115 KeV) electrons during the final two blobs was not greatly depressed from magnetospheric levels, in which case the density pulses were likely observed within the magnetosphere. In fact, energetic electron pitch angle distributions provide even stronger evidence that the blobs were observed within the magnetosphere. Figure 10 shows the energetic particle observations that were discussed but not presented by Saunders [1983]. The ion flux is slightly depressed and the electron flux is greatly depressed from magnetospheric (Region 1) levels inside the blobs (Region 3). The energetic ion and electron fluxes within the blobs are nearly isotropic. The electron flux level within the events falls to levels ranging from 9×10^3 (cm^2–s–sr)$^{-1}$ during the early events to 1×10^3 (cm^2–s–sr)$^{-1}$ during the later events, consistent with the outbound spacecraft's motion through a region of negative radial flux gradient.

Magnetosheath electron fluxes observed after 0550 UT vary over a range of values from background to 2×10^4 (cm^2–s–sr)$^{-1}$ but generally lie within the range of values from 3×10^2 to 4×10^3 (cm^2–s–sr)$^{-1}$. While the average magnetosheath electron flux level is somewhat lower than that in the blobs, magnetosheath flux levels rise to levels comparable to or exceeding those in the blobs during brief intervals centered on 0556, 0605, 0608, and 0623 UT. Scholer et al. [1982] describes these particle bursts as occurring in a magnetosheath layer surrounding FTEs. The comparably low electron flux levels observed during some of the blobs before 0550 UT could be interpreted as observations made in or further radially outward than the layer of energetic electrons described by Scholev et al. [1982] which lies just outside the magnetopause. In this case the

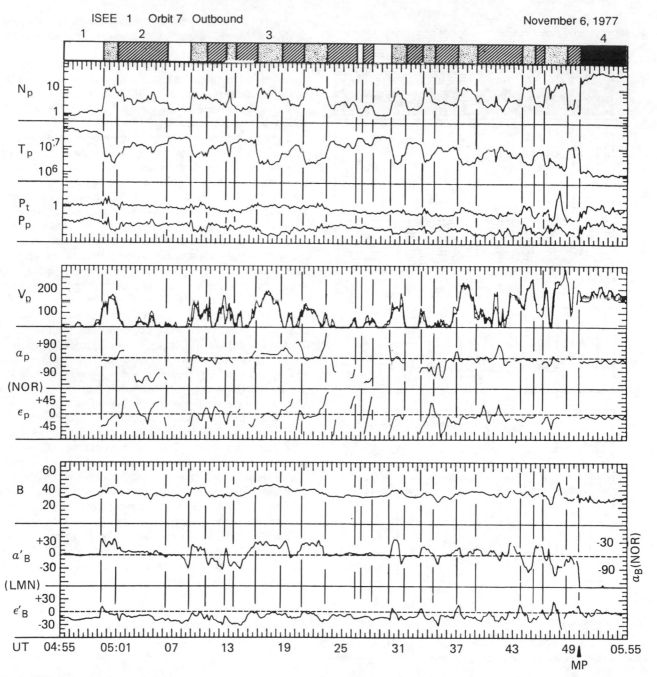

Fig. 9. ISEE-1 plasma and magnetic field observations on November 6, 1977 (adopted from Sckopke et al. [1981]). The upper three panels show the proton density (cm^{-3}), temperature (°K), and pressure (P_p, lower curve) as well as the total pressure $P_t = P_p + B^2/8\pi$, both in units of nPa. The central three panels display the proton flow behavior. The V_p panel shows the bulk speed, α_p measures the flow azimuth in the local tangential plane of the magnetopause ($\alpha_p = 0$ indicates a tailward flow), and ϵ_p measures the flow elevation from this plane ($\epsilon_p > 0$ indicates an outward flow). The lower three panels show the magnetic field as magnitude B, azimuth α_B', and elevation ϵ_B. These angles are defined similarly except that α_B' points toward GSM north. Region 1 is the magnetosphere proper, Region 2 the halo, Region 3 the low-latitude boundary layer, and Region 4 the magnetosheath.

observations are consistent with the magnetopause crossing model. However, the nearly isotropic energetic electron fluxes in the pulses indicate closed, rather than open, magnetic field lines. Mitchell et al. [1987] have also reported trapped distributions of energetic electrons with depressed fluxes within the dawnside LLBL. The nearly isotro-

pic energetic electron pitch angle distributions argue against an interpretation of the blobs in terms of open magnetosheath magnetic field lines or open FTE field lines but rather strongly support the model originally proposed by Sckopke et al. [1981], i.e., one in which the pulsed LLBL lies on closed magnetic field lines. However, it still

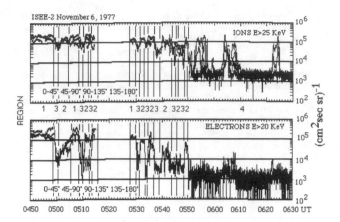

ISEE-2 November 6, 1977

IONS E>25 KeV

REGION

0-45° 45-90° 90-135° 135-180°

1   3 2 1 3232   1 32323 2 3232   4

ELECTRONS E>20 KeV

0-45° 45-90° 90-135° 135-180°

0450   0500   0510   0520   0530   0540   0550   0600   0610   0620   0630 UT

Fig. 10. ISEE-2 energetic particle observations on November 6, 1977 (courtesy of P. Daly and M. Saunders). The two panels show the fluxes of ions with energies greater than 25 KeV and electrons with energies greater than 20 KeV. The region identifier from Figure 9 is reproduced. Although the flux of ions remains relatively high in the dense plasma blobs of Region 3 (variously interpreted as FTEs, LLBL, and magnetosheath intervals), the electron fluxes fall sharply.

isn't clear why blobs on closed magnetic field lines should exhibit greater velocities than those in the nearby magnetosheath.

While the controversy surrounding the events cannot be considered resolved at this time, the differing viewpoints illustrate the need for a simultaneous solar-wind monitor, multi-instrument studies of transient events, and statistical surveys of event structure to determine which features of transient events are typical and which are not. Hapgood and Bryant [1992] have made important first steps in this direction. They order magnetopause crossings with a transition parameter based upon the electron density and temperature. Transient events often fall along the curve determined for gradual transits of the magnetopause, suggesting that the transient events may be no more than abrupt entries into, and exits out of, layers that are continually present at the magnetopause.

Conclusions

This paper discusses possible causes for transient events observed in the outer magnetosphere and described the signatures predicted for each mechanism. We considered ISEE-1/2 observations of transient events in the vicinity of the dayside magnetopause on November 6, 1977. Some aspects of the transient events observed on this day were consistent with boundary waves, but others with FTEs. There is much room for further work. Multiple spacecraft observations, such as those expected during the CLUSTER mission, may help distinguish among possible causes. When the spacecraft are greatly separated, one may be used as a local monitor for the magnetosheath magnetic field orientation and pressure pulses while the others observe transient events within the outer dayside magnetosphere. Using the spacecraft in this manner, one would not be forced to assume the existence of coherent IMF structures over great distances and estimate the times when solar wind features strike the magnetosphere. When the CLUSTER spacecraft are not separated greatly, they may be used to determine the shape and direction of motion of transient events as a function of latitude and local time.

It should also be possible to determine the orientation and direction of motion of the transient events and to use this information to determine their origin. For example, pressure pulses should produce antisunward, moving, concentric ripples in the magnetopause position, whereas magnetic merging produces cylindrical FTEs which move across the magnetopause surface in response to pressure gradients and magnetic curvature forces. Studies using simultaneous IMF observations may help determine if the events move in response to magnetic curvature forces. In particular, it may be possible to determine whether the dawnward or duskward component of the IMF controls the direction of FTE motion. Recent high-time-resolution magnetic field and plasma observations may be used to test the Kelvin–Helmholtz instability criterion over the surface of the magnetopause. If boundary waves are observed in unstable locations but not in stable locations, the waves may be attributed to the instability.

Finally, we note that all of the mechanisms discussed above generate small-scale field-aligned currents in the outer magnetosphere. If the currents remain field aligned, they map along magnetic field lines into the high-latitude dayside ionosphere where they drive transient, localized, convection patterns. Having identified the patterns associated with each mechanism, case and statistical studies of high-latitude electric field, ground magnetometer, radar, and optical observations will yield further information concerning the significance of the various mechanisms proposed as causes for transient events.

Acknowledgements. P. Daly and M. Saunders kindly supplied the ISEE-2 energetic particle observations used in Figure 10. This work was supported by NASA under Task 1 of Space and Naval Warfare Systems Command contract N00039-91-C-0001 to the Applied Physics Laboratory.

References

Chen, L. and A. Hasegawa, A theory of long-period magnetic pulsations, 1. Steady state excitation of field line resonance, *J. Geophys. Res., 79,* 1024–1032, 1974.

Chen, S.-H., M. G. Kivelson, J. T. Gosling, R. J. Walker, and A. J. Lazarus, Anomalous aspects of magnetosheath flow and of the shape and oscillations of the magnetopause during an interval of strongly northward interplanetary magnetic field, *J. Geophys. Res., 98,* 5727–5742, 1993.

Cowley, S. W. H., The causes of convection in the Earth's magnetosphere: A review of developments during the IMF, *Rev. Geophys., 20,* 531–565, 1982.

Cowley, S. W. H., Evidence for the occurrence and importance of reconnection between the earth's magnetic field and the interplanetary magnetic field, in *Magnetic Reconnection in Space and Laboratory Plasmas,* ed. E. W. Hones, Jr., AGU, Washington, D.C., 375–378, 1984.

Cowley, S. W. H. and C. J. Owen, A simple illustrative model of open flux tube motion over the dayside magnetopause, *Planet. Space Sci., 37,* 1461–1475, 1989.

Ding, D. Q., L. C. Lee, and Z. F. Fu, Multiple X line reconnection, 3. A particle simulation of flux transfer events, *J. Geophys. Res., 91,* 13384–13392, 1986.

Ding, D. Q., L. C. Lee, and Z. W. Ma, Different FTE signatures generated by the bursty single X line reconnection and the multiple X line reconnection at the dayside magnetopause, *J. Geophys. Res., 96,* 57–66, 1991.

Dungey, J. W., Electrodynamics of the outer atmosphere, in *Proceedings of the Ionosphere Conference,* p. 225, The Physical Society of London, 1955.

Eastman, T. E. and E. W. Hones, Jr., Characteristics of the magnetospheric boundary layer and magnetopause layer as observed by Imp 6, *J. Geophys. Res., 84,* 2019–2028, 1979.

Elphic, R. C., Multipoint observations of the magnetopause: Results from ISEE and AMPTE, *Adv. Space Sci.*, *8*, 223–238, 1988.

Fairfield, D. H. and N. F. Ness, Magnetic field fluctuations in the earth's magnetosheath, *J. Geophys. Res.*, *75*, 6050–6060, 1970.

Fairfield, D. H., W. Baumjohann, G. Paschmann, H. Lühr, and D. G. Sibeck, Upstream pressure variations associated with the bow shock and their effects on the magnetosphere, *J. Geophys. Res.*, *95*, 3773–3786, 1990.

Farrugia, C. J., R. C. Elphic, D. J. Southwood, and S. W. H. Cowley, Field and flow perturbations outside the reconnected field line region in flux transfer events: Theory, *Planet. Space Sci.*, *35*, 227–240, 1987.

Friis-Christensen, E., M. A. McHenry, C. R. Clauer, and S. Vennerstrøm, Ionospheric traveling convection vortices observed near the polar cleft: A triggered response to sudden changes in the solar wind, *Geophys. Res. Lett.*, *15*, 253–256, 1988.

Fu, Z. F. and L. C. Lee, Simulation of multiple X line reconnection at the dayside magnetopause, *Geophys. Res. Lett.*, *12*, 291–294, 1985.

Goertz, C. K., E. Nielsen, A. Korth, K.-H. Glassmeier, C. Haldoupis, P. Hoeg, and D. Hayward, Observations of a possible ground signature of flux transfer events, *J. Geophys. Res.*, *90*, 4069–4078, 1985.

Gosling, J. T., J. R. Asbridge, S. J. Bame, A. J. Hundhausen, and I. B. Strong, Discontinuities in the solar wind associated with sudden geomagnetic impulses and storm commencements, *J. Geophys. Res.*, *72*, 3357–3363, 1967.

Hapgood, M. A. and D. A. Bryant, Exploring the magnetospheric boundary layer, *Planet. Space Sci.*, *40*, 1432–1459, 1992.

Heikkila, W. J., Comment on Owen and Cowley's analysis of impulsive plasma transport through the magnetopause, *J. Geophys. Res.*, *97*, 1639, 1992.

Heikkila, W. J., T. S. Jørgensen, L. J. Lanzerotti, and C. J. Maclennan, A transient auroral event on the dayside, *J. Geophys. Res.*, *94*, 15277–15290, 1989.

Lee, L. C. and Z. F. Fu, A theory of magnetic flux transfer at the earth's dayside magnetopause, *Geophys. Res. Lett.*, *12*, 105–108, 1985.

Lee, L. C., R. K. Albano, and J. R. Kan, Kelvin–Helmholtz instability in the magnetopause-boundary layer region, *J. Geophys. Res.*, *86*, 54–58, 1981.

Lemaire, J., Impulsive penetration of filamentary plasma elements into the magnetospheres of the Earth and Jupiter, *Planet. Space Sci.*, *25*, 887–890, 1977.

Lerche, I., Validity of the hydromagnetic approach in discussing instability of the magnetospheric boundary, *J. Geophys. Res.*, *71*, 2365–2371, 1966.

Lockwood, M., S. W. H. Cowley, P. E. Sandholt, and R. P. Lepping, The ionospheric signatures of flux transfer events and solar wind dynamic pressure changes, *J. Geophys. Res.*, *95*, 17113–17135, 1990.

Ma, Z. W., J. G. Hawkins, and L. C. Lee, A simulation study of impulsive penetration of solar wind irregularities into the magnetosphere at the dayside magnetopause, *J. Geophys. Res.*, *96*, 15751–15765, 1991.

McHenry, M. A., C. R. Clauer, E. Friis-Christensen, P. T. Newell, and J. D. Kelly, Ground observations of magnetospheric boundary layer phenomena, *J. Geophys. Res.*, *95*, 14995–15005, 1990.

Mitchell, D. G., F. Kutchko, D. J. Williams, T. E. Eastman, L. A. Frank, and C. T. Russell, An extended study of the low-latitude boundary layer on the dawn and dusk flanks of the magnetosphere, *J. Geophys. Res.*, *92*, 7394–7404, 1987.

Miura, A., Anomalous transport by magnetohydrodynamic Kelvin–Helmholtz instabilities in the solar wind-magnetosphere interaction, *J. Geophys. Res.*, *89*, 801–818, 1984.

Miura, A., Simulation of Kelvin–Helmholtz instability at the magnetospheric boundary, *J. Geophys. Res.*, *92*, 3195–3206, 1987.

Ogilvie, K. W. and R. J. Fitzenreiter, The Kelvin–Helmholtz instability at the magnetopause and inner boundary layer surface, *J. Geophys. Res.*, *94*, 15113–15123, 1989.

Ogino, T., R. J. Walker, and M. Ashour-Abdalla, A magnetohydrodynamic simulation of the formation of magnetic flux tubes at the Earth's dayside magnetopause, *Geophys. Res. Lett.*, *16*, 155–158, 1989.

Owen, C. J. and S. W. H. Cowley, Heikkila's mechanism for impulsive plasma transport through the magnetopause: A reexamination, *J. Geophys. Res.*, *96*, 55655574, 1991.

Owen, C. J. and S. W. H. Cowley, Reply, *J. Geophys. Res.*, *97*, 1641–1643, 1992.

Rijnbeek, R. P., S. W. H. Cowley, D. J. Southwood, and C. T. Russell, Observations of "reverse polarity" flux transfer events at the Earth's dayside magnetopause, *Nature, 300*, 23–26, 1982.

Roth, M., On impulsive penetration of solar wind plasmoids into the geomagnetic field, *Planet. Space Sci.*, *40*, 193–201, 1992.

Russell, C. T. and R. C. Elphic, Initial ISEE magnetometer results: Magnetopause observations, *Space Sci. Rev.*, *22*, 681–715, 1978.

Sandholt, P. E., A. Egeland, J. A. Holtet, B. Lybekk, K. Svenes, and S. Asheim, Large- and small-scale dynamics of the polar cusp, *J. Geophys. Res.*, *90*, 4407–4414, 1985.

Saunders, M. A., Recent ISEE observations of the magnetopause and low latitude boundary layer: A review, *J. Geophys.*, *52*, 190–198, 1983.

Schindler, K., On the role of irregularities in plasma entry into the magnetosphere, *J. Geophys. Res.*, *84*, 7257–7266, 1979.

Scholer, M., Magnetic flux transfer at the magnetopause based on single X line bursty reconnection, *Geophys. Res. Lett.*, *15*, 291–294, 1988a.

Scholer, M., Strong core magnetic fields in magnetopause flux transfer events, *Geophys. Res. Lett.*, *15*, 748–751, 1988b.

Scholer, M., Asymmetric time-dependent and stationary magnetic reconnection at the dayside magnetopause, *J. Geophys. Res.*, *94*, 15099–15111, 1989.

Scholer, M., D. Hovestadt, F. M. Ipavich, and G. Gloeckler, Energetic protons, alpha particles, and electrons in magnetic flux transfer events, *J. Geophys. Res.*, *87*, 2169–2175, 1982.

Sckopke, N., Plasma structure near the low-latitude boundary layer: A rebuttal, *J. Geophys. Res.*, *96*, 9815–9820, 1991.

Sckopke, N., G. Paschmann, G. Haerendel, B. U. Ö. Sonnerup, S. J. Bame, T. G. Forbes, E. W. Hones, Jr., and C. T. Russell, Structure of the low-latitude boundary layer, *J. Geophys. Res.*, *86*, 2099–2110, 1981.

Shi, Y., C. C. Wu, and L. C. Lee, A study of multiple X line reconnection at the dayside magnetopause, *Geophys. Res. Lett.*, *15*, 295–298, 1988.

Sibeck, D. G., A model for the transient magnetospheric response to sudden solar wind dynamic pressure variations, *J. Geophys. Res.*, *95*, 3755–3771, 1990.

Sibeck, D. G., Transient events in the outer magnetosphere: Boundary waves or flux transfer events?, *J. Geophys. Res.*, *97*, 4009–4026, 1992.

Sibeck, D. G., W. Baumjohann, R. C. Elphic, D. H. Fairfield, J. F. Fennell, W. B. Gail, L. J. Lanzerotti, R. E. Lopez, H. Lühr, A. T. Y. Lui, C. G. Maclennan, R. W. McEntire, T. A. Potemra, T. J. Rosenberg, and K. Takahashi, The magnetospheric response to 8-min-period strong-amplitude upstream pressure variations, *J. Geophys. Res.*, *94*, 2505–2519, 1989.

Sibeck, D. G., R. P. Lepping, and A. J. Lazarus, Magnetic field line draping in the plasma depletion layer, *J. Geophys. Res.*, *95*, 2433–2440, 1990.

Solodyna, C. V., J. W. Sari, and J. W. Belcher, Plasma field characteristics of directional discontinuities in the interplanetary medium, *J. Geophys. Res.*, *82*, 10–14, 1977.

Sonnerup, B. U. Ö., Magnetopause reconnection rate, *J. Geophys. Res.*, *79*, 1546–1549, 1974.

Southwood, D. J., The hydrodynamic stability of the magnetospheric boundary, *Planet. Space Sci.*, *16*, 587–605, 1968.

Wei, C. Q., L. C. Lee, and A. L. La Belle-Hamer, A simulation study of the vortex structure in the low-latitude boundary layer, *J. Geophys. Res.*, *95*, 20793–20807, 1990.

D. G. Sibeck, The Johns Hopkins University Applied Physics Laboratory, Johns Hopkins Road, Laurel, Maryland 20723-6099.

A Physical Model
of the Geomagnetic Sudden Commencement

TOHRU ARAKI

Department of Geophysics
and
Data Analysis Center for Geomagnetism and Spacemagnetism
Faculty of Science, Kyoto University

A physical model is presented which can explain the global structure of the geomagnetic sudden commencement (SC). Relevant observational results are first summarized in order to construct the model. The most important point is to explain how a simple step-function like increase of the dynamic pressure of the solar wind can produce a complex waveform distribution of SC on the ground. The disturbance field of the SC is decomposed into two sub-fields DL and DP. The DL-field, which shows a monotonic increase of the H-component, is produced by electric currents flowing on the magnetopause and a propagating compressional wave front. The DP-field consists of two successive pulses, the PI (preliminary impulse) and the MI (main impulse), with opposite senses, both of which are produced by twin vortex type ionospheric currents. The ionospheric current for the PI is caused by a dusk-to-dawn electric field transmitted to the polar ionosphere from the compressional wavefront propagating in the dayside magnetosphere. The ionospheric current for the MI of the DP-field is caused by the enhanced dawn-to-dusk convection electric field. The dusk side current vortex of both current systems extends to the dayside equator and produces the equatorial enhancement of PI and MI. The validity of the model is examined from several view points. Finally, results of a numerical synthesis of SC based upon the model are presented.

1. INTRODUCTION

The geomagnetic sudden commencement(SC) is a very clear global phenomenon. It has a clear onset and can be detected almost everywhere within the magnetosphere and almost simultaneously on the earth. The source of SCs is now clearly identified as interplanetary shocks and discontinuities. The global simultaneous occurrence with clear onsets and well identified sources are the main characteristics distinguishing SCs from other geophysical phenomena. We can easily locate SCs and determine their onset. We don't need to worry about identification of the source. In some sense SCs resemble earthquakes, but observations of earthquakes are limited only to the earth's surface and their source can not be studied directly whereas three dimensional observations are possible for SCs, including their source.

Solar Wind Sources of Magnetospheric Ultra-Low-Frequency Waves
Geophysical Monograph 81

Owing to their clear global signatures, SCs continue to attract scientific interest. The almost simultaneous global occurrence of SCs was originally reported by Adams [1892] and Ellis [1892]. The SC phenomena stimulated Chapman and Ferraro [1931, 1932 and 1933] to construct their pioneering theory on the interaction between the solar corpuscular stream and the geomagnetic field. They inferred the existence of a magnetosphere prior to any in situ measurements in the interplanetary space. Gold [1955] noted the rapid arrival of disturbances at earth following solar flares and postulated the existence of an interplanetary shock at the leading edge of a plasma cloud ejected from a solar flare as an explanation for the sharp rise in SCs. Ground studies of SCs made great progress during the IGY (1957-58) following the installation of rapid-run magnetographs. Discovery of the solar wind [Neugebauer and Snyder, 1962] and interplanetary shocks [Sonett et al., 1964] as well as subsequent observations in interplanetary space and the magnetosphere aided in the understanding of the three dimensional structure of SCs.

As described in the following sections, the electric currents responsible for SCs flow over a range of locations from the magnetopause to the inside of the earth. Considering characteristics of the SC phenomena mentioned above, we can effectively use SCs to study the transient response of the magnetosphere, ionosphere and conducting earth system to variations of the solar wind dynamic pressure. When understanding of the transient response is well established, we will be able to use past ground observations of SCs to study solar activity and properties of the solar wind during the pre-satellite era.

2. OBSERVATIONAL FACTS

In this section we summarize those observations necessary to construct a model of the SC phenomena. The lower panel of Figure 1 shows a magnetogram for a typical geomagnetic storm at a low latitude station which started with an SC at 0230 LT. For comparison, a magnetogram for the preceding quiet day is given in the upper panel. As is seen in the lower panel, a sudden increase of the H-component characterizes SCs in low latitudes.

SCs correspond to sudden increases in the solar wind dynamic pressure at interplanetary shocks and discontinuities. The upper panel of Figure 2 shows two hour records of solar wind density and velocity during which an interplanetary shock was observed by ISEE-3 (1817 UT). The shock required about 30 min to reach earth. The lower panel shows an observation of the resulting SC observed by the geosynchronous satellite GOES-2 near local noon at 1850 UT. The inset shows H-component variations at 3 ground stations, Honolulu (Hawaii, geomag. lat. : 21.5°), Wake Is. (west Pacific, 13.2°) and Eusebio (Brazil, 5.4°). The magnetic variation of the SC at geosynchronous orbit near noon is mainly compressional (Hp-component in Fig. 2). Note that the waveform is essentially stepwise at each location of observation but that the rise time increases from approximately 2 min at ISEE-3 in the solar wind to 3 min at GOES-2 in the dayside magnetosphere and more than 4 min on the ground. Maeda et al.[1962] reported that the rise time at Honolulu and Apia (-15.6°) ranges from 1 to 9 min with a maximum occurrence rate of 3-4 min. The amplitude of the SC on the ground is enhanced at dayside equatorial stations (e.g. Eusebio near

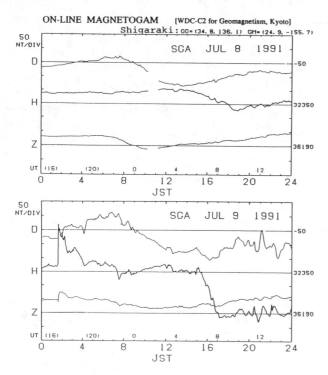

Fig. 1. Lower panel: The ground magnetogram for July 9, 1991 at Shigaraki (geomag. lat. : 24.9°), Japan. An SC occurred at 0130 UT and is followed by a geomagnetic storm. Upper panel: The magnetogram for the previous quiet day for reference.

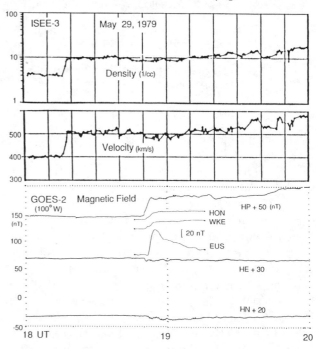

Fig. 2. An interplanetary shock observed by ISEE-3 located $202R_e$ upstream of the earth, and the shock-produced SC observed by a geosynchronous satellite GOES-2 near noon and 3 low latitude ground stations: Honolulu (geomag. lat. : 21.5° and local time :near 7h), Wake Island (13.2°, 6h) and Eusebio (5.4°, 16h). The ISEE-3 data were provided by E. Smith.

15^h LT in Figure 2) as is well known [Sugiura, 1953, Forbush and Vestine, 1955, Maeda and Yamamoto, 1960, Nishida, 1962b, and Rastogi and Sastri, 1974].

It is also well known that there is a linear relationship between the amplitude of the ground SC and the jump in the square root of the solar wind dynamic pressure across the shock or discontinuity [Siscoe et al., 1968, Ogilvie et al., 1968, Smith et al., 1986 and Lepping et al., 1987]. Russell et al.[1992] showed that the inclination of the line depends upon local time of stations used.

The amplitude of SCs at geosynchronous orbit shows a clear local time dependence as shown in the upper panel of Figure 3 [Kokubun, 1983]. The amplitude near midnight geosynchronous orbit is very weak and shows a seasonal variation suggesting effects of the tail current as shown in the middle panel of Figure 3 [Nagano and Araki, 1984]. The lower panel of Figure 3 shows difference in onset time of SC observed by two geosynchronous satellites GOES-2 and -3 versus local time of the satellites on the dayside [Nagano and Araki, 1985]. The onset times are earliest in the range of local times 12-14h. Propagation velocities calculated under the assumption that SCs propagate along the sun-earth line ranges from 300 to 1000 km/s with an average of 600 km/s, consistent with known magnetospheric fast mode velocities.

Wilken et al.[1982] illustrated the propagation of an SC in the magnetosphere using multipoint observations from five satellites in the magnetosphere and one in the interplanetary space as well as ground observations. Figure 4 shows the deformation of the wavefront of the interplanetary shock and SC front in the equatorial plane of the magnetosphere. Propagation of the SC front in the equatorial plane is rather simple. The wavefront in the magnetosphere moves at the fast mode velocity and deforms according to the spatial distribution of that velocity. Namikawa et al.[1964] and Stegelmann and Kenschitzky [1964] calculated the deformation of the wavefront. When the wavefront reaches the ionosphere, a complex interaction between the wave and the conducting earth-ionosphere system begins [Ohnishi and Araki, 1992]. The wave is partly reflected back into the magnetosphere and partly converted into an electromagnetic wave which propagates from the dayside to the nightside in the space between the ionosphere and the earth.

Figure 5 shows an SC observed at 9 stations distributed from auroral to equatorial latitude. In the low latitude region except the dayside equator, the wave-

form is approximately a step-function as seen at Memambetsu (Japan, geomag. lat. : 34.6°), Honolulu (Hawaii, 21.5°), San Juan (Puerto Rico, 29.4°), Guam (south-west Pacific, 4.6°) and Bangui (Central Africa,

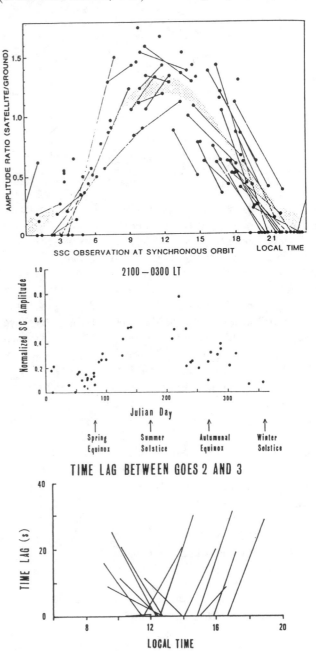

Fig. 3. Upper panel: Local time dependence of the amplitude of SC at geosynchronous orbit [Kokubun, 1983]. Middle panel: Seasonal variation of the amplitude of SC near midnight at geosynchronous orbit [Nagano and Araki, 1986]. Lower panel: Time delay of the SC onset observed by geosynchronous satellites GOES-2 and -3 in the dayside equator [Nagano and Araki, 1985].

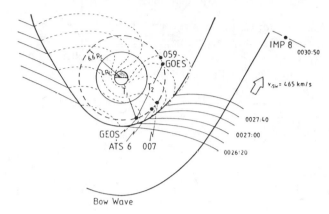

Fig. 4. Deformation of the SC wavefront in the equatorial plane as deduced from multi-satellite and ground observations [Wilken et al., 1982].

4.5°). The waveform at the auroral latitudes consists of two successive pulses. Within the morning side auroral zone (College, Alaska, 65.1°), a sharp positive pulse precedes a broad negative pulse. Within the afternoon auroral zone (Narssarsuaq, Greenland, 70.6°), a negative pulse precedes a positive pulse. The second pulse in both auroral zones recovers gradually in 15-30 min to the initial level. At the dayside equator (Huancayo, Peru, -1.1°) the two pulse structure seen at high latitudes appears again and the waveform is quite similar to that in the afternoon auroral latitude (Narssarssuaq), suggesting a strong coupling between the polar and the equatorial ionosphere. Even at low latitude stations, the two pulse structure with a reduced amplitude appears to be superposed on step-function waveform. The amplitude of the two pulses decreases with decreasing latitude but is enhanced at the dayside equator. Here the first and the second pulse are called the preliminary (PI) and main impulses (MI), respectively. A PI with a negative sense is often called a preliminary reverse impulse (PRI). A periodic oscillation with a period of several minutes is superposed on the two pulse structure at College and Yellowknife. This is called P_{sc} [Saito and Matsushita, 1967].

Figure 6 shows the diurnal variation of the normalized amplitude of MI at Huancayo [Sugiura, 1953] and the latitudinal variation of MI (denoted SSC$^+$) and PRI (denoted SSC^{-+}) at Indian stations [Rastogi and Sastri, 1974]. We see that the rate of the equatorial enhancement is larger for PRI than for MI.

Figure 7 [Araki et al., 1992] shows the fine-scale structure of an SC observed simultaneously in the north-south Alaska Chain of the North American IMS Magnetometer Chain (upper panel) and two chains in the

north-south (middle panel) and east-west (lower panel) directions which are part of the Scandinavian IMS Magnetometer Array [Kueppers et al., 1979]. The plots for the SC in Scandinavia are reproduced from the thesis of Volpers [1984]. The local time is around 3h in Scandinavia and 17h in Alaska.

A sharp negative pulse (PI) in the H-component precedes the MI at the lower latitude side stations (TLK, COL, FYU, AVI and INK) of the Alaska Chain (upper left panel). It changes to a positive pulse at the higher latitude station JOP as shown by a vertical dashed line. The stations CPY and SAH seems to be located in a transition region. Corresponding to this first negative pulse seen at the lower latitude stations of the Alaska Chain in the late afternoon, a positive pulse appears at most of the Scandinavian stations in the early morning. At the two highest latitude stations SOY and MAT a

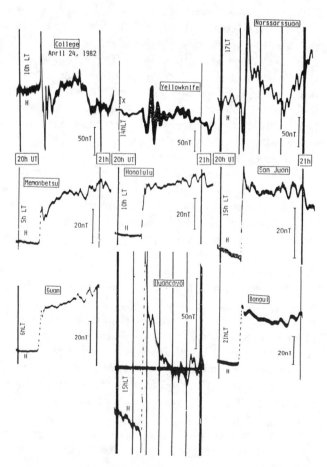

Fig 5. Simultaneous observations of an SC at auroral (upper panels), middle and low latitudes (middle panels), as well as the equator (lower panels). Left and right panels correspond to morning and afternoon, respectively.

and second pulse reverses as one moves from morning to afternoon along the auroral latitude. The first pulse (PI) is positive in the early morning (Scandinavia) and negative in the evening (Alaska) in latitude range from 63° to 68°. The sense of the second pulse (MI) is opposite to that of the first pulse. This two pulse structure seen in Scandinavian and Alaskan chains is consistent with the global waveform distribution shown in Figure 5. Figure 7 also **suggests that both the first and second** pulses change senses in the polar cap.

Figure 8 [Araki, 1977] shows an SC recorded on rapid-run magnetograms at stations from the noon equator to afternoon auroral latitudes. A PRI in the H-component precedes at the auroral latitude (Point Barrow, College

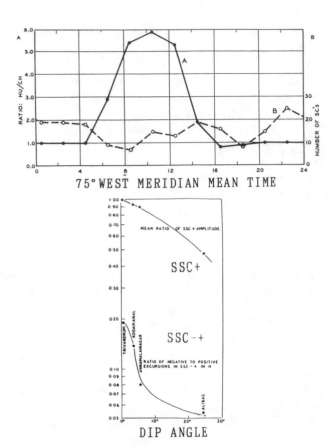

Fig. 6. Upper panel: Diurnal variation of the normalized amplitude of SC at Huancayo near the dip equator [Sugiura, 1953]. Lower panel: Latitudinal variation of the normalized amplitude of MI (denoted as SSC$^+$) and PI (denoted as SSC^{-+}) at Indian stations [Rastogi and Sastri, 1974].

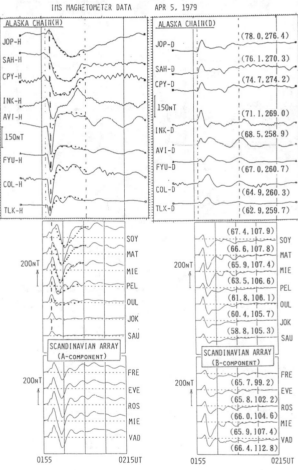

Fig. 7. Fine structure of an SC observed simultaneously along the Alaska chain (top panels) and Scandinavian north-south (middle panels) and east-west (bottom panel) chains in Scandinavia [Araki et al., 1992]. The A- and B-components in the middle and lower panels correspond to H- and D-components, respectively. Numbers in the parentheses show geomagnetic latitude and longitude. The middle and bottom panels were taken from Volper [1984].

small negative pulse begins to appear suggesting that its amplitude grows at the higher latitudes. Note that the geomagnetic latitude of the highest latitude station SOY in Scandinavia is slightly lower than that of AVI of the Alaska Chain. The second pulse (MI) of the H-component shown by dotted curves in the upper and middle left panels is much broader than the initial pulse. It is positive at the lower latitude stations of the Alaska Chain but negative at the higher latitudes. In Scandinavia, the second pulse is negative and its amplitude diminishes to near zero at the lowest latitude stations (SAU and JOK). P_{sc} is superposed on the second pulse at most of the Scandinavian stations and lower latitude stations (COL and TLK) of the Alaska Chain.

Taking into account that the latitude of the 4 highest latitude stations (SOY to PEL) in the Scandinavian north-south chain is equivalent to that of the 4 lowest latitude stations (AVI to TLK) of the Alaska Chain, it will be reasonably concluded that the sense of both first

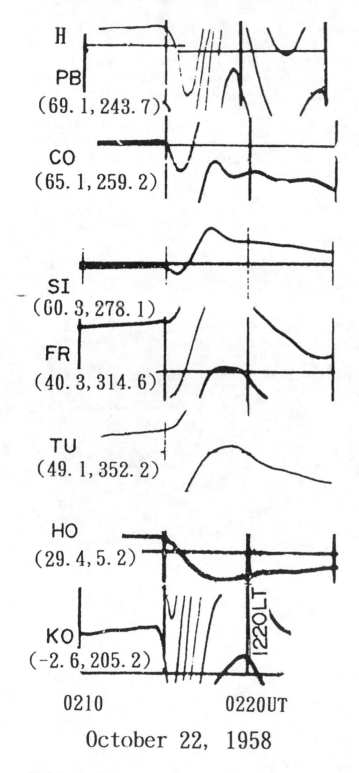

H

PB
(69. 1, 243. 7)

CO
(65. 1, 259. 2)

SI
(60. 3, 278. 1)

FR
(40. 3, 314. 6)

TU
(49. 1, 352. 2)

HO
(29. 4, 5. 2)

KO
(-2. 6, 205. 2)

0210   0220UT

October 22, 1958

Fig. 8. Rapid-run magnetograms showing latitudinal variation of waveforms for an H-component SC. Geomagnetic latitude and longitude are given in the parentheses. The sense of the variation at Honolulu is reversed [Araki, 1977].

and Sitka) in the afternoon and at the noon equator (Koror) but disappears at 3 other stations (Fredericksburg, Tucson and Honolulu) between them.

Matsushita [1962] classified the waveform of SC into 3 categories; SC (normal stepwise SC), $^-$SC (SC preceded by a PRI, called also SC* or SSC*) and SC$^-$ (SC followed by a negative impulse, called also *SC or *SSC) and studied its global distribution. Sano [1962] modified Matsushita's distribution taking into account data from higher latitude stations. Figure 9 shows the results of Matsushita and Sano on the distribution of SC* ($^-$SC) and *SC (SC$^-$). From this Figure we can see that

(1) SC* predominantly appears at afternoon middle and auroral latitudes and the morning side polar cap,

(2) *SC (SC$^-$) predominantly appears in middle and auroral latitudes from morning to noon and the afternoon polar cap,

(3) SC* ($^-$SC) appears in an isolated narrow region along the dayside dip equator.

These results statistically support the case studies shown in Figure 5, 7 and 8.

By checking rapid-run magnetograms from Guam (geomag. lat. : 4.6°), Araki et al.[1985] showed that there are two types of SCs along the nighttime equator. One type of SC has a smooth increase of the H-component. The other type exhibits a small sharp initial positive impulse at the beginning of the smooth increase. The latter type of SC shows a maximum occurrence rate (about 50% of SC) near 3^hLT and corresponds well to PRI at the dayside equator (Huancayo).

Nagata and Abe [1955] analyzed SC* by collecting so called quick-run magnetograms available at that time and proposed the equivalent current system for the preliminary impulse (PI) of SC* which is shown in the upper left panel of Figure 10. Nishida et al. [1966] pointed out the similarity between the equivalent current systems of the SC* and the newly discovered DP-2 variation (the upper middle panel of Figure 10). The upper right panel of Figure 10 is the current system of the PI proposed by Araki et al. [1985] to take into account the observational results described above. It is called DP_{pi}(see section 3). Before drawing this current system it was also confirmed that an ionospheric current flows from day to night at Thule station near the geomagnetic north pole.

Obayashi and Jacobs [1957] decomposed the disturbance field of the SC, $D(SC)$, into two parts, $D_{st}(SC)$ and $D_s(SC)$ (lower panel of Figure 10). The former is obtained by averaging $D(SC)$ longitudinally and so

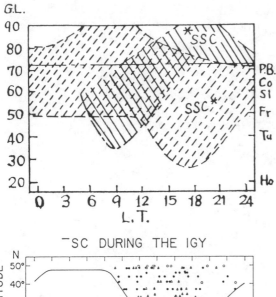

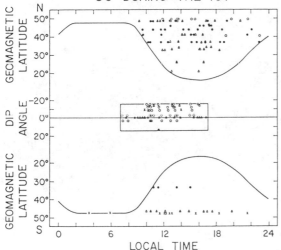

Fig. 9. Distribution of SSC*($^-$SC) and *SSC. Upper panel; Sano [1962], lower panel; Matsushita, [1962].

depends upon only latitude and storm time. The latter is obtained by subtracting $D_{st}(SC)$ from $D(SC)$ and depends also upon local time.

3. Model

Any models of an SC should be consistent with the following observational facts;

(1) The source of the SC in interplanetary space is a simple step-function like increase of the solar wind dynamic pressure.

(2) At dayside geosynchronous orbit a clear stepwise compression of the magnetic field is observed. The amplitude of the SC shows a diurnal variation with a maximum near noon. The compression propagates earthward in the dayside magnetosphere with a relevant HM wave velocity.

(3) Ground observations at high latitudes show that the waveform of the SC in the H-component consists of two successive pulses (PI and MI) with opposite senses. At auroral latitudes in the morning a positive pulse precedes and a negative pulse follows. The sense of the pulses is reversed at afternoon auroral latitudes. The sense is also reversed in the polar cap. The amplitude of the pulses decreases with decreasing latitude.

(4) At middle and low latitudes, the waveform of the H-component becomes more step-function like but the two pulse structure with reduced amplitude is still superposed.

(5) At nightside equatorial latitudes, the waveform of the SC is most step-function like but a small preliminary positive pulse (PPI) may be superposed upon the smooth rise of the SC.

(6) At the dayside equator, the two pulse structure appears again. The waveform is similar to that in high latitude afternoon with a negative PI(PRI). The amplitude of both pulses is enhanced compared with those at low latitudes. The enhancement rate is larger for PRI than MI.

It should be especially stressed that the two pulse structure clearly seen at high latitudes and the dayside equator on the ground is not detected either in interplanetary space or at geosynchronous orbit. This suggests that electric currents flowing in the ionosphere play an important role and one should therefore take into account realistic distributions of the ionospheric conductivity.

Here we decompose the disturbance field of the SC, D_{sc}, into two parts as follows [Araki, 1977 and 1987];

$$D_{sc} = DL + DP \qquad (1)$$

where DL represents a step-function like increase of the H-component dominant in low latitudes and DP represents the two pulse structure dominant at high latitudes. The sub-field, DP, is further decomposed into two parts corresponding to the preliminary sharp impulse (PI) and the following main impulse (MI) as

$$DP = DP_{pi} + DP_{mi} \qquad (2)$$

Figure 11 shows the dependence of the decomposed fields upon local time and latitude. The observed field is interpreted as the superposition of the decomposed fields, and is shown by solid lines.

Let's consider a simple electric circuit which consists of a battery and a lamp (Figure 12(a)). When the switch of this circuit is turned on, the lamp is apparently

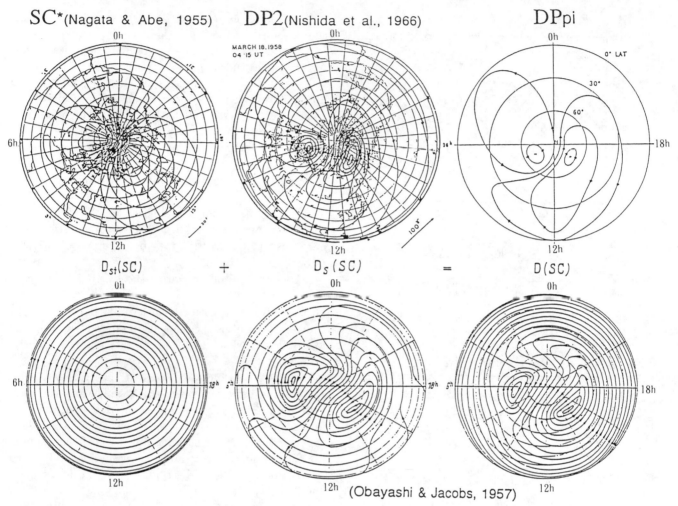

Fig. 10. Equivalent current systems for the PI of SC* [Nagata and Abe, 1955 and Araki et al., 1985], DP-2[Nishida et al., 1966], D_{st}(SC), D_s(SC) and D(SC) [Obayashi and Jacobs, 1957].

lit instantaneously. Actually this is not instantaneous. An electromagnetic wave propagates with the velocity of light (c) from the battery to the lamp. The lamp will be lit after a propagation time l/c (l is length of the circuit). The displacement current (J_D), flowing along the wavefront, connects the conduction current (J) in the wire to form a closed current loop inside of which the magnetic field (b) increases.

Now consider a space plasma with the static magnetic field perpendicular to the plane of the circuit (Figure 12(b)). The displacement current along the wavefront changes to a polarization current (J_p) and the wavefront propagates with the fast mode hydromagnetic wave velocity V_F. The $J_p \times B$ force compresses the plasma in the direction of propagation. This means that the wave is a compressional wave. The propagation time is given by l/V_F in this case.

In the case of an SC, the circuit is replaced by the equatorial magnetopause and the battery is replaced by the electromotive force, ($V \times B$) imposed by the sudden increase of the solar wind dynamic pressure (Figure 12(c)). The dawn-to-dusk magnetopause current (J_M) is enhanced when the dynamic pressure increases and the $J_M \times B$ force due to this current resists the compression. The enhanced magnetopause current (J_M) and the polarization current (J_p) along the compressional wavefront form a current loop inside of which the northward magnetic field is increased. When the wavefront reaches the earth, the H-component of the ground magnetic field begins to increase. This instant corresponds to lighting of the lamp in Figure 12(b). The northward magnetic field keeps increasing during the passage of the wavefront which has a finite thickness, until the compression of the magnetopause ceases. This

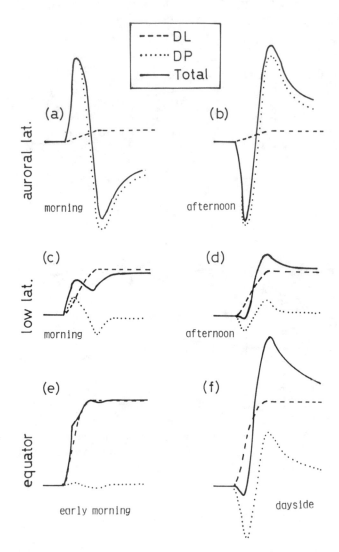

Fig. 11. Decomposition of the SC disturbance field into DP- and DL-sub-fields.

is the DL-field. The $D_{st}(SC)$ proposed by Obayashi and Jacobs [1957] is considered to be the first approximation to the DL-field.

If the compressional wavefront propagates solely within the equatorial plane, it is not necessary to consider other modes of propagation of the SC. The inhomogeneity of the magnetospheric plasma and magnetic field, however, produces another mode (the Alfven mode) which propagates along field lines to the polar ionosphere [Tamao 1964a]. The wave accompanies a field aligned current (FAC, J_F in Figure 12(d)) which flows into the ionosphere on the dusk side and out of the ionosphere on the dawn side. This FAC produces ionospheric currents of the twin vortex type [Tamao 1964b] which are clockwise in the afternoon over the northern

hemisphere and anti-clockwise in the morning, as shown in Figure 12(d). This current system is consistent with that in Figure 10 which represents the DP_{pi}-field.

After the passage of the compressional wavefront toward the magnetotail, the magnetospheric convection has to adjust to the new compressed state of the magnetosphere if the dynamic pressure of the solar wind remains high behind the shock or discontinuity. The convection electric field in the dawn-to-dusk direction has to be enhanced in the compressed magnetosphere. The associated FAC flows into the dawn ionosphere and out from the dusk ionosphere (Figure 12(e)). Then a twin vortex current with opposite sense to the preceding current system appears in the ionosphere, and produces the DP_{mi}-field.

In the twin-vortex type ionospheric current generated by the FAC, the vortex on the afternoon side generally becomes larger than that on the morning side and extends to the dayside equator as is shown in Figure 10. This is due to the existence of the Hall conductivity and day-night conductivity inhomogeneity. The equatorial ionospheric current associated with the afternoon vortex flows westward and eastward for the first (Figure 12(d)) and the second (Figure 12(e)) current systems, respectively, and thus produces a negative impulse (PRI) followed by a positive impulse (MI) at the dayside equator.

4. EXAMINATION OF THE MODEL

The model for SC described above requires that the global twin vortex type ionospheric current forms almost instantaneously when the dusk-to-dawn or dawn-to-dusk electric field is impressed upon the polar ionosphere from the magnetosphere. For example, Figure 8 shows that a PRI appears simultaneously at auroral (Pt. Barrow and College) and equatorial (Koror) latitudes. We therefore have to find a mechanism which enables the instantaneous transmission of a horizontal polar ionospheric electric field to the equator. If we assume an infinite uniform ionosphere with conductivities of the E-region, the equations governing the transmission are diffusion type [Rostoker, 1965] and it takes more than one hour for the polar electric field to be transmitted to the equator. Although the F-region waveguide centered at the altitude of the minimum in the HM wave velocity has been postulated for the horizontal propagation of the Pc-1 geomagnetic pulsation [Manchester, 1966], it can not be applied to SC with longer time scales because the F-region waveguide has a lower cut off frequency around 1 Hz.

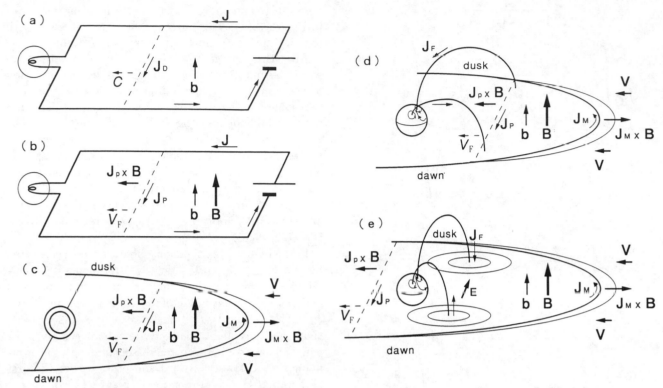

Fig. 12. A model for SC.

Instead, a waveguide formed in the space between the ionosphere and the earth was proposed by Kikuchi et al. [1978], and Kikuchi and Araki [1979b]. The Hall current due to the dawn-dusk electric field impressed in the polar ionosphere produces a dawn-dusk magnetic field in the waveguide below the ionosphere which propagates as the zeroth order TM (transverse magnetic) waveguide mode to the equator carrying the electric field in the ionosphere (Figure 13). The zeroth order mode has no lower cut off frequency. The conduction currents J flowing in the ionosphere and the earth are connected by the vertical displacement current in the space between them so as to accomplish continuity of the current. Since the horizontal scale of the actual electric field is finite, the amplitude decreases as it propagates toward the equator but it will still cause an enhancement of the ionospheric current with the aid of the enhanced Cowling conductivity.

This mechanism for the transmission of the magnetospheric electric field to the equatorial ionosphere through the polar ionosphere and the earth-ionosphere waveguide can be applied not only for SCs but also for other geomagnetic phenomena which are observed simultaneously in the polar and equatorial regions (for example, geomagnetic pulsations and DP-2 variations).

Once the near instantaneous transmission of the polar electric field to the equator is accepted, we can make a static calculation for the global distribution of the ionospheric current produced by the electric field of the DP-field. Tsunomura and Araki [1984] made such a kind of calculation. For this calculation, a realistic con-

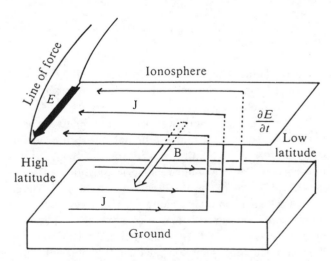

Fig. 13. Transmission of an ionospheric horizontal electric field from high latitudes to the equator [Kikuchi and Araki, 1979b].

ductivity distribution on the spherical thin ionosphere is assumed. The ionospheric current is caused by a pair of field aligned currents (FACs) which flow into the ionosphere with maximum intensity at 18^hLT and $75°$ latitude and flow outward at 06^hLT and the same latitude. The ionospheric current flows out from the footprint of the FAC on duskside, turns first clockwise and then anti-clockwise, and flows into the footprint of the FAC on the dawn side. The ionospheric current flows with westward (eastward) component in the afternoon (morning) at latitudes lower than the FAC footprints and at the dayside (nightside) equator. The current distribution is therefore consistent with the equivalent current system for PI in Figure 10. The intensities of the ionospheric electric fields and currents decrease with decreasing latitude. The electric field decreases sharply at the dayside equator. This is because of a counter polarization produced by electric charges accumulated at the boundary of the enhanced conductivity. In spite of this electric field reduction, the electric current increases sharply enough to explain the equatorial enhancement of the PRI. Depending upon the threshold for the detection of the magnetic field due to this current, the magnetic field variation may be observed only at high latitudes and the dayside equator. This latitudinal variation of the ionospheric current is consistent with the distribution of SC*($^-$SC) in Figure 8 and 9. Takeda [1984] calculated the global distribution of the ionospheric current by taking into account the finite thickness of the ionosphere. His calculation also supports the results described here.

The current system for PI (Figure 10) had been an assumed "equivalent" current system for a long time. Simultaneous magnetic field observations on the ground and above the ionosphere finally proved that the PI is caused by an electric current actually flowing in the ionosphere [Araki et al., 1984].

It was also confirmed that electric current systems with opposite directions are produced when the magnetosphere suddenly expands [Araki and Nagano, 1988]. This supports the validity of the model described above.

5. Numerical synthesis of SC

Based upon the model proposed in Section 3, Osada [1992] synthesized SC disturbance fields using the 3 sub-fields, DP_{pi}, DP_{mi} and DL. Realistic conductivities similar to those of Tsunomura and Araki [1984] were given on a spherical thin shell ionosphere. A pair of field aligned currents (FACs), C1, which cause DP_{pi}-field, flows into (out of) the afternoon (morning) polar

ionosphere symmetrically about the noon meridian as shown in the upper panel of Figure **14.** The footprints of the FAC move along the $75°$ geomagnetic latitude circle from noon to both evening and morning sides and the FAC intensities vary with time. Several minutes (4.5 minutes here) later, the second FAC, C2, responsible for the DP_{mi} begins to flow into (out of) the morning (afternoon) polar ionosphere. The footprints of the second FAC are fixed at 0840 and 1520 local time. The time variation of the intensity and location of the footprints of both FACs is shown in the middle panel of Figure 14. The DL-field is assumed to be a simple increase of the H-component (bottom panel). The magnetic fields produced on the ground by both FACs and ionospheric **currents were calculated using the Biot-Savart law and** were superposed upon the DL-field.

Figure 15 shows the calculated global distribution of H-component waveforms. As expected, the two-pulse structure is clear at high latitudes. A negative pulse precedes in most of the auroral latitude but a positive pulse precedes in a relatively limited morning region ($60°$ and $72°$, 10^hLT). At 10^h and 14^hLT, the sense of the first pulse reverses between auroral latitudes and the polar cap. The two-pulse structure also appears at the dayside (10^hLT) equator. This results from the equatorial enhancement of the DP_{pi} and DP_{mi} current. Figure 15 also suggests possible appearance of a small positive impulse at the equator in the morning (6^hLT).

Thus we can say that the calculated results are consistent with the observations described in the previous sections. It will be possible to obtain better agreement with observations by adjusting the ionospheric conductivities, the location of the FACs, and intensity and time variation of the FACs and the DL-field for particular cases. For a more rigorous estimation of the magnetic field caused by the FAC we might need to take into account the time delay due to wave propagation. We consider, however, that the effect of wave propagation can be neglected as a first approximation because distant FACs do not contribute much to the magnetic field on the ground.

6. Discussion

Ohnishi and Araki [1992] studied the two dimensional interaction between a propagating plane HM wave and the earth-ionosphere system with curvature in order to understand equatorial properties of the DL-field. The ionosphere is assumed to be a thin cylindrical shell with a diurnally varying height integrated Cowling conductivity. Inside the ionosphere, there is a perfectly con-

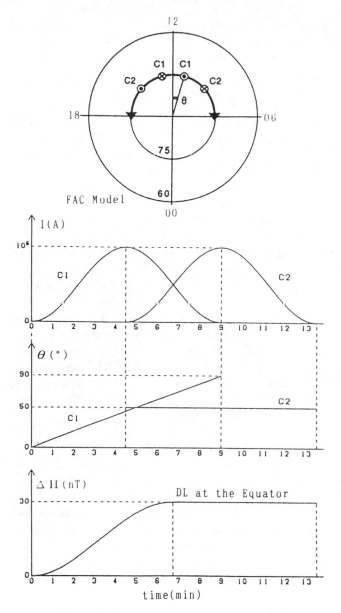

Fig. 14. Synthesis of SC from DP_{pi}-, DP_{mi}- and DL-fields [Osada, 1992].

ducting earth with a radius smaller than the actual radius. An incident plane HM wave propagates from the dayside magnetosphere to the earth with a sharp rise (2 min here) and a much slower decrease of the northward magnetic field. They calculated the time variation of the magnetic field just above the ionosphere (b_a) and on the ground (b_b) and the electric current and charge in the ionosphere. The following 3 points are stressed here;

(1) b_a is larger than b_b (especially on the dayside),

(2) b_b rises gradually and almost simultaneously with the same waveform and amplitude at any local time,

(3) the variation of b_b seems to follow that of b_a but the initial rise of b_b precedes that of b_a in the nightside.

The first point above is supported by MAGSAT observations of SCs [Araki et al., 1984] showing that the averaged amplitude of SC is 1.3 times larger above the ionosphere than on the ground in low latitudes. This enhancement is caused by the westward electric current flowing in the ionosphere. The orbit of MAGSAT was restricted to the dawn-dusk meridian, so the observed ratio will become even larger towards the noon meridian. Note that the ionospheric current flows westward in spite of an increase in the ground H-component. This is because the ionospheric current flows to shield the ground magnetic field increase due to the source current flowing along the magnetopause. The second point is also due to the shielding current flowing in the ionosphere. The third point is caused by difference in propagation speed of HM wave above the ionosphere and electromagnetic wave below the ionosphere. The effect of electromagnetic wave propagation below the ionosphere is observed earlier in the nightside because arrival of the HM wave is delayed.

Contrary to the second point above, there is a small delay (about 1 min) in the onset time of SC observed on the ground. Nishida and Jacobs [1962a] showed that SCs begin at afternoon auroral latitudes and propagate toward lower latitudes and the nightside. Here it should be realized that the observed onset time does not necessarily mean the arrival time of the wavefront. Even if the wavefront arrives, the geomagnetic variation will not be observed if a sufficient source current does not flow. When the source current is in the ionosphere, the ionospheric conductivity will determine threshold level for the detection.

Figure 16 [Kikuchi, 1986] shows an SC observed at Hussafell, Iceland, and Kakioka, Japan. A clear negative impulse appears at Hussafell as expected for high latitudes in the afternoon (around 15^hLT). Almost simultaneously, a negative change occurs in the HF Doppler observation (both 2.5 MHz and 10 MHz) at Akita (around 01^hLT) approximately 300km north of Kakioka. This means that the ionospheric plasma drifts upward due to an eastward electric field originating in the polar ionosphere. The variation of the magnetic field at Kakioka, however, started approximately 1 min later than the onset of the electric field variation at Akita. This might be interpreted to imply that low conductivity in the nighttime ionosphere did not produce a sufficient ionospheric current, although the electric field

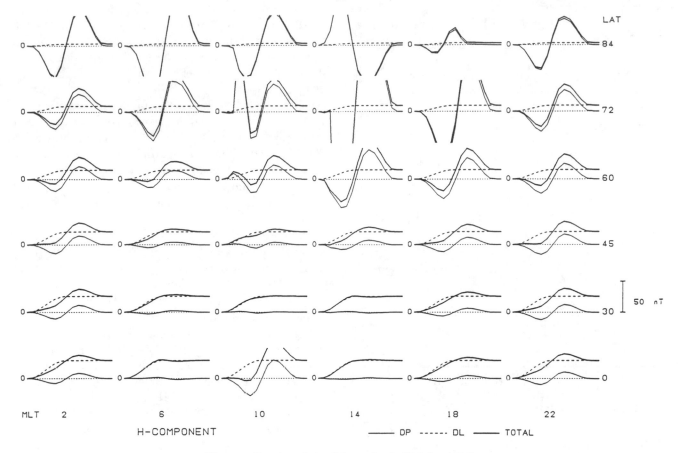

Fig. 15. Results of the SC synthesis [Osada, 1992].

already had arrived.

Further, we have to realize that the observed magnetic field is a superposition of the DL- and DP-fields. Both fields may cancel if they have opposite senses and as a result the onset time may apparently be delayed. Such examples were shown by Araki [1977].

The disappearance of the preliminary reverse impulse (PRI) near the latitude of Honolulu and its isolated reappearance at the dayside equator (Figure 8 and 9) has suggested other mechanisms for the equatorial PRI. Tamao [1964b] suggested that the equatorial PRI is produced by a westward electric current associated with a converted transverse mode HM wave incident directly on the ionosphere. Kikuchi and Araki [1979a] examined whether or not the PRI can be produced by a direct incidence of a HM wave on the equatorial and polar ionosphere. No PRI could be obtained whereas a small PRI-like transient impulse appeared in the polar case.

In the model proposed above, the ionospheric currents with polar origin which are responsible for DP_{pi} and DP_{mi} intensify at the dayside equator (due to en-

hanced Cowling conductivity) and cause the equatorial enhancement of both preliminary reverse impulse and following main impulse. This is the simplest way to interpret the daytime equatorial enhancement of SC. This model may explain why the enhancement rate of PRI is larger than that of MI, because the PRI consists of a pure DP-component which is sharply enhanced at the dayside equator whereas the MI consists of both DP- and DL-components [Araki, 1977]. The amplitude of the DL-field decreases more slowly with increasing latitude. Jacobs and Watanabe [1963] proposed that the westward electric field associated with the compressional wave incident on the equatorial ionosphere causes a downward motion of charged particles which leads to the intensification of the equatorial electrojet through an enhancement of the conductivities. One possible problem for this idea is that it can not explain the observed delay of the peak time of enhanced MI. In the model proposed above, the equatorial enhancement of MI is caused by an enhancement of DP_{mi} which usually achieves its peak value later than the peak of the

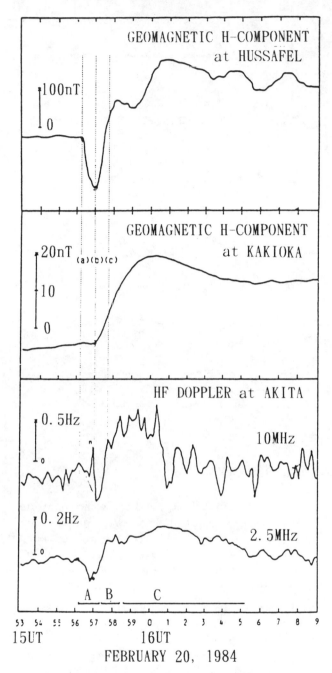

GEOMAGNETIC H-COMPONENT
at HUSSAFEL

100nT

0

GEOMAGNETIC H-COMPONENT
at KAKIOKA

20nT
(a)(b)(c)

10

0

HF DOPPLER at AKITA

0.5Hz

10MHz

0.2Hz

2.5MHz

A   B   C

53 54 55 56 57 58 59 0   1   2   3   4   5   6   7   8   9
15UT   16UT

FEBRUARY 20, 1984

Fig. 16. Simultaneous observations of an SC by the HF Doppler receiver and magnetometers [Kikuchi et al., 1985].

DL-field.

The Cowling conductivity is strongly enhanced in a narrow belt centered along the dayside dip equator. We need to know in more detail what happens when a compressional wave is incident on a medium having such a strong horizontal inhomogeneity. At present, however, no mechanism other than the polar originating iono-

spheric current has been found which can explain the global distribution of the two-pulse SC structure including their equatorial enhancement. The model proposed here is also supported by the HF Doppler observation of ionospheric electric fields during SCs [Kikuchi et al., 1985].

When we analyze responses of ground magnetic fields to the solar wind dynamic pressure, we have to select geomagnetic stations very carefully so as to minimize effects of secondary induced ionospheric currents. Russell et al. [1992] concluded that the H-component observed between 15° and 30° geomagnetic latitude show most healthy response to the dynamic pressure of the solar wind.

As shown in Figures 5 and 7, SC excites damped geomagnetic pulsations (P_{sc}) [e.g., Saito and Matsushita, 1967]. Although it might be possible to consider that PI and MI in the model above are part of P_{sc}, we do not adopt this view. We maintain that the main cause of the P_{sc} is a field line resonance whose period depends strongly upon latitude and occurs in relatively localized regions, while the PI and MI are produced by a global ionospheric current vortex which extends from the polar region to the equator. However, the P_{sc} may be superposed on the PI and MI and modify them.

Wilson and Sugiura [1961] reported that the horizontal vector of SC observed in high latitudes is circularly polarized and concluded that it indicated hydromagnetic waves in the magnetosphere. They used data from stations below auroral latitudes and proposed a polarization rule that the sense of the polarization is anti-clockwise in the morning and clockwise in the afternoon. At geosynchronous orbit, the magnetic vector in the plane perpendicular to the static field rotates clockwise (viewed from north) near dawn and anti-clockwise at dusk [Nagano and Araki, 1984 and Kuwashima et al., 1985]. This is consistent with the polarization rule of Wilson and Sugiura. Using data from the North American IMS Magnetometer Network which include stations in the polar cap, Araki and Allen [1982] demonstrated a latitudinal reversal of the sense of the polarization of SC around 64°-72° corrected latitude.

The model proposed in section 3 assumes that a pair of ionospheric current vortices produced by field aligned currents move from noon to both dawn and dusk sides at high latitudes when the preliminary impulse of SC is excited. The sense of the current rotation in the northern hemisphere is anti-clockwise in the morning and clockwise in the afternoon. Below the moving current vortex, the horizontal magnetic vector changes its direction as shown in Figure 17 [Nagano et al., 1985].

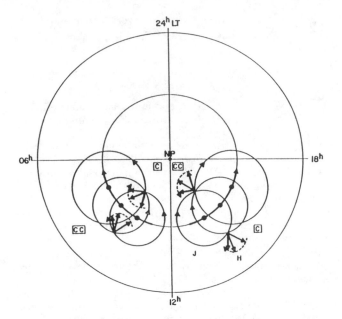

Fig. 17. Rotation of horizontal magnetic vectors under moving twin vortex ionospheric currents [Nagano et al.,1985].

The rotation sense is anti-clockwise in the morning and clockwise in the afternoon under the low latitude side of the current loops, but reverses under the higher latitude side. This polarization sense is consistent with the polarization rule of Wilson and Sugiura [1961] and also with latitudinal polarization reversal found by Araki and Allen [1982].

However, Araki and Allen [1982] also found that half of the 18 events examined did not obey the polarization rule of Wilson and Sugiura. There might be other mechanisms which affect the polarization of SC. Modification by P_{sc} may be one of the candidates.

Electric currents induced in the earth will certainly contribute to the field of SCs. In particular we know that each station has its own inherent sign for the z-component variation during SC [Ivanov, 1964]. We have to study more quantitatively the contribution of the induced current to the H- and D-component variations of the SC. We believe, however, that the main story for the SC phenomena described above will not change.

It is well known that the occurrence rate of substorms increases after an SC [e.g., Kokubun et al., 1977]. Iyemori and Tsunomura [1983] showed that substorm occurrence is most likely 10 min after an SC but that a negative SI does not enhance the probability. They also showed that the occurrence probability of substorms does not depend upon the sense of the IMF-B_z before the SC but that the amplitude of the triggered substorm is larger when it is southward.

Iyemori and Araki [1982] discovered that the equivalent current system for SCs becomes a single vortex type when the IMF is northward. Since the configuration of the magnetosphere depends upon the IMF, it is reasonable to assume that its compression depends upon the IMF conditions. We have to make more detailed studies about the IMF dependence of magnetospheric compressions and expansions.

There have been some debates concerning recently reported bipolar geomagnetic variations in the auroral oval. The bipolar geomagnetic variations have been discussed as ground signatures of flux transfer events (FTE) first reported at the magnetopause by Russell and Elphic [1979] (e.g., Lanzerotti et al., 1986), and subsequently as evidence for the impulsive penetration of plasma through the magnetopause or pressure perturbations at the magnetopause [Friis-Christensen, 1988, Glassmeier et al., 1989 and Sibeck 1993, see review paper by Glassmeier, 1992]. Although the debates have not been concluded, it seems that some of the bipolar events are certainly related to partial compressions of the magnetopause. If so, knowledge accumulated during the study of SC will be useful for interpretation of the bipolar geomagnetic variations. At the same time understanding of the partial compression of the magnetosphere will be effectively used for further study of SC.

There seems to be some confusion in the definition and terminology of geomagnetic sudden commencements [e.g., Joselyn and Tsurutani, 1990]. Because of the frequent occurrence of geomagnetic storms after SCs, they are often called storm sudden commencements (SSC). This is because the IMF might easily become southward or turbulent behind an interplanetary shock or discontinuity which causes an SC. If the IMF remains northward behind the shock or discontinuity, the SC will not be followed by a geomagnetic storm. Such an SC has been called SI (sudden impulse) but there is no difference in the physical mechanisms of SSC and SI. From the present physical point of view, SSC and SI are only phenomenological distinctions which sometimes become necessary from a practical point of view.

There is a possibility, however, that further investigation might distinguish presently unknown subtle differences in the physical mechanisms of SC occurrence and allow them to be classified into several groups. One example is the IMF dependence of SC described previously. In this sense, more detailed studies should be made concerning relationship between the global structure of SC and interplanetary parameters.

7. Concluding remarks

Since SCs are transient phenomena initiated by sudden compressions of the magnetopause, they should be treated as an initial boundary value problem. The complex geometry inside the magnetosphere must play important roles in determining the characteristics of SC. Inhomogeneities and curvature of the medium can not be neglected because spatial scale lengths of SC are comparable to or larger than the size of the magnetosphere. For example, if we take the rise time and averaged propagation speed of an SC as 5 min and 500km/s, respectively, the thickness of the SC wavefront becomes 150,000 km, which is more than twice the size of the dayside magnetosphere. It was already shown that the distribution of the ionospheric conductivity plays an essential role in the interpretation of SC. It is almost impossible, however, to incorporate such a complex geometry or inhomogeneity into time dependent initial boundary value problems. The way to overcome this difficulty is to understand the true physical mechanism of SC by combining solutions of initial boundary value problems with simple geometries and time-independent problems with complex geometry. Our consideration of **the interaction between a plane HM wave and a cylindri**cal ionosphere-earth system is an example of the former and the static calculation of global ionospheric current on the spherical ionosphere with a realistic conductivity distribution and numerical synthesis of SC are examples of the latter. The results show that there are still many aspects of SCs which can be understood as time independent problems.

A few works have treated so far as initial boundary value problems. Here we mention the recent numerical analysis by Lysak and Lee [1992] on the response of three dimensional dipole magnetosphere to pressure pulses at the magnetopause. The results show that the pressure pulses first excite compressional waves which then are converted to shear Alfven waves by inhomogeneity of the magnetosphere. Twin vortex type perturbations appear in the perfectly conducting ionosphere and move azimuthally and latitudinally depending upon the period of the pressure pulse.

In the outer magnetosphere nonlinear effects might be important. Okuzawa [1967] solved a one dimensional piston problem of the magnetopause and discussed intensity variation of finite amplitude compressional waves in a model magnetosphere with density gradient. No other investigations seem to be reported which take nonlinear effects into account.

Acknowledgments. The author is very much grateful to two referees who have contributed to improvement of this paper.

References

Adams, W. G., Comparison of simultaneous magnetic disturbances at several observatories, *Phil. Trans. London, A, 183,* 131-140, 1892.

Araki, T., Global structure of geomagnetic sudden commencements, *Planet. Space Sci., 25,* 373- 384, 1977.

Araki, T. and J. H. Allen, Latitudinal reversal of polarization of the geomagnetic sudden commencement, *J. Geophys. Res., 87,* 5207-5216, 1982.

Araki, T., T. Iyemori and T. Kamei, Sudden commencements observed by MAGSAT above the ionosphere, *J. Geomag. Geoelectr., 36,* 507-520,1984.

Araki, T., J. H. Allen and Y. Araki, Extension of a polar ionospheric current to the nightside equator, *Planet. Space Sci., 33,* 11-16, 1985.

Araki, T., A model of geomagnetic sudden commencement, *Proc. of Symposium on Quantitative Modeling of Magnetosphere-Ionosphere Coupling Processes,* 47-52, March, 1987.

Araki, T. and H. Nagano, Geomagnetic response to sudden expansions of the magnetosphere, *J. Geophys. Res., 93,* 3983-3988, 1988.

Araki, T., H. Shimazu, T. Kamei, and H. Hanado, Scandinavian IMS Magnetometer Array Data and their use for studies of geomagnetic rapid variations, *Proc. NIPR Symp. Upper Atmos. Phys., 5,* 10-20, 1992.

Chapman, S. and V. C. A. Ferraro, A new theory of magnetic storms, *Terr. Mag., 36,* 77-97, 171-186, 1931; *37,* 147-156, 421-429, 1932; *38,* 79-96, 1933.

Ellis, W., On the simultaneity of magnetic variations at different places on occasions of magnetic disturbance, and on the relations between magnetic and earth current phenomena, *Proc. Roy. Soc., 102,* 191, 1892.

Forbush, W. E. and E. H. Vestine, Daytime enhancement of size of sudden commencements and initial phase of magnetic storms at Huancayo, *J. Geophys. Res., 60,* 299-316, 1955.

Friis-Christensen, E., M. A. McHenry, C. R. Clauer, and S. Vennerstrom, Ionospheric traveling convection vortices observed near the polar cleft: A triggered response to sudden changes in the solar wind, *Geophys. Res. Lett., 15,* 253-256, 1988.

Glassmeier, K.-H., Traveling magnetospheric convection twin-vortices: observations and theory, *Ann. Geophysicae, 10,* 547-565, 1992.

Glassmeier, K. -H., M. Hönisch and J. Untiedt, Ground-based and satellite observations of travelling magnetospheric convection vortices, *J. Geophys. Res., 94,* 2520-2528, 1989.

Gold, T., Gas dynamics of cosmic clouds, ed. by H.C. van de Hulst and J.M. Burgers, *North-Holland Publ. Co.,* Amsterdam, 103, 1955.

Ivanov, K. G., Map of the distribution of the sign of the z-component of the ssc field over the Earth's surface, *Geomag. Aeron., 4,* 629-630, 1964.

Iyemori, T. and T. Araki, Single vortex current system in the polar region generated by an interplanetary shock wave, *Geophys. Res. Lett., 9*, 535-538, 1982.

Iyemori, T. and S. Tsunomura, Characteristics of the association between an SC and a substorm onset, *Mem. Natl. Inst. Polar Res., Spec. Issue, No 26*, 139-147, 1983.

Jacobs, J. A. and T. Watanabe, The equatorial enhancement of sudden commencement of geomagnetic storms, *J. Atmos. Terr. Phys., 25*, 267-279, 1963.

Joselyn, J. A. and B. Tsurutani, A note on terminology: Geomagnetic sudden impulses (SIs) and storm sudden commencements (SSCs), *EOS, Trans. AGU, 71*, 1808-1909, 1990.

Kikuchi, T., T. Araki, H. Maeda and K. Maekawa, Transmission of polar electric fields to the equator, *Nature, 273*, 650-651, 1978.

Kikuchi, T. and T. Araki, Transient response of uniform ionosphere and preliminary reverse impulse of geomagnetic storm sudden commencement, *J. Atmos. Terr. Phys., 41*, 917-925, 1979a.

Kikuchi, T. and T. Araki, Horizontal transmission of the polar electric field to the equator, *J. Atmos. Terr. Phys. 41*, 927-936, 1979b.

Kikuchi, T., T. Ishimine and H. Sugiuchi, Local time distribution of HF Doppler frequency deviations associated with storm sudden commencements, *J. Geophys. Res., 90*, 4389- 4393, 1985.

Kikuchi, T., Evidence of transmission of polar electric fields to the low latitude at times of geomagnetic sudden commencement, *J. Geophys. Res., 91*, 3101-3105, 1986.

Kokubun, S., Characteristics of storm sudden commencement at geostationary orbit, *J. Geophys. Res., 88*, 10025-10033, 1983.

Kokubun, S., R. L. McPherron and C. T. Russell, Triggering of substorms by solar wind discontinuities, *J. Geophys. Res., 82*, 74-86, 1977.

Kueppers, F. J., J. Untiedt, W. Baumjohann, K. Lange and A. G. Jones, A two dimensional magnetometer array for ground-based observations of auroral zone electric currents during the International Magnetospheric Study (IMS), *J. Geophys., 46*, 429-450, 1979

Kuwashima, M., S. Tsunomura and H. Fukunishi, SSC associated magnetic variations at the geosynchronous altitude, *J. Atmos. Terr. Phys., 47*, 451-461, 1985.

Lanzerotti, L. J., L. C. Lee, C. G. McLennan, A. Wolfe and L. V. Medford, Possible evidence of flux transfer events in the polar ionosphere, *Geophys. Res. Lett., 13*, 1089-1092, 1986.

Lepping, R. P., A. F.-Vinas, A. Lazarus, M. Sugiura, T. Araki and S. Kokubun, Interplanetary shock-driven waves in the magnetosphere: Propagation delay, *EOS Trans. Am. Geophys. Union, 68*, 1419, 1987.

Lysak, R. and **D.-H.** Lee, Response of the dipole magnetosphere to pressure pulses, *Geophys. Res. Lett., 9*, 937-940, 1992.

Maeda, H. and M. Yamamoto, A note on daytime enhancement of the amplitude of geomagnetic-storm sudden commencements in the equatorial region, *J. Geophys. Res., 65*, 2538-2539, 1960.

Maeda, H., K. Sakurai, T. Ondoh and M. Yamamoto, Solar terrestrial relationships during the IGY and IGC, *Ann. Geophysique, 18*, 305-333, 1962.

Manchester, R. N., Propagation of Pc-1 micropulsations from high to low latitudes, *J. Geophys. Res., 71*, 3749-3754, 1966.

Matsushita, S., On geomagnetic sudden commencements, sudden impulses, and storm durations, *J. Geophys. Res., 67*, 3753-3777, 1962.

Nagano, H. and T. Araki, Polarization of geomagnetic sudden commencements observed by geostationary satellites, *J. Geophys. Res., 89*, 11018-11022, 1984.

Nagano, H., T. Araki, H. Fukunishi and N. Sato, Characteristics of polarization of geomagnetic sudden commencements at geostationary orbit, *Mem. Natl. Inst. Polar Res., Spec. Issue, No.36*, 123-135, 1985.

Nagano, H. and T. Araki, Seasonal variation of amplitude of geomagnetic sudden commencements near midnight at geostationary orbit, *Planet. Space Sci., 34*, 205-217, 1986.

Nagata, T. and S. Abe, Notes on the distribution of SC* in high latitudes, *Rep. Ionosph. Res. Japan, 9*, 33-44, 1955.

Namikawa, T., T. Kitamura, T. Okuzawa and T. Araki, Propagation of weak hydromagnetic discontinuity in the magnetosphere and the sudden commencement of geomagnetic storm, *Rep. Ionos. Space Res. Japan, 18*, 218-227, 1964.

Neugebauer, M. and C. W. Snyder, Solar Plasma Experiment, *Science, 138*, 1095, 1962.

Nishida, A. and J. A. Jacobs, World wide changes in the geomagnetic field, *J. Geophys. Res., 67*, 525-540, 1962a.

Nishida, A. and J. A. Jacobs, Equatorial enhancement of world-wide changes, *J. Geophys. Res., 67*, 4937-4940, 1962b.

Nishida, A., N. Iwasaki and T. Nagata, The origin of fluctuations in the equatorial electrojet; a new type of geomagnetic variations, *Ann. Geophys., 22*, 478-484, 1966.

Nishida, A., Geomagnetic Diagnosis of the Magnetosphere, *Springer-Verlag*, New York, 1978.

Obayashi, T. and J. A. Jacobs, Sudden commencements of geomagnetic storms and atmospheric dynamo action, *J. Geophys. Res., 62*, 589-616, 1957.

Ogilvie, K. W., L. F. Burlaga and T. D. Wilkerson, Plasma observations on Explorer 34, *J. Geophys. Res., 73*, 6809-6824, 1968.

Ohnishi, H. and T. Araki, Two-dimensional interaction between a plane hydromagnetic wave and the earth-ionosphere system with curvature, *Ann. Geophysicae, 10*, 281-287, 1992.

Okuzawa, T., Numerical study on finite amplitude hydromagnetic waves in a model magnetosphere with density plateau, *ISAS Report No.417*, Institute of Space and Aeronautical Science, Tokyo Univ., 215-231, 1967.

Osada, S., Numerical calculation of geomagnetic sudden commencement, *Master Thesis*, Faculty of Science, Kyoto Univ., March 1992.

Rastogi, R. G. and N. S. Sastri, On the occurrence of SSC (-+) at geomagnetic observatories in India, *J. Geomag. Geoelectr., 26*, 529-537, 1974.

Rostoker, G., Propagation of Pi2 micropulsations through the ionosphere, *J. Geophys. Res., 70*, 4388-4390, 1965.

Russell, C. T., and R. C. Elphic, ISEE observations of flux transfer events at the dayside magnetopause, *Geophys. Res. Lett., 6*, 33-36, 1979.

Russell, C. T., M. Ginsky, S. Petrinec and G. Le, The effect of solar wind dynamic pressure changes on low and mid-latitude magnetic records, *Geophys. Res. Lett.*, *19*, 1227-1231, 1992.

Saito, T. and S. Matsushita, Geomagnetic pulsations associated with sudden commencements and sudden impulses, *Planet. Space Sci.*, *15*, 573-587, 1967.

Sano, Y., Morphological studies on sudden commencements of magnetic storms using the rapid- run magnetograms during the IGY, *J. Geomag. Geoelectr.*, *14*, 1-15, 1962.

Sibeck, D. G., W. Baumjohann, and R. E. Lopez, Solar wind dynamic pressure variations and transient magnetospheric signatures, *Geophys. Res. Lett.*, *16*, 13-16, 1989.

Sibeck, D. G., Transient magnetic field signatures at high latitudes, *J. Geophys. Res.*, *98*, 243-256, 1993.

Siscoe, G. L., V. Formisano and A. J. Lazarus, Relation between geomagnetic sudden impulses and solar wind pressure changes–an experimental investigation, *J. Geophys. Res.*, *73*, 4869-4874, 1968.

Smith, E. J., J. A. Slavin, R. D. Zwickl and S. J. Bame, Shocks and storm sudden commencements, in "*Solar Wind-Magnetosphere Coupling*", ed. by Y. Kamide and J. A. Slavin, pp. 345-365, 1986, Tokyo.

Sonett, C. P., D. S. Colburn, L. Davis, Jr., E. J. Smith and P. J. Coleman, Jr., Evidence for a collision-free magnetohydrodynamic shock in interplanetary space, *Physical Rev. Letts.*, *13*, 153-156, 1964.

Stegelmann, E. J. and C. H. von Kenschitzki, On the interpretation of the sudden commencement of geomagnetic storms, *J. Geophys. Res.*, *69*, 139-155, 1964.

Sugiura, M., The solar diurnal variation in the amplitude of sudden commencements of magnetic storms at the geomagnetic equator, *J. Geophys. Res.*, *58*, 558-559, 1953.

Takeda, M., Three dimensional structure of ionospheric currents produced by field aligned currents, *J. Atmos. Terr. Phys.*, *44*, 695-701, 1984.

Tamao, T., Hydromagnetic interpretation of geomagnetic SSC*, *Rep. Ionos. Space Res. Japan*, *18*, 16-31, 1964a.

Tamao, T., The structure of three-dimensional hydromagnetic waves in a uniform cold plasma, *J. Geomag. Geoelect.*, *18*, 89-114, 1964b.

Tsunomura, S. and T. Araki, Numerical analysis of equatorial enhancement of geomagnetic sudden commencement, *Planet. Space Sci.*, *32*, 599-604, 1984.

Volpers, H., Untersuchung und Interpretation zweier durch SSC's angeregter Pc 5- Pulsationsereignisse, gemessen mit einem Magnetometernetz in Nordskandinavien, *Diplomarbeit in Fach PHYSIK angefertigt* in Institut fur Geophysik der Westfalischen Wilhelms Universitat, Mai 1984.

Wilken, B., C. K. Goertz, D. N. Baker, P. R. Higbie and T. A. Fritz, The SSC on July 29, 1977 and its propagation within the magnetosphere, *J. Geophys. Res.*, *87*, 5901-5910, 1982.

Wilson, C. R. and M. Sugiura, Hydromagnetic interpretation of sudden commencement of magnetic storms, *J. Geophys. Res.*, *66*, 4097-4111, 1961.

Tohru Araki, Department of Geophysics and Data Analysis Center for Geomagnetism and Space magnetism,Faculty of Science, Kyoto University, Kyoto 606-01, Japan.

Multiple Brightenings of Poleward-Moving Dayside Auroral Forms

G.J. FASEL, J.I. MINOW, R.W. SMITH, C.S. DEEHR, AND L.C. LEE

Geophysical Institute, University of Alaska-Fairbanks

Auroral forms have been observed moving poleward from the dayside auroral oval into the polar cap. These auroral forms are commonly observed when the IMF $B_z < 0$ and when the auroral oval is expanding or in an expanded condition. In this paper stationary rayed arcs and bands define the instantaneous auroral oval. Poleward-moving auroral forms are defined to be rayed bands that are formed in the dayside auroral oval, separate from it and drift into the polar cap. Recently, Fasel et al. [1992] reported on a new type of poleward-moving auroral forms which display multiple intensity variations during their poleward movement. We present several new examples of this brightening scenario. Recent studies have shown a (1.5-2.0) minute interval between subsequent brightenings. The Alfvén wave travel time from the subsolar magnetopause to the ionosphere is calculated and correlates well with the multiple brightening interval, showing a relationship between the Alfvén wave which carries the field-aligned current and the brightening of the auroral form. Mechanisms capable of launching Alfvén waves are discussed including an extension of new model using both patchy reconnection and multiple x-line reconnection [Lee et al., 1993].

INTRODUCTION

In the region of the cusp, magnetosheath particles penetrate down through the cusp and into the ionosphere. Low energy electrons collide with atomic oxygen to produce a diffuse red band (630.0 nm emissions) which makes up a region close to magnetic noon [Heikkila, 1985]. Sandholt et al. [1989] redefined Heikkila's [1985] definition of the cusp, stating:

"The cusp is a more localized region near noon characterized by low energy precipitation and a minimum in 557.7 nm emission, but also displaying characteristic transient, poleward-moving auroral forms with ∼ 10 kR green line intensity, presumably associated with temporal plasma injections/accelerations connected with the magnetic cusp".

The region around noon surrounding the optical cusp on its equatorward, dawn, and dusk sides also contains low energy electron precipitation producing 630.0 nm emissions and also discrete auroral structures displaying 557.7 nm emissions from higher energy electrons colliding with atomic oxygen [Heikkila,1985]. This boundary region is defined as the cleft (projection of the low latitude boundary layer, LLBL, on the high altitude ionosphere) which combined with the optical cusp, makes up the dayside auroral oval. In this paper, stationary rayed arcs and bands define the instantaneous dayside auroral oval.

In the dayside auroral oval discrete auroral structures have been observed to move poleward into the polar cap [Feldstein and Starkov,1967; Vorobjev et al., 1975, Horowitz and Akasofu, 1977;

Sandholt et al., 1986, 1989,1990]. In this paper poleward-moving forms are defined to be:

1) rayed bands that are formed in the dayside auroral oval, which then separate from the oval and drift into the polar cap, between 9-15 MLT,

2) forms which may move either eastward or westward, dependent upon the IMF B_y component,

3) forms which typically emit 1kR of [OI] 557.7 nm, green line.

The interpretation of these poleward-moving auroral forms has been a hotly debated subject in the past few years. Moving auroral forms are thought to be ionospheric signatures of processes that allow plasma transfer from the solar wind into the earth's magnetosphere via the magnetosheath/magnetopause. Some mechanisms that have been proposed to explain this plasma transfer are [Sandholt et al., 1986, and references therein]: (1) magnetic field line merging, including quasi-steady reconnection and flux transfer events, (2) viscous interaction, (3) gradient drift entry, and (4) impulsive penetration events. While all of the processes listed may be capable of transferring both solar wind particles and momentum into the magnetosphere, magnetic field line merging and solar wind pressure variations [Sibeck, 1990; Lui and Sibeck, 1991] have played major roles in efforts to describe the auroral forms that emerge from the dayside auroral oval and drift into the polar cap.

The field line merging process is thought to be most efficient when the interplanetary magnetic field (IMF) B_z component turns southward. The latitude of the equatorial edge of the dayside auroral oval has been shown to vary with the direction of IMF B_z, the equatorward edge shifting equatorward (poleward) when the IMF B_z component is negative (positive) [Burch, 1973; Akasofu, 1977; Horowitz and Akasofu, 1977; Sandholt et al. 1986, 1988].

Solar Wind Sources of Magnetospheric Ultra-Low-Frequency Waves
Geophysical Monograph 81

Poleward forms that drift poleward into the polar cap have been observed during periods when the oval is expanding equatorward [Vorobjev et al., 1975; Horowitz and Akasofu, 1977]. Vorobjev et al. [1975] noted that the auroral forms are a common feature during equatorward expansions and attributed their existence to the motion of field lines from the dayside to the nightside of the magnetosphere. Horowitz and Akasofu [1977] noted a one-to-one correspondence of luminous poleward-moving forms associated with an equatorward movement of the oval. They also suggested that the auroral forms are linked to the same process that resulted in the equatorward shift of the oval. In a number of cases no such moving auroral forms were seen even though the oval shifted equatorward. The authors suggested their absence may be due to arcs being too dim to detect with their all sky cameras.

Dayside poleward-moving auroral forms are also seen during periods of an expanded dayside oval for both northern and southern hemisphere observations [Sandholt et al., 1986, 1989, 1990; Rairden and Mende,1989; Mende et al., 1990]. During times when the oval is expanding or is already expanded the poleward-moving auroral forms are observed to move away from the dayside auroral oval and into the polar cap and fade [Vorobjev et al., 1975; Horowitz and Akasofu, 1977; Sandholt et al., 1986, 1990; Smith and Lockwood, 1990; Lockwood, 1991; Cowley et al., 1991; Rairden and Mende, 1989]. Observations reviewed by Smith and Lockwood [1990] and Lockwood [1991] were reported to brighten initially in the [OI] 630.0 nm emission on the equatorward side of the optical cusp followed by an intensification in the 557.7 nm emission before poleward motion. These optical events occurred at approximately 8 minute intervals having a lifetime between 2-10 minutes [Sandholt et al., 1986, 1990].

Lui and Sibeck [1991] attributed these poleward-moving auroral forms to changes in the dynamic solar wind pressure. As the dayside magnetopause compresses, plasma in the LLBL becomes energized and flows down field lines into the ionosphere [Lui and Sibeck, 1991]. The equatorward edge of the dayside auroral oval brightens since that is where the LLBL maps to in the ionosphere. According to this model, as the pressure decreases on the dayside magnetopause and increases along the flanks of the magnetosphere and encounters the mantle, a new particle population is energized. The energized plasma then flows down the field lines towards the ionosphere causing the poleward side of the auroral oval to brighten. This would explain the movement of auroral forms through the auroral oval. A fast mode wave is sent towards the magnetotail scattering lobe plasma down field lines into the ionosphere in a poleward succession. This accounts for the poleward movement of the dayside auroral form. Due to a low density of lobe plasma the brightening would die out rather quickly as the auroral form moved away from the dayside auroral oval, poleward into the polar cap.

In contrast observations will be shown of dayside auroral forms that continue to brighten multiple times as they move poleward from the dayside oval. When the auroral forms reach their most poleward position of their brightening sequence as recorded by the optical instruments, they brighten and then fade from view. Most of the events examined display this brightening scenario.

INSTRUMENTATION

The dayside auroral observations were made at Longyearbyen, Svalbard (74.9 N, 114.6 E geomagnetic coordinates, geomagnetic noon 0830 UT) during the boreal winter, when the sun is more than 6 degrees below the horizon for more than two months, permitting optical observations throughout the daytime period. A 5-channel meridian scanning photometer (MSP) [Romick, 1976] records auroral emission intensity as a function of the elevation angle along the magnetic meridian. Each channel is equipped with a 3 angstrom band width filter to isolate an auroral emission. The system employs a filter-tilting mechanism permitting a background subtraction of scattered sky continuum radiation. A mirror rotates once every four seconds to direct light from the meridian to the photometers. Two individual peak and background elevation scans are averaged before writing the data to the record tapes resulting in a time resolution of sixteen seconds for the processed data. The field of view of each photometer is one degree. An all-sky camera (ASC) comprising an image orthicon white light television camera fitted with a fish-eye lens provides real-time video of the aurora over the entire sky.

OBSERVATIONS

31 December 1984

The [OI] 557.7 nm and [OI] 630.0 nm emissions from the dayside auroral oval for 31 Dec 1984 are shown in the MSP greyscale plots in Figure 1. The integrated intensity along the line of sight of the photometer is plotted as a function of elevation angle and Universal Time (UT). Between 0800 and 0900 UT, a typical quiet aurora dominated by the [OI] 630.0 nm emissions is seen poleward of the zenith. The auroral oval begins to shift equatorward at approximately 0905 UT reaching its maximum southern position at 1100 UT. A number of poleward-moving auroral forms are observed during this period, of which five are labelled in Figure 2. The auroral forms appear to form in the auroral oval and drift into the polar cap as has been described earlier [Sandholt et al., 1986, 1989].

In the subsequent discussion we will focus on aspects of the three events labelled c, d, and e in Figure 2 that occur between 0945 and 0957 UT. Each of these auroral forms has a large east-west extent. ASC images of events c, d, and e are displayed in Figure 3 for selected times. Each image is an average of 16 video frames from the original record yielding a 0.5 second integration period per image and is displayed in negative form. On the southern border of each frame in Figure 3 a dark background exists due to the twilight in the horizon. This feature should not be considered as part of the auroral oval. The individual events are labelled with subscripts indicating different stages of evolution (eg. e_n; $n=1,2,3, ...$) in the brightening sequence. From each label a line is drawn to the corresponding auroral form in the ASC sequence. The magnetic orientation is given in the last frame of Figure 3. The brightenings corresponding to the different stages of event e are also identified in the MSP plots in Figure 2. Particular attention was paid to the brightening sequence of event e, however, we begin by looking at the preceding events c and d.

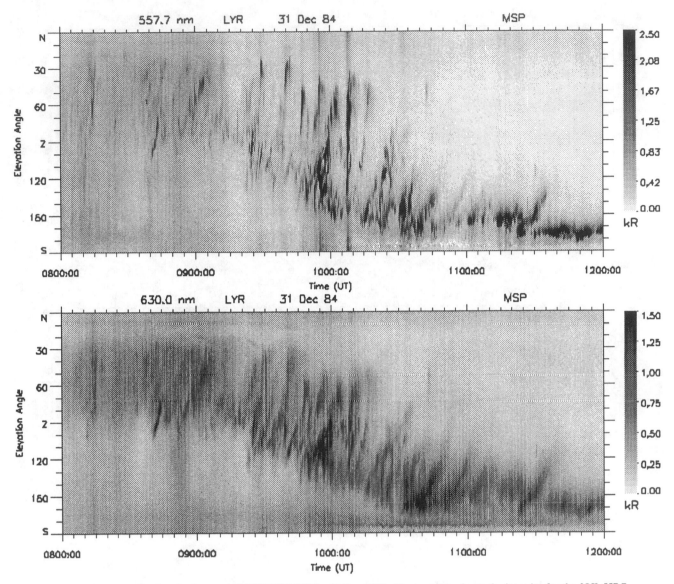

Fig. 1. Greyscale plot for the time period 0800-1200 UT for 31 Dec 1984. The top plate shows the intensity for the [OI] 557.7 nm emissions and the bottom plate displays the [OI] 630.0 nm emissions from the meridian scanning photometer. Intensities for the greyscales are plotted as a function of angle from the horizon and universal time.

The first frame shows the start of the final intensification of event c (c_1) at 0947:00 UT at its northernmost displacement from the oval that can be observed by the ASC and MSP. A separate arc begins to brighten at 0947:09 UT in the auroral oval in the south as event c maximizes in luminosity in the north (c_2). In the third frame, at 0948:13 UT, a third auroral form d_0 just south of c_3 in the zenith is observed. The eastern edge of this auroral form begins to intensify as does an arc in the dayside auroral oval. This intensification of the arc, in the dayside oval, at 0948:13 UT begins with an enhancement of the [OI] 557.7 nm emission as shown in Figure 4. The same initial enhancement of the green emission is present in all of the events labelled in Figure 2. In previous studies the events examined usually had an intensification first in the 630.0 nm emission observed on the equatorward edge

followed by a strong 557.7 nm emission in the moving auroral form [Smith and Lockwood, 1990; Lockwood, 1991]. The arc in the oval continues to brighten, giving rise to stage e_1 at 0949:01 UT. Event d drifts poleward and continues to intensify reaching its maximum at 0949:26 UT (d_3). The intensity of arc e_1 fluctuates after the initial brightening before beginning to dim at 0951:25 UT (e_2). The arc e_3 having a large east-west extent begins to move poleward at 0952 UT. The fourth stage (e_4) of event e shows a localized enhancement in brightness which maximizes at 0953:06 UT (e_5) along with a corresponding brightening in the auroral oval. This is the second intensification of event e. At 0954:27 UT, event e begins to intensify at the east end of the arc (e_6) while simultaneously an arc in the oval is also seen to intensify. This is the third brightening of event e. The brightness

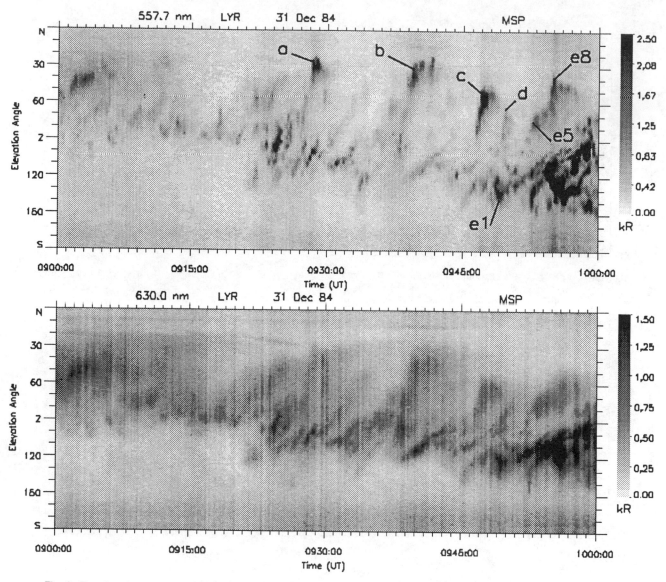

Fig. 2. Top plate shows an expanded MSP plot of the [OI] 557.7 nm emission, while the bottom plate displays the [OI] 630.0 nm emissions, for the time period 0900-1000 UT. Selected events are labelled.

spreads westward through the auroral form reaching its maximum luminosity at 0955:03 UT (e_8). At this point, 0957 UT, event e has reached the poleward most position in the brightening sequence before it begins to fade and finally disappear from our field of view.

27 December 1991

On the 27 December 1991 the dayside oval is just south of the zenith as can be seen from the MSP greyscale plots for the [OI] 557.7 nm and [OI] 630.0 nm emissions, Figure 5. The dayside oval began expanding equatorward around 0735 UT which can be seen from a greyscale plot with an increased time axis (Figure not shown). The expanded time period, 0800 UT- 0900 UT, centered around magnetic noon shows a number of poleward-moving au-

roral forms, especially a little after 0822 UT and 0826 UT. The enhanced density with increasing time indicates a rebrightening of the poleward bound auroral form, just as was seen for the events previously described for 31 December 1984.

The ASC is shown in Figure 6 for selected times for events a,b,c with o designating the dayside auroral oval. The magnetic orientation is the same as given in the last frame of Figure 3. The indexed stages of events a,b,c are shown at intermediate points to show the correspondence between the ASC images and MSP greyscale plots at specific times.

The auroral oval, o, is shown south of the zenith in frame 1. The first four frames show the final stages of event a, (a_1, a_2, a_3, a_4), as it fades from view. Event b (b_1) is seen breaking away from the dayside auroral oval (o) in frame 1 of Figure 6 a

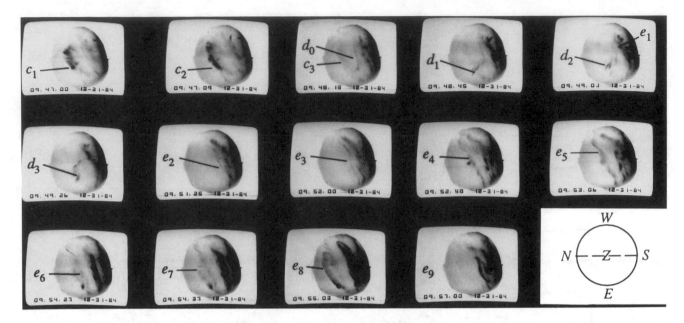

Fig. 3. All sky camera images for the period 0947 to 0957 UT. The magnetic orientation is given in the last frame. Dashed line through the north-south meridian displays the MSP scan. Images are in negative form.

little past 0844 UT. As event b moves poleward away from the dayside oval it continues to intensify along with the oval and split into two rayed bands, events b and c. Event c can be seen in frame 7, $\sim$0846:00 UT and frame 8, $\sim$0846:20 UT, but is not marked. At $\sim$0846:40 UT, frame 9, the two rayed bands, b_9 and c_1, can be seen more clearly.

Both events, b_{11} and c_3 intensify in brightness at $\sim$0847:20 along with an enhancement of the auroral oval. Event b continues to move poleward. At $\sim$0848:00 UTb_{13} and c_5 are seen both in Figures 5 and 6. Until this time it was not possible to see the two arcs in the MSP greyscale, Figure 5. In an expanded MSP greyscale, not shown, with the vertical axis expanded, two distinct arcs can be seen. In frame 15, $\sim$0849 UT, the auroral forms (b_{15} and c_7) fade along with the auroral oval. Two minutes later event b (b_{17}), c (c_{11}), and the auroral oval increase in brightness. The last frame, $\sim$0851:30 UT, shows events b (b_{20}) and c (c_{12}) fading.

12 January 1992

For 12 January 1992 the auroral oval is in an expanded condition. This day displays auroral forms, many which brighten multiple times as the move poleward into the polar cap. Event a which is seen between 0711 UT and 0718 UT in Figure 7, green [OI] 557.7 nm and red [OI] 630.0 nm line emissions, intensifies three times (a_3, a_{10} and a_{17}) after it separates from the auroral oval during its poleward movement.

The ASC, Figure 8, which has the same magnetic orientation as Figure 3 begins with event a separating from the auroral oval (o). At $\sim$0712 UT the auroral form (a_3) begins to intensify as does the auroral oval (o). Event a continues moving poleward. Frames 6 through 9 show event a (a_6, a_7, a_8, a_9) dimming. Another enhancement in brightness occurs at $\sim$0714:10 UT in both the poleward-moving auroral form (a_{10}) and the auroral oval. As the

auroral form continues to move poleward it again begins fading. A third brightening occurs at $\sim$0716:00 UT in both event a (a_{17}) and the auroral oval. After reaching its most poleward position of its brightening history the third brightening event a fades from our field of view, barely visible in frame 25 (a_{25}).

DISCUSSION

Not all poleward-moving auroral forms have the same brightening characteristics. Some have been observed to fade with time [Sandholt et al., 1986, 1989, 1990] while others continue to intensify [Fasel et al., 1992] after they separate from the dayside auroral oval.

Events presented in this paper come from three different dayside magnetic time periods: prenoon, $\sim$ noon, and postnoon. Each event can be generalized as follows:

1) event begins at time of an equatorward shift of, or expanded dayside auroral oval,

2) dayside auroral oval brightens,

3) auroral form breaks away from dayside oval and moves poleward,

4) localized intermittent brightenings occur which in some cases spread either eastward or westward along the arc or rayed band of auroral form,

5) time interval between brightening intensifications is between (1.5 - 2.0) minutes

6) dayside auroral oval brightens about the same time (4) occurs. The events last between 1 to fifteen minutes.

Interpreting poleward-moving auroral forms could provide insight into solar wind/magnetosphere coupling. The transfer of energy from the magnetopause into the ionosphere is still not completely known. Many of the existing models need to be better connected with observations looking at the finer details of

Longyearbyen MSP

31 December 1984

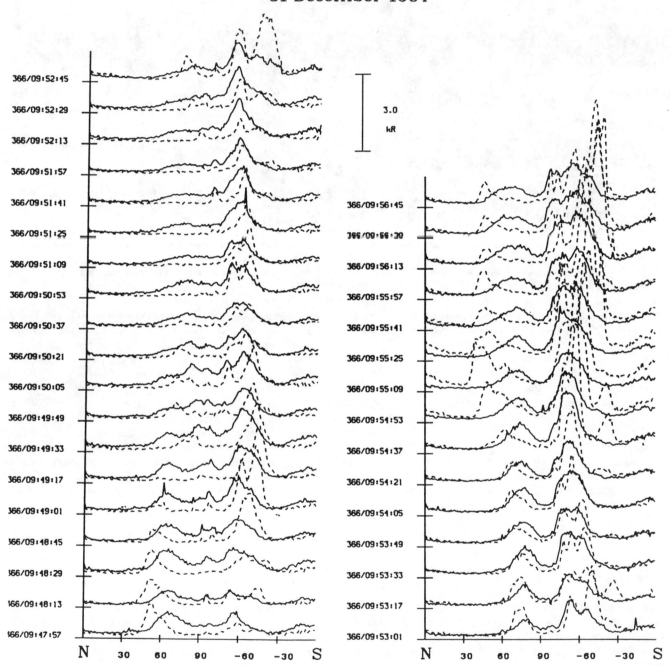

Fig. 4. Stackplot of individual MSP scans for event (e). [OI] 630.0 nm and [OI] 557.7 nm intensities are given by the solid and dashed line, respectively. Vertical axis is labelled with Julian day/hour/minute/second.

auroral forms. Looking at the fine structure of the daytime phenomena such as brightening and motion can give hints to what is happening at the magnetosheath/magnetosphere interface. We know that when the IMF B_z component turns negative, the optical cusp responds by moving equatorwards [Burch, 1973; Horowitz and Akasofu, 1977]. Any brightening and subsequent motion of the poleward-moving auroral forms must be indicative of some process happening at the dayside magnetopause. If brightenings

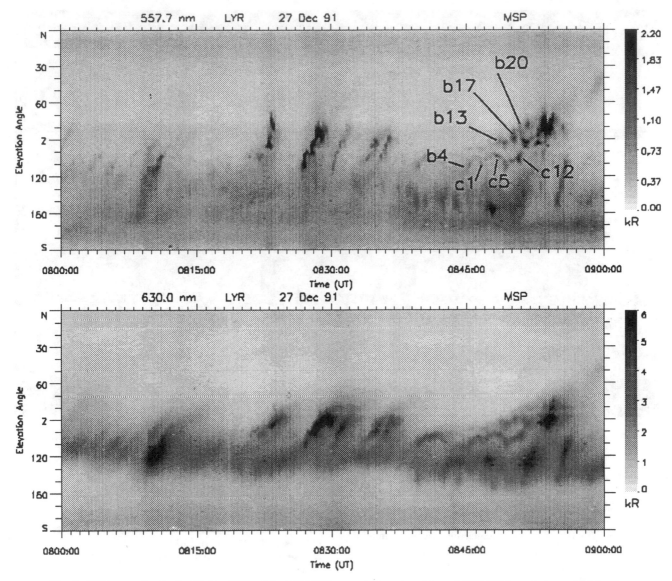

Fig. 5. MSP greyscale plot for 27 Dec 1991, of the [OI] 557.7 nm emissions, top plate, and [OI] 630.0 nm, bottom plate, for the time period 0800-0900 UT. Moving auroral forms *b* and *c* are labeled.

of auroral forms are associated with precipitating electrons (upward Birkeland currents) [Kamide, 1981; Rairden and Mende, 1989], we may conclude that a plasma injection event has taken place along the field lines and that multiple brightenings are associated with multiple injections. Plasma injection events at the magnetosheath interface were used to explain auroral structures observed by Meng and Lundin [1986] in DMSP images of the southern hemisphere dayside oval.

One possible interpretation of the multiple brightening poleward-moving auroral forms is multiple x-line reconnection at the dayside magnetopause [Lee and Fu, 1985]. As illustrated in Figure 9 , geomagnetic field lines may reconnect with magnetosheath field lines at two sites P and Q. At time t_1, reconnection takes place at P with the formation of a magnetic bubble along the

newly open field lines. Alfvén waves may be launched towards the polar ionosphere from the reconnection site P. Electrons can be accelerated to KeV energies as the field-aligned currents associated with the Alfvén waves encounter the high altitude ionosphere, leading to the first brightening of the moving auroral form. At $t_2 > t_1$, reconnection occurs at Q with the formation of a magnetic island and a new magnetic bubble along the open field lines. The second reconnection on the same flux tube launches new Alfvén waves, with associated field-aligned currents, towards the ionosphere leading to a second brightening of the poleward-moving auroral form. Since the geomagnetic field lines have been convected more poleward at t_2, the second brightening should occur more poleward than the first, as observed.

Recently Lee et al. [1993], using a combination of patchy re-

Fig. 6. All sky camera images for 27 Dec 1991. Selected images for the time period 0844:00-0851:30 UT. Magnetic orientation is the same as for Figure 3. Images are in negative form.

connection and multiple x-line reconnection examined the topology of flux tubes and developed a model for the formation of the low latitude boundary layer. This model can be extended to explain multiple brightenings of poleward moving auroral forms [Fasel et al., 1993]. Because reconnection occurs at multiple patches, an open flux tube can be formed having both IMF and geomagnetic field components. Alfvén waves initiated at the merging sites propagate along the flux tube towards the ionosphere, carrying field aligned currents, generate multiple brightenings of the poleward-moving form.

Another possibility for the multiple brightenings is the bouncing of Alfvén waves between the magnetopause and ionosphere after the initial launch of Alfvén waves from the reconnection site. Enhanced brightening of the moving form may occur each time the Alfvén waves are propagated towards the ionosphere. One would expect in this case that each subsequent brightening would be dimmer than the previous one, since the Alfvén waves should lose energy on each bounce.

The travel time, T, for an Alfvén wave from the subsolar magnetopause to the ionosphere can be calculated by

$$T = \int_{s_i}^{s_m} \frac{ds}{V_A(s)} \qquad (1)$$

where ds is the path length integrated from the subsolar magne-

topause, s_m, to the ionosphere, s_i. $V_A(s)$ is the Alfvén velocity

$$V_A(s) = \frac{B}{(\mu_0 \rho)^{\frac{1}{2}}} \qquad (2)$$

where μ_0 is the permeability of free space, B is the magnetic field strength, and ρ is the cold ion mass density. Using a dipole field, a first order approximation for the travel length of the Alfvén wave is

$$l = \int_{s_m}^{s_i} ds \approx 21 R_E \qquad (3)$$

The Alfvén velocity varies from ~ 500 km/s at the equatorial plane to thousands of km/s at a couple of R_E and then dropping back down as the wave approaches the ionosphere, due to the changing magnetic field strength and the plasma density. Using an average Alfvén velocity of 1500 km/s the Alfvén travel time is found to be ~ 89 seconds. Based on EISCAT measurements the Alfvén travel time is found to be ~ 1 minute and using methods by Sampson and Rostoker [1972] the travel time is between (1.4-2.6) minutes [Freeman et al., 1990].

The time between subsequent brightenings found in the poleward moving auroral forms falls within the Alfvén travel time needed for propagation between the subsolar magnetopause and ionosphere.

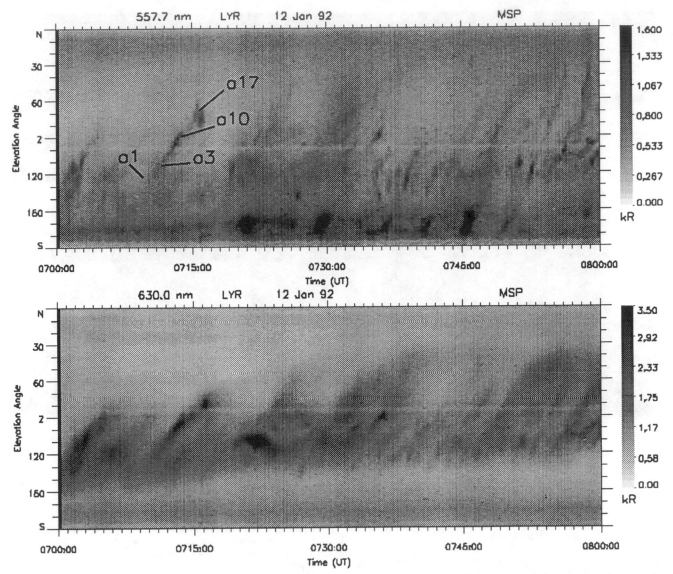

Fig. 7. MSP greyscale plot for 12 Jan 1992, of the [OI] 557.7 nm emissions, top plate, and [OI] 630.0 nm emissions, bottom plate, for the time period 0700-0800 UT. Selected stages for event *a* are labeled.

CONCLUSION

Examples of poleward-moving auroral forms have been selected and discussed from three days in which the auroral oval was expanding or already expanded. The brightening sequence for each auroral form displayed multiple intensifications along the arc or rayed band as they moved poleward. At the most poleward position that can be observed by the ASC and MSP, they intensified again and then faded from view. The time interval brightening intensifications is between (1.5 - 2.0) minutes. Simultaneously with each intensification, an arc in the auroral oval also intensified. Each example had several events which followed the above brightening sequence. These are different from the poleward-moving auroral forms described in earlier studies [Smith and Lockwood, 1990; Lockwood,1991] that brightened at the equatorward edge of the auroral oval and faded as they moved into the polar cap.

Multiple x-line reconnection is a possible mechanism for the multiple brightening sequence observed. Each brightening could be the result of magnetic reconnection on the same flux tube that is convecting antisunward into the polar cap. Patchy reconnection combined with multiple x-line reconnection allows for the formation of open flux tubes consisting of both IMF and geomagnetic fields. Each merging site launches an Alfvén wave carrying field aligned currents towards the ionosphere leading to a brightening of the poleward moving auroral form.

Acknowledgements. Financial support for this research was provided by National Science Foundation, Atmospheric Sciences Section, grants ATM 88-14635, ATM 89-01047, ATM 92-04116, and ATM 91-11509 to the Geophysical Institute of the University of Alaska. The observations on Svalbard were made through a cooperative effort with the Universities of Tromsø under the direction of K. Henriksen of the University of Tromsø. We acknowledge the logistical support and technical assistance of the

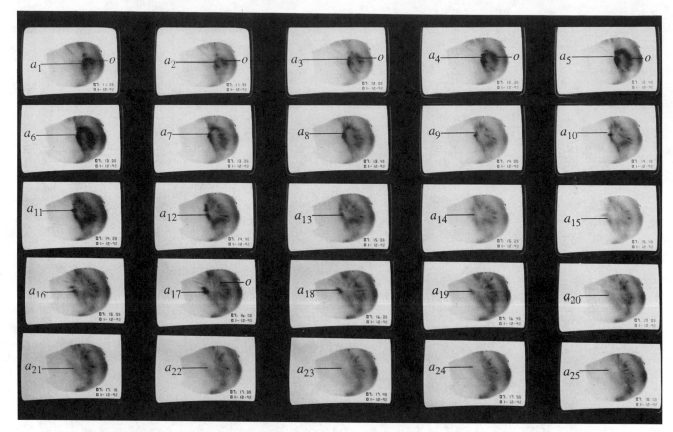

Fig. 8. All sky camera images for 12 Jan 1992. Selected images for the time period 0700:00-0718:10 UT. Magnetic orientation is the same as for Figure 3. Images are in negative form.

Great Norwegian Spitsbergen Coal Company, and the staff of governor's office of Svalbard. The authors would also like to thank S.-I. Akasofu for his discussions and support.

References

Akasofu, S.-I., *Physics of Magnetospheric Substorms*, p. 220, D. Reidel, Boston, Mass., 1977.

Burch, J.L., Rate of erosion of dayside magnetic flux based on a quantitative study of the dependence of polar cusp latitude on the interplanetary magnetic field, *Radio Science, 8*, 955-961, 1973.

Cowley, S.W.H., M.P. Freeman, M. Lockwood, and M. Smith, The ionospheric signature of flux transfer events, in *Proceedings of the CLUSTER workshop on Longyearbyen, ESA SPA-330*, in press, 1991.

Elphic, R.C., M. Lockwood, S.W.H. Cowley, and P.E. Sandholt, Flux transfer events at the magnetopause and in the ionosphere, *Geophys. Res. Lett., 17*, 2241-2244, 1990.

Fasel, G.J., J. Minow, R.W. Smith, C.S. Deehr, and L.C. Lee, Multiple brightenings of transient dayside auroral forms during oval expansions, *Geophys. Res. Lett., 19*, 2429-2432, 1992.

Fasel, G.J., L.C. Lee, and R.W. Smith, A theory of multiple brightening poleward moving dayside auroral forms, submitted to *Geophys. Res. Lett.*, April 1993.

Feldstein, Y.I., and G.V. Starkov, Dynamics of auroral belt and polar geomagnetic disturbances, *Planet. Space Sci., 15*, 209-229, 1967.

Freeman, M.P., C.J. Farrugia, and S.W.H. Cowley, The response of dayside ionospheric convection to the y-component of the magnetosheath magnetic field: a case study, *Planet. Space Sci. 38*, 13-41, 1990.

Friis-Christensen, E., M.A. McHenry, C.R. Clauer, and S. Vennerstrom,

Ionospheric travelling convection vortices observed near the polar cleft: a response to sudden changes in the solar wind, *Geophs. Res. Lett., 15*, 253-256, 1988.

Fukunishi, H., and L. J. Lanzerotti, Hydromagnetic waves in the dayside cusp region and ground signatures of flux transfer events, in *Plasma Waves and Instabilities at Comets and in Magnetospheres, Geophys. Monogr. Ser., 53*, ed. by B. T. Tsurutani and H. Oya, 179-195, AGU, Washington, D. C., 1989.

Goertz, C.K., E. Nielsen, A. North, K.-H. Glassmeier, C. Haldoupis, P. Hoeg, and D. Hayward, Observations of a possible ground signature of flux transfer events, *J. Geophys. Res., 90*, 4069-4078, 1985.

Heikkila, W.J., Definition of the cusp, in *The Polar Cusp*, ed. by J. Holtet and A. Egeland, 387-395, 1985.

Horowitz, J.L., and S.-I. Akasofu, The response of the dayside aurora to sharp northward and southward transitions of the interplanetary magnetic field and to magnetospheric substorms, *J. Geophys. Res., 82*, 2723-2734, 1977.

Kamide, Y., On the empirical relationship between field-aligned currents and auroras, in *Physics of auroral arc formation, Geophys. Monogr. Ser, 25*, ed., by S.-I. Akasofu and J.R. Kan, 192-198, AGU, Washington, D.C., 1981.

Lanzerotti, J.L., L.C. Lee, C.G. Maclennan, A. Wolfe, and L.V. Medford, Possible evidence of flux transfer events in the polar ionosphere, *Geophys. Res. Lett., 13*, 1089-1092, 1986.

Lee, L.C. and Z.F. Fu, A theory of magnetic flux transfer at the earth's magnetopause, *Geophys. Res. Lett., 12*, 105, 1985.

Lee, L.C., Z.W. Ma, Z.F. Fu, and A. Otto, Topology of magnetic flux ropes and formation of fossil flux transfer events, and boundary layer plasmas, *J. Geophys. Res., 98*, 3943-3951, 1993.

Lockwood, M., The excitation of ionospheric convection, *J. of Atm. and*

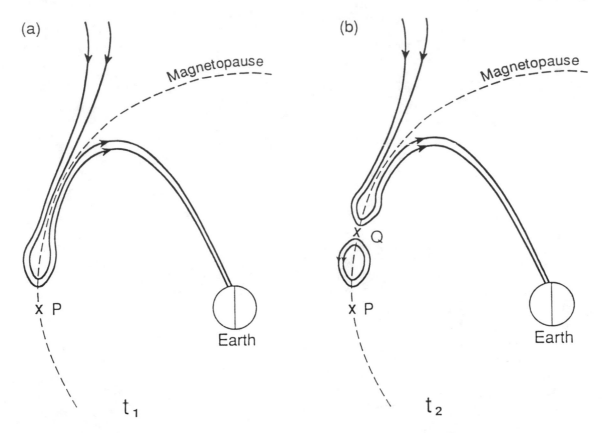

Fig. 9. Multiple reconnections along the same flux tube at times t_1 and t_2. Each reconnection launches Alfvén waves carrying field-aligned currents towards the ionosphere leading to brightening of the poleward-moving auroral form.

Terr. Physics, 53, 177-199, 1991.

Lockwood, M., S.W.H. Cowley, P.E. Sandholt, and R.P. Lepping, The ionospheric signatures of flux transfer events and solar wind dynamic pressure changes, *J. Geophys. Res., 95,* 17113-17135, 1990.

Lockwood, M., P.E. Sandholt, and S.W.H. Cowley, Dayside auroral activity and magnetic flux transfer from the solar wind, *Geophys. Res. Lett., 16,* 33-36, 1989.

Lui, A.T.Y., and D.G. Sibeck, Dayside auroral activities and their implications for impulsive entry processes in the dayside magnetosphere, *J. of Atm. and Terr. Physics, 53,* 219-229, 1991.

Mende, S.B., R.L. Rairden, L.J. Lanzerotti, and C.G. Maclennan, Magnetic impulses and associated optical signatures in the dayside aurora, *Geophys. Res. Lett., 17,* 131-134, 1990.

Meng, C.-I., and R. Lundin, Auroral morphology of the midday oval, *J. Geophys. Res., 91,* 1572,1986.

Oguti, T., T. Yamamoto, K Hayashi, S. Kokubun, A. Egeland, and J.A. Holtet, Dayside auroral activity and related magnetic impulses in the polar cusp, *J. Geomag. Geoelectr., 40,* 387-408, 1988.

Rairden, R.L., and S.B. Mende, Properties of 6300-Å auroral emissions at south pole, *J. Geophys. Res., 94,* 1402-1416, 1989.

Romick, G.J., The detection and study of the visible spectrum of the aurora and airglow, *Methods for Atm. Radiometry, 91,* 63-70, 1976.

Sampson, J.C., and G. Rostoker, Latitude-dependent characteristics of high latitude pc 4 and pc 5 micropulsations, *J. Geophys. Res.,* 6133-6144, 1972.

Sandholt, P.E., C.S. Deehr, A. Egeland, B. Lybekk, R. Viereck, and G.J. Romick, Signatures in the dayside aurora of plasma transfer from the magnetosheath, *J. Geophys. Res. 91,* 10063-10079, 1986.

Sandholt, P.E., IMF control of polar cusp and cleft auroras, *Adv. Space Sci., 9,* 21 - 34, 1988.

Sandholt, P.E., B. Lybekk, A. Egeland, R. Nakamura, and T. Oguti, Midday auroral breakup, *J. Geomag. Geoelectr., 41,* 371-387, 1989.

Sandholt, P.E., M. Lockwood, T. Oguti,, S.W.H. Cowley, K.S.C. Freeman, B. Lybekk, A. Egeland, and D.M. Willis, Midday auroral breakup events and related energy and momentum transfer from the magnetosheath, *J. Geophys. Res., 95,* 1039-1060, 1990.

Sibeck, D.G., A model for the transient magnetospheric response to sudden solar wind dynamic pressure variations, *J. Geophys. Res., 95,* 3755-3771, 1990.

Smith, M.F. and M.L. Lockwood, The pulsating cusp, *Geophys. Res. Lett., 17,* 1069-1072, 1990.

Southwood, D.J., C.J. Farrugia, and M.A. Saunders, What are flux transfer events?, *Planet. Space Sci., 36,* 503-508, 1988.

Vorobjev, V.G., G. Gustafasson, G.V. Starkov, Y.I. Feldstein, and N.F. Shernina, Dynamics of day and night aurora during substorms, *Planet. Space Sci., 23,* 269-278, 1975.

C.S. Deehr, G.J. Fasel, L.C. Lee, J.I. Minow, and R.W. Smith, University of Alaska-Fairbanks, Geophysical Institute, Fairbanks, Ak 99775.

Generation of Region 1 and Mantle Field-Aligned Currents by the Secondary Rotational Discontinuity

Y. LIN AND L. C. LEE

Geophysical Institute and Department of Physics, University of Alaska, Fairbanks

Our simulations of the magnetopause boundary layer in the open field line region indicate that in addition to the rotational discontinuity (RD1) at the magnetopause, there also exists a weaker rotational discontinuity (RD2) in the boundary layer. Associated with this secondary rotational discontinuity (RD2) are a field-aligned current and a kink of magnetic field lines. We suggest that the propagation of this kink field to the cusp ionosphere leads to the generation of the observed near-noon region 1 and mantle field-aligned currents. The secondary rotational discontinuity (RD2) in the dayside boundary layer produces the region 1 field-aligned currents near noon, while the secondary rotational discontinuity in the magnetotail plasma mantle (high-latitude boundary layer) contributes to the mantle field-aligned currents. The dependence of the current direction on the interplanetary magnetic field (IMF) B_y obtained in this model is found to be consistent with observations.

1. INTRODUCTION

Satellite and ground observations in the dayside polar region near noon indicate the existence of a field-aligned current system [e.g., Iijima and Potemra, 1976; McDiarmid et al., 1978, 1979; Wilhjelm et al., 1978; Iijima et al., 1978; Rostoker, 1980; Doyle et al., 1981; Bythrow et al., 1982, 1988; Friis-Christensen et al., 1985; Erlandson et al., 1988]. The intensity of these field-aligned currents increases as the interplanetary magnetic field turns to a southward direction [e.g., Iijima and Potemra, 1976]. Analyses of both satellite and ground magnetic field data over the northern and southern polar regions indicate that the field-aligned current system in each individual event usually consists of a pair of current sheets extending in the east-west direction. The poleward part of the field-aligned current pair is now called the mantle field-aligned currents [e.g., Bythrow et al., 1988; Erlandson et al., 1988] and was traditionally called the cusp field-aligned currents [Iijima and Potemra, 1976]. The equatorward part of the field-aligned current pair merges with the dayside region 1 field-aligned currents. In this paper, we adopt the term "region 1 field-aligned currents near noon" [e.g., Erlandson et al., 1988] to name the equatorward part of the observed current system. Measurements of the magnetic field and charged particles indicate that the mantle field-aligned currents are associated with field lines extending to the tail plasma mantle and the region 1 currents near noon are related to field lines threading the dayside low-latitude boundary layer [Bythrow et al., 1988; Erlandson et al., 1988]. The injection of field-aligned currents to the polar ionosphere may lead to an impulse of magnetic perturbations and subsequent ULF wave trains [Lanzerotti et al., 1986; Lee et al., 1988].

The mantle field-aligned currents and region 1 currents near noon flow in opposite direction [e.g., McDiarmid et al., 1978; Wilhjelm et al., 1978; Doyle et al., 1981; Bythrow et al., 1982; Friis-Christensen et al., 1985; Erlandson et al., 1988; Saunders, 1989]. For IMF $B_y > 0$, the field-aligned currents in the northern (southern) hemisphere are usually observed to exist in the pre-noon (post-noon) region, with the mantle field-aligned currents flowing away from (into) the polar ionosphere and the region 1 currents near noon flowing into (away from) the polar ionosphere. For IMF $B_y < 0$, the field-aligned currents in the northern (southern) hemisphere are usually observed to exist in the post-noon (pre-noon) sector, with the mantle field-aligned currents flowing into (away from) the polar ionosphere and the region 1 currents flowing away from (into) the ionosphere. Figure 1 shows the statistical distribution of the polar field-aligned current system which consists of Region 1, Region 2, and the mantle field-aligned currents [Iijima and Potemra, 1976]. The mantle field-aligned currents are located in the region with latitude 78°-80°. As shown in Figure 1, the mantle field-aligned currents usually flow out of (into) the polar ionosphere when observed in the pre-noon (post-noon) sector. The IMF B_y dependence of the displacement of cusp precipitation region and the location of mantle field-aligned

Solar Wind Sources of Magnetospheric Ultra-Low-Frequency Waves
Geophysical Monograph 81

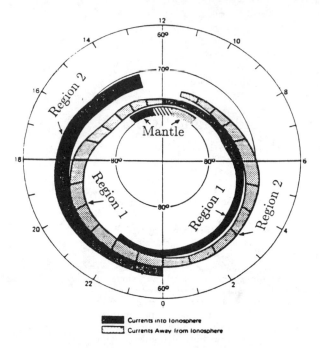

Fig. 1. The observed polar field-aligned current system in the northern hemisphere, which consists of Region 1, Region 2, and the poleward part of the cusp field-aligned currents [from Iijima and Potemra, 1976]. The mantle field-aligned currents are located in the region with latitude 78°-80°.

currents was reviewed by Cowley et al. [1991].

Several mechanisms for the origin of the region 1 and mantle field-aligned currents have been proposed [e.g., Lee and Kan, 1979; D'Angelo, 1980; Cowley, 1981; Primdahl and Spangslev, 1981; Lee et al., 1985; Siscoe and Sanchez, 1987; Saunders, 1989; Lundin et al., 1992]. In particular, Lee et al. [1985] suggested that the field-aligned currents are caused by the leakage of the field-aligned currents associated with the rotational discontinuities at the dayside and tail magnetopause. The open magnetopause has been frequently identified as a rotational discontinuity [e.g., Paschmann et al., 1979; Sanchez et al., 1990], which can be considered as a localized large-amplitude Alfvén wave with twisted field lines. Because of the variation of magnetic field and plasma conditions across the magnetopause, magnetic tension of the kink field lines associated with rotational discontinuity may be released and propagate as Alfvén waves to the polar ionosphere. In Lee et al.'s [1985] mechanism, the leakage of the field-aligned currents in the rotational discontinuity at the dayside magnetopause leads to the generation of the region 1 field-aligned currents near noon, and that from the nightside magnetopause contributes to the mantle field-aligned currents. The study showed that the direction of the field-aligned currents at the magnetopause rotational discontinuity is controlled by IMF B_y. Siscoe and Sanchez [1987] also suggested that the field-aligned currents associated with the rotational discontinuity at the tail magnetopause feed the mantle field-aligned currents.

On the other hand, Saunders [1989] proposed that the mantle field-aligned currents arise from the release of magnetic tension in reconnected open field lines moving over the dayside magnetopause. In his mechanism, the mantle field-aligned current is generated from the twisted field lines at the dayside boundary layer.

Recently, Lin and Lee [1993] studied the evolution of the dayside magnetopause boundary layer after the onset of magnetic reconnection. They found that apart from the rotational discontinuity (RD1) at the magnetopause, several other MHD discontinuities (shocks) and expansion waves are present in the boundary layer. For the general cases in which the magnetic fields in the magnetosheath and magnetosphere are not anti-parallel, there exists a secondary rotational discontinuity (RD2) in the boundary layer earthward of the primary rotational discontinuity (RD1) at the magnetopause. However, they did not relate RD2 to the observed field-aligned currents. The boundary layer structure has also been studied by Heyn et al. [1988], Biernat et al. [1989], Scholer [1989], and Shi and Lee [1990].

After the magnetosheath and magnetospheric fields reconnect at the dayside magnetopause, the Maxwell stress accelerates plasma and drags the new open field lines toward the geomagnetic tail. The new open field lines at the tail magnetopause then evolve to form the plasma mantle [Coroniti and Kennel, 1979; Swift and Lee, 1982; Siscoe and Sanchez, 1987; Sanchez et al., 1990]. Siscoe and Sanchez [1987] studied the structure of the tail mantle for the limiting case in which the plasma density in the lobe is zero.

In this paper, we extend our earlier simulations [Lin et al., 1992; Lin and Lee, 1993] to include the structure of the plasma mantle. For the tail lobe with a finite plasma density, a secondary rotational discontinuity (RD2) is found to exist in the plasma mantle, similar to the dayside boundary layer. We examine the field-aligned currents associated with secondary rotational discontinuities in the dayside boundary layer and in the tail mantle. The relaxation of the twisted magnetic fields in RD2 may cause the injection of field-aligned currents into the cusp ionosphere. We suggest that the field-aligned currents associated with RD2 in the dayside (tail) boundary layer may result in the generation of the observed region 1 (mantle) field-aligned currents.

In Lee et al.'s [1985] mechanism the field-aligned currents are generated by RD1 **at the magnetopause, while in our** present model the field-aligned currents are generated by RD2 in the boundary layer. In Saunders' [1989] mechanism the mantle field-aligned currents are generated in the dayside boundary layer, while in our present model the mantle currents are generated in the tail mantle region. In all three models, the observed field-aligned currents are generated by the same mechanism, i.e., the relaxation of magnetic tension in twisted fields.

Note that our model is closely related to magnetic reconnection at the dayside magnetopause. When the IMF has a strong northward component, the polar cap "NBZ" current system is often observed. The NBZ current system may be generated by reconnection at the tail magnetopause or by the distorted plasma sheet [see Iijima and Shibaji, 1987 and references therein].

2. SIMULATION MODELS

Simulations of the dayside boundary layer have been presented by Lin and Lee [1993]. In this paper, we briefly describe the simulation result of dayside boundary layer and extend our simulations to the tail plasma mantle. The field-aligned currents in the boundary layer are then examined.

In our study, an initial current sheet at the magnetopause separates two uniform plasma regions: the magnetosheath with low magnetic field strength (B) and high plasma mass density (ρ) and the magnetosphere with high magnetic field and low plasma density. Across the dayside magnetopause, these two regions have antiparallel magnetic field components in the z (north-south) direction and dawn-dusk magnetic field components in the y direction. The normal of the magnetopause current sheet is in the x direction pointing to the Sun. Note that the GSM coordinate system is used in this paper. Initially, total pressure (plasma thermal pressure plus magnetic pressure) is assumed constant everywhere, and plasma flow velocity is assumed zero everywhere. As a result of the dayside reconnection, a non-zero normal component of magnetic field, $B_n = -B_x$, is present at the magnetopause current sheet. The dayside boundary layer is formed through the evolution of the open magnetopause current sheet. In our 1-D simulations of the dayside boundary layer, all the dependent variables are functions of the spatial coordinate x and the time t only. The time t can be related to the coordinate z by $z \simeq \bar{v}_z t$, where $\bar{v}_z$ is an effective convection velocity. Thus the 1-D simulation results can be used to determine the 2-D reconnection layer in the xz plane [Lin et al., 1992].

Initially, the dayside magnetopause current sheet is located at $x = x_c = 0$. The magnetosheath is in the region with $x > 0$ and the magnetosphere in $x < 0$. Let the subscript "s" denote the physical quantities in magnetosheath, and "m" the quantities in the magnetosphere. The initial profiles of the z-component magnetic field (B_z), magnetic field strength, and plasma temperature (T) are given by

$$B_z(x) = \frac{1}{2}(B_{zs} + B_{zm}) + \frac{1}{2}(B_{zs} - B_{zm})\tanh(x/\delta) \quad (1)$$

$$B(x) = \frac{1}{2}(B_s + B_m) + \frac{1}{2}(B_s - B_m)\tanh(x/\delta) \quad (2)$$

$$T(x) = \frac{1}{2}(T_s + T_m) + \frac{1}{2}(T_s - T_m)\tanh(x/\delta) \quad (3)$$

where δ is the half-width of the initial current sheet. The profile of B_y is then determined by $B_y(x) = [B(x)^2 - B_z(x)^2 - B_x^2]^{1/2}$. The profile of plasma pressure P is determined by the total pressure balance

$$P(x) + B(x)^2/2\mu_0 = P_m + B_m^2/2\mu_0 \quad (4)$$

In our simulations, $B_s < B_m$, $\rho_s > \rho_m$, and $T_s < T_m$ are used. A free boundary condition with $\partial/\partial x = 0$ is used for each physical quantity at $x = \pm L_x/2$, where L_x is the length of simulation domain. In our simulations, the spatial grid size used is $\Delta x = 0.2\delta$, and the number of **grid** points is 2000.

On the other hand, the open tail magnetopause also separates two uniform plasma regions: the magnetosheath and the magnetospheric lobe. The lobe magnetic field is approximately in the x direction and y axis is directed from dawn to dusk. The normal of the current sheet is in the z direction pointing northward. A large plasma flow in the tail magnetosheath ($v_x \neq 0$) is included in our simulation. The evolution of the open magnetopause current sheet in the magnetotail leads to the formation of plasma mantle. In the simulation of the mantle structure the dependent variables are functions of z and t. The time t can be related to the coordinate x by $x \simeq \bar{v}_x t$, where $\bar{v}_x$ is the effective convection velocity ($\bar{v}_x < 0$). Therefore the 1-D simulation results can be used to determine the 2-D reconnection layer in the xz plane.

The initial profiles of magnetic field, temperature and pressure are constructed in the same way as those for the dayside boundary layer. In addition, $T_s > T_m$ in the magnetotail [e.g., Hardy et al., 1975; Rosenbauer et al., 1975; Newell et al., 1991], while $T_s < T_m$ across the dayside boundary. The initial profile of the x-component of plasma flow velocity across the tail magnetopause is given by

$$v_x(z) = \frac{1}{2}(v_{xs} + v_{xm}) + \frac{1}{2}(v_{xs} - v_{xm})\tanh[z/\delta] \quad (5)$$

For simplicity we ignore the plasma flow in the lobe and set $v_{xm} = 0$.

A resistive MHD code and a hybrid code, as described by Lin and Lee [1993], have been used in our simulations. In the MHD simulations, the magnetic field is expressed in units of B_m, the plasma density in ρ_m, the temperature in T_m, the plasma thermal pressure in P_m, and the spatial coordinate in δ. The velocity is expressed by the Alfvén speed in the magnetosphere, $v_{Am} \equiv B_m/\sqrt{\mu_0\rho_m}$, the time is expressed by $t_{Am} \equiv \delta/v_{Am}$, the current density $\mathbf{J}$ is expressed in units of $J_m \equiv B_m/(\mu_0\delta)$, and the resistivity is expressed in units of $\eta_m \equiv v_{Am}\delta/\mu_0$.

3. FIELD-ALIGNED CURRENTS IN THE DAYSIDE BOUNDARY LAYER AND PLASMA MANTLE

We first show the simulation results of a case for the dayside boundary layer. In this case, $B_n = -0.25B_{tm}$, $B_{ts} = 0.9B_{tm}$, $\beta_m = 0.2$, $\beta_s = 0.46$, $\rho_s = 10\rho_m$, $T_s = 0.128T_m$, $\theta_m = 0°$, and $\theta_s = 120°$, where B_t is the magnitude of tangential magnetic field, and $\theta = \tan^{-1}(B_y/B_z)$ is the polar angle. From the MHD simulation, it is found that there exist five discontinuities and expansion waves in the resulting magnetopause boundary layer. These five discontinuities and waves from the magnetosheath side to the magnetospheric side are, in turn, a time-dependent intermediate shock (TDIS1) at the magnetopause, a slow expansion wave (SE), a contact discontinuity (CD), a slow shock (SS), and a secondary time-dependent intermediate shock (TDIS2) in the boundary layer. The upper left plot of Figure 2 shows the position of these five discontinuities and waves as a function of time t.

The right panels of Figure 2 shows spatial profiles of the tangential magnetic field components (B_y and B_z), field-aligned current density ($J_\parallel$), plasma density, temperature, and tangential components of plasma flow velocity (v_y and v_z) obtained at $t = 743 t_{Am}$ in the MHD simulation of Case 1. The discontinuity ranging from state "a" to state "b" is the time-dependent intermediate shock TDIS1. Behind TDIS1 is the slow expansion wave SE whose upstream quantities are marked by "b" and downstream by "e". The magnetic field strength increases and plasma density and temperature decrease across the slow expansion wave. The discontinuity bounding the boundary layer from the magnetospheric side is the secondary time-dependent intermediate

shock TDIS2 whose upstream quantities are marked by "i" and downstream by "h". Across the slow shock SS from state "g" to state "f", the magnetic field strength decreases and plasma density increases. The contact discontinuity CD exists at the center of the boundary layer. Through the contact discontinuity from "d" to "e", the plasma density changes while the magnetic field is conserved.

The lower left plot of Figure 2 shows the corresponding hodogram of tangential magnetic field at $t = 743 t_{Am}$. The magnetic field changes direction across the time-dependent intermediate shocks. It is seen that the tangential magnetic field rotates from point "a" to point "b" across TDIS1 by $\sim 104^\circ$, and the tangential field rotates across TDIS2 from point "i" to point "h" by $\sim 16^\circ$.

The width of time-dependent intermediate shock increases with time, while the strength decreases. As $t \to \infty$, the time-dependent intermediate shocks evolve to a structure with an infinite width, across which the magnetic field and plasma density are conserved.

In the resistive MHD model, the kinetic effects of ions in collisionless plasma are not included. In the hybrid simulations, the results shown in Figure 2 obtained from the resistive MHD model are modified by the ion kinetic effects. The time-dependent intermediate shock TDIS1 evolves to a pri-

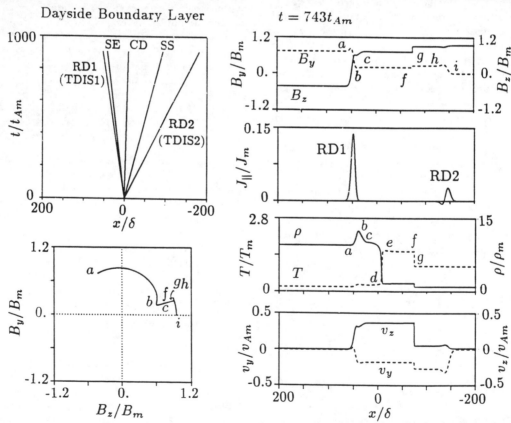

Fig. 2. MHD simulation results of the dayside boundary layer. The upper left plot shows the positions of discontinuities and expansion wave as a function of time, and the lower left plot shows the hodogram of tangential magnetic field. The right column shows the spatial profiles of magnetic field and plasma quantities at $t = 743 t_{Am}$. The field-aligned current density $J_\parallel$ is concentrated at the rotational discontinuities RD1 and RD2.

mary rotational discontinuity (RD1) at the magnetopause in a short time (< 20 ion gyroperiods, or $< 1min$). The width of this rotational discontinuity is $\sim 300km$. The time-dependent intermediate shock TDIS2 evolves to a secondary rotational discontinuity (RD2) in the boundary layer in $\sim 20sec$. The width of RD2 is $\sim 500km$. As the time-dependent intermediate shock evolves to the corresponding rotational discontinuity, the magnetic field strength downstream of the time-dependent intermediate shock increases to approximately the upstream value. In addition, due to the mixing of magnetosheath and magnetospheric plasmas along the magnetic field lines, the contact discontinuity does not exist in the boundary layer, and the slow expansion wave and slow shock are modified [Lin and Lee, 1993]. Hereafter we will call the time-dependent intermediate shock at the magnetopause RD1 and that in the boundary layer RD2.

It is seen from Figure 2 that the field-aligned current density ($J_{\parallel}$) is concentrated in the two time-dependent intermediate shocks (rotational discontinuities). In this case with $B_{ys} > 0$, the field-aligned currents are positive. The rotation angle of magnetic field across RD2 (or TDIS2) in the boundary layer is found to be $\sim 20\%$ of that in RD1 (or TDIS1). The presence of field-aligned currents in the magnetopause-boundary layer is associated with the kink of magnetic field in the two rotational discontinuities.

In the tail magnetosheath, a large anti-sunward plasma flow along the magnetopause is observed [e.g., Sanchez et al., 1990; Gosling et al., 1991]. Across the magnetopause boundary, the plasma density and temperature decrease from the magnetosheath to the lobe, while the magnetic field strength increases [e.g., Hardy et al., 1975; Rosenbauer et al., 1975]. Figure 3 shows the MHD simulation results of Case 2 for the structure of plasma mantle. In Case 2, $B_n = -0.25B_{tm}$, $B_{ts} = 0.7B_{tm}$, $\beta_m = 0.04$, $\beta_s = 1.0$, $\rho_s = 10\rho_m$, $T_s = 1.4T_m$, $\theta_m = 0^o$, and $\theta_s = 80^o$, where $\theta = tan^{-1}(B_y/B_x)$ in the magnetotail. The shear plasma flow velocity in the magnetosheath is chosen as $v_{xs} = -v_{As}$, where v_{As} is the magnitude of tangential Alfvén velocity in the magnetosheath. Similar to Case 1, a time-dependent intermediate shock (TDIS1) exists at the magnetopause and a secondary time-dependent intermediate shock (TDIS2) is present in the boundary layer. Between these two time-dependent intermediate shocks are a slow expansion wave (SE), a contact discontinuity (CD), and a slow shock (SS). It is seen from Case 1 and Case 2 that for $B_{ys} > 0$, the field-aligned currents in the tail mantle are negative, while field-aligned currents in the dayside boundary layer are positive.

In the hybrid simulation of Case 2, the steady rotational discontinuities RD1 and RD2 are also formed very quickly from the time-dependent intermediate shocks. The tangential magnetic field rotates about 70° across RD1 and 10° across RD2. The rotation angle of magnetic field across RD2 is $\sim 20\%$ of that in RD1.

As shown above, a secondary rotational discontinuity RD2

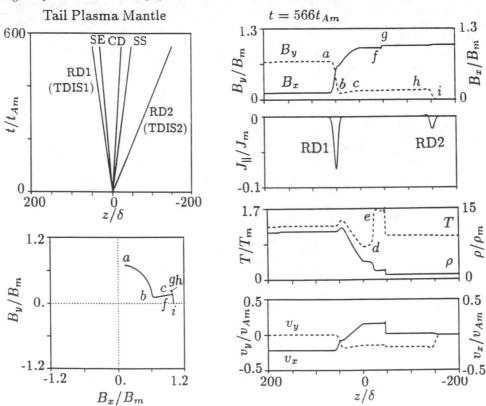

Fig. 3. Same as Figure 2 except for the nightside plasma mantle.

exists in the dayside boundary layer and nightside plasma mantle. Associated with this secondary rotational discontinuity are a kink of magnetic field lines and a field-aligned current. We suggest that the kink magnetic field associated with RD2 may propagate as an Alfvén wave along the magnetic field to the polar ionosphere, contributing to the observed field-aligned currents. The secondary rotational discontinuity (RD2) in the dayside boundary layer produces the region 1 field-aligned currents near noon, while RD2 in the tail plasma mantle generates the mantle field-aligned currents. In the following, we show that the secondary rotational discontinuity RD2 can provide the observed field-aligned currents with sufficient current density and proper flow direction.

Based on the MHD formulation, the density of field-aligned current in a rotational discontinuity in the dayside magnetopause boundary layer can be written as

$$J_{\|}(x) = \mathbf{J} \cdot \mathbf{B}/B = (B_t^2/\mu_0 B)d\theta/dx \qquad (6)$$

Let L_{RD} be the thickness of the rotational discontinuity. The field-aligned current density can be estimated as

$$J_{\|} \simeq (B_t^2/\mu_0 B)\Delta\theta/L_{RD} \qquad (7)$$

where $\Delta\theta$ is the rotation angle of tangential magnetic field and the magnitude of the tangential magnetic field B_t is a constant across the rotational discontinuity. It is seen that the current density is proportional to the total rotation angle $\Delta\theta$ for a rotational discontinuity with a constant width L_{RD}. On the other hand, for the rotational discontinuity in the plasma mantle,

$$J_{\|} \simeq -(B_t^2/\mu_0 B)\Delta\theta/L_{RD} \qquad (8)$$

Our simulations indicate that the resulting rotational discontinuities and the associated field-aligned currents are controlled by the quantity $(\theta_s - \theta_m)$, the magnetosheath to magnetospheric density ratio ρ_s/ρ_m, and the magnetic field ratio B_{ts}/B_{tm}. The quantity $(\theta_s - \theta_m)$ is the angle between the tangential magnetic fields in the magnetosheath and in the magnetosphere. Let $\Delta\theta_1$ and $\Delta\theta_2$ denote, respectively, the rotation angle of tangential magnetic field across RD1 and RD2. The top plot of Figure 4 plots the rotation angle $\Delta\theta_2$ across RD2 in the dayside boundary layer obtained from our simulations as a function of $(\theta_s - \theta_m)$. In this plot, we use the same parameters as in Case 1 except that $(\theta_s - \theta_m)$ is allowed to vary. It is seen that $|\Delta\theta_2|$ of RD2 has a maximum $\sim 17^o$ when $(\theta_s - \theta_m) \sim \pm 110^o$, and $\Delta\theta_2 = 0^o$ when $(\theta_s - \theta_m) = 0^o$ and $\pm 180^o$. The rotation angle of tangential magnetic field across RD1 (not shown) is $\Delta\theta_1 = (\theta_s - \theta_m) - \Delta\theta_2$. Note that $|\Delta\theta_1|$ has a maximum value of 180° at $(\theta_s - \theta_m) = \pm 180^o$. Also shown in the top plot of Figure 4 is the corresponding value of the field-aligned current density $J_{\|2}$ of RD2 in the dayside boundary layer, which is expressed in units of

$J_0 \equiv B_m/(\mu_0 L_{RD2})$, where L_{RD2} is the thickness of RD2. For $\Delta\theta_2 \sim 17^o$, $B \sim B_t \sim 50nT$, and $L_{RD2} \sim 500km$, we obtain $J_0 \sim 8 \times 10^{-8} A/m^2$ and $J_{\|2} \sim 2.4 \times 10^{-8} A/m^2$.

In the earlier model of Lee et al. [1985], the field-aligned currents at RD1 are responsible for the generation of the observed region 1 and mantle currents. Our simulation results show that the flow direction of field-aligned currents in the primary rotational discontinuity RD1 is the same as that associated with RD2 in the boundary layer.

We have also calculated the rotation angle $\Delta\theta_2$ in the tail mantle as a function of $(\theta_s - \theta_m)$. It is found that the results are nearly the same as those shown in the top plot of Figure 4, except that the sign of $J_{\|2}$ is reversed.

The bottom plot of Figure 4 shows the rotation angles $\Delta\theta_1$ and $\Delta\theta_2$ and the corresponding current density $J_{\|2}$ in RD2 as a function of ρ_s/ρ_m. In this calculation, we assume $(\theta_s - \theta_m) = 120^o$ and $T_s/T_m = (\rho_m/\rho_s)^{1/2}$ for the dayside magnetopause. Since the total pressure balance is used across the initial current layer, the ratio B_{ts}/B_{tm} decreases as ρ_s/ρ_m increases. It is seen from the figure that as ρ_s/ρ_m increases from 1 to 50, $\Delta\theta_2$ decreases from 60° to 4.5^o. Consequently, the quantity $J_{\|2}/J_0$ in RD2 decreases from ~ 1 to ~ 0.08, corresponding to a field-aligned current density from $\sim 8.4 \times 10^{-8} A/m^2$ to $6.7 \times 10^{-9} A/m^2$. We have also obtained the results for the tail plasma mantle. For $(\theta_s - \theta_m) = 80^o$, the obtained angle $\Delta\theta_2$ decreases from

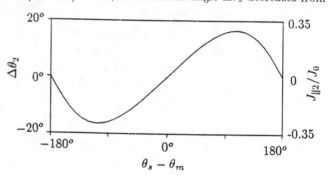

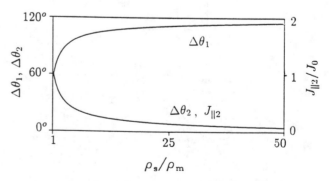

Fig. 4. The upper plot shows the rotation angle of tangential magnetic field $\Delta\theta_2$ across RD2 in the dayside boundary layer as a function of $(\theta_s - \theta_m)$, which is the angle between the tangential magnetic fields in the magnetosheath and in the magnetosphere. The bottom plot shows the rotation angles $\Delta\theta_1$ and $\Delta\theta_2$ as a function of the density ratio ρ_s/ρ_m. Also shown are the corresponding values of the current density $J_{\|2}$ in RD2.

40° to 3° and the quantity $J_{\|2}/J_0$ in RD2 decreases from 0.7 to 0.05 as ρ_s/ρ_m increases from 1 to 50.

As the secondary rotational discontinuity (RD2) in the dayside boundary layer propagates away from the reconnection region, the Maxwell stress in RD2 may be released due to the variation of magnetic field and plasma conditions along the dayside boundary, The relaxation of the kink fields results in the injection of field-aligned currents into the cusp ionosphere. For $\rho_s/\rho_m = 10$ and $0^o < |\theta_s - \theta_m| < 180^o$, the typical rotation angle of magnetic field in RD2 is $0^o < |\Delta\theta_2| < 17^o$. Taking $B_m \sim 50nT$, the current density in RD2 has the magnitude $0 < |J_{\|2}| < 2.4 \times 10^{-8} A/m^2$, with an average current density $\sim 1.4 \times 10^{-8} A/m^2$. From the conservation of field-aligned current flux, the field-aligned current density $J_{\|i}$ at the ionospheric altitude can be estimated as $J_{\|i} \sim J_{\|2}(B_i/B_m)$, where $B_i \sim 5 \times 10^{-4} nT$ is the magnetic field strength in the polar ionosphere. From the satellite and ground observations, the average current density of the region 1 field-aligned currents is $|J_{\|i}| \sim 1.3$-$3 \times 10^{-6} A/m^2$ [e.g., Iijima and Potemra, 1976; Bythrow et al., 1988], corresponding to $|J_{\|2}| \sim 1.3$-$3 \times 10^{-9} A/m^2$ in RD2. Thus the field-aligned currents in RD2 in the dayside boundary layer can account for the region 1 field-aligned currents observed in the cusp ionosphere.

Similarly, for the magnetotail with $\rho_s/\rho_m = 10$, the typical rotation angle of magnetic field in RD2 in the plasma mantle is $0^o < |\Delta\theta_2| < 12^o$. Taking $B_m \sim 30nT$, the field-aligned current density is in the range $0 < |J_{\|2}| < 1 \times 10^{-8} A/m^2$, with an average current density $\sim 5.7 \times$ $10^{-9} A/m^2$ in the plasma mantle. Our estimate indicates that the rotational discontinuity RD2 in the tail boundary layer can provide a sufficient current density for the observed mantle field-aligned currents.

We now discuss the direction of the field-aligned currents associated with RD2 in our model as a function of the sign of IMF B_y. Figure 5 shows a sketch of the global view of the field-aligned currents for IMF $B_y > 0$ and IMF $B_y < 0$ in the northern hemisphere. The view of the figure is from the dayside magnetosheath above the equatorial plane. For a positive IMF B_y, the interplanetary magnetic flux penetrates into the earth's magnetosphere mainly from the dawn side of the northern magnetosphere and leaves from the dusk side of the southern magnetosphere [e.g., Cowley, 1981; Akasofu and Roederer, 1984]. Thus the region 1 and mantle field-aligned current density in the northern hemisphere is expected to be larger in the pre-noon region than in the post-noon region. For IMF $B_y > 0$, we have $0^o < \theta_s < 180^o$. Since $\theta_m \sim 0^o$, we obtain $0^o < (\theta_s - \theta_m) < 180^o$. It is seen from Figure 4 that $J_\| > 0$ in the dayside boundary layer. Therefore our simulation results indicate that in the northern hemisphere, the field-aligned currents associated with RD2 in the dayside boundary layer flow predominantly into the pre-noon region of the polar ionosphere if IMF $B_y > 0$. On the other hand, for IMF $B_y > 0$, $J_\| < 0$ in the plasma mantle and thus the field-aligned currents from the plasma mantle flow away from the pre-noon region of the polar ionosphere.

For IMF $B_y < 0$, the field-aligned current density in the

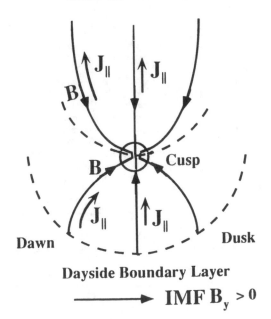

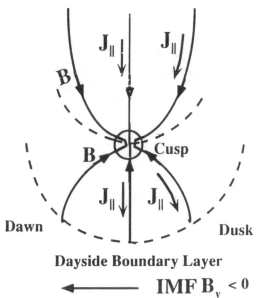

Fig. 5. Global view of the cusp field-aligned currents in the northern hemisphere for IMF $B_y > 0$ and IMF $B_y < 0$ based on our model. The view is from the dayside magnetosheath above the equatorial plane. The lower dashed line is located in the equatorial plane and the upper dashed line separates the dayside and nightside magnetic field lines.

northern hemisphere is expected to be larger in the post-noon region than in the pre-noon region. Therefore the field-aligned currents associated with RD2 in the dayside boundary layer (plasma mantle) flow away from (into) the post-noon region of the polar ionosphere.

In the southern hemisphere, the flow directions and locations of the mantle and region 1 field-aligned currents associated with RD2 are systematically reversed.

Our study indicates that the flow direction of field-aligned currents from RD2 in the dayside boundary layer (tail plasma mantle) is consistent with that of the the observed region 1 (mantle) field-aligned currents near noon. For all values of the IMF B_y, the mantle field-aligned current is expected to be distributed as that shown in Figure 1.

Note that in earlier models of Lee et al. [1985] and Siscoe and Sanchez [1987], the region 1 and mantle field-aligned currents are associated with the kink magnetic fields in the rotational discontinuity RD1 at the magnetopause. However, the rotation angle $|\Delta\theta_1|$ has a maximum value of 180° and the current intensity $|J_\parallel|$ has a maximum when $(\theta_s - \theta_m) = \pm180^o$, corresponding to a southward IMF with $B_y \sim 0$. This is not consistent with the observation of region 1 and mantle field-aligned currents, which shows a minimum intensity $|J_\parallel|$ when IMF B_y is very small [McDiarmid et al., 1978]. On the other hand, in our present RD2 model the current intensity $|J_{\parallel 2}|$ in RD2, as shown in the upper plot of Figure 4, equals zero when $(\theta_s - \theta_m) = \pm180^o$ (or IMF $B_y \sim 0$).

In summary, we have carried out a numerical simulation for the dayside and nightside magnetopause boundary layers in the open field line region. On the basis of the numerical simulations, we suggest that the propagation of the kink field associated with the secondary rotational discontinuity RD2 to the cusp ionosphere leads to the generation of the observed region 1 and mantle field-aligned currents. In our model, the rotational discontinuity RD2 in the dayside boundary layer produces the region 1 field-aligned currents near noon, whereas RD2 from the tail plasma mantle contributes to the observed mantle field-aligned currents.

Acknowledgments. This work was supported by NASA SPTP grant NAG 5-1504 and NSF grant ATM-91-11509 to the University of Alaska.

References

Akasofu, S.-I. and M. Roederer, Dependence of the polar geometry on the IMF, *Plant. Space, Sci., 32,* 111, 1984

Biernat, H. K., M. F. Heyn, R. P. Rijnbeek, V. S. Semenov, and C. J. Farrugia, The structure of reconnection layers: Application to the Earth's magnetopause, *j. Geophys. Res., 94,* 287, 1989.

Bythrow, P. F., T. A. Potemra, and R. A. Hoffman, Observations of field-aligned currents, particles, and plasma drift in the polar cusps near solstice, *J. Geophys. Res., 87,* 5131, 1982.

Bythrow, P. F., T. A. Potemra, R. E. Erlandson, and L. J. Zanetti, Birkeland currents and charged particles in the high-latitude prenoon region: A new interpretation, *J. Geophys. Res., 93,* 9791, 1988.

Coroniti, F. V. and C. F. Kennel, Magnetospheric reconnection, substorms, and energetic particle acceleration, in *Particle Acceleration in Planetary Magnetospheres,* edit by J. Arons, C. Max and C. Mckee, pp. 169-178, American Institute of Physics, New York, 1979.

Cowley, S. W. H., Magnetospheric asymmetries associated with the y-component of the IMF, *Planet. Space. Sci., 29,* 79, 1981.

Cowley, S. W. H., J. P. Morelli, and M. Lockwood, Dependence of convective flows and particle precipitation in the high-latitude dayside ionosphere on the x and y components of the interplanetary magnetic field, *J. Geophys. Res., 96,* 5557, 1991.

D'Angelo, N., Field-aligned currents and large scale magnetospheric electric fields, *Ann. Geophys., 36,* 31, 1980.

Doyle, M. A., F. J. Rich, W. J. Burke, and M. Smiddy, Field-aligned currents and electric fields observed in the region of the dayside cusp, *J. Geophys. Res., 86,* 5656, 1981.

Erlandson, R. E., L. J. Zanetti, T. A. Potemra, P. F. Bythrow, and R. Lundin, IMF B_y dependence of region 1 Birkeland currents near noon, *J. Geophys. Res., 93,* 9804, 1988.

Friis-Christensen, E., Y. Kamide, A. D. Richmond, and S. Matsushita, Interplanetary magnetic field control of high-latitude electric fields and currents determined from Greenland magnetometer data, *J. Geophys. Res., 90,* 1325, 1985.

Gosling, J. T., M. F. Thomsen, S. J. Bame, R. C. Elphic, and C. T. Russell, Observations of reconnection of interplanetary and lobe magnetic field lines at the high-latitude magnetopause, *J. Geophys. Res., 96,* 14097, 1991.

Hardy, D. A., H. K. Hills, and J. W. Freeman, A new plasma region in the distant magnetic tail, *Geophys. Res. Lett., 2,* 169, 1975.

Heyn, M. F., H. K. Biernat, R. P. Rijnbeek, and V. S. Semenov, The structure of reconnection layer, *J. Plasma Phys., 40,* 235, 1988.

Iijima, T. and T. A. Potemra, Field-aligned currents in the dayside cusp observed by Triad, *J. Geophys. Res., 81,* 5971, 1976.

Iijima, T. and T. Shibaji, Global characteristics of northward IMF-associated (NBZ) field-aligned currents, *J. Geophys. Res., 92,* 2408, 1987.

Iijima, T., R. Fujii, T. A. Potemra, and N. A. Saflekos, Field-aligned currents in the south polar cusp and their relationship to the interplanetary magnetic field, *J. Geophys. Res., 83,* 5595, 1978.

Lanzerotti, L. J., L. C. Lee, C. G. Maclennan, A. Wolfe, and L. V. Medford, Possible evidence of flux transfer events in the polar ionosphere, *Geophys. Res. Lett., 13,* 1089, 1986.

Lee, L. C. and J. R. Kan, A unified kinetic model of the tangential magnetopause structure, *J. Geophys. Res., 84,* 6417, 1979.

Lee, L. C., J. R. Kan, and S.-I. Akasofu, On the origin of the cusp field-aligned currents, *J. Geophys., 57,* 217, 1985.

Lee, L. C., Y. Shi, and L. J. Lanzerotti, A mechanism for the generation of cusp region hydromagnetic waves, *J. Geophys. Res., 93,* 7578, 1988.

Lin, Y., L. C. Lee, and C. F. Kennel, The role of intermediate shocks in magnetic reconnection, *Geophys. Res. Lett., 19,* 229, 1992.

Lin, Y. and L. C. Lee, Structure of the dayside reconnection layer in resistive MHD and hybrid models, *J. Geophys. Res., 98,* 3919, 1993.

Lundin, R., M. Yamauchi, and J. Woch, A new theory and observations of the large-scale field-aligned currents near local noon, *EOS, 73,* 263, 1992.

McDiarmid, I. B., J. R. Burrows, and M. D. Wilson, Magnetic field perturbations in the dayside cleft and their relationship to the IMF, *J. Geophys. Res., 83,* 5753, 1978.

McDiarmid, I. B., J. R. Burrows, and M. D. Wilson, Large-scale magnetic field perturbations and particle measurements at 1400 km on the dayside, *J. Geophys. Res., 84,* 1431, 1979.

Newell, P. T., W. J. Burke, C.-I. Meng, E. R. Sanchez, and M. E. Greenspan, Identification and observations of the plasma mantle at low altitude, *J. Geophys. Res.*, *96*, 35, 1991.

Paschmann, G., B. U. Ö. Sonnerup, Papamastorakis, N. Sckopke, G. Haerendel, S. J. Bame, J. R. Asbridge, J. T. Gosling, C. T. Russell, and R. C. Elphic, Plasma acceleration at the Earth's magnetopause: Evidence for reconnection, *Nature*, *282*, 243, 1979.

Primdahl, F. and F. Spangslev, Cusp region and auroral zone field-aligned currents, *Ann. Geophys.*, *37*, 529, 1981.

Rostoker, G., Magnetospheric and ionospheric currents in the polar cusp and their dependence on the B_y component of the interplanetary magnetic fields, *J. Geophys. Res.*, *85*, 4167, 1980.

Sanchez, E. R., G. L. Siscoe, J. T. Gosling, and E. W. Hones, Jr., Observations of rotational discontinuity-slow expansion fan structure of the magnetotail boundary, *J. Geophys. Res.*, *95*, 61, 1990.

Saunders, M. A., Origin of the cusp Birkeland currents, *Geophys. Res. Lett.*, *16*, 151, 1989.

Scholer, M., Asymmetric time-dependent and stationary magnetic reconnection at the dayside magnetopause, *J. Geophys. Res.*, *94*, 15099, 1989.

Shi, Y. and L. C. Lee, Structure of the reconnection layer at the dayside magnetopause, *Planet. Space Sci.*, *38*, 437, 1990.

Siscoe, G. L. and E. Sanchez, An MHD model for the complete open magnetotail boundary, *J. Geophys. Res.*, *92*, 7405, 1987.

Swift, D. W. and L. C. Lee, The magnetotail boundary and energy transfer process, *Geophys. Res. Lett.*, *9*, 527, 1982.

Wilhjelm, J., E. Friis-Christensen, and T. A. Potemra, The relationship between ionospheric and field-aligned currents in the dayside cusp, *J. Geophys. Res.*, *83*, 5586, 1978.

L. C. Lee and Y. Lin, Geophysical Institute and Department of Physics, University of Alaska, Fairbanks, AK 99775.

Control of the Generation of Field-Aligned Currents and Transverse ULF Waves by the Magnetic Helicity Input

YAN SONG, ROBERT L. LYSAK AND NAIGUO LIN

School of Physics and Astronomy
University of Minnesota

When the solar wind impinges on the magnetopause, part of the solar wind kinetic and magnetic energy is converted into the free energy stored in the shear, torsional and linked magnetic structure. The magnetic helicity inherent in the magnetic structure can be transformed into twist helicity carried by field-aligned current filaments directly through localized, time-dependent reconnection. The twisted magnetic structure gives the observational characteristics of localized and transient magnetic field structures, which have been called flux transfer events (FTEs) [Russell and Elphic, 1978] or transient magnetic field events (TMFEs) [Kawano et al., 1992]. The twisted magnetic structure radiates away as ULF waves carrying the field-aligned filaments. The occurrence rate and the polarization of the TMFEs (or FTEs) gives information on the generation of the field-aligned current and the transverse ULF waves.

The direct generation of localized field-aligned current filaments and transverse ULF waves through dynamic reconnection is called the direct dynamo effect [Song and Lysak, 1992], which is determined by the dynamic pressure of the solar wind, magnetic shear and the velocity shear between the solar wind and the magnetopause. Based on a calculation of the normalized twist helicity, the relationship between the spatial distribution of the transient magnetic field events and the magnetic local time (MLT) at the dayside magnetopause is evaluated. The dependence of the polarization of the transient magnetic field events on the IMF B_y are investigated. The results are compared with the observations of the TMFE.

INTRODUCTION

Transient magnetic field variations near the dayside magnetopause have been observed with satellites and explained by patchy reconnection in terms of flux transfer events (FTE) [e.g., Russell and Elphic, 1978], impulsive penetration [Lemaire, 1977; Heikkila, 1982; Lundin, 1988] and solar wind pressure pulses [Sibeck et al., 1989]. The magnetic field variation at the dayside magnetopause often shows a bipolar perturbation in the radial component and a one-sided pulse in the other components. When the duration of the magnetic field perturbation is longer than about one minute and the amplitude of the magnetic perturbation is larger than a few nanoTesla, such bipolar signatures have been named transient magnetic field events (TMFEs) by Kawano et al. [1992] or FTEs by Russell and Elphic [1978]. For simplicity, we will continue to use the term "TMFE" to represent the transient magnetic field perturbation at the dayside magnetopause.

The obvious dependence of the occurrence of the TMFEs with southward IMF has been shown, e.g., in statistical survey of FTEs [e.g., Berchem and Russell, 1984; Rijnbeek et al.,

Solar Wind Sources of Magnetospheric Ultra-Low-Frequency Waves
Geophysical Monograph 81

1984] and of TMFEs [Kawato et al., 1992]. The strong southward IMF dependence implies that the TMFEs and the FTEs are reconnection-related phenomena.

The observational results of the TMFEs show [Kawano et al., 1992] a reversal in the polarity around noon: the left-handed events are in the prenoon sector and the right-handed events are in the postnoon sector. The polarity is mixed near local noon. The prenoon-postnoon dependence has been shown in the statistical studies of the FTE observations [e.g., Figure 8(c), of the paper by Berchem and Russell, 1984], although the dependence was not clearly reported in the paper. The latitude dependence has also been reported by Berchem and Russell [1984] and Kawano et al. [1992]. Standard-type and reverse-type events are mainly observed in the northern and southern hemisphere, respectively. The occurrence rate of the TMFEs increases with increasing latitude.

Large-scale field-aligned current sheets, often called Birkeland currents, have been detected by satellites in the mid-70s [Zmuda and Armstrong, 1974; Iijima and Potemra, 1976]. There are three distinct regions of Birkeland current system, i.e., region 1, region 2 and the cusp currents. The dayside region 1 and the cusp current system are generated by the physical processes occurring on the dayside magnetopause and the boundary layer, such as reconnection and viscous interaction [e.g., Sato and Iijima, 1979]. In the northern hemisphere, the region 1 currents flow into the ionosphere in the morning sector and away from the ionosphere in the evening sector. For $B_y > 0$ ($B_y < 0$) the region 1 current flowing into (away from) the ionosphere has more extension across noon and its intensity is larger on the dawn side (dusk side).

We note that the polarization of ULF waves at high latitude [Samson et al., 1971], the direction of the region I current [Iijima and Potemra, 1976] and the polarity of the TMFEs are correlated. They have a consistent MLT dependence: the TMFE is left-handed (right-handed) polarized on the dawn (dusk) side, which corresponds to a flow direction towards (away from) the ionosphere of the region 1 field-aligned current in the northern hemisphere. This polarization pattern is consistent with that of the ULF waves at the magnetopause [see section 3]. The maximum values of both the region 1 current intensity and the occurrence rate of the TMFE are located between about 9:00-10:00 MLT on the morning side and 14:00-15:00 MLT on the afternoon side [e. g., Iijima and Potemra, 1978; Kawano et al., 1992].

In fact, the above three phenomena can be considered to be the result of the direct dynamo effect. The direct dynamo effect at the magnetopause has been introduced by Song and Lysak [1992]. Basically, dynamo processes may be classified as indirect and direct dynamos, which are based on purely dissipative and inductive transport processes, respectively. In the direct dynamo process, localized, time-dependent reconnection converts the helicity contained in the sheared magnetic structure into twist helicity and shows bipolar magnetic signatures at the magnetopause, which is recognized as a TMFE. TMFEs are a source of ULF waves which radiate away from the reconnection site, carrying field-aligned currents.

In the previous models, generation of field-aligned current has been simply connected with the large scale convection caused by magnetic drag [Dungey, 1961] or viscous drag [Axford and Hines 1961]. The classical reconnection model [Dungey, 1961], which is based on the traditional MHD theory, intrinsically excludes the inductive dynamo effect [Song and Lysak, 1992] and the relationship between the reconnection rate and the dynamical properties of the solar wind has not been clearly given. The relationship between reconnection and the generation of ULF waves is also unclear. Although the addition of the meso-scale anomalous viscosity caused by the Kelvin-Helmholtz instability can provide a sufficiently large viscous drag [e.g., Miura, 1984], this anomalous viscosity itself is unable to explain the strong dependence of the current generation on the IMF. On the other hand, the dynamic effect of the solar wind has been emphasized by the impulsive penetration model [Lemaire, 1977; Heikkila, 1982; Lundin, 1988] and solar wind pressure pulses [Sibeck et al., 1989], but the electromagnetic coupling aspect has not been clearly considered.

Traditionally, the field-aligned gradient of the field aligned current $J_{\parallel}$ is evaluated by $\partial J_{\parallel}/\partial s = -\nabla \cdot \mathbf{J}_{\perp}$, where $\mathbf{J}_{\perp}$ is composed of diamagnetic and polarization currents. However, the mechanism of the generation of field aligned current is not clearly given by the divergence-free condition, even in the more elaborate expressions for the field-aligned gradient of the field-

aligned current given by Hasegawa and Sato [1979] and Vasyliunas [1984]. It is unclear from $\nabla \cdot \mathbf{J} = 0$ condition whether the kinetic and magnetic energy extracted by the polarization processes is balanced by Joule heating or part of the kinetic and magnetic energy is directly converted into the magnetic energy stored in the compressed or twisted magnetic field. It is also unclear whether the generation of the field aligned current is a large-scale or a meso-scale phenomenon. The rich physical processes of field aligned current generation are simply and ambiguously hidden behind the "current divergence free" ($\nabla \cdot \mathbf{J} = 0$) statement.

A statistical description of the direct generation of the field-aligned current in a turbulent plasma is given by the MHD anomalous transport theory based on the mean-field magnetohydrodynamics [e.g., Krause and Radler, 1980; Moffatt, 1978], which describes the relationship between the average $\mathbf{E}$ field and the spectra of the energy and helicity. Ohm's law for the mean current and electric field becomes [e.g., Krause and Radler, 1980]:

$$\frac{\mathbf{J}_0}{c^2/4\pi\beta} = \mathbf{E}_0 + \frac{1}{c}\mathbf{U}_0 \times \mathbf{B}_0 + \frac{1}{c}\alpha \mathbf{B}_0. \qquad (1)$$

where magnetic and velocity fields are separated into mean and fluctuating parts with $\mathbf{U} = \mathbf{U}_0 + \delta\mathbf{u}$ and $\mathbf{B} = \mathbf{B}_0 + \delta\mathbf{B}$, and $(\alpha\mathbf{B}_0 - \beta\nabla\times\mathbf{B}_0) = <\delta\mathbf{u}\times\delta\mathbf{B}>$. The so-called α-effect gives a time-irreversible, field-aligned electric field in the case of lack of mirror symmetry of the magnetofluid. From the "circuit" viewpoint, the α-effect provides an average time-dependent impedance, which describes the direct generation of an average $J_{\parallel}$. It has been pointed out [Song and Lysak, 1992] that non-disspative reconnection can occur if the time-average impedance caused by the α and β effect breaks up the "frozen in" condition in the reconnection region. By increasing the electromagnetic energy of the transverse $\delta B_{\perp}$ and $\delta u_{\perp}$ fields, the Poynting flux ($\int \frac{B_0^2}{4\pi}\mathbf{U}_{in} \cdot d\mathbf{S}$) is basically balanced by the induced electromagnetic energy in the form of the field-aligned currents ($\int |<\frac{\alpha}{c}\mathbf{J}_0 \cdot \mathbf{B}_0>| \, dV$), where the $\mathbf{U}_{in}$ is the inflow velocity.

In the MHD meso-scale, plasma "inductors" and "capacitors" can be formed by localized and time-dependent deformation and reconnection of small flux tubes. The ability of magnetization of a plasma mainly comes from the meso-scale transverse part of the polarization currents, which are much larger than the current loops caused by the cyclotron gyration of the charged particles. Topologically, the properties of the interacting flux tubes can be described by the magnetic helicity.

The magnetic helicity, $K = \int_V \mathbf{A} \cdot \mathbf{B} \, dV$, is a measurement of the net linkage of magnetic field lines (see, for example, Moffatt [1969, 1978]; Berger and Field [1984]), which obeys a conservation law [e.g., Jensen and Chu, 1984]: $\partial/\partial t (\mathbf{A} \cdot \mathbf{B}) + c\nabla \cdot (\phi \mathbf{B} + \mathbf{E} \times \mathbf{A}) = -2c(\mathbf{E} \cdot \mathbf{B})$, where $\mathbf{E} = -\nabla\phi - (1/c)\partial\mathbf{A}/\partial t$. ϕ and $\mathbf{A}$ are the electric potential, and magnetic vector potential respectively. Gauge-invariant expressions for the magnetic helicity have been given by Berger and Field [1984] and Jensen and Chu [1984]. During magnetic reconnection, the relative magnetic helicity is basically conserved [e.g., Berger, 1982; 1984]; therefore, the twist helicity after reconnection can be determined by the helicity inherent in the shear, linked and torsional magnetic structure before reconnection.

The evaluation of twist helicity due to the shear flow has been presented in solar physics research [cf. Berger, 1991] to understand how random motions in the photosphere or convection zone of a star can generate currents and twisted flux tubes in the corona plasma. The twist helicity due to the shear magnetic fields for FTEs in the earth's dayside magnetopause has been evaluated [Song and Lysak, 1989a; Wright and Berger, 1989]. However, only a kinematic approach was used in the evaluation. It has been shown [Sonnerup, 1987] that a field-aligned current filament can be generated when an elbow-shaped magnetic flux tube continuously cuts through the magnetopause magnetic flux by magnetic drag. In this model, the dynamics due to the magnetic tension force has been emphasized; however, the drag force due to shear flow and the dynamics due to the impinging of the solar wind have not been discussed.

In this paper, a normalized twist helicity input is determined by the dynamic pressure of the solar wind and the drag forces caused by magnetic shear and velocity shear between the solar wind and the magnetopause. The evaluation is executed for the case, where the threshold conditions for the dynamic reconnection and twist helicity input are satisfied. To a first approximation, the occurrence rate of the TMFEs is proportional to the input of the twist helicity, and the sign of the helicity gives the polarization of the TMFEs. Thus, the relationship between the spatial distribution of the transient magnetic field events and the magnetic local time (MLT) at the dayside magnetopause is obtained. The dependence of the polarization of the transient magnetic field events on the IMF B_y are investigated. The results explain some of the basic character of the observations of the TMFEs and the region 1 current.

In section 3, the relationship between the normalized twist helicity input rate and the occurrence rate as well as the polarization of the TMFEs is examined. Section 4 gives a brief discussion. Modeling results which distinguish the effects between pressure perturbations and magnetic reconnection by including magnetic helicity injection are presented by Lysak et al. [1993] in this monograph.

2. DYNAMIC RECONNECTION

The relative motion of magnetofluids, such as when the solar wind impinges the earth's magnetopause or the rising magnetofluids driven by the buoyancy force in the earth's liquid core interact with the azimuthal magnetic flux, converts the kinetic and magnetic energy of the motion of the magnetofluids into the electromagnetic energy stored in the distorted magnetic structure. The helicity inherent in the sheared magnetic field and velocity field can be converted into twist helicity through localized and time-dependent reconnection. In general, the direct generation of field-aligned current and transverse Alfvén waves is the natural result of the interaction when there is a lack of mirror symmetry of the magnetofluids.

When reconnection connects different components of the magnetic field, the electromagnetic energy can be directly deposited in the twisted magnetic field carrying the field-aligned current rather than being purely dissipated [Song and Lysak, 1989b]. Unlike classical reconnection theory, where reconnection is just a dissipative process, dynamical reconnection changes a conservative system into a dissipative system and becomes an important part of the process of reorganizing the magnetofluids into new states [Song and Lysak, 1989a, b].

Figure 1 shows schematically the comparison of the reconnection in the indirect (non-inductive) dynamo case (Fig. 1a) and in the direct (inductive) dynamo case (Fig. 1b). In the former case (Fig. 1a), the Poynting flux of electromagnetic energy flowing into the reconnection region is mainly balanced by Joule heating and bulk kinetic energy of plasma flow. It is analogous with a resistive circuit with voltage V=IR, where I and R are the current and the resistance, respectively. In the latter case (Fig. 1b), twist helical waves or small flux tubes are formed, carrying the current and vorticity filaments forming meso-scale inductors and capacitors (L-C). The voltage is balanced by the product of the current I and the impedance Z

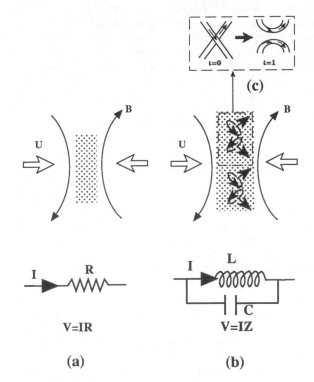

Fig. 1. Schematic diagrams showing the reconnection regions for (a) an indirect (non-inductive) dynamo case where the Ohm's law is $E \approx J/\sigma$, which is similar to a circuit with voltage V=IR, where I and R are the current and the resistance, respectively. (b) a direct (inductive) dynamo case where the E field is connected to the current filament (J) and the vorticity (ω) generation, which is similar to a inductor/capacitor (L/C) circuit with V=IZ, where Z is the impedance provided by the L/C elements. (c) Schematical illustration of physical processes forming L/C loops.

given by the L-C elements. Figure 1c gives a schematic illustration of the physical process of forming L-C loops in the direct dynamo case.

The formation and interaction of the small "inductors" and "capacitors" not only provides an enhanced impedance, but also leads to the possible generation of transversely polarized Alfvén waves and/or large scale twisted magnetic flux tubes accompanied with field-aligned current filaments ($J_\parallel$) by the dynamic reconnection process. The twist of the magnetic flux tube or polarization of the transverse Alfvén waves are determined by the helicity input, giving the TMFE signatures.

Since the formation of the L-C elements is the result of the solar-wind/magnetosphere interaction and depends on the solar wind parameters, dynamic reconnection must have a driven nature. Based on the formation, conversion and propagation of the L-C elements mentioned above, the threshold for the inductive dynamo effect and the reconnection rate in the direct dynamo case will be evaluated in a future paper. In the next section, the input of the twist helicity is evaluated for the case where the threshold conditions are satisfied.

3. The Evaluation Of The Helicity

When the solar wind impinges on the magnetopause, both the velocity shear and the magnetic shear can convert part of the solar wind kinetic and magnetic energy into the stretched and sheared magnetic structure. Specifically, when the solar wind parameters satisfy the threshold condition, the magnetic field at the magnetopause is effectively distorted by the solar wind.

Roughly speaking, the amplitude of the TMFEs is proportional to the quantity of the twist helicity. The occurrence rate of the TMFEs is, therefore, roughly proportional to the injection of the twist helicity. The sign of the injected twist helicity corresponds to the rotational polarity of the TMFEs, the polarization of ULF waves and the direction of the region 1 current.

The relation between helicity injection of different signs and the polarization of ULF waves is illustrated in Figure 2 which shows the case of the morning sector. The positive twist helicity injection corresponds to left-handed rotation ("standard") TMFEs at northern latitudes, and right-handed rotation ("reverse") TMFEs at southern latitudes [see, Kawano et al., 1992]. The positive helicity twists the magnetic flux tube so that the produced field aligned currents flow away from the equatorial region in both northern and southern hemisphere. As the structure of twisted field lines propagates towards the ionospheres, an observer on plane A in the northern hemisphere will observe the vector of the transverse variation, **b**, sweeping points 1, 2, and 3 in sequence, i. e. a left handed polarization. However, in the southern hemisphere, on plane B, the twisted structure propagates southward, thus the wave vector sweeps points 1', 2', and 3' in sequence, and also produces left-handed polarization. In the afternoon sector, the opposite helicity injection occurs and the wave polarization reverses. This polarization pattern at the magnetopause is consistent with the polarization of the ULF waves which has long been observed [see, for example, a review by Lanzerotti and Southwood, 1979].

By dynamic reconnection, the twist helicity K_{tw} can be generated by releasing the mutual helicity K_l due to magnetic field

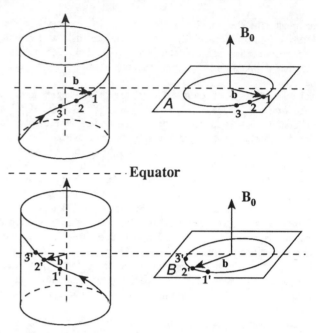

Fig. 2. Schematic diagram illustrating the relation between the helicity injection and the polarization of ULF waves (see text for detail).

linkage and by the flow of the relative helicity through the open boundary [Berger, 1984].

(i) The Generation of Twist Due to Magnetic Shear:

For simplicity, consider two flux tubes with one crossing. The mutual (linkage) helicity is $K_l = \pm \Phi^2$, where Φ is the magnetic flux carried by each of the tubes and the sign is determined by the right hand rule [see, e.g., Berger and Field, 1984]. Assume that uniformly twisted tubes are generated after reconnection. By helicity conservation, the twist helicity K_{tw} for each tube comes from the released mutual helicity K_l

$$K_{tw} = K_l = T\Phi^2 = \pm \frac{1}{2}\Phi^2, \qquad (2)$$

where T is the twist number [e.g., Berger and Field, 1984; Song and Lysak, 1989a].

(ii) The Rate of Twisting Due To Velocity Shear:

In general, the torsional magnetic structure formed by the velocity shear contains helicity. Figure 3a describes schematically how the helicity is injected into the flux tube by the drag of the open flux tube after reconnection. Figure 3b shows schematically that viscosity drags the closed flux tube and generates a helical magnetic structure; then, reconnection converts the helicity inherent in the torsional flux tube into twist helicity. For simplicity, consider only a segment of the flux tube [Figure 3c], which has three footpoints on the two planes S_1 and S_2, i.e., D_1^-, D_2^+, D_3^+, where "+" and "−" correspond to $B_z > 0$ and $B_z < 0$ respectively. The normal direction of the plane S_1 and S_2 is $z > 0$, and the twist helicity will be injected into the segments (I) and (II), which are located in the $z < 0$ region of the plane

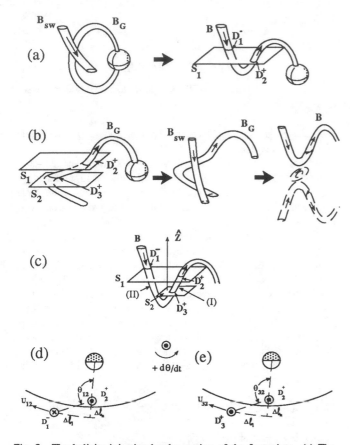

$$F_B = \frac{1}{2\pi}((\omega_1^- + \omega_2^+) - 2\frac{d\theta_{12}}{dt})\,\Phi^2 \qquad (3)$$

$$F_V = \frac{1}{2\pi}((-\omega_3^+ + \omega_2^+) - 2\frac{d\theta_{32}}{dt})\,\Phi^2 \qquad (4)$$

respectively. In equations (3) and (4), ω_i is the spin rate of the ith footpoint and $d\theta_{ij}/dt$ is the time derivative of the relative angle θ_{ij} of the ith and jth footpoints of the flux tube. The positive direction for ω_i and $d\theta_{ij}/dt$ is the right-hand helical direction relative to the normal direction of the planes S_1 and S_2 (see, Figure 3d and 3e). The $d\theta_{ij}/dt$ can be roughly estimated by $u_{ij}/\Delta l_n$ (Figure 3d and 3e), where u_{ij} is the relative velocity between the tubes i and j, and Δl_n is the distance between two tubes. $\Theta_{ij} = \theta_{ij}/2\pi$ is called the winding number [Wright and Berger, 1989]; therefore, $d\Theta_{ij}/dt \approx u_{ij}/(2\pi\Delta l_n)$. During a time period Δt, the change of the winding number $\Delta\Theta_{ij} = u_{ij}\Delta t/(2\pi\Delta l_n)$.

Considering only the relative motion between the footpoints, the helicity flow across the boundary is $F = F_B = F_V \approx 2(d\Theta/dt)\Phi^2 = (u/(\pi\Delta l_n))\Phi^2$ (see, Fig. 3c and d). If the ratio $u/\Delta l_n$ is the same for the case (a) (Figure 3a) and (b) (Figure 3b), there is no difference between the two cases. In general, the term $(u/(\pi\Delta l_n))\Phi$ can be considered as an addition of magnetic flux. For example [Sonnerup, 1987], if an elbow-shaped flux tube with velocity u cuts the magnetic flux in between the bent flux tube with width $(u/(2\pi\Delta l_n))\Phi$, the same result will be obtained. From the viewpoint of the helicity input, the three processes, i.e., (i) drag by the reconnected tube; (ii) viscous drag first, then reconnection; (iii) the reconnected tube cutting the magnetic flux; are equivalent. Since the helicity is conserved on the reconnection time scale, the twist helicity can be calculated from the initial and final magnetic structure, which is independent of the order of the physical processes.

(iii) Distortion Level:

Figure 4a shows the distortion of a flux tube by the perturbed part of the impinging solar wind. We have

$$\frac{B_\perp}{B_0} \approx \frac{\Delta y}{\Delta z} \qquad (5)$$

where B_0 is magnetic field at the magnetopause. We assume that the ratio between the perturbed part of the dynamic pressure and the average dynamic pressure is a constant, i.e., $\gamma = \delta W_{sw}/W_{sw} = \delta(1/2\rho_{sw} u_{sw}^2)/(1/2\rho_{sw} u_{sw}^2)$ and γ is a constant.

When the threshold conditions for an effective distortion are satisfied, the kinetic energy of the solar wind is nearly balanced by the magnetic energy in the distorted field. When the distortion level is high, i.e., $B_\perp/B_0$ is large, the energy balance is

$$\delta W_{swy}\Delta y_{sw}\cdot S \approx \frac{B_\perp^2}{8\pi}\Delta y \cdot S \qquad (6)$$

where S and the Δy_{sw} are the area and length of the perturbed part of the solar wind, respectively, and $S = \Delta x \Delta z$. $\delta W_{swy} = \delta(1/2\rho_{sw} u_y^2)$, where u_y is the y component of the solar wind velocity. Combining Equations (5) and (6) for fixed values

Fig. 3. The helicity injection by the motion of the footpoints. (a) The helicity is injected into a flux tube by the drag of the open flux tubes, which come from velocity shear and magnetic shear. In this case, reconnection occurs first. (b) the helicity injection is caused by dragging a closed flux tube. The drag force comes from velocity shear and/or magnetic shear. In this case, reconnection occurs after dragging. (c) A segment of the flux tube, intersected by two planes S_1 and S_2 which divide the tube into four parts. The self-rotation and circling about each other of the footpoints D_1^- and D_2^+ inject helicity into the segments (I) and (II), which corresponds to case (a). The self-rotation and circling about each other of the footpoints D_3^+ and D_2^+ inject helicity into the segment (II), which corresponds to case (b). (d) and (e) show schematically the relative circle between the footpoints. The z-direction points out of the paper. The positive direction for $d\theta_{12}/dt$ and $d\theta_{32}/dt$ is the right-handed helical direction relative to the z, which is showing in the box with the dashed line. u_{12} and u_{32} are the relative velocity in the tangential direction between D_1^- and D_2^+ and between D_3^+ and D_2^+ respectively, where $d\theta_{12}/dt \approx u_{12}/\Delta l_n$ and $d\theta_{32}/dt \approx u_{32}/\Delta l_n$.

S_1. The motion caused by magnetic or viscous drag is easily analyzed as the self-rotation of the footpoints and circling about each other of the footpoints. The input rate of the magnetic helicity can be evaluated by the helicity flux across the boundaries per unit time. For the magnetic drag case, the helicity flux F_B caused by the motion of the footpoints D_1^- and D_2^+ flows into segments I and II (Figure 3d). The helicity flux across the plane boundary for the viscous drag case F_V flows into segment II (Figure 3e). F_B and F_V can be expressed as [cf. Berger, 1984]:

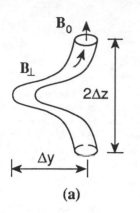

(a)

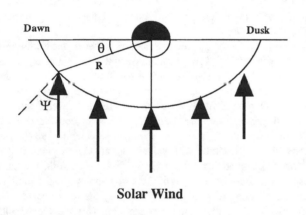

Solar Wind

(b)

Fig. 4. (a) Schematic view of the distortion of a flux tube by the perturbation of the impinging solar wind. Here $B_\perp/B_0 \approx \Delta y/\Delta z$. (b) Schematic diagram of the relationship between the angle Ψ and angle θ, where Ψ is the angle between the magnetopause normal and the direction of the solar wind velocity. R is the radial distance of the magnetopause and θ is the angle measured from dawn (6:00 MLT).

of Δz and Δy_{sw}, we have

$$\frac{\frac{1}{2}\rho_{sw} u_y{}^2}{B_0{}^2/8\pi} \propto \Delta y^3 \qquad (7)$$

If ρ_{sw} and B_0 are constants, the distortion length $\Delta y \propto u_y{}^{2/3}$.

When the distortion level is low, we have the energy equipartition of the perturbed magnetic and velocity fields, i.e., $\delta u_y \approx B_\perp/\sqrt{4\pi\rho}$. The energy balance gives

$$\delta W_{swy}\Delta y_{sw}\cdot S \sim \frac{B_\perp B_0}{4\pi}\Delta y \cdot S \qquad (8)$$

In this case, we have $\Delta y \propto u_y$.

When the solar wind impinges on the magnetopause, the solar wind velocity **u** can be decomposed into the tangential component u_l and the normal component u_n, where $u_l = u\sin\Psi$, $u_n = u\cos\Psi$, and Ψ is the angle between the magnetopause normal and the direction of the solar wind velocity [Figure 4b]. The distortion of the magnetopause in the tangential and normal

directions are

$$\Delta l \propto u_l{}^m = (u\sin\Psi)^m \qquad (9)$$

$$\Delta l_n \propto u_n{}^m = (u\cos\Psi)^m \qquad (10)$$

where m ranges from 1 to 2/3 as the distortion level increases from small to large.

(iv) The Relationship Between The Normalized Helicity and Magnetic Local Time:

We assume the occurrence rate of the TMFEs and FTEs is approximately proportional to the twist helicity input during a time period $\Delta\tau$. $\Delta\tau$ is larger than the reconnection time scale and smaller than $\Delta\tau_A$, the time scale of the twist propagation across the generator region at the Alfvén velocity.

The twist helicity caused by releasing the mutual helicity is (see, equation (2)) $K_{tw1} = \pm(1/2)\Phi^2$, where the "+" and "−" correspond to the $B_y>0$ and $B_y<0$ cases, respectively. The area of the cross section of the magnetic flux tube is $\sim\Delta l_n\Delta l_t$, where Δl_t is the longitudinal length of the cross section of the flux tube. If Δl_t does not change, we have $\Phi \propto \Delta l_n$; therefore, $\Phi^2 \propto \Delta l_n{}^2$. The drag by the solar wind dynamic pressure (magnetic or viscous) causes the change of the winding number. The magnetic flux is assumed unchanged. We have $K_{tw2} = 2\Theta_{12}\Phi^2$, where $\Theta_{12} = u_{12}\Delta t/(2\pi\Delta l_n) = \Delta l_{12}/(2\pi\Delta l_n)$. The total twist helicity is roughly

$$K_{total} = K_{tw1}+K_{tw2} = (\pm\frac{1}{2}+2\Theta_{12})\Phi^2 = (\pm\frac{1}{2}+\frac{\Delta l_{12}}{\pi\Delta l_n})\Phi^2 \quad (11)$$

Combining equations (9)-(11), we have

$$K_{total} \propto \cos^{2m}\Psi \left(\frac{\tan^m\Psi}{\pi} \pm \frac{1}{2}\right). \qquad (12)$$

Based on the calculation by Holzer and Slavin [1978], the relationship between the magnetic local time and the angle Ψ is:

$$\Psi = \text{atan}(-\frac{dy}{dx}) = \text{atan}(\frac{\cos\theta}{\varepsilon + \sin\theta}) \qquad (13)$$

where the magnetopause surface is approximated by an ellipse with the earth at one focus [Holzer and Slavin, 1978] and ε is the eccentricity of the ellipse, $\varepsilon \sim 0.4$. Magnetic local time is $t = 6 + (\theta/180)\times12$. Based on the equations (12) and (13), the relationship between the normalized twist helicity and the magnetic local time t is shown in Figure 5. The curves in the upper and lower part of Figure 5 are the normalized helicity input vs. MLT for $B_y>0$ and $B_y<0$ respectively. The solid lines are for the $m = 2/3$ case, the dashed lines correspond to $m = 1$. Only if the twist helicity input is larger than a certain value will the magnetic perturbation be observed and defined as an event, such as a TMFE. For example, we might assume that when the normalized helicity input is larger than ±0.35 in our calculation (see the two horizontal lines in the Figure 5), it would be classified as a measurable event.

The result of the calculation shows that:

(i) The helicity changes sign around noon, with the positive helicity in the dawn side and the negative helicity in the dusk

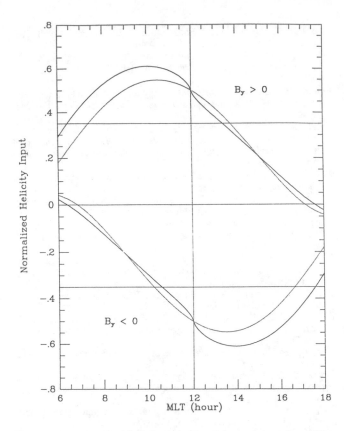

Fig. 5. The normalized helicity input by the magnetic and velocity shear versus MLT. The four curves are for $B_y>0$ (upper) and $B_y<0$ (lower), respectively. The solid lines are for the $m = 2/3$ case, the dashed lines correspond to $m = 1$. The normalized helicity input on the two horizontal lines are ± 0.35.

side, which correspond to the left-handed and right-handed rotational polarity of the TMFEs [Kawano et al., 1992] and the left-handed and right-handed polarization of the ULF waves. They also correspond to the flow direction towards and away from the ionosphere of the region 1 field-aligned current in the northern hemisphere. For $B_y>0$ ($B_y<0$) case, the positive (negative) helicity is dominant in the dawn (dusk) side.

(ii) The maximum normalized helicity input is located at 9:00 to 11:00 MLT and 13:00 to 15:00 MLT ($t_{peak} \sim 10:00$ and 14:00 MLT). This is consistent with observations of the peak time period for the occurrence rate of the TMFEs and the FTEs [see, e.g., Kawano et al., 1992]. It is also roughly consistent with observations of the MLT distribution of the region 1 current intensity [e.g., Iijima and Potemra, 1978]. The reason for this distribution can be explained intuitively by an analysis of the helicity input. The helicity K can be considered as a product of Φ_1 and Φ_2 for two winding tubes. The reconnected flux Φ_1 depends roughly on the normal component of the impinging force driven by the solar wind dynamic pressure, i.e., $\Phi_1 \propto u_n{}^m \propto (cos\,\Psi)^m$. The tangential component of the impinging force of the solar wind drags the flux tube to form a shear structure and $\Phi_2 \propto u_l{}^m \propto (sin\,\Psi)^m$. Thus, $K \propto (<u_n\,u_l>)^m \propto (cos\Psi sin\Psi)^m$, where $m \sim 1$. The term

$(cos\,\Psi sin\Psi)^m$ has a maximum value when $\Psi = \pi/4$, which corresponds to $t_{peak} \sim 9:00$ and 15:00 MLT. Since the helicity input due to the magnetic shear is proportional to the normal component of the impinging force of the solar wind, it is largest around the subsolar point. The helicity input by the magnetic shear moves the peak time towards noon.

(iii) The sign of the normalized helicity input is mixed near local noon from 10:30 to 13:30 MLT, which is consistent with the observation of TMFEs. This can be simply explained as the addition of the helicities. The helicity input caused by the velocity shear is nearly symmetric around noon, and has a positive value on the dawn side and a negative value on the dusk side. Its absolute value increases from noon to the dawn or dusk side. The helicity input caused by the magnetic shear depends on the sign of B_y, which is positive for the $B_y>0$ case, and negative for the $B_y<0$ case. The absolute value of the helicity caused by magnetic shear decreases from noon to the dawn or dusk sides. Therefore, the existence of the mixed region shows the effect of the magnetic shear. This result may also explain that for $B_y>0$ ($B_y<0$) case, the region 1 current flowing into (away from) the ionosphere has more extension across noon [Iijima and Potemra, 1978].

4. CONCLUSION AND DISCUSSION

Transverse ULF waves and field-aligned current filaments can be generated by the direct dynamo process through dynamic reconnection, giving signatures of the transient magnetic field perturbation at the magnetopause, such as the TMFEs. The direct dynamo related to the dynamic reconnection is a meso-scale physical process, which is directly driven by the solar wind.

The normalized twist helicity due to the velocity shear, magnetic shear and the impinging force of the solar wind is evaluated. The results of the calculation show good agreement with the observations of the TMFEs [Kawano et al., 1992] and the FTEs [e.g., Berchem and Russell, 1984]. The results also support the observations of the direction of the region 1 current and the polarization of the ULF wave.

In our calculation, the force to form the helical tube comes from the solar wind dynamic pressure, which is a function of the angle Ψ. A more thorough calculation including the resultant of the forces of the solar wind dynamic pressure and the magnetic tension of the elbow-shape of the flux tube will give also the latitude distribution of the TMFEs.

Since the direct dynamo effect depends on the solar-wind dynamic pressure, viscous drag and the magnetic topology of the solar-wind/magnetosphere system, the reconnection and viscous drag must have a driven nature. When the magnetopause is distorted by the impinging of the solar wind, the magnetic structures caused by dynamic reconnection can provide the viscosity [cf. Crooker, 1989]. The viscous drag of the solar wind can directly increase the twist helicity and cause the magnetic flux tubes to have further reconnection. The magnetic shear, velocity shear and the dynamic pressure effect are intrinsically connected in the formation of twisted, helical flux tubes. Thus reconnection, viscous interaction and impulsive penetration have an intrinsic connection and describe different aspects of a direct dynamo process.

Acknowledgments. The authors are grateful to M. J. Engebretson and G. Le for the helpful discussions on the magnetosheath data. We also thank two referees for their helpful comments. This work was supported by NASA grant NAGW-2690 and by NSF grant ATM-9111791 to the University of Minnesota.

REFERENCES

Axford, W. L., and C. O. Hines, A unifying theory of high latitude geophysical phenomena and geomagnetic storms, *Can. J. Phys., 39,* 1433, 1961.

Berchem, J., and C. T. Russell, Flux transfer events on the magnetopause: spatial distribution and controlling factors, *J. Geophys. Res., 89,* 6689, 1984.

Berger, M. A., Rapid reconnection and the conservation of magnetic helicity (abstract), *Bull. Am. Astro. Soc., 14,* 978, 1982.

Berger, M. A., Rigorous new limits on magnetic helicity dissipation in the solar corona, *Geophysical and Astrophysical Fluid Dynamics, 30,* 79, 1984.

Berger, M. A., Generation of coronal magnetic fields by random surface motions. I. mean square twist and current density, *Astron. and Astrophys., 252,* 369, 1991.

Berger, M. A., and G. B. Field, The topological properties of magnetic helicity, *J. Fluid Mech., 147,* 133, 1984.

Crooker, N. U., Imbedded open flux tubes and "viscous interaction" in the low latitude boundary layer, in *Physics of Magnetic Flux Ropes,* C. T. Russell, and E. R. Priest (eds.), p489, AGU monograph 58, 1989.

Dungey, J. W., Interplanetary magnetic field and the auroral zones, *Phys. Rev. Lett, 6,* 47, 1961.

Hasegawa, A., and T. Sato, Generation of field aligned current during substorm, in *Dynamics of the Magnetosphere,* S. -I. Akasofu (ed.), p529, D. Reidel, Dordrecht-Holland, 1979.

Heikkila, W. J., Impulsive plasma transport through the magnetopause, *Geophys. Res. Lett., 9,* 159, 1982.

Holzer, R.E., and J. A. Slavin, Magnetic flux transfer associated with expansions and contractions of the dayside magnetosphere, *J. Geophys. Res., 83,* 3831, 1978.

Iijima, T., and T. A. Potemra, Field-aligned currents in the dayside cusp observed by Triad, *J. Geophs. Res., 81,* 5971, 1976.

Iijima, T., and T. A. Potemra, Large-scale characteristics of field-aligned currents associated with substorms, *J. Geophys. Res., 83,* 599, 1978.

Jensen, T., and M. S. Chu, Current drive and helicity injection, *Phys. Fluids, 27,* 2881, 1984.

Kawano, H., S. Kokubun and K. Takahashi, Survey of Transient Magnetic Field Events in the Dayside Magnetosphere, *J. Geophys. Res., 97,* 10677, 1992.

Krause, F. and K. H. Radler, *Mean-field Magnetohydrodynamics and Dynamo Theory,* Oxford, Pergamon Press, 12, 1980.

Lanzerotti, L. J., and D. J. Southwood, Hydromagnetic waves, *Solar System Plasma Physics,*

eds. by C. F. Kennel, L. J. Lanzerotti, and E. N. Parker, Vol. 3, 109, North Holland, Amsterdam, 1979.

Lemaire, J., Impulsive penetration of filamentary plasma elements into the magnetospheres of the earth and Jupiter, *Planet. Space Sci., 25,* 887, 1977.

Lundin, R., On the magnetospheric boundary layer and solar wind energy transfer into the magnetosphere, *Space Sci. Rev., 48,* 263, 1988.

Lysak, R. L., Y. Song, and D. H. Lee, Generation of ULF waves by fluctuations in the magnetopause position, *Solar Wind Sources of Magnetospheric ULF Pulsations,* M. J. Engebretson (eds.), AGU Monograph xx, p. xxx, 1993.

Moffatt, H. K., The degree of knottedness of tangled vortex lines, *J. Fluid Mech., 35,* 117, 1969.

Moffatt, H. K., *Magnetic Field Generation in Electrically Conducting Fluids,* Cambridge University Press, New York, 1978.

Miura, A., Anomalous transport by magnetohydrodynamic Kelvin-Helmholtz instabilities in the solar wind-magnetosphere interaction, *J. Geophys. Res., 89,* 801, 1984.

Rijnbeek, R. P., S. W. H. Cowley, D. J. Southwood, and C. T. Russell, A survey of dayside flux transfer events observed by ISEE 1 and 2 magnetometers, *J. Geophys. Res.,89,* 786, 1984.

Russell, C. T., and R. C. Elphic, Initial ISEE magnetometer results: magnetopause observations, *Space Sci. Rev., 22,* 681, 1978.

Samson, J. C., J. A. Jacobs, and G. Rostoker, Latitude dependent characteristics of long period geomagnetic pulsations, *J. Geophys. Res., 76,* 3675, 1971.

Sato, T., and T. Iijima, Primary source of large-scale Birkeland currents, *Space Sci. Res., 24,* 347, 1979.

Sibeck, D. G., W. Baumjohann, and R. E. Lopez, Solar wind dynamic pressure variations and transient magnetospheric signatures, *Geophys. Res. Lett., 16,* 13, 1989.

Song, Y. and R. L. Lysak, Evaluation of twist helicity in FTE flux tubes, *J. Geophys. Res., 94,* 5273, 1989(a).

Song, Y. and R. L. Lysak, Dynamo effect of 3-d time-dependent reconnection in the dayside magnetopause, *Geophys. Res. Lett., 8,* 913, 1989(b).

Song, Y. and R. L. Lysak, Solar-wind/magnetospheric dynamos: MHD scale collective entry of the solar wind energy, momentum and mass into the magnetosphere, *Substorms I,* p. 149, Kiruna, Sweden, 1992.

Sonnerup, B. U. O., On the stress balance in flux transfer events, *J. Geophys. Res., 92,* 8613, 1987.

Vasyliunas, V. M., Fundamentals of current description, in *Magnetospheric Currents,* T. A. Potemra (eds.), AGU monograph 28, p63, 1984.

Wright, A. N., and M. A. Berger, The effect of reconnection upon the linkage and interior structure of the magnetic flux tubes, *J. Geophys. Res., 94,* 1295, 1989.

Zmuda, A. J. and J. C. Armstrong, The diurnal variation of the region with vector magnetic field changes associated with field-aligned currents, *J. Geophys. Res., 79,* 2501, 1974.

Yan Song, Robert L. Lysak and Naiguo Lin, School of Physics and Astronomy, University of Minnesota, 116 Church Street SE, Minneapolis, MN 55455

Ground Signatures of Travelling Convection Vortices

HERMANN LÜHR AND WIEBKE BLAWERT

Institut für Geophysik und MeteorologieTechnische Universität Braunschweig, Germany

Processes in the solar wind can excite ULF waves in the magnetosphere in many ways. The cusp/cleft is a region especially receptive to perturbations originating at the magnetopause. One particular kind of signature at high latitudes are localized ionospheric travelling convection vortices. In several cases solar wind dynamic pressure variations have been identified as the cause for this phenomenon. Early studies suggested - without being able to prove it - that the vortices are the ionospheric manifestation of oppositely directed field-aligned currents connected to a source region at the magnetopause which moves rapidly tailward. Here we provide observational evidence by combining EISCAT and magnetometer data that the magnetic deflections on the ground are generated by ionospheric Hall current vortices. The existence and the shape of the associated field--aligned currents could also be confirmed by satellite and ground-based observations. A rather detailed picture of the temporal and spatial development of a travelling convection vortex event could be drawn using a widely spread network of magnetometer stations spanning the local time sectors 05 to 15 MLT. Convection vortices show up both sides of noon. They all travel tailward. Their intensity increases during the first four minutes and then starts to fade away. There is a clear prenoon/afternoon asymmetry with the largest perturbations developing between 09 and 10 MLT. These features derived from a single case study have been compared to two statistical analyses. Both studies agree very well with our findings. Furthermore, it can be concluded that similar systems of vortices are also generated in the southern hemisphere.

INTRODUCTION

The high latitude polar ionosphere exhibits a large variety of different phenomena. The entire dayside magnetopause can be shown to map into the relatively small area of the polar cusp. Any modification of the magnetopause configuration is expected to give rise to an ionospheric response explaining the many different signatures in this region. Ground-based magnetic field observations integrate contributions of ionospheric currents from an area of more than 100 km in radius. Due to the field line mapping geometry the source regions at the magnetopause might thus be widely

Solar Wind Sources of Magnetospheric Ultra-Low-Frequency Waves
Geophysical Monograph 81

distributed or even two different phenomena may be mixed in one observation. All this has to be kept in mind when looking at magnetograms from this region.

The most direct response to changes in the solar wind is seen at the equatorward boundary of the dayside cusp around magnetic local noon. In this region irregular magnetic pulsations and transient events are often observed. Kleymenova et al. [1985] found that these pulsations with a period around 8 minutes occur for both interplanetary magnetic field (IMF) polarities, $B_z > 0$ and $B_z < 0$, and that they show a clear occurrence peak in the prenoon sector between 0600 and 1100 magnetic local time (MLT).

Possible processes at the magnetopause responsible for the magnetic field variations observed in the cusp region include flux transfer events (FTEs). Lee [1986] and Southwood et al. [1988] have proposed two different models of FTE field -aligned current systems. The earlier model consists of a central core of field-aligned current which spreads radially in the ionosphere and closes through sheaths of field-aligned currents at some distance. The second system consists of two oppositely directed field-aligned currents on the flanks of the FTE flux tube. The predicted motion of the flux tubes is perpendicular to the line connecting the vortex centers. Con2siderable effort has been made to relate ionospheric signatures to one or the other FTE models [Goertz et al., 1985; Lanzerotti et al., 1986; Elphic et al., 1990].

Another mechanism that might give rise to transient variations at cusp latitudes is a changing solar wind dynamic pressure. Sibeck et al. [1989a] and Sibeck [1990] have proposed a qualitative model in which dynamic pressure variations generate magnetic signatures near the magnetopause which resemble those reported as being typical for FTEs. Pressure pulses create a disturbance on the magnetopause that propagates tailward. Inside the magnetosphere the magnetic perturbation of the radial component becomes according to FTE terminology "standard" ("reverse") type at northern (southern) latitudes, respectively [Sibeck, 1990; Kawano et al., 1992]. Up to now it is not clear what fraction of transient events is due to solar wind pressure pulses, and it should also be remembered that the two processes, patchy reconnection and pressure pulses need not be mutually exclusive [Sibeck et al., 1989b; Elphic et al., 1990].

The ionospheric signatures of pressure pulse induced perturbations were first published by Friis-Christensen et al. [1988]. Subsequently, several authors have related transient magnetic variations at high latitudes to pressure variations in the solar wind [e.g. Sibeck et al., 1989a; Potemra et al. 1989; Lühr et al., 1990]. The purpose of this paper is to summarize observational results of transient magnetic field events which have been obtained in the vicinity of the dayside cleft. In particular, we report on travelling medium scale ionospheric current systems. Pioneering papers of this phenomenon have been published by Friis-Christensen et al. [1988] and Glaβmeier et al. [1989]. Both groups of authors employed dense arrays of ground-based magnetometers to monitor the ionospheric currents.

They report on pairs of oppositely rotating cells of current vortices showing up preferably in the prenoon hours around 09 magnetic local time (MLT). These vortices were found to move rapidly westward at speeds of several kilometers per second approximately along lines of constant L-shells around 72° invariant latitude.

Figure 1 shows a reproduction of the principal figures of the pioneering papers [Friis-Christensen et al., 1988; Glaβmeier et al. 1989]. In both cases we see a pair of oppositely rotating vortices. For the generation of these plots the total horizontal magnetic perturbation vectors have been rotated by 90°. Glaβmeier et al. used a clockwise rotation which gives the direction of equivalent ionospheric currents. Friis-Christensen et al. preferred to rotate counter-clockwise hence displaying the equivalent plasma convection pattern.

GREENLAND CHAIN MAGNETIC PERTURBATIONS
PLOTTED AS EQUIVALENT CONVECTION

28 JUNE 1986
10:06-10:21 UT

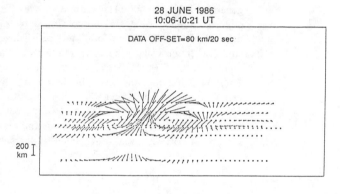

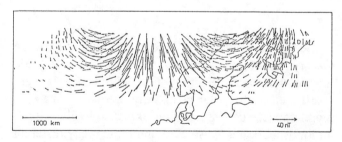

Fig. 1.   Ionospheric patterns of convection vortices deduced from magnetic field observations. The arrows point in opposite directions in the graphs. In the top panel (from Friis-Christensen et al., [1988]) the plasma convection has been plotted while in the bottom panel (from Glaβmeier et al. [1989]) equivalent currents are shown. The spatial scales in both panels are the same.

Vectors observed simultaneously along a meridional chain are plotted in a vertical column spaced according to their latitudinal separation. Readings of the next point in time are plotted in the same way but somewhat offset to the right. In case of a stationary current system moving overhead of the observer this kind of display gives a correct spatial impression of the current pattern. The velocity of the system can be estimated from the propagation delay between stations being separated in the east/west and/or north/south direction.

The two events displayed in Figure 1, although studied independently, show remarkable similarities. In both cases the vortices exhibit the same sense of rotation, and the vortex centers are observed near 73° inv. latitude overhead Greenland [Friis-Christensen et al., 1988] or north of Scandinavia, i.e. north of 67° inv. latitude [Glaßmeier et al. 1989]. The vortex centers are aligned along an east/west line and separated by about 1000 km. Both groups of authors consider a generation mechanism at the dayside magnetopause. The vortices are suggested as being driven by a pair of oppositely directed field-aligned currents connecting the source region at the magnetopause to the polar ionosphere.

In what follows we will try to find observational evidence for the suggestions and assumptions made in the pioneering papers. In the next section we investigate whether the observed magnetic field variations are consistent with a vortex pattern generated by field-aligned currents. Subsequently the special features of these field-aligned currents are studied. Furthermore, we show for a single case the temporal and spatial evolution of such vortex systems on a global scale. Features derived from this case study are compared with statistical analysis published elsewhere.

THE IONOSPHERIC VORTEX PATTERN

From Figure 1 we have seen that the magnetic perturbations observed during the kind of event studied here have been interpreted in terms of travelling vortices. In general, however, it is not possible to determine the ionospheric current configuration unambiguously from ground-based magnetometers alone. Before taking measurements from other facilities into account we will give a short theoretical description of the fields and currents involved. A rather extensive theo-

retical study of magnetic field perturbations generated by field-aligned current filaments has been given by McHenry and Clauer [1987]. We will restrict ourselves to simpler configurations. Let us assume a field-aligned current carrying a total current, I, equally distributed over the cross section, A, flowing vertically into a homogeneously conducting ionosphere. At a distance 2d apart from the downward current we place an identical upward current. Figure 2 shows the geometry used in our calculations. The electric potential distribution outside the cross section, A, is that of two opposite point charges

$$\Phi_e = C \left(\ln\sqrt{(y-d)^2 + x^2} - \ln\sqrt{(y+d)^2 + x^2} \right) \quad (1)$$

where the factor $C = I/2\pi\Sigma_p$ and Σ_p is the Pedersen conductance. Assuming a circular cross section with radius r, we get for the potential inside the area A

$$\Phi_i = C \frac{1}{2r^2}[r^2 - x^2 - (y+d)^2]$$
$$+ C \ln \frac{1}{r} \sqrt{(y-d)^2 + x^2} \quad (2a)$$

and

$$\Phi_i = -C \frac{1}{2r^2}[r^2 - x^2 - (y-d)^2]$$
$$- C \ln \frac{1}{r} \sqrt{(y+d)^2 + x^2} \quad (2b)$$

for the footprints of the downward and upward field-aligned current filament, respectively.

The electric field can be calculated as $\mathbf{E} = -\text{grad } \phi$, hence we get for the outside

$$E_y = sign \frac{I}{2\pi\Sigma_p} \left(\frac{y+d}{(y+d)^2 + x^2} - \frac{y-d}{(y-d)^2 + x^2} \right) \quad (3b)$$

where sign is 1 for y > 0 and -1 for y < 0. Inside we get

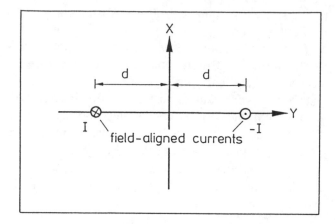

Fig. 2. Sketch of the current configuration used for the calculation of the potential distribution. Vertical currents flow into and out of the conducting ionosphere.

$$E_x = C \left(\frac{x}{r^2} - \frac{x}{(y-d)^2+x^2} \right) \qquad (4a)$$

and

$$E_x = -C \left(\frac{x}{r^2} - \frac{x}{(y+d)^2+x^2} \right) \qquad (4c)$$

$$E_y = -C \left(\frac{y-d}{r^2} - \frac{y+d}{(y+d)^2+x^2} \right) \qquad (4d)$$

for the downward and upward current, respectively.

The electric field outlined above drives both Hall an Pedersen currents in the ionosphere

$$J_H = \Sigma_H \frac{1}{B} B \times E \qquad (5a)$$

$$J_p = \Sigma_p E \qquad (5b)$$

where Σ_H and Σ_P are height-integrated Hall and Pedersen conductivities, respectively, and B is the ambient magnetic field pointing vertically downward in our case.

According to Fukushima's theorem [Fukushima, 1969] only the source-free part of the ionospheric current contributes to the magnetic variations on the ground. Since we derived our electric field from a scalar potential, curl $E = 0$ is satisfied by definition. This implies that curl $J_P = 0$ and div $J_H = 0$, hence the magnetic perturbations are generated solely by the Hall current. If we can show that the ground-based magnetic field variations are generated by Hall currents, our model configuration is a valid description of the real current configuration.

Figure 3 shows the distribution of equivalent currents derived from our simple model. For the parameters we used the following values: field-aligned current, I = 200 kA, separation of upward and downward current, 2d = 800 km and ratio between ionospheric conductances, $\Sigma_H/\Sigma_P = 2$. Strongest currents appear halfway between the field-aligned currents, and they point southward if there is a downward field-aligned current in the west and an upward in the east. Our derived current pattern is in good agreement with the observations shown in Figure 1, hence the presented model could be a suitable description of the real current configuration.

In order to check the above hypothesis we have studied in some detail a travelling convection vortices

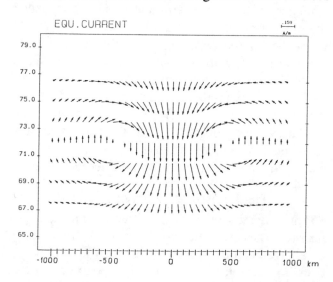

Fig. 3. Calculated distribution of equivalent currents. The centers of the field-aligned currents have been placed on both sides 400 km apart from the origin. Southward equivalent currents result from downward currents in the west and upward in the east.

event which has been observed by a variety of instruments [Lühr et al., 1993]. The observations were made during the coordinated AMPTE-EISCAT campaign in 1984. The AMPTE/IRM and UKS spacecraft were monitoring the solar wind while EISCAT was sampling the F-region of the high latitude polar ionosphere employing the special program UK-Polar. In addition to these data ground magnetic recordings have been taken into account.

Figure 4 shows a map of Scandinavia showing the magnetometer stations and displaying furthermore the two dwell positions of the Tromsö radar beam during the Polar program. In the subsequent figures readings from the western beam are marked by open circles (o) and from the eastern beam by crosses (x). The positions of the first four range gates are labled by tick marks.

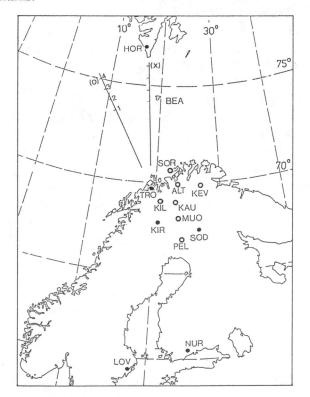

Fig. 4. Map of Scandinavia showing the locations of the various magnetometer stations. The sites of the EISCAT Magnetometer Cross are marked by open circles. In addition readings from magnetometers on Bear Island (BEA) and Hornsund (HOR) are used in this study. Also included are the two beam positions of the EISCAT program UK-POLAR and the first four range gates.

Magnetic variations recorded during this event are shown in Figure 5. Both the northward, H, and eastward, D, components along a meridional chain are plotted for half an hour centered on the event at 0635 UT, 27 October 1984. The northernmost station HOR exhibits in the H component a strong positive deflection while all stations south of it vary in antiphase. In the D component we also see a prominent positive peak but here all stations on the profile are in phase, and the H and D variations are about 90° out of phase. From the recordings of the two stations at the flanks, KIL and KEV (cf. Fig. 4) we can deduce a westward propagation of the current system amounting to $0.15°s^{-1}$. The bottom traces of Figure 5 show recordings of the Japanese station Syowa, Antarctica, which is magnetically conjugate to a place in the center of Iceland. In H the signature is about 4 min delayed. This is consistent with the derived propagation speed of $0.15°s^{-1}$ and the separation of 34° in magnetic longitude.

The event presented here shows all the features used by Glaßmeier et al. [1984] to identify their 85 events,

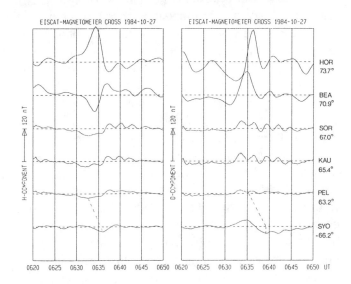

Fig. 5. Magnetic field recordings along a north/south profile and at Antarctica (SYO) taken during the passage of a TCV. In the right margin the inv. latitudes of the stations are indicated. The H component variations at HOR are in anti-phase with that of the sites in the south, in D all stations are in phase. Such a signature implies a position of the vortex centers somewhere between HOR and BEA.

namely (1) transient character, (2) H and D component in quadrature, and (3) westward propagation, hence we accept it as a travelling convection vortex (TCV) event.

Simultaneous observations obtained by EISCAT are shown in Figure 6. For the same period of half an hour line-of-sight (LOS) velocities and ion temperatures at the four gates are plotted. From the patterns of circles and crosses in these graphes one can see that data are taken over 2.5 min alternating between the two beam directions. Rather striking in this figure is the sudden and short-lived plasma flow burst at 0635 UT reaching LOS velocities up to 1500 m s^{-1}. Five minutes later there is another albeit smaller burst. Simultaneously with the peaks in velocity the ion temperature jumps up. This coincidence suggests a frictional heating of the ions which implies that the plasma is driven by an electric field rather than by neutral winds. Comparing the phase of the velocity signals it is obvious that features like zero crossings or extrema show up earlier at southern gates than at northern. The delay between adjacent gates is about 15 s. Taking into account the gate separation of 67 km and the westward propagation speed of 5 km s^{-1} at this latitude we obtain a line of constant phase running from southwest to northeast tilted at an angle of about 40° with respect to geomagnetic north. The second burst shows the same phase characteristic.

Solar wind parameters observed by AMPTE-IRM are shown in Figure 7. From top to bottom the bulk velocity, the electron density, magnetic field magnitude, azimuth and elevation in geocentral solar magnetospheric (GSM) coordinates are plotted over the same period of half an hour as before. Remarkable in this figure is the density peak at 0630 UT. The electron count rate doubles for about 1 min, goes down before it rises again. The full extent of the second peak is unknown because of a data gap at that time. Because of the expanded scale the solar wind speed looks rather variable, but it stays constant on average. The magnetic field is pointing almost radially towards the sun with a slight northward elevation (theta). All field components show fairly strong fluctuations. This indicates that IRM was engulfed by a region of up-streaming ions reflected at the bow shock. Such signatures are typical for radial field geometries. Simultaneously with the first density enhancement the field magnitude goes up to 39 nT and the elevation angle

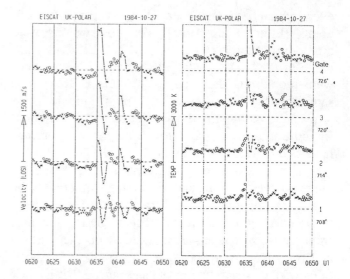

Fig. 6. Ionospheric plasma drifts and ion temperatures obtained by EISCAT in the F region. Observations from the eastern beam are marked by (x) and from the western by (o). Remarkable are the sudden increases in ion temperature and in line-of-sight (LOS) velocity at both 0635 and 0640 UT.

theta becomes larger, exceeding 30°. As a consequence IRM is for a short while disconnected from the bow shock which is indicated by the fading of the upstream waves. After the passage of the high density regions all parameters return to their initial values.

We assume as has been done before by Sibeck et al. [1989b] that the solar wind pressure pulse resulting from the enhanced density is the cause for the TCV observed on the ground. Evidence for this relation is provided (1) by the delay time of 6 min between the density peaks at IRM and the flow bursts in the ionosphere and (2) by the coincidence of the two enhancements as well in the solar wind densities as in the ionospheric flows. Etemadi et al. [1988] reported delay times between solar wind features observed by IRM and the response of the high latitude ionosphere. According to their Table 2 a delay of 6.8 min is found for a solar wind speed of 450 km s^{-1}. Magnetic field merging at the magnetopause does not seem to be important for this event, since IMF B$_z$ was clearly positive.

For a more direct comparison between the magnetic deflections and the EISCAT plasma flow observations we choose a presentation as used in Figure 1. Assuming a time stationary current pattern moving overhe-

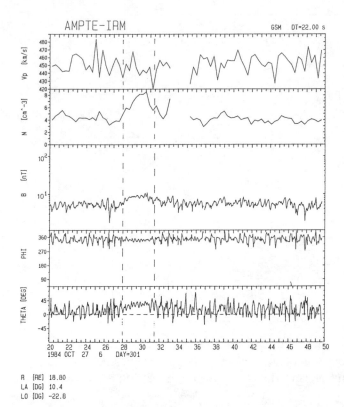

AMPTE-IRM   GSM DT=22.00 s

R   [RE] 18.80
LA [DG] 10.4
LO [DG] -22.8

Fig. 7. Solar wind data obtained by the AMPTE/IRM spacecraft for the period 0620 to 0650 UT. From top to bottom are shown: Solar wind speed and density, magnetic field magnitude, azimuth, and elevation in GSM coordinates. Of special interest for this case are the density enhancements at 0630 and 0634 UT.

ad the observer temporal variations can be interpreted as spatial gradients. Figure 8 shows the observations from 0630 to 0640 UT plotted as a snapshot of the convection vortices taken at 0637:40 UT. The map of Scandinavia has been included in the figure in order to get an idea of the spatial scale of the vortices. Equivalent currents have been deduced from the magnetic field recordings by rotating the horizontal deflection vector by 90° in a clockwise direction. This procedure gives meaningful results, if the spatial extent of the current structure is much larger than the height of the current layer. Equivalent current vectors are marked by a solid dot at their origin.

The determination of full plasma drift vectors is more difficult, since only LOS-velocities are available. In spite of this limitation certain conclusions can be drawn. At times when the radar records no LOS velocity the plasma drift must have been perpendicular to the beam direction. Such situations have been marked in Figure 8 by circles with short lines indicating the possible flow directions. We furthermore made use of the westward movement of the pattern. Each time the radar beam swings from east to west it overtakes the pattern and samples the same region another time but now at a somewhat different angle. This situation occurred twice during the displayed period, at 0633 and 0638 UT. Drift vectors obtained in this way are plotted as arrows with a tick at their origin. In a third approach we assumed that the plasma drift is nearly parallel to the radar line-of-sight whenever the maxima in velocity and ion temperature coincides. This happened shortly after 0635 UT. These vectors have been plotted without ticks at their ends.

The pattern of equivalent currents in Figure 8 deduced from magnetic field recordings gives a clear picture of a pair of vortices with opposite senses of rotation. The centers of both vortices are drifting along latitudes halfway between HOR and BEA, and they are separated by about 1000 km in east/west direction.

There is a remarkable similarity between the event described here and the two shown in Figure 1. In all three cases we observe the same sense of rotation of the two vortices, they all show a westward propagation of a few kilometers per second, they appear in the local time sector around 09 MLT, and in all cases the vortex centers are separated by about 1000 km.

Plasma drift vectors are quite sparse in Figure 8 but they are in good agreement with the magnetic field measurements. The drift arrows point in directions opposite to the currents indicating that the magnetic deflections on the ground are caused by Hall currents. The plasma motion in the F region monitored by EISCAT at about 300 km altitude is dominated by the $(\mathbf{E} \times \mathbf{B})$-drift

$$v = \frac{1}{B^2} \, \mathbf{E} \times \mathbf{B} \tag{6}$$

Combining Eqs. (5b) and (6) we get

$$v = -\frac{1}{B\Sigma_H} \, \mathbf{J}_H \tag{7}$$

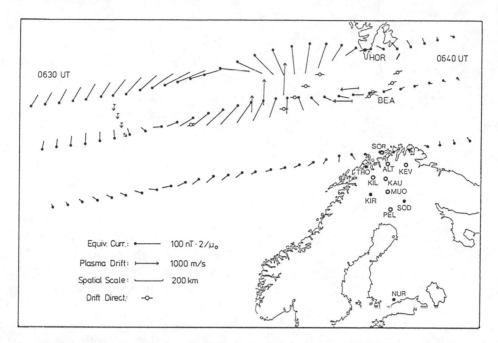

Fig. 8. Equivalent current and plasma drift pattern of the travelling vortex event on 27 Oct. 1984. Current densities have been estimated from magnetic field recordings and plasma drift velocities are obtained by EISCAT. Currents and plasma flows exhibit opposite directions.

The ionospheric Hall current is carried by electrons, therefore it is expected to be anti-parallel to the F region plasma drift. From the combination of radar and magnetometer data we have shown unambiguously that TCVs are accompanied by Hall current vortices. Precisely this kind of Hall currents was predicted by our simple model introduced above. Friis-Christensen et al. [1988] and Glaßmeier et al. [1989] assumed the same three dimensional current system without being able to prove it.

The magnetic signature at SYO, Antarctica, shows inverse variations in the D component compared to northern hemisphere stations. A similar behaviour has been reported by Lühr et al. [1990] during another event. The comparison of H and D components suggests that conjugate vortices in the two hemispheres have opposite senses of rotation. Since the geomagnetic field has almost opposite directions near the north and south pole, the opposite rotation implies according to Eq. (6) similar electric field signatures in both ionospheres. The easiest way to achieve the similarity at these widely separated places is to assume a common source near the equatorial plane on field lines connecting the conjugate locations.

THE FIELD-ALIGNED CURRENTS

Our simple current model shown in Figure 2 requires field-aligned currents flowing into and out of the ionosphere. All before mentioned authors dealing with TCVs also assumed the double vortex structure being an indication of a pair of localized antiparallel field-aligned currents connected to the ionosphere. In the following we will try to provide observational evidence for the existence of these currents and take a first glance at their characteristics.

According to Fukushima's theorem the field-aligned/Pedersen current circuit is invisible from ground magnetometers at high latitudes. At mid latitude stations, however, a magnetic effect is expected. For the event on 27 October 1984 we know the footprints of the field-aligned currents rather well (centers of vortices, 72.5° inv. lat., cf. Fig. 8). We assumed a line current flowing along a dipolar field line into the western cell and another equally strong one out of the eastern cell. Current closures are assumed in the ionosphere and in the equatorial plane. The magnetic effect on the ground can be calculated for this model by using Biot-Savart's law

$$B = \frac{\mu_o I}{4\pi} \int \frac{r \times dl}{r^3} \qquad (8)$$

where **r** is the vector pointing from the observer to the current element d**l**, and d**l** is running all around the circuit. We did these calculations for a place at 50° geomagnetic latitude. In the lower part of Figure 9 the computed magnetic effect of a pair of field-aligned currents fitting the configuration of the 27 Oct. event, i.e. intersecting the ionosphere at the centers of the vortices and drifting westward at a speed of $0.15°$ s^{-1}, is shown. The north component, B_x, exhibits a unipolar negative deflection while in the east component, B_y, we expect a bipolar negative/positive transition. The time of closest approach to the field-aligned current filaments is marked by the minimum in B_x and the zero crossing in B_y.

The upper part of Figure 9 shows the normal run magnetogram of the magnetic observatory Niemegk near Berlin at 49° inv. latitude. The traces of the X and Y component around 0635 UT closely resemble, although being small, the predicted variations. Both, the shape of the observed deflections and their duration are in good agreement with the predictions. Furthermore, the time of closest approach estimated from the Niemegk recordings is 0636 UT which fits excellently the time of passage of the vortices north of Scandinavia. Due to this convincing evidence we accept that the variations at Niemegk are caused by the field-aligned currents. By normalizing the computed amplitudes to the observations one can scale the 4total current in the circuit. The resulting value is I = 350 kA. It has to be kept in mind, however, that the horizontal components of ground-based magnetic field recordings are enhanced by the induction effect in the ground. For this location and this period range an overestimate of 10% to 15% seems reasonable, hence we may assume a total current, I = 300 kA, for the 27 October 1984 event.

The above observation provides evidence for the existence of field-aligned currents and also offers some information on the current strength but it does not say anything about the current density and the shape of the current filaments. To get these details one needs a satellite flying through the flux tubes. The only in-situ observation of this kind we know of was reported by Vogelsang et al. [1993]. These aut-

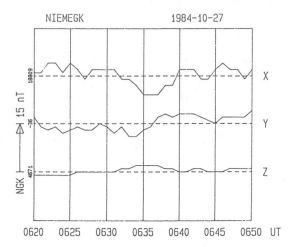

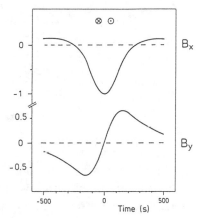

Fig. 9. Mid-latitude magnetogram from the observatory Niemegk. Below are the expected magnetic field deflections at this station caused by a pair of anti-parallel field-aligned currents drifting at a speed of 5 km s^{-1} from east to west along cusp latitudes.

hors scanned the Viking data set and found one event where they had a coincident field-aligned current crossing in space and passage of a TCV on the ground in the vicinity of the satellite footprint.

The upper part of Figure 10 shows electric and magnetic field recordings of Viking from 0610 to 0620 UT on 21 April 1986. The steep magnetic field gradients around 0613 and 0616 UT have been interpreted in terms of current tube crossings. Using Ampère's law it is possible to calculate the current density

$$curl\ B = \mu_o\ j \qquad (9)$$

Solving this for a field-aligned current filament and

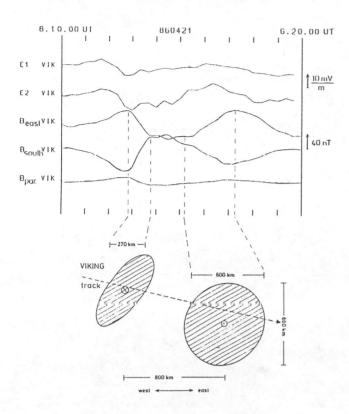

Fig. 10. Electric and magnetic field observations by Viking over a period of 10 min. Below are cross-sections of field-aligned currents deduced from the above measurements. The dimensions are scaled to ionospheric heights.

assuming a moving time stationary current system one gets:

$$j_r = \frac{1}{\mu_o \, v_\perp} \frac{\partial B_\perp}{\partial t} \qquad (10)$$

where $v_\perp$ is the component of the relative velocity between the flux tube and the spacecraft perpendicular to the field line. The spacecraft velocity can be computed from the orbital elements, and the propagation speed of the vortex structure has been estimated from ground observations to be about $0.16°s^{-1}$. At the position of Viking (3 R_E and 69° lat.) the relative velocity of the spacecraft amounts to $v_\perp = 16$ km s^{-1} pointing eastward. Mapped down to the ionosphere this corresponds to 5.3 km s^{-1}.

Vogelsang et al. [1993] reported a field-aligned current density of 1 µA m^{-2} and 0.4 µA m^{-2} for the downward and upward currents, respectively, at ionospheric heights. Beside the current density they also could make predictions on the shape of the current filaments. The east/west extent of the two current tubes can be estimated from the duration of the field gradients around 0613 and 0616 UT and from the relative velocity. Between the two field slopes there is a period of no field change. This has been interpreted as a current free gap between the filaments. From the rate of field change when approaching and when departing from the flux tubes the authors estimated the north/south extent. Furthermore they used the magnetic field direction at Viking when entering and exiting the current filaments to determine the tangent of the structure. The lower part of Figure 10 shows the shape of the current tubes resulting from the in-situ observations. The total current was reported to be I = 130 kA and found to be equally strong in both tubes. Here again we see a tilt of the major axes of about 40° running from southwest to northeast.

Figure 11 is a snapshot taken at 0614 UT. The derived field-aligned currents properly scaled have been overlayed on a map. In addition equivalent currents deduced from ground-based magnetic observations have been plotted on the map. Both spaceborn and ground-based observations fit nicely together and provide strong evidence that the field-aligned current filaments are part of the TCV identified on the ground.

The observations presented provide convincing evidence that the travelling vortices are an ionospheric manifestation of rapidly moving pairs of oppositely directed field-aligned currents, as suggested already in the pioneering papers [Friis-Christensen et al., 1988; Glaßmeier et al., 1989].

SPATIAL AND TEMPORAL DEVELOPMENT OF CONVECTION VORTICES

Up to now we have presented TCV observations which were obtained by magnetometer networks of rather limited longitudinal extent. In all cases a time stationary current system moving overhead the stations was assumed. To be able to distinguish between temporal and spatial variations it is important to have a number of meridional chains of magnetometers covering several hours in local time but still dense enough to avoid spatial aliasing.

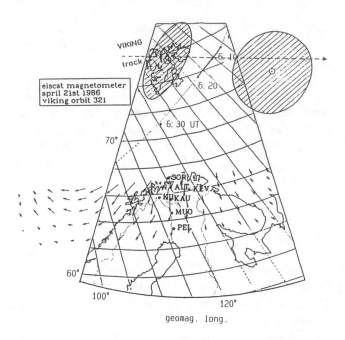

Fig. 11. A synoptic view of the field-aligned currents and ionospheric currents. The results from Viking measurements and the equivalent currents deduced from ground observations are overlaid on a map for one point in time.

For the event on 9 October 1984, 1245 UT we have been able to combine ground observations from a large number of magnetometers with data from three satellites in the magnetosphere and one in the solar wind. Figure 12 shows a map containing all the 28 ground stations included in this study. There are meridional chains over Scandinavia, east coast of Greenland including Iceland, west coast of Greenland and several pairs of stations in Canada. The magnetic local time coverage extends from 05 to 15 MLT. Magnetic local noon is indicated on the map and it runs north/south across Greenland. Footprints of the three magnetospheric satellites are marked by open circles. Their location has been mapped along field lines, using the Tsyganenko 1989 model, down to the ionosphere. The AMPTE/CCE spacecraft was located at about 9 R_E radial distance and 10° north of the magnetic equator. Both GOES 5 and GOES 6 are geostationary spacecraft and their magnetic local times run approximately 4.8 and 6.3 hours behind universal time, respectively.

As mentioned in the previous sections TCVs are believed to be triggered by solar wind pressure variations. It is therefore appropriate to start the description of observations in the solar wind and then proceed earthward. Figure 13 shows magnetic field measurements from the four satellites. AMPTE/IRM located at 12 R_E close to local noon is sampling the magnetosheath before and after the event. Probably due to a transient solar wind pressure pulse the bow shock is swept past the location of this satellite, and IRM was engulfed by the solar wind for 1 min. Solar wind parameters taken during this minute are electron density, $n_e = 2.5$ cm^{-3}, bulk velocity, $v = 720$ km s^{-1} and magnetic field in GSM, $B_x = 0.8$ nT, $B_y = -5.2$ nT, $B_z = 5.6$ nT. We note that the solar wind speed was high on that day and the IMF was predominantly perpendicular to the Earth/Sun line. Shown in Figure 13 is the GSM B_z component of the magnetic field. Within the solar wind we have a clear positive B_z, and also in the sheath B_z stays positive over the whole period displayed. Reconnection processes seem not to be important for this event.

In case of the three other satellites shown in Figure 13 the component parallel to the background field is plotted, hence we see changes in field magnitude. As a consequence of the solar wind pressure pulse we observe at all locations a magnetospheric compression, first an overshoot and than a persistant enhance-

Fig. 12. Map of all the stations used in analysing the 9 Oct. 1984 event. Footprints of the three magnetospheric satellites are marked by open circles. Magnetic noon runs north/south across Greenland .

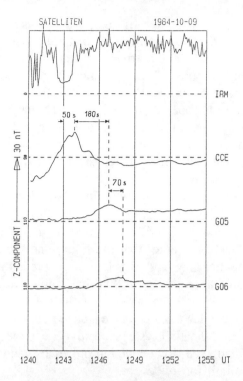

Fig. 13. Magnetic field recordings of four satellites. IRM is located in the magnetosheath except for the minute around 1243 UT when the bow shock is swept earthward across the spacecraft. The three spacecraft in the magnetosphere experience a magnetic field compression progressively delayed towards earlier local times.

ment. The peak in field strength, however, appears delayed toward earlier local times and also the amplitude becomes progressively smaller with distance from local noon. From the delay at the various satellites we can estimate a propagation speed of the disturbance. Assuming that the pressure front hits the magnetopause first at the subsolar point and then propagates tailward we obtain a velocity, $v_p = 0.28°s^{-1}$, for the segment between CCE and GOES 5 and $v_p = 0.33°s^{-1}$ from GOES 5 to GOES 6.

On the ground we also observed clear signatures of a TCV. All stations marked by a large dot in Figure 12 show magnetic variations consistent with the notion of drifting vortices. Only a few places in the very north are not effected by this event. Rather than presenting magnetograms of the many stations we show in Figure 14 equivalent current plots obtained along four meridional chains. From top to bottom observa-

tions from east Greenland, west Greenland, central Canada, and northern Scandinavia are shown. Along the east Greenland chain including Iceland we see before 1246 UT a clockwise rotating current vortex and thereafter another one with an opposite sense of rotation. After 1250 UT the vortex related perturbations have faded away. At the west Greenland chain we also have a clockwise vortex first, but the change to the anti-clockwise occurs only after 1249 UT. In central Canada there is evidence for a clockwise vortex showing up after 1247 UT.

A quite different picture emerges from the Scandinavian and Svalbard stations. Here we had to reverse the time axis to obtain a correct spatial impression of the current vortex. This is indicative of a feature travelling eastward. Since here the local time is about 1520 MLT for this event, an eastward motion is consistent with a tailward propagation of a perturbation at the magnetopause. Around 1249 UT we see a clear current vortex rotating anti-clockwise. Before 1247 UT there are some indications of a vortex with an opposite sense.

In the previous sections we interpreted these kinds of perturbations in terms of travelling vortices. The series of equivalent current plots in Figure 15 nicely confirms this view. At 1246 UT the east Greenland chain records a southward current, three minutes later it has reached the west coast of Greenland. Similarily, there are northeast currents observed at Greenland west around 1244 UT, and 3 minutes later Fort Churchill (FCC) records the onset of similar currents. These delays can be used to compute a propagation speed. For the motion from the east to the west coast of Greenland which are separated by about 40° in longitude we obtain $v_p = 0.33°s^{-1}$ and further to Fort Churchill we get $v_p = 0.38°s^{-1}$. These numbers have to be compared with the propagation speed determined in space. On average the ground observations give slightly higher results. The much better time resolution of the spacecraft data, however, favours the velocity obtained in space. Both sets of readings suggest that features speed up on their way downtail.

Another quantity that can be read from the plots in Figure 14 is the current density. The strongest currents are observed on the west coast of Greenland around 1248 UT. Equivalent current densities reach

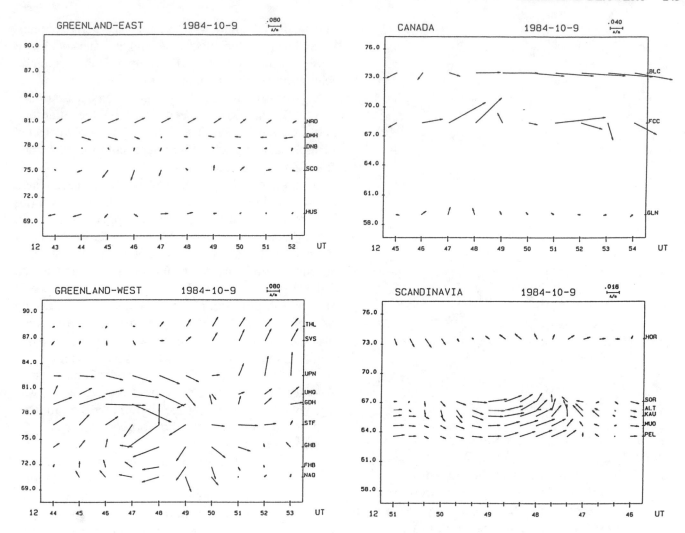

Fig. 14. Equivalent current patterns deduced from four meridional chains of magnetometers. The ordinates show invariant latitudes. Note the different scales for the current density.

values up to 0.24 Am⁻¹. At the east coast of Greenland strongest currents are observed at 1246 UT but they reach only one-third of the above value. At Fort Churchill peak values are achieved at 1248 UT amounting to 0.1 Am⁻¹ (note the change in scale). Rather weak currents of only 0.02 Am⁻¹ have been recorded over Scandinavia.

In order to see the full dynamic of the spatial and temporal development of the current vortices, it would be ideal to plot the spatial distribution over the whole area for each point in time. In Figure 15 we sketched the distribution of currents for two times. The upper part of the figure shows the situation right at the beginning of the event. At 1245 UT we have clear evidence for a clockwise current vortex over Green-

land. Such a vortex is expected to be associated with downward field-aligned currents. This picture contains only a part of the whole story, because the field-aligned return current is missing but this is the part of the circuit actually observed. We may well have missed a vortex to the west and/or to the east, since we have neither observations from Baffin Island nor from the north Atlantic. Four minutes later, shown in the lower part of Figure 15, the clockwise vortex has moved to the west and there are two additional ones with opposite sense of rotation. On the afternoon side there is another pair of vortices, that is travelling eastward.

Comparing the upper and lower parts of Figure 15 we can confirm the aforementioned drift of the vortices, away from local noon. All previous studies

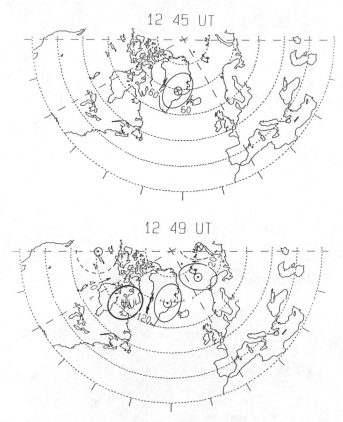

Fig. 15. Sketches of the current vortices at two points in time. During the initial phase, shown in the upper panel, only one vortex can be identified unambiguously. Four minutes later during the main phase five vortices are decernable. The numbers by the arrows denote the equivalent current density in mA m⁻¹.

using magnetometer arrays confined to a narrow local time sector have assumed, however, a time stationary current system moving overhead the observer. For the event presented here we see that there is considerable temporal change between the two parts of Figure 15 being separated by 4 min. The lower sketch coincides approximately with the main phase of the event. The initial vortex has doubled in current strength (the numbers in Fig. 15 give the mean equivalent current density in mA m⁻¹) and similarly all the other vortices are strongest at this time. About 4 min after their forming the vortices reach their climax and then start to fade away. During the event presented here the peak current density is observed west of Greenland, i.e. around 10 MLT, and it points southward. All previously presented TCV events in this paper show strong southward equivalent currents and were obser-

ved in the 09 to 10 MLT sector. The activity maximum between the clockwise and counter-clockwise vortex around 0930 MLT seems to be a common feature of TCVs. The northward current further to the west over central Canada is significantly less (J = 40 mA m⁻¹), and finally the currents on the afternoon side are even weaker. The vortices were only detected in the magnetograms after the event was identified from Greenland recordings.

In summary: The 9 Oct. 1984 is another case where a solar wind pressure pulse observed in space is followed by a TCV in the auroral ionosphere. Here again the IMF B_z was clearly positive. The perturbation travelled tailward away from noon. The propagation speeds determined in the magnetosphere and in the ionosphere are essentially the same. The ionospheric current vortices form close to local noon and develop on their way to the flanks. The strongest currents are observed 4 min after onset between 09 and 10 MLT pointing southward. By that time several vortices are observable both on the morning and afternoon side. Another 4 min later the perturbations have faded away everywhere. Multiplying the lifetime by the propagation speed suggests that the vortices travel about 6 hours in local time.

Drawing together all the observations presented we think TCVs develop a number of common features. Figure 16 shows a sketch of the distribution of vortices versus local time during their main phase. The heaviness of the ellipses is intended as a qualitative measure of the ionospheric current density. In the center of the vortices the direction of the associated field-aligned currents has been depicted. Dotted circles indicate upward currents and crossed circles downward currents. The number of vortices shown is not meant to be exhaustive. There might well be additional but weaker ones which have not been identified observationally so far.

The east/west extension of the vortices is of the order of 2 to 3 hours. On their way downtail the events seem to speed up, hence their apparent diameter increases with distance from local noon. The major axis of the vortices is tilted with respect to the magnetic meridian such that observers in the south see features earlier than those in the north. The wavelets above and below the vortices labled H and D indicate the kind of magnetic signature expected during TCV events in the different time sectors at stations north or

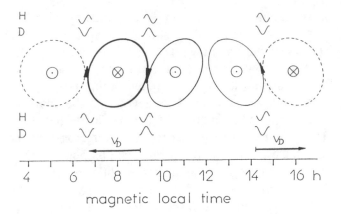

magnetic local time

Fig. 16. Sketch of the distribution of vortices during the main phase of a TCV event. The dotted and crossed circles in the centers denote currents out of and into the ionosphere, respectively. The wavelets labled H and D represent typical ground signatures north and south of the vortex centers for various local times.

south of the vortex centers. The H and D signals are in quadrature and H is in anti-phase above and below the centers. Right at the latitude of the centers there is only a variation in D. By comparing the observed signatures with the depicted wavelets it should be possible to determine the location of an individual station relative to the vortex system. In the following section we will use this sketch of vortices as a reference pattern for the assessment of observations published elsewhere.

STATISTICAL SURVEYS

The temporal and spatial development of a travelling convection vortices event outlined in the previous section has been deduced from a single case study. In order to assess the validity of the features presented here we have compared them with statistical studies. The two papers containing the most comprehensive statistics on this kind of events are Glaßmeier et al. [1989] (here after referred to as Paper G) and Lanzerotti et al. [1991] (here after referred to as Paper L).

Paper G used as mentioned before observations of a magnetometer array in Scandinavia, while Paper L employed data from two conjugate stations at about 74° inv. latitude, Iqaluit (IQ) on the south coast of Baffin Island and South Pole (SP) Station, Antarctica. Although Paper L presents a somewhat different interpretation for their transient events, we will show that

their findings nicely fit the characteristics of TCVs. Using single station observations they had no chance to make unambiguous statements on ionospheric features involving temporal and spatial variations.

The two groups of authors used somewhat different selection criteria. In Paper G events were identified by visual inspection of the magnetograms. Transient isolated variations exhibiting about 90° phase difference between H and D were selected.

For Paper L a computer-based event selection code was used. The magnetograms were searched for extrema and then a time window of 8 min centered on the extremum was taken for further tests. The deflections in the window had to exceed 40 nT in H and D and 50 nT in the Z component. Cases where the amplitude criteria were satisfied for at least two of the components were accepted. Special attention was paid to variations in the vertical, Z, component, that had to exceed the adjacent deflections by a factor of 2.

Both papers analysed the local time and seasonal dependence of their events. Figure 17 shows the occurrence rate of events versus local time detected at Scandinavia and Iqaluit according to Papers G and L, respectively. In both cases there is a clear occurrence peak in the prenoon hours around 09 MLT. At Iqaluit we see a secondary peak at 14 MLT, while there are no events reported from Scandinavia during the afternoon hours.

Figure 17 clearly demonstrates that TCVs are a daytime phenomenon, consistent with our assumption of a generation at the dayside magnetopause. Accepting solar wind/magnetopause interactions as the cause for the events would imply an equal distribution over the daytime hours. To explain the modulation of occurrence rates and the difference between Scandinavia and Iqaluit we have to recall the selection criteria. Paper L used an amplitude criterion, and Paper G also favoured large events, since small ones are hard to distinguish from other variations. In this context Figure 17 provides a local time distribution of deflection amplitudes. Such an interpretation is consistent with our sketch in Figure 16. Largest amplitudes are predicted in the vicinity of the middle vortex on the morning flank. This is precisely the time sector when most of the events have been detected. The considerably lower detection frequency during the afternoon hours also confirms the finding of the prenoon/after-

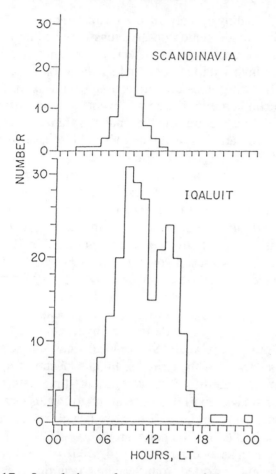

Fig. 17. Local time of occurrence of the magnetic impulse events in Scandinavia as reported by Paper G and at Iqaluit, Baffin Island by Paper L (from Lanzerotti et al. [1991]).

noon asymmetry, and the location of the afternoon occurrence peak coincides with the local time sector of the stronger afternoon vortex. The complete lack of afternoon events in the Scandinavian sample is probably due to the distance from the vortex centers, about 5° in latitude.

Concerning the seasonal distributions, Papers G and L reach at somewhat different conclusions, but looking at their graphs a preference for occurrence during the months around equinox seems to emerge. This could be related to the tilt of the solar rotation axis. Around the equinoxes the Earth/Sun line reaches highest solar latitudes. Another possibility would be the tilt between GSM z and GSE z directions which also maximizes at equinox. Russell and McPherron [1973] convincingly showed that the semiannual peak

in geomagnetic activity can be explained by it. To test this hypothesis the data should be sorted by IMF sectors (toward and away). If there is a preference for TCV to occur in one sector during spring, the opposite one is expected to be favoured during fall.

A good part of Paper L is devoted to the comparison of ground signatures at the conjugate locations. Their Figure 9 shows that for a large majority of the conjugate events the northward, H, component varies in phase in the two hemispheres while the eastward, D, component varies in antiphase. These phase relations suggest that corresponding vortices in the two hemispheres have opposite senses of rotation and have, because of the opposite direction of the ambient magnetic field, similar electric field patterns at conjugate locations.

The interpretation of the vertical component recorded by ground-based magnetometers is generally difficult, because of the induction effects in the ground. We do not want to pay too much attention to this component but there are some qualitative results which seem to be noteworthy. Figure 18 shows the local time dependence of positive or negative deflections in the vertical, V, component separately for the two stations IQ and SP. There are distinct distributions for positive and negative polarities. A positive V component corresponds to a downward deflection in the north and an upward in the southern hemisphere. At IQ we hence find downward deflections during the morning hours and upward in the afternoon. At SP we have opposite deflections. The signatures seen in the northern hemisphere component may well have been produced by clockwise current vortices in the morning and counterclockwise currents during the afternoon hours. In the southern hemisphere an opposite sense of rotation is required. Using the same arguments as before that due to the amplitude criterion the occurrence statistics can be reinterpreted in terms of an amplitude distribution versus local time, Figure 18 implies that there is an absolutely dominating clockwise current vortex on the morningside around 09 MLT. The dominating vortex in the afternoon peaks around 14 MLT and exhibits a counter-clockwise sense. These findings are fully consistent with our proposed vortex pattern in Figure 16 and provide additional evidence for the prenoon/afternoon assymetry. The deflections at South Pole confirm the above

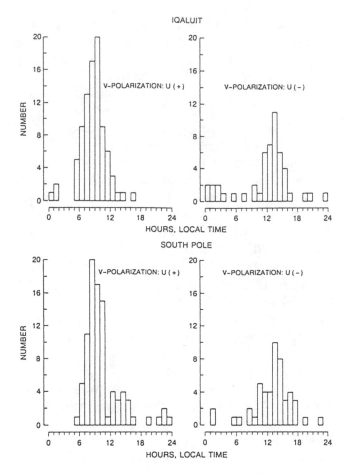

Fig. 18. Magnetic impulse events sorted according to station and deflection in the vertical, V, component. A positive V points downward in the northern hemisphere and upward in the southern (from Lanzerotti et al. [1991]).

statement that conjugate vortices show in principle the same behaviour except that they exhibit opposite senses of rotation.

A final question to be discussed here is the latitude at which the centers of the vortices typically move. From Figure 16 we can read that the magnetic signatures are different for stations being located north or south of the vortex centers. The H component changes phase when crossing the central latitude but D stays in phase all along a meridional profile. According to the model calculations shown in Figure 3 strongest deflections in D are expected halfway between the vortex centers. Figure 19 shows the type of variations as reported by Paper L for the H component at South Pole station. During the morning hours around 09

MLT a plus/minus transition clearly dominates while in the afternoon around 14 MLT the opposite polarity change prevails. Since the H deflections are the same in both hemispheres, as mentioned before, we may compare the observations at SP with our sketch in Figure 16 valid for the northern hemisphere. A plus/-minus change in H around 09 MLT is predicted for a station poleward of the centers. On the otherhand a switch from minus to plus in the afternoon suggests a location equatorward of the centers. This result implies that the morning side vortices occur on lower latitudes than the afternoon ones.

Paper L contains further data suitable to test this hypothesis, namely the polarization ellipses of the horizontal components. According to their Figure 12 (not shown here) at IQ the major axis of the ellipse points towards north around 08 MLT, rotates gradually towards west and points almost due west at 14 MLT. Comparison with our Figure 16 shows that a strong positive H deflection at 08 MLT requires a station well north of the vortex centers. As local time progresses the centers the vortices are located further poleward causing an increase of D and a decrease of H deflection. At 14 MLT IQ seems to be on the same latitude as the vortex centers. Here the H deflection vanishes and only the negative D prevails. The observed variation of the major axis versus local time obviously is another confirmation of our previous finding that the vortices in the afternoon are located at higher latitudes. This could be another reason why no afternoon events have been identified in Scandinavia

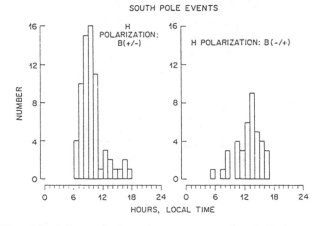

Fig. 19. Magnetic impulse events at South Pole sorted according to type of polarization in the H component (from Lanzerotti et al. [1991]).

by Paper G. The invariant latitudes of Iqaluit and South Pole have been quoted by Paper L as 74°N and 74.8°S, repectively.

All the statistical results presented here are in agreement with the pattern of vortices sketched in Figure 16 and some of the analyses adds further information to it. The sequence of vortices shown there seems to be a reasonable description of a TCV event during its main phase.

CONCLUSIONS AND SUMMARY

In this paper we tried to pull together as many published observations on travelling convection vortices as possible. The result of this effort is a rather comprehensive picture of the ionospheric current circuits associated with TCVs. Many of the features of these events have been predicted already by the pioneering papers. Both, Friis-Christensen et al. [1988] and Glaßmeier et al. [1989] interpreted the observed ground signatures as being generated by westward drifting Hall current vortices. Only by combining EISCAT measurements and magnetic field recordings were Lühr et al. [1993] able to show unambiguously the Hall character of the equivalent currents.

The two early papers also interpreted their transient variation in the ground-based magnetic recordings as the ionospheric manifestation of a pair of oppositely directed field-aligned currents moving rapidly to the west. Observational evidence for the existence of field-aligned current filaments in the centers of the vortices has been given by Vogelsang et al. [1993] and Lühr et al. [1993]. Although employing completely different kinds of data both groups of authors came up with similar gross features of the current filaments. Total field-aligned currents have been found to amount to a few hundred Ampères and the separation of the current centers was about 800 km in both cases. Field-aligned current densities determined in-situ were of the order of 1 µA m^{-2}. The same magnitude was already estimated by Glaßmeier and Heppner [1992] from magnetic field recording on the ground.

Knowing the field-aligned current and the ionospheric Pedersen conductivity one can calculate the electric power converted to heat in the ionosphere. For the 27 October 1984 event we estimated a total field-aligned current of 300 kA. Employing the potential difference of 25 kV, as emerging from a model by Lühr et al. [1993] for this event, we obtain a power input of 7.5· 10^9 W into the ionosphere which is a remarkable amount for a quiet time event. The density of Joule heating can be computed as h = $\Sigma_p E^2$. Using numbers obtained by the above mentioned model calculation (Σ_p = 4 S and E = 57 mV m^{-1}) we get h = 13 mW m^{-2} in the center of the event.

Really new concepts concerning the evolution of TCVs have evolved from simultaneous observations covering almost the whole dayside oval. Important for this kind of observation is the existence of a number of meridional magnetometer chains separated by not more than two hours in local time to avoid spatial aliasing. With such a net of observing stations we unambiguously could determine the propagation of the perturbations from noon to the night side over both flanks. The speed determined by spacecraft in the magnetosphere agrees rather well with the estimates obtained on the ground. Features seem to speed up on their way down tail. Typical propagation speeds are 0.1° to 0.3°s^{-1} amounting to 3 to 10 km s^{-1} in the ionosphere or 130 to 400 km s^{-1} at the magnetopause. There is a clear correlation between solar wind velocity and propagation speed.

The perturbation amplitude increases during the first four minutes or so on its way to the flanks. That is why strongest events are generally observed 2 to 3 hours away from noon. After about four more minutes the perturbations have faded away. At the time of peak intensity at least five vortices can be identified.

The diameter of a single vortex is limited to 2 to 3 hours in local time. Multiplying the life time of 8 min by an average propagation speed of 0.15° s^{-1} a vortex will travel twice its diameter. Using a single meridional chain of stations it is therefore not possible to observe more than two vortices. Which of the pairs will be detected depends on the local time of the chain. The coincidence of propagation distance and diameter is the reason why in some publications the term twin vortex events is used.

The scale size of the disturbance at the magnetopause can be estimated by mapping the ionospheric dimensions along the field lines into the equatorial plane. A segment of 2 hours in local time has a length of about 6 R$_E$ at the magnetopause. This size is much greater than the bulge on the magnetopause generated by flux transfer events. In a statistical study Rijnbeek et al. [1984] found a diameter of the flux tube of the order of 1 R_E on average.

Convection vortices in the northern ionosphere have mirror images in the south. This feature already suggested by Lühr et al. [1990; 1993] has been confirmed statistically by Lanzerotti et al. [1991]. Using data only from a single station in each hemisphere a large fraction of the events passed the criterion for conjugacy. All the statistical features derived independently at the two widely separated locations are so similar that there is no doubt of close communication between the two sites. If one had a more extended array of magnetometers at one of the common footpoints, one could allow for a certain change of the field line mapping geometry. Such a configuration would probably identify a much larger fraction of conjugate events, since the field models used to determine conjugate sites reflect the situation for quiet conditions, but TCVs are dynamic phenomena.

In many of the published case studies a solar wind pressure pulse of a few minute duration could be identified as the cause for the TCV event. Even after one of the AMPTE Lithium releases in the solar wind a TCV was observed. An addmixture of only 40 g ionized Li^+ vapor to the flowing solar wind was obviously sufficient to cause magnetic deflections on the ground up to 300 nT which are among the largest ever reported in connection with TCVs [Lühr et al., 1990]. Despite the clear relation between pressure pulses and TCVs we do not want to exclude other excitation mechanisms, e.g. rapid changes of the IMF.

All case studies presented here which include solar wind data exhibited a positive IMF B_z component during the event. Reconnection seems therefore not to be important for these cases. Going, however, through unpublished events and taking the recent statistical study by Konik et al. [1993] into account there are as many events during times of negative B_z. In our oppinion the IMF B_z component is not a controlling factor for the occurrence of TCVs. For this and other reasons we do not think that classical FTEs [e.g. Rijnbeek et al., 1984] can be the cause for TCVs. The scale size of an FTE would be about 200 km in the ionosphere. It is hard to imagine that current systems of this size moving through the ionosphere at about 10 km s^{-1} produce much of a magnetic signal on the ground. The kind of current circuit in the ionosphere discussed here differs considerably from the "ionospheric signatures of FTEs" as proposed by, e.g. Lee [1986] and Southwood et al. [1988]. According to the model of Lee [1986] a single vortex of Hall currents is expected. Southwood et al. propose a double vortex system in the ionosphere, but the expected propagation is perpendicular to the line connecting the centers. All events studied here exhibited propagations along the line of centers.

In this article almost nothing has been said about processes near the magnetopause, the probable source region of TCVs. The high magnetic latitude (73° inv. lat. on average) for example is much in favour with this idea. Furthermore, the apparently high degree of conjugacy of TCV events argues for a common source for both hemispheres close to the equatorial plane. Publication of in-situ observations in the outer magnetosphere related to TCVs are still rare. Only recently Kawano et al. [1992] presented a rather comprehensive survey of transient events in the AMPTE/CCE data set from the dayside. Their analysis revealed a number of interesting features of the disturbances at the magnetopause, but the magnetospheric current system they deduce from their observations does not fit the field-aligned current pattern seen in the ionosphere. It has to be stressed here, however, that current estimates from single spacecraft observations are always ambiguous.

There have been several attemtps to model the response of the magnetospheric/ionospheric system to solar wind pressure pulses impinging on the magnetopause (e.g. Kivelson and Southwood [1991]; Lysak and Lee [1992]; Glaßmeier [1992]). Although these models differ in some respect they all predict two pairs of vortices, one on either side of noon. The expected sense of rotation in the morning sector is the same as that of the pair between 05 and 08 MLT in Figure 16 and for the afternoon sector it is identical to the pair observed between 13 and 16 MLT. However, the strongest signals around 09 MLT which are most obvious in all studies dealing with TCVs are not covered by these models. Also the apparent asymmetry of the deflection amplitude between prenoon and afternoon cannot be explained by them. We may speculate here that the cause for a TCV event is for example a solar wind pressure pulse but in addition there seems to be a selective magnetospheric process which is the cause of the strong vortices around 09 MLT and for the prenoon/afternoon asymmetry.

In summary we may state: the ionospheric features of travelling convection vortices are well documented by now, but the processes taking place at the magnetopause and in the magnetosphere are far less understood. It really needs more studies and more observations in the magnetosphere, preferably from multi-spacecraft missions, to shed further light on the generation processes of TCVs.

Acknowledgements. The authors are indepted to the many researchers who generously supplied us with data for the 9 Oct. 1984 event. In particular we would like to thank E. Friis-Christensen from the Danish Meteorological Instiute for the Greenland data, G.J. van Beek from the Geological Survey of Canada for the Canadian recordings, N. Sato from the Institute of Polar Research, Tokyo for the data from Husafell, T. Araki from WDC C2 for Geomagnetism, Kyoto for the data from Syowa, J. Jankowski from the Polish Academy of Science for the data from Hornsund, T.A. Potemra from the Johns Hopkins University/APL for the AMPTE/CCE data and K. Takahashi JHU/APL for supplying GOES 5 and 6 data. We are grateful to K.-H. Glaßmeier and P. Weidelt for their fruitful discussions during the evaluation of the paper. One of us (WB) was supported by grants from the Deutsche Forschungsgemeinschaft.

REFERENCES

Elphic, R.C., M. Lockwood, S.W.H. Cowley, and P.E. Sandholt, Flux transfer events at the magnetopause and in the ionosphere, *Geophys. Res .Lett., 17,* 2241, 1990.

Etemadi, A., S.W.H. Cowley, M. Lockwood, B.J.I. Bromage, D.M. Willis and H. Lühr, The dependence of highlatitude dayside ionospheric flows on the northsouth component of the IMF: A high time resolution correlation analysis using EISCA "Polar" and AMPTE-UKS and IRM data, *Planet. Space Sci., 36,* 471, 1988.

Friis-Christensen, E., M.A. McHenry, C.R. Clauer, and S. Vennerstrom, Ionospheric travelling convection vortices observed near the polar cleft: A triggered response to sudden changes in the solar wind, *Geophys. Res. Lett., 15,* 253, 1988.

Fukushima, N., Equivalence in ground geomagnetic effect of Chapman-Vestine's and Birkeland-Alfvén's electric current-systems for polar magnetic storms, *Rep. Ionos. Space Res., 90,* 4069, 1965.

Glaßmeier, K.-H., Travelling magnetospheric convection twin-vortices: Observations and theory, *Ann. Geophys., 10,* 547, 1992.

Glaßmeier, K.-H., C. Heppner, Travelling magnetospheric convection twin-vortices: another case study, global characteristics, and a model, *J. Geophys. Res., 97,* 3977, 1992.

Glaßmeier, K.-H., M. Hönisch, and J. Untiedt, Groundbased and

satellite observations of travelling magnetospheric convection twin-vortices, *J. Geophys. Res., 94,* 2520, 1989.

Goertz, C.K., E. Nielsen, A. Korth, K.-H. Glaßmeier, C. Haldoupis, P. Hoeg, and D. Hayward, Observations of a possible ground signature of flux transfer events, *J. Geophys. Res., 90,* 4069, 1985.

Kawano, H., S. Kokubun, and K. Takahashi, Survey of transient magnetic field events in the dayside magnetosphere, *J. Geophys. Res., 97,* 10677, 1992.

Kivelson, M.G. and D.J. Southwood, Ionospheric travelling vortex generation by solar wind buffeting of the magnetosphere, *J. Geophys. Res., 96,* 1661, 1991.

Kleymenova, N.G., O.V. Bolshakova, V.A. Troitskaya, and E. Friis-Christensen, Two forms of long period geomagnetic pulsations near the equatorial of the dayside polar cusp, *Geomagnetism and Aeronomy, 25,* 139, 1985.

Konik, R.M., L.J. Lanzerotti, A. Wolfe, D. Venkatesan, and C.G. Maclennan, Cusp-latitude magnetic ionosphere events, 2.Interplanetary magnetic field and solar wind conditions, J. Geophys. Res., (submitted) 1993.

Lanzerotti, L.J., L.C. Lee, C.G. Maclennan, A. Wolfe, and L.V. Medford, Possible evidence of flux transfer events in the polar ionosphere, *Geophys, Res. Lett., 13,* 1089, 1986.

Lanzerotti, L.J., R.M. Konik, A. Wolfe, D. Venkatesan, and C.G. Maclennan, Cusp-latitude magnetic impulse events, 1. Occurrence statistics, *J. Geophys. Res., 96,* 14009, 1991.

Lee, L.C., Magnetic flux transfer at the Earth's magnetopause, in *Solar Wind-Magnetosphere Coupling,* Eds. Y. Kamide and J. Slavin, *Terra, Tokyo,* 297, 1986.

Lühr, H., W. Baumjohann, and T.A. Potemra, The AMPTE Lithium releases in the solar wind: A possible trigger for geomagnetic pulsations, *Geophys. Res. Lett. 17,* 2301, 1990.

Lühr, H., W. Blawert, and H. Todd, The ionospheric plasma flow and current patterns of travelling convection vortices: A case study, J. atmos. terr. Phys. (in press) 1993.

Lysak, R.L. and D.-H. Lee, Response of the dipole magnetosphere to pressure pulses, *Geophys. Res. Lett., 19,* 937, 1992.

McHenry, M.A. and C.R. Clauer, Modeled ground magnetic signatures of flux transfer events, *J. Geophys. Res., 92,* 11231, 1987.

Potemra, T.A., H. Lühr, L.J. Zanetti, K. Takahashi, R.E. Erland son, G.T. Marklund, L.P. Block, L.G. Blomberg, and R.L. Lepping, Multi-satellite and ground-based observations of transient ULF waves, *J. Geophys. Res., 94,* 2543, 1989.

Rijnbeek, R.P., S.W.H. Cowley, D.J. Southwood, and C.T. Rus sell, A survey of dayside flux transfer events observed by ISEE 1 and 2 magnetometers, *J. Geophys. Res., 89,* 786, 1984.

Russell, C.T. and R.L. McPherron, Semiannual variation of geo magnetic activity, *J. Geophys. Res., 78,* 92, 1973.

Sibeck, D.G., A model for the transient magnetospheric response to sudden solar wind dynamic pressure variations, *J. Geophys. Res., 95,* 3755, 1990.

Sibeck, D.G., W. Baumjohann, R.C. Elphic, D.H. Fairfield, J.F Fennell, W.B. Gail, L.J. Lanzerotti, R.E. Lopez, H. Lühr, A.T.Y. Lui, C.G. Maclennan, R.W. McEntire, T.A. Potemra,

T.J. Rosenberg, and K. Takahashi, The magnetospheric response to 8-minute-period strong-amplitude upstream pressure variations, *J. Geophys. Res., 94,* 2505, 1989a.

Sibeck, D.G., W. Baumjohann, and R.E. Lopez, Solar wind dynamic pressure variations and transient magnetospheric signatures, *Geophys. Res. Lett., 16,* 13, 1989b.

Southwood, D.J., C.F. Farrugia, and M.A. Sounders, What are flux transfer events?, *Planet. Space Sci., 36,* 503,1988.

Vogelsang, H., H. Lühr, H. Voelker, J. Woch, T. Bösinger, T.A. Potemra, and P.-A. Lindqvist, An ionospheric travelling convection vortex event observed by groundbased magnetometers and by Viking, Geophys. Res. Lett., (submitted) 1993.

H. Lühr and W. Blawert, Institut für Geophysik und Meteorologie, Technische Universität, Mendelssohnstrasse 3, D-38106 Braunschweig, Germany.

The Coupling of Solar Wind Energy to MHD Cavity Modes, Waveguide Modes, and Field Line Resonances in the Earth's Magnetosphere

J. C. SAMSON AND R. RANKIN

Canadian Network for Space Research and Department of Physics
University of Alberta, Edmonton, Alberta, Canada

Field line resonances associated with the coupling of compressional energy in magnetohydrodynamic waves to shear Alfvén waves are commonly observed in the magnetosphere. These resonances often have quantized frequencies with 1.3, 1.9, 2.6, and 3.4 mHz being very common. We shall look at possible magnetohydrodynamic cavity or waveguide modes which might produce the quantized frequencies that are observed, and outline a model for the coupling of the compressional energy in the waveguide modes to field line resonances. Experimental evidence from ground based magnetometer and radar measurements indicates that transient phenomena such as pressure pulses from the solar wind may be the most likely source of energy for these modes. We shall also show observational evidence for large velocities, up to 200 km/s, for the velocity fields associated with the resonances in the equatorial plane. These large shear flows associated with the resonances might be nonlinearly unstable to Kelvin-Helmholtz instabilities, leading to vortex structures within the resonance. We discuss the possibility of Kelvin Helmholtz instabilities in large amplitude resonances, as well as other possible nonlinear effects such as localized tearing mode associated with large field aligned currents. We show 3-dimensional, magnetohydrodynamic computer simulations of the nonlinear evolution of the field line resonances and the formation of the vortex structures.

INTRODUCTION

The nonuniform magnetoplasmas in the Earth's magnetosphere, coupled with the boundary of the magnetosphere, give a structure which should easily be excited to give natural, almost monochromatic modes of oscillations of magnetohydrodynamic (MHD) waves. These natural modes are due to cavity or waveguide modes of compressional Alfvén waves reflected at turning points in the inner magnetosphere and at an outer boundary, possibly the magnetopause. The possibility of the existence of these cavity modes was first pointed out by Kivelson et al. [1984] and Kivelson and Southwood [1985]. These compressional cavity modes should have discrete, quantized frequencies, which we will loosely call normal modes. They can be excited by a variety of sources including pressure pulses from the solar wind, transient dayside reconnection, Kelvin Helmholtz instabilities (KHI) in the low latitude boundary layer and substorm associated plasma instabilities. These quantized, compressional modes are coupled to shear Alfvén resonances, or field line resonances (FLRs) on dipole field lines with matching frequencies. These FLRs are very narrow in the radial direction (typically 0.3 to 0.6 R_E in the equatorial plane) but extend over many 10's to 100's of degrees in the azimuthal direction. The resonances can grow to very large amplitudes and nonlinear effects can become very important.

Nonlinear effects are manifest only in a three dimensional geometry, and lead to the further azimuthal structuring of the FLR. In the equatorial plane, the nonlinear KHI within the resonance leads to the formation of vortices with azimuthal scale sizes of a number of R_E. Near the ionosphere, where field aligned currents in the resonances are large, localized tearing mode instabilities can lead to small scale vortex structures with azimuthal scale sizes of one km or so. In fact, the FLRs have large enough potential drops to accelerate auroral electrons to several keV, and thus the small scale vortex structures might be seen in auroral emissions.

A remarkable aspect of this MHD model of the waveguide modes and FLRs in the magnetosphere is that an impressively simple and parsimonious setting leads to an enormous richness in the prediction of a variety of structures in the magnetosphere. The model requires only three elements, a nonuniform magnetoplasma, a boundary for this plasma, and a source of energy. The quantized normal modes, the quantized FLRs, and the nonlinear vortex structures associated with the resonances evolve naturally without any further assumptions or constraints.

Solar Wind Sources of Magnetospheric Ultra-Low-Frequency Waves
Geophysical Monograph 81

In this review we shall outline the linear theory of MHD cavity or waveguide modes in the magnetosphere. We shall show experimental observations, in the high latitude ionosphere and in the magnetosphere, indicating that these modes do indeed exist. We shall then discuss nonlinear phenomena associated with large amplitude FLRs, show some computer models of these phenomena, and discuss experimental observations which might confirm the existence of nonlinear effects.

THE LINEAR MODEL, CAVITY MODES, WAVEGUIDES, AND FIELD LINE RESONANCES

Theory

We assume a Cartesian geometry with gradients only in the x-direction (the radial direction in the magnetosphere) and choose $\vec{B}_0$ in the z-direction. Linearization of the MHD equations gives the vector equation

$$\rho_0 \ddot{\vec{\xi}} = \nabla\left(\vec{\xi} \cdot \nabla p_0 + \gamma p_0 \nabla \cdot \vec{\xi}\right) + \frac{1}{\mu_0}\left\{\nabla \times \left[\nabla \times \left(\vec{\xi} \times \vec{B}_0\right)\right]\right\} \times \vec{B}_0$$
$$+ \frac{1}{\mu_0}\left(\nabla \times \vec{B}_0\right) \times \left[\nabla \times \left(\vec{\xi} \times \vec{B}_0\right)\right]$$

(1)

where $\vec{\xi}$ is the plasma displacement, p_0 is the zeroth order pressure and ρ_0 is the density. The components of the equation can be written in the form

$$\frac{dP_T}{dx} = \rho_0\left[\omega^2 - k_z^2 V_A^2(x)\right]\xi_x(x)$$

(2)

$$\frac{d\xi_x(x)}{dx} = -\frac{G(x)}{\omega^2 - k_z^2 V_A^2(x)}\frac{P_T}{\rho_0}$$

(3)

where $P_T = p + B_0 b_\| / \mu_0$ is the total perturbation pressure, the plasma displacement is given by

$$\vec{\xi} = \vec{\xi}(x)e^{i\left(k_y y + k_z z - \omega t\right)},$$

(4)

and

$$G(x) = \frac{\omega^2\left[\omega^2 - \left(k_y^2 + k_z^2\right)V^2\right] + k_z^2\left(k_y^2 + k_z^2\right)V_S^2 V_A^2}{\omega^2 V^2 - k_z^2 V_S^2 V_A^2}$$

(5)

where V_S is the sound speed, V_A is the Alfvén speed, and $V^2 = V_A^2 + V_S^2$. Solving for ξ_x gives the equation

$$\frac{d^2\xi_x}{dx^2} + \frac{F'}{F}\frac{d\xi_x}{dx} + G\xi_x = 0$$

(6)

where

$$F = \frac{\rho_0\left(\omega^2 - k_z^2 V_A^2\right)}{G}.$$

(7)

This equation has two turning points, x_t ($G(x_t) = 0$), and two resonances at positions, x_r ($F(x_r) = 0$), corresponding to the shear Alfvén resonance ($\omega^2 - k_z^2 V_A^2 = 0$) and the cusp resonance ($\omega^2 V^2 - k_z^2 V_S^2 V_A^2 = 0$).

The basic configuration we are considering is given in Figures 1 and 2. The energy input is from a pressure pulse from the solar wind, which leads to a distortion moving antisunward along the magnetopause. Compressional MHD waves set up cavity or waveguide modes between the outer boundary, possibly the magnetopause, and the surface of the turning point. In this schematic the waveguide is in the morning sector. The compressional waveguide energy is coupled to FLRs earthward of the surface of the turning point.

If we assume that a Cartesian geometry is appropriate near the equatorial plane, then x is in the radial direction, y is the azimuthal direction, and z is magnetic field aligned. For one frequency component, ω, of the incident, broad band compressional pulse, there is a turning point on field lines where

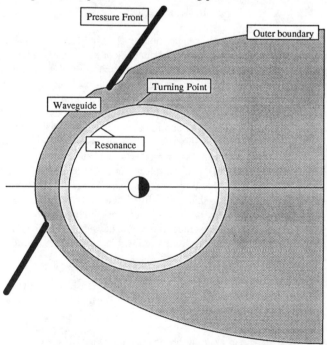

Fig. 1. A schematic of the MHD waveguide in the equatorial plane of the magnetosphere.

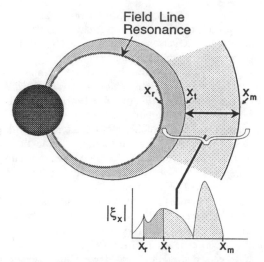

Fig. 2. A cross section of the MHD waveguide in the local dawn meridian. A schematic for the radial displacement for n=2 is shown in the inset at the bottom.

$G(x_T) = \omega^2/V_A^2(x_T) - k_y^2 - k_z^2 = 0$. In the region between the magnetopause and the turning point we use the WKB solution of equation (6), given by

$$\xi_x(x) = F^{-1/2}(x)G^{-1/4}(x)\exp\left[i\int^x G^{1/2}(x')dx'\right] \qquad (8)$$

In essence, $G^{1/2}(x) = k_x(x)$ can be considered to be the "spatially dependent wavenumber" in the x-direction. Positive k_x gives a reflected (outgoing) wave and negative k_x give an incident wave. If the resonance is far earthward of the turning point, then the wave is completely reflected at the turning point, with a $\pi/2$ phase shift in order to match the Airy function solution which is valid near the turning point. At the magnetopause, $x = x_m$, the boundary condition is $\xi_x(x_m) = 0$ due to the large increase in density and decrease in Alfvén velocity at the magnetopause. Consequently, for fixed k_y and k_z, the system will have quantized eigenfrequencies, ω_n, given by solutions of the phase integral equation.

$$\Phi_n = \int_{x_t}^{x_m} G^{1/2}(x,\omega_n)dx - (n-1/4)\pi = 0. \qquad (9)$$

Samson et al. [1992a] have integrated equations (2) and (3) numerically and found that the numerical solutions give eigenfrequencies which are within 3% of the WKB frequencies.

One problem with this scenario is that it is difficult to see why k_z and especially k_y should have fixed values. The parameter k_z might be determined by the length of field lines from ionosphere to ionosphere, but a fixed k_y requires an azimuthally symmetric cavity or boundaries at the azimuthal ends of the cavity. Bearing in mind the basic configuration of the magnetosphere near the Earth (see e.g. Figure 1) it is very

difficult to see how an assumption of azimuthal symmetry could be valid. Similarly it is not clear where any further azimuthal boundaries might be. Samson et al. [1992a] resolved this dilemma by pointing out that it might be better to consider a waveguide for the MHD modes, with antisunward propagation when the source of energy originated from pressure pulses in the solar wind. They pointed out that the value of Φ_n is very insensitive to the value of k_y (and k_z) if the Alfvén velocity falls off rapidly with radial distance in the waveguide. Consequently the modes at discrete frequencies are recovered.

The coupling of the compressional fast mode and the shear Alfvén wave at the resonance is relatively well understood, at least in the linear regime, with abundant publications on this topic [see e.g. Chen and Hasegawa, 1974, Southwood, 1974, and Tamao, 1966]. Near the shear Alfvén resonance, equation (6) becomes the modified Bessel equation of zero order [Southwood, 1974], with the approximate form

$$\frac{d^2\xi_x}{dx^2} + \frac{1}{x-x_r+i\gamma}\frac{d\xi_x}{dx} - k_y^2\xi_x = 0 \qquad (10)$$

where $\gamma = \mathrm{Im}\,\varepsilon(x_r)/(d\,\mathrm{Re}\varepsilon/dx)_{x=x_r}$,

$$\mathrm{Re}\varepsilon = \frac{\omega^2}{V_A^2} - k_z^2 \qquad (11)$$

and the imaginary component of ε takes into account losses. The solution is

$$\xi_x = \xi_0 \ln\left(k_y(x-x_r+i\gamma)\right) \qquad (12)$$

A schematic of the approximate solution for $|\xi_x(x)|$, including the field in the region between the turning point and the outer boundary, is given in the inset at the bottom of Figure 2.

The solution for the radial component of the electric field is

$$E_x(x) = \frac{ik_yE_0}{\left(\mathrm{Re}\varepsilon - k_y^2\right)(x-x_r+i\gamma)}. \qquad (13)$$

Near the resonance there is a large peak in the radial component of the electric field, and a 180° phase shift across this peak [Walker et al., 1979; Walker, 1980]. If losses are relatively small, then the fields in the vicinity of the resonance can become quite large, leaving the assumption of linear theory in some doubt. We shall address this issue later when we look at nonlinear effects. The large fields in the shear Alfvén resonance or FLR are likely to be the most observable manifestations of the cavity modes in the magnetosphere, particularly for observations made at ionospheric altitudes.

The coupling of the compressional modes should lead to FLRs

with quantized frequencies ω_n. Since the effective length of field lines decrease in an earthward direction (k_z increases) , in the equatorial plane the radial distance to the FLR will decrease as ω_n increases. In the ionosphere, the frequency will increase with decreasing latitude.

There are numerous published examples of numerical or computational models of linear cavity modes and FLRs in the magnetosphere. Allan et al. [1986] used Laplace transform techniques to compute the wave fields in a time dependent model of cavity modes in a magnetosphere having field lines with a cylindrical geometry. They showed that an impulsive energy input could lead to a discrete set of frequencies and shear Alfvén resonances for given azimuthal wave numbers. Inhester [1987] looked at a time dependent numerical model using a Cartesian or box configuration. He also included the effects of electron inertia in his equations. Zhu and Kivelson [1988] used a box model and Laplace transform methods to find numerical solutions for the cavity modes. Lee and Lysak [1991] have looked at more realistic dipole configurations. All of these numerical models produced coupled discrete cavity modes and shear Alfvén resonances.

To illustrate the form of the coupled waveguide or cavity mode solutions, we shall use the results given by Harrold and Samson [1992]. Harrold and Samson integrated equations (2) and (3) numerically in order to determine the modes which would evolve from a pressure pulse from the solar wind. Their configuration was very similar to that in Figure 1, but they used a Cartesian box model to find the wave fields. They chose a reflecting boundary at the bow shock in order to accommodate the very low frequencies of observed discrete resonances which might be associated with waveguide or cavity modes [Samson et al., 1992a]. Reflection at the bow shock occurs because the solar wind flow velocity exceeds the Alfvén velocity.

Harrold and Samson's numerical solutions are shown in Figure 3. The compressional waveguide or cavity modes are clearest in the plot of $|\xi_x|$, and the shear Alfvén resonances in the plot of $|E_x|$. A surface wave labeled by S is the eigenmode of a KHI at the low latitude boundary layer (LLBL). A slow mode wave reflecting from the bow shock (BS) and LLBL is labeled SL. Other modes below about 1.3 mHz are evanescent with no propagation in the x-direction. The coupled waveguide or cavity modes can be seen in the discrete modes labeled 0 (1.3mHz), 1 (1.9mHz), 2 (2.5 mHz), 3 (3.4 mHz), and 4 (4.2mHz). The FLRs (the peaks in the plot of $|E_x|$) are embedded in the continuum spectrum, but have much larger amplitudes. These amplitudes are somewhat arbitrary as an imaginary component is added to k_z in order to take into account ionospheric losses. These peaks correspond to the solution for E_x given in equation (13).

Observations

In this review we shall outline only recent observations which are relevant to the cavity or waveguide model we are discussing. A more comprehensive review of earlier observations of FLRs and MHD waves is given by Samson [1991]. There are only a few observations of possible compressional oscillations associated with cavity or waveguide modes in the magnetosphere. Kivelson et al. [1984] attributed a quasimonochromatic compressional wave seen by the ISEE 1 satellite to a cavity mode. Warnecke et al. [1990] noted FLRs at about 3.1 mHz on the ground which were presumably being driven by compressional oscillations at 3.1 mHz seen at the same time by the AMPTE IRM satellite.

Observations of FLRs which might be associated with cavity or waveguide modes are far more numerous. McDiarmid and Allan [1990] showed that some coherent radar pulsation data could be explained by FLRs driven by cavity modes. Warnecke et al. [1990] noted FLRs at about 1.8 mHz and 2.4 mHz, in addition to the 3.1 mHz resonance mentioned above. These resonances were presumably driven by a sudden impulse. Mitchell et al. [1990] describe very large amplitude FLRs seen in particle and electric field data from the ISEE 1 and 2 satellites when they traversed the equatorial dawn flank magnetosphere. These resonances had frequencies of about 2.0-2.2 mHz at 10-11 R_E, increasing to about 3 mHz at 9 R_E. Mitchell et al. pointed out that the frequencies changed "incrementally" rather than continuously as the ISEE 1 satellite moved radially. This is exactly what would be expected if the resonances were coupled to discrete waveguide or cavity modes. In addition, the calculated velocity fields of these FLRs were very large (Figure 4) ranging between 100 and 200 km/s. As we shall show later these large velocity fields indicate that the resonances might evolve nonlinearly.

Recently, multiple, discrete FLRs have been seen in the F-region drift velocities measured by the Johns Hopkins University/Applied Physics Laboratory radar at Goose Bay in Canada [Ruohoniemi et al., 1991; Samson et al., 1991a; Samson et al. 1992a; Samson et al., 1992b; Walker et al., 1992]. Samson et al. [1991a, 1992b] showed that these multiple resonances are also seen in magnetometer data from the Canadian CANOPUS array. These resonances tend to occur predominantly in the local morning sector and have frequencies of about 1.3, 1.9, 2.6, and 3.4 mHz. In fact these frequencies are remarkably stable changing by no more than 5-10% over intervals as long as 3-4 hours [Walker et al., 1992] and from day to day and month to month [Samson et al., 1991a]. These discrete resonances have been attributed to cavity modes or waveguide modes in the morningside magnetosphere.

In order to illustrate the features of these resonances we will look in detail at one interval, 0400-0800 UT, January 11, 1989 (see also Ruohoniemi et al. [1991], Samson et al.[1992a], Walker et al. [1992]). Time series plots of the Doppler velocities measured on beam 10 are shown in Figure 5. The wave trains of the resonances are quite evident as is their impulsive nature, and the decrease in frequency with increasing latitude. Figure 6 shows power spectra of the time series of the Doppler velocities in the interval from 0400 to 0600 UT. Four discrete spectral peaks are clearly evident, a 1.3 mHz peak near the poleward border of the field of view at about 71.5°, a 1.8-1.9 mHz peak near 70.5°, a 2.7 mHz peak near 69.75°, and a 3.3-3.4 mHz peak at the southern border of the field of view at about 69.0°. These discrete latitudinally localized spectral peaks are exactly what

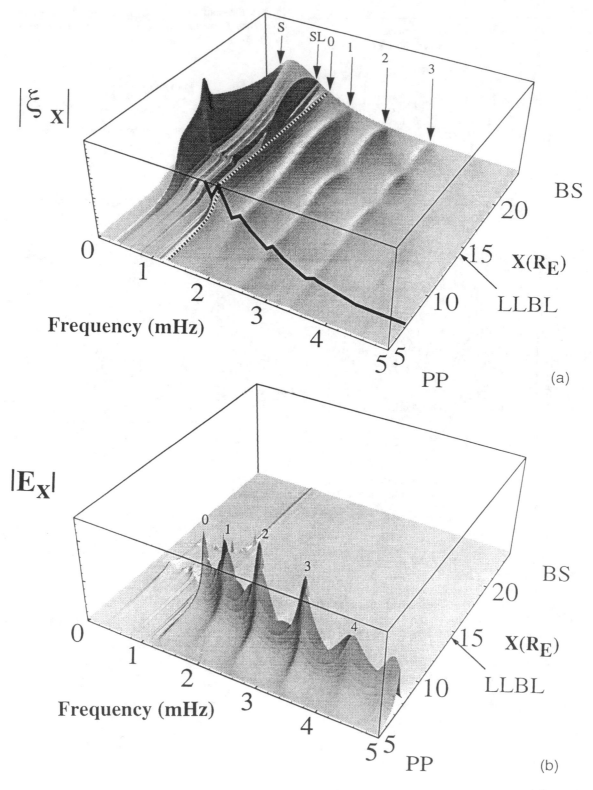

Fig. 3. Numerical solutions of equation (5) for the radial components of the displacement field and the electric field of the cavity or waveguide mode. The solid line is the locus of the points at the peak of ξ_x (equation (12)) at the position of the continuum spectrum. The position of the plasmapause is labeled PP. Other terms are explained in the text.

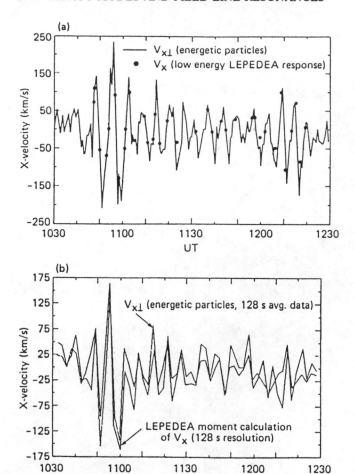

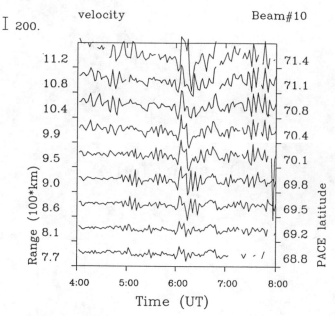

whereby the configuration of the waveguide or cavity might be determined from the radar measurements of the discrete FLRs. They assumed an Alfvén velocity profile of the form

$$v_A(x) = v_p \left(\frac{x_p}{x} \right)^l \tag{14}$$

where x_p is the position of the plasmapause and v_p is the Alfvén

Fig. 4. Fluid velocity fields of FLRs which were estimated from energetic particle and LEPEDEA data from the ISEE-1 satellite [Mitchell et al., 1990]. The X-direction is toward the sun (GSM coordinates) and is approximately in the azimuthal direction in the early morning sector where these data were recorded.

1989/01/11

Fig. 5. Line of sight Doppler velocities measured on beam 10 of the Goose Bay HF-radar in the interval from 0400-0800 UT, 11 January, 1989. The velocity scale is given by the bar at the top left, and has the range 200 m/s. Positive Doppler is toward the radar. The PACE coordinates are described by Baker and Wing [1989].

would be expected if they were due to FLRs driven by cavity or waveguide modes. (Compare Figure 6 with Figure 3.)

Figure 7 shows the time-dependent frequency and amplitude of the 2.7 mHz oscillations at 69.7° in the interval from 0415 UT to 0745UT [Walker et al., 1992]. The amplitude and phase were calculated by using the analytic signal representation of the time series. Note the impulsive nature of the amplitude envelope, with a strong enhancement near 0615 UT (see also Figure 5) and the extremely stable frequency of the oscillations in the intervals with substantial amplitudes. Plots of the latitudinal changes in the phase and power of the 2.7 mHz oscillations in the wave packet between 0545-0645 UT are shown in Figure 8. The amplitude is very localized in latitude, with a peak at 69.7°. A pronounced phase change occurs at the latitude of the peak, evolving from about 120° near the beginning of the packet to 180° at the end. The 180° phase change and the pronounced latitudinal peak are all characteristic of the predictions of steady-state models of FLRs (see equation (13)).

Samson et al. [1991a, 1992a] have described a method

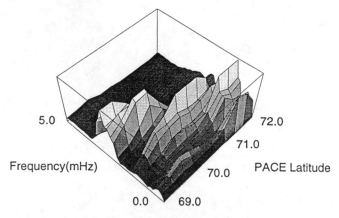

Fig. 6. Power spectra of the velocity data (Figure 5) in the interval from 0400 to 0600 UT. The power spectra at each latitude are normalized to have a maximum of 1.0.

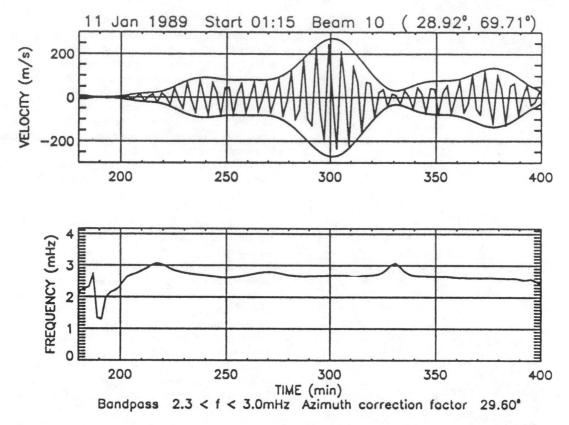

Fig. 7. Top: the filtered (2.3-3.0 mHz) Doppler velocity data from beam 10 (Figure 5), at 69.7°, and in the interval from 0415 to 0755 UT. The amplitude envelope has been estimated by using the analytic signal representation [Walker et al., 1992]. Bottom: the time dependent frequency estimated from the analytic signal representation. The velocity data have been scaled to correspond to equivalent east-west flows [Walker et al., 1992].

velocity just outside the plasmapause. They determined the position of the outer boundary, x_m, and the parameter, l (equation(14)), by minimizing the objective function

$$\sum_{n=1}^{4} \Phi_n^2 \left(2\pi f_n(obs)\right) \qquad (15)$$

with respect to x_m and l. The phase integral, Φ_n, is given in equation (9) and $f_n(obs)$=1.3,1.9,2.6, and 3.4 mHz. They assumed that the position of the turning point, x_t, was near the resonance and that the resonances were fundamental mode oscillations. The minimization yielded $l = 3.0 \pm 0.1$, and $x_m = 14.8 \pm 0.2$ R$_E$. The estimated value of x_m is very close to the position of the magnetopause on the dawn flank [Sibeck et al., 1991] suggesting that the magnetopause might be the outer boundary of the waveguide. Similarly the value of l suggests that much of the waveguide is in a region with dipolelike field lines and very gradual changes in density. All these parameters are compatible with the schematic shown in Figure 1, with the outer boundary at the magnetopause.   However, there is one problem with this model. The predicted Alfvén velocities near

the magnetopause are very low, near 40 km/s, and would require plasma densities almost an order of magnitude larger than values estimated from satellite measurements [Moore et al., 1987]. Our data indicate that the average mass density in the cavity would have to be about 40 amu/cc, suggesting the possible presence of O$^+$. These densities are, however, compatible with densities determined from FLRs [Poulter et al., 1984; Walker et al., 1992]. Poulter et al. suggested a plasma composition with greater than 50% O$^+$. The low Alfvén velocities led Harrold and Samson [1992] to look at other configurations for the waveguide, with reflection at the bow shock.

In fact, Mitchell et al. [1991], Warnecke et al. [1990], and Walker et al. [1992] have noted similar problems with the densities in trying to model the frequencies of observed FLRs. In all cases the relatively low frequencies of the FLRs suggested that equatorial mass densities might be considerably larger than accepted values. Walker et al. [1992] calculated that mass densities in the equatorial plane at about 10 R$_E$ would have to be several tens of amu/cc in order to predict the observed frequencies. Mitchell et al. [1991] have suggested that the extra mass might come from oxygen in the equatorial magnetosphere. Clearly this problem needs further consideration.

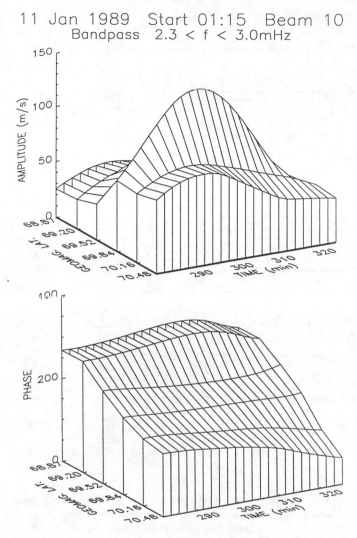

11 Jan 1989 Start 01:15 Beam 10
Bandpass 2.3 < f < 3.0mHz

Fig. 8. Time dependent and latitude dependent amplitudes and phases of the 2.7 mHz pulsations in the interval from 0545-0645 UT in Figure 7 [Walker et al., 1992].

NONLINEAR EFFECTS

Two of the nonlinear processes which might be associated with FLRs are shown in Figure 9. The diagrams indicate different phases of the wave cycle for a fundamental mode FLR. The nonlinear effects in FLRs are three dimensional in nature and lead to the evolution of azimuthal structures with scale sizes much smaller than the azimuthal wavelength of the resonance. The top diagram in Figure 9 indicates the phase of the wave cycle when the field aligned currents are maximum, just above the auroral ionosphere (see Greenwald and Walker [1980], Figure 1). Greenwald and Walker [1980] and Walker et al. [1992] have shown that field aligned currents in resonances can be of the order of 1 to 5 $\mu A/m^2$ above the auroral ionosphere. During the phase with upward field aligned currents in the resonance, the configuration of the field aligned currents and

electric fields in the resonance is very similar to the configurations found in auroral arcs. Seyler [1990] has shown that small scale arcs evolve through the nonlinear development of azimuthal structures through collisionless tearing and reconnection. Consequently we suspect that if the resonance has small scale latitudinal structure, on the order of a number of kilometers, then finite electron inertia effects could lead to localized tearing mode instabilities and the nonlinear development of azimuthal structures with small scale (several kilometers) vortices within the resonance. Unfortunately, we do not yet have any measurements of resonances with latitudinal resolutions less than about 40 km (see Figure 8 and Walker et al. [1992]).

The bottom diagram in Figure 9 shows the wave cycle advanced by 90° with respect to the phase in the top diagram. The fundamental mode has electric and velocity field nodes at the ionosphere, and antinodes at the equator. During this part of the cycle, the electric and velocity fields reach a temporal maximum at the equator. The electric field has a 180° phase shift across the resonance (see equation (13)), and so the velocity field of the resonance has a strong shear, with flow in the azimuthal direction. If the velocity fields are large enough, then this shear can lead to nonlinear KHI with growth rates large enough to allow the instability to evolve in only a fraction of the resonance period. The measured fluid velocities of the resonances in the auroral ionosphere are of the order of 1 km/s, [Samson et al., 1992b; Walker et al 1992], and using the

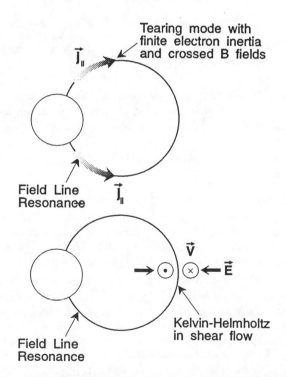

Fig. 9. Nonlinear phenomena which might occur in the fundamental mode of a FLR. Top: nonlinear tearing mode associated with large field aligned currents near the auroral ionosphere. Bottom: 90° later in the wave cycle. Nonlinear KHI in the equatorial plane of the resonance.

resonance model given by Walker [1980] we estimate that maximum velocities in the equatorial plane are often greater than 100 km/s. The resonances seen by Mitchell et al. [1990] (see Figure 4) had velocity fields that had maxima of almost 200 km/s.

In order for this nonlinear KHI to evolve substantially in the resonance it is necessary for the effective e-folding time for growth to be much less than half a wave period. If growth is too slow, the velocity fields of the resonance reverse after a half cycle, suppressing any vortices which are forming. In addition, the Alfvén travel time from the equator to the ionosphere and back is one half wave period. If the ionosphere is conducting, the reflected Alfvén waves can suppress the growth of the KHI when they reach the equatorial plane. There does not yet appear to be an adequate theory for these nonlinear effects in resonances, so we give here only a very simple analytic treatment of the KHI. Computer models of the fully nonlinear 3-D MHD evolution of the instability will follow.

To determine the growth rates we use the linear approximation for the KHI, and the velocity fields in the resonance. In practice, this growth rate changes in time as the velocity field in the resonance changes during the wave period. If the resonance can be modeled as a discontinuous profile in the x-direction, with the shear at the position of the resonance, then the growth rate ω_i, is

$$(2\omega_i)^2 = \left[\vec{\kappa} \cdot (\vec{v}_2 - \vec{v}_1)\right]^2 - \left[2\vec{\kappa} \cdot \vec{v}_A\right]^2 \qquad (16)$$

where we have assumed that the density and magnetic field change slowly in the radial direction at the equator. Subscripts 1 and 2 indicate respectively the earthward and anti-earthward side of the resonance. Maximum growth occurs for the wavevector of the KHI, $\vec{\kappa}$, in the azimuthal direction. In addition, maximum growth occurs when $\kappa d \approx 0.6$, where d is the thickness of the shear layer [Walker, 1981]. Then the maximum growth rate is

$$\omega_i = \kappa v = 0.6 v/d \qquad (17)$$

where $v = |\vec{v}_1| = |\vec{v}_2|$.

The measured widths of resonances in the ionosphere are on the order of 40km. Mapped to the equatorial plane the width is about 0.3 R_E. The velocity fields of the resonances in the equatorial plane can be as large as 200 km/s so we choose 100 km/s as a reasonable amplitude. To compute an effective growth rate or approximate time averaged growth rate we use $v=50$ km/s. Using these values for v and d we find that $\omega_i = 0.015$ s^{-1} and the e-folding time is about 70 s. This value is in reasonable agreement with the values determined from the computational model we discuss below.

The frequencies of the FLRs seen in the auroral ionosphere vary from 1 and 4 mHz. The lowest frequency resonance has a half period of 500 s, and consequently there are 7 e-folding times in this half cycle. This time is more than adequate for the KHI

vortices to evolve out of background MHD noise. The scale size of the vortices in the azimuthal direction in the equatorial plane is approximately $(2\pi / 0.6)d$ or about 3 R_E. This maps to 400 km in the auroral ionosphere, which is compatible with the scale sizes of auroral surges and omega bands [Samson et al., 1992b].

A computational model of the nonlinear evolution of FLRs gives a far more realistic picture of this process. Rankin et al. [1993] used a three dimensional, MHD code to simulate the nonlinear evolution of the KHI in the equatorial plane of a FLR. They used a Cartesian geometry and initialized the system with a FLR with a maximum velocity field of 100 km/s, a radial scale size of roughly 0.3 R_E and a period of 1400 s. They found that the nonlinear evolution of the resonance in the equatorial plane leads to strong vortex cells from the nonlinear KHI, and that these vortex structures propagate to the ionosphere. As the KHI evolves, the largely azimuthal flow (v_y) of the resonance is distorted giving substantial radial flows (v_x) as the vortices develop. Figure 10 shows the evolution of the maximum in v_x in the equatorial plane of the resonance. The evolution of the nonlinear KHI is very clear with v_x reaching almost 60-70 km/s at 300s (the Alfvén speed is 280 km/s). The effective e-folding time before reaching nonlinear saturation is about 90 s. This compares very well with the 70 s estimated from our simplified analytic model. Figures 11 and 12 show the evolution of the vorticity in the equatorial plane. By 250 s the KHI vortex is large enough to greatly distort the FLR. As the vortex cells propagate to the ionosphere, they further distort the FLR, and eventually lead to its break-up.

DISCUSSION AND CONCLUSIONS

The model of the MHD waveguide, the computational models of nonlinear effects in the FLRs driven by the waveguide modes and the observations of the discrete low frequency (1-4 mHz) FLRs which we have presented in this review suggest that

Growth of Vx at Equator

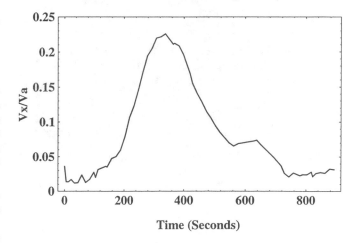

Fig. 10. The temporal evolution of the radial component of the velocity field of a large amplitude FLR at the equator [Rankin et al., 1993]. The radial velocity field increases as a nonlinear KHI evolves.

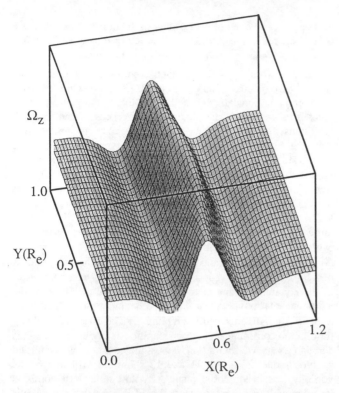

Fig. 11. The magnetic field aligned component of the vorticity field, Ω_z, of a FLR in the equatorial plane [Rankin et al., 1993]. The fields correspond to the time of 100 s in Figure 10. Note the slight kink in the central maximum of the vorticity as the KHI begins to evolve.

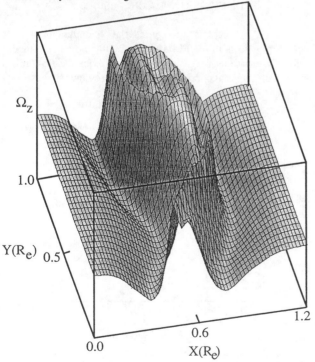

Fig. 12. The magnetic field aligned component of the vorticity in the FLR at 250 s.

studies of ULF plasma waves in the magnetosphere still have many exciting frontiers to explore. In some ways we really have taken only a first glimpse at these phenomena, and much remains to be done, observationally, computationally, and theoretically. Perhaps the most exciting aspect of the MHD waveguide model is the economy of assumption (there are only three) and the richness of the model's predictive power. The richness of the predictive power also leads to many unanswered questions. We shall address a few of these questions here.

One of the major unsolved problems with this waveguide model is the extreme stability of the frequencies of the FLRs (see Figure 7 and Samson et al. [1991a], Samson et al. [1992a,b], Walker et al., [1992]). Presumably the frequencies of the waveguide modes would be sensitive to the position of the magnetopause, or the bow shock in the model of Harrold and Samson [1992]. Other scenarios are possible, including cavity modes in the magnetotail. Walker et al. [1992] concluded after modeling modes in the magnetotail that the waveguide mode in Figure 1 still was the most likely scenario. The waveguide mode predicts the observed antisunward propagation and the observations of resonances on the dayside and in the morning (e.g. Mitchell et al.[1990]) and late night sector. Nevertheless, many more observations of the discrete low frequency resonances, and possibly the compressional component of the waveguide mode, are needed before we can be sure that this waveguide model is appropriate. Most observations have been during relatively quiet times. It would be interesting to see if the frequencies of the resonances changed during more active intervals.

Another problem is the large plasma densities required to explain the very low frequencies of the observed resonances [Mitchell et al., 1990; Warnecke et al., 1990; Walker et al., 1992b] and the estimated low Alfvén velocities in the waveguide model of Samson et al. [1992a]. The question is "Is there hidden cold dark matter in the magnetosphere?" Mitchell et al. [1990] have suggested that a "hidden" population of oxygen might be present in equatorial regions. Clearly this problem needs to be addressed.

Energy at discrete frequencies from the waveguide (or any possible cavity modes) should also propagate into the magnetotail. The effects of these waves would possibly be most evident on field lines threading the plasmasheet boundary layer, where there is a strong Alfvén velocity gradient in the Y-direction (geocentric coordinates). Goertz and Smith [1989] have noted that there is strong resonance absorption in the plasma sheet boundary layer, though they suggested that compressional energy propagated from the tail lobes whereas we have compressional energy propagating nearer the center of the tail. The Alfvén gradient in the plasma sheet boundary layer mapped to the ionosphere would show increasing Alfvén velocity with latitude. This is the opposite trend from that in the dipolar field lines, where the discrete FLRs have been observed. Consequently waveguide energy coupled to shear Alfvén waves in the plasma sheet boundary layer would be observed in the ionosphere as discrete monochromatic oscillations, but with a latitudinal phase shift which is the opposite of that for FLRs on dipolar field lines (see the discussion in Samson et al. [1992b]).

In radar data, the signature of the coupling to shear Alfvén waves in the plasma sheet boundary layer would be equatorward moving bands, not the poleward moving bands seen on dipolar field lines outside the plasmapause [Walker et al., 1979].

In terms of the dissipative and nonlinear effects associated with the FLRs, there are an enormous number of problems to be addressed, many beyond the scope of this review. The large field aligned currents (1-5 $\mu A/m^2$) and potentials (2-5 keV at the ionosphere) indicate that FLRs might lead to the formation of auroral arcs through a variety of processes, including the generation of electrostatic ion cyclotron waves and finite electron inertia and ion gyroradius effects [Samson et, 1991a; Samson et al., 1992b]. Correlated optical observations of auroral arcs and radar observation of FLRs in the auroral ionosphere are needed. These observations would also help in determining whether the nonlinear processes in the resonances, including tearing mode and KHI, occur and whether any of these processes play a role in the formation of auroral structures.

It is clear that more computer simulations of the nonlinear evolution of FLRs are needed. It would be interesting to repeat the simulations done by Seyler [1990], but in a configuration which included a large amplitude FLR. The modeling of the nonlinear KHI needs to be extended by having a fully 3-D MHD simulation of the resonance driven by an external source of compressional Alfvén waves. A variety of possibilities exist in the driven scenario. If ionospheric losses are small enough, the resonance will narrow with time, possibly reaching a width where finite electron inertia effects or finite ion gyroradius effects can lead to linear (or nonlinear) mode conversion to electron inertia or kinetic Alfvén waves. The electron inertia or kinetic Alfvén waves can then limit the growth of the resonance as they propagate energy away from the resonance. If these kinetic effects are not important, then there is every possibility that nonlinear KHIs play a role in the evolution and dissipation of the resonance. The combination of possibilities is clearly complicated, and further computational studies are needed.

Turning finally to theory, it is clear that we have a good understanding of the theory needed to describe the linear cavity mode or waveguide model, including FLRs. But, what about theories for the nonlinear evolution of the FLRs? To our knowledge, there are no published theories which directly address the nonlinear aspects of the FLRs which we have discussed in this paper. Much needs to be done.

Acknowledgments. Research for this project has been supported in part through the Natural Sciences and Engineering Research Council of Canada and in part through the Canadian Networks of Centers of Excellence Program. We would like to thank B. G. Harrold and A.D.M. Walker for a variety of suggestions regarding MHD waves in the magnetosphere.

REFERENCES

Allan, W., S.P. White, and E.M. Poulter, Impulse excited hydromagnetic cavity and field line resonances in the magnetosphere, *Planet. Space Sci., 34*, 371-385, 1986.

Baker, G., and S. Wing, A new magnetic coordinate system for conjugate studies of high latitudes, *J. Geophys. Res., 94*, 9139-9144, 1989.

Chen, L., and A. Hasegawa, A theory of long period magnetic pulsations, 1, Steady state exitation of field line resonance, *J. Geophys. Res., 79*, 1024-1032, 1974.

Greenwald, R.A., and A.D.M. Walker, Energetics of long period hydromagnetic waves, *Geophys. Res. Lett., 7*, 745-748, 1980.

Harrold, B.G., C.K. Goertz, R.A. Smith, and P.J. Hansen, Resonant Alfvén wave heating of the plasma sheet boundary layer, *J. Geophys. Res., 95*, 15039-15046, 1990.

Harrold, B.G. and J.C. Samson, Standing ULF modes of the magnetosphere: A theory, *Geophys. Res. Lett., 19*, 1811-1814, 1992.

Inhester, B., Numerical modeling of hydromagnetic wave coupling in the magnetosphere, *J. Geophys. Res., 92*, 4751-4756, 1987.

Kivelson, M.G., J. Etcheto, and J.B. Trotignon, Global compressional oscillations of the terrestrial magnetosphere: The evidence and a model, *J. Geophys. Res., 89*, 9851-9856, 1984.

Kivelson, M.G., and D.J. Southwood, Resonant ULF waves: A new interpretation, *Geophys. Res. Lett., 12*, 49-52, 1985.

Lee, D.-H., and R.L. Lysak, Impulse excitation of ULF waves in the three-dimensional dipole model: The initial results, *J. Geophys. Res., 96*, 3479-3486, 1991.

McDiarmid, D.R., and W. Allan, Simulation and analysis of auroral radar signatures generated by a magnetospheric cavity mode, *J. Geophys. Res., 95*, 20911-20922, 1990.

Mitchell, D.G., M.J. Engebretson, D.J. Williams, C.A. Cattell, and R. Lundin, Pc 5 pulsations in the outer dawn magnetosphere seen by ISEE 1 and 2, *J. Geophys. Res., 95*, 967-975, 1990.

Moore, T.E., D.L. Gallagher, J.L. Horwitz, and R.H. Comfort, MHD wave breaking in the outer plasmasphere, *Geophys. Res. Lett., 14*, 1007-1010, 1987.

Poulter, E.M., W. Allan, J.G. Keys, and E. Nielsen, Plasmatrough ion mass densities determined from ULF pulsation eigenperiods, *Planet. Space. Sci., 32*, 1069-1078, 1984.

Rankin, R., B.G. Harrold, J.C. Samson, and P. Frycz, The nonlinear evolution of field line resonances in the Earth's magnetosphere, *J. Geophys. Res.,* In press, 1993.

Ruohoniemi, J.M., R.A. Greenwald, K.B. Baker, and J.C. Samson, HF radar observations of Pc 5 field line resonances in the midnight/early morning MLT sector, *J. Geophys. Res., 96*, 15697-15710, 1991.

Samson, J.C., Geomagnetic pulsations and plasma waves in the Earth's magnetosphere, in *Geomagnetism, Vol. 4*, edited by J.A. Jacobs, pp. 481-592, Academic, San Diego, 1991.

Samson, J.C., R.A. Greenwald, J.M. Ruohoniemi, T.J. Hughes, and D.D. Wallis, Magnetometer and radar observations of MHD cavity modes in the Earth's magnetosphere, *Can. J. Phys., 69*, 929-937, 1991a.

Samson, J.C., B.G. Harrold, J.M. Ruohoniemi, R.A. Greenwald, and A.D.M. Walker, Field line resonances associated with MHD waveguides in the Earth's magnetosphere, *Geophys. Res. Lett., 19*, 441-444, 1992a.

Samson, J.C., T.J. Hughes, F. Creutzberg, D.D. Wallis, R.A. Greenwald, and J.M. Ruohoniemi, Observations of a detached, discrete arc in association with field line resonances, *J. Geophys. Res., 96*, 15683-15695, 1991b.

Samson, J.C., D.D. Wallis, T.J. Hughes, F. Creutzberg, J.M. Ruohoniemi, and R.A. Greenwald, Substorm intensifications and field line resonances in the nightside magnetosphere, *J. Geophys. Res., 97*, 8495-8518, 1992b.

Seyler, A mathematical model of the structure and evolution of small scale discrete auroral arcs, *J. Geophys. Res., 95*, 17199-17216, 1990.

Sibeck, D.G., R.E. Lopez, and E.C. Roelof, Solar wind control of the magnetopause shape location and motion, *J. Geophys. Res., 96*, 5489-5496, 1991.

Southwood, D.J., Some features of field line resonances in the magnetosphere, *Planet. Space Sci., 22*, 483-491, 1974.

Tamao, T., Transmission and coupling resonances of hydromagnetic disturbances in the non-uniform, earth's magnetosphere, *Sci. Rep. Tohoku Univ., Ser 5, 17*, 43- 72, 1966.

Walker, A.D.M., Modeling of Pc5 pulsation structure in the magnetosphere, *Planet. Space Sci., 28*, 213-223, 1980.

Walker, A.D.M., The Kelvin-Helmholtz instability in the low latitude boundary layer, *Planet Space Sci., 29*, 1119-1133, 1981.

Walker, A.D.M. R.A. Greenwald, W.F. Stuart, and C.A. Green, Stare auroral radar observations of Pc 5 geomagnetic pulsations, *J. Geophys. Res., 84,* 3373-3388, 1979.

Walker, A.D.M., J.M. Ruohoniemi, K.B. Baker, R.A. Greenwald, and J.C. Samson, Spatial and temporal behavior of ULF pulsations observed by the Goose Bay HF radar, *J. Geophys. Res., 97,* 12187-12202, 1992.

Walker, A.D.M., J.M. Ruohoniemi, K.B. Baker, R.A. Greenwald, and J.C. Samson, Spectral properties of magnetotail oscillations as a source of Pc 5 pulsations, *Adv. Space Res.,* In press, 1993.

Warnecke, J., H. Luhr, and K. Takahashi, Observational features of field line resonances excited by solar wind pressure variations on 4 September, 1984, *Planet. Space Sci., 38,* 1517-1531, 1990.

Zhu, X., and M.G. Kivelson, Analytic formulation and quantitative solutions of the coupled ULF wave problem, *J. Geophys. Res., 93,* 8602-8612, 1988.

J.C. Samson and R. Rankin, Canadian Network for Space Research, Dept. of Physics, CS-005 Biological Sciences Bldg., University of Alberta, Edmonton, Alberta T6G 2E9, Canada.

Generation of ULF Magnetic Pulsations in Response to Sudden Variations in Solar Wind Dynamic Pressure

G. I. KOROTOVA

IZMIRAN, Troitsk, Moscow Region, 142092 Russia

D. G. SIBECK

The Johns Hopkins University Applied Physics Laboratory, Laurel, MD

We present simultaneous solar wind, magnetospheric, and ground magnetometer observations to study the magnetospheric response to a series of sudden sharp variations in the solar wind dynamic pressure. In addition to an impulsive response, each variation in the pressure excited a packet of damped, long-period (3–15 min) pulsations. The sense of the pulsations reversed near local noon and with increasing geomagnetic latitude. One packet with coherent oscillations was observed over a 30–50° range of longitudes. We consider a possible explanation in terms of cavity resonances.

INTRODUCTION

The pressure variations associated with features both intrinsic to the solar wind (tangential discontinuities and interplanetary shocks) and generated at the bow shock (foreshock pressure pulses) continually buffet the magnetosphere. They compress the magnetopause, enhance magnetospheric magnetic field strengths, and excite resonant azimuthal magnetic field oscillations [Sugiura et al., 1968; Kaufmann and Walker, 1974; Nopper et al., 1982]. Low-latitude ground magnetometers record sudden impulse/sudden storm commencement (SI/SSC) signatures as simple step function increases (or decreases) in the H component, whereas high-latitude ground magnetometers often record bipolar signatures in the H component [Matsushita, 1957]. The sense of the bipolar signature depends upon local time; positive/negative signatures generally occur prior to local noon and negative/positive signatures after local noon. Recent observations with high time and spatial resolution allow the latter signatures to be interpreted as antisunward-moving convection vortices [Friis-Christensen et al., 1988].

SI and SSC also excite damped resonant oscillations with periods in the range from 5–15 min in high-latitude ground magnetograms [Wilson and Sugiura, 1961; Saito and Matsushita, 1967]. Kivelson and Southwood [1985] noted that observed resonances are often dominated by discrete frequencies and suggested that an impulsive source might excite magnetohydrodynamic (MHD) cavity mode oscillations at discrete eigenfrequencies. Samson et al. [1992] recently reported oscillations at discrete periods of 5, 6.3, 8.8, and 12.8 min observed by the Goose Bay radar at high latitudes during local morning.

Previous work provided a wealth of detail concerning the statistical patterns of signatures seen by ground magnetometers but was often hampered by the lack of any solar wind or magnetospheric magnetic field observations. Glassmeier and Heppner [1992] presented a case study comparing the geosynchronous magnetic field strength to global H component variations at ground stations but without simultaneous solar wind observations.

In this paper, we will present the high-time-resolution solar wind plasma and magnetic field observations, multipoint magnetospheric magnetic field observations, and global ground magnetometer observations necessary to understand the magnetospheric response to a series of abrupt step-function solar wind dynamic pressure variations which occurred on November 24, 1986. We will emphasize the relationship between the solar wind pressure variations and the transient and quasi-periodic signatures which they excite in the ionosphere. In particular, we will present evidence for sudden sharp variations in the solar wind dynamic pressure and argue that they are a likely source for widespread resonances with discrete frequencies.

SOLAR WIND OBSERVATIONS

During the 3-hr period of interest from 0900–1200 UT on November 24, 1986, IMP-8 was located in the solar wind and moved

Solar Wind Sources of Magnetospheric Ultra-Low-Frequency Waves
Geophysical Monograph 81

from GSE (x, y, z) = (20.9, −27.5, −5.5) to (22.4, −26.2, −4.6) R_E, as shown in Figure 1. In Figure 2 the panels show, from top to bottom, the solar wind speed, density, dynamic pressure, magnetic field strength, latitude, and longitude in GSM coordinates. The solar wind speed generally varied between 450 and 500 km s^{-1}. Prior to 0925 UT and after 1105 UT, the solar wind density was low (~5–10 cm^{-3}), and the IMF strength was high (~16–20 nT). During the intervening period, IMP-8 detected an embedded region of high densities (20–30 cm^{-3}) and lower magnetic field strengths (8–12 nT). Magnetic field and density variations during this period were generally out of phase, indicating a rough balance between solar wind thermal and magnetic

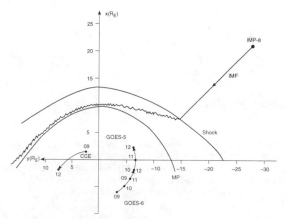

Fig. 1 The locations of IMP-8, AMPTE CCE, and GOES-5 and -6 in a geocentric, solar-ecliptic X-Y projection.

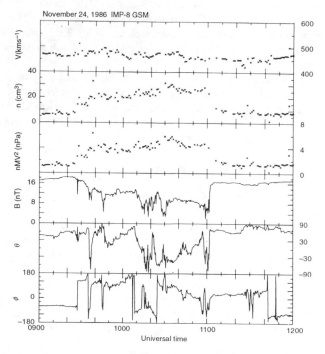

Fig. 2 IMP-8 solar wind observations. From top to bottom: the solar wind velocity, density, dynamic pressure, magnetic field strength, latitude, and longitude in GSM coordinates. Latitude 0°, longitude 0° indicates the sunward direction, and longitude 90° is duskward.

pressures. Solar wind dynamic pressure variations during the interval from 0925–1105 UT mainly corresponded to variations in the solar wind density and included step function increases at 0925–0927, 0934–0937, 1005–1008, and 1025–1030 UT, as well as a step function decrease at 1101–1103 UT. There were other numerous, less abrupt fluctuations in the solar wind dynamic pressure. During the period from 0925 to 1100 UT, the solar wind dynamic pressure varied from 4–6 nPa, about twice its typical value. The IMF was generally northward except during the interval from 1010–1100 UT, and generally pointed either duskward and antisunward or dawnward and sunward, as illustrated in Figure 1.

MAGNETOSPHERIC OBSERVATIONS

During the period of interest, geosynchronous satellites GOES-5 and -6 moved through the post-midnight magnetosphere. As shown in Figure 1, GOES-6 moved from 0200 to 0500 LT, and GOES-5 moved from 0400 to 0700 LT. Figure 3 shows GOES-5 and -6 magnetospheric magnetic field observations in N, E, and P coordinates, where P is parallel to the spacecraft spin axis, E points earthward along the radius connecting the satellite and earth, and eastward-pointing N completes the triad. A series of magnetospheric magnetic field compressions were observed at both satellites. Inspection of the N component shows most clearly that GOES-5 observed the features about 1-2 min prior to GOES-6, consistent with the antisunward propagation expected for features travelling with the solar wind and magnetosheath flow. Although there was no detailed correspondence between each variation in the solar wind dynamic pressure and the magnetospheric magnetic field strength, there were abrupt magnetospheric magnetic field strength increases at 0925, 0935–0940, 1000–1005, and 1032–1035 UT, and the GOES-5 field strength fell abruptly at 1105–1110 UT. The amplitude of the field strength increase at 0925 UT was ~5 nT at GOES-5, but only ~3 nT at GOES-6. The times of the prominant magnetospheric magnetic field strength increases and decreases correspond approximately to those of the solar wind pressure increases and decreases. That effect results from the location of IMP-8 upstream of the dawn bow shock and the often typical spiral IMF orientation illustrated in Figure 1. The field strength increases at 0935-0940, 1000–1005, and 1030 UT were followed by azimuthal magnetic field oscillations in the E component at both satellites. Azimuthal oscillations which began again at about 1050 UT, following a drop in the GOES-5 magnetic field strength, intensified at the time of the 1105–1110 UT field strength decrease at GOES-5 and then decayed. The oscillations from 0935–0945 and 1055–1110 UT reached a peak amplitude of 3–4 nT and had periods of from 3–4 min, whereas those from 1030-1050 UT reached amplitudes of ~1-2 nT and had periods of 45–60s.

As Figure 1 showed, the CCE satellite moved from a radial distance of 3.0 R_E and a magnetic local time of 16.0 to a radial distance of 7.2 R_E and a magnetic local time of 19.0 during the period of interest. The satellite was therefore located in the inner dusk magnetosphere. Figure 4 shows CCE observations, detrended by removing a cubic fit to the observations during the interval from 0918-1120 UT. It is interesting to compare the variations in the total magnetic field strength observed at the CCE to those seen at GOES-5 and -6. Whereas the CCE observes an ~10-nT compression at the time of the SC, GOES-5 and -6 observe much smaller increases in the

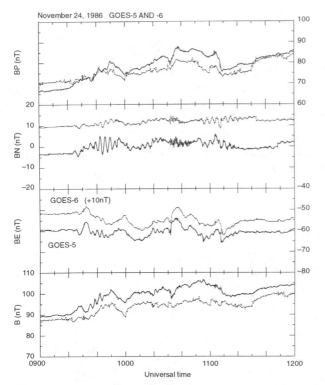

Fig. 3 GOES-5 and -6 magnetic field observations in N, E, P coordinates (P is parallel to spacecraft spin axis, E points earthward along radius connecting satellite and earth, and N completes the triad).

dawn magnetosphere. Furthermore, there is at most only a general correspondence among enhanced magnetic field strengths at the CCE (Figure 4), enhanced magnetic field strengths at GOES-5 and -6 (Figure 3), and enhanced solar wind dynamic pressures (Figure 2). Clearly the signatures produced by variations in the solar wind dynamic pressure depend strongly upon the location of the observing satellite. Unlike GOES-5 and -6 on the dawn side of the magnetosphere, CCE observes little evidence for resonant azimuthal (B_x) oscillations in response to variations in the solar wind dynamic pressure applied to the magnetosphere.

GROUND OBSERVATIONS

For purposes of this study, we have assembled digital records of ground magnetograms from the 27 mid- and high-latitude northern hemisphere observatories indicated in Figure 5. The panels in Figure 5 illustrate variations in the X components at each of these stations. It is clear that the SC at 0925 UT marked the onset of a disturbed period at all stations. However, the response varied greatly from station to station. Considering first the initial impulsive response, we note positive/negative variations at pre-noon stations in Greenland and eastern North America, but negative/positive variations at stations near and just after local noon. Stations in central North America recorded slight negative, then strong positive, variations. The response on the nightside was more complex. Note that the amplitude of the dayside signatures decays rapidly with decreasing latitude: the responses at GLL and OTT are much less than those at BLC and FCC, the response at STJ is much less than those at THL, GDH, and NAQ,

and the response at BOX is much less than those at MMK, ABK, and SOD.

To study global characteristics of pulsations, we applied a band-pass filter in the range from 3–20 min to the X traces of the ground magnetograms. Figure 6a shows pulsations observed at auroral latitudes. We can define three discrete packets of pulsations during this interval: 0923–1020, 1030–1100, and 1100–1120 UT. Comparing Figure 6a with Figure 2 indicates that the first packet began during a period of strongly northward IMF, the second during a period of southward IMF, and the third at a transition between northward and southward IMF. Each packet begins with an impulsive oscillation which then decays with time. The oscillations were coherent over a 30–50° longitude range. The first packet of pulsations followed the 0923 UT SC and was characterized by regular 8–10 min period oscillations in the morning (FCC) and afternoon (DIK) sectors of magnetosphere. The second packet followed the 1030 UT increase in solar wind dynamic pressure and magnetospheric compression. It was marked by several irregular pulse-like variations. The third packet followed the 1100 UT peak in solar wind dynamic pressure and magnetospheric compression. Three cycles of a 7–8 min period oscillation were detected by nearly all stations. Figure 6b shows responses at high latitudes. Here again, the pulsations were observed in packets, but these did not coincide in time with the packets in the auroral zone and were more localized in latitude and longitude. A packet from 0923–1000 UT with a 9-min period was observed at CBB and BLC only and was therefore confined to about 5° in latitude and 20° in longitude. A second packet with period 13–15 min was noted at BLF from 1000–1120 UT. A third packet of

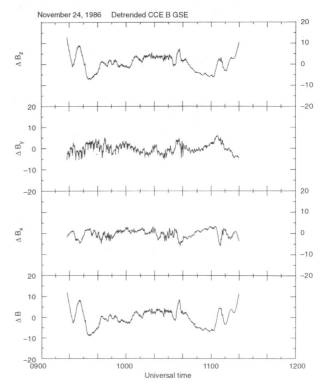

Fig. 4 Detrended CCE magnetic field observations in GSE coordinates.

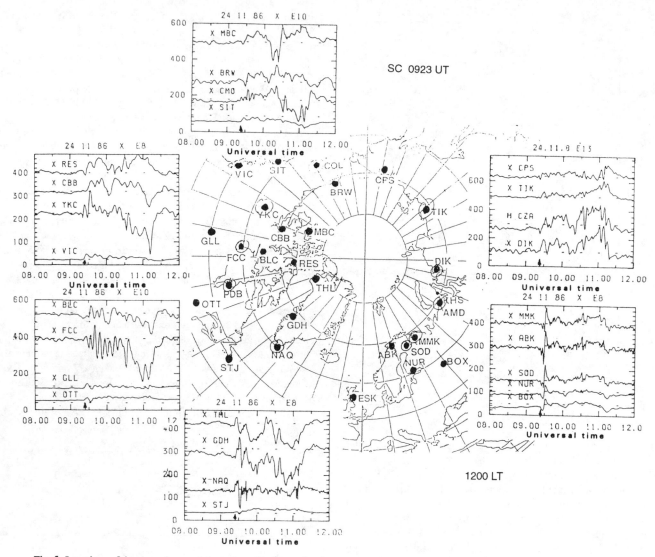

Fig. 5. Locations of the ground magnetometers used in this study. Separate panels present the north/south components of the magnetic field at various stations.

pulsations with period 12 min was observed at MBC from 1010–1040 UT.

The polarization characteristics of the pulsations in the horizontal plane varied with latitude and local time. Polarization in the auroral zone tended to be elliptical. During the first packet, counterclockwise rotation was observed in the morning sector, clockwise rotation in the afternoon sector, and mixed in the midnight sector. Figure 7a shows the pattern of auroral zone polarization observed during this packet as a function of local time. No clear pattern versus local time could be discerned during the second packet of auroral zone pulsations. Counterclockwise rotations were observed during the third packet of auroral zone pulsations.

At higher latitudes, the sense of rotation was primarily clockwise in the midnight and morning sectors (MBC, RES, CBB, BLC), but counterclockwise just prior to local noon (GDH, THL). However, counterclockwise polarization was observed from 1027–1056 UT at

MBC and 1042–1105 UT at BLC. Furthermore, from 1050–1100 and 1110–1120 UT, clockwise polarization was observed at THL. The sense of polarization at all mid-latitude stations was counterclockwise until 1020 UT. After that, clockwise polarizations were observed at post-noon stations, i.e., from 1020–1050 UT at ESK and NUR, and from 1020–1120 UT at BOX. Figure 7b presents hodographs for the first packet of pulsations at high-, auroral-, and middle-latitude stations at pre-noon local times for the interval from 0923–1020 UT, following the SC. The sense of polarization at high-latitude stations was righthanded, opposite that at auroral and middle latitudes.

To study the spectral characteristics of the SC-associated pulsations, we calculated amplitudes of the Fourier spectra for the interval from 0900–1200 UT. Figure 8 shows examples of X-component spectra. Many spectral maxima appear in the range of periods from 3 to 18 min. The dominant periods were a function of latitude and

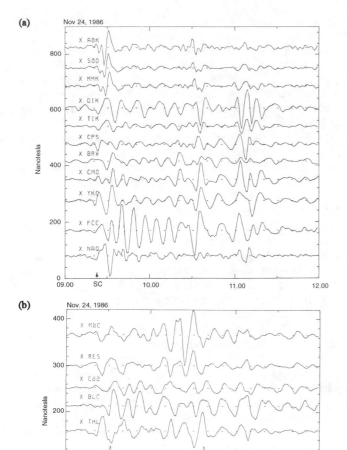

(a) Nov 24, 1986

(b) Nov. 24, 1986

Fig. 6 The north-south component of ground magnetograms band-passed filtered in the range of periods from 3–20 min at (a) auroral latitudes (60 < λ < 70°) and (b) high latitudes (λ > 75°).

ground magnetometer observations to study the global magnetospheric response to sudden variations in the solar wind dynamic pressure on November 24, 1986. Our results confirm and extend those obtained by previous authors.

We first used simultaneous solar wind and magnetospheric observations to demonstrate the expected correspondence between varia-

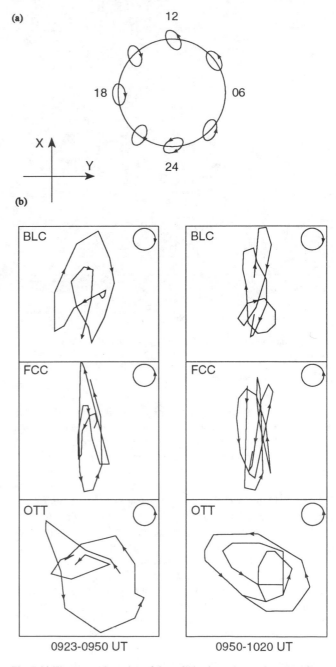

Fig. 7 (a) The sense of rotation of the polarizations observed at auroral zone latitudes as a function of local time. (b) Hodograms of the horizontal magnetic field vectors observed at high (BLC), auroral (FCC), and middle (OTT) latitude stations in the morning sector of magnetosphere during the period from 0923–1020 UT.

local time. At middle and subauroral latitudes, the main spectral maximum corresponds to a 13.8-min period at NUR and BOX, but to 12 min at other locations. Periods of 9.5, 7.5, and 3.4 min were common. At some stations, periods of 18, 6.7, and 5.2 min also occurred.

At auroral latitudes, the primary spectral peak corresponded to a period of 9.5 min in the afternoon sector (ABK, SOD, MMK), to 12 min in the evening sector (DIK), and to 13.8 min in the midnight and morning sectors (YKC, BRW, CMO). A secondary spectral peak occurred at 7.7 min in the afternoon sector and at 8.2 min in the morning sector. The latter peak became dominant at FCC. Peaks at 18, 10, 6.9, 3.7 min were also observed. At high latitudes, the main spectral peaks occurred at 11.2 min at MBC, RES, and CBB but at 13.8 min at BLC. Another peak at 9.5 min was widespread. Periods of 18, 7.8, 6.7, 5.3, and 3.3 min were noted at some stations.

SUMMARY AND CONCLUSIONS

We have used simultaneous solar wind, magnetospheric, and

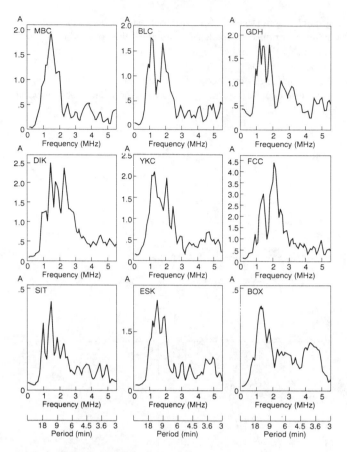

Fig. 8 Results of a spectral analysis of the X-components of nine ground magnetograms during the period from 0900–1200 UT. MBC, BLC, and GDH are at high latitudes; DIK, YKC, and FCC are at auroral latitudes; and SIT, ESK, and BOX at mid-latitudes. The amplitudes are in arbitrary units.

tions in the solar wind dynamic pressure, antisunward moving compressions in the dawnside magnetospheric magnetic field, and impulsive signatures at high geomagnetic latitudes. However, duskside magnetospheric magnetic field strength variations observed by the CCE did not greatly resemble those on the dawnside obtained by the GOES satellites. This may suggest that the scale lengths of the relevant solar wind features seen at pre-noon local times by IMP-8 did not extend far past the earth-sun line. Alternatively, we have previously suggested that the IMF orientation controls the nature of the solar wind interaction with the bow shock and therefore the solar wind dynamic pressure applied to the magnetosphere [Sibeck et al., 1989]. As shown in Figure 1, the fact that the pre-noon magnetopause lay behind the quasi-parallel bow shock and the post-noon magnetopause behind the quasi-perpendicular bow shock suggests that the differing magnetic field responses from CCE and GOES may indeed depend upon the bow shock interaction.

Each compression of the magnetospheric magnetic field caused an initial bipolar variation in high-latitude dayside ground magnetometer X-components: north/south before local noon and south/north after local noon. A similar pattern was reported by Matsushita [1957]. We showed that the amplitude of the ground response decays rapidly equatorward from a peak in the vicinity of the auroral oval.

Increases in solar wind dynamic pressure also initiated elliptically polarized oscillations in the auroral oval which were polarized counterclockwise before local noon and clockwise after local noon, consistent with results previously reported by Wilson and Sugiura [1961]. The amplitudes of these oscillations were greater on the dawnside than on the duskside, consistent with previous statistical studies as well as simultaneous magnetospheric observations by GOES-5 and -6 and the CCE.

The sense of polarization for the resonant oscillations reported here reverses with latitude, as in the case of the impulsive events by Araki and Allen [1982]. Thus it seems likely that the impulsive and resonant events have common origins but that the impulsive events are even more rapidly damped. The packets were localized in latitudinal and longitudinal extent. Although the dominant periods ranged from 3 to 18 min and were a function of local time, secondary peaks with similar periods were frequently observed over a wide range of latitudes and local times. We found no evidence indicating that the transient ground events and quasi-periodic oscillations were initiated solely during periods of southward IMF, and therefore we conclude that they are unrelated to magnetic merging.

The bipolar ground magnetometer signature observed in the auroral oval at the start of each wave packet greatly resembles those seen during antisunward moving convection vortices [e.g., Friis-Christensen et al., 1988]. For example, the bipolar signatures reach peak amplitudes at auroral zone stations and appear to be associated with transient compressions of the magnetospheric magnetic field and SI/SC signatures at low latitudes. Yet Glassmeier [1992] argues that the ground signatures of travelling convection vortices are quite localized and therefore differ from the large scale convection patterns that correspond to the global SI/SC response. Since several studies now indicate that the traveling convection vortices may be observed over a wide range of local times, the discrepancy may well result from the use of low-time-resolution data in the older SI/SC studies and an assumption that the transient events were observed simultaneously over a wide range of local times. The travelling convection vortices should be distinguished from the vortices discussed by Saunders et al. [1992], which swell (but do not move) in the antisunward direction during periods of southward IMF.

Variations in the solar wind dynamic pressure launch fast-mode waves into the magnetosphere. Hasegawa et al. [1983] showed that these waves may couple to resonant transverse oscillations over a wide range of local times and latitudes. The periods of the resulting oscillations would depend on the observation site. On the other hand, the fast-mode waves may initiate resonant-cavity-mode oscillations at fixed frequencies observed over a wide range of latitudes and local times. Our results indicate that a wide variety of frequencies were excited, but some common frequencies were observed by a number of stations. It does not seem to us to be too surprising that none of the common frequencies we observed correspond to those so clearly defined by Samson et al. [1992], because our observations were made under relatively unusual conditions, namely solar wind pressures twice as great as on average. Perhaps the observations we report provide evidence for both resonant-cavity-mode oscillations and local-azimuthal oscillations.

Acknowledgements. Work at APL was supported by NASA under Task 1 of Space and Naval Warfare Systems Command contract N00039-91-C-5001 to the Navy. IMP-8 observations were provided by the NSSDC.

REFERENCES

Araki, T. and J. H. Allen, Latitudinal reversal of polarization of the geomagnetic sudden commencement, J. *Geophys. Res.*, *87*, 5207-5216, 1982.

Friis-Christensen, E., M. A. McHenry, C. R. Clauer, and S. Vennerstrom, Ionospheric travelling convection vortices observed near the polar cleft: A triggered response to sudden changes in the solar wind, *Geophys. Res. Lett.*, *15*, 253-256, 1988.

Glassmeier, K.-H., Traveling magnetospheric convection twin-vortices: Observations and theory, *Ann. Geophys.*, *10*, 547-565, 1992.

Glassmeier, K.-H. and C. Heppner, Traveling magnetospheric convection twin vortices: Another case study, global characteristics, and a model, *J. Geophys. Res.*, *97*, 3977–3992, 1992.

Hasegawa, A., K. H. Tsui, and A. S. Assis, A theory of long period magnetic pulsations, 3. Local field line oscillations, *Geophys. Res. Lett.*, *10*, 765-767, 1983.

Kaufmann, R. L. and D. N. Walker, Hydromagnetic waves excited during an ssc, *J. Geophys. Res.*, *79*, 5187-5195, 1974.

Kivelson, M. G. and D. J. Southwood, Resonant ULF waves: A new interpretation, *Geophys. Res. Lett.*, *12*, 49-52, 1985.

Matsushita, S., On sudden commencements of magnetic storms at higher latitudes, *J. Geophys. Res.*, *62*, 162-166, 1957.

Nopper, R. W., Jr., W. J. Hughes, C. G. Maclennon, and R. L. McPherron, Impulse-excited pulsations during the July 29, 1977 event, *J. Geophys. Res.*, *87*, 5911-5916, 1982.

Saito, T. and S. Matsushita, Geomagnetic pulsations associated with sudden commencements and sudden impulses, *Planet. Space Sci.*, *15*, 573-587, 1967.

Samson, J. C., B. G. Harrold, J. M. Ruohoniemi, R. A. Greenwald, and A. D. M. Walker, Field-line resonances associated with MHD waveguides in the magnetosphere, *Geophys. Res. Lett.*, *19*, 441-444, 1992.

Saunders, M. A., M. P. Freeman, D. J. Southwood, S. W. H. Cowley, M. Lockwood, J. C. Samson, C. J. Farrugia, and T. J. Hughes, Dayside ionospheric convection changes in response to long-period interplanetary magnetic field oscillations: Determination of the ionospheric phase velocity, *J. Geophys. Res.*, *97*, 19373-19380, 1992.

Sibeck, D. G., W. Baumjohann, R. C. Elphic, D. H. Fairfield, J. F. Fennel, W. B. Gail, L. J. Lanzerotti, R. E. Lopez, H. Lühr, A. T. Y. Lui, C. G. Maclennan, R. W. McEntire, T. A. Potemra, T. J. Rosenberg, and K. Takahashi, The magnetospheric response to 8-min-period strong-amplitude upstream pressure variations, *J. Geophys. Res.*, *94*, 2505-2519, 1989.

Sugiura, M., T. L. Skillman, B. G. Ledley, and J. P. Heppner, Propagation of the sudden commencement of July 8, 1966 to the magnetotail, *J. Geophys. Res.*, *73*, 6699-6709, 1968.

Wilson, C. R. and M. Sugiura, Hydromagnetic interpretation of sudden commencements of magnetic storms, *J. Geophys. Res.*, *66*, 4097-4111, 1961.

D. G. Sibeck, The Johns Hopkins University Applied Physics Laboratory, Johns Hopkins Road, Laurel, Maryland 20723-6099.

Generation of ULF Waves by Fluctuations in the Magnetopause Position

ROBERT L. LYSAK AND YAN SONG

School of Physics and Astronomy
University of Minnesota

DONG-HUN LEE

Department of Astronomy and Space Science, Kyunghee University

The position of the magnetopause fluctuates in response to perturbations in the solar wind dynamic pressure and the interplanetary magnetic field. These fluctuations can generate ULF waves in the magnetosphere, with the character of these waves depending on whether the frequency of the fluctuations is greater than or less than field line resonant frequencies in the magnetosphere. These variations have been studied in the three-dimensional dipole model [Lee and Lysak, 1991; Lysak and Lee, 1992]. This model has recently been improved by the addition of an ionosphere with finite conductivity and by the imposition of an open tail boundary condition which allows wave energy to pass out of the dayside magnetosphere. The finite ionospheric conductivity has the effect of damping pulsations on low L-shells more than outer shells, since the Alfvén waves interact more frequently with the ionosphere on these shorter field lines. This gives resonant ionospheric signatures which are more localized toward the magnetopause. In order to introduce effects associated with the interplanetary magnetic field, we have adopted a model involving a B_y-dependent magnetic helicity injection. This model produces a dawn-dusk asymmetry in the observations of vortices, with the vortices being stronger on the dawn (dusk) side of the northern (southern) hemisphere for $B_y > 0$ and on the dusk (dawn) side for $B_y < 0$.

INTRODUCTION

ULF waves in the magnetosphere can be driven either from internal sources such as MHD instabilities or wave-particle interactions or from external forcing due to the solar wind-magnetopause interaction. It is the purpose of this paper to investigate the latter of these sources. Oscillations in the location of the magnetopause can be produced due to a varying equilibrium stand-off distance caused by variations in the magnetosheath pressure. Such variations may be directly related to phenomena in the solar wind, such as interplanetary shocks, or may be produced by magnetosheath instabilities, as is discussed in other contributions to this monograph. In either case, pressure enhancements (decreases) can cause the magnetopause to move inward (outward) in order to maintain the overall pressure balance.

The magnetopause position can also vary due to the effect of magnetic reconnection. Reconnected flux tubes are transported tailward by the solar wind flow, perhaps leading to the forma-

tion of flux transfer events (FTEs) [Russell and Elphic, 1978]. Such flux tubes can also perturb the magnetopause, causing the magnetospheric plasma to flow around the obstacle [Papamastorakis et al., 1989; Sonnerup et al., 1992]. The localized reconnection process injects magnetic helicity across the magnetopause [Song and Lysak, 1989a; Wright and Berger, 1989], leading to a dynamo effect generating field-aligned currents [Song and Lysak, 1989b]. Such perturbations radiate away from the reconnection site as ULF waves [Song et al., 1993].

From the ground, the most distinct signature of these processes is the observation of the so-called traveling convection vortices (TCVs) which have been observed by a number of groups [Lanzerotti et al., 1987; Friis-Christensen et al., 1988; Glassmeier et al., 1989]. These vortices generally consist of two or more convection vortices of alternating polarity, corresponding to alternating field-aligned current filaments. Although the simplest of these structures is a dual vortex pattern, more complicated structures are often observed [McHenry et al., 1987, 1990]. Lanzerotti et al. [1991] have shown that a triple current system is often observed, with the central current being the strongest.

In this paper, we will present modeling results which will attempt to distinguish between the effects of pressure perturbations and magnetic reconnection by including magnetic helicity injection. This model is based on a three-dimensional model of MHD wave propagation [Lee and Lysak, 1991; Lysak and Lee, 1992]. This model has been extended to include the presence of an ionosphere with finite conductivity, as opposed to the infinite conductivity model presented earlier. In addition, we have implemented a new boundary condition corresponding to the direct injection of magnetic helicity.

MODEL DESCRIPTION

The basic equations of the model are the cold plasma MHD equations for the magnetic field and the perpendicular components of the electric field (here written in cgs units):

$$\frac{\partial \mathbf{B}}{\partial t} = -c \ \nabla \times \mathbf{E} \tag{1}$$

$$\frac{\partial \mathbf{E}_\perp}{\partial t} = \frac{V_A^2}{c} (\nabla \times \mathbf{B})_\perp \tag{2}$$

These equations are cast into dipole coordinates (μ, ν, ϕ), where μ is the field-aligned coordinate, $\nu = -1/L$ is in the outward normal direction, and ϕ is the usual azimuthal angle. Details of the two-dimensional model on which this model is based are presented in Lee and Lysak [1989], and the three-dimensional extensions have been discussed by Lee and Lysak [1991] and Lysak and Lee [1992].

Two fundamental changes have been made compared to previous versions of this model. First, as is well known [Scholer, 1970; Mallinckrodt and Carlson, 1978; Goertz and Boswell, 1979], the conducting ionosphere reflects Alfvén waves incident upon it with a reflection coefficient given by:

$$R = \frac{E_{ref}}{E_{inc}} = \frac{\Sigma_A - \Sigma_P}{\Sigma_A + \Sigma_P} \tag{3}$$

where $\Sigma_A = c^2/4\pi V_A$ (cgs units) or $\Sigma_A = 1/\mu_0 V_A$ (SI units) is called the Alfvén conductivity and Σ_P is the ionospheric Pedersen conductivity. Note that for infinite Pedersen conductivity this reflection coefficient is -1, indicating that the reflected and incident electric fields exactly cancel, i.e., the total electric field vanishes. The reflection coefficient given in Equation (3) has been used to calculate the ionospheric electric field in the present code, as opposed to the previous papers in which $\mathbf{E} = 0$ at the ionosphere was assumed. It should be noted that the reflection coefficient given by (3) is only valid for uniform conductivities; for conductivity profiles with gradients, the coupled Ohm's Law and current continuity equations must be solved explicitly [Glassmeier, 1983].

A second model improvement has been implemented which decreases the necessary computation time while not significantly affecting the results. This has been to eliminate the nightside magnetosphere from the model, effectively allowing the wave energy to escape down the geomagnetic tail. This has been done by adopting a zero reflection coefficient as given by Equation (3) on the dawn and dusk meridians. Comparisons of runs with this boundary condition with the runs shown in Lysak and Lee [1992] indicate no significant difference.

One final difference concerns a new type of diagnostic used.

Previously, we have presented snapshots of the ionospheric signatures of runs at particular times. While this gives an idea of the general structure of the waves and is useful for studies of magnetospheric eigenmodes [e.g., Lee and Lysak, 1993], it does not compare directly with ground observations. Friis-Christensen et al. [1988] and Glassmeier et al. [1989] have presented the data from arrays of magnetometers in a vector format in which each successive time was offset by a particular amount, providing a picture of the vortex structure under the assumption that the structure was static and moving over the magnetometer array at a fixed rate. Since each grid point in our model is in effect a magnetometer, a magnetometer chain can be constructed at each local time in the model and displayed in the same format as the true observations. By comparing these different types of displays, a more complete picture of the vortex structure can be deduced.

The runs discussed below are driven in a similar manner to those presented in Lysak and Lee [1992]. It is assumed that a pulse of enhanced solar wind dynamic pressure is incident on the magnetosphere. Thus, the magnetopause is compressed inward when the pressure pulse hits, and relaxes back outward as the pulse passes by. This motion of the magnetopause is implemented in the code by applying a boundary condition on the azimuthal electric field E_ϕ, which corresponds to a velocity in the radial (ν) direction. The pulse has the form of an anti-sunward propagating plane wave in the solar wind in which the wave front is perpendicular to the sun-earth line. Thus, the front interacts with the magnetosphere at noon magnetic local time first, and then propagates toward both the dawn and dusk, with its amplitude diminishing as it reaches the terminator. The wavelength of all runs is assumed to be 6 R_E in the solar wind, and the amplitude of perturbation corresponds to a magnetopause displacement of 0.1 R_E at the sub-solar point. Two periods of oscillation will be considered: a fast case with 60 second period, and a slower case with a period of 240 seconds.

IONOSPHERIC CONDUCTIVITY EFFECTS

In order to assess the effect of finite ionospheric conductivity, we performed runs in which the ionospheric conductivity was set to a constant value of 1 mho. This value is rather low for a dayside conductivity, but serves to emphasize the effect of ionospheric damping. According to Equation (3), this implies that a fraction of the wave energy is lost every time the Alfvén wave interacts with the ionosphere. Figure 1 shows a snapshot of the perpendicular components of the ionospheric magnetic field for a run with a 60 second pulse. As noted in Lysak and Lee [1992], this period is fast enough to excite field line resonances in the magnetosphere. For reference, the 60 second period corresponds to the fundamental, third harmonic and fifth harmonic at latitudes of 64°, 69°, and 71°, respectively (only odd harmonics are excited due to the symmetry of the driving perturbation). Figure 1 shows enhancements in the toroidal field corresponding to the third and fifth harmonics; however, structure at the fundamental is absent. This is in contrast to the case with infinite conductivity, shown in the bottom panel of Figure 4 of Lysak and Lee [1992], in which the fundamental was excited.

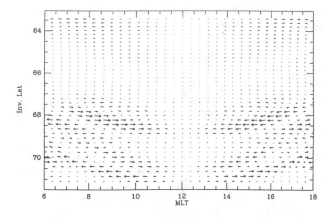

Fig. 1. Magnetic field vectors at the ionosphere boundary of the simulation at a time of 240 seconds for a run with a driving pulse with 60 second period and an ionospheric conductivity of 1 mho. Vectors are normalized such that a 50 nT field is represented by a vector with length equal to the horizontal spacing of the points. Note that in this figure, as well as Figures 3 and 6, the view is up along the field line in the northern hemisphere ionosphere

This difference is due to the variation in the Alfvén travel time on the various field lines. At lower latitudes (and therefore lower L-shells) the Alfvén travel time is much shorter than on the outer L-shells, and thus the wave interacts more often with the ionosphere. Each interaction causes a loss of wave energy according to Equation (3), and so the wave energy is damped more quickly on the inner shells than the outer ones.

A different view of these runs is given in Figure 2. In this figure, the magnetic perturbations at the ionosphere is given as a function of time and invariant latitude for the grid points at 9.75 MLT. Each successive time value is displaced toward the east (i.e., to the right) since the motion of the vortex on the morning side is to the west. The top panel gives the results for the infinite conductivity case, while the bottom panel is for a conductivity of 1 mho. A number of things can be seen from this plot. First of all, in both runs there is a transient period of about 120 seconds during which the waves are unstructured and reflect the direct driving of the system by the input pulse. Note that with a period of 60 seconds and a wavelength of 6 R_E, the pulse requires 100 seconds to travel the 10 R_E distance from the sub-solar point to the terminator. Since the entire pulse is 60 seconds in duration, the driving pulse is totally out of the system by 160 seconds, and the compressional phase of the pulse, which lasts 30 seconds into the run, leaves the system by 130 seconds.

After the driving pulse leaves the system, the wave structure organizes itself into resonant field lines where the amplitude is larger and non-resonant field lines with lower amplitudes. In the infinite conductivity run, there is still noticeable wave power on nearly all field lines; however, in the finite conductivity run, the non-resonant field lines have decayed away while toroidal oscillations on the resonant shells damp more slowly. Note also a characteristic poleward motion of the nodes and antinodes of the wave. This dispersion is due to the difference in Alfvén travel time along each of the field lines.

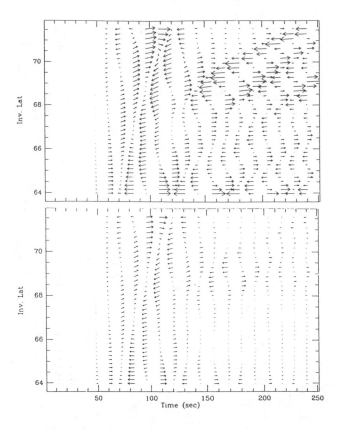

Fig. 2. Time history of the magnetic field at the ionosphere boundary for a set of grid points at 9.75 hours MLT. The top panel gives results for a run with infinite ionospheric conductivity, while the bottom panel is a run with 1 mho conductivity. Vectors are again normalized to 50 nT equal to the horizontal spacing of the points. In this figure, as well as Figures 4, 5, and 7, the view is looking down on the northern hemisphere ionosphere.

Let us now turn to longer period perturbations, and assume a 240 second period, which is typical for traveling convection vortices. This period is longer than the fundamental global eigenperiod for our parameters and corresponds to the field line resonance frequency near the outer part of the simulation, at a latitude of 70° [Lysak and Lee, 1992]. Thus, the ionospheric signatures for these runs do not penetrate far into the magnetosphere and remain near the magnetopause.

Figure 3 shows snapshots of the ionospheric magnetic perturbations at times of 160, 240 and 400 seconds. Note that these snapshots are separated by half the total period of the wave. A few facts should be noted in looking at these plots. First, this is the northern hemisphere ionosphere. Secondly, the figure is oriented from the point of view of an observer looking back up the field line. Thus, a clockwise rotation of the magnetic field vectors corresponds to an upward field-aligned current, and counter-clockwise rotations are downward field-aligned currents. In other words, the background magnetic field vector comes out of the page, i.e., in the downward direction for the northern hemisphere. It should also be noted that the vectors in this figure have been normalized so that the distance between

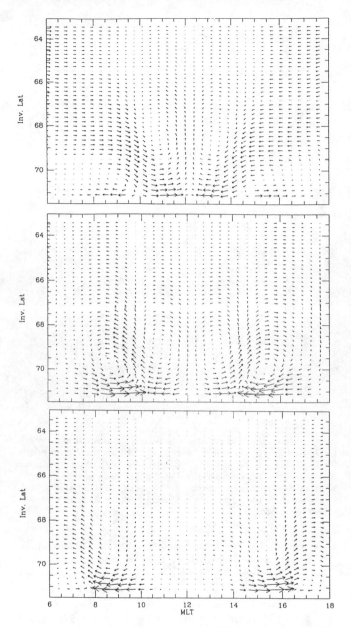

Fig. 3. Ionospheric magnetic perturbations at 3 times for a run with 240 second driving period. The top panel is at 160 seconds, the middle panel is 280 seconds, and the bottom panel is at 400 seconds. Vectors are normalized to 5 nT.

adjacent grid points in 5 nT, as opposed to 10 nT in the corresponding figure (Figure 2) of Lysak and Lee [1992].

At 160 seconds into this run (top panel), incompletely formed vortices can be seen with centers at 70.5° latitude and magnetic local times of 9 and 15 hours, and new vortices are forming at about 68° and 11 and 13 hours MLT. The leading vortices, those at 9 and 15 MLT, have a current which is downward on the dawn side and upward at dusk, while the 11 and 13 MLT vortices have the opposite polarity. At 280 seconds (top panel), the leading vortices have moved across the terminator, while

the second set of vortices is now at 8.5 and 15.5 MLT. In addition, these vortices have moved poleward to about 70° latitude. By this time, a third vortex pair can be seen, with polarity the same as the leading vortex structure. By 400 seconds (bottom panel) the second pair of vortices is almost out of the system, and the third pair is very weak. At this late time in the run, the driving pulse has left the system, and the damping due to the finite ionospheric conductivity has caused the vortex structures to decay.

This current system can be understood either from a mechanical or an electrodynamic point of view. The input at the magnetopause consists of a radially inward velocity followed by a radially outward velocity. Note that the inward velocity region is on the antisunward side of the pulse, i.e., at earlier local times at dawn and later local times at dusk. In this inward region, the magnetic field is compressed, leading to an enhancement of the magnetic pressure which diverts the flow to both sides azimuthally. Similarly, in the region of outward flow the magnetic pressure is decreased, which draws the flow in azimuthally. Thus, the flow between the inward and outward velocity regions is driven azimuthally toward noon on both the dawn and dusk sides. The magnetic field lines are frozen into this flow; the resulting motions twist the field lines in a sense which gives rise to a downward current at dawn and upward at dusk. Weaker, oppositely directed currents lead and follow the entire structure by a similar argument.

From an electrodynamic point of view, the inward and outward flows are accompanied by dawnward and duskward directed azimuthal electric fields. Thus, the electric field has a positive divergence on the dawn side and negative divergence at dusk; therefore, the region between the inward and outward flow has positive space charge at dawn and negative charge at dusk. These charges discharge along the magnetic field lines, again producing downward currents on the dawn side and upward currents at dusk, preceded and followed by currents of opposite polarity. The resulting flows and currents are indicated schematically in Figure 4.

While these considerations give a rough idea as to the formation of the field-aligned currents, the positions of the resulting signatures in the ionosphere are influenced by the combined propagation of the compressional and transverse waves through the magnetosphere. The compressional signal propagates inward to the magnetopause until it reaches the position of the field line resonance, which is at 70° for this case as mentioned above. It is at this point that the ionospheric vortices form.

Another view of this run is given in Figure 5, which shows the magnetic field vectors at MLT=9 (top panel) and MLT=15 (bottom panel) plotted as a function of time. In this figure, the latitude scale has been reversed, so that the point of view is now looking down on the ionosphere from above. Thus clockwise (counterclockwise) rotation of the vectors corresponds to downward (upward) current. Note also that the time scale on the afternoon figure has been reversed, since the motion of these vortices is from west to east. The triple vortex structure can be clearly seen in this figure. In each case, the central current provides the most complete vortex pattern with the lead-

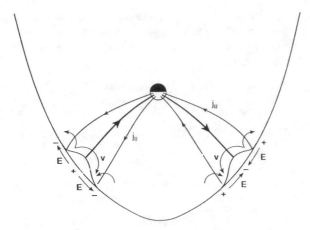

Fig. 4. Schematic illustration of the effect of compression of the magnetosphere, showing the plasma flows, electric fields, and current systems produced by this interaction.

ing and trailing vortices being not as well defined. The sense of this central current is downward (upward) in the morning (afternoon). In other words, the central currents are in the Region 1 sense, while the leading and trailing currents are

opposite. This is consistent with the model current systems proposed by Lanzerotti et al. [1991] and Lühr and Blawert [1993].

Figure 6 shows results from another run in which the magnetopause was compressed and stayed compressed, i.e., the solar wind dynamic pressure increases from one value to a higher value. Again morning (top panel) and afternoon (bottom panel) histories are shown. As might be expected from the above considerations, the magnetopause compression gives rise to vortices with currents going up and then down on the morning side, and vice versa in the afternoon. The opposite pattern should hold for an expansion of the magnetopause. Thus, the triple vortex pattern in Figure 5 can be considered to be the result of a compression followed by an expansion. The trailing current associated with the compression combines with the leading current of the expansion to make the central current stronger.

INJECTION OF MAGNETIC HELICITY

The above models all referred to a case in which the magnetopause was compressed without any change in the linkage of magnetic field lines, i.e., without magnetic reconnection. Indeed, the interplanetary magnetic field (IMF) does not play any role in these models. While statistical studies of the IMF

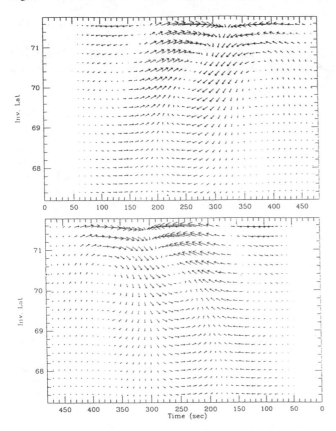

Fig. 5. Time history of the ionospheric magnetic field for the run of Figure 3 at 9 MLT (top panel) and 15 MLT (bottom panel). Vectors are normalized to 5 nT. Note that the time scale for the 15 MLT case is plotted in reverse to account for the eastward propagation of events in the afternoon.

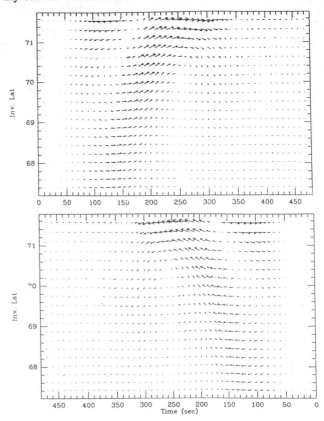

Fig. 6. Time histories of the ionospheric magnetic field at 9 MLT (top) and 15 MLT (bottom) for a run similar to that of Figures 3 and 4 but with only a compressional phase and no relaxation back to the initial state. Vectors are again normalized to 5 nT, and the 15 MLT time scale is reversed.

direction in association with these events are not common, it seems reasonable to ask the question of whether the IMF orientation has any effect on the production of ULF waves in general and traveling vortices in particular. Magnetic reconnection is likely to produce variations in the magnetopause position by changing closed flux tubes into open ones and transporting them tailward as a flux transfer event (FTE).

FTEs are open flux tubes with excess total pressure within them, which is balanced by the twist of the magnetic field around them [e.g., Farrugia et al., 1988]. As these structures propagate around the magnetopause, they can act as a pressure perturbation on the magnetopause, and the resulting flow around the obstacle can be analyzed [Papamastorakis et al., 1989; Sonnerup et al., 1992]. In addition to distorting the magnetopause, propagating FTE structures cause continuing reconnection to occur as they cut through the interplanetary B_y component [Sonnerup, 1987]. This reconnection adds toroidal flux to the flux tube, resulting in a generation of field-aligned current through the tube. This process can be described as a current dynamo effect [Song and Lysak, 1989b; see also Song et al., this monograph] associated with the localized reconnection.

This current dynamo can be described in terms of an injection of magnetic helicity across the magnetopause. Magnetic helicity is defined in terms of the magnetic field and vector potential:

$$K = \int dV \ \mathbf{A} \cdot \mathbf{B} \qquad (4)$$

where the integral is over a volume bounded by magnetic surfaces. In the presence of finite B_y, it has been shown [Song and Lysak, 1989a; Wright and Berger, 1989] that the reconnection of a single flux tube gives rise to a transfer of helicity given by $\Delta K = \pm(1/2)\Phi^2$, where Φ is the amount of flux which is reconnected. The sign in this relation is equal to the sign of the IMF B_y component. Maxwell's equations imply a conservation law for the magnetic helicity [Moffatt, 1978]:

$$\frac{\partial}{\partial t}(\mathbf{A} \cdot \mathbf{B}) + c \, \nabla \cdot (\mathbf{E} \times \mathbf{A}) = -2c \, \mathbf{E} \cdot \mathbf{B} \qquad (5)$$

Note that the right hand side of this equation vanishes for ideal MHD. Note that Equation (5) implies that the quantity $c\,\mathbf{E} \times \mathbf{A}$ is the flux of magnetic helicity. Thus, we can treat the transfer of helicity across the magnetopause by considering this helicity flux.

Putting this in terms of the dipole coordinates used in the model, we need to consider the helicity flux in the $\hat{\nu}$ direction, i.e., the direction across L-shells. Since the parallel electric field is assumed to be zero, this implies that the relevant term in the helicity flux is $cE_\phi A_\mu$. Note that the parallel component of the vector potential A_μ is responsible for twisting the flux tube, and is related to the field-aligned current by $\nabla^2 A_\mu = -(4\pi/c)j_\mu$. Thus, the injection of magnetic helicity across the magnetopause can be modeled by imposing a boundary condition on A_μ along with the azimuthal electric field E_ϕ considered in the previous runs.

The magnitude of A_μ can be estimated by simple considerations of the geometry of the reconnection. Consider a geomagnetic flux tube with a field B_0 in the $\hat{\mu}$ direction which reconnects for a distance δz along its length. Then the total helicity transferred can be written as:

$$\Delta K = \int dV \ \Delta A_\mu B_0 = \delta z \ \Delta A_\mu \Phi \qquad (6)$$

where Φ is the reconnected flux $\int \mathbf{B}_0 \cdot d\mathbf{S}$. Equating (6) to the relation $\Delta K = \pm(1/2)\Phi^2$ yields:

$$A_\mu = \pm \frac{\Phi}{2 \, \delta z} \qquad (7)$$

Now consider the flux passing through the magnetopause, which must be equal to the total reconnected flux:

$$\Phi = \delta y \ \delta z \ B_n \qquad (8)$$

where B_n is the normal magnetic field and δy is the distance of the reconnection patch in the azimuthal direction. Flux conservation yields the usual condition for the normal magnetic field:

$$B_n = \frac{v_n}{V_A} B_0 \qquad (9)$$

which may be written in terms of the azimuthal electric field:

$$B_n = \frac{c}{V_A} E_\phi \qquad (10)$$

Combining Equations (7), (8), and (10) yields a relationship between the parallel vector potential and the azimuthal electric field:

$$A_\mu = \pm \frac{c \, E_\phi}{2 V_A} \delta y \qquad (11)$$

Note that since the $\hat{\nu}$ direction is oriented out of the magnetosphere, the ν component of the helicity flux should have the opposite sign to the total helicity transferred, i.e., the opposite sign to B_y. Therefore, the plus sign in Equation (11) corresponds to $B_y < 0$, and vice versa. Note as well that the argument above has assumed that reconnection is 100% efficient. Since this is unlikely to be true, we have multiplied the result of Equation (11) by an efficiency factor of 0.20.

We have performed runs adding an input of vector potential according to Equation (11), with all other parameters being the same as the 240 second run shown in Figures 3 and 5. First, a run was performed in which positive helicity was input, corresponding to $B_y > 0$, without any perturbation of the magnetopause position. The results of this run are shown in Figure 7. Note that the injection of positive magnetic helicity leads to a primary current system which is parallel to the geomagnetic field, i.e., downward in the northern hemisphere and upward in the southern hemisphere, on both the dawn and dusk sides of the magnetosphere; negative helicity input gives rise to currents from north to south (Song and Lysak, 1989a). Note also in this figure that the current vortices have moved inward from the magnetopause itself, and occur near the 70° location of the field line resonance.

Next, the helicity injection and the motion of the magneto-

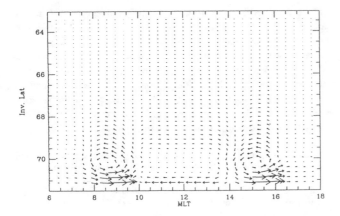

Fig. 7. Ionospheric magnetic perturbations at 280 seconds for a run in which magnetic helicity is injected without a compression of the magnetopause. Note the symmetry between the vortices observed on dawn and dusk sides.

pause were combined into a single run. Figure 8 shows snapshots for the northern ionosphere (top panel) and the southern ionosphere (bottom panel) at a time of 280 seconds for the $B_y > 0$ case. These panels should be compared with the middle panel of Figure 3. It is clear from these figures that the vortex system appears on the dawn side in the northern hemisphere and the dusk side in the southern hemisphere. Note in these figures the background magnetic field is upward out of the plane of the figure. Therefore, the dawn vortex in the top panel has a downward current sense while the dusk vortex in the south has an upward current. To lowest order, these currents can be understood as a superposition of the current system associated with the pressure fluctuation, which is down at dawn and up at dusk (in the north), and the current associated with the helicity injection which is down in both local time sectors in the north. Thus, the currents in the northern dusk side tend to be canceled out, destroying the vortex structure there. Corresponding arguments indicate that vortices should be observed in the dusk southern hemisphere for $B_y > 0$, and that for $B_y < 0$, vortices should be observed in the dusk northern ionosphere and the dawn southern ionosphere. Thus, this model implies that there should be a strong dependence of the local time of vortex observation on the sign of the interplanetary B_y component.

DISCUSSION

The results of the runs presented above indicate some of the complexities which can arise in describing the interaction of externally driven ULF waves in the magnetosphere. One new result presented here is the use of a finite ionospheric conductivity in the models. The major effect of this is to introduce a loss mechanism for wave energy into the model. This has simplified the interpretation of the results since wave energy does not propagate around the system indefinitely as in the case of infinite conductivity. This effect is most clearly seen on lower L-shells, which damp ULF waves more rapidly because of more frequent encounters with the dissipative ionosphere.

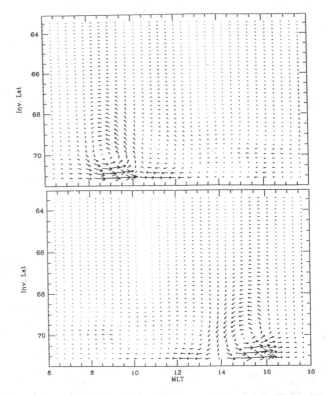

Fig. 8. Ionospheric magnetic perturbations at 280 seconds for runs in which magnetic helicity is injected for the $B_y > 0$ case according to the discussion leading to Equation (11). Top panel shows the northern ionosphere, bottom panel the southern ionosphere. Note that the background magnetic field comes out of the plane of the figure in both panels; thus, the counterclockwise vortex in the top panel corresponds to a downward current, while the counterclockwise vortex in the bottom panel is an upward current. Vectors are normalized to 10 nT.

For higher frequency waves such as the 60 second period runs presented here, the damping effect does not destroy the field line resonant structure, except at these lower L-shells. For lower frequencies, the vortex structure becomes more clear since extraneous waves are damped from the system. Our assumed ionospheric conductivity has been picked to be a rather low value of 1 mho in order to emphasize the effect of ionospheric damping on the system.

This has allowed the sense of the current vortices produced by fluctuations in the magnetopause position to be clearly identified. A tailward-propagating compression at the magnetopause leads to an upward current followed by a downward current vortex on the dawn side. The opposite sense of current occurs on the dusk side. Similarly, a rarefaction produces an downward current followed by a upward current at dawn. Thus, an isolated pressure enhancement in the solar wind produces a strong downward current filament preceded and followed by weaker upward filaments at dawn, while at dusk the central current is upward. This model is consistent with one shown in Figure 4 of Lanzerotti et al. [1991] and implicit in the works of others on this topic. Note that these statements about upward and downward currents hold for both the northern and southern ionospheres.

An important aspect of the pressure driven structures is that the model predicts an anti-symmetry between the dawn and dusk current systems. This implies that it will be important to make measurements of these vortex systems from global arrays of magnetometers which can span the noon sector. Preliminary results of this sort are presented by Lühr and Blawert [1993]. It is to be expected that results from the newly installed MACCS array of J. Hughes and M. Engebretson will provide interesting new results of this type. Another aspect clearly seen in the model is that the vortex structures appear first on more equatorward field lines and then appear to propagate somewhat poleward, as can be seen in Figure 3. At a single local time, however, the first vortex is seen more poleward and later ones more equatorward, as can be seen in Figure 5. Again, two-dimensional arrays of magnetometers are much more effective than single longitudinal chains for distinguishing this type of motion. It should be emphasized that in any case, this latitudinal motion is much smaller than the longitudinal motion of the vortices.

Another new feature presented here is the direct injection of magnetic helicity into the magnetosphere. This effect is expected to be the result of localized reconnection of the geomagnetic field with an interplanetary field which has a finite B_y component. Such localized reconnection gives rise to a current dynamo effect which twists flux tubes in a B_y-dependent manner. This helicity injection has the effect of breaking the symmetry between dawn and dusk sides, since the current systems associated with a tailward-moving dynamo consist of a strong downward current filament for $B_y > 0$ in the northern hemisphere, and the reverse for $B_y < 0$, on both the dawn and dusk side. In this case, the sense of the currents are reversed between the northern and southern hemisphere. Thus, the combination of the dynamic pressure driven motions of the magnetopause and helicity injection due to B_y dependent reconnection leads to a tendency for vortices to be observed in the northern dawn and southern dusk ionospheres for $B_y > 0$, and in the northern dusk and southern dawn for negative B_y. Again, global magnetometer arrays, coupled with observations of the IMF, could determine whether this effect is observed.

It is worth noting that this result depends only on the sense of the currents generated by the two different effects, and does not include much of the complex nonlinear physics involved in the localized reconnection process. In particular, the effect of the magnetic curvature force in pulling reconnected flux tubes toward the dawn and dusk sectors is neglected in this description. It is interesting to note, however, that this force has the same effect as the results presented here, e.g., for $B_y > 0$ the northern part of a reconnected flux tube will be dragged toward the dawn by this force. A more complete analysis of the forces acting on reconnected flux tubes as a function of the IMF is given elsewhere in this volume by Song et al. [1993].

In summary, the three-dimensional dipole model has proven useful in understanding the dynamics of the interaction of the solar wind with the magnetosphere. While the model has limitations, notably that it is a linear model and that it does not include zero-order distortions from the dipole geometry, it still can provide useful input for observers attempting to interpret data from magnetometer arrays. The results clearly suggest that global observations, spanning the noon sector, will be most useful in understanding the nature of the solar wind-magnetosphere-ionosphere interactions on these intermediate scales. Hopefully future observations coupled with more complete models will provide a ground-based means of diagnosing the state of the dynamics of the magnetopause.

Acknowledgments. This work has benefited from discussions with many of our colleagues at this Chapman Conference, in particular H. Lühr and T. Potemra. This work was supported in part by NSF grant ATM-9111791 and NASA grants NAGW-2653 and NAGW-2690, by a research grant from Kyunghee University, and by a grant of Cray-2 time from the Minnesota Supercomputer Institute.

REFERENCES

Farrugia, C. J., R. P. Rijnbeek, M. A. Saunders, D. J. Southwood, D. J. Rodgers, M. F. Smith, C. P. Chaloner, D. S. Hall, P. J. Christiansen, and L. J. C. Woolliscroft, A multi-instrument study and flux transfer event structure, *J. Geophys. Res., 93,* 14465, 1988.

Friis-Christensen, E., M. A. McHenry, C. R. Clauer, and S. Vennerstrǿm, Ionospheric traveling convection vortices observed near the polar cleft: a triggered response to sudden changes in the solar wind, *Geophys. Res. Lett., 15,* 253, 1988.

Glassmeier, K.-H., Reflection of MHD waves in the Pc4-5 period range at ionospheres with nonuniform conductivity distributions, *Geophys. Res. Lett., 10,* 678, 1983.

Glassmeier, K.-H., M. Hoenisch, and J. Untiedt, Ground-based and satellite observations of traveling magnetospheric convection twin-vortices, *J. Geophys. Res., 94,* 2520, 1989.

Goertz, C. K. and R. W. Boswell, Magnetosphere-ionosphere coupling, *J. Geophys. Res., 84,* 7239, 1979.

Lanzerotti, L. J., R. D. Hunsucker, D. Rice, L.-C. Lee, A. Wolfe, C. G. Maclennan, and L. V. Medford, Ionosphere and ground-based response to field-aligned currents near the magnetospheric cusp region, *J. Geophys. Res., 92,* 7739, 1987.

Lanzerotti, L. J., R. M. Konik, A. Wolfe, D. Venkatesan, and C. G. Maclennan, Cusp latitude magnetic impulse events, 1, occurrence, statistics, *J. Geophys. Res., 96,* 14,009, 1991.

Lee, D.-H., and R. L. Lysak, Magnetospheric ULF wave coupling in the dipole model: the impulsive excitation, *J. Geophys. Res., 94,* 17,097, 1989.

Lee, D.-H., and R. L. Lysak, Impulsive excitation of ULF waves in the three-dimensional dipole model: initial results, *J. Geophys. Res., 96,* 3479, 1991.

Lee, D.-H., and R. L. Lysak, Numerical simulation studies of ULF waves in the dipole model, *Solar Wind Sources of Magnetospheric ULF Pulsations,* this monograph, 1993.

Lühr, H., and W. Blawert, Ground signatures of traveling convection vortices, *Solar Wind Sources of Magnetospheric ULF Pulsations,* this monograph, 1993.

Lysak, R. L., and D.-H. Lee, The response of the dipole magnetosphere to pressure pulses, *Geophys. Res. Lett., 19,* 937, 1992.

Mallinckrodt, A. J., and C. W. Carlson, Relations between transverse electric fields and field-aligned currents, *J. Geophys. Res., 83,* 1426, 1978.

McHenry, M. A., and C. R. Clauer, Modeled ground magnetic signatures of flux transfer events, *J. Geophys. Res., 92,* 11231, 1987.

McHenry, M. A., C. R. Clauer, E. Friis-Christensen, P. T. Newell, and J. D. Kelly, Ground observations of magnetospheric boundary layer phenomena, *J. Geophys. Res., 95,* 14,995, 1990.

Moffatt, H. K., *Magnetic Field Generation in Electrically Conducting Fluids,* Cambridge University Press, Cambridge, 1978.

Papamastorakis, I., G. Paschmann, W. Baumjohann, B. U. O. Sonnerup, and H. Lühr, Orientation, motion, and other properties of flux transfer event structures on September 4, 1984, *J. Geophys. Res., 94*, 8852, 1989.

Russell, C. T., and R. C. Elphic, Initial ISEE magnetometer results: magnetopause observations, *Space Sci. Rev., 22*, 681, 1978.

Scholer, M., On the motion of artificial ion clouds in the magnetosphere, *Planet. Space Sci., 18*, 977, 1970.

Song, Y. and R. L. Lysak, Evaluation of twist helicity in FTE flux tubes, *J. Geophys. Res., 94*, 5273, 1989a.

Song, Y. and R. L. Lysak, Dynamo effect of 3-d time-dependent reconnection in the dayside magnetopause, *Geophys. Res. Lett., 16*, 911, 1989b.

Song, Y., R. L. Lysak, and N. Lin, Control of the generation of field-aligned current and ULF waves by the magnetic helicity input, *Solar Wind Sources of Magnetospheric ULF Pulsations*, this monograph, 1993.

Sonnerup, B. U. O., On the stress balance in flux transfer events, *J. Geophys. Res., 92*, 8613, 1987.

Sonnerup, B. U. O., L.-N. Hau, and D. W. Walthour, On steady field-aligned adiabatic flow, *J. Geophys. Res., 97*, 12,015, 1992.

Wright, A. N., and M. A. Berger, The effect of reconnection upon the linkage and interior structure of magnetic flux tubes, *J. Geophys. Res., 94*, 1295, 1989.

Dong-hun Lee, Department of Astronomy and Space Science, Kyunghee University, Yongin, Kyungki, Korea 449-701

Robert L. Lysak and Yan Song, School of Physics and Astronomy, University of Minnesota, 116 Church Street SE, Minneapolis, MN 55455

Magnetotelluric Sounding of the Crust and Hydrodynamic Monitoring of the Magnetosphere With the Use of ULF Waves

Vyatcheslav A. Pilipenko and Evgeniy N. Fedorov

Institute of the Physics of the Earth, Moscow, Russia

The problem of unification of magnetospheric diagnostics and magnetotelluric sounding with the use of ULF waves is reviewed. Some fundamental problems of magnetotellurics which cannot be resolved without a detailed knowledge of the MHD wave propagation from solar wind sources through the magnetosphere and the ionosphere are discussed. A summary is given of the various methods to determine experimentally the ULF spatial structure parameters - resonance frequency, its meridional gradient and width of the resonance. Recent experimental results considered demonstrate the promising prospects of these methods.

INTRODUCTION

The practical applications of ground-based observations of ULF waves are mainly related to hydromagnetic diagnostics (HD) of magnetospheric plasma and with magnetotelluric sounding (MTS) of the Earth's crust. These two approaches are being developed practically independently, although the problem of their unification has already been put forward [Gugliel'mi, 1989b]. Typical daytime ULF pulsations are excited by solar wind waves and irregularities. The process of MHD wave energy transport from an extra-magnetospheric source to the ground is inevitably related to the transformation of a compressional wave into an Alfven field line oscillation. The fundamentals of this process form the basis of resonance theory [Chen and Hasegawa, 1974; Southwood, 1974]. This theory predicts a specific spatial structure of the ULF field near a resonance field line. On the other hand, for magnetotelluric models the proper choice of the structure and the type of a primary wave impinging on the crust is of particular importance. But the resonance structure of the ULF field, verified by numerous experiments, and its possible significance for the MTS fundamentals have not been considered so far. In this review we would like to draw attention to the problem of "what MTS specialists should know about the magnetospheric field line resonances". Also, we will show that original MTS data, which include the records of magnetic and electric components of the ULF telluric field, can be used by the magnetospheric community for the development of new methods of HD. Hence, magnetotellurics, as well as diagnostics of the magnetosphere, could be enriched by the "coupling" of these two approaches to the study of ULF waves. Because of the author's ignorance in the field of magnetotellurics, the list of references on this topic does not pretend to be comprehensive (a more complete set of references can be found in the book by Kaufman and Keller [1981]).

MAGNETOTELLURICS AND MAGNETOSPHERIC FIELD LINE RESONANCE

In common MTS methods of geophysical prospecting with the use of ULF pulsations details of their generation mechanisms and of MHD wave propagation and transformation in the magnetosphere and ionosphere are not considered. Below we will analyze the situations when the peculiarities of disregarded processes, in particular, resonant properties of the magnetosphere and mode conversion in the ionosphere, could put in question some principles of interpretation of magnetotelluric data.

Tikhonov-Cagniard model

The basic physical principles of MTS are rather clear and are well explained in many reviews and books (e.g. Dmitriev and Berdichevsky, [1979]; Kaufman and Keller, [1981]). The horizontal components of the magnetotelluric field, E_t and H_t, recorded at the same point of the Earth's surface, are related through an impedance condition. The surface impedance, in turn, is determined by the distribution of the crustal conductivity σ along depth z. The magnetotelluric problem consists in the measurement of a surface impedance and the restoration of a dependence $\sigma(z)$ from the parametric dependence of impedance on frequency. The basis of MTS is the Tikhonov-Cagniard (TC) model based on the following physical considerations. In the atmospheric gap a ULF field has a quasi-static character. While

Solar Wind Sources of Magnetospheric Ultra-Low-Frequency Waves
Geophysical Monograph 81

penetrating a highly conductive Earth (the condition of "high conductivity" will be given below), the ULF electromagnetic field at the Earth's surface is characterized by an impedance coinciding with the impedance of a plane vertically propagating wave. Thus, for convenience the TC model is often referred to as the model of a plane vertically incident wave, bearing in mind a certain incorrectness of this definition. In general, the transverse components of a magnetotelluric field near the Earth's surface are related by the TC operator $\hat{Z}$, which parametrically depends on frequency f and is determined by a distribution $\sigma(z)$:

$$\mathbf{E}_t = \hat{Z}(\mathbf{H}_t) \tag{1}$$

In the case of a stratified medium the operator $\hat{Z}$ reduces to an integro-differential transformation, i.e.

$$\mathbf{E}_t = \int G(\mathbf{r} - \mathbf{r}')\left[\mathbf{n} \times \mathbf{H}_t(\mathbf{r}')\right]d\mathbf{r}' \tag{2}$$

where $\mathbf{n}$ is the normal to the Earth's surface, and the integration is being made over Earth's surface. The kernel G of an integral operator can be determined through the spectral impedance $Z^{(h)}(\omega, \kappa)$ [Dmitriev and Berdichevsky, 1979]. In the case of a vertically incident plane wave the relationship (2) reduces to an extremely simple form

$$\mathbf{E}_t = -Z_o\left[\mathbf{n} \times \mathbf{H}_t\right] \tag{3}$$

Equation (3) is known in the theory of radiowave propagation over highly conductive surfaces as the Leontovich boundary condition. In magnetotellurics, Z_O is the TC impedance, which is a functional of a conductivity. If a field is homogeneous over horizontal scales which are greater than the skin-depth d (i.e., $(kd)^2 \ll 1$, where $d=(2\pi\omega\sigma)^{-1/2}$), then the spectral impedance $Z^{(h)}(\omega, k)$ coincides with the TC impedance Z_O [Wait, 1954]. In fact, the TC model has a much wider range of applicability than follows from Wait's criteria [Price, 1962]. When spatial variations of horizontal field components are linear, then the integral operator (2) degenerates to a matrix operator. The action of this operator is reduced to a simple multiplication by Z_O, as in equation (3). Hence, if variations of $\mathbf{H}_t(\mathbf{r})$ at the spatial scales not less than 3d are linear, then a formal definition of TC impedance is possible. On the basis of this important conclusion the applicability of the TC model for interpretation of MTS data seems nearly universal.

Directional analysis of H and E modes

However, for a layer with low conductivity the condition of the validity of the TC model may be violated. In this case the layer's impedance would be determined not only by the conductivity, but by the spatial structure of an incident field as well. As a first approximation of local ULF field horizontal structure, a model of an inhomogeneous plane wave can be adopted

$$(E, H) \sim \exp(-i\omega t + i\mathbf{kr}) \tag{4}$$

where $\mathbf{k}=\mathbf{k}'+i\mathbf{k}'$ denotes the horizontal component of a complex wave vector, and $\omega=2\pi f$. The theory of MTS, based on the plane inhomogeneous wave conception [Dmitriev, 1970; Tikhonov et al., 1974], includes some other physical effects. The ionospheric currents excited by a magnetospheric MHD wave can induce electromagnetic disturbances in the Earth's crust via two mechanisms. The first one, actually assumed above, is electromagnetic induction. Since the atmospheric gap has some electric conductivity, a galvanic mechanism, i.e. the penetration of some part of the ionospheric current through the atmosphere into the Earth, is also possible. The existence of these two mechanisms is formally revealed in the splitting of Maxwell's equations into two independent sub-systems. One corresponds to the H (magnetic) mode (the above consideration has assumed the presence of this mode only), and the other to the E (electric) mode. These modes have different partial spectral impedances, $Z^{(h)}(\omega, k)$ and $Z^{(e)}(\omega, k)$. Only when the condition of strong skin effect $(kd)^2 \ll 1$ (i.e. the Leontovich boundary condition) is fulfilled do the impedances of these partial modes coincide $Z^{(h)} \approx Z^{(e)} \approx Z_O$. In the general case, when an E-mode is present, the TC operator Z comprises not only an integral term, as in the equation (2), but also an integro-differential term

$$\mathbf{E}_t(\mathbf{r}') = \int G(\mathbf{r} - \mathbf{r}')\mathbf{H}_t(\mathbf{r}')d\mathbf{r}' + \int \Delta G(\mathbf{r} - \mathbf{r}')\hat{L}(\mathbf{H}_t)d\mathbf{r}' \tag{5}$$

where $\hat{L}$ denotes some second-order differential operator [Dmitriev and Berdichevsky, 1979]. Chetaev [1985] developed practical algorithms for splitting an original ULF field into partial H and E modes. These algorithms require simultaneous 6-component data at one point or 5-component ($\mathbf{E}_t$, $\mathbf{H}_t$, H_z) measurements at two nearby points. As a result, the partial surface impedances $Z^{(h)}$, $Z^{(e)}$ and the wave vector $\mathbf{k}$ can be determined experimentally. There are geoelectrical structures where an effective skin-depth in the Pc3-4 frequency range is comparable to a horizontal field scale (about 100-200 km at mid-latitudes). In these cases the partial impedances become dependent on wave propagation characteristics. Then, for wave packets with certain |k|, the condition of total reflection from an underlying layer can be fulfilled. In that case the presence of the layer's boundary is revealed by the characteristics of the wave, and partial impedances must have a pronounced peculiarity near critical values. This interpretation of ULF magnetotelluric data forms the basis for a new scheme of MTS suggested by Chetaev [1985]. The new scheme of MTS (also known as directional analysis of ULF field) was verified experimentally in the Ukraine and gave a reasonable profile of upper mantle.

Possible input of electric mode in the ULF pulsation structure

As expression (5) shows, the presence of an E-mode modifies

equation (3) for the spectral components of a ULF telluric field, so the impedance relationships should contain additional terms with a difference of spectral impedances, $\Delta Z = (Z^{(h)} - Z^{(e)})/k^2$. In classical magnetotellurics the input of an E-mode is considered to be negligible and the terms $\sim \Delta Z$ are ignored [Berdichevsky et al., 1971]. On the other hand, Tikhonov et al. [1974]; Chetaev [1985], and Savin et al. [1991] suggested that many experimentally observed paradoxes of MTS (non-orthogonality of E and H ellipses, large magnitudes of vertical currents in bore-holes, opposite rotation of E and H polarization ellipses, etc.) cannot be resolved without regard for E-mode existence. They supported their conclusions by the experimental application of directional analysis algorithms for the analysis of daytime Pc3 pulsations [Chetaev, 1985; Savin et al., 1991]. In their experiments with 6-component telluric field registration, large vertical electric currents in bore holes reaching 10^{-8} A/m^2 were detected. The authors ascribed these intense currents to an incident ULF field, but not to the influence of geoelectrical inhomogeneities. Consequently, the E-mode appeared to be an essential input to the ULF field structure. The assumption of an electric mode excitation in the Earth-ionosphere waveguide was made also by Kikuchi and Araki [1979]. They invoked the $(TH)_0$ mode for the interpretation of preliminary reverse impulses of SSC, observed at near-equatorial latitudes.

The ambiguity with the possible occurrence of an electric mode cannot be resolved within the framework of magnetotellurics theory only. The essential help can be obtained from the analysis of the problem of MHD wave transmission from the magnetosphere through the ionosphere. Theoretical calculations of magnetospheric ULF wave incidence upon the ionosphere [Hughes and Southwood, 1976; Alperovich and Fedorov, 1984b] show a very low effectiveness of E-mode excitation. In essence, the rate of E-mode generation by magnetospheric processes should be proportional to the rate of magnetospheric vertical current penetration into a low-conductivity atmosphere. For all theoretical models and for a wide range of ionospheric and pulsation parameters the vertical component of the ULF electric field in the atmosphere does not exceed several V/m. Direct measurements of the vertical component of the electric field in the air have revealed variations of E_z in the ULF frequency range with amplitudes less than 10 V/m [Chetaev et al., 1975; Chetaev et al., 1977]. As follows from estimates [Berdichevsky et al., 1971], the E-mode input into the pulsation structure should be taken into account in MTS only when the magnitude of the vertical electric field component in the air exceeds 10^2 V/m.

So far there is still some uncertainty concerning the possible input of the E-mode in the structure of the geomagnetic pulsation field and its role in magnetotellurics, in particular, in the interpretation of bore-hole observation data. The decisive solution of this problem would require specialized 6-component observations of ULF pulsations, including the atmospheric vertical electric field registration. Reliable numerical techniques which would enable one to separate the vertical electric fields and currents, induced by lateral inhomogeneities and by an incident ULF wave, correspondingly, should be developed.

Resonant structure of ULF field

The schemes of MTS described above were based on a phenomenological conception of the spatial structure of a ULF field, whereas the physical mechanisms governing ULF wave propagation were not considered. Let us discuss now why the "magnetospheric" aspect of the physics of ULF waves might be essential for the problems of MTS data interpretation.

MHD disturbances from remote parts of the magnetosphere (e.g. solar wind disturbances, magnetopause surface waves, magnetosheath turbulence, etc.) propagate inside an inhomogeneous magnetosphere and, through a mode transformation, excite standing Alfven oscillations along the Earth's magnetic field lines. In their turn, the Alfven waves impinging the ionosphere are, in most cases, the sources of ULF geomagnetic pulsations (Pc3-5, Pi2) observed on the ground. The process of mode transformation is most effective in the vicinity of the resonant geomagnetic shells where the frequency, f, of the external source coincides with the local frequency $f_R(L)$ of Alfven field line oscillations. The mathematical description of the spatial field structure of the field perturbation $\mathbf{b}$ in the magnetosphere near resonant shells according to the qualitative theory of differential equations [Kivelson and Southwood, 1986; Krylov and Fedorov, 1976; Krylov et al., 1981] can be expressed in the form of an asymptotic expansion:

$$\mathbf{b} = \alpha w^{-1} + \beta \ln(w)$$
$$\alpha = \sum \alpha_n w^n, \beta = \sum \beta_n w^n, \qquad (n = 0, 1, ...) \tag{6}$$

where $\mathbf{b}, \alpha, \beta$ are vector functions of the form $\mathbf{b} = (b_\varphi, b_\nu)$, $w = \nu - \nu_R(f) + i\varepsilon_m$, ν is the coordinate of a magnetic shell, φ is the azimuthal coordinate, $\nu_R(f)$ is the point where $f_R = f$, and ε_m is the width of a resonance region in the magnetosphere. For the determination of coefficients α_n and β_n a recurrent system of equations can be obtained. For the dipole geometry these recurrent equations give [Lifshitz and Fedorov, 1986]

$$b_\varphi(\nu, f) = b_0(f) h_\varphi^{-1} \{ w^{-1} + ... + C_0 \ln(m[h_\nu / h_\varphi] w) + ... \}$$
$$b_\nu(\nu, f) = b_0(f) h_\varphi^{-1} \{ im(h_\nu / h_\varphi) \ln(m[h_\nu / h_\varphi] w) + ... \} \tag{7}$$

where m denotes the azimuthal wave number, and h_φ and h_ν are Lame's coefficients. The expression (7) is, in fact, the generalization of the original simple "box model" of the magnetospheric resonator [Southwood, 1974]. The leading terms of the asymptotic expansion (7) and the expressions given by the "box field" model coincide, with the expense of some correction (the coefficient C_0 vanishes in the simple geometries with a constant curvature radius, as in [Southwood, 1974; Radoski, 1974]).

The leading term in the asymptotic decomposition (7) which

describes the resonant singularity of the b_φ component near a resonant shell, i.e. when $|\nu - \nu_R(f)| \le \varepsilon_m$, can be presented in the form

$$b_\varphi(\nu, f) = \frac{i\varepsilon_m}{\nu - \nu_R(f) + i\varepsilon_m} b_o(f) \qquad (8)$$

Based on the equation (8) the meridional structure of the azimuthal component of the ULF field can be qualitatively represented as the combination of a "source" spectrum (term $\sim b_o(f)$) and a magnetospheric resonance response (term $\sim [\nu - \nu_R(f) + i\varepsilon_m]^{-1}$). The "source" part is related to a disturbance transported by a large-scale fast compressional wave and it has a weak dependence on the ν coordinate. But, the resonant magnetospheric response related to the Alfven wave excitation is strongly localized and it causes rapid variations of amplitude and phase when a resonant shell is crossed. The radial component b_ν has a weaker logarithmic singularity near the resonance, so that the resonant behavior of this component would barely be noticeable.

Upon transmission through the ionosphere, the spatial structure of large scale ($|mv\ h| < 1$) oscillations keeps the same form, if the $\pi/2$ rotation is taken into account: $b_\nu \rightarrow D$, $b_\varphi \rightarrow H$ [Hughes and Southwood, 1976; Leonovich and Mazur, 1991]. The width ε of the resonance peak, as observed on the ground, is smoother as compared with that above the ionosphere, namely $\varepsilon = \varepsilon_m + h$ (h is the height of the ionospheric E-layer) [Alperovich et al., 1991].

The leading term (8) which describes amplitude and phase characteristics of H component on the ground can also be represented in the form [Gugliel'mi, 1989a]

$$H(x, f) = \frac{h_R(f)}{1 + i\zeta} \qquad (9)$$

where $\zeta = (x - x_R)/\varepsilon$ denotes the normalized distance from the resonant point $x_R(f)$, $h_R(f)$ is the amplitude of the pulsation at the resonant point, and x is the coordinate of a magnetic shell, as measured along a geomagnetic meridian on the ground.

Distortions of MTS curves

When the applicability condition of the TC model breaks down the impedances are determined not only by the geoelectrical properties of an underlying crust but also by the spatial structure of an incident ULF wave. Simple estimates show that the validity of the TC impedance can be guaranteed only for a crust with a resistivity lower than $10^3 \Omega m$. As an example, the numerical calculations of the ground ULF field excited by an oscillating linear current (e.g. auroral electrojet) with periods 10-600 sec in the dayside ionosphere, performed by Alperovich and Fedorov [1984a], can be mentioned. Their calculations show that for a value of the Earth's resistivity of $\rho = 10^5\ \Omega m$ the apparent impedances differ 2-5 times from the TC impedance up to

distances 3×10^3 km from the current. For $\rho = 10^4\ \Omega m$ impedances differ by 1.5 - 2 times up to $R < 1.5 \times 10^3$ km, and only for $\rho = 10^3\ \Omega m$ are the apparent impedances close to the TC impedance.

However, the resonant structure (8,9) of ULF pulsations cannot be approximated either by a plane vertically incident wave, or by a plane inhomogeneous wave, or by in-phase oscillating ionospheric currents. Hence, the magnetospheric resonant effects may cause distortions of a standard MTS curve near a local resonant frequency which could be misinterpreted as a false feature of the Earth's crust structure. This situation is illustrated in Figure 1, taken from Alperovich et al. [1991]. As an example, a 4-layer geoelectrical structure with the following parameters was chosen: $\rho_1 = 30\ \Omega m$, $h_1 = 3$ km; $\rho_2 = 3 \times 10^3\ \Omega m$, $h_2 = 50$ km; $\rho_3 = 3 \times 10^2\ \Omega m$, $h_3 = 50$ km; $\rho_4 = 3 \times 10^{-2}\ \Omega m$, $h_4 = \infty$. The numerical calculations of the field structure and corresponding impedances were made for a vertically incident plane wave (solid line) and for a resonant structure (7) at latitudes $\Phi = 60^o$ (dashed line) and $\Phi = 55^o$ (dotted line). The comparison of the synthetic apparent resistance curves, $\rho_a(T)$, for the classical TC model and for the resonance structure of the incident field shows additional extrema near the local resonance periods (100 sec and 46 sec, correspondingly). Deflection of the two curves reaches 30%. Distortion of the MTS curves by the resonant effects would be especially pronounced above low-conductivity layers.

Most experimental studies of the role of resonance effects in the formation of a local meridional ULF structure have verified the predictions of the resonance theory. Hence, the unambiguous application of standard MTS methods requires a preliminary HD analysis of magnetospheric resonance frequencies in a region under study. This fact demonstrates the necessity of unification of the two approaches to the use of geomagnetic pulsations, HD and MTS [Gugliel'mi, 1990; Pilipenko, 1990]. Below, we will consider some examples which illustrate the fruitfulness of such a combined approach and indicate new opportunities for HD.

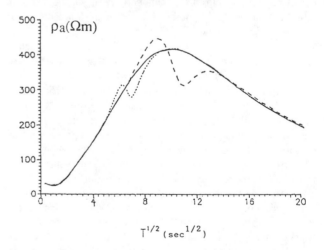

Fig 1. The apparent resistance curves $\rho_a(T)$ above a 4-layer geoelectrical structure: for the classical TC model, i.e. vertically incident plane wave (solid line), and for the resonance structure of the incident field, calculated for latitude $\Phi = 60^o$, T=100 s (dashed line), and for $\Phi = 55^o$, T=46 sec (dotted line) .

HYDROMAGNETIC DIAGNOSTICS OF THE MAGNETOSPHERIC PLASMA BY THE DATA OF GROUND BASED OBSERVATIONS

The above discussion shows that, for correct MTS or HD, effective and simple methods of operative determination of the resonant frequency distribution in a given region should be elaborated. With this aim in mind, we will subsequently analyze various ground-based methods of determination of f_R and its meridional gradient. Frequencies of field line Alfven oscillations vary considerably in space (from 10 sec at low latitudes to 10 min at high latitudes), and in time. Therefore, a nearly real-time monitoring of Alfven resonant frequencies is desirable for a productive MTS. Below, we will review the methods of f_R determination with the use of the same magnetotelluric data.

Gradient method

Despite the seemingly simplicity of the resonance model [Southwood, 1974; Chen and Hasegawa, 1974], a theoretically predicted amplitude and phase meridional structure corresponds well to the experimental local structure of various types of ULF waves at sub-auroral and middle latitudes. It was found that the occurrence region of Alfven resonance spreads to rather low latitudes, $\Phi \leq 20^0$.

The principal problem of the experimental determination of a resonant field line frequency $f_R(L)$ is, that for most events, when the spectral content of ULF pulsations is examined, the input of the resonant magnetospheric response and the input of a source are comparable. So, in most cases, a spectral peak does not necessarily correspond to a local resonant frequency, and the width of a spectral peak cannot be directly used to determine the Q-factor of the magnetospheric resonator. This ambiguity can be resolved with the help of the experimental methods described below, in particular the gradient method, proposed by Baransky et al. [1985]. Precision measurements of the gradient of spectral amplitude (desirably phase also) on a small base make it possible to remove the influence of the source spectrum, so that relatively weak resonant effects can be observed.

The simple relationships stemming from function (9) properties enable one to estimate the resonant frequency of the field line between the stations, the width of a resonant peak ε, and an Alfven frequency gradient in the magnetosphere. The ratio between amplitude spectra and the difference of phase spectra of ULF H-components, recorded in points x_1 and x_2 have the following form:

$$G(f) = \frac{|H(x_1,f)|}{|H(x_2,f)|} = \left(\frac{1+\zeta_2^2}{1+\zeta_1^2}\right)^{1/2}$$

$$\Delta\varphi(f) = \arctan\left(\frac{\zeta_2-\zeta_1}{1+\zeta_1\zeta_2}\right)$$

(10)

A schematic of the model functions G(f) and $\Delta\varphi(f)$ is shown in Fig. 2. The important features of these functions are: a) $G(f_A)=1$

for frequency $f_A=f_R(x_C)$, where the point $x_C =(x_1+x_2)/2$ is located at the midpoint between the stations; b) G(f) reaches extreme values G_+ and G_- at frequencies f_+ and f_- , which correspond to the points $x_R(f_\pm) = x_C \pm (\varepsilon^2 +(\Delta x/2)^2]^{1/2}$, where $\Delta x=x_1 - x_2$; c) $G_+ G_- = 1$ and $G_+ - G_- = \Delta x/\varepsilon$; d) $\Delta\varphi(f)$ reaches an extreme value $\Delta\varphi^*=2\arctan(\Delta x/2\varepsilon)$ at resonant frequency f_A. These relationships constitute, in essence, the practical background of the gradient method and they have been used in one or another form elsewhere [Baransky et al., 1985,89; Kurchashov et al., 1987; Green et al., 1993].

The meridional gradient of the Alfven frequency can be roughly estimated from the data of gradient measurements as

$$\frac{\partial f_R}{\partial x} \cong \left(\frac{\Delta f}{2}\right)\left(\varepsilon^2 + \left(\frac{\Delta x}{2}\right)^2\right)^{-1/2}$$

(11)

where $\Delta f= f_+ - f_-$. The gradient method makes it possible to restore a smooth $f_R(L)$ profile in a limited interval of latitudes, using the possibility to estimate ε and 3 resonant frequencies, f_R, f_+, f_- [Baransky et al., 1989]. When amplitude measurements of the H-component at a meridional net of stations (not less than 3) are available another modification of the gradient method for the restoration of a continuous $f_R(L)$ profile can be used [Gugliel'mi et al., 1989].

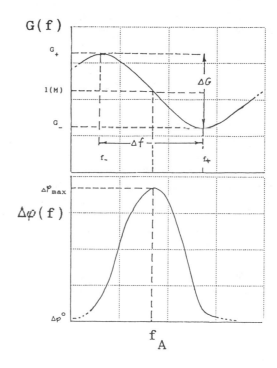

Fig 2. Schematic spectral plots as predicted from the resonance theory: ratio of amplitude spectra, G(f) (top); and phase difference $\Delta\varphi(f)$ between two nearby stations (bottom).

Amplitude-phase gradient method

One of the modifications of the gradient method uses simultaneous information about amplitude and phase at two points [Gugliel'mi et al., 1989]. Assuming that the dependence of the complex amplitude $H(f,x)$ has the form of (9) and using the measurements of $H(f,x)$ at two points x_1 and x_2, a system of two linear equations with complex coefficients and three unknown parameters x_R, ε, h_R can be obtained. As a result, one can obtain the dependence of the resonance position $x_R(f)$ and the resonance width on frequency

$$\frac{x_R(f)-x_1}{\Delta x} = \frac{1-G\cos(\Delta\varphi)}{G^2-2G\cos(\Delta\varphi)+1}$$
$$\frac{\varepsilon}{\Delta x} = \frac{G\sin(\Delta\varphi)}{G^2-2G\cos(\Delta\varphi)+1} \qquad (12)$$

The meridional gradient of Alfven frequency can be estimated, using this method, knowing the slope of $G(f)$ at resonant frequency by the relation

$$\frac{\partial f_R(x=x_R)}{\partial x} \cong \left(\frac{\partial G(f=f_A)}{\partial f}\right)^{-1}\frac{1-\cos(\Delta\varphi)}{\Delta x} \qquad (13)$$

A continuous distribution of resonant frequency as a function of x can be determined, after reversing the experimentally determined dependence (12): $x_R(f) \Leftrightarrow f_R(x)$.

Modification of the gradient method in the presence of a lateral geoelectrical inhomogeneity

The ideas developed for the gradient method and its modifications work well when both stations are situated in areas with similar geoelectrical conditions. The presence of a lateral geoelectrical inhomogeneity, especially if the condition of a strong skin-effect is violated, may substantially distort the resonant structure of pulsations. A geoelectrical structure will change the magnetic component that is oriented across the extent of the structure. A magnetometer situated in a region with higher conductivity will have an abnormally high magnetic field and an additional positive phase shift.

A rigorous modification of the gradient method for the case of a crust's geoelectrical inhomogeneity would require numerical calculations of a complicated self-consistent problem. But, as a first approximation, elimination of geological influence can be done with the use of the following simple method [Green et al., 1993]. Let us suppose that the influence of a geoelectrical inhomogeneity can be expressed as some coefficient M, which the ratio $G(f)$ is multiplied by, and an additive phase shift, $\Delta\varphi^0(f)$. Then the experimentally measured functions G' and $\Delta\varphi'(f)$ will be presented in the form: $G'=MG$ and $\Delta\varphi'(f)=\Delta\varphi(f)+\Delta\varphi^0(f)$. Further,

let us suppose that a weak dependence of the unknown coefficients $M(f)$ and $\Delta\varphi^0(f)$ on frequency for a limited frequency band near a resonant frequency can be neglected. Then the coefficient M can be found experimentally from the following simple relationship resulting from the properties of function $G(f)$:

$$M = \left(G'_-G'_+\right)^{1/2} \qquad (14)$$

So, the amplitude ratio (10) and phase difference (11) will be shifted to the new levels M and $\Delta\varphi^0(f)$ (Fig. 2a,b).

POLARIZATION METHODS OF THE ULF RESONANT STRUCTURE STUDY

The remarkable asymmetry of resonant properties between various components of the ULF field mentioned above suggests that resonance effects would manifest themselves not only in the spatial structure, but in the polarization properties, too. On the basis of this idea a number of polarization methods of f_R determination were suggested. These methods can supplement the gradient method and can even be used for the diagnostics of the $f_R(L)$ distribution from single-station data. Here we are using the concept of polarization methods in a very limited sense, only as the study of the amplitude and phase relationships between two different components of the ULF field in a limited frequency band. More sophisticated methods of polarization analysis of multi-channel time series (e.g. used by Samson [1986]) are not considered.

Polarization of horizontal magnetic components

Polarization of horizontal magnetic components may reflect a spatial structure of ULF waves. As the equations (7) show, the resonant response of the magnetosphere is characterized by a pronounced asymmetry of the H and D components. At the same time a source spectrum is imposed on both components in the same way. Hence, even when a resonance response is masked by a source spectrum, the ratio $H(f)/D(f)$ should reveal a maximum at a resonant frequency [Baransky et al., 1990]. This conclusion is illustrated by Figure 3 from [Vellante et al., 1993], where the results of f_R determination by this method at stations Niemegk (L=2.7) and l'Aquila (L=1.6), situated along the same meridian, are shown. The spectral peaks in ratio of H and D components delineate several subsequent resonant harmonics at the mid-latitude station Niemegk and the fundamental mode at the low-latitude station l'Aquila.

Useful information about pulsation structure is hidden in polarization parameters: the ellipticity and the orientation of the main axis. Keeping the leading asymptotic terms in the solution of MHD equations (2), the ratio between the transverse magnetic components in a dipole geomagnetic field has been obtained

$$b_v / b_\varphi = -imY\ln(mY) \qquad (15)$$

Here $Y=(s+i\eta)$, $s=(L/L_R-1)(4-3/L)^{-1/2}$ is the normalized distance

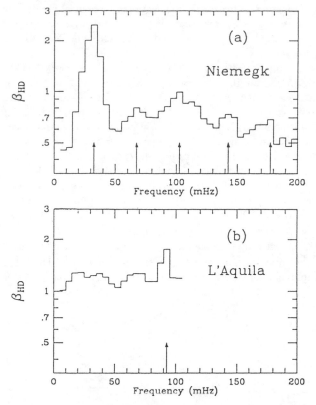

Fig. 3. Averaged ratio, β_{HD}, between the spectral densities of H and D components for Niemegk (a) and l'Aquila (b) stations. Arrows mark the spectral peaks, probably corresponding to resonant frequencies.

from the resonant shell L_R, and $\eta=(\varepsilon/L)(4-3/L)^{-1/2}$ is the normalized width of a resonance peak. The coefficient $(4-3/L)^{1/2}$ arises due to the difference between the scaling factors in the radial and azimuthal directions during the projection onto the ionosphere. Similar to [Hasegawa and Lanzerotti, 1978] from equation (15) the parameters p and q can be obtained which are related with the ellipticity and the orientation angle of the main axis. The behavior of these parameters near the resonant shell, i.e. when $|s|< |\eta|<1$, is described by the following relations

$$b_v / b_\varphi = p + iq$$
$$p = m\eta \ln|m\eta|, \quad q = ms\ln|m\eta| + m\eta Arg(mY)$$
(16)

If the dissipation in the system is low, so $\|\ln|m\eta\||>>|m\eta|$, then the expression (16) for q(s) predicts that when crossing over the resonant shell, i.e $s\rightarrow0$, the polarization of the pulsation must change. When the Q-factor of the magnetospheric resonator is low the effect of a polarization switch may happen only at some shell, shifted from the resonant one. In this case the polarization features of horizontal magnetic components are difficult to use for the practical determination of a resonant frequency, because this shift depends on two poorly known parameters, the azimuthal wave

number m and the scale of magnetic field inhomogeneity. So, in contrast with the expectations of resonance theory [Chen and Hasegawa, 1974; Southwood, 1974], the switch of polarization has turned out to be a less reliable method for the determination of resonant frequency in the real magnetosphere, especially at low latitudes. Nonetheless, the polarization parameters may give additional information about the direction of an azimuthal phase velocity and the sign of a radial gradient of Alfven velocity in the magnetosphere. For that purpose it's enough to use only the signs of ellipticity and orientation angle .

Resonant properties of the vertical magnetic component

The vertical component of the ULF magnetic field on the ground is a sensitive indicator of inhomogeneity of both the ULF field and the crustal conductivity [Southwood and Hughes, 1978]. So, use of H_z component data might be helpful for the examination of resonant features of the geomagnetic pulsation spatial structure. In a case of strong skin effect the relationship between vertical and horizontal magnetic components of the ULF field is given by the following formula [Wait, 1954; Gugliel'mi et al., 1989a,b]

$$H_z(x,f) = \frac{ic}{2\pi f}\left\{Z_0 \text{div}\mathbf{H}_t + \mathbf{H}_t \nabla Z_0\right\}$$
(17)

For regions with homogeneous conductivity of the Earth's crust, estimates show that for typical Pc3-5 and Pi2 pulsations the meridional component of the ULF field gradient predominates in equation (17). Under these conditions, as follows from (9,17), the resonant effects in the behavior of the H_z component might be even more pronounced than in the H component [Green et al., 1991].

The complex value of the surface impedance Z_0 in equation (17) can be determined experimentally or it can be removed from consideration with the help of impedance relation (3). Thus the relationship between components H_z and E_y is:

$$H_z / E_y = -(c / 2\pi f \varepsilon)\left\{1+i\zeta\right\}^{-1}$$
(18)

Qualitative plots of the amplitude curve $|H_z/E_y|\sim(1+\zeta^2)^{-1}$ and the phase curve $\psi(f)=ArgH_z-ArgE_y=\pi-\arctan\zeta$ are shown in Figure 4. At the frequency of an Alfven resonance, $f=f_R$, the ratio $|H_z/E_y|$ has a local maximum, and the components H_z and E_y must be out of phase. The slope of the curve $\psi(f)$ at the frequency $f=f_R$ is determined by the resonance width e and by the Alfven frequency gradient. So, an experimental plot of $\psi(f)$ can be used to estimate the gradient of an Alfven frequency in the magnetosphere using the data from a single ground-based station

$$\frac{\partial f_R\left(x = x_R\right)}{\partial x} \cong \left\{\frac{\varepsilon\partial\psi\left(f = f_R\right)}{\partial f}\right\}^{-1}$$
(19)

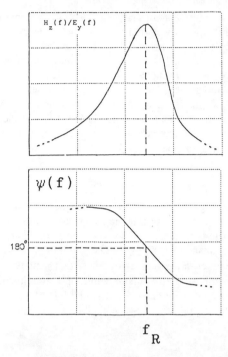

Fig 4. Schematic spectral plots of the relationships between the spectra of H_z and E_y components near the resonance frequency f_R: ratio of amplitudes (top), and phase difference,(bottom).

Through the use of H_z and electric components of the ULF field it is possible to reconstruct the functional dependence $f_R(x)$ in the vicinity of the observation point, with the data from a single station [Gugliel'mi, 1989b, Gugliel'mi et al., 1989b]. From equation (18) the distance between the observation point (x=0) and the resonant shell $x_R(f)$ as well as the width of the resonance ε can be determined by

$$x_R = (c/\omega)|E_y/H_z|\sin\psi$$
$$\varepsilon = -(c/\omega)|E_y/H_z|\cos\psi \qquad (20)$$

Reversing the dependence (20), $x_R(f) \Leftrightarrow f_R(x)$, the variation of the resonant frequency $f_R(x)$ can be determined in the vicinity of the observation point.

EXPERIMENTAL VERIFICATION OF GRADIENT AND POLARIZATION METHODS

Some theoretical predictions listed above can be verified using data from a meridional array of magnetometers at L=1.5 [Green et al., 1993]. The measurements of ULF magnetic fields were performed with triaxial ring-core fluxgate magnetometers, and the horizontal electric components E_x and E_y were measured using grounded dipoles. Figure 5 demonstrates the spectral curve, calculated with the use of cross-spectral analysis, for the pair of stations TM and AK with a baseline Δx=250 km. Let us compare

the properties of these functions with the theoretical predictions given above. The frequency band ~ 70-80 mHz around the spectral peak in the H amplitude spectrum (Fig. 5a) will be examined.

The plots $G(f)=H^{(TM)}/H^{(AK)}$ (Fig. 5b) and $\Delta\varphi(f)=\varphi^{(AK)}-\varphi^{(TM)}$ (Fig. 5c) show in this spectral domain the specific behavior predicted by the resonance theory. But these curves are shifted from the levels indicated in the theoretical plots (Fig. 2a,b): $G(f)>1$ and $\Delta\varphi(f)$ is below 0. This shift is caused by a pronounced difference of the geoelectrical properties of the Earth's surface at the two stations. The surface impedances calculated for TM and AK indicate a magnitude of apparent resistivity, ρ, as $\cong 10^2$ Ωm and $\cong 10^3 - 10^4$ Ωm, respectively. The value for M has been calculated using equation (14) of the modified gradient method. For this event the new reference level is M=1.73 (Fig. 5b). The new reference level for the phase shift, $\Delta\varphi^0(f)$, can be approximated by the boundary of phase excursions in the phase curve, and for this event $\Delta\varphi^0(f)=-30^0$ (Fig. 5c). Thus, the amplitude and phase curves allow us to identify the resonant

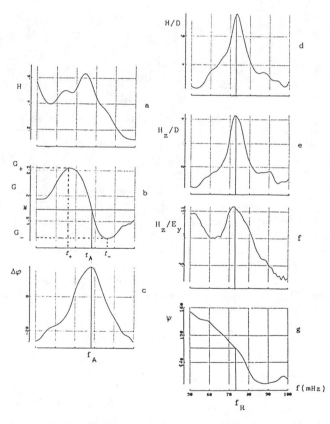

Fig. 5. Spectral plots of the TM and AK stations for the interval 2:00 - 2:40 UT, August 17, 1990:
a) the amplitude spectrum of the H-component at TM;
b) the amplitude gradient $G(f)=H^{(TM)}/H^{(AK)}$;
c) the phase difference $\Delta\varphi(f)=\varphi^{(AK)}-\varphi^{(TM)}$;
d) the ratio H/D at TM;
e) the ratio H_z/D at TM;
f) the ratio H_z/E_y at TM;
g) the phase difference $\psi(f)=Arg(H_z)-Arg(E_y)$ at TM.

frequency, f_A, as 79 mHz, for the field line intersecting the Earth midway between TM and AK.

The selective amplification at the resonant frequency of the H-component as compared to the D-component is stressed by the $H(f)/D(f)$ ratio (Fig. 5d). The same effect is more evident in the $H_z(f)/D(f)$ ratio (Fig. 5e). This effect is absent for other non-resonant peaks that occur in the spectrum of the ULF signal.

The ratio of the H_z and E_z amplitudes, from station TM, has a clear local maximum near the resonant frequency (Fig. 5f), in agreement with theoretical predictions (Fig. 4a).

The slope of the phase curve $\psi(f)$ (Fig. 5g) between the H_z and the E_y components in a resonant frequency band corresponds to the position of a resonant field line inside the plasmasphere, where $df_R/dx < 0$. The value $\psi(f=f_R) \sim 150^o$ differs somewhat from the theoretically predicted 180° (Fig. 4b). It seems that this shift is caused by the influence of geoelectrical inhomogeneity.

For a comparison of the gradient method to the polarization method only the data of TM station may be used, because in this case, the approximation of a strong skin effect is valid. The amplitude ratio and the phase shift between the H components at two stations (Fig. 5b,c) give the value of resonant frequency of a magnetic shell between TM and AK, as f_A=79 mHz. Polarization methods (Fig. 5d,e,f) indicate the resonance frequency to be about f_R=74 mHz for a field line over the station at TM. Because the station TM is situated to the north of station AK, the relation $f_R < f_A$ holds, which corresponds to the expected meridional structure of resonance frequencies at these latitudes. So, the gradient and the polarization methods give consistent results of the resonance frequency determination, keeping in mind the simplified approach to account for the crust's lateral inhomogeneity.

Using the spectral plots of the horizontal gradients, $G(f)$ and $\Delta\varphi(f)$, (Fig. 5b,c), estimates of the width of the resonance region and the local gradient of Alfven frequency along the meridian have been determined for the region between TM and AK. The amplitude ratio $G(f)$ and the relationships (10,11) gave estimates of ε=300 km and df_R/dx=-0.032 mHz/km. The same parameters were estimated by the amplitude-phase gradient method with the use of relationship (13) and ε=270 km and df_R/dx=-0.04 mHz/km were obtained. So, both methods give consistent estimates of the Alfven frequency gradient along the meridian in the magnetosphere. This experimentally determined value for the gradient of resonant frequency can be used to account for the observed difference between resonant frequencies f_R(TM) and f_A, f_R(TM)=79mHz- 0.04 mHz/km ×125km=74mHz.

The most reliable estimates of the resonant frequency, resonance width and the gradient of Alfven frequency in the experiment are f_R=62-79 mHz, ε=300-400 km, and $|df_R/dx|$=0.03-0.04 mHz/km. [Green et al., 1993]. The obtained gradients correspond to the characteristic scale of Alfven frequency inhomogeneity at ionospheric heights $a=(|dlnf_R/dx|)^{-1} = (2-4)\times10^3$ km. The Q-factor of the magnetospheric resonator at these latitudes can be estimated then as $Q=(2\varepsilon_m/a) = 3-5$. These values are much lower then the predictions using the existing models of ionospheric Joule dissipation. The reason for this discrepancy is related to the fact that the approximation of a "thin" ionosphere, used in these models, is not valid at low latitudes. With the use of a proper model the values of Q, estimated by the above method, could be used to determine the ionospheric conductivity.

CONCLUSION

In this review we have attempted to draw the attention of geophysicists to the necessity to merge the theory of magnetospheric MHD waves and MTS methods. The disregarding of any of these aspects of the physics of ULF waves may lead to significant errors. For example, the influence of lateral geoelectrical inhomogeneity may obscure the magnetospheric resonance effects in the ground-based structure of ULF pulsations, which makes HD impossible. In their turn, the resonant effects cause distortions of MTS curves that may be misinterpreted as some features of a geoelectrical structure.

Simple modification of the gradient method described above make it possible to exclude to a zero-order approximation the influence of geology and to reveal the resonance effects. But this method requires a theoretical justification. For that purpose numerical calculations of the problem of resonant ULF field reflection from laterally inhomogeneous geoelectrical structure should be attempted.

The new methods of HD which use multi-component structure of a magnetotelluric field seem rather promising. The first attempt to apply these methods for the local Alfven frequency determination showed a good correspondence between them and the gradient methods. All the potentialities of the combined application of the gradient and the polarization methods to restore the meridional structure of a plasma density in the magnetosphere are still not realized.

Nowadays, the idea to use ground-based observations of ULF waves for the monitoring of solar wind parameters is intensively studied [e. g. Engebretson et al., 1987]. Usually, for that purpose the frequency and the amplitude of ULF waves are used which are supposed to be controlled by a solar wind source. But, in practice the observed ULF signals are distorted by the magnetospheric resonant effects. So, for an effective diagnostic of solar wind sources reliable methods of separation of resonant features from source properties are needed. The basic physical principles of these methods have been outlined in this paper.

Acknowledgments. We acknowledge stimulating discussions of the theoretical and observational aspects of the considered problems with L. Baransky, A. Green, W. Worthington, L. Van'jan, and useful suggestions made by both referees.

REFERENCES

Alperovich, L.S., and E.N. Fedorov, The role of the finite earth's conductivity in a spatial distribution of geomagnetic pulsations, Izv. AN SSSR, Fizika Zemli, 11, 90, 1984a.

Alperovich, L.S. and E.N. Fedorov, The influence of the ionosphere on the propagation of hydromagnetic wave beams, Izv. Vuzov (Radiofizika), 27, 1238, 1984b.

Alperovich L.S., E.N. Fedorov, and T.B. Os'makova, About peculiarities of a telluric field near resonant magnetic shell, Izv. AN SSSR, Fizika Zemli, 7, 60, 1991.

Baransky, L.N., Yu.E. Borovkov, M.B. Gokhberg, S.M. Krylov, and V.A.

Troitskaya, High resolution method of direct measurement of the magnetic field line's eigen frequencies, Planet. Space Sci., 33, 1369, 1985.

Baransky, L.N., S.P. Belokris, Yu.E. Borovkov, M.B. Gokhberg, E.N. Fedorov, and C.A. Green, Restoration of the meridional structure of geomagnetic pulsation fields from gradient measurements, Planet.. Space Sci., 37, 859, 1989.

Baransky, L.N., S.P. Belokris, Yu.E. Borovkov, and C.A. Green, Two simple methods for the determination of the resonance frequencies of magnetic field lines, Planet. Space Sci., 38, 1573, 1990.

Berdichevsky, M.N., L.L. Van'jan, and V.I. Dmitriev, About possibility to neglect the vertical currents during magnetotelluric sounding, Izv. AN SSSR, Fizika Zemli, 5, 69, 1971.

Chen, L., and A. Hasegawa, A theory of long-period magnetic pulsations. 1. Steady state excitation of field line resonance, Geophys. Res., 79, 1024, 1974.

Chetaev, D.N. Directional analysis of magnetotelluric observations, "Nauka", Moscow, 228pp (in Russian), 1985.

Chetaev, D.N., E.N. Fedorov, S.M. Krylov et al. On the vertical electric component of the geomagnetic pulsation field, Planet. Space Sci., 23, 311, 1975.

Chetaev, D.N., V.A. Morgunov, S.V. Shamanin, and V.P. Lependin, Doubling period effect in the Ez component of the geomagnetic pulsation field, Planet. Space Sci., 25, 205, 1977.

Dmitriev, V.I., Impedance of multi-layered media for an inhomogeneous plane wave, Izv. AN SSSR, Fizika Zemli, 7, 63, 1970.

Dmitriev, V.I., and M.N. Berdichevsky, The fundamental model of magnetotelluric sounding, Proc. IEEE, 67(7), 1034, 1979.

Engebretson, M.J., L.J. Zanetti, T.A. Potemra, W. Baumjohann, H. Luehr, and M.H. Acuna, Simultaneous observation of Pc3-4 pulsations in the solar wind and in the Earth's magnetosphere, Geophys. Res., 92, 10053, 1987.

Green, A.W., E.W. Worthington, L.N. Baransky, E.N. Fedorov, N.A. Kurneva, V.A. Pilipenko, A.A. Bektemirov, and E. Philipov, Alfven field line resonances at low latitude (L=1,5). J. Geophys. Res., (in press), 1993.

Green A.W., E.W. Worthington, V.A. Pilipenko, G.B. Herzog, and Kurneva, Influence of magnetospheric Alfven resonance on spectra of Pc3-4 wave packets at middle latitudes, Geomagn. and Aeronomy, 31, 619, 1991.

Gugliel'mi, A.V., Diagnostics of the plasma in the magnetosphere by means of measurement of spectrum of Alfven oscillations. Planet. Space Sci., 37, 1011, 1989a.

Gugliel'mi, A.V., Hydromagnetic diagnostics and geoelectric exploration, Uspekhi Fizitcheskikh Nauk, 158, 605, 1989b.

Gugliel'mi, A.V., N.A. Zolotukhina, and I.P. Kharchenko, Hydromagnetic diagnostics by the ULF observation at geomagnetic meridian. Geomagn. and Aeronomy, 29, 926, 1989a.

Gugliel'mi, A.V., M.B. Gokhberg, and V.F. Ruban, Hydromagnetic diagnostics and geoelectric prospecting on the basis of single observatory, Doklady AN SSSR, 308, 578, 1989b.

Hasegawa, A., and L.J. Lanzerotti, On the orientation of hydromagnetic waves in the magnetosphere, Rev. Geophys. Space Phys., 16, 263, 1978.

Hughes, W.J. and D.J. Southwood, The screening of micropulsation signals by the atmosphere and ionosphere, J. Geophys. Res., 81, 3234, 1976.

Kaufman, A.A., and G.V. Keller, The magnetotelluric sounding method, Methods in Geochemistry and Geophysics, vol.15, New York, Elsevier, 595 pp., 1981.

Kikuchi, T., and T. Araki, Horizontal transmission of the polar electric field to the equator, J. Atmos. Terr. Phys., 41, 927, 1979.

Kivelson, M.G., and D.J. Southwood, Coupling of global magnetospheric MHD eigenmodes to field line resonances, J. Geophys. Res., 91, 4345, 1986.

Krylov, A.L., and E.N. Fedorov, About eigen oscillations of bounded cold plasma volume, Dokl. AN SSSR, 231, 68, 1976.

Krylov, A.L., A.E. Lifshitz, and E.N. Fedorov, About resonant properties of the magnetosphere, Fizika Zemli, 6, 49, 1981.

Kurchashov, Yu.P., Ya.S. Nikomarov, V.A. Pilipenko, and A. Best, Local meridional structure of mid-latitude geomagnetic pulsations, Ann. Geophysicae, 5A, 147, 1987.

Leonovich, A.S., and V.A. Mazur, An electromagnetic field, induced in the ionosphere and atmosphere and on the Earth's surface by low frequency Alfven oscillations of the magnetosphere: General theory, Planet. Space Sci., 39, 221, 1991.

Lifshitz, A.E. and E.N. Fedorov, Hydromagnetic oscillations of the magnetosphere-ionosphere resonator, Doklady AN SSSR, 287, 90, 1986.

Pilipenko, V.A., ULF Waves on the ground and in space, J. Atmos. and Terr. Phys., 52, 1193, 1990.

Price, A., The theory of magnetotelluric method when the source field is considered, J. Geophys. Res., 67, 1907, 1962.

Radoski, H.R., A theory of latitude dependent geomagnetic micropulsations: the asymptotic fields, J. Geophys. Res., 79, 596, 1974.

Samson, J.C., Pure states, polarized waves, and principal components in the spectra of multiple geophysical time series, Geophys. J. R. Astr. Soc., 72, 647, 1983.

Savin, M.G., V.M. Nikiforov, and V.M. Kharakhinov, About anomalies of vertical electric component of telluric field at Northern Sakhalin, Izv. AN SSSR, Fizika Zemli, 2, 100, 1991.

Southwood, D.J., Some features of field line resonances in the magnetosphere, Planet. Space Sci., 22, 483, 1974.

Southwood, D.J., and W.J. Hughes, Source induced vertical components in geomagnetic pulsation signals, Planet. Space Sci., 726, 15, 1978.

Tikhonov, A.N., D.N. Chetaev, V.A. Morgunov et al., Concerning studies of the Earth's crust, Doklady AN SSSR, 217, 1063, 1974.

Vellante M., Villante U., De Lauretis M., R. Core, A. Best, D. Lenners, and V.A. Pilipenko, Simultaneous geomagnetic pulsation observations at two latitudes: resonant mode characteristics Ann. Geophysicae, (in press), 1993.

Wait, J.R., On the relation between telluric currents and the Earth's magnetic field, Geophysics, 19(2), 281, 1954.

V.A. Pilipenko and E.N. Fedorov, Institute of the Physics of the Earth, Bolshaya Gruzinskaya 10, Moscow, 123810, Russia.

Numerical Studies on ULF Wave Structures in the Dipole Model

DONG-HUN LEE

Department of Astronomy and Space Science, Kyung Hee University

ROBERT L. LYSAK

School of Physics and Astronomy, University of Minnesota

Propagation of ULF waves inside the magnetospheric cavity depends on the local Alfven speed and polarization as well as ambient field geometry. We have studied the dynamical properties of ULF waves in the dipole model. We investigate two-dimensional wave structures on a meridional plane for different frequencies and azimuthal wavenumbers. An impulse is assumed at the magnetopause in the numerical model and an inverse Fourier transform is applied to obtain a single frequency component. The amplitude of compressional oscillations is found to be largest near the equatorial region and become significantly attenuated near the polar region. We suggest that the dipole geometry may be responsible for such localization. Each structure is compared with previous theoretical and numerical results. The general effects of dipole geometry on wave propagation will be discussed in detail.

INTRODUCTION

The role of global compressional modes for the impulsive excitation of ULF waves in the magnetosphere has been recently emphasized. After Kivelson and Southwood [1985; 1986] suggested a theory of global cavity modes, many numerical studies [Allan et al., 1986a; Inhester, 1987; Krauss-Varban and Patel, 1988; Lee and Lysak, 1989; 1990; 1991a,b; Zhu and Kivelson, 1988; Fujita and Patel, 1992] have been successfully performed to improve the understanding of hydromagnetic wave properties when the system is confined by some boundaries and excited by a stimulus at the magnetopause. Both theoretical and numerical studies indicate that field line resonances (FLR) can be strongly excited where the global mode frequency is matched to any transverse standing mode. Initially, a straight-field system, the box model [Kivelson and Southwood, 1985; 1986], was introduced to study the qualitative properties of the global mode coupling problem. More recently, many of the geometric effects not contained in the box model have been examined by considering curved field lines. For instance, in the box model, the FLR occur on the inner shells of the magnetospheric cavity. However, only a small fraction of the compressional wave energy may reach such low-latitude magnetic shells

Solar Wind Sources of Magnetospheric Ultra-Low-Frequency Waves
Geophysical Monograph 81
Copyright 1994 by the American Geophysical Union.

since the coupling location of FLR in the straight fields is always closer to the inside of the magnetosphere than the cutoff region. Although no harmonic coupling exists in the one-dimensional box model, it has been shown in the dipole model [Lee and Lysak, 1989] that coupling can arise at multiple regions inside the cavity wherever the compressional wave frequency is harmonically matched to the transverse modes. This supports the idea that FLR may be easily excited at various magnetic shells inside the magnetosphere. Analytical studies [e.g., Mond et al., 1988; Chen and Cowley, 1988; Wright; 1992] in a dipole field and in a general curved-field system have been consistent with these numerical results. Lee and Lysak [1990] also showed that dipole geometry plays a crucial role in understanding the wave propagation and coupling, which strongly depends on the local Alfven speed distribution and azimuthal wavenumber. In order to understand the dynamical observational features of ULF waves, it is important to visualize the structure of each mode both across and along field lines. The transverse mode structure has been well investigated even in a dipole field [Cummings et al., 1969; Chiu, 1987; Lee and Lysak, 1989] since it is a field-aligned one-dimensional standing wave. However, only a few studies [e.g. Allan et al., 1986; Zhu and Kivelson, 1988] have been made on the compressional mode structure. These studies mostly investigated the structure across the field lines. The effect of field-aligned inhomogeneity was first introduced by Southwood and Kivelson [1986], representing the two-dimensional characteristics of a straight field system. Lee and Lysak [1989; hereafter referred as P1] presented a global compressional mode structure on the two-dimensional meridional plane which shows a complicated structure due to the curved boundaries. They also proved that a global cavity and its frequency strongly depend on the azimuthal wavenumber [Lee and Lysak, 1990]; as the azimuthal wave number becomes higher, the compressional wave energy becomes more localized near the magnetopause. In fact, these effects mainly originate in geometry rather than inhomogeneity. Recently, Fujita and Patel [1992] showed that their numerical results from eigenmode analysis were consistent with the previous results in the time-dependent model of Lee and Lysak [1989; 1990]. In this paper, we present various mode structures on the dipole meridian for different frequencies and azimuthal wavenumbers. We will compare them with our previous results using the WKB approximation [Lee and Lysak, 1990]. The compressional energy distribution will be also compared with recent observational results. Finally, the overall dynamics of compressional wave propagation inside the dipole magnetosphere will be discussed.

WAVE PROPAGATION IN A DIPOLE FIELD

The effect of a dipole field becomes important since the curvature of the ambient field implies that transverse and compressional waves follow different ray paths. In addition, the inner boundary of wave propagation inside the magnetosphere significantly depends on geometry as well as on the local Alfven speed [Lee and Lysak, 1990]. In this model, the equatorial density is given by

$$n_{eq}(L) = n_o \left(\frac{r_{mp}}{r}\right)^3 \tag{1}$$

and the density variation along each field line is given by

$$n(r, L) = n_{eq}(L) \left(\frac{R_E L}{r}\right)^6 \tag{2}$$

where $r_{mp} = 10 R_E$ and $n_o = 1 \, cm^{-3}$. If we consider the magnetospheric cavity to be a region between an inner boundary defined by the turning points of compressional wave propagation and an outer boundary which is usually the magnetopause, the WKB approximation may be used to determine the inner boundary [Lee and Lysak, 1990]. The cutoff frequency ω_{min} is obtained by noting that the compressional wave frequency ω_c has to satisfy

$$\omega_c^2 > \omega_{min}^2(r, L) = k_\phi^2 V_A^2 \tag{3}$$

where k_ϕ is the azimuthal wavenumber; $k_\phi = 2\pi/\lambda_\phi = m/r$ where r is the distance to the dipole axis. Figure 1 shows the equatorial cutoff frequency for various azimuthal wavenumbers. As pointed out in Lee and Lysak [1990], as m increases, the cutoff points move toward the magnetopause and the region of propagation is confined to lower latitude. Figure 2 represents the cutoff region on the meridional plane for two global modes when $m=3$ and 10, respectively. As the azimuthal wavenumber becomes higher and the frequency becomes lower, the compres-

sional waves are restricted to the outer boundary and propagation is allowed only in the lower latitude region. We will confirm this theoretical prediction with the following numerical experiments.

MODEL

The cold MHD wave equations as given in P1 are solved in time and space where the meridian is represented by 94 x 41 grid points in the parallel ($\hat{\mu}$) and radial ($\hat{\nu}$) directions, respectively. We excited the system with a compressional impulse at the magnetopause. The details of the input and Alfven speed distribution is presented in P1. All time histories are recorded and transformed through an FFT. In order to investigate the structure of a single mode, we applied an inverse Fourier transform after filtering all other frequencies except for the selected mode frequency. Thus, we obtained the two-dimensional structure of single mode oscillations on the meridional plane. Two different azimuthal wavenumbers were selected ($m=3$ and 10) which enabled us to compare the results with our previous simulation studies [P1; Lee and Lysak, 1990]. We investigate the first two global mode structures in this paper.

MODE STRUCTURES

Figure 3 shows the spectrum of compressional waves for $m=10$. The first global mode is found near $f=0.042$ Hz and the second near $f=0.056$ Hz, respectively. The spectrum of $m=3$ is shown in Plate 1 in P1 and the first two modes are $f=0.027$, 0.060 Hz in the compressional component. Hereafter, we will focus on the structures of the two lowest compressional modes for each case. Figures 4(a) and 4(b) show the global mode structure for $m=3$ at $f=0.027$ and $f=0.060$ Hz, respectively. The dipole meridian is mapped into a rectangle, in which the upper and lower boundaries are the northern and southern ionospheres and vertical lines are magnetic shells. Therefore, the left boundary in the following figures corresponds to the plasmapause ($L=5$) and the right end is the magnetopause ($L=10$). We averaged all amplitudes for more than ten periods of each mode and thus each figure represents the actual energy distribution of compressional oscillations on the meridian. It is evident that the $f=0.027$ mode has one node and $f=0.060$ has two nodes along the equator, while there is no qualitative difference along the field lines. In addition, the region of the $f=0.060$ oscillations is found to be more confined to the lower latitude region in comparison with the $f=0.027$ structure. This is consistent with the results of the condition (3) and Figure 2. Figure 5 represents an averaged mode structure for $m=10$. The oscillations becomes more concentrated into the outer equatorial region compared to $m=3$. The structure of the first mode ($f=0.042$) in Figure 5(a), however, is extended deeper into the inner magnetosphere, while the $f=0.056$ mode is strongly restricted to the outer shells. It is interesting to note that the $f=0.056$ mode has a harmonic structure along the field lines rather than across magnetic shells in contrast with the other structures. This is easy to understand since global modes in the dipole meridian should consist of fully two-dimensional

0.12
0.10
0.08
0.06
0.04
0.02
0.00
5
6
7
8
9
10
L(Re)
ω min (Hz)
15
13
11
9
7
5
3
m = 1

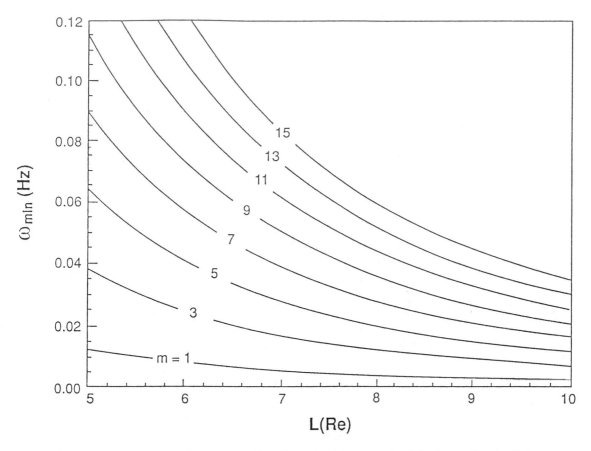

Fig. 1. The equatorial cutoff frequency ω_{min} for various azimuthal wavenumber. [after Lee and Lysak, 1990]

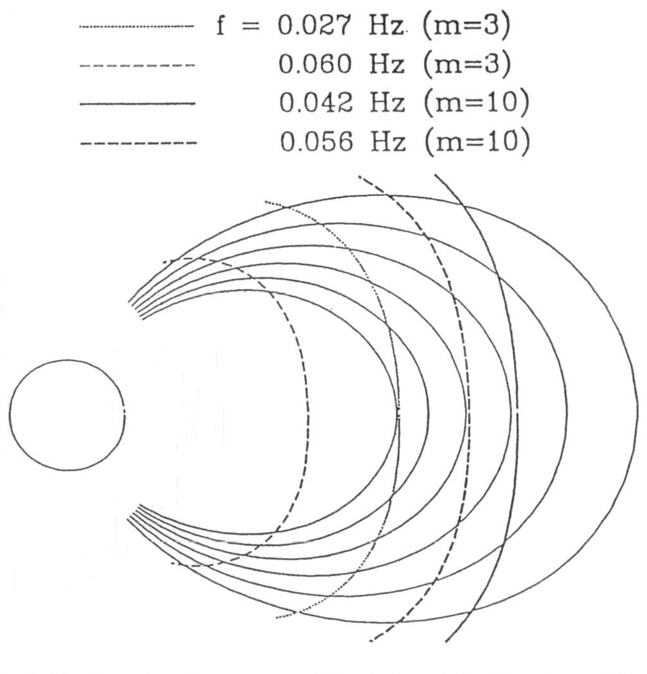

f = 0.027 Hz (m=3)
0.060 Hz (m=3)
0.042 Hz (m=10)
0.056 Hz (m=10)

Fig. 2. The inner boundary on a meridional plane defined by the cutoff condition (3).

harmonic oscillations, as distinguished from any straight field model [P1]. In fact, the structure and propagation properties of compressional waves show a good agreement between the theory of the cutoff condition and the simulation results. In addition, it should be mentioned that the mode structures of the toroidal component for the above cases are found to be well consistent with our previous results [P1; Lee and Lysak, 1990] associated with the coupling problem.

Discussion

Several points may be emphasized with respect to the above results. First, our results suggest that the inner boundary must exist and depends on the frequency and the azimuthal wavenumber in the WKB approximation. This may be verified by statistical observational studies on compressional wave energy distribution for various frequencies and m values. The inner boundary in ULF studies also becomes important when we consider the effect of boundary interference [Lee and Lysak, 1991a]. For instance, even a narrow-banded energy source may result in a relatively strong broad-banded spectrum as long as the inner boundary is located near the magnetopause since the inner boundary continuously reflects incoming waves which will destructively interfere with a driving source. Secondly, this model assumes a monotonic Alfven speed profile inside the magnetosphere, which is not the case if the plasmasphere is

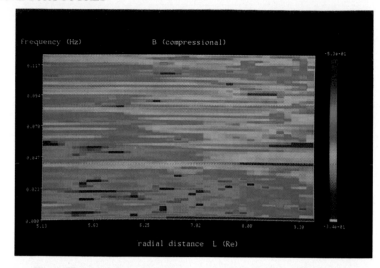

Fig. 3. The wave spectrum of a compressional magnetic field where m is assumed to be 10.

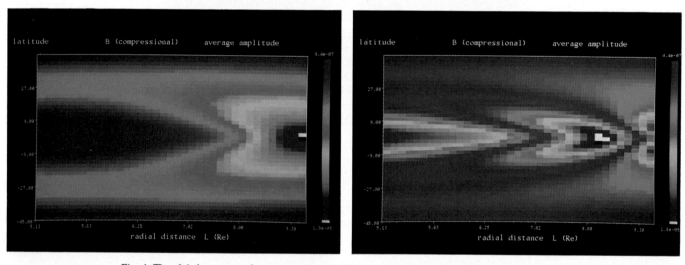

Fig. 4. The global structure of a compressional mode for $m=3$: (a) $f=0.027$ Hz and (b) $f=0.060$ Hz.

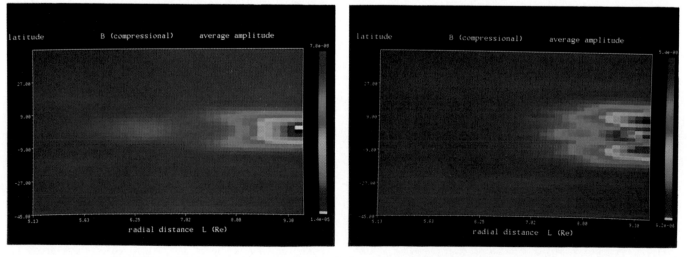

Fig. 5. The global structure of a compressional mode for $m=10$: (a) $f=0.042$ Hz and (b) $f=0.056$ Hz.

included. If the Alfven speed varies non-monotonically across the magnetic fields, the curves of cutoff frequency in Figure 1 will have a dip rather than being simply decreasing toward the magnetopause. This implies that the region of propagation in Figure 2 will have an additional allowed region of compressional oscillations like a "island" in the inner damping region and waves may stay inside such local island. These effects have been investigated in the box-model and the cylindrical model [Allan et al., 1986b; Zhu and Kivelson, 1989]. They will have to be further studied in the dipole model by assuming a different density profile in the future. Finally, it should be pointed out that the characteristics of above structures are well consistent with ground-based and satellite observations of Pc 3-4 waves [e.g. Yumoto et al., 1985]. Yumoto et al. [1985], for instance, showed that after external sources excite Pc 3-4 waves near the magnetopause, they are likely to propagate into the lower-latitude equatorial region. Figure 4 and 5 both suggest that the compressional waves mainly have maximum amplitude near the equator and they can not propagate easily in the high latitude region. Thus, our results may explain this feature as a result of dipole geometric effects.

Acknowledgments . This work was supported in part by NSF grant ATM-9111791 and NASA grant NAGW-2653 and by '92 research fund from Kyung Hee university, Korea. Computing costs were supported by a grant of Cray time from the Minnesota Supercomputer Institute.

REFERENCES

Allan, W., S. P. White, and E. M. Poulter, Impulse-excited hydromagnetic cavity and field-line resonances in the magnetosphere, *Planet. Space Sci., 34*, 371, 1986a.

Allan, W., E. M. Poulter, and S. P. White, Hydromagnetic wave coupling in the magnetosphere-Plasmapause effects on impulse-excited resonances *Planet. Space Sci., 34*, 1189, 1986b.

Chen, L., and S. C. Cowley, On field line resonances of hydromagnetic Alfven waves in dipole magnetic field, *Geophys. Res. Lett., 16*, 895, 1989.

Chiu, Y. T., Resonant Alfven waves on auroral field lines, *J. Geophys. Res., 92*, 3402, 1987.

Cummings, W. D., R. J. O'Sullivan and P. J. Coleman, Jr., Standing Alfven waves in the magnetosphere, *J. Geophys. Res., 74*, 778, 1969.

Fujita, S., and V. L. Patel, Eigenmode analysis of coupled magnetohydrodynamic oscillations in the magnetosphere, *J. Geophys. Res., 97*, 13777, 1992.

Inhester, B., Numerical modeling of hydromagnetic wave coupling in the magnetosphere, *J. Geophys. Res., 92*, 4751, 1987.

Kivelson, M. G., and D. J. Southwood, Resonant ULF waves: A new interpretation, *Geophys. Res. Lett., 12*, 49, 1985.

Kivelson, M. G., and D. J. Southwood, Coupling of global magnetospheric MHD eigenmodes to field line resonances, *J. Geophys. Res., 91*, 4345, 1986.

Krauss-Varban, D., and V. L. Patel, Numerical analysis of the coupled hydromagnetic wave equations in the magnetosphere, *J. Geophys. Res., 93*, 9721, 1988.

Lee, D. H., and R. L. Lysak, Magnetospheric ULF wave coupling in the dipole model: the impulsive excitation, *J. Geophys. Res., 94*, 17097, 1989.

Lee, D. H., and R. L. Lysak, Effects of azimuthal asymmetry on ULF waves in the dipole magnetosphere, *Geophys. Res. Lett., 17*, 53, 1990.

Lee, D. H., and R. L. Lysak, Impulsive excitation of ULF waves in the three-dimensional dipole model: the initial results, *J. Geophys. Res., 96*, 3479, 1991a.

Lee, D. H., and R. L. Lysak, Monochromatic ULF wave excitation in the dipole magnetosphere, *J. Geophys. Res., 96*, 5811, 1991b.

Mond, M., E. Hameiri and P. N. Hu, Coupling of MHD waves in inhomogeneous magnetic field configurations, *J. Geophys. Res., 95*, 89, 1990.

Southwood, D. J., and M. G. Kivelson, The effect of parallel inhomogeneity on magnetospheric wave coupling, *J. Geophys. Res., 91*, 6871, 1986.

Wright, A. N., Asymptotic and time-dependent solutions of magnetic pulsations in realistic magnetic field geometries, *J. Geophys. Res., 97*, 6439, 1992.

Yumoto, K., T. Saito, S.-I. Akasofu, B. T. Tsurutani and E. J. Smith, Propagation mechanism of daytime Pc 3-4 pulsations observed at synchronous orbit and multiple ground-based stations, *J. Geophys. Res., 90*, 6439, 1985.

Zhu, X. M., and M. G. Kivelson, Analytic formulation and quantitative solutions of the coupled ULF wave problem, *J. Geophys. Res., 93*, 8602, 1988.

Zhu, X. M., and M. G. Kivelson, Global mode ULF pulsations in a magnetosphere with a nonmonotonic Alfven velocity profile, *J. Geophys. Res., 94*, 1479, 1989.

Dong-Hun Lee, Department of Astronomy and Space Science, Kyung Hee University, Yongin, Kyunggi, Korea, 449-701.

Robert L. Lysak, School of Physics and Astronomy, University of Minnesota, 116 Church Street SE, Minneapolis, MN 55455

observed nearly every day at L = 1.83 (42° geomagnetic), becomes increasingly difficult to find using the crossphase technique at L = 1.70 (40° geomagnetic) and L=1.58 (37°; Worthington, personal communciation). The spectral power division technique Ḣ/Ḋ seems to be more successful at these lower latitudes. Since the crossphase technique evaluates the phase difference associated with the frequency difference between adjacent uncoupled field lines, no resonance signature would be detected if both stations were located symmetrically around the peak of the eigenfrequency-latitude curve. The power division method would be unaffected. There is therefore indirect evidence of a turning point in the latitudinal variation of toroidal mode eigenfrequency, just below 40° geomagnetic latitude. Furthermore, while there is often evidence of resonances at higher latitudes, beyond the station array, resonance characteristics at latitudes $L \leq 1.5$ have not yet been observed in the meridional amplitude and polarization data.

The second trend is that while at least three harmonics are observed in almost all the crossphase spectra recorded at L = 2.78 (53° geomagnetic), harmonic structure has never been observed at L = 1.83 (42° geomagnetic). Poulter et al. [1988, 1990] showed that at low latitudes ionospheric mass loading effects result in non-integer multiple harmonic spacing. However, the disappearance of higher harmonics in low latitude experimental data is probably related to the enhanced Joule dissipation experienced by the higher order wavemodes.

SUMMARY AND CONCLUSIONS

One aim of recent coordinated multipoint campaigns has been to develop reliable diagnostic techniques for monitoring solar-terrestrial interaction processes. Low latitude ground measurements of ULF plasma waves are now approaching this capability. Simple methods now exist which allow the temporal evolution of magnetospheric plasma processes to be monitored, via the signature of field line resonances. The use of multistation ground magnetometer arrays provides further spatial information on these processes. Recent experimental campaigns have therefore been able to provide measurements of parameters previously only available through modeling. Table 1 summarizes some of the recently measured properties of field line resonances in the plasmasphere. Only fundamental mode eigenfrequencies are shown. Equatorial plasma densities ρ_{eq} are calculated using a model of the form r^{-m}. This is not valid below about 40° latitude [Poulter et al., 1990]. Present outstanding problems include the determination of the low latitude limit of resonances, and verification of the mechanism by which apparent standing field line oscillations around 15-25 mHz are frequently observed at widely spaced low latitude sites.

Acknowledgments. This work was supported by the Australian Research Council and the University of Newcastle. C.L.W. received financial assistance from a Commonwealth Postgraduate Research Award, and C.W.S.Z., Q.F., and S.H.L. from University of Newcastle Postgraduate Research Scholarships, during this study. The KUJ field station was operated by T. Kitamura and O. Saka, Physics Department, Kyushu University, Fukuoka, Japan. The assistance of volunteer operators at the Australian field stations is also acknowledged.

REFERENCES

Allan, W., and E. M. Poulter, ULF waves-their relationship to the structure of the Earth's magnetosphere, *Rep. Prog. Phys., 55,* 533-598, 1992.

Anderson, B. J., M. J. Engebretson, and L. J. Zanetti, Distortion effects in spacecraft observations of MHD toroidal standing waves: theory and observations, *J. Geophys. Res., 94,* 13425-13445, 1989.

Ansari, I. A., and B. J. Fraser, A multistation study of low latitude Pc 3 geomagnetic pulsations, *Planet. Space Sci., 34,* 519-536, 1986.

Baransky, L. N., Y. E. Borovkov, M. B. Gokhberg, S. M. Krylov, and V. A. Troitskaya, High resolution method of direct measurement of the magnetic field lines' eigen frequencies, *Planet. Space Sci., 33,* 1369-1374, 1985.

Baransky, L. N., S. P. Belokris, Y. E. Borovkov, M. B. Gokhberg, E. N. Fedorov, and C. A. Green, Restoration of the meridional structure of geomagnetic pulsation fields from gradient measurements, *Planet. Space Sci., 38,* 859-864, 1989.

Baransky, L. N., S. P. Belokris, Y. E. Borovkov, and C. A. Green, Two simple methods for the determination of the resonance frequencies of magnetic field lines, *Planet. Space Sci., 38,* 1573-1576, 1990.

Chen, L., and S. W. H. Cowley, On field line resonances of hydromagnetic Alfvén waves in a dipole magnetic field, *Geophys. Res Lett., 16,* 895-897, 1989.

Cummings, W. D., R. J. O'Sullivan, and P. J. Coleman, Standing Alfvén waves in the magnetosphere, *J. Geophys. Res., 74,* 778-793, 1969.

Engebretson, M. J., L. J. Zanetti, T. A. Potemra, and M. H. Acuna, Harmonically structured ULF pulsations observed by the AMPTE CCE magnetic field experiment, *Geophys. Res. Lett., 13,* 905-908, 1986.

Fedorov, E. N., B. N. Belenkaya, M. B. Gokhberg, S. P. Belokris, L. N. Baransky, and C. A. Green, Magnetospheric plasma density diagnosis from gradient measurements of geomagnetic pulsations, *Planet. Space Sci., 38,* 269-272, 1990.

Fraser, B. J., R. L. McPherron, and C. T. Russell, Radial Alfvén velocity profiles in the magnetosphere and their relation to ULF field line resonances, *Adv. Space Res., 8,* 49-52, 1988.

Fraser, B. J., P. W. McNabb, F. W. Menk., and C. L. Waters, A personal computer induction magnetometer system for recording geomagnetic pulsations, *ANARE Res. Notes, 80,* 83-92, 1991.

Gough, H., and D. Orr, The effect of damping on geomagnetic pulsation amplitude and phase at ground observatories, *Planet Space Sci., 38,* 619-628, 1984.

Gough, H., and D. Orr, Observations of forced oscillations of the magnetosphere by a geostationary satellite and an extensive ground magnetometer array, *Planet. Space Sci., 34,* 863-877, 1986.

Gough, H., D. Orr, and U. Wedeken, Ground observations of geomagnetic pulsations during a quiet magnetospheric interval correlated with satellite plasma measurements, *J. Geophys., 52,* 92-101, 1984.

Green, C. A., The role of ground arrays of magnetometers in the study of pulsation resonance regions, *Planet. Space Sci., 30,* 1199-1208, 1982.

Green, A. W., E. W. Worthington, L. N. Baransky, E. N. Fedorov, N. A. Kurneva, V. A. Pilipenko, D. N. Shvetzov, A. A. Bektemirov, and G. V. Philipov, Alfvén field line resonances at low latitudes (L — 1.5), *J. Geophys. Res.,* in press, 1993.

TABLE 1. Summary of daytime plasmaspheric field line resonance properties

L value	ω_R (mHz)	$\partial\omega_R/\partial\theta$ (mHz/°)	Width (km)	ρ_{eq} (cm^{-3})	Reference
3.6	13		120-160	3.2×10^2	Baranksy et al.[1989], Federov et al. [1990]
3.6	10	0.3			Kurchashov et al. [1989]
3.2	19	2.2	120-160	4.0×10^2	Baransky et al. [1989], Federov et al. [1990]
3.0	24	3.4			Baransky et al. [1990]
2.9	18	2.1			Gough and Orr [1986]
2.8	15 ± 2, 15 ± 5	1.5		2.0×10^3	Waters et al. [1991b], Waters [1992]
2.3	30-35				Vellante et al. [1993]
2.3	23			3.4×10^3	Worthington [1992]
2.0	30-37	3.0			Newcastle Group
1.9	28		250		Ziesolleck et al. [1993]
1.8	50 ± 2, 45 ± 5	5	175 ± 10		Waters et al. [1991], Waters [1992]
1.8	56	2.6			Hattingh and Sutcliffe [1987]
1.7	40-50				Newcastle Group
1.6	85-95				Vellante et al. [1993]
1.6	66	3.5	300	1.2×10^4	Green et al. [1993]
1.55	73				Green et al. [1993]
1.45	69	0			Hattingh and Sutcliffe [1987]
1.20	56	4.8		1.2×10^5	Yumoto et al. [1993]
1.12	45				Yumoto et al. [1993]

Hanson, H. W., D. C. Webb, and D. Beamish, A high resolution study of continuous pulsations in the European sector, *Planet. Space Sci.*, 27, 1371-1382, 1979.

Hattingh, S. K. F., and P. R. Sutcliffe, Pc 3 pulsation eigenperiod determination at low latitudes, *J. Geophys. Res.*, 92, 12433-12436, 1987.

Hughes, W. J., and D. J. Southwood, The screening of micropulsation signals by the atmosphere and ionosphere, *J. Geophys. Res.*, 81, 3234-3240, 1976a.

Hughes, W. J., and D. J. Southwood, An illustration of modification of geomagnetic pulsation structure by the ionosphere, *J. Geophys. Res.*, 81, 3241-3247, 1976b.

Hughes, W. J., R. L. McPherron and J. N. Barfield, Geomagnetic pulsations observed simultaneously on three geostationary satellites, *J.*

Geophys. Res., 83, 1109-1116, 1978.

Hughes, W. J., H. J. Singer, M. Smiddy, J. R. Wygant, and R. R. Anderson, Field aligned plasma distributions derived from ULF field line resonant oscillations, Abstract, Suppl. EOS Trans. AGU, 72, 426, 1991.

Kivelson, M. G., and D. J. Southwood, Resonant ULF waves: a new interpretation, *Geophys. Res. Lett.*, 12, 49-52, 1985.

Kivelson, M. G., and D. J. Southwood, Coupling of global magnetospheric MHD eigenmodes to field line resonances, *J. Geophys. Res.*, 91, 4345-4351, 1986.

Kurchashov, Yu. P., Ya. S. Nikomarov, V. A. Pilipenko, and A. Best, Field line resonance effects in local meridional structure of mid-latitude geomagnetic pulsations, *Ann. Geophys.*, 5A, 147-154, 1987.

Lanzerotti, L. J., L. V. Medford, C. G. MacLennan, T. Hasegawa, M.

H. Acuna, and S. R. Dolce, Polarization characteristics of hydromagnetic waves at low geomagnetic latitudes, *J. Geophys. Res., 86*, 5500-5506, 1981.

Lathuillere, C., F. Glangeaud, J. L. Lacoume, and G. Lejeune, Relationship between ionospheric electric field and ground magnetic pulsations in the Pc 3 domain at mid latitude, *J. Geophys. Res., 86*, 7669-7678, 1981.

Lefeuvre, F., M. Parrat and D. Jones, Magnetospheric wave characteristics derived from cross-spectral analysis, in: *Proc. Conf. Achievements of the IMS, ESA-SP 217*, 723-728, 1984.

Lin, N. G., L. J. Cahill, M. J. Engebretson, M. Sugiura, and R. L. Arnoldy, Dayside pulsation events near the plasmapause, *Planet. Space Sci., 34*, 155-181, 1986.

Menk, F. W., Spectral structure of mid-latitude Pc 3-4 geomagnetic pulsations, *J. Geomag. Geoelectr., 40*, 33-61, 1988.

Miletits, J. Cz., J. Verö, J. Szendröi, P. Ivanova, A. Best, and M. Kivinen, Pulsation periods at mid-latitudes - a seven-station study, *Planet. Space Sci., 38*, 85-95, 1990.

Mond, M., E. Hameiri and P. N. Hu, Coupling of magnetohydrodynamic waves in inhomogeneous magnetic field configurations, *J. Geophys. Res., 95*, 89-95, 1990.

Obayashi, T., and J. A. Jacobs, Geomagnetic pulsations and the Earth's outer atmosphere, *Geophys. J.R. Astr. Soc., 1*, 53-63, 1958.

Ochadlick, A. R., Time series and correlation of pulsations observed simultaneously by two aircraft, *Geophys. Res. Lett., 17*, 1889-1892, 1990.

Orr, D, Magnetospheric hydromagnetic waves: their eigenperiods, amplitudes and phase variations; a tutorial introduction, *J. Geophys., 55*, 76-84, 1984.

Orr, D., and H. W. Hanson, Geomagnetic pulsation phase patterns over an extended latitudinal array, *J. Atmos. Terr. Phys., 43*, 899-910, 1981.

Poulter, E. M., and W. Allan, Transient ULF pulsation decay rates observed by ground based magnetometers: the contribution of spatial intregration, *Planet. Space Sci., 33*, 607-616, 1985.

Poulter, E. M., W. Allan, G. J. Bailey, and R. J. Moffett, On the diurnal period variation of mid-latitude ULF pulsations, *Planet. Space Sci., 32*, 727-734, 1984.

Poulter, E. M., W. Allan, and G. J. Bailey, ULF pulsation eigenperiods within the plasmasphere, *Planet. Space Sci., 36*, 185-196, 1988.

Poulter, E. M., W. Allan, and G. J. Bailey, The effect of density inhomogeneity on standing Alfvén wave structure, *Planet. Space Sci., 38*, 665-673, 1990.

Radoski, H. R., A note on the problem of hydromagnetic resonances in the magnetosphere, *Planet. Space Sci., 19*, 1012-1013, 1971.

Saka, O., and J. S. Kim, Spatial phase structure of low-latitude Pc 3-4 pulsations, *Planet. Space Sci., 33*, 1073-1079, 1985.

Samson, J. C., The spectral matrix, eigenvalues and principal components in the analysis of multi-channel geophysical data, *Ann. Geophys., 1*, 115-119, 1983.

Samson, J. C., Geomagnetic pulsations and plasma waves in the Earth's magnetosphere, in: *Geomagnetism* Vol. 4., Academic Press, New York, 1991.

Singer, H. J., W. J. Hughes, and C. T. Russell, Standing hydromagnetic waves observed by ISEE1 and 2: radial extent and harmonic, *J. Geophys. Res., 87*, 3519-3529, 1982.

Takahashi, K., and R. L. McPherron, Harmonic structure of Pc 3-4 pulsations, *J. Geophys. Res., 87*, 1504-1516, 1982.

Takahashi, K., and R. L. McPherron, Standing hydromagnetic oscillations in the magnetosphere, *Planet. Space Sci., 32*, 1343-1359, 1984.

Takahashi, K., R. L. McPherron, and W. J. Hughes, Multispacecraft observations of the harmonic structure of Pc 3-4 magnetic pulsations, *J. Geophys. Res., 89*, 6758-6774, 1984.

Takahashi, K., B. J. Anderson, and R. J. Strangeway, AMPTE CCE observations of Pc 3-4 pulsations at L = 2-6, *J. Geophys. Res., 95*, 17179-17186, 1990.

Troitskaya, V. A., Micropulsations and the state of the magnetosphere, in: *Solar-terrestrial physics*, eds. J.W. King and W.S. Newman, Academic Press, New York, 1967.

Troitskaya, V. A., and A. V. Gul'elmi, Geomagnetic micropulsations and diagnostics of the magnetosphere, *Space Sci. Rev., 7*, 689-768, 1967.

Vellante, M., U. Villante, R. Core, A. Best, D. Lenners, and V. A. Pilipenko, Simultaneous geomagnetic pulsation observations at two latitudes: resonant mode characteristics, *Ann. Geophys.*, in press, 1993.

Verö, J., Experimental aspects of low latitude pulsations - a review, *J. Geophys., 60*, 106-119, 1986.

Walker, A. D. M., R. A. Greenwald, W. F. Stuart, and C. A. Green, STARE auroral radar observations of Pc 5 geomagnetic pulsations, *J. Geophys. Res., 84*, 3373-3388, 1979.

Warner, M. R., and D. Orr, Time of flight calculations for high latitude geomagnetic pulsations, *Planet. Space Sci., 27*, 679-689, 1979.

Waters, C. L., Low latitude geomagnetic field line resonance, PhD thesis, Dept. of Physics, University of Newcastle, Australia, 1992.

Waters, C. L., F. W. Menk, B. J. Fraser, and P. M. Ostwald, Phase structure of low latitude Pc 3-4 pulsations, *Planet. Space Sci., 39*, 569-582, 1991a.

Waters, C. L., F. W. Menk, and B. J. Fraser, The resonance structure of low latitude Pc 3 geomagnetic pulsations, *Geophys. Res. Lett., 18*, 2293-2296, 1991b.

Worthington, E. W., Field line resonances and plasma diagnostics of the Earth's magnetosphere, PhD thesis, Dept. of Geophysics, Colorado School of Mines, U.S. Department of the Interior, 1992.

Yavorsky, B., and A. Detlaf, *Handbook of physics*, English translation, Mir Publishers, Moscow, p 123, 1975.

Yumoto, K., Generation and propagation mechanisms of low-latitude magnetic pulsations - a review, *J. Geophys., 60*, 79-105, 1986.

Zanetti, L. J., T. A. Potemra, R. E. Erlandson, M. J. Engebretson, and M. H. Acuna, Geomagnetic field-line resonant harmonics measured by the Viking and AMPTE/CCE magnetic field experiments, *Geophys. Res. Lett., 14*, 427-430, 1987.

Zelwer, R., and H. F. Morrison, Spatial characteristics of mid-latitude geomagnetic micropulsations, *J. Geophys. Res., 77*, 674-694, 1972.

Zhu, X., and M. G. Kivelson, Analytic formulation and quantitative solutions of the coupled ULF wave problem, *J. Geophys. Res., 93*, 8602-8612, 1988.

Ziesolleck, C. W. S., B. J. Fraser, F. W. Menk, and P. W. McNabb, Spatial characteristics of low latitude Pc 3-4 geomagnetic pulsations, *J. Geophys. Res., 98*, 197-207, 1993.

Q. Feng, B. J. Fraser, S. H. Lee, P. W. McNabb, F. W. Menk, C. L. Waters, and C. W. S. Ziesolleck, Department of Physics, The University of Newcastle, Callaghan, N.S.W., Australia 2308.

Simultaneous Observation of Pc 3,4 Pulsations
in the Magnetosphere and at Multiple Ground Stations

T. J. ODERA

Physics Department, Maseno University College

D. VAN SWOL AND C. T. RUSSELL

Institute of Geophysics and Planetary Physics
University of California, Los Angeles, CA

Measurements obtained by the ISEE-1,2 spacecraft in the magnetosphere, and at fourteen ground stations of the IGS array of magnetometers were examined at the times of magnetic conjugacy. Simultaneous observations show that whenever compressional waves appear in the magnetosphere Pc 3,4 pulsations are seen at all ground stations, from high latitude (L ~ 7) to low latitude (L ~ 2). The amplitude of the Pc 3,4 pulsations on the ground generally decreases with the decrease in latitude. The one occurrence of quasi-continuous transverse waves in the magnetosphere was not detected at any of the ground stations. These results give further evidence for the importance of compressional waves in transporting Pc 3,4 wave energy through the magnetosphere. However, because of the bias of our selection criteria to long duration events, these observations do not provide information on the relative occurrence rates of transverse and compressional waves in the magnetosphere.

INTRODUCTION

The properties of Pc 3,4 pulsations observed on the surface of the earth have been consistently associated with three solar wind parameters, namely the solar wind velocity, interplanetary magnetic field (IMF) direction and magnitude. An obvious inference is that it is the solar wind which drives the magnetosphere, causing the Pc 3,4 pulsations observed on the ground.

Recent reviews, notably by Yumoto [1985], Odera [1986], and Vero [1986] have summarized statistical evidence for the control of the occurrence probabilities, amplitude and frequency of the Pc 3,4 pulsations observed on the ground by the solar wind velocity and by the IMF orientation and magnitude, respectively. In particular Greenstadt et al. [1980] described the conditions for the control of the occurrence probability and amplitude of Pc 3,4 pulsations in terms of cone angle, Θ_{XB} (the angle between the IMF and the earth-sun-line). Accordingly, when the cone angle is small large amplitude waves can illuminate the magnetopause, enhancing the transfer of pulsations into the magnetosphere. Conversely, large cone angle

Solar Wind Sources of Magnetospheric Ultra-Low-Frequency Waves
Geophysical Monograph 81
Copyright 1994 by the American Geophysical Union.

suppresses pulsation transfer.

The control of frequency by the IMF magnitude was first presented by Troitskaya et al. (1971) in the context of the empirical relation, $T_{sec} = 160/B_{nT}$ (or $f_{mHz} = 6.25\ B_{nT}$). The empirical relation has been confirmed by numerous ground-based studies e.g. Vero and Hollo [1978], Green et al. [1983], and Odera and Stuart [1985]; and by satellite studies, e.g. Russell and Hoppe [1981] and Yumoto et al. [1984].

However, contradictory results have been obtained regarding the control by the solar wind parameters of Pc 3,4 waves recorded on board geostationary satellites. The results presented by Arthur and McPherron [1977] indicated that the occurrence of Pc 3 pulsations at synchronous orbit was not particularly dependent on the IMF conditions. In particular they found no relation between the pulsation frequency and the IMF magnitude, but there was a clear, although weak, relationship between the angle of the IMF from the sun-earth line and the amplitude of the pulsations. Takahashi et al. [1984] indicated that pulsation events exhibiting the harmonic structure (i.e., high-harmonic standing waves) at geostationary orbit show a weak negative correlation between the IMF cone angle Θ_{XB} and the power of pulsations in the Pc 3,4 range of frequency. They also demonstrated that these pulsations show a weak negative correlation between pulsation frequencies and the IMF magnitudes.

It was suggested by Yumoto et al. [1984] that the above

contradictory results could be explained considering the existence of two types of waves, namely standing Alfven waves with larger amplitudes on local field lines and compressional waves with smaller amplitudes propagating from outside the magnetosphere. Both types of waves had been experimentally observed, and theoretically discussed by Yumoto and Saito [1983]. It appeared from the work of Yumoto et al. [1984] that there is a good correlation between the IMF magnitude and frequency of compressional waves in the magnetosphere and the Pc 3,4 pulsations on the ground. Specifically, the authors found a better frequency-IMF magnitude relation for the compressional waves at synchronous orbit than for the Pc 3,4 pulsations, which they observed simultaneously at a low latitude ground station.

Further observational study of the relation between the properties of the compressional waves in the magnetosphere and those of the Pc 3,4 pulsations on the ground has been made recently by Odera et al. [1991] who principally examined a single wave event seen at ISEE-2 and on the ground. The study confirms that Pc 3, 4 wave energy is most readily transported through the magnetosphere by compressional fluctuations. The purpose of this paper is to extend Odera's paper by using more data, both in space and on the ground. We make a direct comparison between the properties of four compressional and one transverse wave events at the ISEE-1,2 spacecraft and those of the Pc 3,4 pulsations observed simultaneously at fourteen (14) ground stations of the IGS array of magnetometers.

OBSERVATION AND ANALYSIS

Data

Both ground and satellite data used in this study were obtained at the times of magnetic 'conjugacy', i.e., when the foot point of the field line traversed by the ISEE-1,2 spacecraft was close to the IGS chain of magnetometers. We will refer to the magnetic conjugacy simply as the footprint of the spacecraft on the ground.

Ground data were obtained from fourteen stations of the network of magnetometers in the northwestern Europe, operated during the International Magnetospheric Study (IMS) by the Geomagnetism Unit of the Institute of Geological Sciences (IGS) now British Geological Survey (BGS), [Stuart, 1971]. Figure 1 is a map illustrating the disposition of the northern hemisphere sites that were occupied during the IMS. Overlaid on the map are L-shells calculated for 1978.0 epoch at 120 km altitude from the main field model of Barraclough et al. [1975]. This distance

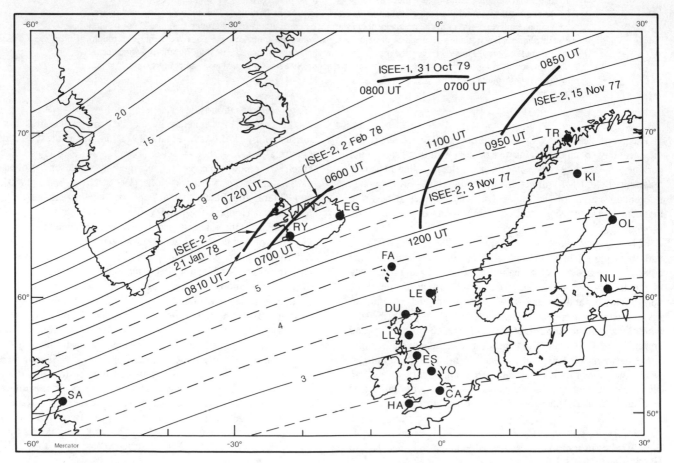

Fig. 1.   Map of north-western Europe showing the location of the IGS magnetometer stations and L-value contours and the path of the ISEE-1,-2 spacecraft mapped along field lines.

is approximately, but not exactly, the distance of the equator crossing of dipole field lines emanating from the ground station. The thick lines on the map are the conjugate points of the ISEE-1,2 spacecraft. In Table 1 is a list of the sites, their geographical and magnetic coordinates, and the corresponding L-values. The stations are listed in the order of decreasing geomagnetic latitudes.

The configuration of the IGS array of magnetometers is described in detail elsewhere by Green [1981]. In brief, the basic configuration, in Europe, is of two lines, approximately along geomagnetic meridians, about 2 hours apart in local time with a connecting azimuthal line at mid-latitude (L ~ 3.3). The eastern meridional line through Scandinavia covers L-value of ~ 3.3 - 6.3 and the western line through the UK to Iceland, L ~ 2.4 - 6.5. In addition there is a station at St. Anthony in Newfoundland, about 4 hours to the west of the UK line. It should be remembered, at this point, that the local time (LT) for the UK line is essentially universal time (UT).

The instrument at each site was a rubidium vapor magnetometer, with an essentially flat frequency response and a noise level typically ≤ 0.1 nT, recording magnetic variations in three orthogonal directions (magnetic NE, NW and Z). The three components can easily be converted to the conventional components, H, D and Z. The sensitivity of the magnetometer was usually in the range of 0.04 - 0.16 nT. The magnetometers

TABLE 1. IGS Magnetometer Stations

Stations	Identifier	Dipole Coordinates Lat.	Long.	Geographic Coords. Lat.	Long.	L. value
Reykjavik	Ry	69.8	72.5	64.2	-21.7	6.20
Eidar	Eg	69.5	81.9	65.4	-14.4	6.20
Tromso	Tr	66.9	117.6	69.7	18.9	6.30
Faroe	Fa	65.0	86.0	62.0	- 6.8	4.46
Kiruna	Ki	65.0	116.5	67.8	20.4	5.42
Lerwick	Le	62.2	89.9	60.1	- 1.2	3.75
St. Anthony	Sa	62.1	20.3	51.4	-55.6	4.19
Oulu	Ol	61.7	117.9	65.1	25.5	4.43
Durness	Du	61.5	85.0	58.6	- 4.8	3.53
Loch Laggan	Ll	59.9	84.2	56.9	- 4.4	3.23
Eskdalemuir	Es	58.1	84.3	55.3	- 3.2	2.94
Nurmijarvi	Nu	57.6	113.7	60.5	24.7	3.39
York	Yo	56.4	85.6	53.9	- 1.1	2.71
Cambridge	Ca	54.6	85.7	52.2	0.1	2.48

sampled at 2.5 sec intervals, their timing accuracy controlled by crystal clocks, and recorded digitally on cassette tapes [Riddick et al., 1975].

The satellite data were obtained from ISEE-1 and 2. The two spacecraft have identical fluxgate magnetometers which are described in detail by Russell [1978]. The data for this paper were sampled at the rate of four times per second and averaged to the same resolution as the ground data, with five seconds averaging overlapped by 2.5 seconds.

The ISEE-1,2 orbit was highly eccentric. With their initial apogee of very nearly 23 R_E, the satellites crossed low L-values very rapidly. The two spacecraft followed each other at a controllable distance of separation. The orbital plane for the satellites was nearly equatorial, and the spacecraft was located in the morning side of the magnetosphere during the four events we observed in late 1977 and early 1978. For the event observed in October 1979, the spacecraft orbit was nearly at 40° magnetic north, in the morning side of the magnetosphere. Figure 2 shows the tracks of the satellites for (a) the meridian and (b) L-value and local time, projections of orbits on which the five Pc 3,4 events in the magnetosphere were observed. Each track is marked with a number, corresponding chronologically to the Pc 3,4 waves detected on the spacecraft.

Identification of Pc 3,4 wave events

The examination of ground records formed an integral part of the exercise for event identification. This exercise was carried out only for the cases of the spacecraft footprint being close to the IGS ground stations. We identified 150 intervals during which the ISEE-1,2 footprint was very close to the IGS chain of magnetometers. Only the intervals when the spacecraft were in the dayside magnetosphere were scrutinized. For these 50 cases we visually scanned through both spacecraft and ground analog records to select the intervals with good data, in terms of continuity and low noise.

In the last stage we looked for suitable Pc 3,4 events, first in the magnetosphere and then on the ground. Further we proceeded in the opposite sense, i.e. we looked for the events first on the ground, and then in the magnetosphere. A suitable event was chosen on the basis of continuous oscillations of at least ten cycles with an amplitude well above the noise level of the recording instrument within a segment of 40 minutes of data. To confirm the suitability of the event, power spectral analysis was performed on the data, and an event was finally described as a peak in the spectrum if it was sufficiently higher than the background field. Although more than one peak could occur in a single segment, a dominant peak within the Pc 3,4 frequency range was always most certainly preferred. Once an event was finally identified, the spectral matrix in the radial-east-north, REN, coordinate system (described below) was averaged over the half-power band-width of each event (e.g. the shaded region in Figure 3) yielding a single spectral matrix for each event, from which the spectral amplitude and weighted frequency was determined. The whole exercise led to the identification of five (5) Pc 3,4 wave events both in the magnetosphere and on the ground. Table 2 shows the list of the five Pc 3,4 events observed in the magnetosphere when the footprint of the spacecraft was close to the IGS chain of magnetometers. Event number 3 has been described in detail by Odera et al. [1991]. It is to be noted that, for the Pc 3,4 events which were observed at many stations over a large area from high latitude to low there were similar wave events observed simultaneously at ISEE-1,2 in the magnetosphere, but the converse is not true.

Figure 4 represents the high time resolution plots of magnetic field fluctuations obtained from ISEE-2 on February 2, 1978 from 0610-0650 UT interval, together with the magnetic field record obtained from seven ground stations, which were situated close to the region conjugate to the ISEE spacecraft. The data were filtered with a band-pass filter of 20-50 mHz. The top three traces show the ISEE-2 magnetic field variations plotted in the REN coordinate system for which the components are: B_R (radially outward from the center of the earth, B_E (magnetically eastward), and B_N (perpendicular to the other two components in the plane containing the earth's magnetic dipole axis and the radial vector). The seven traces below show the ground data, with each trace representing the H-component of the magnetic field variations at each station. Only the H-component is shown, rather than the D-component, to illustrate the waveform since, in this case, more of the wave power was in the H-component than in the D-component. The stations are arranged in the order of decreasing magnetic latitude. The amplitude scales are indicated and the temporal resolution of all the traces are identical.

For the REN coordinate system in which the ISEE-2 data are

presented, the B_R, B_E and B_N-components approximately give radially transverse, azimuthally transverse and compressional modes, respectively near the magnetic equator. We categorize the wave event at the ISEE-2 spacecraft as compressional [see Odera et al., 1990], although there is a comparable amount of power in the azimuthally transverse mode. This point will become clear in the next paragraphs.

During the interval shown in Figure 4, the ISEE-2 spacecraft was inbound and only about 2° north of the magnetic equator, in the morning sector of the magnetosphere, as shown in Figure 2 by the line marked (4). There is a Pc 3 (T ~ 28 sec.) wave event on the ISEE-2 spacecraft record, with a 'switch on' at about 0627 UT. The activity is enhanced for a 10 minute interval (0630-0640 UT). Thereafter the activity remains low and irregular. A similar wave event (not shown here) was also detected on the ISEE-1 spacecraft.

The ground stations show a similar waveform to the one in the magnetosphere. In particular, the 'switch on' of activity at 0627 UT is clearer at all ground stations than in the

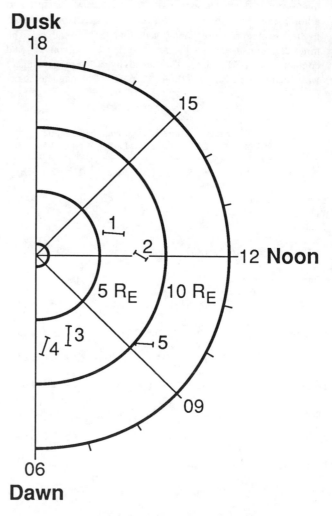

Fig. 2. The tracks of the ISEE-1,-2 orbits projected on (a) the meridional plane (magnetic latitude vs radial distance) and (b) equatorial plane (L-value vs local time).

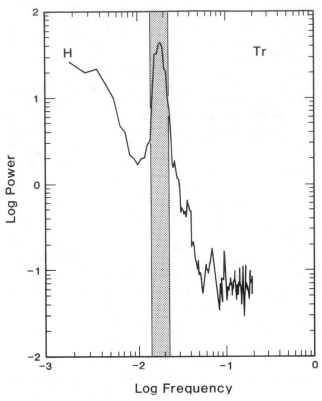

Fig. 3. An example of power spectrum from which a Pc 3,4 event was selected.

waveform (presented in the H-component), but their amplitude consistently decreases with a decrease in the magnetic latitude. The amplitudes of the D-components (not presented here) are comparable to the H-components and their latitudinal trend is the same.

A more careful look at the satellite record in Figure 4 indicates that the amplitudes of both B_E and B_N-components are comparable, at least during the interval of enhanced oscillations. This suggests that, although the wave may be azimuthally polarized, there is a considerable compressional wave power. This point will be shown more clearly by power spectral analysis presented in the later paragraphs.

The compressional Pc 3,4 waves in the magnetosphere and the Pc 3,4 pulsations on the ground are not only qualitatively similar but also quantitatively well related. Correspondence between the compressional waves detected at the ISEE-2 spacecraft on February 2, 1978 and the H-component of the Pc 3,4 pulsations observed simultaneously at many stations on the ground was further examined using cross-correlation analysis. The analysis was performed at a segment of 10 minutes, over the 0630-0640 interval during the enhanced oscillations, both in the magnetosphere and on the ground. The correlation analysis was also performed between the ground stations themselves, with Lerwick being the reference station. Figure 5 shows the results of the analysis: (a) the correlation coefficient between the compressional component (B_N) of the waves at the ISEE-2 spacecraft and the H component of the Pc 3,4 pulsations in the ground as a function of L-value and (b) the correlation coefficient between the H-components of the ground data, as a function of L-value, where unity represents perfect correlation.

Examination of Figures 5(a,b) indicates that (a) the coefficient of correlation is high and is essentially above 0.7, from low latitude to high latitude stations; the coefficient does not change with L-values. (b) The coefficient of correlation is also high between the ground stations themselves; but the coefficient is

magnetosphere. Similarly there is a clear enhancement of activity at the 0630-0640 UT interval. Unfortunately, the lack of simultaneous solar wind data does not allow us to determine whether the solar wind was responsible for the enhancement of the wave activity. The ground stations show variations in their

Table 2. Pc 3,4 waves observed at ISEE-1,2

Event No.	Date	Time Interval (UT)	Polarization	Amplitude (nT)	Wt. freq. mHz	Period	Type	Kp. index (sec)
1	Nov. 3, 1977	1100-1140	Transverse	0.7	17.6	56.8	Pc 4	1⁻
2	*Nov. 15, 1977	0840-0920	Compressional	1.23 [1.43]	35.2 [14.3]	28.5 [70]	Pc 3 [Pc 4]	4⁻
3	Jan. 21, 1978	0730-0810	Compressional	0.33	18.6	53.7	Pc 4	0
4	Feb. 2, 1978	0610-0650	Compressional	0.19	35.3	28.3	Pc 3	3⁻
5	*Oct. 31, 1979	0700-0740	Compressional	0.12 [0.17]	26.5 [13.3]	37.7 [75.2]	Pc 3 [Pc 4]	1⁺

* Two peaks, and the second peak-value in square bracket.

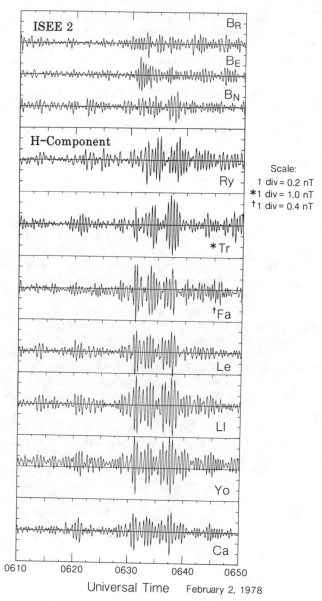

Scale:
1 div = 0.2 nT
*1 div = 1.0 nT
†1 div = 0.4 nT

Fig. 4. The wave form of Pc3,4 pulsations observed simultaneously, both in the magnetosphere (ISEE-2 spacecraft) and on the ground (IGS stations) on February 2, 1978 (0610-0650 UT).

latitude dependent, being largest (about 0.9) at low latitude stations, and smallest (about 0.5) at high latitude stations. The low or high latitude, in this case, is determined with respect to Lerwick, which was chosen to be the reference station for the purpose of cross-correlation analysis. Lerwick was chosen because it lies nearly in the mid-latitude of the IGS array of magnetometers, and also because it has the highest correlation coefficient in (a) above. The results of the analysis imply that the compressional Pc 3,4 waves in the magnetosphere and the Pc 3,4 pulsations on the ground are very similar in waveform and that the similarity is even higher between the ground stations themselves, at low latitude.

In order to examine the satellite-ground comparison of frequency, power spectra of the magnetic field data, from both the ISEE-2 spacecraft and the ground stations were calculated using Fast Fourier Transform (FFT) method. The power spectra were calculated from segments of 10 minutes of data, during the enhancement of the wave activity i.e. from 0630 to 0640 UT. The data were detrended, without removing the average field, before the power spectra were calculated; no filter was applied in this exercise. Figure 6 shows examples of power spectra for both satellite and ground data. The upper left-hand panel shows the power spectra for each of the three components of the magnetic field at the ISEE-2 spacecraft. The spectra are plotted on the same horizontal and vertical axis. The spectral peaks for all the components are sharp and are at the weighted frequency of 35.3 mHz. The right-hand panels are a series of seven spectra arranged in the order of decreasing latitude. Each panel shows the power in the H-component of the ground magnetic field at the respective stations. The panels are plotted in the same frequency and power scales. The horizontal line drawn across each panel marks the zero-log power level, the vertical dashed line marks the position of the weighted frequency calculated from the power spectral matrix at the ISEE-2 spacecraft.

Examination of Figure 6 indicates that there is similarity between the frequency at the ISEE-2 satellite and that at all ground stations. Also the power level at the ground stations

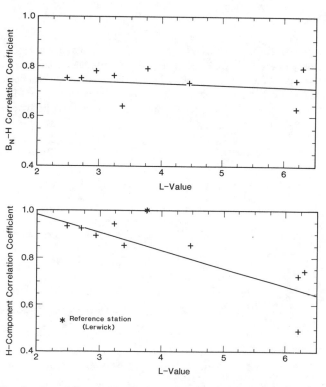

Fig. 5. Correlation coefficicient as a function of L-value for, (a) the ISEE-2 satellite (B_N-component) and the ground (H-component) data, (b) the H-components of theground data, with Lerwick as a reference station.

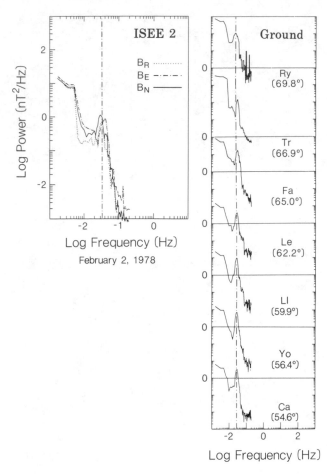

Fig. 6. Power spectra of the Pc3,4 wave events, both in the magnetosphere and on the ground.

generally decreases with decrease in latitude. Further, the compressional power is larger than, although comparable to, the azimuthal power and the power in the radial component is small.

From the direct comparison of the wave-like events, observed at the ISEE- 1,2 satellites, with the simultaneous magnetic field data on the ground, we found good satellite-ground correlation, such as the one described above, on four orbits. Such events were simultaneously observed in the magnetosphere and on the ground over a large area from high latitude to low, and where the footprint of the spacecraft was close to the IGS chain of magnetometers. However, there was one exceptional case involving a Pc 4 wave event which was observed as a monochromatic transverse wave in the magnetosphere, but was not recorded at any of the ground stations. This event is described in detail below in a similar format as the one presented above.

Figure 7 shows the time-amplitude records of the Pc 4 wave event detected at the ISEE-1 spacecraft and the magnetic field fluctuations recorded simultaneously at six ground stations on November 3, 1977 from 1100-1140 UT. The data were filtered with a band-pass filter of 10-25 mHz. The top three traces show the ISEE-1 magnetic field variations plotted in the REN coordinate system. [There was no corresponding data from the ISEE-2 spacecraft in this interval]. The bottom six traces show the simultaneous ground data, with each trace representing the H-component of the magnetic field variations from the ground stations, the stations are arranged in the order of decreasing latitude. The scales are identical at all stations, except at Tromso where the signal was strongest. The temporal resolutions of all the traces are identical.

During the interval shown in Figure 7 the ISEE-1 was inbound in the afternoon sector of the magnetosphere; the orbital plane of the spacecraft was only 1° north of magnetic equator as shown in Figure 2 by the line marked (1).

Examination of Figure 7 indicates that there is a strong transverse Pc 4 (T ~ 56.6 sec.) wave event at the ISEE-1 spacecraft. From the REN coordinate system in which the

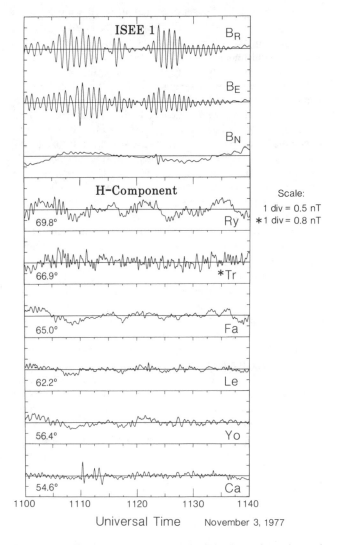

Fig. 7. Amplitude-time plot of magnetic field fluctuations observed simultaneously at the ISEE-1 spacecraft and at six ground stations on November 3, 1977 (1100-1140 UT).

satellite data are displayed, it is clear that the wave is both in the azimuthal and radial transverse modes; there is very little or no power in the compressional mode. We categorize this event as essentially in the transverse mode. We note that Engebretson et al. [1988; 1992] have classified similar events as radially polarized. They may be a different class of waves than the more purely azimuthal wave. The activity is very strong and monochromatic in the magnetosphere. Even before the filter was applied the oscillations are outstandingly monochromatic throughout the interval. The maximum amplitude calculated from the power spectral trace is 0.7 nT and the weighted frequency, calculated from the half-power spectral band, is 17.6 mHz (T = 56.6 sec.). The activity 'switches on and off', but maintains a single frequency of oscillations throughout the interval. Unfortunately, because of lack of simultaneous solar wind data, we cannot determine whether the solar wind directly produced these oscillations.

The ground data show no correlation at all with the wave event detected on board the ISEE-1 spacecraft. At all stations on the ground there is very little or no activity in the bandwidth of the Pc 4 pulsations observed at the ISEE-1 spacecraft, and no clear switch 'on' or "off" can be detected on the records.

The absence of correspondence between ISEE-1 and ground data for this event was further shown by power spectral analysis. The power spectra were calculated for 40 minutes of data, covering the whole interval from 1100-1140 UT, both in space and on the ground. Figure 8 shows the spectra in the azimuthal, radial and compressional components of the magnetic field fluctuations in the magnetosphere and in the H-component of the field fluctuations on the ground. The power and frequency scales are indicated. The horizontal line for the zero-log-power is marked on the spectra. The vertical dashed line for the weighted frequency calculated from the spectral trace at the ISEE-1 spacecraft is also marked on both satellite and ground spectra. The satellite data show strong power in the azimuthal and radial components. These are the transverse components of the Pc 4 waves detected by the ISEE-1 spacecraft. The spectral peaks are sharp and are at the weighted frequency of 17.6 mHz. On the ground, there is virtually no power at all in the bandwidth of the Pc 4 wave event observed in space.

For the five Pc 3,4 events that we have examined in the magnetosphere there are two classes of wave modes, namely compressional and transverse wave modes. We have demonstrated that the compressional waves in the magnetosphere are observed simultaneously as Pc 3,4 pulsations over a large area on the ground, from high latitude to low. In contrast the transverse waves in the magnetosphere have virtually no corresponding signals at stations on the ground.

Statistical relation between the Pc 3,4 pulsations amplitude and L-value

For each of the five events listed in Table 2 a spectral amplitude was calculated from the spectral matrix, i.e., the power summed up over the three sensors of the spacecraft and ground magnetometers. Table 2 shows the amplitudes of the waves on the spacecraft only. The variation of amplitude with

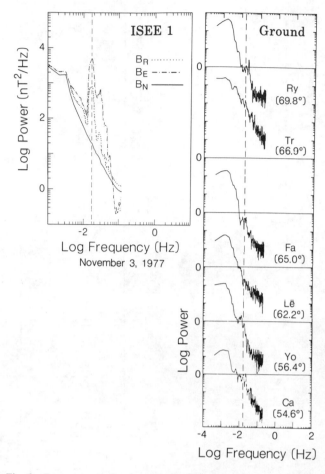

Fig. 8. Power spectra for the magnetic field fluctuations at the ISEE-1 spacecraft and at six ground stations.

latitude (L-value) of the ground data is illustrated in Figure 9, which shows a family of curves of the amplitude L-value variation for the five events.

Examination of Figure 9, indicates specifically that events numbers 2 and 4 have higher amplitudes than the others and that they have a steeper increase of their amplitudes at high latitudes. Generally there is a gradual increase of amplitude at the low L-values (L < 3.75), but the increase is rapid at high L-value (L > 3.75). The change of variation seem to occur at the mid-latitude (L ~ 3.75). This trend is valid only for the four events described here as compressional wave modes (there was no data for the event number 5, for the stations at high latitudes, L > 4.5). For the transverse event, number 1, the situation is very different. The amplitude is small (very nearly at the noise level) and remains low across the station chain, from low L-value to high.

The frequency of each event on the ground did not change with L-value. For each, the maximum range of frequency across the chain of stations was less than 10%; this is within the experimental error. In other words, variation of frequency with L-value was very small and sufficiently covered by experimental

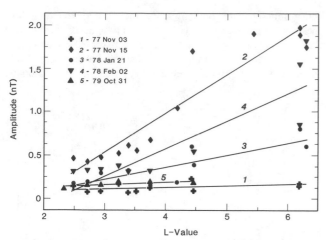

Fig. 9. A family of curves showing the relation between the amplitude of the Pc3,4 pulsations on the ground and L-values. Amplitudes are obtained by integrating under the peak of the spectrum shown in Fig. 3,

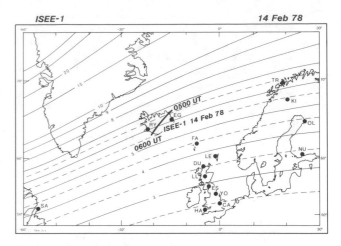

Fig. 10. Conjugate path of ISEE-1 on February 14, 1978.

errors. One important implication of this result is that the frequency of the Pc 3,4 waves observed at the ISEE-1,2 spacecraft was very nearly the same as the frequency of the Pc 3,4 pulsation observed simultaneously on the ground, from low latitudes to high.

Short-lived oscillations

One of the criteria for event selection as discussed above was the persistence of the oscillation at the satellites. However as noted by Singer et al. [1982], the passage of the two spacecraft through the standing field line resonances may produce short-lived (at the satellite) oscillations which may appear at quite different times at the two spacecraft, but which occur at the same L-value. The waves persist over the 10 minutes or longer between the crossing of the same L-shell by ISEE-1 and -2 but in the satellite frame they appear to be quite brief in duration. Thus, we would not include such events in our study. This may, in fact, explain why our study apparently favored compressional events over transverse events.

To ascertain whether we had biassed our study in this manner, we examined a subset of the original conjugate passes. Figure 10 shows the feet of the field liner passing through ISEE on 14 February 1978 on a pass crossing Iceland between Eidar and Reykjavik. Data are not available from Reykjavik on this day. Figure 11 shows the compressional and east-west component of the waves seen at ISEE-1 and -2. The data are shifted by 10 minutes so that Figure 11 represents data obtained at the same L-value. We see good agreement in the east-west component, especially at 0510 at ISEE-2 (0520 at ISEE-1). There is little wave power in the compressional component in the Pc 3,4 band. This observation is very similar to those made by Singer et al. [1982].

Figure 12 shows that there is continuous wave power observed on the ground. Since it is continuous on the ground whereas it is time limited in space, there is low coherence between ground and space. Figure 13 shows that the wave frequencies seen on the ground are very similar to those seen in

space especially at the same L-value (Eidar and Tromso). However, the low latitude stations see very little power at this frequency and sense instead principally a lower frequency wave. This behavior is consistent with the observations discussed above on November 3, 1977, except that on this day very little wave power was observed at any of the ground stations.

In short, then our event selection procedure has a bias toward selecting compressional events because these events are coherent

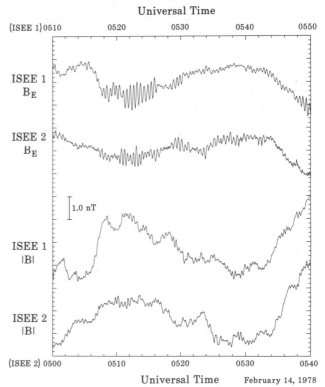

Fig. 11. East-west and compressional components of the magnetic field observed by ISEE-1 and -2 on February 14, 1978 conjugate pass. ISEE-1 and ISEE-2 data have been aligned at times of constant L-value crossing.

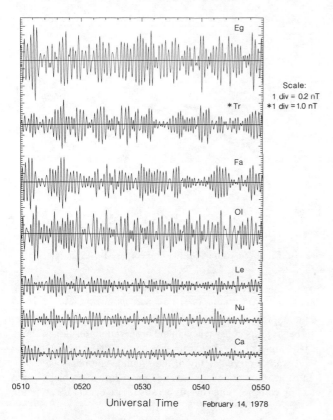

Scale:
1 div = 0.2 nT
*1 div = 1.0 nT

Fig. 12. Horizontal component of the magnetic field measured at the ground stations of the IGS array on February 14, 1978.

time resolution data, both in space and on the ground. This does not mean that the waves were in phase at all sites. The similarity in waveform between the compressional waves at the ISEE-2 spacecraft and the H-component of the magnetic field fluctuations at the ground stations is very clear in the 0630-0640

over a larger region of the magnetosphere. However, our conclusion that there are two quite distinct classes of waves, a compressional class and a transverse resonant class is not affected by our bias. Thus, one should not use the relative occurrence rates of observations of these two wave types in our study to be representative of their overall occurrence in the magnetosphere.

DISCUSSION AND SUMMARY

The primary purpose of this paper is to compare directly the compressional and transverse waves in the magnetosphere with the Pc 3,4 pulsations on the ground, with a view to identifying the pulsations (on the ground) which come from the same wave events in space. We emphasize the similarity in spectra and waveform between the compressional wave mode in the magnetosphere and the Pc 3,4 pulsations recorded simultaneously at many stations on the ground, from high latitude to low. However, we find no correspondence between the transverse waves in the magnetosphere and the Pc 3,4 pulsations on the ground. Figures 4-8 summarize these most important features of the pulsations observed simultaneously in the two regions.

Simultaneity in the context of this paper refers to the fact that the wave events were identified within the same segment of high

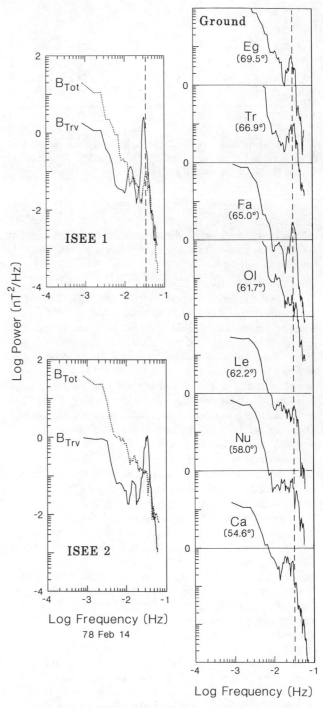

Fig. 13. Power spectra at ISEE-1 and -2 and the ground stations during the conjugate pass on February 14, 1978.

UT interval as shown quantitatively in Figure 4, when the amplitude is enhanced in the two regions. There is an onset of comparatively large amplitude oscillations at about 0630 UT; the oscillations continue to about 0640 UT, when the amplitude decays and remains small throughout the remaining interval. Unfortunately, the lack of simultaneous solar wind data for this event does not allow us to determine if some features in the solar wind were responsible for the enhancement. The evidence of similarity in waveform is given quantitatively in Figure 5 which shows the high correlation coefficient not only between the satellite and ground data alone, but also between the ground stations themselves.

We found good correspondence between the compressional waves in the magnetosphere and the Pc 3,4 pulsations on the ground for four events listed in Table 2. There were simultaneous solar wind data available only for events numbers 2 and 5, on November 15, 1977 and October 31, 1979 respectively. For event number 2 the predicted period (T_{pr}) is 31.0 sec., and the measured period (T_m) is 28.5 sec. The predicted period was calculated on basis of the empirical relations, $T = 160/B$, first presented by Troitskaya et al. [1971]. The cone angle for the event θ_{XB} is 28° and the solar wind speed is V_{sw} 630 km/sec. For the event number 5 the predicted period T_{pr} is 29 sec., and the measured period is 37.7 sec.; the cone angle, θ_{XB} is 71°, and the solar wind speed is 386.5 km/sec. It appears from these results that event number 2 is dependent on the solar wind parameters while event number 5 is not. But the situation is mixed for event number 2 because there are two dominant peaks for the event. The 70 sec. peak (plotted in Figure 9), which was the strongest in the magnetosphere and on the ground, could not possibly be driven by the IMF.

For the transverse waves (see Figure 7) the situation is very different from that of the compressional waves. Only one event of the transverse type was observed in the magnetosphere at the ISEE-1 spacecraft. While there are relatively large amplitude monochromatic oscillations in space, there is hardly anything on the ground (at all stations) which corresponds to the oscillations in space. The magnetic field fluctuations on the ground are very small and irregular. The peak frequency of these fluctuations at Ry, Le and Ca coincidentally matches that at the ISEE-1 spacecraft (see Figure 8), but the spectral amplitude at these stations is very small above the background power level.

It is to be noted that this event should have been the most likely to be detected on the ground by the station chains. From Figure 1, the event occurred right in the middle, between the UK and Scandinavian sub-chains of the IGS magnetometers array. The footprint of the ISEE-1 was very close, both to the UK and Scandinavian stations.

The similarity in spectra between the compressional waves in the magnetosphere and the Pc 3,4 pulsations on the ground is illustrated qualitatively by Figure 6. The spectral line of the weighted frequency (35.3 mHz) calculated for the compressional wave event at the ISEE-2 spacecraft passes very close to the spectral peaks at all the ground stations. Such similarity in spectra between the compressional waves in the magnetosphere

and the Pc 3,4 pulsations at two low latitude ground stations (L = 1.3 -1.8) has been studied indirectly by Yumoto et al. [1984]. The authors found that the statistical relationship between frequency and the IMF magnitude had a similar trend, both in space and on the ground, but that the frequency-magnitude relation was better in statistical sense, in the outer magnetosphere than at the low latitude ground stations. Our data demonstrate directly that the frequency of the compressional Pc 3,4 waves in the magnetosphere is very close to the frequency of the Pc 3,4 pulsations on the ground, from high latitude to low.

In summary, it appears from the above considerations that whenever compressional waves appear in the magnetosphere, Pc 3,4 pulsations are recorded at all ground stations, with a constant frequency, from high latitude to low. The one occurrence of transvere waves in the magnetosphere was not detected at any of the ground stations.

There are two interesting questions which can be related to the above results. What is the excitation mechanism for the waves observed here and how can the compressional waves reach the ground while the transverse waves cannot? A conclusive answer to these questions is beyond the scope of this paper. We only give a brief outline of theories which may help resolve these quesions in future.

Before we give the outline, it necessary to note that the magnetic activity during all five events reported here was generally quiet, except for two events on November 15, 1977 and February 2, 1978 when the magnetic activity was (slightly) disturbed. The conditions for the magnetic activity for the five events are shown in Table 2. According to the statistical relation between the plasmapause position with the Kp index, presented early by Orr and Webb [1975] and recently by Horwitz et al. [1990], many IGS stations from mid-latitude to low latitude (L ~ 4.0 - 2.5) were in the plasmasphere, even during the most disturbed day on November 15, 1977. The stations at the high latitudes (L ~ 4.0 - 6.5) were possibly in the plasma trough during all events, except on January 21, 1978 when it was extremely quiet (magnetically) and all stations in the IGS array of magnetometers were comfortably within the plasmasphere.

Now we return to the question regarding excitation mechanism. There are two ways in which the waves observed here could be excited: one is by the upstream wave source, and another is by global (cavity) mode waves. The way the upstream wave source can stimulate compressional waves in the magnetosphere has been discussed by Yumoto et al. [1985]. Accordingly the upstream waves can propagate through the earth's bow shock, across the magnetosheath, and into the magnetosphere without significant changes in spectra, and can be observed as compressional Pc 3,4 waves at synchronous orbits (L = 6.67). The compressional Pc 3,4 waves, which can propagate further into the inner magnetosphere (L < 6.67) can couple with various hydromagnetic oscillations in the plasmasphere, and finally can be observed as the low-latitude Pc 3,4 pulsations on the ground.

The global mode of the magnetosphere described by Zhu and Kivelson [1989] can also contribute to the compressional waves of the Pc 3,4 frequency range at low latitude regions of the magnetosphere. Because of the rapid decrease of plasma density at the plasmapause, the variation of the Alfven velocity naturally separates the magnetosphere into inner (low latitude) and outer (high latitude) parts. Low frequency wave perturbations are found to be trapped in the Alfven velocity valley and the wave amplitudes in the inner magnetosphere are greatly enhanced.

The global mode involves the magnetospheric cavity which can be regarded as a resonant cavity for low frequency waves. The low frequency waves supported in this cavity can be excited by either external or internal sources depending on whether they are outer cavity waves or inner cavity waves, respectively. The outer cavity waves can be excited by perturbations in the magnetopause, but the inner cavity waves require an internal source of excitation and do not respond to external perturbation. Between the two extremes (outer and inner cavity waves) there are 'tunneling' waves with comparable amplitude in both inner and outer parts of the magnetosphere. The tunnel effect means that if some of the wave energy levels are approximately equal in the two cavities, the wave functions will 'tunnel' through the velocity barrier and have finite amplitude on both sides of the barrier. The tunneling waves are particularly relevant to the generation of low latitude waves because they allow perturbations at the magnetopause to excite oscillations of the inner cavity by tunneling through the Alfven velocity barrier. The energy of the global mode may be coupled to the shear Alfven waves in the same manner as in the original field line mechanism.

Which of the two excitations mechanism described above contributed to the waves observed here is a difficult question which can only be answered in part but not in whole. The upstream wave mechanism involves the control of the solar wind parameters, especially the IMF magnitude and orientation. It is not possible to test the solar wind control because of lack of simultaneous solar wind data for most of the events. For the two events with simultaneous solar wind data we have argued that one (e.g. November 15, 1977 event) is partly dependent on the solar wind parameters, and another (e.g. October 31, 1979 event) is not. It is possible that both events are not related to the upstream waves.

The global mode theory predicts a frequency that is the same everywhere in the magnetosphere, possibly consistent with our observation here. But most of the wave events were observed at generally quiet time magnetically and we did not notice any great impulsive change in the field magnitude to drive the cavity. Also the condition for 'tunneling' waves seems to be very rare, and we have no means of testing it. Hence it remains unclear how the low latitude pulsations could appear.

In summary we note that compressional fluctuations are important in delivering Pc 3, 4 wave energy through the magnetosphere. The way the compressional waves can reach the ground can be visualized in the frame work of the average conductivity (σ) of the earth and the polarization of the waves themselves. In this context we ignore the effect of atmosphere and ionosphere on the waves. In fact the ionosphere has no effect on compressional waves, since the compressional waves do not carry field aligned currents. We apply the boundary conditions for the waves. In an ideal situation when we consider the earth to be perfectly conducting ($\sigma = \infty$) and the compressional wave incident (bn_1) to the earth's surface, there is no transmitted wave (bn_3) into the ground, because the reflected wave (bn_2) is equal and opposite to the incident wave; i.e. $|bn_1| = |-bn_2|$. This is compatible with the usual statement that 'compressional' waves are perfectly reflected, and are not seen on the ground. Transverse polarized waves have no normal component on the ground i.e. $b_n = 0$; therefore, it doesn't matter whether the earth is perfectly conducting or not.

In a real situation the earth does not behave as a perfect conductor. The conductivity of the earth is finite and can be set to be $\sigma >> \epsilon\omega$. At the reflection point a small perturbation, $bn_3 = |bn_1| - |-bn_2|$ exists in the ground for the (fast mode) compressional waves. We expect small amplitude perturbations on the ground, possibly consistent with our few tenths of nanotesla (nT) of amplitudes we have recorded on the ground stations. In the future, it remains experimentally important to carry out quantitative estimates of realistic conductivities (σ) at various stations on the ground, in order to verify the order of magnitude of the wave amplitude that can be detected on the ground.

Acknowledgements. We are grateful to the British Geological Survey (BGS) for permission to use the IGS data. In particular we are thankful to W. F. Stuart, C. A. Green and T. Harris, whose assistance enabled us to prepare the IGS data. We are also grateful to M. G. Kivelson for her useful comments. This paper was prepared with the assistance of IGPP/UCLA, the National Aeronautics and Space Administration (NASA) under research grant NAGW-2027, and the National Science Foundation (NSF) under grant NSF ATM 88-15711.

REFERENCES

Arthur, C. W. and R. L. McPherron, Interplanetary magnetic conditons associated with synchronous orbit observations of Pc 3 magnetic pulsations, *J. Geophys. Res., 82,* 5138, 1977.

Barraclough, D. R., J. M. Harwood, B. R. Leaton, and S. R. C. Malin, A model of the geomagnetic field at epoch 1975, *Geophys. J. R. Astro. Soc., 43,* 645, 1975.

Engebretson, M.J., L.J. Zanetti, T.A. Potemra, D.M. Klumpar, R.J. Strangeway, and M.H. Acuna, Observations of intense ULF pulsation activity near the geomagnetic equator during quiet times, *J. Geophys. Res., 93,* 12795, 1988.

Engebretson, M.J., D.L. Murr, K.N. Erickson, R.J. Strangeway, D.M. Klumpar, S.A. Fuselier, L.J. Zanetti, and T.A. Potemra, The spatial extent of radial magnetic pulsation events observed in the dayside near synchrous orbit, *J. Geophys. Res., 97,* 13741, 1992.

Green, C. A., Continuous magnetic pulsations on the IGS array of magnetometers, *J. Atmos. Terr Phys., 43,* 833, 1981.

Green, C. A., T. J. Odera, and W. F. Stuart, The relationship between the strength of the IMF and the frequency of magnetic pulsations on the ground and in the solar wind, *Planet. Space Sci., 31,* 559, 1983.

Greenstadt, E. W., R. L. McPherron, and K. Takahashi, Solar wind control of daytime, midperiod geomagnetic pulsations, *J. Geomag. Geoelectr., 32,* Suppl. II, SII 89, 1980.

Horwitz, J. L., R. H. Comfort, and C. R. Chappel, A statistical characterization of plasmasphere density structure and boundary locations, *J. Geophys. Res., 95,* 7937, 1990.

Odera, T. J., and W. F. Stuart, The relationship between the IMF magnitude and the frequency of Pc 3,4 pulsations on the ground: Simultaneous events, *Planet. Space Sci., 33,* 387, 1985.

Odera, T. J., Solar wind controlled pulsations: A review, *Rev. Geophys., 24,* 55, 1986.

Odera, T. J., D. Van Swol, C. T. Russell, and C. A. Green, Simultaneous observation of Pc 3,4 pulsations in the magnetosphere and on the ground, *Geophys. Res. Lett., 18,* 1671-1674, 1991.

Orr, D., and D. C. Webb, Statistical studies of geomagnetic pulsations with peiods between 10 and 70 sec and their relationship to the plasmapause region, *Planet. Space, Sci., 23,* 1169, 1975.

Riddick, J. C., J. Brown, and A. J. Forbes, A low power moveable observatory unit for magnetometer array application, *Inst. Geol. Sci. Geomag. Unit, Int. Rep. no. 17,* 1975.

Russell, C. T., The ISEE-1 and -2 fluxgate magnetometers, *IEEE Trans. Geosci. Electron, GE-16,* 239, 1978.

Russell, C. T., and M. M. Hoppe, The dependence of upstream wave peiods on the interplanetary magnetic field strength, *Geophys. Res. Lett., 8,* 615, 1981.

Singer, H. J., W. J. Hughes, and C. T. Russell, Standing hydromagnetic waves observed by ISEE-1 and -2: Radial extent and harmonic, *J. Geophys. Res., 87,* 3519-3529, 1982.

Stuart, W. F., The high resolution magnetic stations operated by IGS, *Geomagnetism Unit Report, No. 8,* 1971.

Takahashi, K., R. L. McPherron, and T. Terasawa, Dependence of the spectrum of Pc 3-4 pulsations on the interplanetary magnetic field, *J. Geophys. Res., 89,* 2770-2780, 1984.

Troitskaya, V. A., T. A. Plysova-Bakounina, and V. A. Gul'elmi, Relationship between Pc2-4 pulsations and interplanetary magnetic field, *Dokd. Akad. Nauk, USSR, 197,* 1312, 1971.

Vero, J. and L. Hollo, Connection between interplanetary magnetic field and geomagnetic pulsations, *J. Atmos. Terr. Phys., 40,* 857, 1978.

Vero, J., Experimental aspects of low latitude pulsations − A review, *J. Geophys. Res., 60,* 106-119, 1986.

Yumoto, K., and T. Saito, Relation of compressional HM waves at GEOS 2 to low-latitude Pc 3 magnetic pulsations, *J. Geophy. Res., 88,* 10,041, 1983.

Yumoto, K., T. Saito, B. T. Tsurutani, E. J. Smith, and S. I. Akasofu, Relation between the IMF magnitude and Pc 3 magnetic pulsations in the magnetosphere, *J. Geophy. Res., 89,* 9731, 1984.

Yumoto, K., T. Saito, S. I. Akasofu, B. T. Tsurutani, and E. J. Smith, Propagation mechanism of daytime 3-4 pulsations observed at synchronous orbit and multiple ground-based stations, *J. Geophys. Res., 90,* 6439, 1985.

Yumoto, K., Low-frequency upstream wave as a probable source of low-latitude Pc 3,4 magnetic pulsations, *Planet. Space Sci., 33,* 239, 1985.

Zhu, X., and M. G. Kivelson, Global mode ULF pulsations in a magnetosphere with a nonmonotonic Alfven velocity profile, *J. Geophys. Res., 94,* 1479, 1989.

T. J. Odera, Physis Department, Maseno College University, Private Bag, Maseno, Kenya, Africa.

D. Van Swol and C.T. Russell, Institute of Geophysics and Planetary Physics, University of California, Los Angeles, CA 90024-1567.

Pc3 Pulsations in the Cusp

JOHN V. OLSON

Geophysical Institute, University of Alaska, Fairbanks, AK

BRIAN J. FRASER

Physics Department, University of Newcastle, Newcastle 2308, NSW, Australia

The presence of enhanced Pc3 pulsation activity is one of the salient features of the dayside cusp/boundary layer system as observed by ground-based magnetometers. Using the results of earlier studies which correlated optical and magnetic signatures of the cusp and nearby boundary layers, we are able to differentiate the Pc3 signals present in those two structures. We have studied the Pc3 spectrum beneath the boundary layers and the cusp using data from pairs of stations in order to infer the local coherence of the pulsations. The first pair of stations used were Cape Parry and Sachs Harbor in Canada which are separated by approximately 150 km. The second pair are the near conjugate stations of Longyearbyen, Svalbard and Davis, Antarctica. Pc3 detected in the boundary layers are characterized by large amplitude and broad-band features and show little or no inter-station coherence. The Pc3 detected in the central cusp tend to be narrow-band packets which show good correlation, even between hemispheres. These observations are consistent with those of Engebretson [1986b, 1991a]. The lack of coherence between the two northern stations implies a short correlation length for the precipitating beams producing the Pc3 variations in the boundary layers. The correlation of central cusp Pc3 implies coherent illumination of the cusp region which leads to the conclusion that these pulsations have their source upstream of the magnetopause.

INTRODUCTION

An observer located beneath the ionospheric footprint of the polar cusp can look up at phenomena whose sources are tens of earth radii away spanning regions which include the bow-shock, magnetosheath, magnetopause and boundary layers. An ensemble of waves and turbulent flows have access to the ionosphere via the cusp. These include the solar wind variations including those generated in the foreshock upstream of the bow shock, radiation from the parallel and perpendicular shocks, magnetosheath turbulence and waves, magnetopause boundary variations including the variations generated by flux-transfer events, pressure variations, the Kelvin-Helmholtz instability etc., and the waves and particle variations which take place in the boundary layers just inside the magnetopause. The ionospheric cusp is the focus of these phenomena and ground observations are comprised of their superposition. The fact that experimental investigations at the cusp may lead to a better understanding of the flow of magnetic flux, energy and momentum from the solar wind to the magnetosphere has heightened interest in such investigations and their interpretations.

Virtually every observation made at cusp latitudes on the ground or on satellites as they pass through the cusp region of the magnetosphere has witnessed a broad spectrum of magnetic fluctuations. (See reviews by D'Angelo [1977], Fraser-Smith [1982], Troitskaya [1985], Arnoldy et al. [1988], and Glassmeier, [1989]). Good correlations between solar wind parameters and the amplitude of ground pulsations have been found by many investigators. While the cusp region spectrum extends from essentially dc up into the vlf and beyond, this paper will concern itself with the signals in the intermediate ULF frequency range from approximately 10 to 50 mHz (nominally the Pc3 band) which occur at the cusp.

It has been known for some time that the amplitude of Pc3 pulsations measured at the ground peaks at cusp latitudes on the dayside [Troitskaya, 1985]. Also, satellite investigations have shown that Pc3 pulsations are detected throughout the magnetosphere [Takahashi and McPherron, 1982, Anderson et al., 1990, Engebretson et al., 1986a, Fraser et al., 1988], at the boundaries and in regions outside the magnetosphere (see Odera, [1986] and Yumoto [1986] for reviews). Others, including Kato et al. [1985], Tonegawa et al. [1985], Engebretson [1986b] and Olson [1986], have investigated the details of the cusp spectrum. Their studies show a broad spectrum of wave energy in the cusp data with imbedded packets of Pc3 waves. Engebretson et al. [1986, 1987] found that ground Pc3 could be separated into two classes: an apparently spatially incoherent portion which appears to be generated in the overhead ionosphere by fluctuating precipitation of kilovolt electrons; and a second, spatially coherent component which correlates with solar wind parameters. In particular, Engebretson [1986b] found a direct correspondence of the frequency, f, of the coherent Pc3 pulses with the magnitude of the interplanetary magnetic field (IMF). This observation was confirmed and extended by Slawinski et al. [1988] and Wolfe et al. [1990] and currently the relationship found shows $f \sim 5.8|B|$. Other correlations have been found also. Wolfe et al. [1987] and Yumoto et al. [1987] using multivariate analysis found the solar wind speed to be the most important solar wind parameter in predicting pulsation energies at the ground.

The association of pulsation occurrence and amplitude with certain solar wind parameters has been investigated by many authors using ground data from a wide range in latitudes. The earliest reports were by Gringauz et al. [1971] who found correlation of Pc3-4 amplitudes with solar wind density fluctuations. Later Vinogradov and Parkhomov [1974] reported correlations between Pc3-4 amplitudes and solar wind velocity. Further studies by Gul'elmi [1974] Greenstadt and Olson [1976, 1977, 1979], Webb and Orr [1976], Singer et al. [1977], and Vero and Hollo [1978] show consistent correlation between solar wind magnetic field amplitude and θ_{XB}, the so-called "cone angle" at almost all latitudes. The cone angle is the angle the IMF field vector makes with the earth-sun line. We will return to this correlation as an important component in understanding the origins of magnetospheric Pc3s later in this paper.

In this paper we will show examples of cusp spectra and the Pc3 band variations they contain using data we have taken at several sites including Cape Parry and Sachs Harbor in northern Canada and Svalbard, Norway and its conjugate site at Davis, Antarctica (see Table 1). Through correlations of the cusp spectrum with optical instruments and the solar wind data, as well as the cross-correlation of magnetometer data between stations, we will review the nature of cusp Pc3 and set the stage for an interpretation of their sources.

THE ULF SPECTRUM AT THE CUSP

In Figure 1 we show the spectrum of ULF waves taken using an induction magnetometer during a 24-hour interval on 9 January 1992 at Longyearbyen, Svalbard, Norway. The spectrum covers the frequency band between dc and 50 mHz, the instrument Nyquist frequency. Longyearbyen, located at a geomagnetic latitude of 74.9°, rotates from the polar cap beneath the auroral oval at about 0400 UT passes local geomagnetic noon at approximately 0830 UT and moves beneath the auroral oval back into the polar cap at approximately 1300 UT. The spectral features observed between

TABLE 1. Station Locations

| Station Name | Geographic | | Geomagnetic | | Magnetic |
	Latitude	Longitude	Latitude	Longitude	Noon (UT)
Cape Parry	70.2	235.3	74.6	-83.4	22:55
Sachs Harbor	72.0	235.9	76.3	-86.1	22:45
Longyearbyen	78.2	15.4	74.9	114.5	09:10
Davis	-68.6	78.0	-74.3	101.5	09:50

0400 UT and 1300 UT are now known to be associated with the ionospheric footprint of the cusp and boundary layers [McHarg and Olson 1992]. The lower panel of Figure 1 shows the total power in the horizontal (X and Y) induction magnetometer signals while the middle panel shows the right-hand polarized power and the top panel shows the left-hand polarized power. Note that near 0400 UT there is a relatively sudden increase in power at all frequencies with a maximum in the lowest frequency band near 1 mHz. As time progresses, the spectral amplitudes diminish until noon (approximately 09:10 UT, marked by the arrowhead) when a relative minimum in the power above 10 mHz is observed and the maximum power has moved upward to near 5 mHz in an arch-like form. Notice that the signals in the 1-5 mHz band between 0400 and 0800 UT are right hand polarized. After approximately 0900 UT, as we pass local magnetic noon, the spectral power above 10 mHz again increases while the location of the maximum spectral power decreases in frequency returning to near 1 mHz by 1200 UT. Finally by 1300 UT the spectrum has become quiet again.

The interpretation of the spectral features described above becomes clear when the data are correlated with data from a meridian scanning photometer located at the same site. In the northern hemisphere, Svalbard has the unique property that at local noon in the wintertime it is located behind the terminator allowing optical measurements to be made as we move past the cusp. The features observed in the magnetometer spectrum correlate well with the station location relative to the cusp and boundary layers as inferred from optical data. The broadband portion of the spectrum located between 0400 and 0800 UT and between 0900 and 1200 UT represent the signatures of the boundary layers. The region of relatively limited signal in the Pc3 band and in which the Pc5 arch reaches its highest frequency is associated with the central cusp. This important feature of the cusp Pc5 spectrum was reported by McHarg and Olson [1992] and Olson and Fraser [1992] and will be the subject of analysis in a separate report. Above 10 mHz, (nominally the Pc3 band) it can be seen that the regions beneath the boundary layers show the greatest activity. However, in examples shown below, the central cusp region often shows Pc3 band activity as well. It is the similarities and differences between the cusp and boundary-layer Pc3 signals that are the subject of this study.

INTERSTATION CORRELATIONS

The boundary-layer and cusp Pc3 signals can be discriminated by investigating interstation correlations. In this section we will show the results of the analysis of two cusp stations located approximately 150 km apart in the northern hemisphere and then inspect the data from two stations in opposite hemispheres which are approximately conjugate magnetically.

As one reviews magnetometer data from cusp regions it should be kept in mind that the state of the overhead ionosphere has an important effect on the magnetic signatures observed. In the northern hemisphere only Longyearbyen, Svalbard offers the opportunity to observe the winter cusp with a dark ionosphere. In the southern hemisphere the South Pole station shares this attribute. In the future there may be more sites available as unmanned observatories are established in Antarctica. The ionosphere is sunlit

Longyearbyen Induction Coils
9 Jan 1991

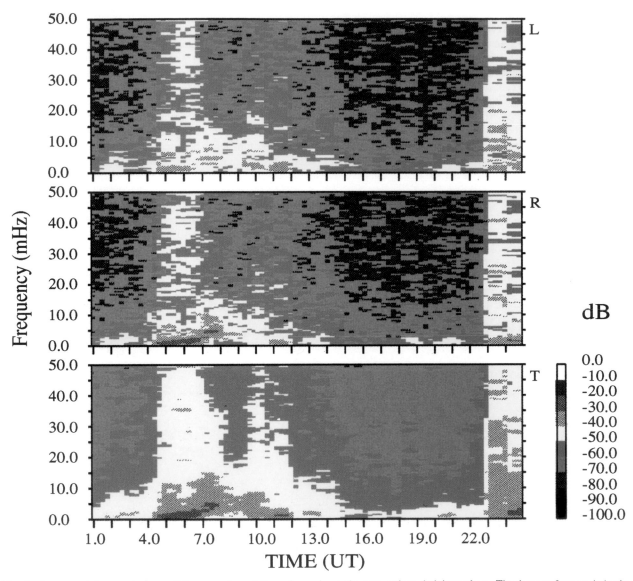

Fig. 1. A representative spectrogram of the magnetic variations observed near the cusp under a dark ionosphere. The data are from an induction magnetometer at Longyearbyen, Svalbard. The bottom panel shows the total horizontal power, the middle panel shows the power of the right-hand polarized components, and the top panel shows the power of the left-hand polarized components. Note the sudden increase in broadband signal power near 0400UT. This marks the passage of the station beneath the ionospheric footprint of the magnetospheric boundary layers. The decrease in broadband power near 1200UT marks the exit of the station from beneath the boundary layers. The region near 0900, local noon, represents the signals present in the central cusp. Note that there the broadband Pc3 amplitudes are diminished while the Pc5 spectrum has risen to a maximum near 5mHz.

during cusp passages at all other stations included here. When the ionosphere is sunlit, the central cusp tends to show more activity than identified in Figure 1 above and the identification of cusp and boundary layer becomes somewhat more difficult using the magnetometer data alone.

CAPE PARRY AND SACHS HARBOR, NWT, CANADA

During the period from 1982 through 1985 induction magnetometer data were taken at nominal cusp sites at Cape Parry and Sachs Harbor, NWT, Canada. The location of these two stations is given in Table 1.

In Figure 2 we show spectrograms of the total horizontal power in the induction magnetometer data from Cape Parry and Sachs Harbor data for 20 May 1985. We have selected the time between 1300UT and 2400UT which contains local noon (marked by arrowheads) and the signatures of the cusp and boundary layers for these stations. Both stations show a similar spectrum in that Pc5 activity begins near 1700UT and a broadband increase in the ULF signals shows an increase between 1800UT and 1900UT which we identify as the passage beneath the oval and the boundary layer near the cusp. Between 2000UT and 2200UT the stations are both beneath the central cusp and just after 2200UT, with the broadband increase in ULF power, the stations rotate back beneath the oval and pass beyond the boundary layer just before 2400UT. Note that significant Pc3 power is observed not only during the broadband increases identified as boundary layers but also as an isolated region in the spectrum above 30mHz during the passage beneath the central cusp (2000UT - 2200UT).

In order to investigate the coherence of the signals detected between the two sites we will present the results of two techniques. The first is based upon the traditional coherence estimate as described in standard textbooks (for example Jenkins and Watts, [1968]). In Figure 3 we present the coherence function plotted as a function of frequency for four intervals selected from the data in

Figure 2. The estimates were computed using the Welch method. The expected mean and variance are 0.14±0.37. This means that random fluctuations would show a mean coherence of 0.14 and that values could be expected to vary about that value by ±0.37. From this we would consider any value above 0.5 to represent coherent variations between the two data sets.

The solid lines in Figure 3 represent the estimates of coherence between the X axes at Cape Parry and the dot-dash lines represent the coherence between the Y axes. In scanning the coherence from interval to interval one sees a general increase in coherence across the band as the stations proceed under the cusp near noon (22:45 - 22:55 UT). Missing from this analysis is a representation of the power contained in these coherent signals. Also, one sees that the coherence function is dependent on the orientation of the sensors, a defect which can lead to difficulty in interpretation. To circumvent these limitations we turn to a more modern estimator of coherence.

The data shown in Figure 2 were taken from the X and Y induction coils at each site. When combined the data from the two stations provide four time series which can be scanned for coherent variations. To scan the data for coherent signals, the four-channel data stream composed of the induction magnetometer data between 1300UT and 2400UT were pure-state filtered using an algorithm developed by Samson and Olson [1981]. Briefly,

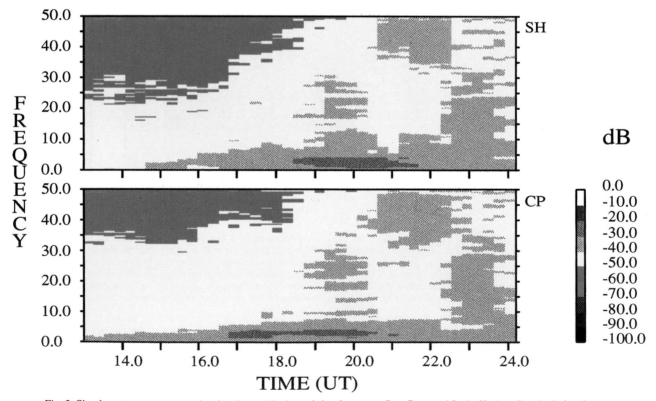

Fig. 2. Simultaneous spectrograms showing the total horizontal signal power at Cape Parry and Sachs Harbor, Canada during the cusp passage on 20 May 1985. Magnetic noon at Cape Parry: 22:55 UT; at Sachs Harbor: 22:45 UT.

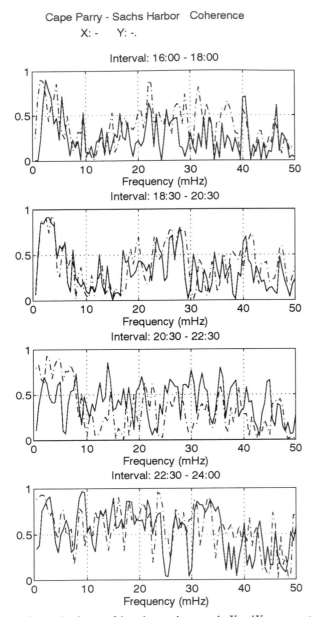

Cape Parry - Sachs Harbor   Coherence
X: -   Y: -.

Interval: 16:00 - 18:00

Interval: 18:30 - 20:30

Interval: 20:30 - 22:30

Interval: 22:30 - 24:00

Fig. 3. Spectral estimates of the coherence between the X and Y components of the induction magnetometer data taken from Cape Parry and Sachs Harbor on 20 May 1985. Note that as the stations move toward local noon (near 22:50 UT) the coherence levels increase across the band. These estimates are characterized by a variance of 0.34 and estimates greater than 0.5 may be considered significant.

this process identifies the signals which are coherent, that is, do not resemble the spectrum of incoherent noise. When applied to the four-channel data stream composed from the data from two stations, the filter passes signals which are coherent at one or the other or both stations. The spectrograms of the filtered signals from the two sites are shown in Figure 4. Here we have weighted the signal power by the esimate of coherence given by the pure-state estimator. It should be noted that this estimator is invariant to sensor orientation.

There are several features to note as one compares the spectrograms of the raw data (Figure 2) with those of the filtered data (Figure 4). First, note that the broadband Pc3 power associated with the boundary layer has been reduced by more than 20dB in most cases while the Pc3 signal identified with the central cusp has survived the process. This indicates that even over distances of the order of 100km, the boundary layer Pc3 spectrum is not coherent while the signals in the central cusp tend to be coherent. The residual signal power can be understood when one considers that magnetometers are effectively wide field of view instruments. It is not surprising that some signal survives the filter process. Engebretson et al. [1990] has made similar observations which show low coherence between various measurements of cusp region Pc3. We will return to the implications of these observations in our discussion.

It is also clear from both coherence analyses (Figures 3 and 4) that the strongest coherent signal is that which occurs in the 1-5mHz, Pc5 band. The Pc5 band contains both pulses and wavetrains. The correlation of the pulses between sites shows poleward motion of the current filaments responsible for the variations [Olson, 1989]. This feature is probably the most consistent indicator of the cusp and, as will be seen below, is a common, coherent feature of the data from conjugate sites [Olson and Fraser, 1992]. Because of the convenience afforded by the spectrogram format of the pure-filtered data to identify coherent signals we will rely solely upon that technique in the following analysis.

SVALBARD - DAVIS CONJUGATE OBSERVATIONS

The observatory sites at Longyearbyen, Svalbard Norway and Davis in Antarctica are nearly conjugate magnetically. Although they are separated in local time by approximately 1 hour, the cusp spectra observed are very similar. The data we will report on here are the first overlapping data sequence which was available to us. While only covering three days, the data show useful information concerning the coherence of signals in the ULF band. We will be analyzing data from January 1991. It should be kept in mind that since these two stations are located at high latitudes in opposite hemispheres they lie beneath very different overhead ionospheres during this period. During January the ionosphere above Longyearbyen is dark and the conductivity present is due primarily to particle precipitation. At Davis, the ionosphere is sunlit and therefore fully conducting in all regions primarily due to solar ultraviolet radiation.

21 JANUARY 1991

In order to look for coherent signals between these two stations, we have combined the four channels of data from the two stations and passed them through the data-adaptive pure-state filter as described above. The result is shown in Figure 5. In this figure, the total power is shown for the signals which showed significant correlation between the four data channels. One sees that in the Pc3 band there is a region several hours wide where Pc3 signals can be identified. Within this region, the maximum Pc3 signal

Cape Parry, Sachs Harbor Induction Mags
Total Power 20 May 1985 4 stn pf

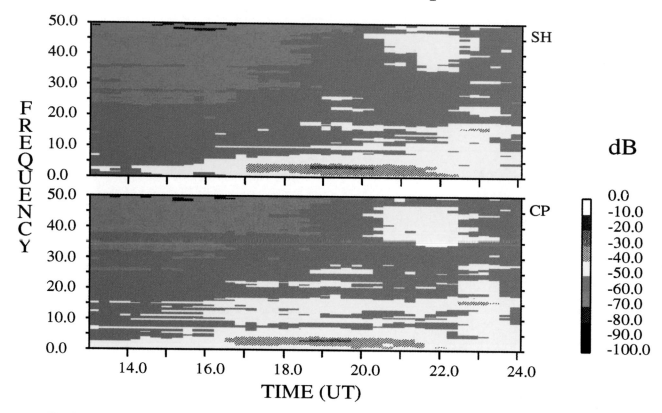

Fig. 4. Simultaneous spectrograms showing the horizontal signal power from Figure 2 after pure-state filtering. The pure-state filter produces multivariate coherence estimates which are invariant to sensor orientation. Note that only the central cusp Pc3 near 2130UT and the Pc5 signals are present after filtering indicating coherence between the two stations.

power occurs between 0900 UT and 1100 UT., centered on local noon. Note also that the Pc3 signals elsewhere in the band have been reduced by more than 10dB. We interpret this to indicate that the Pc3 observed in the central cusp, near local noon, are coherent between the two stations while the Pc3 signals observed away from noon, under the footprint of the boundary layers are not coherent.

22 JANUARY 1991

Figure 6 shows the pure-state filtered power for 22 January 1991. Here, although there is some residual power in the Pc3 band at Davis, there is no indication of correlated power in the spectrogram from Longyearbyen. This is consistent with the theme we are developing. The Pc3 energy seen in the footprint of the boundary layers is not coherent between these two sites. Unfortunately during this day there was little or no Pc3 energy in the central cusp region which found its way to the stations so no conclusion can be drawn concerning its coherence.

23 JANUARY 1991

Figure 7 shows the pure-state filtered power for 23 January 1991. Here the boundary layer signatures are present both in the prenoon and postnoon sectors. This is perhaps most easily seen in the Longyearbyen data with the prenoon boundary layer present between 0600 and 0800 UT and the afternoon boundary layer present between 1000 and 1200 UT. The Davis spectra show increases in Pc3 signal power at these times. Recall that Davis lies below a sunlit ionosphere so we do not expect the contrast between boundary layer and central cusp spectral powers to be as large as it is at Longyearbyen. The only Pc3 band signal to show coherence lies near 0700 UT, in the morningside footprint of the boundary layer. This is one of the few indications of coherence for Pc3 pulsations in the boundary layers. While unusual, it is not unexpected as we will discuss below.

DISCUSSION

We have shown data from stations in the same hemisphere

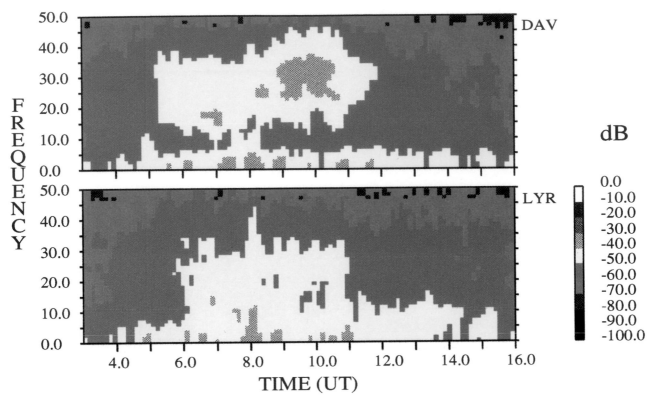

Fig. 5. Simultaneous spectrograms showing the horizontal signal power from Davis and Longyearbyen on 21 January 1991 after pure-state filtering. Note that only the central cusp Pc3 near 0900UT and the Pc5 signals are present after filtering indicating coherence between the two stations.

separated by only 150 kilometers as well as data from nearly conjugate stations in opposite hemispheres. In the example shown in Figures 2 and 3, the most coherent signals seen in the Pc3 band are those which can be associated with the central cusp. The large amplitude, broadband Pc3 signals in the boundary layers do not seem to be as coherent, either locally or between conjugate hemispheres. However, there is variability in the data which allows some qualification of these conclusions. While we do not have sufficient data to allow us to define all of the parameters necessary to explain these variations, we believe one can begin to build a picture of the cusp and boundary layers which can explain the data we have presented.

We take as a beginning assumption that the Pc3 variations observed beneath the footprint of the boundary layers are produced locally by modulated particle precipitation. Very strong evidence for this assumption was given by Engebretson et al. [1991a] in an analysis of Pc3 band variations observed in optical and magnetometer data. They concluded that the variations observed were produced through the periodic modulation of ionospheric conductivities by precipitating ~1 keV electrons.

Proceeding from this evidence, we next assume that the region of the ionosphere which is illuminated by the precipitating particles is limited to scale sizes on the order of 100 km or less. With this assumption one can begin to understand the lack of correlation seen in the boundary layer spectra presented above. That is to say, one would not expect to see correlations between stations separated by more than 100 km in the same hemisphere since differing particle populations are modulating the overhead ionosphere differently. However, for conjugate sites, it is possible although unlikely that two stations might be beneath a region illuminated by the same boundary layer beam and see coherent variations. This would explain the data from 23 January 1991 in which some correlation was seen in the conjugate Pc3 data from the boundary layers.

In the central cusp region when Pc3 signals are present, they appear to be correlated in the data from stations in the same hemisphere and for conjugate stations. These signals often appear as band-limited packets whose frequencies are correlated with the magnitude of the IMF. From the magnetometer data the central cusp region appears to be several hours wide so it is possible for two stations separated by a few hundred kilometers, or by an hour in local time, to be beneath the central cusp. For the two sites to observe correlated signals we must assume that the cusp is

LYR - DAV Induction Mags
Total Power 22 Jan 1991, 4 stn pf

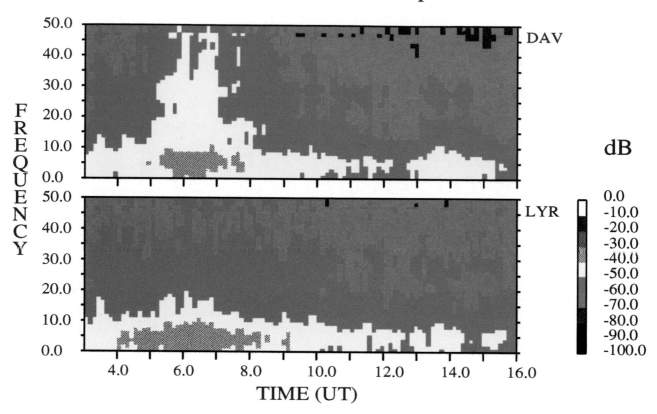

Fig. 6. Simultaneous spectrograms showing the horizontal signal power from Davis and Longyearbyen on 22 January 1991 after pure-state filtering. Note that while there seems to be coherent Pc5 signals present, no Pc3 signals show coherence between the two stations.

coherently illuminated by the Pc3 signal itself. The reason for this is that if one considers the cusp as a current source attempting to radiate Pc3 into the earth-ionospheric cavity then one must assume coherence in the current across the central cusp. This is because the frequency of the lowest mode which can propagate in the earth-ionospheric wave guide is that of the Schumann resonance which is observed to be near 8 Hz. The cutoff for the F2 region waveguide is of the order of 0.5 Hz [Manchester, 1968]. Signals in the Pc3 band occur in the frequency range from 20 to 50 mHz, one to two orders of magnitude below these cutoff frequencies. These signals would be heavily attenuated and would not be expected to be observed over large distances at the ground. The fact that coherent signals are seen over some distance seems to imply the coherence of the cusp variations across the ionospheric footprint of the central cusp. Within this description it can be understood why Pc3 are confined very closely to the cusp and have been used as cusp identifiers in ground magnetometer data.

If the cusp ionosphere is viewed as a coherent current source re-radiating some of the Pc3 energy incident upon it, then it may be an important source of magnetospheric Pc3 energy. The magnetospheric wave guide, that is that portion of the dayside

magnetosphere bounded by the magnetopause, ionosphere and plasmasphere on the dayside, has cutoff frequencies in the range of a few mHz [Samson et al., 1991]. Thus the Pc3 band signals present at the cusp can be radiated into this cavity. This may help explain the ubiquitous presence of Pc3 signals in the dayside magnetosphere.

The existence and efficacy of a cusp radiator filling the magnetospheric cavity with ULF wave energy in the Pc3 band can be tested through correlation studies of magnetospheric Pc3 with solar wind IMF properties. The changes in the magnitude of magnetospheric Pc3 pulsations as a function of IMF cone angle have been studied by Engebretson et al. [1986a]. They show a sudden onset of harmonically structured Pc3 signals in the magnetosphere (see their Figure 3) at the position of the AMPTE/CCE satellite near 0500 UT when the spacecraft was at approximately 9 Re. Inspection of IMF data showed that shortly after 0500 UT the cone angle decreased from near 90° to values near 30°. In a later paper, Engebretson et al. [1991a] suggested that the cusp could be viewed as a region of fluctuating current whose amplitude is modulated by the level of precipitation forming what they termed an "ionospheric transistor". We are initiating a study to model the propagation of Pc3

LYR - DAV Induction Mags
Total Power 23 Jan 1991, 4 stn pf

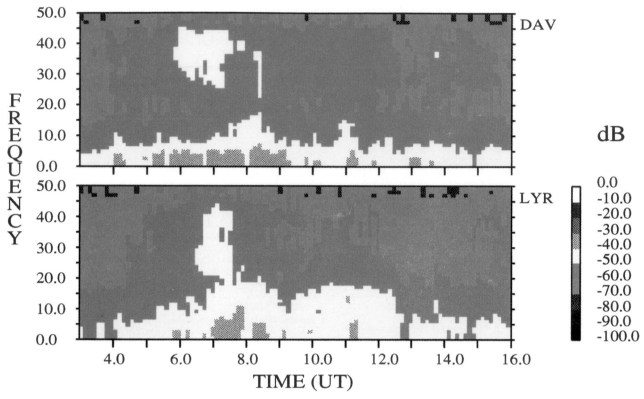

Fig. 7. Simultaneous spectrograms showing the horizontal signal power from Davis and Longyearbyen on 23 January 1991 after pure-state filtering. Note that while there seems to be coherent Pc5 signals present, the only Pc3 correlation seems to be associated with the pre-noon boundary layer.

energy into the magnetospheric wave guide.

Using the observed correlation of cusp Pc3 with solar wind IMF we believe the correlations we have seen imply a global source of the central cusp Pc3 located upstream of the magnetopause. There are two principal correlations between Pc3-4 pulsations and solar wind quantities which have been found. The first is the correlation of Pc3 amplitudes with small cone angles in the solar wind and the second is the direct correlation of the frequency of narrow-band cusp Pc3 with the magnitude of the IMF. Since the magnetopause is bathed with the plasma which passes through a very small region near the nose of the shock it hasOm been presumed that the nature of the shock near the nose may be important in the description of the pulsation spectrum observed at the earth. For small solar wind cone angles the shock at the nose should assume the parallel-shock geometry with an increase in downstream turbulence. It is also through this region of the shock that narrow-band ion-cyclotron waves generated in the foreshock region would travel to reach the ionosphere. Engebretson et al. [1991b] have reported on a study of the character of turbulence in the magnetosheath and its correlation with IMF cone angle. They find no evidence for increased narrow-band pulsations during times of low cone angle. However, they did

note increased, broadband power with minimum frequency at or near that of the upstream waves. Clearly, more studies are required to begin to settle this question.

Acknowledgements This study represents one aspect of the research carried out under grants from the National Science Foundation: ATM 8416865, ATM 9112834 and ATM 9213361. Dr. Fraser was also supported in part by grants from the Australia Research Council, the Australian Science Advisory Committee and the University of Newcastle. The authors gratefully acknowledge the support of the Great Norwegian Spitsbergen Coal Company and the Norwegian Polar Institute. Observations at Davis were carried out under the direction of Dr. G. Burns and Dr. R. Morris of the Australian Antarctic Division. We would also like to acknowledge useful discussions with M. G. McHarg, M. Engebretson and B. Anderson. We also acknowledge several useful suggestions by the referees.

REFERENCES

Anderson, B. J., M. J. Engebretson, S. P. Rounds, L. J. Zanetti, and T. A. Potemra, A statistical study of Pc 3-5 pulsations observed by the AMPTE/CCE magnetic fields experiment, 1, Occurrence distributions, *J. Geophys. Res.*, *95*, 10495, 1990.

Arnoldy, R. L., L. J. Cahill, Jr., M. J. Engebretson, L. J. Lanzerotti, and A. Wolfe, Review of hydromagnetic wave studies in the Antarctic, *Rev. Geophys.*, *26*, 181, 1988.

D'Angelo, N., Plasma waves and instabilities in the polar cusp: A review, *Rev. Geophys.*, *79*, 299, 1977.

Engebretson, M. J., L. J. Zanetti, T. A. Potemra, and M. H. Acuna, Harmonically structured ULF pulsations observed by the AMPTE/CCE magnetic field experiment, *Geophys. Res. Lett.*, *13*, 905, 1986a.

Engebretson, M. J., C.-I. Meng, R. L. Arnoldy, and L. J. Cahill, Jr., Pc3 pulsations observed near the south polar cusp, *J. Geophys. Res.*, *91* 8909, 1986b.

Engebretson, M. J., B. J. Anderson, L. J. Cahill, Jr., R. L. Arnoldy, T. J. Rosenberg, D. L. Carpenter, W. B. Gail, and R. H. Eather, Ionospheric signatures of cusp latitude Pc 3 pulsations, *J. Geophys. Res.*, *95*, 2447, 1990.

Engebretson, M. J., L. H. Cahill, Jr., R. L. Arnoldy, B. J. Anderson, T. J. Rosenberg, D. L. Carpenter, U. S. Inan, and R. H. Eather, The role of the ionosphere in coupling upstream ULF wave power into the dayside magnetosphere, *J. Geophys. Res.*, *96*, 1527, 1991a.

Engebretson, M. J., N. Lin, W. Baumjohann, H. Luehr, B. J. Anderson, L. J. Zanetti, T. A. Potemra, R. L. McPherron, and M. G. Kivelson, A comparison of ULF fluctuations in the solar wind, magnetosheath, and dayside magnetosphere 1. magnetosheath morphology, *J. Geophys. Res.*, *96*, 3441, 1991b.

Fraser, B. J., R. McPherron, and C. T. Russell, Radial Alfven velocity profiles in the magnetosphere and their relation to ULF wave field line resonances, *Adv. Space Res.*, *8*, 49, 1988.

Fraser-Smith, A. C., ULF/lower-ELF electromagnetic field measurements in the polar caps, *Rev. Geophys.*, *20*, 497, 1982.

Glassmeier, K. H., ULF pulsations in the polar cusp and cap, in *Electromagnetic Coupling in the Polar Clefts and Caps*, P. E. Sandholt and A. Egeland, eds., Kluwer, 1989.

Gringauz, K. I., E. K. Solomatina, V. A. Troitskaya and R. V. Shchepetnov, Variations of solar wind flux observed by several spacecraft and related pulsations of the earth's electromagnetic field, *J. Geophys. Res.*, *76*, 1065, 1971.

Greenstadt, E. W. and J. V. Olson, Pc 3,4 activity and interplanetary field orientation, *J. Geophys. Res.*, *81*, 5911, 1976.

Greenstadt, E. W. and J. V. Olson, A contribution to ULF activity in the Pc 3-4 range correlated with IMF radial orientation, *J. Geophys. Res.82*, 4991, 1977.

Greenstadt, E. W. and J. V. Olson, Geomagnetic pulsation signals and hourly distributions of IMF orientation, *J. Geophys. Res.*, *84*, 1493, 1979.

Gul'elmi, A. V. and O. V. Bol'shakova, Diagnostics of the interplanetary medium by measurements of pulsations, *Space Sci. Rev.*, *16*, 331, 1974.

Jenkins, G. M., and D. G. Watts, Spectral analysis and its applications, Holden Day, 1968.

Kato, Y., Y. Tonegawa, and K. Tomomura, Dynamic spectral study of Pc 3-5 magnetic pulsations obseved in the north polar cusp regions, *Natl. Inst. Pol. Res. Japan, Special Issue*, *36*, 58, 1985.

Lin, N., M. J. Engebretson, R. L. McPherron, M. G. Kivelson, W. Baumjohann, H. Luehr, T. A. Potemra, B. J. Anderson, and L. J. Zanetti, A comparison of ULF fluctuations in the solar wind, magnetosheath and dayside magnetosphere 2. Field and plasma conditions in the magnetosheath, *J. Geophys. Res.*, *96*, 3455, 1991.

Manchester, R. N., Correlation of Pc1 pulsations at spaced stations, *J. Geophys. Res.*, *73*, 3549, 1968.

McHarg, M. G. and J. V. Olson, Correlated optical and ULF magnetic observations of the winter cusp - boundary layer system, *Geophys. Res. Lett.*, *19*, 817, 1992.

Newell, P. T., and C.-I. Meng, The cusp and the cleft/boundary layer: Low altitude identification and statistical local time variation, *J. Geophys. Res.*, *93*, 14549, 1988.

Odera, T. J., Solar wind controlled pulsations, a review, *Rev. Geophys.*, *24*, 55, 1986.

Olson, J. V., ULF signatures of the polar cusp, *J. Geophys. Res.*, *91*, 10055, 1986.

Olson, J. V., Poleward propagation of pulsations near the cusp, *Planet. Space Sci.*, *37*, 775, 1989.

Olson, J. V., and B. J. Fraser, Conjugate ULF pulsations in the cusp, in press, *COSPAR Adv. in Space Res.*, 1992.

Samson, J. C., and J. V. Olson, Data-adaptive polarization filters for multichannel geophysical data, *Geophysics*, *46*, 1423, 1981.

Samson, J. C., R. A. Greenwald, J. M. Ruohoniemi, T. J. Hughes, and D. D. Wallis, Magnetometer and radar observations of magnetohydrodynamic cavity modes in the earth's magnetosphere, *Can. J. Phys.*, *69*, 929, 1991.

Singer, H. J., C. T. Russell, M. G. Kivelson, E. W. Greenstadt and J. V. Olson, Evidence for the control of Pc 3,4 magnetic pulsations by the solar wind velocity *Geophys. Res. Lett.*, *4*, 377, 1977.

Slawinski, R., D. Venkatesan, A. Wolfe, L. J. Lanzerotti, and C. G. Maclennan, Transmission of solar wind hydromagnetic energy into the terrestrial magnetosphere, *Geophys. Res. Lett.*, *15*, 1275, 1988.

Takahashi, K. and R. L. McPherron, Harmonic structure of Pc 3-4 pulsations, *J. Geophys. Res.*, *87*, 1504, 1982.

Tonegawa, Y., H. Fukunishi, L. J. Lanzerotti, C. G. Maclennan, L. V. Medford, and D. H. Carpenter, Studies of the energy source for hydromagnetic waves at auroral latitudes, *Mem. Natl. Inst. Pol. Res. Japan*, *38*, 73, 1985.

Troitskaya, V. A., ULF wave investigations in the dayside cusp, *Adv. Space Res.*, *5*, 219, 1985.

Tsurutani, B. T., E. J. Smith, R. M. Thorne, R. R. Anderson, D. A. Gurnett, G. K. Parks, C. S. Lin and C. T. Russell, Wave-particle interactions at the magnetopause: contributions to the dayside aurora, *Geophys. Res. Lett.*, *8*, 183, 1981.

Vero, J. and L. Hollo, Connections between interplanetary magnetic field and geomagnetic pulsations, *J. Atmos. Terres. Phys.*, *40*, 857, 1978.

Vinogradov, P. A. and Parkhomov, MHD waves in the solar wind - a possible source of geomagnetic Pc 3 pulsations, *Geomag. Aeron.*, *15*, 109, 1974.

Webb, D. and D. Orr, Geomagnetic pulsations (5-50 mHz) and the interplanetary magnetic field, *J. Geophys. Res.*, *81*, 5941, 1976.

Wolfe, A., E. Kamen, L. J. Lanzerotti, C. G. Maclennan, J. F. Bamber, and D. Venkatesan, ULF geomagnetic power at cusp latitudes in response to upstream solar wind conditions, *J. Geophys. Res.*, *92*, 168, 1987.

Wolfe, A., D. Venkatesan, R. Slawinski and C. G. Maclennan, Conjugate area study of Pc3 pulsations near the cusp, *J. Geophys. Res.*, *95*, 10,695, 1990.

Yumoto, K., Generation and propagation mechanisms of low-latitude magnetic pulsations - A review, *J. Geophys.*, *60*, 79, 1986.

Yumoto, K., A. Wolfe, T. Teresawa, E. L., Kamen, and L. J. Lanzerotti, Dependence of Pc3 magnetic energy spectra at South Pole on upstream solar wind parameters, *J. Geophys. Res.*, *92*, 12,437, 1987.

J.V. Olson, Geophysical Institute, University of Alaska, Fairbanks, AK 99775-0800.

B.J. Fraser, Physics Department, University of Newcastle, Newcastle 2308, NSW, Australia.

Global Cavity Mode-Like and Localized Field-Line Pc 3-4 Oscillations Stimulated by Interplanetary Impulses (Si/Sc): Initial Results From the 210° MM Magnetic Observations

K. Yumoto, A. Isono, K. Shiokawa, H. Matsuoka and Y. Tanaka

Solar-Terrestrial Environment Laboratory, Nagoya University, Japan

F. W. Menk and B. J. Fraser

Department of Physics, University of Newcastle, Newcastle, Australia

210° Magnetic Meridian Magnetic Observation Group

In order to investigate whether or not global cavity-mode and localized field-line oscillations can be excited in the real inner magnetosphere by interplanetary impulses (sc/si), we have analyzed magnetic data from the 210° magnetic meridian chain stations recorded during July 1990 - September 1991. It is found that two types of Pc 3-4 magnetic pulsations were stimulated at low latitudes just after 13 sc/si events. Most of the Pc 3-4 pulsations were standing field-line oscillations, having maximum power densities at L=1.57-1.60 and/or at higher latitudes and in-phase and out-of-phase relations of H- and D-component variations, respectively, at the magnetic conjugate points. Only 4 global cavity mode-like oscillations, showing a discrete spectrum, larger amplitude at lower latitudes (L<1.6) and phase delays among the latitudinally separated stations, were stimulated by large sudden commencements and impulses within the dayside magnetosphere from 09:00 to 15:00 LT during higher levels of magnetic activity (Kp≥7+). The duration of the Pc 3-4 pulsations stimulated by the sc/si events is found to be less than 20 min.

Introduction

It is more than forty years since sinusoidal magnetospheric oscillations, first recorded almost a hundred years before and known as magnetic pulsations, were attributed to magnetohydrodynamic (MHD) waves in the Earth's magnetosphere. Most magnetic pulsations having a period longer than 10 sec in the magnetosphere are generally believed to be field-line oscillations excited by "external" and "internal" source waves through resonance mechanisms [see review paper of Yumoto, 1988]. There are four likely main external sources of magnetospheric MHD waves in the 2-100 mHz range in the dayside magnetosphere: upstream waves excited by ion cyclotron resonance instability of reflected ion beams in the Earth's foreshock, surface waves on the magnetospheric boundary, quasi-period dynamic pressure oscillations in the solar wind, and sudden impulses caused by interplanetary shocks, discontinuities, and dayside reconnection. These external source waves can drive

Solar Wind Sources of Magnetospheric Ultra-Low-Frequency Waves
Geophysical Monograph 81
Copyright 1994 by the American Geophysical Union.

dayside Pc 3 (T=10-45 s), Pc 3-5 (10-600 s), sc/si-associated (Psc/si; T>1 s), and FTE-associated pulsations. Inhomogeneities in the magnetosphere are regarded as important in the mode coupling of the source waves with low-frequency MHD waves, i.e., torsional Alfvén and magnetospheric (and/or plasmaspheric) cavity oscillations [see e.g. references of Yumoto et al., 1990].

Three types of spectra are predicted by the MHD simulation for externally excited pulsations [cf. Yumoto et al., 1990]. One is a discrete shear Alfvén spectrum of a standing field-line oscillation, excited by narrow-band external wave sources through the field line resonance mechanism [Chen and Hasegawa, 1974; Southwood, 1974]. The second is the continuous shear Alfvén spectrum of standing field-line oscillations with spatially varying frequencies, stimulated by impulsive source waves having larger azimuthal wave number. The third is the discrete spectrum of quantized "global" or compressional cavity-mode oscillations with spatially constant eigenfrequencies [cf. Yumoto and Saito, 1983; Kivelson and Southwood, 1986; Allan et al., 1986; Lee and Lysak, 1990]. The standing field-line oscillations are characterized by a maximum amplitude at the resonance points

MAGNETOSPHERIC MHD WAVES EXCITED BY SUDDEN IMPULSE

(1) FIELD LINE OSCILLATION (ALFVÉN MODE)
EIGEN FREQUENCY; LATITUDINAL DEPENDENCE
AMPLITUDE; MAXIMUM AT RESONANCE POINT
PHASE; H, IN-PHASE, D, OUT-OF-PHASE

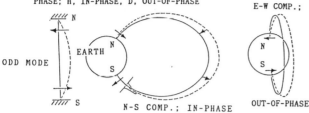

ODD MODE

E-W COMP.;

N-S COMP.; IN-PHASE   OUT-OF-PHASE

(2) CAVITY MODE OSCILLATION
EIGEN FREQUENCY; DEPENDENT ON SIZE OF SPHERE
AMPLITUDE; GLOBAL PROFILE
PHASE; UNRESOLVED

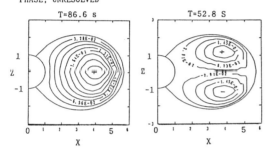

Fig. 1. Two possible MHD oscillations excited by interplanetary sudden impulses. (1) Standing field-line oscillation (Alfvén mode). A schematic illustration of odd mode oscillation. (2) Cavity mode oscillation. Examples of fundamental and second harmonic oscillations in the plasmasphere obtained by Itonaga et al. [1992].

with in-phase and out-of-phase relations of H- and D-component variations at the magnetic conjugate points as shown in the upper panel of Fig. 1. The cavity mode-like oscillations are expected to have a broad amplitude profile without clear magnetic conjugacy of polarization fields [see the bottom panel; e.g. Itonaga et al., 1992].

Psc/si magnetic pulsations are caused by sudden changes in the solar wind dynamic pressure, and are expected to be a good source of global mode MHD waves. However, there has been little observational confirmation of such cavity modes. Except for a few studies on sc-associated pulsations observed on the radar and on the ground magnetometer network [see e.g. Yeoman and Lester, 1990, and references therein] and low-latitude Pi 2 pulsations [Yumoto et al., 1989; Yeoman and Orr, 1989; Sutcliffe and Yumoto, 1991; Takahashi et al., 1992], no satellite or ground-based observations of the global cavity mode-like oscillations or the coupling between cavity and shear Alfvén modes have been reported.

In the present paper using the 210° magnetic meridian chain data [Yumoto et al., 1992], we investigate whether or not such cavity mode-like oscillations in the Pc 3-4 frequency band can be stimulated at low latitudes by interplanetary impulses (sc/si).

GEOMAGNETIC FIELD DATA

Imaging the Earth's magnetosphere using ground-based magnetometer arrays is still one of the major techniques for investigating the dynamical feature of the solar wind-magnetosphere interactions. Magnetic field data from coordinated ground stations make it possible to study magnetospheric processes by separating temporal changes and spatial variations of the phenomena, to clarify global and latitudinal structures and propagation characteristics of magnetic variations from higher to equatorial latitudes, and to understand the global generation mechanisms of the magnetospheric phenomena.

Magnetic observations along the 210° and 250° magnetic meridians are being conducted by the Solar-Terrestrial Environment Laboratory (STEL), Nagoya University, for the STEP period from 1990 to 1997 [Yumoto et al., 1992]. The project is carried out in collaboration with and/or with support from the following institutes and organizations; Tohoku Univ. (THU), Tohoku Inst. of Technol. (TIT), Tokai Univ. (TKU), and Kakioka Magnetic Obs. (KMO) in Japan, Univ. of Newcastle(UNC), Electr. Res. Lab. (DSTO), CSIRO Tropical Ecosystems Res. Centre (TERC), Weipa North State School (WNSS), IPS Radio and Space Services at Learmonth (IPS), Birdsville Police Station (POB), Dalby Agriculture College (DAC), Australia Antarctic Division (AAD) in Australia, Pacific Tsunami Warning Center (PTWC), Guam Magnetic Obs. of U.S. Geological Survey (USGS), and Univ. of Alaska, Fairbanks (UAF) in USA, Paradise Wewak Hotel (PWH) and Univ. of Papua New Guinea (UPNG) in PNG, Biak station of Natl. Inst. of Aeronautics and Space (LAPAN) in Indonesia, and three institutes (IKIR, IKFIA, IFZ) of Academy of Science in Russia.

In July, 1990, we installed the first fluxgate magnetometer systems of the array at Moshiri (MSR), Chichijima (CBI) and Kagoshima (KAG) in Japan, and Adelaide (ADE), Birdsville (BSV) and Weipa (WEP) in Australia. Figure 2 shows these 210° magnetic meridian stations in geomagnetic coordinates. The BSV site (L=1.57) is located near the magnetic conjugate point of MSR (1.60) in Japan. The CBI (L=1.14), WEP (1.18) and ADE (2.13) sites are located near the meridian of the conjugate stations. The KAG site (1.22) is situated ~2° north of the conjugate point of Darwin (1.18) which is located ~12° west in geomagnetic longitude of the WEP station. At Ewa Beach (EWA; L=1.18, Λ=269.05°), Hawaii, the same magnetometer system was installed in January, 1991.

In June, 1991, the fluxgate magnetometer observations were also started at Onagawa (ONW, L=1.38) in Japan, at Wewak (WWK: 1.06) in Papua New Guinea, and at Guam (GUA: 1.03) in USA. We completed installations of the magnetometer systems at Dalby (DAL: 1.58), Darwin (DAR: 1.18), and Learmonth (LMT: 1.47) in Australia in the summer of 1991, and Biak (BIK: 1.05) in Indonesia in May, 1992. We are extending the 210° magnetic meridian chain to northern high latitudes in Siberia in cooperation with IKFIA, IKIR, and IFZ of Academy of Science in Russia. The magnetometer systems were installed at St.

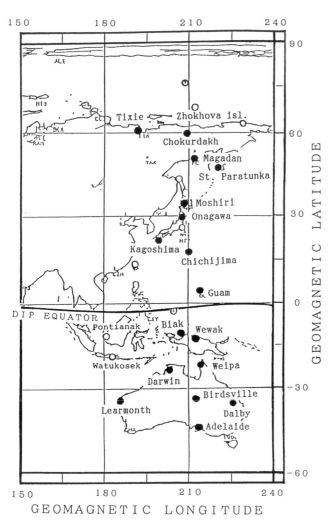

Fig. 2. Map showing the locations of the 210° magnetic meridian chain stations in geomagnetic coordinates (see Table 1). Reproduction of Fig.1, Yumoto et al.[1992].

Paratunka (PTK, L=2.11), Magadan (MGD: 2.85), Chokurdakh (CHD: 5.50), and Tixie (TIK; L=5.83, Λ =196.50°). The installation at Zhokhova island (ZHI: 9.05) is planned for 1993. The conjugate magnetic observations at Kotzebue in Alaska and Macquarie island in Australia around the 250° magnetic meridian will start in June 1993.

Table 1 summarizes the station names, the geographic and geomagnetic coordinates, and the L values of proposed observation sites including established stations, where additional instruments with a high time resolution will be installed during the STEP period. The corrected geomagnetic coordinates and L values are calculated for 100 km altitude for each of the stations on January 1, 1991 using the IGRF-85 model. Commencement months and abbreviations of institutes and organizations, which support or collaborate with STE Lab's magnetic observation team, are given in the last two columns in Table 1.

Magnetic variation data (ΔH, ΔD, ΔZ, dH/dt, dD/dt, dZ/dt) from all stations, except for CBI, are obtained using ring-core fluxgate magnetometers with identical logging systems (DCR-3, KOSMO Ltd.) and time signal generators as shown in detail in Yumoto et al. [1992]. The resolutions of ordinary analogue outputs V_O (ΔH, ΔD, ΔZ) in the 0-2.5 Hz frequency range are +/-300 nT, +/-1000nT, and +/-2000nT/ +/-10 volt for low, middle, and high-latitude stations, respectively. The time-derivative components (V_{TD}; dH/dt, dD/dt, dZ/dt) are obtained by putting an analogue circuit at the output terminals of the ordinary components (V_O). The V_{TD} outputs in the frequency range of 0.0-0.1 Hz exhibit essentially the same frequency response as an induction magnetometer. The noise level of the magnetometer system is lower than 0.1 nT rms equivalent. The six magnetic signals (ΔH, ΔD, ΔZ, dH/dt, dD/dt, dZ/dt) and time pulses (1 min, 1hr, 24 hr) are registered on a digital cassette tape using the digital data logger with sampling rate of 1 sec and 16 bit resolution of 0.012 nT, 0.039 nT, and 0.078nT/LSB at the low, middle and high-latitude stations, respectively. Each cassette tape holds 21 days of data. Fluxgate magnetometer data from the CBI station of the Kakioka Magnetic Observatory, are obtained by the same logging system. The time pulses (1 min, 1 hr, 24 hr) from the time signal generator are also recorded on the digital cassette tape to check the crystal clock inside the data logger. The time signal generator is maintained accurate to within +/- 25 ms by automatic comparisons with standard radio transmissions from WWVH (Maui, Hawaii), JJY (Koganei, Japan), WWV (Boulder, USA) and Omega.

Data from higher latitude stations, which are now deployed in the 210° meridian, were not available for this study. Thus we have analyzed magnetic data from 6 low-latitude stations at L≤2.1 in the next section.

DATA ANALYSES

In order to find low-latitude Pc 3-4 magnetic pulsations stimulated by magnetospheric impulses, we selected 13 sc and si events observed at 6 stations (MSR, CBI, KAG, WEP, BSV, ADE) along the 210° magnetic meridian during the local daytime of 05-19 LT from July 21, 1990 to September 8, 1991. Figure 3 shows one example, on March 24, 1991, of the selected sc/si events in the H-component magnetograms from the meridional chain stations. The geomagnetic storm with a sudden commencement (sc) of ΔH~202 nT at the Kakioka Magnetic Observatory (KAK; not shown in Fig. 3) began at 03:42 UT, developed rapidly over the next two hours, and ΔH reached a minimum value of -180 nT at KAK. The magnitude of the initial impulse (sc) is the largest ever recorded at Kakioka. A sudden impulse (si) occurred again at 05:17 UT, and the H magnetic variation developed slowly in the inner magnetosphere at CBI and WEP (L<1.2). Another geomagnetic storm began with a sudden impulse (si) at 20:31 UT on March 24, developed over the next eight hours, and reached a ΔH minimum of about -300 nT at KAK around 04 UT on March 25, 1992.

Power spectral densities for 20-min intervals just before, just after and +20-min after the sc/si onsets were calculated to identify spectral peaks at the separated low-latitude stations along the 210° magnetic meridian, and to clarify latitudinal variation in the common spectral components in the Pc 3-4 frequency range. The

Station Name	Abbr.	Geographic		Geomagnetic		L	Associ. Inst.	Beginning
		Lat, deg	Long, deg	Lat, deg	Long, deg			
Zhokhova isl.	ZHI	76.24	152.74	70.59	210.61	9.05	IKFIA,IFZ	
Tixie	TIK	71.58	129.00	65.53	196.50	5.83	IKFIA,IFZ	92/8-
Chokurdakh	CHD	70.62	147.89	64.75	211.78	5.50	IKFIA,IFZ	92/8-
Magadan	MGD	59.97	150.86	53.70	218.34	2.85	IKIR	92/8-
St. Paratunka	PTK	52.94	158.25	46.49	225.60	2.11	IKIR	92/8-
Moshiri	MSR	44.37	142.27	37.76	212.96	1.60	STEL	90/7-
Onagawa	ONW	38.43	141.47	31.79	212.25	1.38	THU	91/6-
Kagoshima	KAG	31.48	130.72	25.23	201.99	1.22	STEL	90/7-
Chichijima	CBI	27.15	142.30	20.65	212.74	1.14	KMO	90/7-
Guam	GUA	13.58	144.87	9.02	215.18	1.03	USGS	91/6-
Biak	BIK	-1.08	136.05	-12.02	206.94	1.05	LAPAN	92/5-
Wewak	WWK	-3.55	143.62	-14.08	215.00	1.06	PWH,UPNG	91/6-
Darwin	DAR	-12.40	130.90	-23.22	202.42	1.18	TERC,UNC	91/8-
Weipa	WEP	-12.68	141.88	-23.06	214.07	1.18	WNSS,UNC	90/7-
Learmonth	LMT	-22.22	114.10	-34.36	184.64	1.47	IPS,UNC	91/8-
Birdsville	BSV	-25.83	139.33	-37.08	212.86	1.57	POB	90/7-
Dalby	DAL	-27.18	151.20	-37.30	226.53	1.58	DAC,UNC	91/8-
Adelaide	ADE	-34.67	138.65	-46.72	213.34	2.13	DSTO	90/7-
Kotzebue	KOT	66.88	197.40	64.63	249.25	5.45	UAF	
Ewa Beach	EWA	21.32	202.00	22.72	269.05	1.18	PTWC/USGS	91/1-
American Samoa	ASA	-14.28	170.70	-20.67	244.81	1.14		
Macquarie isl.	MCQ	-54.50	158.95	-64.77	247.60	5.50	AAD	92/11-

Table 1. Geographic and corrected geomagnetic coordinates, L values, supporting institutes for the 210° and 250° magnetic meridian chain stations, and commencement months. Geomagnetic parameters are calculated by using the IGRF-1985 model for 100 km altitude of each station on January 1, 1991.

power spectral densities for 20 min data were calculated by means of the Fast Fourier Transform method. Then, we examined wave characteristics to clarify which of cavity mode-like and field-line oscillations can be stimulated predominantly after the impulsive sc/si events in the real inner magnetosphere.

Table 2 summarizes the calculated power density (nT^2/Hz) of 20-min intervals just before, just after, and +20-min after each sc/si impulse observed at the MSR station, and magnetic activity (Kp). It is found that the 13 sc/si events have 40 identical eigen periods (T) among the six low-latitude stations. The ssc events in Table 2 are listed in NOAA's "Solar-Geophysical Data Prompt Report". The arrows in the table indicate that all identical power spectral densities in the Pc 3-4 frequency range, except for two components of T=26.9 sec at 20:31 UT on March 24, 1991 and T=19.7 sec at 06:50 UT on June 13, 1991, increase just after the onset of sc/si events, and most component decreases after 20 min. of the sc/si onsets.

Figure 4 shows a typical example of power spectra for standing field-line oscillations stimulated at low latitudes by a sudden impulse on August 26, 1990. The left and right panels indicate superimposed power spectral of H-component magnetic variations observed at the six low-latitude stations during the period of 05:25-05:45 UT just before, and in the interval of 05:50-06:10 UT just after the sc event, respectively. Power spectral levels of pulsations after the sc event became slightly higher than those

before the event. Three oscillations with identical discrete spectral components T=51.3 s, 34.1 s, and 16.5 s period (vertical lines in the right panel), are found to be excited just after the sc onset. It is noteworthy that the short-period components at 16-20 sec are visible in the spectra of MSR and BSV before the sc event, and were most effectively stimulated by the sc around the conjugate points of Moshiri and Birdsville (L=1.57-1.60), as identified by the shaded area in the figure. This oscillation corresponds to the localized short-period (10-20 sec) Pc 3 magnetic pulsation observed at low latitudes, i.e., a torsional Alfvén resonance oscillation around 38° magnetic latitude as illustrated in Fig. 7 of Yumoto et al. [1992].

The other example of power spectral densities of 20-min interval pulsation data at the 6 chain stations is shown in Fig. 5, to examine if a cavity mode-like oscillation can be stimulated after the impulsive sc event. The left, middle, and right panels show superimposed power spectra of H-component magnetic variations observed during the periods of 03:20-03:40 UT just before the sc onset, and in the intervals of 03:45-04:05 UT and 04:05-04:25 UT just after the event, respectively. Spectral levels of pulsations after the sc event became 10^3 times greater than those before the event. Three identical spectral components T=64.4 s, 39.6 s, and 32.3 s period, are found to be excited just after the onset of the sc event at different latitudes in the inner magnetosphere. The local standing field-line oscillation with frequency near 45 mHz and

March 24, 1991 (H-COMPONENT)

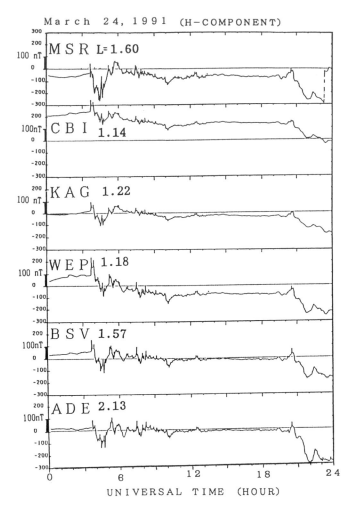

Fig. 3. H-component amplitude-time records of the March 24, 1991 ssc storm event observed simultaneously at the 210° magnetic meridian chain stations of MSR, CBI, and KAG in the northern hemisphere and WEP, BSV, and ADE in the southern hemisphere. Reproduction of Fig. 10, Yumoto et al. [1992].

maximum amplitude at MSR and BSV (L=1.57-1.60) was also activated after the sc event. It is noteworthy that spectral peaks around T=64.4 s (15.5 mHz), which did not exist before the sc but appeared after the event, show exactly the same frequency at the L=1.1-2.1 meridional stations and similar power levels of > 10^3 (nT^2/Hz). This component is indicated by a solid vertical line in the middle panel. Higher frequency components at 39.6 s (25 mHz) and 32.3 s (31 mHz) were also stimulated after the sc onset. The right panel shows another interesting characteristic that these global Pc 3-4 pulsations having the same period at different locations continued for about 20 min., and then disappeared concurrently. The identical periods at different latitudes and larger amplitudes at lower latitudes are quite different features than those of standing field-line oscillations. These can be explained by invoking a cavity mode oscillation in the plasmasphere or in the magnetosphere [see references of Yumoto et al., 1990; Lee and Lysak, 1990; Itonaga et al., 1992;

Fujita and Glassmeier, 1993a and b].

Figure 6 shows amplitude-time records of the H component signal at the 6 stations, which have been bandpass filtered in the 50-100 sec period range for the interval 03:30-04:20 UT of the March 24, 1991 event in Fig. 5. It is noteworthy that the duration of cavity mode-like oscillation (T=64.4 s) is about 20 min., which may be associated with duration of the driving source mechanism in the solar wind and/or with the characteristic time constants for the trapped wave energy in the plasmasphere to escape from the plasmapause or dissipate in the ionosphere. Recently, Fujita and Glassmeier [1993a and b] discussed cavity-mode oscillations allowing for energy flow across the outer boundary, i.e. the plasmapause was taken to be a inperfect refractor. Further theoretical and observational studies on cavity mode-like oscillations are needed to understand the physical meaning of the duration of 20 min.

In order to investigate the latitudinal variation of amplitude of the stimulated pulsations, we have plotted the power spectral peak densities, which are normalized by each maximum power density among the 6 stations, as a function of magnetic latitude of the chain stations. The left panel of Fig. 7 shows the latitudinal variation of normalized power spectral densities observed on August 26, 1990 (see Fig. 4), indicating maximum powers located around 38 magnetic latitude and/or at higher latitudes. These power density profiles can be explained by standing field-line oscillations excited in the inner magnetosphere. On the other hand, the March 24, 1991 event (see the middle panel of Fig.5) shows larger power densities at equatorial and lower latitudes of |Φ|≤30°. The power density profile of the 64.4 s period component cannot be interpreted by a "normal" localized standing field-line oscillation, but can be explained by a cavity mode-like oscillation stimulated in the inner magnetosphere (and/or in the plasmasphere). It is also found that there are dips in the power energy distribution of four cavity-like oscillations (see in Fig. 9) around 25°N and 35°S magnetic latitudes as shown in the right panel. This may be explained by an existence of a peculiar region, e.g. a discontinuity of plasma distribution around L = 1.6 where a standing field-line resonance can be stimulated easily, and which can be a "node" of cavity-like oscillations in the plasmasphere. Another possibility is an effect from an anomalous geological electrical conductivity structure, because the KAG station is located near the sea coast. Further statistical comparisons of characteristics of magnetic variations at KAG with those at other stations, and theoretical and simulation studies are needed to separate the anomalous electromagnetic response due to the coastal effect, and to understand the dips in the power energy distribution in the right panel of Fig. 7.

We also investigated phase relationships of these stimulated oscillations. After bandpass filtering in the 15.5-18.0 sec period range for the interval 06:01-06:06 UT on August 26, 1990, and in the 52.0-74.0 sec period range for the interval 03:45-03:55 UT on March 24, 1991, we plotted the amplitude-time records of H and D components at the six stations in the left and right panels of Fig. 8, respectively. The cutoff power level of the Finite Impulse Response digital bandpass filter [Rabiner, 1971] is -60dB. The left panel shows in-phase relation of H-component and out-of-

Fig. 5 and in the right panels of Figs. 7 and 8.

(4) Most of the Pc 3-4 oscillations stimulated at low latitudes by sc/si events have a duration shorter than 20 min. (see Table 2).

The results (1) and (2) suggest that the solar wind impulses can readily stimulate a standing field-line oscillation in the inner magnetosphere. The result (3) indicates a possibility of cavity resonance oscillations in the Pc 3-4 frequency band at low latitudes, which are not easily stimulated by the external impulsive sources (sc/si) but tend to be excited within the daytime magnetosphere only by great sudden commencements and impulses. The question (4), why most of the stimulated oscillations can appear only in the interval less than 20 min., will be clarified by further theoretical [e.g. Fujita and Glassmeier, 1993a and b] and observational analyses in near future.

Future studies on sc/si-stimulated Pc 3-4 pulsations using the 210° magnetic meridian stations will be done to get unique information on (1) Energy transfer (or wave propagation) processes from high latitude through middle and low latitudes to the equatorial region, (2) Coupling processes between the external sources and the stimulated pulsations at auroral, middle, low and equatorial latitudes, and (3) Conjugacy of fine-scale structures of the stimulated oscillations observed in the northern and southern hemisphere.

Acknowledgements. The authors would like to express their sincere thanks to Prof. H. Oya, Tohoku Univ., and Prof. S. Kokubun, Univ. of Tokyo, for their ceaseless supports for the coordinated magnetic observation project along the 210° magnetic meridian. The magnetic observations are conducted by STE Lab., Nagoya Univ. in cooperations with Tohoku Univ. (T. Saito, T. Takahashi), Tohoku Inst. of Technol. (M. Seto, Y. Kitamura), Kakioka Magnetic Obs. (S. Tsunomura), Tokai Univ. (T. Sakurai) in Japan, Univ. of Newcastle (B.J. Fraser, F.W. Menk), Electr. Res. Lab., DSTO (K.J.W. Lynn), CSIRO Tropical Ecosystems Research Centre (L. Corbett), and IPS Radio and Space Services, Learmonth Solar Observatory (J. Kennewell) in Australia, U.S. Geolog. Survey (A.W. Green), and Univ. of Alaska (S.-I. Akasofu, J. Olson) in USA, Univ. of Papua New Guinea (D. Yeboah-Amankwah) in PNG, LAPAN (W. Sinambela, O. Sobary) in Indonesia, and IFZ (V.A. Pilipenko), IKIR (E. Vershinin, V.F. Osinin), IKFIA (G. Krymskij, S.I. Solovyev), and IZMIRAN in Russia. Fluxgate magnetometer data from the Chichijima station are obtained by courtesy of the Kakioka Magnetic Observatory. Magnetic observations at Adelaide, Birdsville, Dalby and Weipa in Australia, Guam and Ewa Beach in U.S.A, and Wewak in PNG are kindly maintained by John Dunkely of DSTO, Owen Harms of Birdsville Police Station, M. McGilchrist and Dalby Agriculture College, Michael Sbrizzi of Weipa North State School, P. Hattori of Guam U.S. Geolog. Survey, M. Blackford and F. Takenouchi of Pacific Tsunami Warning Centre, and S. Kawabata of Paradise Wewak Hotel, respectively. The 210° magnetic meridian project is also supported by M. Nishino, T. Katho, M. Satho, Y. Katho, M. Sera, Y. Ikegami, and K. Hidaka of STE Lab., Nagoya Univ. Thanks are due to the Ministry of Education, Science, and Culture of Japan for financial support as the Grantsin-Aid for Oversea Scientific Survey (02041039, 03041061, 04044077), for General Scientific Research (02402015), for Developmental Scientific Research (02504002) and for the International STEP project.

REFERENCES

Allan, W., E. M. Poulter, and S. P. White, Hydromagnetic wave coupling in the magnetosphere-plasmasphere effects on impulse-excited resonance, *Planet. Space Sci., 12,* 1189-1200, 1986.

Chen, L., and A. Hasegawa, A theory of long-period magnetic pulsations, 1. Steady state excitation of field line resonance, *J. Geophys. Res., 79,*1024-1032, 1974.

Fujita, S. and K.-H. Glassmeier, Cavity-mode magnetohydrodynamic oscillations with energy flow across the outermost L-shell, *J. Geomag. Geoelectr.,* in press 1993a.

Fujita, S. and K.-H. Glassmeier, Characteristics of the magnetohydrodynamics oscillations trapped in the imperfect cavity: Implications of theoretical results to observations, *J. Geomag. Geoelectr.,* inpress 1993b.

Lee, D.-H., and R. L. Lysak, Effects of azimutal asymmetry on ULF waves in the dipole magnetosphere, *Geophys. Res. Lett., 17,* 53-56, 1990.

Itonaga M., T.-I. Kitamura, O. Saka, H. Tachihara, M. Shinohara, and A. Yoshikawa, Discrete spectral structure of low-latitude and equatorial Pi 2 pulsation, *J. Geomag. Geoelectr., 44,* 253-259, 1992.

Kivelson, M. G., and D. J. Southwood, Coupling of global magnetospheric MHD eigenmodes to field line resonances, *J. Geophys. Res., 91,* 4345-4351, 1986.

Rabiner, L. R., Technique for designing finite-duration impulse-response digital filters, *IEEE Trans. Commum.,* COM-19, April 1971.

Southwood, D. J., Some features of field line resonances in the magnetosphere, *Planet. Space Sci., 22,* 483-491, 1974.

Sutcliffe, P., and K. Yumoto, On the cavity mode nature of low latitude Pi 2 pulsations, *J. Geophys. Res., 96,* 1543-1551, 1991.

Takahashi, K., S.-I. Ohtani, and K. Yumoto, AMPTE CCE observations of Pi 2 pulsations in the inner magnetosphere, *Geophys. Res. Lett., 19,* 1447-1450, 1992.

Yeoman T. K., and D. Orr, Phase and spectral power of mid-latitude Pi 2 pulsations: Evidence for a plasmaspheric cavity resonance, *Planet.. Space Sci., 37,* 1367-1383, 1989.

Yeoman T. K., and M. Lester, Characteristics of MHD waves associated with storm sudden commencements observed by SABRE and ground magnetometers, *Planet. Space Sci., 38,* 603-616, 1990.

Yumoto, K., External and internal sources of low-frequency MHD waves in the magnetosphere - A review, *J. Geomag. Geoelectr., 40,* 293-311, 1988,

Yumoto, K., and T. Saito, Relation of compressional HM waves at GOES 2 to low-latitude Pc 3 magnetic pulsations, *J. Geophys. Res., 88,* 10041-10052, 1983.

Yumoto, K., K. Takahashi, T. Saito, F.W. Menk, B.J. Fraser, T.A. Potemra, and L.J. Zanetti, Some aspects of the relation between Pi 1-2 magnetic pulsations observed at L=1.3-2.1 on the ground and substorm-associated magnetic field variations in the near-Earth magnetotail observed by AMPTE/CCE, *J. Geophys. Res., 94,* 3611-3618, 1989.

Yumoto, K., S. Watanabe, and H. Oya, MHD responses of a model magnetosphere to magnetopause perturbations, *Proc. Res. Inst. Atmos., Nagoya Univ., 37,* 17-36, 1990.

Yumoto, K. Y. Tanaka, T. Oguti, K. Shiokawa, Y. Yoshimura, A. Isono, B. J. Fraser, F.W. Menk, J.W. Lynn, M. Seto, and 210° MM Magnetic Observation Group, Globally coordinated magnetic observations along 210° magnetic meridian during STEP period: 1. Preliminary results of low-latitude Pc 3's, *J. Geomag. Geoelectr., 44,* 261-276, 1992.

A. Isono, H. Matsuoka, K. Shiokawa, Y. Tanaka, and K. Yumoto, Solar-Terrestrial Environment Laboratory, Nagoya University, Toyokawa, Aichi 442, Japan

B. J. Fraser and F. W. Menk, Department of Physics, University of Newcastle, Newcastle, N.S.W. 2308, Australia

Studies of the Occurrence and Properties of Pc 3-4 Magnetic and Auroral Pulsations at South Pole, Antarctica

M. J. Engebretson,[1] J. R. Beck,[1] R. L. Rairden,[2] S. B. Mende,[2]
R. L. Arnoldy,[3] L. J. Cahill, Jr.,[4] and T. J. Rosenberg[5]

South Pole Station, Antarctica is uniquely situated for studies of the high latitude winter ionosphere both because of the extended darkness its geographical location allows and because it is located near the nominal latitude of the foot of the dayside cusp/cleft region during periods of moderate geomagnetic activity. In this study we use data from magnetometers, all-sky imagers, and zenith-viewing photometers at South Pole Station to compare optical and magnetic signatures of Pc 3-4 pulsations as a function of the proximity of the dayside auroral oval to the observing station. We find that the optical and magnetic pulsations are closely linked: narrowband Pc 3-4 optical pulsations occur only when similarly narrowband Pc 3-4 magnetic pulsations are also present. Both statistically (using photometer data) and on a case-by-case basis (using all-sky imager data), we find that modulations in 427.8 nm auroral light at Pc 3-4 frequencies occur during times when the dayside auroral oval is at least slightly poleward of the observing station. These observations place the location of the optical Pc 3-4 pulsations under either the boundary layer or the outer edge of the magnetosphere, but not under plasma mantle field lines. In addition, we have so far found no evidence of these optical pulsations in the cusp itself. Comparison of optical pulsation waveforms indicates the existence of two different precipitation regimes. When South Pole Station lies under field lines close to the magnetospheric boundary, the optical pulsations appear to be embedded in the background level, with excursions roughly symmetric both above and below this level. When the station lies under field lines situated deeper in the magnetosphere, the optical pulsations occur in clearly separated "packets" with levels clearly above background levels. One explanation for this difference is that electrons precipitating in the first case come from the magnetospheric boundary layer, and share in the variability of this region, while those in the second case originate as trapped outer magnetospheric electrons which are destabilized by packets of traveling compressional Pc 3-4 waves. In either case the magnetic and optical pulsations are a consequence of variations which impinge on the dayside magnetosphere from a highly disturbed magnetosheath downstream from a quasi-parallel subsolar bow shock.

Introduction

A large fraction of magnetospheric Pc 3-4 pulsations (7 - 100 mHz) is regarded now as having its ultimate source at or upstream of the Earth's bow shock. These pulsations are thought to be created as part of the solar wind — shock interaction associated with quasi-parallel shock processes, and their occurrence in the magnetosphere is directly related to the cone angle of the interplanetary magnetic field, which also governs the locations at which quasi-parallel shock geometry will occur on the Earth's bow shock [Greenstadt, 1972, Russell et al., 1983; Odera, 1986, Greenstadt, 1991]. Both in the

[1] Department of Physics, Augsburg College, Minneapolis, Minnesota.
[2] Lockheed Palo Alto Research Laboratory, Palo Alto, California.
[3] University of New Hampshire, Durham.
[4] University of Minnesota, Minneapolis.
[5] University of Maryland, College Park.

Solar Wind Sources of Magnetospheric Ultra-Low-Frequency Waves
Geophysical Monograph 81

magnetosphere and on the ground the occurrence of Pc 3-4 magnetic pulsations is associated with IMF cone angles below 45° [Russell et al., 1983; Yumoto et al., 1984, 1985; Engebretson et al., 1986], suggesting that waves generated at or near the subsolar region of the Earth's bow shock are convected by magnetosheath flow past the Earth's magnetosphere.

A continuing question, however, concerns the means by which this wave activity is transmitted into the Earth's magnetosphere and thence to the ground. In particular, it is not clear whether the Pc 3-4 pulsations observed at cusp/cleft latitudes reach these high latitudes via equatorial entry of compressional waves, followed by mode coupling to transverse waves [e.g., Russell et al., 1983], or by more direct entry along outer magnetospheric magnetic field lines to these latitudes [Engebretson et al., 1991a].

Recent studies of Pc 3-4 magnetic pulsations observed at South Pole Station, Antarctica (~74° invariant latitude) within several hours of local magnetic noon have shown that observed narrowband pulsations have a frequency consistent with an upstream driving source, and that they are often accompanied by roughly simultaneous pulsations in 427.8 nm wavelength auroral light, as detected by a zenith-viewing photometer [Engebretson et al., 1990, 1991a]. These observations suggested that a modulation of electron precipitation at

Pc3-4 frequencies, the cause of the pulsations in auroral light, might be involved in producing at least some of the magnetic pulsation activity seen on the ground.

We have undertaken additional studies of pulsation data from search coil magnetometers and photometers at South Pole, and have also compared these with data from an all-sky auroral imaging camera at South Pole in an attempt to determine what part of the auroral region might be responsible for these optical pulsations. Magnetic field data are recorded continually at South Pole throughout the year at 10 samples/s [Engebretson et al., 1986], and auroral emissions at wavelengths of 427.8 nm (in the N2+ first negative band) and 630.0 nm (the OI(1D) line) are routinely recorded at South Pole during local winter (May to September) by both all-sky imagers, recording one image per minute [Rairden and Mende, 1989; Rairden et al., 1990], and zenith-viewing photometers with 55° full angle field of view, recorded each second [Wu and Rosenberg, 1992]. Because the 630.0 nm emission is from a metastable state with a lifetime of 110 s, rapid pulsations are seldom evident in optical signals at this wavelength. Such pulsations are often evident in the 427.8 nm photometer data, however, in association with magnetic pulsations of the same frequency.

Data Base and Sample Events

In order to develop a large data base for statistical studies, we produced color Fourier spectrograms of magnetometer, photometer, and ELF/VLF receiver data from South Pole for 94 days during the 1990 austral winter (May through early August, 1990). Each time series was differenced before FFT calculation in order to whiten the spectra. The spectrogram for June 16, 1990 (Figure 1) shows typical features of the events of interest for this study. This figure shows color-coded power levels from 1000 to 2000 UT June 16, 1990 as a function of frequency (vertical axis) and time (horizontal axis). The day was magnetically quiet, with three-hour K_p indices of 2+, 1, 1+, 1-, 1, 1, 1+, and 1-. The time interval chosen spans local magnetic noon (1530 UT) at the station. A band-limited region of enhanced power between ~15 and ~35 mHz is evident in the magnetic field signals (XBB and YBB) from 1000 to 2000 UT. Power at similar frequencies is enhanced in the 427.8 nm photometer signal (PHO2) from 1100 to 1630 UT, and in the 1 kHz to 2 kHz ELF/VLF emission band (VLF2) from 1240 to 1400 and 1430 to 1630 UT. Much weaker modulations are also evident in the 630.0 nm photometer signal (PHO1) during the largest modulations at 427.8 nm. High frequency "tails" on the brightest optical features indicate that these pulsations are not pure sinusoids. No pulsation signal is evident in the 30 MHz riometer signal (RIO3).

IMP 8 satellite measurements indicated that the IMF magnitude varied from 3.6 to 4.5 nT during this interval. We thus infer an upstream wave frequency of 21-27 mHz [see, for example, Engebretson et al., 1990], consistent with the frequency of the waves observed at South Pole. The IMF cone angle was below 45° (favorable for propagation of upstream waves to Earth) during nearly all of this ten-hour interval.

We will show in Figures 2 and 3 two one-hour intervals of roughly simultaneous magnetic and optical pulsations observed on this day, and in Figures 4 and 5 the corresponding all-sky images showing the location of the dayside auroral oval relative to South Pole Station during these times. These intervals indicate two typical classes of optical pulsation activity, which we suggest are related to different magnetospheric regions equatorward of the boundary between open and closed field lines.

Figure 2 shows magnetic field, auroral light, ELF/VLF emissions, and riometer absorption, based on one-second samples, for a one hour period from 1300 to 1400 UT on this day. Packets of simultaneous Pc 3-4 pulsations are evident in the magnetic field traces (XBB and YBB) and the 427.8 nm photometer trace (PHO2) throughout the hour. No temporal correlations are evident during this hour between individual magnetic field variations and optical variations. Although good temporal correlations were occasionally observed for individual wave packets, the considerably different fields of view of the magnetometer and photometer help make excellent correlations the exception rather than the rule. The degree of modulation in brightness at 427.8 nm shown here, approximately 30-50%, was somewhat above the average level of ~20% observed in this study. The 630.0 nm signal exhibits minor fluctuations in phase with the 427.8 nm pulses, but they clearly do not dominate the signal at this wavelength.

Clear but modest-amplitude modulations in 1-2 kHz ELF/VLF emissions at Pc 3-4 frequencies also occurred throughout most of the hour. By analogy with ion cyclotron waves, electron cyclotron waves are generated at ELF/VLF frequencies proportional to the magnitude of the magnetic field at the locations of wave generation, which are presumably at or near the minimum point(s) along individual field lines. The observation that the emissions at higher ELF/VLF frequencies were not modulated (not shown) suggests an origin of the modulations in the outer magnetosphere (see Tsurutani and Smith [1977] for a summary of observed dayside ELF/VLF frequencies). We will report further on these modulations and their sources in a subsequent paper. There is no evident correlation between the riometer absorption and either individual Pc 3 pulsations or wave packets, with the possible exception of a broad increase near 1342 UT.

Many of the Pc 3-4 fluctuations in the 427.8 nm signal shown in Figure 2, including most before 1320 UT, appear to be sharper downward than upward; i.e., they appear to be negative pulses. This polarity is common (but pulses of both polarities are observed) during the entire 1100-1330 UT interval, and appears to be typical for events when auroral activity at 630.0 nm is only slightly poleward of the station.

From ~1430 (by which time the 630.0 nm activity was 60° poleward from the zenith) until ~1640, signals in 427.8 nm were as shown in Figure 3, which shows one hour's data from 1500 to 1600 UT in the same format as Figure 2. Magnetic pulsations during this hour were similar to those in the previous figure, but sometimes less sinusoidal. Optical pulsations at 427.8 nm were uniformly positive, often occurred in bursts, and appeared well above background values. The considerably smaller signals at 630.0 nm also indicated proportionally large increases above background values. The correlation between wave packets in magnetic and optical pulsations was slightly better between 1430 and 1640 UT (for example, near 1510 UT), but there were still few one-to-one correlations of individual magnetic and optical pulses.

Although the background signal at 427.8 nm was roughly similar during both intervals, the background signal at 630.0 nm was nearly

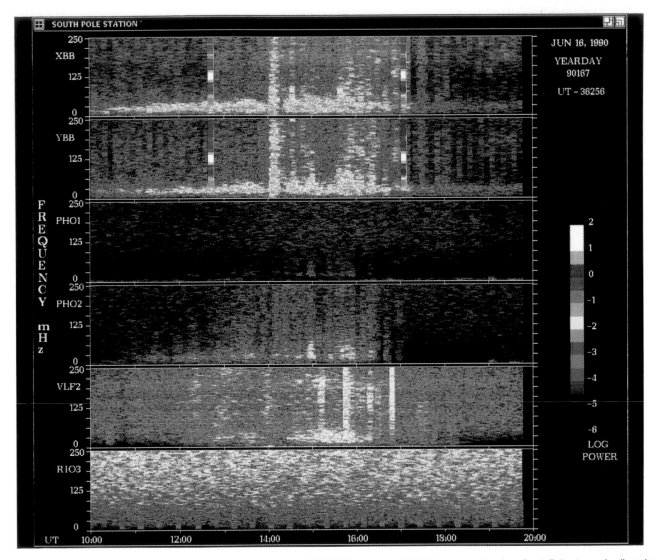

Fig. 1: Color Fourier spectrogram of magnetometer, auroral photometer, ELF/VLF receiver, and 30-MHz riometer data from South Pole, Antarctica (invariant latitude ~74°, local magnetic noon ~1530 UT) from 1000 to 2000 UT June 16, 1990. Color-coded power levels are shown as a function of frequency (vertical axis) and time (horizontal axis). Power levels in each panel are normalized to fit into the available color bar, but follow the logarithmic scale shown at the right.

a factor of 5 lower during the second interval. Comparison of these signals may be used to infer relative hardness of electron precipitation. We infer that during this second interval the pulsating precipitation was harder than in the first case. This is confirmed by the riometer data for this interval, which shows clear increases in absorption associated with pulsation bursts in the 427.8 nm signal, and by the prompt response in the 630.0 nm signal (indicating emission predominantly at lower altitudes). These observations all suggest that during this time South Pole was located under field lines deeper in the magnetosphere. (This is also consistent with the all-sky imager data shown below.) Further, a comparison of fluctuations in the ratio of 630.0 nm to 427.8 nm emission near 1302 and 1558 UT indicates a reversal in sense: in the first case we infer softer precipitation during pulses, and during the second case harder precipitation.

Figures 4 and 5 show all-sky camera images at 630.0 nm (left) and 427.8 nm (right) oriented such that the poleward direction is up for both images. Intensities are shown using a quasi-logarithmic color scale. The lower two levels are tinted red and blue to qualitatively indicate wavelength channel (630.0 and 427.8 nm, respectively). The white circle near the center of each image denotes the boundary of the field of view of the photometers. As a consistency check, we compared the integrated signal from the region inside this circle to the photometer data. The resulting 1-min resolution time series matched the temporal behavior detected by the photometers.

The all-sky images on this day showed that auroral activity at 630.0 nm was moving from zenith poleward throughout the period from 1100 to 1500 UT, but that there was little overhead intensity at 427.8 nm wavelength at any time, although emissions at this wavelength sometimes did show a fluctuation in overall brightness.

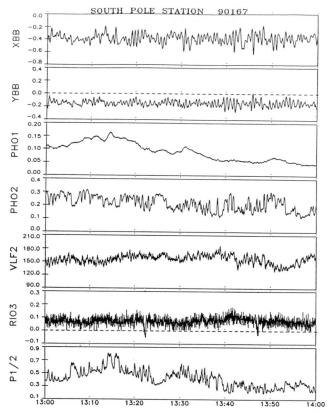

Fig. 2: Time series plot of data from South Pole, Antarctica, from 1300 to 1400 UT June 16, 1990. From top to bottom the panels display search coil magnetometer data (XBB and YBB, in nT/s), zenith-viewing photometer data (PHO1 = 630.0 nm, PHO2 = 427.8 nm, in kilorayleighs), ELF/VLF signals in the 1 kHz to 2 kHz band (in microvolts/m), riometer absorption (in dB) and the ratio of the auroral brightnesses at 630.0 and 427.8 nm.

Unfortunately the sampling rate of 1 min did not allow us to identify the spatial region or regions responsible for the pulsations. The lack of emission features at 427.8 nm associated with the oval observed at 630.0 nm suggests that the oval was produced by very soft electron precipitation.

Figure 4 shows the location of the oval at 1325 UT. The brightest part of the oval was 55 to 60° poleward from the zenith, but its equatorward edge (as defined by the 630.0 nm image) was in the photometers' field of view, and was the source of at least a large fraction of the 630.0 nm photometer signal. In contrast, the 427.8 nm image showed little or no spatial structure during this period other than a localized bright region at the poleward edge of the bright oval.

Figure 5 shows that by 1510 UT the auroral oval had thinned; although its brightest features were still 55 to 60° poleward from the zenith, the field of view of the photometers did not include any part of the oval. The 427.8 nm image was even weaker at this time, and was almost featureless. Consistent with Figure 3, the integrated signal level was low during the 1.8-s integration time used for this image, but fluctuated by approximately a factor of two during the preceding minutes.

DMSP satellite observations of precipitating particles for this day were consistent with the all-sky data (P. T. Newell, personal communi-

nication, 1993). During the four southern hemisphere dayside passes between 0950 and 1450 UT the auroral oval was poleward of 74° invariant latitude, and the level of precipitation at 74° invariant latitude was very low during the 1450 UT pass. Because of its sun-synchronous orbit, however, the DMSP satellites never fly overhead of South Pole, and do not reach high geomagnetic latitudes after ~1500 UT. Thus the DMSP data provide only indirect or qualified support for the boundary identifications made here.

The images presented here are typical of those seen throughout the dayside observation region during times of optical Pc 3-4 pulsations, in that both the background levels and the modulations are subvisual. In addition, the region of brightest (and mostly red) auroral emission was poleward of the observing station during all times both magnetic and optical pulsations were observed. Optical data at times indicated very soft overhead precipitation, characteristic of the cusp, near local magnetic noon. In the cases studied in detail so far, only magnetic pulsations were observed during these times. In no case studied (between 1000 and 2000 UT) did optical pulsations occur when the brightest auroral region was equatorward.

STATISTICAL STUDY

Our earlier event studies [Engebretson et al., 1990, 1991a] of Pc 3-4 pulsations in various data sets (magnetic field, auroral emissions, and ELF/VLF emissions) at South Pole had suggested that these signals all occurred roughly together, but with magnetic pulsations much more likely to occur than modulations in auroral light or ELF/

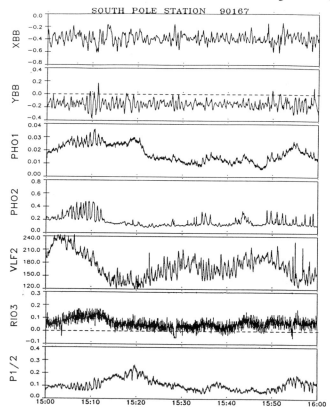

Fig. 3: Time series plot of data from South Pole, Antarctica, from 1500 to 1600 UT June 16, 1990, as in Figure 2.

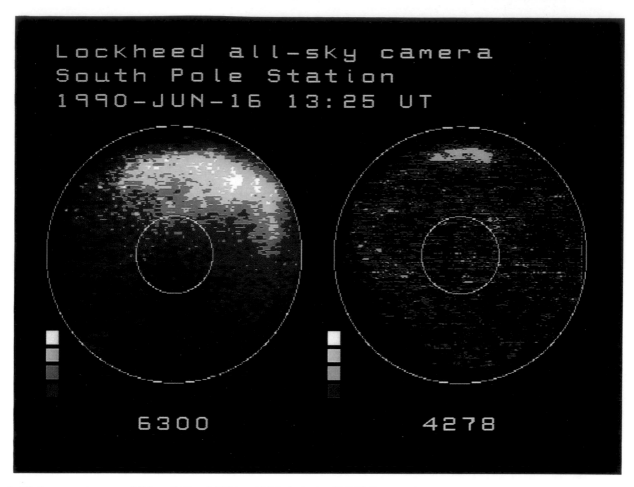

Fig. 4: All-sky camera images at 630.0 nm (left) and 427.8 nm (right) obtained at 1325 UT June 16, 1990, oriented such that the poleward direction is up for both images. The four color chips indicate roughly logarithmic emission rate levels of > 1000 Rayleighs (white), between 600 and 1000 R (yellow), between 300 and 600 R (orange/aqua), between 100 and 300 R (red/blue), and < 100 R (black). The lower two levels are tinted red and blue to qualitatively indicate wavelength channel (630.0 and 427.8 nm, respectively). The white circle near the center of each image denotes the boundary of the field of view of the photometers.

VLF emissions. In order to study these differences further, we used the spectrograms to compile a data base of pulsation occurrence as a function of time and magnetic activity. Pulsation events (in magnetometer or photometer signals) were visually identified from the spectrograms if in a given hour there was narrowband power at least 2 color levels (> 10 dB) above background (i.e., above power levels at frequencies above and below the narrowband feature). Note that this criterion gives a relative power threshold, not an absolute one. In practice, however, pulsations with amplitude below 0.2 nT peak to peak amplitude were too weak to satisfy our selection criterion.

Figure 6 shows the occurrence of Pc 3-4 pulsations, as a function of universal time (from 1000 to 2000) and magnetic local time (bins are centered on times from 0700 to 1600 MLT). Both magnetic and optical pulsations in this data set are clearly more frequent before local magnetic noon (1530 UT) than after. The distribution of optical pulsations is consistent with the earlier results of Wu and Rosenberg [1992], who used a computer-based identification of optical pulsation activity in 10-min intervals, without regard to spectral character, between period limits of 10 and 40 s. It is significant that the occurrences of optical pulsations in this study in every case formed a subset of the occurrences of magnetic pulsations; i.e., in no case during these 94 days did we observe a narrowband optical pulsation without simultaneously observing a narrowband magnetic pulsation (to within the few minutes resolution of the spectrograms). Given the wide spatial range of a ground-based magnetometer and the much narrower field of view of the photometers, these data are evidence for a very close connection between the two types of pulsations. It is consistent with our data that all or nearly all of the observed magnetic pulsations may be associated with modulated electron precipitation. Detailed calculations based on an assumed Maxwellian spectrum with characteristic energy consistent with the optical observations indicate that the observed precipitation will, under the influence of the 10 to 25 mV/m ionospheric electric fields typical near the dayside oval [Friis-Christensen et al., 1985], cause changes in ionospheric currents which will in turn generate magnetic pulsation signals of the magnitudes observed in this study (D. L. Detrick, T. J. Rosenberg, and Z. Wang, unpublished manuscript, 1993).

Figures 7 and 8 provide plots of occurrence rates of narrowband

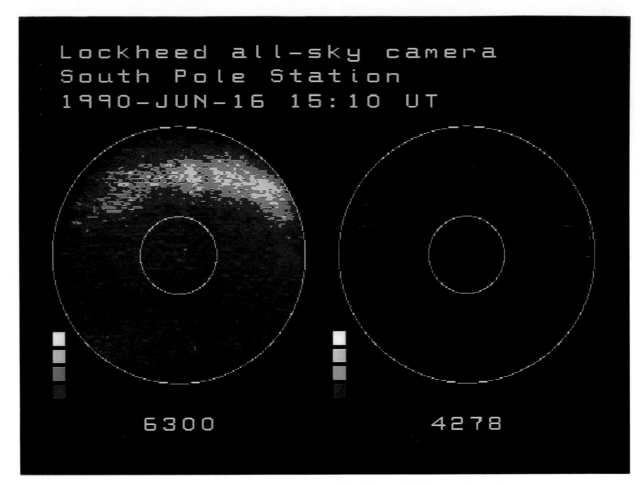

Fig. 5: All-sky camera images obtained at 1510 UT June 16, 1990, as in Figure 4.

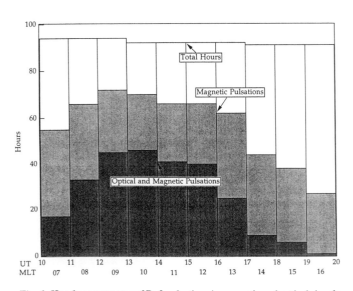

Fig. 6: Hourly occurrences of Pc 3 pulsations in magnetic and optical signals at South Pole Station, Antarctica, during a 94-day period May-July 1990. There were no cases in which optical Pc 3 pulsations were observed without magnetic pulsations.

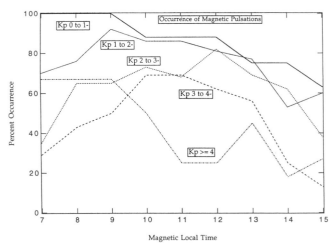

Fig. 7: Percentage occurrence rates of Pc 3 magnetic pulsations at South Pole Station as a function of local time, for several ranges of K_p. Values shown are the percentage of the total days available during the interval studied when magnetic pulsations were observed.

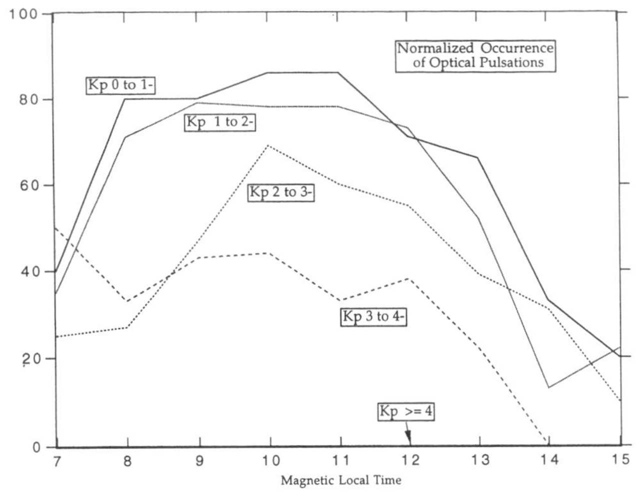

Fig. 8: Normalized occurrence rates of simultaneous Pc 3-4 magnetic and optical pulsations at South Pole Station, plotted as a function of local time for several ranges of K_p. Values shown are the percentage of the times optical pulsations are observed, relative to the times magnetic pulsations are observed.

magnetic and optical pulsations as a function of K_p. The magnetic data (Figure 7) show a maximum somewhat before local noon for all levels of activity, and a tendency toward higher occurrence levels when K_p is lower. Figure 8 shows that optical pulsations are much more common at pre-noon local times, and diminish greatly in occurrence with higher K_p values. The local time distributions are in agreement with the findings of Wu and Rosenberg [1992] except for the lowest K_p bin: the Wu and Rosenberg result showed an approximately symmetric distribution peaked at noon for $K_p < 1$, in contrast to the prenoon distributions at all higher K_p levels. More significantly, the restriction to narrowband modulations in our study apparently greatly reduced the occurrence of optical pulsations at high K_p levels relative to the Wu and Rosenberg results.

It is well known from many earlier studies that the auroral oval expands equatorward when K_p values increase, and that the ionospheric foot of the magnetospheric boundary (comprising the cusp and boundary layer regions) is approximately above or slightly poleward of South Pole Station for low values of K_p. The upstream wave source of Pc 3-4 pulsations is physically unrelated to geomagnetic disturbances of the sort parameterized by K_p, however, and on geometrical grounds should, and does, exhibit only a weak dependence on K_p: very quiet (or very disturbed) geomagnetic conditions correspond to a more strongly positive (or negative) IMF B_z component, respectively, which is inconsistent with the low IMF cone angle favorable for Pc 3-4 pulsations. The sharp drop in occurrence of Pc 3-4 optical pulsations as K_p increases thus provides independent evidence that an optical pulsation source region rather localized in latitude, probably the boundary layer and/or the outer edge of the magnetosphere, has moved equatorward away from the station. Our inference that the two types of optical pulsations observed represent precipitation in or near the boundary layer and deeper within the magnetosphere is also consistent with this statistically observed latitudinal variation: In each of the cases of harder optical precipitation at Pc 3-4 frequencies we have identified, K_p was at or below

1, consistent with a poleward retreat of the dayside oval.

Although it is tempting to suggest that the local time asymmetry in occurrence is related to upstream foreshock geometry (the dominance of the gardenhose IMF orientation), it is important to remember that Pc 3 pulsations appear to reach the magnetosphere only if they are convected earthward from the subsolar bow shock [Greenstadt, 1972, 1991]. Waves from the dawn side of the bow shock cannot cross the magnetosheath stream flow quickly enough to reach the dayside magnetopause. The larger probability of pulsation occurrence in the morning sector may, however, be related to the local time asymmetry of the auroral oval noted by McDiarmid et al. [1974]. McDiarmid et al. showed, based on ~1000 satellite crossings of the dayside auroral zone for $K_p \leq 3$, that the 35-keV electron trapping boundary exhibited a significant asymmetry in invariant latitude across local noon, with highest invariant latitude of the trapping boundary near 0900 local time, or three hours before local magnetic noon. An equatorward expansion of such an asymmetric oval with increasing K_p would be consistent with our observations, if the Pc 3-4 optical pulsations originated at or equatorward/earthward of the boundary layer. We are unaware, however, of any more recent studies confirming this asymmetry.

DISCUSSION and SUMMARY

We have found that narrowband modulations in auroral emissions at Pc 3-4 frequencies occur only when Pc 3-4 magnetic pulsations are present, and exhibit a strong preference, both statistically and on a case-by-case basis, for times when the auroral oval is poleward of South Pole. Under these conditions South Pole is under either the boundary layer, the cusp, or regions of closed field lines slightly deeper within the magnetosphere. The optical data indicate that the pulsations occur in a particle precipitation regime equatorward of the cusp and cleft (the regions of the softest, most intense dayside precipitation).

We have also found evidence that the pulsation signals are somewhat different depending on the proximity of the dayside auroral oval to the observing station. During times when part of the dayside auroral oval is slightly poleward and within the field of view of the photometers, as in the case shown in Figure 2, the optical signals appear to be simple modulations above and below a somewhat irregular baseline. At times when the oval is farther poleward, such as shown in Figure 3, we see bursts of harder precipitation above a more level background. It appears that the type of optical signal observed by the zenith-viewing photometers may be useful in determining what region of the magnetosphere is overhead.

In addition, we have noted that modulations of ELF/VLF signals also are linked to Pc 3-4 magnetic pulsations, although auroral and ELF/VLF emissions do not exhibit a clear temporal relationship to each other. In particular, we have at times observed Pc 3-4 optical signals when there is no evident ELF/VLF modulation. The Coroniti and Kennel [1970] nonlinear pitch angle modulation mechanism is widely accepted as a means by which compressional ULF waves can cause modulations of electron precipitation. Because this mechanism invokes the growth of ELF/VLF emissions as an intermediate step, however, one would expect in every case to find ELF/VLF modulations when optical Pc 3-4 pulsations are observed. That this is not always the case indicates that there may be an additional mechanism for modulated electron precipitation.

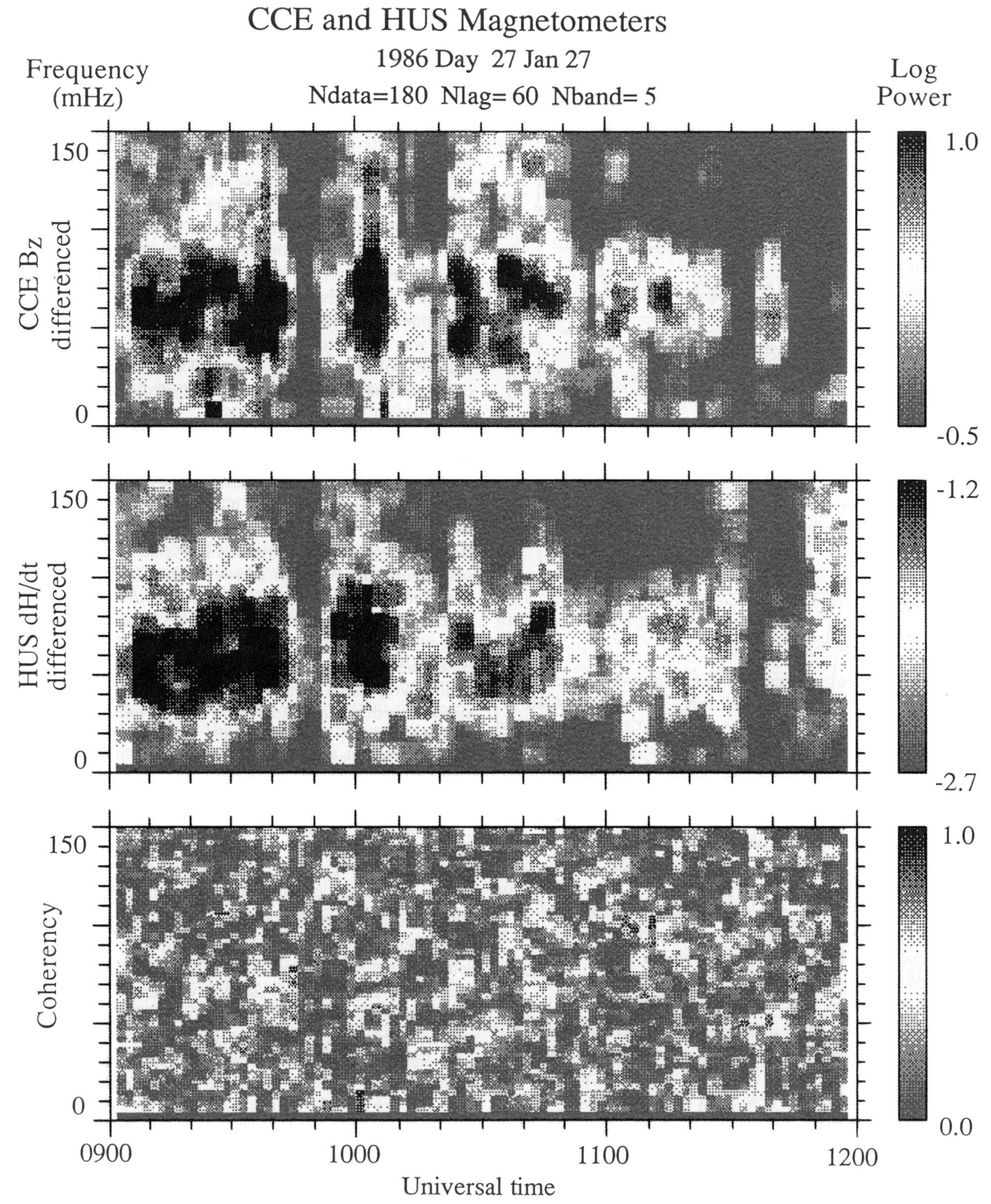

CCE and HUS Magnetometers
1986 Day 27 Jan 27
Ndata=180 Nlag= 60 Nband= 5

Fig. 6. Dynamic auto spectra and coherency for B_z at CCE and H at Husafell, Iceland, for a 3-hour interval including the Pc 3 event discussed in the text. The Husafell data were acquired with a search coil magnetometer. The data were time differenced prior to spectral estimate, in order to remove strong low frequency ($f < 20$ mHz) power.

line the fluctuation level of both the magnetic field and particle flux in the dayside magnetosheath increases [Engebretson et al., 1991b]. Provided that the total pressure of these fluctuations is nonzero, part of the fluctuation energy will be transmitted into the magnetosheath as fast mode waves. The magnetic field polarization of Pc 3 pulsations observed by CCE is consistent with propagation of fast mode waves.

We then need to consider how the waves might propagate in the magnetosphere and from the magnetosphere to the ground. For this discussion we use a simple rectangular coordinate system instead of a cylindrical system but adhere to the same notation for the coordinates: $\mathbf{e}_\rho$ for the latitudinal direction, $\mathbf{e}_\phi$ for the longitudinal direction, and $\mathbf{e}_z$ along the ambient magnetic field. In a simplest situation the fast mode waves transmitted from the subsolar magnetopause will have a wave vector nearly parallel to $\mathbf{e}_\rho$. In that case the magnetic field perturbations of the transmitted wave will be strongly compressional. In addition, from model magnetospheric plasma density distributions such as that of Moore et al. [1987], one expects that the phase velocity of the fast mode waves is at a minimum at the magnetic equator and that as a consequence the amplitude of the waves will be greatest at the equator [Takahashi and Anderson, 1992].

The fast mode waves, however, cannot propagate in a purely compressional mode throughout the magnetosphere. Due to inhomogenieties in the plasma density and ambient magnetic field, fast mode waves inevitably couple with Alfvén waves, and we need to consider the consequence of the coupling. The coupling is particularly strong at locations where the resonance condition $\omega = k_z V_A$ is satisfied between the fast mode waves and Alfvén waves [Chen and Hasegawa, 1974; Southwood, 1974], where ω and k_z are the frequency and the parallel wave number for the source wave, and V_A is the local Alfvén velocity. When the coupling occurs the magnetic field perturbation becomes strongly polarized in the azimuthal (ϕ) direction. Unfortunately, we cannot determine whether such coupling occurred at the location of CCE, because the resonantly excited Alfvén waves likely had a standing structure along the geomagnetic field line, and, if the standing wave was an odd mode, there was no transverse magnetic field perturbation at the equator. To proceed, then, let us assume that the coupling occurred and that the coupled waves propagated to the ionosphere.

We next discuss the effect of the ionosphere on MHD waves based on previous theoretical studies by Nishida [1964; 1978] and Hughes and Southwood [1976]. We quote the results from Nishida [1978], who used a simple system consisting of a uniform magnetosphere, an infinitely thin ionosphere, a uniform atmosphere, a perfectly conducting ionosphere, and an ambient magnetic field which is perpendicular to the ionosphere (see Figure 7). In accordance with the discussion given above, we assume that the wave field varies as $g(z)e^{i(k_\rho \rho - \omega t)}$. In this case the magnetic field perturbation in the magnetosphere consists of the latitudinal (ρ) and field aligned (z) components for the fast mode waves, while it

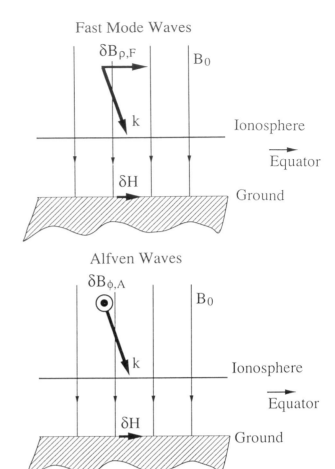

Fig. 7. Geometrical relationship among the ambient magnetic field, the wave vector, and magnetic field perturbation for Alfvén or fast mode waves propagating from the magnetosphere to the ground through the ionosphere.

consists of only the azimuthal (ϕ) component for the Alfvén waves.

One important fact noted by Nishida [1964] is that in the atmosphere there is no vertical current so the horizontal magnetic field perturbation there (and on the Earth's surface) has no ϕ component regardless of the mode of waves incident on the ionosphere. This fact can give a simple explanation to why the Pc 3 pulsations observed on the ground had a perturbation mainly in the H component.

The amplitude of pulsation signals detected on the ground depends on the ionospheric conductivities and on the temporal and spatial scales of the incident waves. According to Eq. (168) of Nishida [1978], the ground-level amplitude is

$$\delta H = \delta B_\rho(z=0) = \frac{2\left\{\delta B_{\rho,\mathrm{F}} - \left(\dfrac{\Sigma_\mathrm{H}}{\Sigma_\mathrm{P}}\right)\delta B_{\phi,\mathrm{A}}\right\}e^{-|k_\rho|d}}{\left\{1 - \dfrac{i}{2}\left(1 + \dfrac{\Sigma_\mathrm{H}^2}{\Sigma_\mathrm{P}^2}\right)\dfrac{\mu_0 \Sigma_\mathrm{P}\omega}{|k_\rho|}\left(1 - e^{-2|k_\rho|d}\right)\right\}} \quad (1)$$

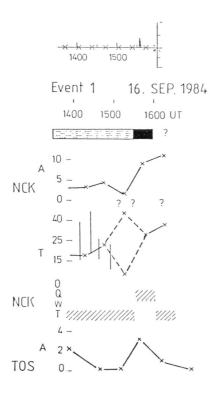

Event 1   16. SEP. 1984

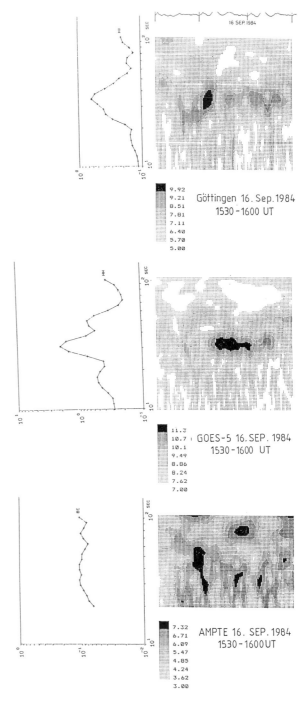

Fig. 4.a., Plot of parameters of the upstream waves and of surface pulsations. Event 1, Sept. 16, 1984, 1330–1600 UT. From top to bottom: presence (black bar) or absence (open bar) of upstream waves (gaps indicate lacking data); Pc 3 amplitude at the NCK observatory (hand-scaled); average period at NCK, full lines indicate reliable, dashed less reliable data, or the presence of two spectral peaks, thin vertical lines indicate the period range of upstream waves (if it could be determined with some reliability); NCK regularity, increasing upwards, for explanation of the letters, see text; averaged TOS amplitudes (30 min intervals). The note '?' indicates cases when upstream waves are present, but the period range is not well determined.

gle was favourable (20°) for the propagation of UP waves, as in most cases studied.

Slight UP wave activity was observed by AMPTE/IRM after 0630 UT, otherwise the solar wind was quiet.

NCK Pc 3 activity was low except for the interval 0630–0700 UT, when it increased from 2-4 to 15 units. The period of the then regular pulsations corresponded to the actual B-value. An impulsive pulsation event (IPE) at 0629 UT corresponded to an UP-switch-on at 0625 UT, another at 0649 UT perhaps to a small, sharp UP-event at 0645 UT.

At TOS, the increase of the Pc 3 activity was less spectacular, but the IPE-s at 0629 and 0649 UT can be seen there, too.

The array observed L-dependent periods (17 to 39 s) and the IPE-s can be seen everywhere. GOES 5 recorded an increase at 0629 UT only in the Pc 4-5 range.

Fig. 4.b., High-resolution dynamic spectra from Göttingen (H north component) and from the GOES 5 (H normal component which points eastward) and AMPTE/CCE (H east component) satellites, together with the GTT analog record (top). Left from the dynamic spectra, corresponding amplitude spectra are plotted. Symbols for log-amplitude are given for each spectrum. GOES 5 was located at 74.5° west longitude in geosynchronous orbit (6.67 Earth radii) in the Earth's equatorial plane. AMPTE/CCE travelled from 13.2 to 13.4 magnetic local time in the equatorial plane at L=9.1.

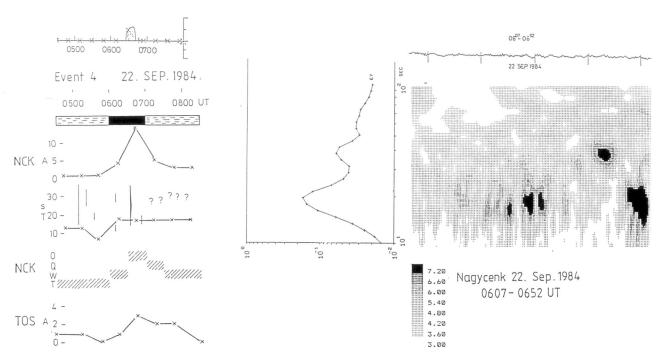

Fig. 5. The same as Fig. 4., but for Event 4, Sept. 22., 1984, 0430–0830 UT (Dynamic spectrum from NCK).

Event 9, October 6, 1984, 0830-1400 UT (Fig. 6)

Geomagnetically quiet (Kp 2), V = 410 km/s, B = 16 nT, cone angle 35°: a quiet interval with high interplanetary magnetic field. During this interval, UP conditions altered several times with nUP or with intervals when the situation was not clear (no satellite data).

Clear UP-intervals coincided with maxima of the NCK Pc 3 activity in each case. During UP, amplitudes were 20 to 25 units, in contrast to 10 units otherwise. These pulsations were very regular (O), or less regular toward the end of the event. Periods were longer than expected from |B|, namely 20 s instead of an expected 10 s.

TOS amplitudes were similarly affected by changes of UP (amplitudes 3 to 4 units) and nUP (1 to 2 units) situations. There were several onsets of the UP-activity. The first of them at 0825 UT was followed by an IPE at 0828 UT, further ones at 1033 and 1126 UT, by less clear IPE's at 1037 and 1129 UT, and finally one at 1223 UT, and finally by a clear IPE again at 1227 UT. The switch-offs of the UP-activity at 1158 and 1255 UT were accompanied by Pc 4–5 impulses (in the latter case at 1301).

The array observed prominent L-dependent periods (13 to 36 s) during a significant part of this event (0900-1400 UT). Some L-independent period activity (16 s) was also observed. The SOD dynamic spectrum chosen as an example here reflects the UP-nUP changes (e.g. UP 1035 to about 1050 UT).

5. RESULTS AND CONCLUSIONS

A direct comparison of the upstream wave source using data from the AMPTE/IRM satellite and ground-based pulsation data from mid-latitude and polar observatories led to the following main results.

1. If upstream waves are present in the spectra of the foreshock interplanetary magnetic field, the amplitudes of the Pc 3 pulsations increase at all stations studied by a factor of about 2 to 3, the higher value is more likely.

2. The amplification occurs at the mid-latitude station NCK (L~2) in the Pc 3 (and in the Pc 5) range, at the auroral zone station TOS in the Pc 3 and Pc 5 ranges, and at the polar cap station NYA in the Pc 3–Pc 4 ranges, thus toward higher latitude the period range of the amplification extends to longer and longer periods, roughly corresponding to the eigenperiods of the field lines. The amplification is less clear at higher latitude stations.

3. Together with the larger amplitudes, the 'regularity' of the pulsations increases, too, i.e. spectral peaks get narrower if upstream waves are present. Previous results have shown that more regular pulsations are connected to field line resonances, thus it seems that upstream waves excite more effectively field line resonances.

4. In spite of the presumed connection between upstream waves and field line resonances, no direct proof was found for a better correlation between B-values and observed periods in the absence of upstream waves when the correlation B-T

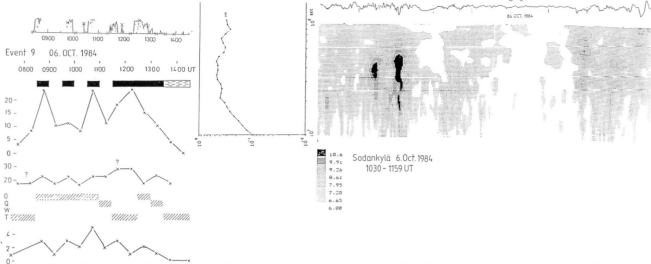

Fig. 6. The same as Fig. 4., but for Event 9, Oct. 06, 1984, 0730–1430 UT (Dynamic spectrum from Sodankylä).

is less disturbed by field line resonances. The present data did not confirm such a suggestion, even if the correlation B-T was slightly better for nUP-conditions.

A sample of 9 selected events were studied in details. These events were selected to include cases with continuous upstream wave activity, without upstream waves, and — in the most cases — with alternating UP and nUP conditions. The studies led to the following picture of the development of pulsations during an UP-event:

1. At the mid-latitude observatory NCK, pulsation amplitudes (and regularity) responded quickly (delay around 4 min) to changes in the UP-nUP-conditions, increasing at nUP-UP transitions, and decreasing at UP-nUP transitions. The changes at the auroral and polar stations were less convincing. Indications of UP onsets at higher latitude stations are unambiguous, but of more complicated forms (e.g. double peaks).

2. A study of all events studied from the array has shown that periods of the pulsations corresponded in many cases to those expected from the relation 160/|B|, and in a few cases even the change of the period with changing B could be observed. Nevertheless, there were events, too, when the periods did not correspond to this relation. Such events are discussed in the present paper, too. The connection between B and T for these events is discussed in details by Verő et al. [1993].

3. Satellites in the magnetosphere, GOES 5 and AMPTE/CCE indicated less clearly UP onsets, in some cases there was no indication at the satellites. Impulses were variable in frequency, too.

4. Data of the Central- and Northern-European array were available for the majority of the events. No connection was found between the occurrence or absence of the upstream waves and the pulsations with L-dependent (magnetospheric) or L-independent (solar wind) periods. Two of the events (Events 1 and 9) occurred during UP-conditions with L-independent periods, two including Event 4 during

UP with L-dependent periods, while one, which occurred during nUP-conditions, contained both L-dependent and L-independent periods, and one during nUP, included L-dependent periods, three events had mixed character.

5. The end of UP-intervals coincide in about half of the cases with longer (50-200 s) period impulses, both at surface stations and in the magnetosphere. (Long period activity is quite common around pulsation switch-offs [Verő and Tátrallyay, 1973]).

We concluded that upstream wave activity in the solar wind fore-shock region is a prerequisite for high amplitude pulsations at any latitude. Such events are supposed to be generally more regular (narrower peak in the spectrum), nevertheless, a direct connection between upstream waves and field line resonances could not be confirmed. However, at higher latitudes (auroral zone and polar cap) the range of pulsation periods which are amplified during upstream wave activity, extends towards longer and longer periods.

The low correlation between IMF B-values and pulsation periods is surprising in this sample, since L-dependent periods were found relatively seldom in the sample of measurements at the array. This lack of correlation is supposed to be a random effect, since more extended statistical studies show the expected correspondance between B and T.

Acknowledgements. The investigations were supported by a Hungarian state grant, OTKA No. 1171. The authors are indebted to the collegues who helped them to obtain pulsation records from observatories and satellites. A grant for this purpose from the Soros foundation is acknowledged, too.

References

Asbridge, J.R., S.J. Bame and I.B. Strong, Outward flow of protons from the earth's bow shock, *J. Geophys. Res.*, 73, 5777-5782, 1968.

Couzens, D.A., Interplanetary Medium Data Book Suppl. 3A (1977-1985), NSSDC-World Data Center A 86-049, Greenbelt,

1986.

Etemadi, A., S.W.H. Cowley, M. Lockwood, B.J.I. Bromage, D.M. Willis and H. Lühr, The dependence of high-latitude dayside ionospheric flows on the north-south component of the IMF: A high time resolution correlation analysis using EISCAT, Polar and AMPTE UKS and IRM data, *Planet. Space Sci.*, 36, 471, 1988.

Fairfield, D.H., Bow shock associated waves observed in the far upstream interplanetary medium, *J. Geophys. Res.*, 74, 3541-3553, 1969.

Gosling, J.T., J.R. Asbridge, S.J. Bame, G. Paschmann and N. Sckopke, Observations of two distinct populations of bow shock ions in the upstream solar wind, *Geophys. Res. Lett.*, 5, 957-960, 1978.

Greenstadt, E.W., I.M. Green, G.T. Inouye, A.J. Hundhausen, S.J. Bame and I.B. Strong, Correlated magnetic field and plasma observations of the earth's bow shock, *J. Geophys. Res.*, 73, 51-60, 1968.

Gul'elmi, A.V., Diagnostics of the magnetosphere by means of pulsations, *Space Sci. Rev.*, 16, 331-345, 1974.

Mann, G. and H. Lühr, Comparison of low frequency magnetic field fluctuations upstream of earth's bow shock with a strong Alfvénic turbulence model, *Ann. Geophysicae*, 9, 681-689, 1991.

Miletits, J. Cz., J. Verő, J. Szendrői, P. Ivanova, A. Best and M. Kivinen, Pulsation periods at mid-latitudes — a seven station study, *Planet. Space Sci.*, 38, 85-97, 1990.

Paschmann, G., N. Sckopke, S.J. Bame, J.R. Asbridge, J.T. Gosling, C.T. Russel and E.W. Greenstadt, Association of low frequency waves with suprathermal ions in the upstream solar wind, *Geophys. Res. Lett.*, 6, 209-212, 1979.

Russel, C.T. and N.M. Hoppe, The dependence of upstream wave periods on the interplanetary magnetic field, *Geophys. Res. Lett.*, 8, 615-617, 1981.

Scarf, F.L., R.W. Fredericks, L.A. Frank, C.T. Russel, P.J. Coleman, Jr. and M. Neugebauer, Direct correlations of large amplitude waves with suprathermal protons in the upstream solar wind, *J. Geophys. Res.*, 75, 7316-7322, 1970.

Veldkamp, J., A giant geomagnetic pulsation, *J. Atm. Terr. Phys.*, 17, 320-324, 1960.

Verő, J., Experimental aspects of low-latitude pulsations — A review, *J. Geophys.*, 60, 106-120, 1986.

Verő, J. and M. Tátrallyay, Changes of geomagnetic micropulsations following sudden impulses. *J. Atm. Terr. Phys.*, 35, 1507-1515, 1973.

Verő, J., L. Holló, A. Egeland and A. Brekke, Connections between high- and middle-latitude pulsations. *J. Atm. Terr. Phys.*, 52, 789-796, 1990.

Verő, J. and J.Cz. Miletits, Impulsive pulsation events and pulsation beats. *J. Atm. Terr. Phys.*, 1993 (in press).

J. Verő and B. Zieger, Geodetic and Geophysical Research Institute of the Hungarian Academy of Sciences, POB 5, H–9401 Sopron, Hungary.

H. Lühr, Institut für Geophysik und Meteorologie der Technischen Universität Carolo-Wilhelmina, Mendelssohnstrasse 3, D–3300 Braunschweig, Germany

Hydromagnetic Frequency Spectra in the
High Latitude Quiet Magnetosphere

A. WOLFE, R. D. KELMAN AND S. E. WARREN

NYCTC of CUNY, Brooklyn, NY

C. G. MACLENNAN AND L. J. LANZEROTTI

AT&T Bell Laboratories, Murray Hill, NJ

Hydromagnetic frequency spectral levels are characterized under quiet magnetosphere conditions at high (~74°) geomagnetic latitudes. Magnetometer data recorded at the conjugate stations South Pole, Antarctica, and Iqaluit, Canada, (L~13) are analyzed for the eight quietest days of 1989. For each of the days the daily sum planetary index $\Sigma Kp < 6$. Hourly power spectral analyses are performed on all ground magnetic field components to determine total power levels and spectral slopes over both a wide period band (20-2500s) and a narrow band (20-100s). Spectral slopes less than -3 are found to occur. Such slopes are steeper than those reported from earlier studies of the low latitude quiet magnetosphere. The diurnal and conjugate nature of these results is presented. In addition, variations in hourly solar wind plasma parameters and interplanetary magnetic field values as recorded on the IMP-8 spacecraft are studied for their associations with variations in the ground-based observations.

INTRODUCTION

The goal of this work is to characterize hydromagnetic activity at high latitudes during geomagnetically quiet times. Geomagnetic quiet conditions have previously been studied two decades ago by Lanzerotti and Robbins [1973] using ground data from low to mid latitude stations. More recently Lanzerotti et al. [1990] examined the natural electromagnetic environment spanning ten decades of frequency for a month during austral winter conditions at McMurdo Station, Antarctica, in the southern polar cap region. In addition, magnetometer data sets that were not restricted to geomagnetic conditions or that were reported for specialized reasons were studied more frequently this past decade from low and high latitude stations without preselection of quiet times [DeLauretis et al., 1991; Olson, 1986; Wolfe and Meloni, 1981; Wolfe et al. 1990] and see review papers by Arnoldy et al., [1988] and Fraser-Smith, [1982]. However, the characteristics of the quiet magnetosphere at high latitudes have not been previously reported in the literature as a separate ULF study. Our study addresses this omission. We examine in this paper high resolution magnetometer data (H, D, and V components; defined

conventionally as the south-north, west-east, and vertical directions, respectively) recorded at the near-conjugate stations South Pole (SP), Antarctica, and Iqaluit (IQ), Canada, both near the dayside cusp/cleft region (L~13).

DATA AND METHODOLOGY

Hourly power spectral analyses are computed using data from all three components of the vector magnetic field data, which are sampled at 10s intervals at SP (after smoothing and decimation of the original 1s data samples) and 5s intervals at IQ. Spectral slopes and integrated powers (magnetic energy densities) are then calculated in both broad (20-2500s) and narrow (20-100s) period bands. The diurnal and conjugate nature of these results is reported as well as their relationship to solar wind plasma and interplanetary magnetic field conditions. Hourly averages of the solar wind conditions are obtained from measurements on the IMP 8 spacecraft.

Geomagnetic quiet days were selected for the year 1989 (a year near the solar sunspot maximum) using the condition that the daily sum planetary index ΣKp was less than six. This restrictive condition was satisfied on only eight days of the year. Table 1 lists the date and ΣKp index for each of these eight quietest days used in this study. This initial study could later be augmented since conjugate station data are available from mid-July 1985. We choose to begin the analysis with our most recent data as a first sample.

Solar Wind Sources of Magnetospheric Ultra-Low-Frequency Waves
Geophysical Monograph 81

Table 1. Quiet Days 1989		
Day of Year	Date	ΣK_p
183	July 3	5^-
184	July 4	5^-
196	July 16	3
216	Aug. 5	4
285	Oct. 13	2
286	Oct. 14	1^+
328	Nov. 25	3^+
343	Dec. 10	5^-

Magnetometer data time series for August 5 are shown in Fig. 1 for both stations as an illustration of the data during a defined quiet day. Most of the day is seen to be extremely quiet except for large variations during the first couple of hours UT prior to local midnight. Magnetic local time LT~ UT −4h at IQ and SP.

RESULTS

Hourly power spectral analyses are then calculated for each of the eight days of data. An example of these spectra (hour 5 UT on Aug. 5) is shown in Fig. 2a. These spectra are selected to show the steepest power law slopes measured from IQ data on this day (S_1 is defined as the spectral slope in the 20-2500s period band and S_2 is the slope in the 20-100s band). A spectral power-law slope equal to −5.4 is found in the narrow band for the H-component. The corresponding South Pole spectra also show steep slopes, in contrast to earlier mid-latitude quiet-day studies which revealed flatter slopes, between −1 and −2 [Lanzerotti and Robbins, 1973]. Flatter spectra were also found at high latitudes, as evidenced by the spectra in Fig. 2b. This figure, for 14 UT on Aug. 5, shows the flattest slopes found on this day at IQ.

It is instructive to examine the nature of the difference between the steepest and flattest spectral slopes observed at IQ as shown in Figs. 2a and 2b. The log power spectral densities are near a value of 2 nT^2/Hz at a frequency of 10^{-2} Hz for IQ-H and D at both 5 and 14 UT (near local midnight and local noon, respectively). However, the power "fall-offs" at higher frequencies are very different during these two UT hours. The power levels decrease by less than one decade at 14 UT (from ~2 to ~1 near local noon), whereas they decrease by about three decades near local midnight at 5 UT (from ~2 to ~−1). These differences are due to significant enhancements in the power in the Pc3-4 range (period $\tau \sim 20-100$ sec) measured during the 14 UT hour.

Spectral slopes for each hour on all eight days are calculated in the two period bands at both stations. Figure 3 shows the diurnal variation of the maximum and minimum spectral slopes measured for the H-component at each station for all eight days. The top

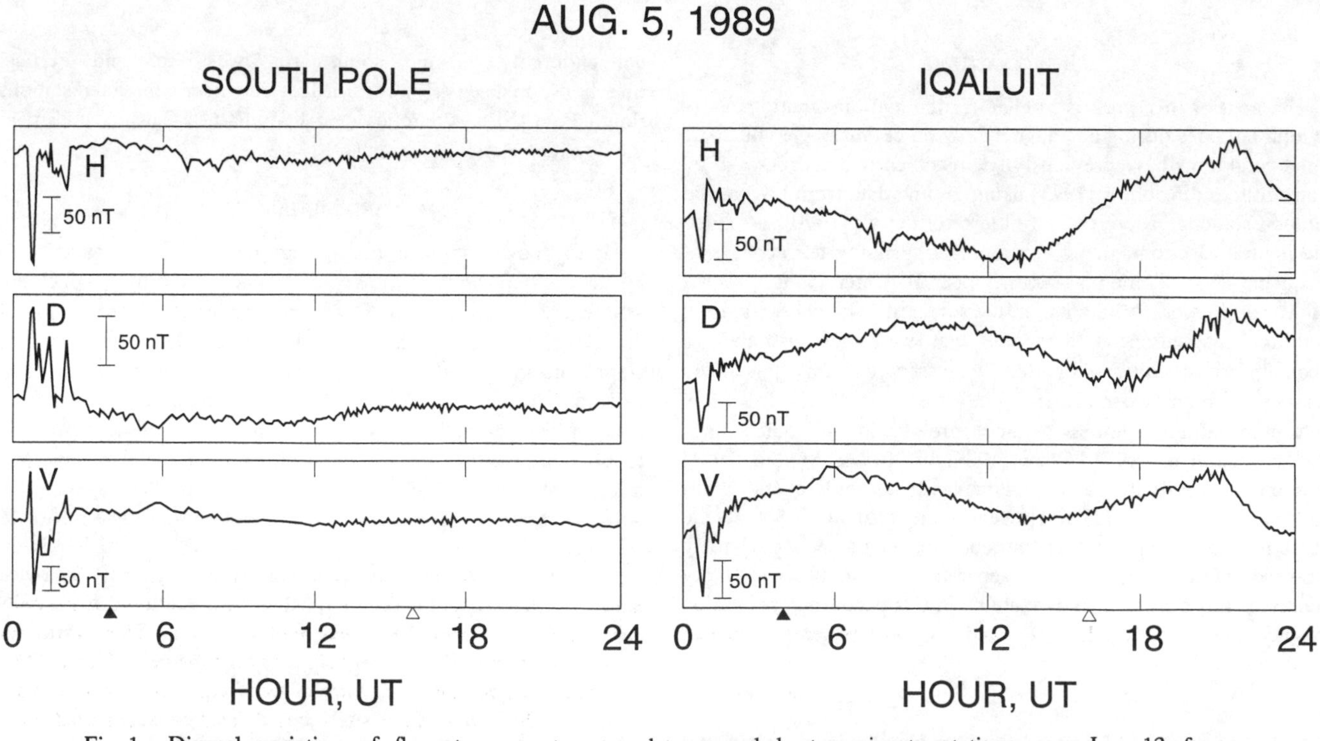

AUG. 5, 1989

SOUTH POLE

IQALUIT

Fig. 1. Diurnal variation of fluxgate magnetometer data recorded at conjugate stations near L = 13 for a geomagnetically quiet day. Local midnight (4 UT) and local noon (16 UT) are indicated by the solid and open triangles respectively.

HOUR 5 UT, 5 AUG. 1989

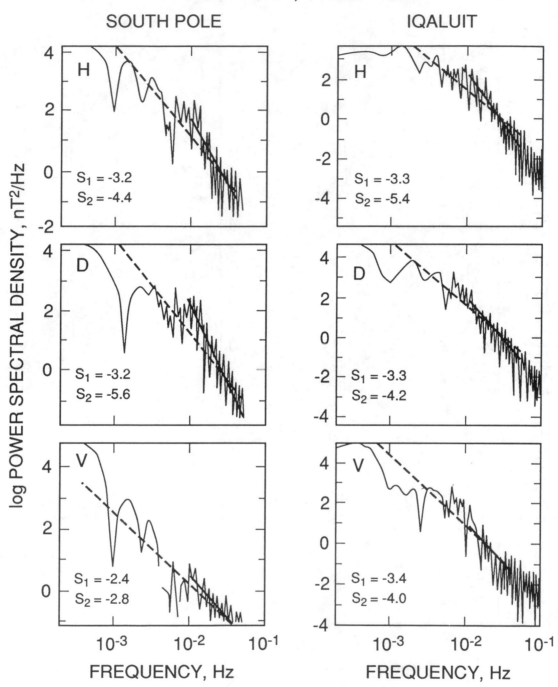

Fig. 2a. Power spectra selected to show the steepest slopes at Iqaluit. The slope S_1 is calculated over a wide period band (20-2500s) and the slope S_2 is calculated over a narrower period band (20-100s).

panel shows that the broad band maximum slopes generally vary between values of -1 and -2 while the broad band minimum slopes generally vary between values of -3 and -4 at both SP and IQ. The bottom panel shows that the narrow band maximum slopes lie generally between 0 and -2 while the narrow band minimum slopes vary considerably between -2 and -6 at both stations. One standard deviation in the slope values is calculated to be $\sigma \sim 0.1$ (broad band) and $\sigma \sim 0.2$ (narrow band).

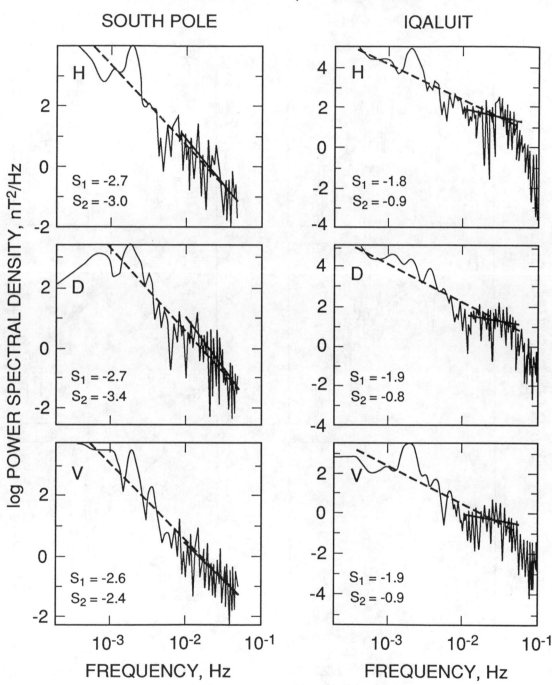

Fig. 2b. Power spectra selected to show the flattest slopes at Iqaluit.

Therefore, slopes less than −3 at high latitudes are commonly observed for quiet day spectra.

Magnetic energy densities (integrated power spectra) for each hour on all eight days are also calculated in the two period bands at both stations. The diurnal variation of the maximum and minimum log powers observed in each hour for the eight days are shown for the H-component in Fig. 4. Large variations in the integrated power are seen at both stations, except for the minimum power in the narrow period band shown in the bottom panel.

In order to explore this last finding further, we plot and compare

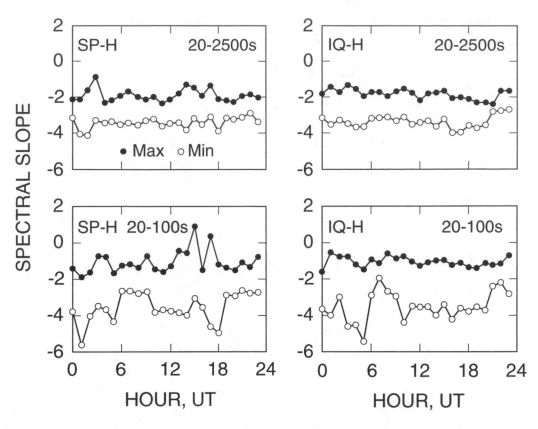

Fig. 3. Diurnal variation of maxima and minima spectral slopes over both period bands, during the eight geomagnetically quietest days of 1989.

the narrow band power minima in all three components at both stations (Fig. 5). Variations in the log power minima are seen to be generally less than 15% in all components at both stations. This observation then characterizes high latitude quiet day spectral "background" power levels. Furthermore, significantly higher minimum power levels are observed in the H and D components, with few exceptions, at IQ than at SP over these eight days.

The "conjugate" nature of the integrated power levels in the two bands at the IQ-SP station pair are shown in Fig. 6. Hourly integrated power values of the H-component for the eight quiet days are shown for the broad band (upper panel) and narrow band (lower panel). A significant relationship is seen between the IQ and SP powers in the narrower band (correlation coefficient R=0.64; significant beyond the 99.9% confidence level for the number of points), but a weaker relationship is found in the broad band.

The power and spectral slope data were examined as a function of local time for each of the eight individual quiet days. Plotted in Figure 7 are the integrated (20-100s band) power spectra for the H-component at each station on each day. On day 183 (Figure 7a) the power levels are quite similar in conjugate regions for all hours, indicating the night time magnetic disturbances, except for the local day hours, from ~1000-1900 UT, when they are larger at the northern station. During the following day, there are enhancements of integrated power levels during local day in both

hemispheres, with essentially equal power levels measured. There is a tendency for the integrated power levels to be higher at IQ than at SP for all of the other quiet days, except day 343 (Figure 7h). On this day, the local day time integrated power levels are enhanced more at SP than at IQ.

The diurnal variations of the power law slopes for the H-component for each of the days are shown in Figure 8. In general, there is a tendency for the times of greater power at a given station (corresponding panel of Figure 7) to be accompanied by steeper slopes. This is quite clear in the case of the local day power during day 183, for example (Figures 7a and 8a). However, the local day power law slopes on day 343 from 12-18 UT (Figures 7h and 8h) do not follow this trend.

Solar wind plasma and interplanetary magnetic field conditions were both recorded only on one day (August 5) of the eight days studied. Changes in the solar wind dynamic pressure are found to be highly associated with changes in the log powers (Fig. 9a) and spectral slopes (Fig. 9b) as observed on the ground. The calculated quantity NV^2 represents hourly averages of the dynamic pressure, where N is the solar wind density (cm^{-3}) and V is the solar wind speed ($km\ s^{-1}$).

A power increase of about one decade in the H-component on the ground can be associated with an approximately twenty percent increase of the solar wind pressure (Fig. 9a). This relationship is found to be significant for both period bands at both stations.

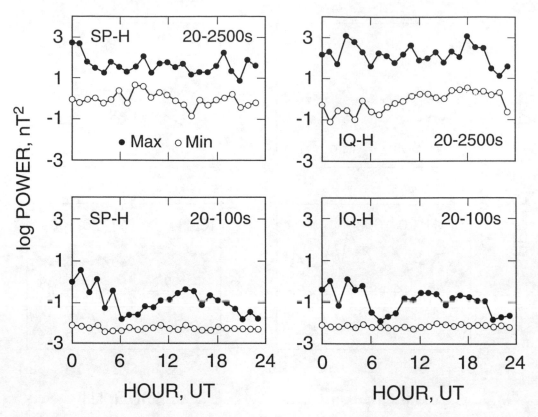

Fig. 4. The top panel shows the diurnal variation of maxima and minima log powers of the H-component (wide band) at the high latitude stations. The bottom panel shows the corresponding variation within the Pc3-4 period range.

Steeper spectral slopes in the vertical component at SP are also found to be associated with dynamic pressure increases (Fig. 9b). Such a relationship was not seen at IQ. Significant correlations between the ground data and the IMF data were not found for this day.

SUMMARY AND DISCUSSION

Power spectral slopes and magnetic energy densities in two period bands are calculated for hydromagnetic waves in the high latitude quiet magnetosphere. These spectra and spectral features are further examined for their diurnal variation, conjugate nature and relationship to solar wind parameters. Several aspects of these results are found to characterize quiet times. We observe that spectral slopes with power law slopes steeper than −3 are commonly found to occur. Such values are much less than those reported for quiet time, low to mid latitude power spectra [Lanzerotti and Robbins, 1973] and for interplanetary magnetic field observations [Burlaga et al., 1989; Montgomery et al., 1987; Siscoe et al., 1968; and Tu et al.; 1984] where slopes between −1.5 and −2.0 were found. Recently, DeLauretis et al. [1991] reported slopes between −3 and −4 at low and high latitude stations for average spectra calculated without preselection of geomagnetic quiet conditions. Therefore, our present work finds new results which extend the previous hydromagnetic wave studies to the high latitude quiet magnetosphere.

Hydromagnetic power "background" levels are found from the high latitude spectra. The power minima for each UT hour of the eight quiet days is shown in Fig. 5 to have little variation in the Pc3-4 frequency range (20-100s period). Similarities in conjugate regions are also noted by the high correlation in the Pc3-4 hourly powers shown for the H-component in Fig. 6.

In contrast, considerable diurnal variation of magnetic power levels occurs on each day. Lower power values are found to occur at South Pole than at Iqaluit. The diurnal variation of the integrated power levels (20-100s band) and the changes in the hemisphere with dominant power during these quiet days are probably caused by the sunlit ionosphere conditions (compare days 183 and 343, Figures 7a,h). The IQ ionosphere undergoes only a small diurnal sunlight cycle in the northern summer, while the SP ionosphere is sunlit throughout the entire day 343 (although the sun is low on the horizon). On most of the other days, the ionosphere at SP has some sunlight at some local times, but the sun always has a very large zenith angle.

The situation on day 184, as compared to day 183, is most interesting in terms of the daytime integrated power conditions. There is obviously much more conjugacy on the former day than on the latter, and this bears further study. A ready speculation is that even though day 183 was geomagnetically quiet, the stations were not at the ends of a conjugate flux tube. After a day of prolonged quiet, however, they were more conjugate and observed

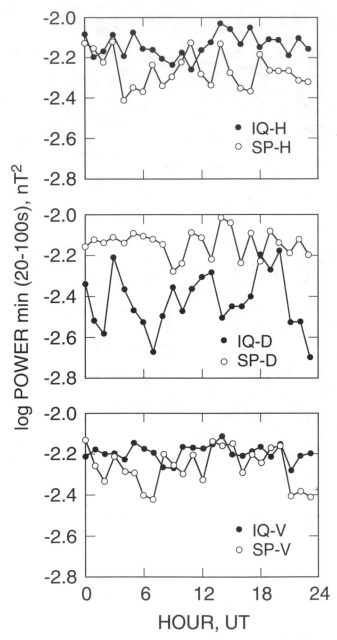

Fig. 5. Diurnal variation of H, D and V components of the minima log power (20-100 sec) at South Pole and Iqaluit. Note that variations are generally less than 15%.

similar levels of solar wind-produced hydromagnetic activity, even though the SP ionosphere was in darkness. Unfortunately, solar wind data are not available to examine on these days. However, lower latitude magnetometer data, for example from the Bell Laboratories station at Tuckerton, NJ, in the same local time sector, might provide information on the solar-wind produced wave activity on these days and therefore information on the difference between the conjugate points.

It is also interesting to note that on the quiet days with enhanced

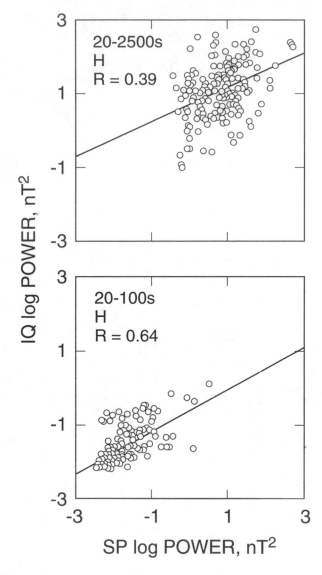

Fig. 6. Relationship between South Pole and Iqaluit for H-component log powers (20-2500 sec – top panel and 20-100 sec – bottom panel). A significant correlation is obtained for the narrow period band.

local day integrated power levels that there are decreases in these levels near local noon (e.g., days 183, 184, 216, 343; Figures 7a,b,d,h). This same decrease was reported and commented upon in the early study by Lanzerotti and Robbins [1973] and is undoubtedly related to the flow of the solar wind across the magnetopause. For example, the nose of the magnetopause is a stagnation point for the solar wind and could not be a location for the onset of the Kelvin-Helmholz instability.

Solar wind dynamic pressure values are found to be associated with integrated power levels and spectral slope values at both stations for the day when interplanetary data are available. The integrated power levels increase and the spectral slopes steepen

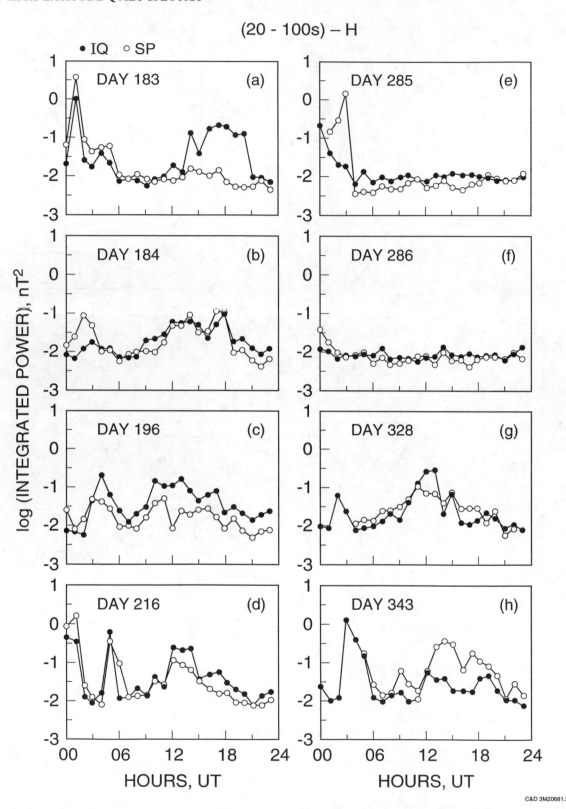

Fig. 7. Local time dependence of the integrated power levels (20-100s band) for each of the eight quiet days at both stations.

(20 - 100s) – H

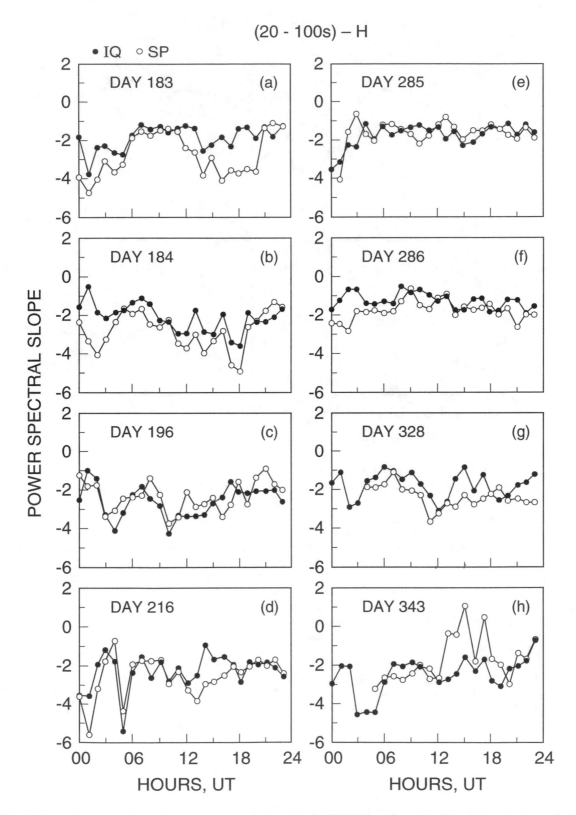

Fig. 8. Local time dependence of the power spectral slopes for the 20-100 band for each of the eight quiet days at both stations.

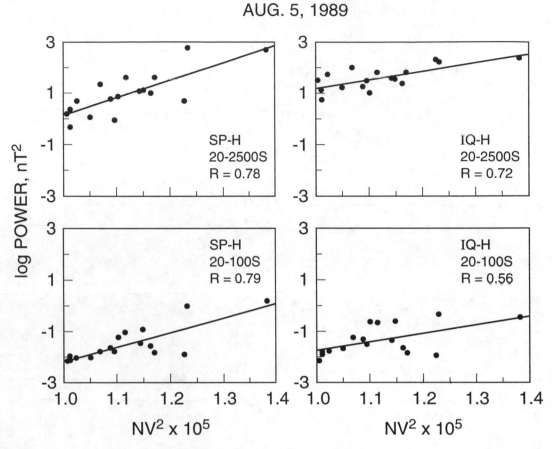

Fig. 9a. Scatter plots of log power in both period bands for the H-component at both stations and hourly averages of solar wind dynamic pressure. High correlations are found.

with increasing solar wind pressure on an hourly time scale (Figs. 9a,b). Upstream dynamic pressure variations have previously been reported to be a driving source for magnetospheric oscillations with periods in the range from 200 to 600s [Sibeck et al., 1989]. Changes in pressure cause motion of the magnetopause which results in compressional oscillations (~500s) in the magnetosphere. Sudden changes in NV^2 have been known for a long time to trigger long period Pc5 wave activity [Wilson and Sugiura, 1961; Kaufmann and Walker, 1974]. More recently Junginger and Baumjohann [1988] from electric field observations on GEOS 2 and Southwood and Kivelson [1990] reported on long period oscillations in association with solar wind dynamic pressure changes although Lin et al. [1991] reported that such pressure changes have little effect on Pc3-4 in the outer magnetosphere. Our present results not only support the above works but they extend the idea of solar wind dynamic pressure variations as a source of broad-band Pc5 type variations to near cusp/cleft latitudes during geomagnetic quiet times. However, ground magnetometer-IMF data relationships do not reveal significantly high correlations. Shorter time scales should be examined in future studies to reveal possible IMF control of hydromagnetic background levels observed on the surface of the earth.

Acknowledgments. The work at NYCTC by A.W. has been supported in part by NSF grants DPP 89-21094 and the GEM program ATM-91-12522. R.D. Kelman and S.E. Warren were supported by a "Research Experiences for Undergraduates" supplement to the former award. We are also grateful to the staff of the NSSDC/WDC for providing IMP8 hourly averages of solar wind and IMF parameters.

REFERENCES

Arnoldy, R. L., L. J. Cahill Jr., M. J. Engebretson, L. J. Lanzerotti and A. Wolfe, Review of hydromagnetic wave studies in the Antarctic, *Rev. Geophys.* 26, 181-207, 1988.
Burlaga, L. F., W. H. Mish and D. A. Roberts, Large-scale fluctuations in the solar wind at 1 AU: 1978-1982, *J. Geophys. Res.* 94, 177-184, 1989.
DeLauretis, M., U. Villante, M. Vellante and A. Wolfe, An analysis of power spectral indices in the micropulsation frequency range at different ground stations, *Planet Space Sci., 39*, 975-982, 1991.
Fraser-Smith, A. C., ULF/Lower-ELF electromagnetic field measurements in the polar caps, *Rev. Geophys. Space Phys.* 20, 497-512, 1982.
Junginger, H. and W. Baumjohann, Dayside long-period magnetospheric pulsations: Solar wind dependence, *J. Geophys. Res., 93*, 877-883, 1988.
Kaufmann, R. L. and D. N. Walker, Hydromagnetic waves excited during an SSC, *J. Geophys. Res., 79*, 5187-5195, 1974.
Lanzerotti, L. J., C. G. Maclennan and A. C. Fraser-Smith, Background

AUG. 5, 1989

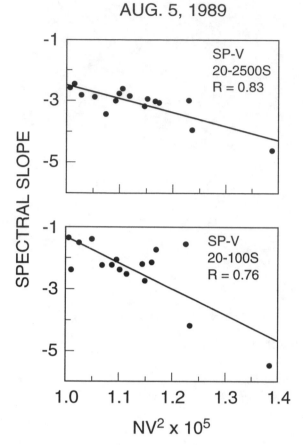

Fig. 9b. Scatter plots of spectral slope for the vertical component at SP and solar wind pressure.

magnetic spectra: ~10^{-5} to ~10^5 Hz, *Geophys. Res. Lett. 17*, 1593-1596, 1990.

Lanzerotti, L. J. and M. F. Robbins, ULF geomagnetic power near L = 4, 1, quiet day power spectra at conjugate points during December solstice, *J. Geophys. Res., 78*, 19, 3816-3827, 1973.

Lin, N., M. J. Engebretson, R. L. McPherron, M. G. Kivelson, W. Baumjohann, H. Luehr, T. A. Potemra, B. J. Anderson and L. J. Zanetti, A comparison of ULF fluctuations in the solar wind, magnetosheath and dayside magnetosphere, 2, field and plasma conditions in the magnetosheath, *J. Geophys. Res., 96*, 3455-3464, 1991.

Montgomery, D., M. R. Brown and W. H. Matthaeus, Density fluctuation spectra in magnetohydrodynamic turbulence, *J. Geophys. Res., 92*, 282-284, 1987.

Olson, J. V., ULF signature of the polar cusp, *J. Geophys. Res., 91*, 10055-10062, 1986.

Sibeck, D. G., W. Baumjohann, R. C. Elphic, D. H. Fairfield, J. F. Fennell, W. B. Gail, L. J. Lanzerotti, R. E. Lopez, H. Luehr, A. T. Y. Lui, C. G. Maclennan, R. W. McEntire, T. A. Potemra, T. J. Rosenberg and K. Takahashi, The magnetopheric response to 8-minute period strong-amplitude upstream pressure variations, *J. Geophys. Res., 94*, 2505-2519, 1989.

Siscoe, G. L., L. Davis, Jr., P. J. Coleman, Jr., E. J. Smith and D. E. Jones, Power spectra and discontinuities of the interplanetary magnetic field: Mariner 4, *J. Geophys. Res., 73*, 61-82, 1968.

Southwood, D. J. and M. G. Kivelson, The magnetohydrodynamic response of the magnetospheric cavity to changes in solar wind pressure, *J. Geophys. Res., 95*, 2301-2309, 1990.

Tu, C., Z. Pu, and F. Wei, The power spectrum of interplanetary Alfvénic fluctuations: derivation of the governing equation and its solution, *J. Geophys. Res., 89*, 9695-9702, 1984.

Wilson, C. R., and M. Sugiura, Hydromagnetic interpretation of sudden commencements of magnetic storms, *J. Geophys. Res., 66*, 4097-4111, 1961.

Wolfe, A. and A. Meloni, ULF geomagnetic power near L = 4 6. Relationship to upstream solar wind quantities, *J. Geophys. Res., 86*, 7507-7512, 1981.

Wolfe, A., D. Venkatesan, R. Slawinski and C. G. Maclennan, A conjugate area study of Pc3 pulsations near cusp latitudes, *J. Geophys. Res., 95*, 10695-10698, 1990.

R. D. Kelman, S. E. Warren and A. Wolfe, NYCTC of CUNY, 300 Jay Street, Brooklyn, NY 11201.

C. G. Maclennan and L. J. Lanzerotti, AT&T Bell Laboratories, 600 Mountain Avenue, Murray Hill, NJ 07974.

Electric and Magnetic Field Fluctuations at High Latitudes in the Dayside Ionosphere During Southward IMF

E. M. BASINSKA,[1] W. J. BURKE,[2] N. C. MAYNARD,[2]
W. J. HUGHES,[1] D. J. KNUDSEN,[3] AND J. A. SLAVIN[4]

Low-altitude satellites provide snapshot-like images of ionospheric cusp/cleft regions that reflect processes taking place near the dayside magnetopause. These regions are characterized by rapid fluctuations in both electric and magnetic fields which are collocated with intense fluxes of low-energy, magnetosheath-like particles and often exhibit very narrow, spike-like features. The present study concentrates on the analysis of the electric and magnetic field fluctuations detected during approximately seventy crossings through the dayside ionosphere by the DE-2 satellite, when the IMF had a southward component. The data span three hours (10–13) in magnetic local time. The electric field fluctuations usually increased sharply at the low-latitude edge of the enhanced soft electron precipitation. Fluctuations with scale sizes greater than 10 km are consistent with steady field-aligned currents closing by ionospheric Pedersen currents. For fluctuations with smaller spatial/temporal scale sizes there is evidence for hydromagnetic wave activity. Here the ratios of electric to magnetic field amplitudes indicate the superposition of Alfvén waves and stationary field-aligned currents.

INTRODUCTION

Over the past decade impressive gains have been made in understanding the large-scale morphology of the dayside ionosphere at high latitudes and its electrical coupling to the magnetosphere and the magnetosheath. At ionospheric altitudes coupling manifests itself through convective electric-field, field-aligned current (FAC) and particle-precipitation patterns. This region also hosts a number of small-scale structures on whose particle and field characteristics we report here, based on observations taken during about seventy dayside ionospheric crossings by the Dynamics Explorer-2 (DE-2). For all these passes interplanetary magnetic field (IMF) measurements indicate that IMF B_z was directed southward.

During periods of southward IMF, midday convection patterns show large east-west flows that depend on the polarity of IMF B_y [Heppner and Maynard, 1987].

[1]Center for Space Physics, Boston University, Boston, Massachusetts
[2]Geophysics Directorate, Hanscom Air Force Base, Massachusetts
[3]Hertzberg Institute for Astrophysics, Ottawa, Canada
[4]Goddard Space Flight Center, Greenbelt, Maryland

Solar Wind Sources of Magnetospheric Ultra-Low-Frequency Waves
Geophysical Monograph 81

Analyses of magnetic deflections in the midday sector reveal large-scale FACs that are directly associated with cusp and mantle precipitation [Bythrow et al., 1988; Erlandson et al., 1988]. Depending on the kinds of measured dayside precipitation, we refer to the FACs as "cusp" or "mantle" currents. Precipitation regions on the dayside are traditionally designated as either "cusp" or "cleft" according to the spectral characteristics of precipitating electrons [Newell and Meng, 1988]. Newell et al. [1991] showed that cleft precipitation may be drawn from sources in the low-latitude boundary layer (LLBL) and the boundary plasma sheet (BPS). Cusp precipitation divides into that of the cusp proper and the plasma mantle (PM). The most probable locations to encounter each of these regions have been mapped out by Newell and Meng [1992]. The DE-2 ion detectors have a much smaller geometric factor than the DMSP sensors, but to the degree possible, we use the criteria of Newell et al. [1991] to identify the various particle-precipitation regions sampled by the DE-2 satellite.

Low-altitude satellites typically cross the few degrees of magnetic latitude of cusp/mantle precipitation in less than two minutes. In addition to intense, highly-structured fluxes of magnetosheath-like particles, large-scale azimuthal plasma flows, and dayside current systems, satellite and sounding rocket measurements indicate the presence of intense, small-scale electric and magnetic field fluctuations [Maynard and Johnstone, 1974; Maynard, 1985]. These are attributable either to small-scale field-aligned currents, manifesting quasi-stationary coupling between high and low altitudes, or to transient phenomena propagating by means of hydrodynamic waves.

Interpretations of single-satellite observations as resulting from spatial or temporal phenomena are often ambiguous. It is possible, however, to address this problem by comparing simultaneously measured electric and magnetic fields. Ratios of electric to magnetic field fluctuation amplitudes, as well as phase shifts between them, differ for spatial and temporal signals [Knudsen et al., 1990, 1992]. Ratios of these fields measured on relatively large scales (> 50 km) are consistent with anticipated Pedersen conductivities indicative of quasi-stationary structures; at smaller scale sizes the effective conductance decreases [Berthelier et al., 1989; Forget et al., 1991; Ishii et al., 1992]. This could be caused both by quasi-stationary structures and by temporal phenomena. Use of the techniques developed in these references should indicate whether significant fractions of the spectral power in the fluctuations may be attributed to travelling and/or reflected hydromagnetic waves. Estimates of the Poynting flux direction may also be used to determine whether the energy sources of the fluctuations are located above or below the observing spacecraft.

The following section which presents DE-2 measurements is divided into two parts. The first part gives examples of: (1) the location of the electromagnetic fluctuations in global magnetic coordinates and with respect to large-scale convection, FAC and particle precipitation patterns, and (2) their spectral characteristics. In the second part we divide the data into three frequency bands and compare their statistical characteristics with respect to IMF variability and estimated Alfvén and Pedersen impedances. In the final section we summarize the main observational results and discuss possible energy sources that could lead to the observed electric and magnetic field fluctuations.

OBSERVATIONS

The DE-2 satellite was launched into a 90° inclination orbit in August 1981. We have analyzed field and particle measurements from about seventy crossings of the high-latitude, dayside ionosphere when the IMF had a southward component, mainly during the first three months of the satellite's operation. The experiment complement on DE-2 includes the vector electric field instrument (VEFI) [Maynard et al., 1981], the three-axis, vector magnetometer (MAG-B) [Farthing et al., 1981], and the low-altitude plasma instrument (LAPI) [Winningham et al., 1981]. Measurements from both MAG-B and VEFI were sampled at rates of 16 s^{-1} in each field component. LAPI compiled spectra for electrons and ions with energies between 5 eV and 32 keV, in 32 bins, at 15 different pitch angles, once every sec-

ond. LAPI also contains Geiger-Mueller (GMT) counters to detect fluxes of electrons with energies > 35 keV. They were mounted on a scan platform and measured fluxes at pitch angles of 0° and 90°.

Figure 1 gives an overview of data selected for this study. It shows traces of the DE-2 orbital segments, on an invariant latitude (IL) versus magnetic local time (MLT) grid, during which the intensity of electric field fluctuations exceeded 10 mV/m on time scales shorter than 5 seconds. As illustrated in combined particle and field measurements presented below, the electric field fluctuations are collocated with intense fluxes of magnetosheath-like particles. Allowing for an equatorward migration of cusp/cleft precipitation during periods of southward IMF [Burch et al., 1973], the IL-MLT distribution in Figure 1 roughly corresponds to the regions designated by Newell and Meng [1992] as most probable for detecting LLBL and cusp precipitation. Our criteria excluded regions with mantle-like particles most of the time. The remainder of this section has two parts. The first presents two illustrative examples of the DE-2 fluctuation measurements. We then examine their frequency responses to IMF variability and to characteristic impedances.

Illustrative Examples

Plate 1 shows more than three minutes of data acquired during two DE-2 passes on 19 October 1981. The electric and magnetic field measurements are presented in a satellite-centered coordinate system. X is positive along the direction of satellite motion; Y is positive in the upward direction, and Z is perpendicular to the satellite track. In both cases the satellite travelled poleward across the northern midday auroral ionosphere. The top panels display E_x, the component of the electric field along the satellite velocity. In these orbits, positive values of E_x correspond to poleward electric fields which drive westward plasma convection. The ΔB_z traces, in the second panels, represent differences between measured and MAGSAT model magnetic

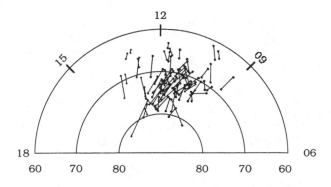

Fig. 1. Distribution of observed electric field fluctuations with amplitudes > 10 mV/m plotted on an invariant latitude versus magnetic local time grid.

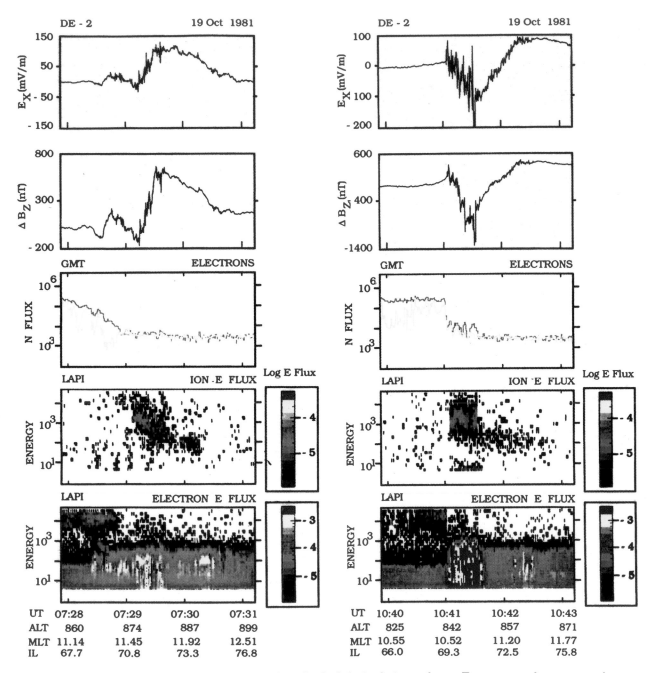

Plate 1. Two DE-2 passes through the dayside, high-latitude ionosphere. From top to bottom panels show: (1) electric field component along satellite trajectory, (2) magnetic field component transverse to satellite trajectory, (3) number flux of > 35 keV electrons at pitch angles of 0° (yellow) and 90° (blue), (4) energy flux of precipitating ions with energies in the range 5 eV - 32 keV, (5) energy flux of precipitating electrons with energies in the range 5 eV - 32 keV.

field components [Langel et al., 1980] in the direction transverse to the satellite trajectory. ΔB_z is positive towards the east. Positive slopes in ΔB_z correspond to downward FACs. Note that in this coordinate system, contributions of E_x and ΔB_z to the field-aligned component of the Poynting flux $S_y = \mu_o(E_x \Delta B_z - E_z \Delta B_x)$,

is toward the ionosphere if their variations correlate positively. Since E_z was not measured by DE-2, the contribution of the second term is unknown.

The third panel of Plate 1 contains the directional fluxes of electrons with energies > 35 keV at pitch angles of 90° (blue) and 0° (yellow). The bottom two panels contain directional, differential energy fluxes of downcoming ions and electrons with energies in the 5 eV to 32 keV range at pitch angles centered near 44° and 7°, respectively.

In the two examples we see that the electric and magnetic fields have both long (minute) and short (≤ 3 second) time variations, corresponding to the large-scale convection/FAC systems and to the fluctuations that are the topic of this report. We consider the field fluctuations first with respect to particle precipitation characteristics, and then with respect to the convection/FAC systems. At magnetic latitudes less than $\sim 69°$ the amplitudes of electric and magnetic field fluctuations were small. Here the GMT detectors measured anisotropic fluxes of > 35 keV electrons (panel 3) and significant energy fluxes of electrons with energies near 10 keV, characteristic of dayside, central plasma sheet (CPS) precipitation [Newell et al., 1991]. Low-energy (< 80 eV) electrons detected at this time are atmospheric photoelectrons from the conjugate ionosphere. The intensity of field variations increased significantly as the satellite entered regions of enhanced, soft (< 1 keV) electron precipitation, while fluxes of energetic electrons decreased dramatically and became nearly isotropic.

The precipitation characteristics of the ions in the two cases have differences and similarities. In the example on the right side of Plate 1, the onsets of ion and soft-electron precipitation are nearly simultaneous with the decrease in fluxes of energetic electrons and the appearance of electric and magnetic field fluctuations. In the other example ion precipitation was displaced poleward of other phenomena by about 1° IL. Here the ions show an energy-versus-latitude dispersion characteristic of cusp precipitation. In both cases, poleward of the most intense fluxes of keV ions, there are regions, 1° to 3° wide in IL, of low energy (~ 100 eV) ions. This is a characteristic signature of mantle precipitation [Newell et al., 1991]. The largest amplitude variations in the electric and magnetic fields were detected where number fluxes of magnetosheath-like ions and electrons were also most intense. Fluctuations decreased significantly in the region of mantle precipitation and vanished in the polar cap. Similar dependence of the intensity of small-scale electric and magnetic field variations on location within dayside boundary layers is a persistent feature of most DE-2 dayside passes studied by us so far.

Signatures of large-scale convection and FAC systems can be discerned in the electric and magnetic field measurements shown in the top two panels. On these scales the variations in E_x and ΔB_z appear quite similar within the individual orbits, but dissimilar between orbits. The orbit-to-orbit variations are quite easily explained in terms of the large-scale convection pattern. During the earlier pass, DE-2 crossed the westward reaches of the afternoon convection cell; while during the later pass it crossed the eastward part of the morning cell.

Multiple FAC sheets were detected during the first satellite crossing starting at 07:28 UT. The most intense field fluctuations were observed in the region of downward FAC, between 0729:12 and 0729:32 UT. At this time the current's polarity changed sign, and the intensity of fluctuations decreased significantly. Comparing the locations of these FACs with ion precipitation characteristics, we interpret the downward current (7:29:15–7:29:35 UT) and poleward upward current (7:29:35–7:30:40 UT) as the cusp and mantle FAC systems [Bythrow et al., 1988; Erlandson et al., 1988].

During the second pass, starting at approximately 10:40 UT, the most intense field fluctuations were again observed in the equatorward part of the magnetosheath-like particle precipitation. This time they are associated with a region of an upward FAC crossed between 10:41:05 and 10:41:32 UT. After that time DE-2 moved into the region of a downward current, collocated with mantle-like particles. The amplitudes of field fluctuations measured within the mantle were lower by more than an order of magnitude. The critical observational point is that the electric and magnetic field fluctuations are closely associated with the cusp/mantle current system, with the larger-amplitude fluctuations in the cusp current, independent of its polarity.

Plate 2 shows the fluctuations in the form of dynamic power spectra for electric (upper panels) and magnetic (lower panels) fields. The logarithm of the spectral power is color coded. During the second satellite pass, powers in both electric and magnetic field fluctuations exceeded those measured during the first pass by almost two orders of magnitude. The magnetic power appears concentrated below 2 Hz, while relatively-intense electric field variations extended to higher frequencies. This corresponds to magnetic field spectra being steeper than those of the electric field. To illustrate this more quantitatively, Figure 2 shows power spectra averaged over the dayside passes shown in Plate 2. Spectra were calculated for the 64 seconds of data starting at 7:28:45 UT (Figure 2a), and 10:41:00 UT (Figure 2b). On average, the amplitude ratios of electric to magnetic field variations tend to increase with increasing frequency. The increase, however, is not a monotonic function of frequency, as the electric and magnetic field spectra have different local minima and maxima.

Statistical Characteristics

The data presented in the previous subsection show that electric and magnetic fluctuations are large in the region of cusp/mantle precipitation and dayside, large-scale, field-aligned currents. Since some fraction of the magnetic field lines associated with this regions are believed to connect to the interplanetary medium, we have examined the variability of the IMF with respect to fluctuations detected in the ionosphere. The power spectral densities in electric field fluctuations were calculated for

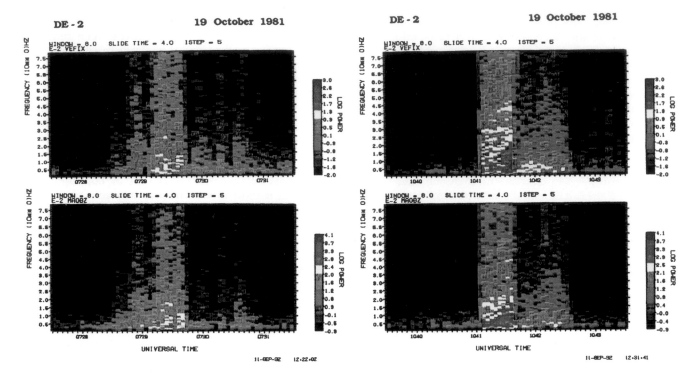

Plate 2. Dynamic power spectral densities in electric field ($P_E(f)$ - top panel) and magnetic field ($P_B(f)$ - bottom panel) fluctuations. $P_E(f)$ is measured in $(\text{mV/m})^2/\text{Hz}$, and $P_B(f)$ in $(\text{nT})^2/\text{Hz}$.

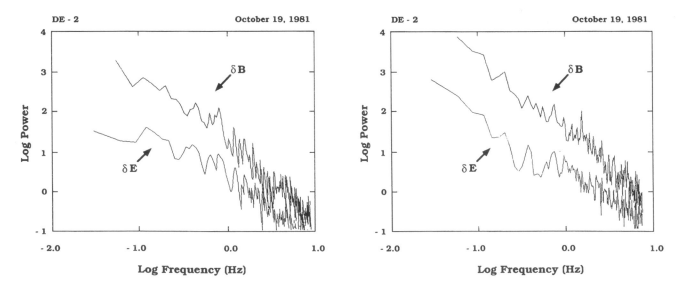

Fig. 2a. Power in electric ($P_E(f)$) and magnetic ($P_B(f)$) field fluctuations calculated for 64 seconds of data starting at 7:28:45 UT. $P_E(f)$ is measured in $(\text{mV/m})^2/\text{Hz}$, and $P_B(f)$ in $(\text{nT})^2/\text{Hz}$.

Fig. 2b. Power in electric ($P_E(f)$) and magnetic ($P_B(f)$) field fluctuations calculated for 64 seconds of data starting at 10:41:00 UT. $P_E(f)$ is measured in $(\text{mV/m})^2/\text{Hz}$, and $P_B(f)$ in $(\text{nT})^2/\text{Hz}$.

all orbital segments presented in Figure 1. Lengths of these segments were selected, as was mentioned before, based on the presence of electric field fluctuations with amplitudes exceeding 10 mV/m on time scales shorter than 5 seconds. Our selection criteria included regions with cusp/cleft particle characteristics, and most of the time excluded regions classified as mantle. Since most of the intervals fell into a range of 32–64 seconds, we used a 1024 point FFT spectra, filling in the difference between the number of real data points and 1024 points with zeroes. We have compared power in electric field fluctuations with the hourly values of the rms variations of the magnitude of the interplanetary magnetic field published by Couzens and King [1986]. Figure 3 compares electric field power, integrated over three frequency bands with the changes in the IMF. This figure shows that for frequencies below 2 Hz most intense field fluctuations tend to be measured at times when the rms variations in the IMF magnitude are large, while points of low intensity cluster at low rms values. Little correlation is seen at frequencies above 2 Hz. Similar results are found for magnetic field fluctuations.

To help understand the nature of the observed field fluctuations, following Knudsen et al. [1990], we introduce a measure of frequency-dependent impedance Z_{obs}

$$Z_{obs}(f) = \mu_0 \left(\frac{P_E(f)}{P_B(f)} \right)^{\frac{1}{2}} \qquad (1)$$

where $P_E(f)$ and $P_B(f)$ correspond to power spectral densities of electric and magnetic field measurements, respectively. There may be two extremes represented in the data: first, encounters with spatial FAC sheets that close by Pedersen currents in the lower ionosphere and second, encounters with travelling Alfvén waves.

One could also encounter superposition of waves impinging onto the ionosphere and those reflected from it. In case of quasi-stationary field-aligned currents, the measured impedance is determined by the inverse of the integrated Pedersen conductivity, $1/\Sigma_P$, while for travelling wave it equals the local Alfvén impedance. Superposition of incident and reflected Alfvén waves results in a range of impedance values and non-zero phase shifts between electric and magnetic field spectra [Knudsen et al., 1990, 1992; Basinska et al., 1992]. A priori we expect the rate of Pedersen encounters to be highest for low-frequency structures and the rate of travelling wave encounters to be highest for high-frequency structures. This reflects the time required for a low-altitude satellite to cross cusp/cleft latitudes relative to the wave periods. Longer period hydromagnetic waves will almost always appear as stationary FACs, while Pc-1 waves (in a frequency range 0.1–1.0 Hz) undergo many oscillations in the minute or so of electric/magnetic field sampling of cusp/cleft field lines.

Figure 4 displays values of Z_{obs} for three frequency ranges plotted as functions of calculated Pedersen impedances $1/\Sigma_P$. The measured impedance was calculated according to the formula (1) for each frequency band, where P_E and P_B corresponded to integrated power of electric and magnetic field, respectively. The height-integrated Pedersen conductivities are assumed to be caused primarily by solar illumination and were calculated according to the formula given by Wallis and Budzinski [1981]. The three panels show impedances measured in the frequency bands 0.1–0.5 Hz (left), 0.5–2.0 Hz (center), and 2.0–7.5 Hz (right). Solid lines correspond to $Z_{obs} = 1/\Sigma_P$, the values that would be measured if the conductivity were correctly calculated and if the satellite crossed sets of stationary FACs. The

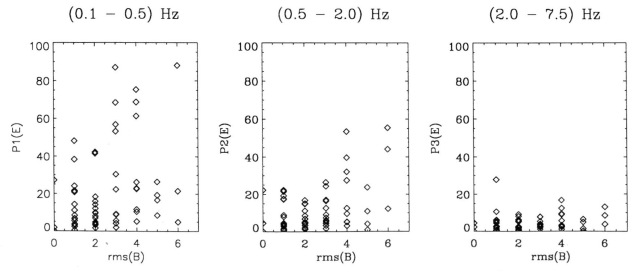

Fig. 3. Scatter plots of integrated power in electric field fluctuations P_E in $(\text{mV/m})^2$ over three bands in frequency versus rms fluctuations in IMF magnitude (measured in nT).

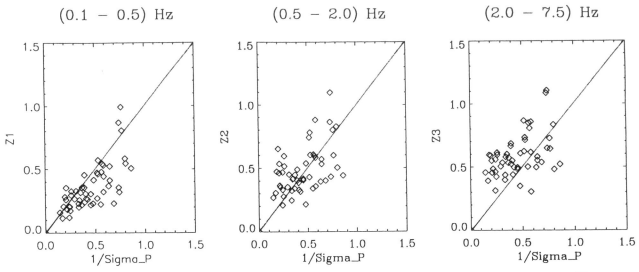

Fig. 4.   Measured impedance Z_{obs} (Ohms) versus inverse of integrated Pedersen conductivity $1/\Sigma_P$ (Ohms) due to solar illumination.

lower boundaries of the scatter clouds increase with increasing frequency from ~ 0.1 Ohms for the lowest frequency band to ~ 0.3 Ohms in the highest. Also, the relative locations of the scattered points with respect to the $Z_{obs} = 1/\Sigma_P$ line changes from band to band. At frequencies of 0.1–0.5 Hz, most data points fall below this line. They scatter about equally above and below the line in the frequency range 0.5–2.0 Hz, and are mostly above the line at highest frequencies. From the distribution of points in the lowest frequency band we conclude that the values of Σ_P contained in Figure 4 are probably underestimates of the actual height-integrated conductivities. Values of ionospheric conductivities calculated to a formula published by Robinson and Vondrak [1984] are somewhat higher than those shown in Figure 4. However, since validity of their equation is restricted to solar zenith angles below 85°, eliminating a large fraction of our data points, we decided to compare our measurements with Σ_P calculated according to the formula published by Wallis and Budzinski [1981].

Scatter plots of measured impedances versus local Alfvén impedances, $Z_A = \mu_o V_A$, are shown in the three panels of Figure 5. Alfvén speeds were calculated using measured magnetic fields B and average plasma densities n from the Langmuir probe data, and by assuming that O^+ is the dominant species at DE-2 altitudes. Again, the impedances associated with the scatter clouds move up with increasing frequency. For lowest frequencies all but two of the data points fall below the $Z_{obs} = Z_A$ line. For the highest-frequency fluctuations, the satellite measurements tend to approach the local Alfvén impedance. Similar results from low- and mid-altitude measurements were previously reported by Gurnett et al. [1984], Berthelier et al. [1989], Forget et al. [1991], and Ishii et al. [1992]. Fluctuations in

the lowest frequency band, allowing for a systematic error in conductivity estimates, appear to be caused by quasi-stationary field-aligned currents closing through the ionosphere. Increase of the measured impedance in higher frequency bands indicate a presence of travelling and/or reflected hydromagnetic waves in addition to quasi-stationary structures.

Figure 6 presents calculated values of the observed impedance Z and phase differences $\Delta\Phi$ between the electric and magnetic field measurements for the event shown on the right side of Plate 1. There is a distinct phase shift of approximately 45° at a frequency of about 0.6 Hz. This phase shift is associated with an increase in impedance which reaches 0.4 Ohms. A simultaneous increase in Z and presence of a non-zero phase-shift could be a result of superposition of incident and reflected Alfvén waves, as suggested by Knudsen et al. [1990]. Observed values are in agreement with predictions of a simple model of ionospheric reflection [Basinska et al., 1992].

SUMMARY AND DISCUSSION

In this paper we have examined fluctuations in electric and magnetic fields measured, during periods of southward IMF by the DE-2 satellite as it passed over the midday, high-latitude ionosphere. The main observational results are as follows:

1. Large-amplitude electric and magnetic field fluctuations are persistent features of the cusp/cleft region that collocate with fluxes of magnetosheath-like particles. At sub cusp/cleft latitudes and in the polar cap the fluctuation amplitudes are small. Also, fluctuations encountered within the mantle region, although still relatively large, can be up to two orders of magnitude higher than within cusp/cleft region (see Plates 1 and 2).

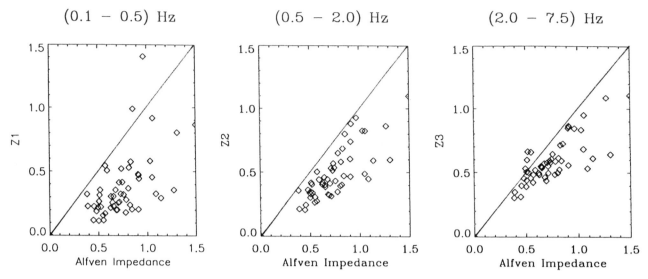

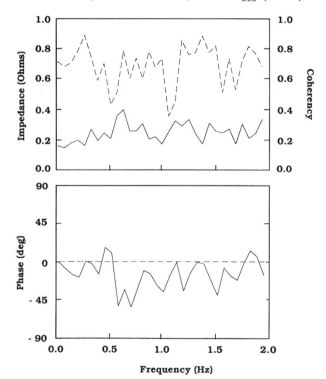

Fig. 5. Measured impedance Z_{obs} (Ohms) versus local Alfvén impedance $Z_A = \mu_0 V_A$ (Ohms).

Fig. 6. Measured impedance Z_{obs} (Ohms) shown as solid line, coherence shown as dashed line, and phase shift (degrees) for electric and magnetic field measurements shown on the right side of Plate 1. During this pass phase shifts between E_x and ΔB_z deviate up to $-50°$ simultaneously with an increase in $Z(f)$, suggesting that reflected Alfvén waves may be present in the data.

2. The most intense fluctuations occur at the lowest latitudes of cusp/cleft precipitation coincident with the equatorward part of the cusp/mantle FAC system,

independent of the current's polarity (see Plate 1).

3. The power in electric and magnetic field fluctuations tends to increase with increasing variability in the magnitude of the IMF.

4. The ratios of measured electric to magnetic field amplitudes increase with increasing frequency. At frequencies from 0.1 to 0.5 Hz, measured impedances are approximately equal to estimated values of $1/\Sigma_P$, allowing for systematic errors in calculated values of Pedersen conductivities. For higher-frequency fluctuations, the ratios of electric to magnetic field amplitudes are generally higher than $1/\Sigma_P$, approaching values of the local Alfvén impedance. Our interpretation of these results calls for quasi-stationary field-aligned currents dominating field fluctuations measured below 0.5 Hz. At higher frequencies measured values of impedance and phase-shifts indicate the presence of travelling and/or reflected hydromagnetic waves in addition to stationary structures.

Recent work attempting to identify possible solar-wind sources for Pc-3, 4 and 5 pulsations commonly observed by means of ground magnetometers represent a major topic of this Chapman conference. Electric and magnetic field variations measured by DE-2, having periods between 0.1 and 10 s, fall in the Pc-1 range. Ground measurements indicate that the occurrence of Pc-1 pulsations maximizes beneath the cusp/cleft ionosphere [Hansen et al., 1991, 1992]. There is also evidence that some dayside Pc-1 pulsations are associated with Pc-5 signals [Arnoldy et al., 1988] resulting from magnetopause processes such as flux transfer events, solar-wind pressure pulses, and/or perturbations due to Kelvin Helmholtz instability. Song et al. [1989], reported that wave power in the Pc-1 frequency band, as measured by ISEE near the magnetopause, is higher under IMF B_z south conditions.

A variety of sources may be responsible for low-frequency electric and magnetic field fluctuations observed during DE-2 crossings of the dayside regions. Among them are: (i) broadband magnetosheath turbulence, (ii) quasi-stationary, small-scale, field-aligned currents, (iii) multiple reflections of Alfvén waves between the ionosphere and the magnetopause, (iv) ionospheric resonant cavity modes, and (v) resonant interactions of background plasma fluctuations with magnetosheath ions.

(i) Broadband ULF turbulence is commonly detected in the magnetosheath by high-latitude spacecraft. Hydromagnetic waves that are transmitted through the magnetopause propagate towards the ionosphere along magnetic field lines threading the dayside boundary layers. The highest frequency of the observed spectrum is determined by a minimum value of ion cyclotron frequency along the propagation path. Assuming that within the exterior cusp region $B \approx 10$ nT, the cyclotron frequency of hydrogen would be approximately 1 Hz. Waves below that frequency may reach ionospheric regions and give rise to temporal components for signals detected by DE-2.

(ii) Forget et al. [1991] interpreted Aureol-3 electric and magnetic field measurements in terms of stationary, small-scale, field-aligned currents that close in the ionospheric F-layer. In this case, effective height-integrated conductivities are less than would be calculated integrating across the entire ionosphere, corresponding to an increase in the observed impedance. Frequency dependent effective conductivities reflect scale-dependent mapping of stationary electric field structures between F and E regions [Farley, 1959, 1960]. Processes described by Farley become significant for electric field structures with spatial sizes of order 1 km and less. Since the Nyquist frequency of DE-2 measurements is 8 Hz, corresponding to scale-sizes of approximately 1 km, the mechanism described by Farley [1959, 1960] and Forget et al. [1991] should not contribute in a significant way to our measurements.

(iii) Multiple reflections of Alfvén waves between the ionosphere and the magnetopause were invoked by Basinska et al. [1992] to explain a train of almost linearly polarized, phase-shifted electric and magnetic field fluctuations detected by the DE-2 satellite at the poleward boundary of the cusp with northward IMF. Linking the observed electromagnetic oscillations with the dynamics of merging, they described them as transient oscillations in the frame of convecting plasma caused by the sudden change of the flow at the magnetopause. In the fixed frame of the boundary they form a standing wave pattern similar to those formed at the foot of a waterfall. A very similar scenario was examined by Mallinckrodt and Carlson [1978], except they considered plasma convecting through the electric field structure associated with an auroral arc. Similar phenomenon could be expected for merging with southward IMF giving rise to fluctuations at the boundary between open and closed field lines.

(iv) An analogous mechanism of multiple reflections of hydromagnetic waves is at the base of the model of the ionospheric resonant cavity discussed by Lysak [1991]. However, in this model the reflecting layer above the satellite is located at the altitude of 1–2 R_E [R. L. Lysak, private communication, 1993] due to the significant gradient in Alfvén velocity between the ionosphere and the lower magnetosphere. As resonant cavity modes maximize at frequencies around 0.1–1.0 Hz, they may contribute to fluctuations detected on board of the DE-2 satellite.

(v) An ion-cyclotron instability due to resonant interactions of background plasma fluctuations with hot anisotropic plasmas has been suggested as a mechanism for the excitation of Pc-1 waves observed in a vicinity of the ionospheric cusp/cleft. For example, Cole et al. [1982] used this mechanism to explain trains of irregular pulsations with rising periods (IPRPs) that usually last for few minutes and appear as hydromagnetic whistlers with increasing periods on dynamic spectra [Maltseva et al., 1976; Morris et al., 1982]. In this case the large-amplitude Pc-1 waves result from a cyclotron instability of the energetic magnetosheath ions trapped on closed flux tubes of the LLBL that are filled with a cold plasma from the ionosphere. The rising-period effect would come from increasing flux-tube volumes (decreasing equatorial magnetic fields) in the region of the resonant interactions.

If ion cyclotron resonant interactions are responsible for a temporal part of electric and magnetic field fluctuations observed by DE-2, then highest fluctuations should be associated with the most intense fluxes of magnetosheath ions. Indeed, this is the case for most DE-2 passes studied by us so far, as illustrated in Plates 1 and 2. Intense fluctuations in Pc 1–2 frequency range occurred mainly in regions of LLBL and cusp precipitation. Fluctuations persisted within mantle region, but at a decreased level. In what follows we expand on this possible explanation for the small-scale electric and magnetic field variations detected at low altitudes within the cusp/cleft region.

In general, Pc-1 waves can propagate through a cold, magnetized plasma in either the Alfvénic or magnetosonic modes. Gyroresonant interactions of energetic ions with left-hand Alfvén waves and with right-hand magnetosonic waves differ significantly. Resonant interactions with Alfvén waves require that the ions and waves travel in opposite directions. Resonance with magnetosonic waves occurs when the ions and waves travel in the same direction, with the ions overtaking the waves. Ion energies required for resonance with Alfvén-mode waves are less than those required for resonance with waves propagating in the magnetosonic mode [Arnoldy et al., 1988]. Waves detected propagating toward the ionosphere require free-energy sources that reside in energetic ion populations and are located above the spacecraft. The resonant ions must move upward if the waves are Alfvénic and downward if they are magnetosonic.

Anisotropic ion distributions are sources of free-energy for growth of waves with frequencies less than local ion gyrofrequency [Kennel and Petschek, 1966]. Specific resonance requirements and ion distributions normally observed within the dayside boundary layers lead, under an assumption of nearly parallel propagation, to a prediction of the dominant polarization modes that should be observed within these regions.

1. Near the equatorial plane of the LLBL, locally trapped ion distributions are frequently observed. Such an anisotropy provides free-energy for the growth of Alfvén waves. Waves reaching the northern ionosphere are amplified through interactions with ions travelling toward the southern ionosphere.

2. Along recently-merged magnetic field lines threading the equatorward boundary of the cusp, ion distributions have a significant field-aligned component [Hill and Reiff, 1977]. Ions in the high-energy part of the distribution should pass through regions where resonant excitation of magnetosonic waves may occur.

3. The distribution functions of ions on field lines further removed from the ionospheric projection of the merging line contain growing fractions of particles reflected by the Earth's magnetic field. Owing to the velocity filter effect [Reiff et al., 1977] the average energy of precipitating ions decreases with distance from the merging line. There are fewer ions with energies sufficient to support the growth of magnetosonic (right-hand polarized) waves. Here waves reaching the ionosphere should mostly be Alfvénic (left-hand polarized) that grow by pitch-angle scattering some fraction of the mirrored ions toward the nearly-empty atmospheric loss cone.

Thus, if resonant interactions with background plasmas is the source of the waves we observe, then a properly instrumented satellite moving across the LLBL, cusp and mantle should detect Pc-1 fluctuations whose dominant polarity changes from left to right then back to left. Waves observed by Morris et al. [1982] and discussed by Cole et al. [1982] were mostly left-hand polarized, consistent with expectations for a LLBL source. The fact that Arnoldy et al. [1988] were unable to distinguish a dominant polarity for the Pc-1 waves may suggest that what they detected at ground level is consistent with a dual-source explanation, where right-hand polarized waves are driven by resonant interactions with precipitating, magnetosheath ions within the cusp, while left-hand polarized waves grow due to interactions with mirroring ions with upward loss-cone distributions. We advise some degree of caution in interpreting such measurements. The lack of a dominant polarity does not necessarily indicate a superposition of left- and right-hand polarized waves, since in case of highly oblique wave propagation their polarization is almost linear. Also wave coupling through the ionospheric wave guide may lead to a presence of right-hand polarized waves at ionospheric altitudes, even though only left-hand polarized waves are present above the ionospheric wave guide [Fujita and Tamao, 1988; Fujita, 1988; Popecki et al., 1992].

Acknowledgements. We would like to thank Dr. J. D. Winningham for providing access to DE-2 particle data. One of us (EMB) would like to thank Dr. Gary Erickson for many stimulating discussions and his help in preparing camera-ready copy of this article. This work was supported in part by U.S. Air Force contract F19628-90-K-0003 and NASA grant NAGW-2856 with Boston University, and by AFOSR Task 2311G5.

REFERENCES

Arnoldy, R. L., M. J. Engebretson, and L. J. Cahill Jr., Bursts of Pc 1-2 near the ionospheric footprint of the cusp and their relationship to flux transfer events, *J. Geophys. Res.*, *93*, 1007–1016, 1988.

Basinska, E. M., W. J. Burke, N. C. Maynard, W. J. Hughes, J. D. Winningham, and W. B. Hanson, Small-scale electrodynamics of the cusp with northward interplanetary magnetic field, *J. Geophys. Res.*, *97*, 6369–6379, 1992.

Berthelier, A., J. C. Cerisier, J. J. Berthelier, J. M. Bosqued and R. A. Kovrazkhin, The electrodynamic signature of short scale field aligned currents, and associated turbulence in the cusp and dayside auroral zone, in *Electromagnetic Coupling in the Polar Clefts and Caps*, edited by P. E. Sandholt and A. Egeland, pp. 299–310, Kluwer Acad. Press, Dordrecht, 1989.

Burch, J. L., Rate of erosion of dayside magnetic flux based on a quantitative study of the dependence of polar cusp latitude on the interplanetary magnetic field, *Radio Sci.*, *8*, 955–961, 1973.

Bythrow, P. F., T. A. Potemra, R. E. Erlandson, L. J. Zanetti, and D. M. Klumpar, Birkeland currents and charged particles in the high-latitude prenoon region: a new interpretation, *J. Geophys. Res.*, *93*, 9791–9803, 1988.

Cole, K. D., R. J. Morris, E. T. Maltseva, V. A. Troitskaya and O. A. Pokhotelov, The relationship of the boundary layer of the magnetosphere to IPRP events, *Planet. Space Sci.*, *30*, 129–136, 1982.

Couzens, D. A., and King, J. H., *Interplanetary Medium Data Book-Suppl. 3A*; NSSDC/WDC-A-R&S 86-04a, 1986.

Erlandson, R. E., L. J. Zanetti, T. A. Potemra, P. F. Bythrow, and R. Lundin, IMF B_y dependence of region 1 Birkeland currents near noon, *J. Geophys. Res.*, *93*, 9804–9814, 1988.

Farley, D. T., Jr., A theory of electrostatic fields in a horizontally stratified ionosphere subject to a vertical magnetic field, *J. Geophys. Res.*, *64*, 1225–1233, 1959.

Farley, D. T., Jr., A theory of electrostatic fields in the ionosphere at nonpolar geomagnetic latitudes, *J. Geophys. Res.*, *65*, 869–877, 1960.

Farthing, W. H., M. Sugiura, B. G. Ledley and L. J. Cahill Jr., Magnetic field observations on DE-A and -B, *Space Sci. Instru.*, *5*, 551, 1981.

Forget, B., J.-C. Cerisier, A. Berthelier, and J.-J. Berthelier, Ionospheric closure of small-scale Birkeland currents, *J.*

Geophys. Res., 96, 1843–1847, 1991.

Fujita, S., Duct propagation of hydromagnetic waves in the upper ionosphere, 2. Dispersion characteristics and loss mechanism, *J. Geophys. Res., 93,* 14,674–14,682, 1988.

Fujita, S., and T. Tamao, Duct propagation of hydromagnetic waves in the upper ionosphere, 1. Electromagnetic field disturbances in high latitudes associated with localized incidence of a shear Alfvén wave, *J. Geophys. Res., 93,* 14,665–14,673, 1988.

Gurnett, D. A., R. L. Huff, J. D. Menietti, J. L. Burch, J. D. Winningham, and S. D. Shawhan, Correlated low-frequency electric and magnetic noise along the auroral field lines, *J. Geophys. Res., 89,* 8971–8985, 1984.

Hansen, H. J., B. J. Fraser, F. W. Menk, Y.-D. Hu, P. T. Newell, and C.-I. Meng, High latitude unstructured Pc-1 emissions generated in the vicinity of the dayside auroral oval, *Planet. Space Sci., 39,* 709–719, 1991.

Hansen, H. J., B. J. Fraser, F. W. Menk, Y.-D. Hu, P. T. Newell, C.-I. Meng, and R. J. Morris, High-latitude Pc 1 bursts arising in the dayside boundary layer region, *J. Geophys. Res., 97,* 3993–4008, 1992.

Heppner J. P., and N. C. Maynard, Empirical high-latitude electric field models, *J. Geophys. Res., 92,* 4467–4489, 1987.

Hill, T. W., and P. H. Reiff, Evidence of magnetospheric cusp proton acceleration by magnetic merging at the dayside magnetopause, *J. Geophys. Res., 82,* 3623–3628, 1977.

Ishii, M., M. Sugiura, T. Iyemori and J. A. Slavin, Correlation between magnetic and electric field perturbations in the field-aligned current regions deduced from DE 2 observations, *J. Geophys. Res., 97,* 13,877–13,887, 1992.

Kennel, C. F. and H. C. Petschek, Limit on stably trapped particle fluxes, *J. Geophys. Res., 71,* 1–28, 1966.

Knudsen, D. J., M. C. Kelley, G. D. Earle, J. F. Vickrey, and M. Boehm, Distinguishing Alfvén waves from quasi-static field structures associated with discrete aurora: sounding rocket and HILAT satellite measurements, *Geophys. Res. Lett., 17,* 921–924, 1990.

Knudsen, D. J., M. C. Kelley, and J. F. Vickrey, Alfvén waves in the auroral ionosphere: a numerical model compared with measurements, *J. Geophys. Res., 97,* 77–90, 1992.

Langel, R. A., R. H. Estes, G. D. Mead, E. B. Fabiano, and E. R. Lancaster, Initial geomagnetic field model from MAGSAT vector data, *Geophys. Res. Lett., 7,* 793–796, 1980.

Lysak, R. L., Feedback instability of the ionospheric resonant cavity, *J. Geophys. Res., 96,* 1553–1568, 1991.

Maltseva, E. T., V. A. Troitskaya, and F. Z. Feygin, Intervals of pulsations with rising periods (IPRP) in polar caps, *Planet. Space Sci., 24,* 673–678, 1976.

Mallinckrodt, A. J., and C. W. Carlson, Relations between transverse electric fields and field-aligned currents, *J. Geophys. Res., 83,* 1426–1432, 1978.

Maynard, N. C., E. A. Bielecki, and H. F. Burdick, Instrumentation for vector electric field measurements from DE-B, *Space Sci. Instru., 5,* 523–534, 1981.

Maynard, N. C., Structure in the dc and ac electric fields associated with the dayside cusp region, in *The Polar Cusp,* edited by J. A. Holtet and A. Egeland, pp. 305–322, D. Reidel, Hingham, MA, 1985.

Maynard, N. C., and A. D. Johnstone, High-latitude day side electric field and particle measurements, *J. Geophys. Res., 79,* 3111–3123, 1974.

Morris, R. J., K. D. Cole, E. T. Maltseva, and V. A. Troitskaya, Hydromagnetic "whistles" at the dayside cusp: IPRP events, *Planet. Space Sci., 30,* 113–127, 1982.

Newell, P. T. and C.-I. Meng, The cusp and cleft/boundary layer; low-altitude identification and statistical local time variation, *J. Geophys. Res., 93,* 14,549–14,556, 1988.

Newell, P. T., W. J. Burke, C.-I. Meng, E. R. Sanchez, and M. E. Greenspan, Identification and observations of the plasma at low altitude, *J. Geophys. Res., 96,* 35–45, 1991.

Newell, P. T., and C.-I. Meng, Mapping the dayside ionosphere to the magnetosphere according to particle precipitation characteristics, *Geophys. Res. Lett., 19,* 609–612, 1992.

Popecki, M. A., R. L. Arnoldy, M. J. Engebretson, and L. J. Cahill, High-latitude ground observations of Pc-1/2 micropulsations, submitted to *J. Geophys. Res.,* 1992.

Reiff, P. H., T. W. Hill, and J. L. Burch, Solar wind plasma injection at the dayside magnetospheric cusp, *J. Geophys. Res., 82,* 479–491, 1977.

Robinson, R. M., and R. R. Vondrak, Measurements of *E* region ionization and conductivity produced by solar illumination at high latitudes, *J. Geophys. Res., 89,* 3951–3956, 1984.

Song, P., C. T. Russell, N. Lin, J. Strangeway, J. T. Gosling, M. Thomson, T. A. Fritz, D. G. Mitchell, and R. R. Anderson, Wave and particle properties of the subsolar magnetopause, in *Physics of Space Plasmas, SPI Conference Proceedings, No. 9,* edited by T. Chang, G. B. Crew, and J. R. Jasperse, pp. 463–476, Scientific Publishers, Cambridge, MA, 1989.

Wallis, D. D., and E. E. Budzinski, Empirical models of height integrated conductivities, *J. Geophys. Res., 86,* 125–137, 1981.

Winningham, J. D., J. L. Burch, N. Eaker, V. A. Blevins, and R. A. Hoffman, The low altitude plasma instrument (LAPI), *Space Sci. Instru., 5,* 465–475, 1981.

E. M. Basinska and W. J. Hughes, Center for Space Physics, Boston University, 725 Commonwealth Avenue, Boston, MA 02215.

W. J. Burke and N. C. Maynard, Geophysics Directorate, PL/GPSG, Hanscom AFB, MA 01731.

D. J. Knudsen, Hertzberg Institute for Astrophysics, Ottawa, Canada K1A 0R6.

J. A. Slavin, GSFC, Code 696, Greenbelt, MD 20771.

Pc 1 Waves Generated by a Magnetospheric Compression During the Recovery Phase of a Geomagnetic Storm

R. E. Erlandson and L. J. Zanetti

The Johns Hopkins University Applied Physics Laboratory, Johns Hopkins Road, Laurel, MD

M. J. Engebretson

Augsburg College, Department of Physics, Minneapolis, Minnesota

R. Arnoldy

University of New Hampshire, Institute of Earth, Oceans, and Space, Durham, New Hampshire

T. Bösinger and K. Mursula

University of Oulu, Department of Physics, Oulu, Finland

A multipoint ground-satellite observation of a Pc 1 wave event is used to investigate electromagnetic ion cyclotron (EMIC) wave generation during a magnetospheric compression. The event occurred on September 15, 1986, from approximately 0400 to 0900 UT. Viking satellite observations were acquired by the magnetic and electric field experiments from 0622 to 0637 UT near 60° invariant latitude, 1130 magnetic local time, and an altitude of 13,500 km. The ground magnetic field observations were acquired throughout the event using the Finnish ground-station chain (Rovaniemi, Ivalo, Kilpisjarvi) and Sondre Stromfjord, South Pole, McMurdo, and Siple. The event occurred during the recovery phase of a large geomagnetic storm, where D_{st} reached -180 nT. There was a transient increase in D_{st} during the recovery phase of the storm. The wave event was observed during this transient increase in D_{st}, which is interpreted as a signature of a magnetospheric compression. The correlation between the D_{st} index and EMIC waves is used to investigate the association between magnetospheric compressions during the recovery phase of a geomagnetic storm and EMIC wave generation. Viking satellite and Finnish ground-station observations are used to compare spectral amplitudes in the magnetosphere and on the ground and to estimate ionospheric attenuation of Pc 1 waves.

Introduction

Low-frequency fluctuations in the Pc 1 frequency range (0.2–5 Hz) observed on the ground are believed to be due to the electromagnetic ion cyclotron (EMIC) instability generated by temperature anisotropies of protons in the energy range from a few keV to a few hundred keV [Cornwall, 1965; Kennel and Petschek, 1966; Liemohn, 1967]. The EMIC instability was later confirmed as the source of Pc 1 pulsations using simultaneous wave and particle data recorded at geostationary orbit [Mauk and McPherron, 1980]. In a statistical study using 7500 hours of data acquired by the AMPTE/CCE satellite it was found that EMIC waves are generated near the magnetic equator at all local times from $L \sim 3.5$ to $L > 9$ [Anderson et al., 1992]. EMIC waves occur most often in the outer magnetosphere ($L > 7$) and in the magnetic local time (MLT) sector from 1200 to 1500 hours [Erlandson et al., 1990; Anderson et al., 1992].

The occurrence of some types of Pc 1 wave activity have been found to be associated with geomagnetic storms. This association, however, depends on the type or morphology of the waves (see Jacobs et al. [1964] and Fukunishi et al. [1981] for a discussion of Pc 1 wave classification) and the phase of the storm. For example, IPDP (intervals of pulsations of diminishing periods) tend to occur during the main phase of geomagnetic storms in the afternoon–evening sector [Heacock, 1971; Barfield and McPherron, 1972; Bossen et al., 1976]. Periodically structured Pc 1 pulsations tend to occur during the recovery phase of geomagnetic storms [Wentworth, 1964; Plyasova-Bakounina and Matveyeva, 1968; Heacock and Kivinen, 1972]. On the other hand, unstructured Pc 1 pulsations, dominant at high latitudes, do not appear to be associated with geomag-

Solar Wind Sources of Magnetospheric Ultra-Low-Frequency Waves
Geophysical Monograph 81

netic storms [Kuwashima et al., 1981].

The generation of Pc 1 waves has also been associated with sudden impulses and sudden storm commencements [Troitskaya, 1961; Teply and Wentworth, 1962; Heacock and Hessler, 1965; Saito and Matsushita, 1967]. Olson and Lee [1983] have shown that wave generation during a magneto-spheric compression can be explained by the corresponding increase in proton temperature anisotropy; they also found that the maximum growth rate occurs near noon, just inside the magnetopause. A further test on the effect of compressions was performed by Anderson and Hamilton [1993], who found that 47% of sudden magnetic field increases greater than 10 nT in the 8 to 16 MLT sector at $L > 6$ resulted in EMIC wave generation. A number of studies have also investigated the spectral structure of Pc 1 waves during magnetospheric compressions [Troitskaya et al., 1968; Hirasawa, 1981; Kangas et al., 1986]. Kangas et al. [1986] found that the Pc 1 spectral structure could vary significantly from one event to the next. It was speculated that this variability is the result of the particular state of the magnetosphere at the time of the compression.

The purpose of this paper is to investigate the generation of Pc 1 waves by a magnetospheric compression that occurs during the recovery phase of a geomagnetic storm. The event is studied using multipoint ground-satellite and Viking satellite observations acquired on September 15, 1986.

INSTRUMENTATION

Data used in this investigation were obtained from the magnetic and electric field experiments on the Viking satellite and from ground-station magnetometers. The field line projection to a 100-km altitude of the Viking satellite and location of ground stations used are shown in Figure 1. In that figure the locations of ground stations in the Southern Hemisphere have been mapped to their conjugate point in the Northern Hemisphere. The ground stations located in the Northern Hemisphere include Rovaniemi (ROV), Ivalo (IVA), Kilpisjarvi (KIL), and Sondre Stromfjord (SSF). The ground stations located in the Southern Hemisphere include Siple (SIP), South Pole (SP), and McMurdo (MCM). Search coil magnetometers were used on the ground and were sampled at a rate of 10 Hz (SSF, SIP, SP, and MCM) and at 5 Hz (ROV, IVA, and KIL).

For reference, the location of $L = 4$ and 12 MLT (at 0630 UT) are indicated by the dashed lines in Figure 1.

The geographic and geomagnetic locations of the ground stations are listed in Table 1. The Finnish chain was at approximately 0900 MLT at a time of 0630 UT, resulting in a separation of 2.5 h in MLT with the Viking satellite.

OBSERVATIONS AND DISCUSSION

The wave event discussed in this paper occurred during the recovery phase of a geomagnetic storm. In addition, the event occurred during solar minimum at the boundary between solar cycles 21 and 22 when the sunspot number was 0. The D_{st} index during this storm is shown in Figure 2. The main phase of the geomagnetic storm occurred on September 12, 1986, and the recovery phase lasted for 3 to 4 days. At 0400 UT on

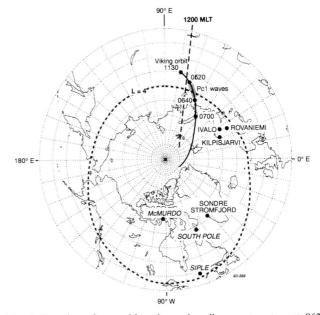

Fig. 1. Location of ground-based search coil magnetometers at 0630 UT and Viking's ionospheric footprint from 0620 to 0720 UT. The box along the Viking trajectory indicates the location of the wave event at Viking. The stations located in the Southern Hemisphere (Siple, South Pole, and McMurdo) have been projected to their conjugate point in the Northern Hemisphere and are shown in italics.

TABLE 1. Ground Station Geographic and Geomagnetic Locations

Station	Geographic latitude (deg)	Geographic longitude (deg)	Geomagnetic latitude (deg)	Geomagnetic longitude (deg)	L Value	MLT (h) 630 UT
Rovaniemi	66.8	25.9	63.2	107.4	4.8	9.00
Ivalo	68.7	27.3	65.0	209.9	5.5	9.18
Kilpisjarvi	69.0	20.9	65.7	105.3	6.0	8.87
S. Stromfjord	66.0	−52.0	73.5	40.5	13.3	4.55
Siple	−76.0	−83.0	−61.4	3.1	4.2	2.05
South Pole	−90.0	N/A	−74.2	18.6	14.7	3.08
McMurdo	−78.0	165.0	−80.6	−32.0	>15.0	23.72

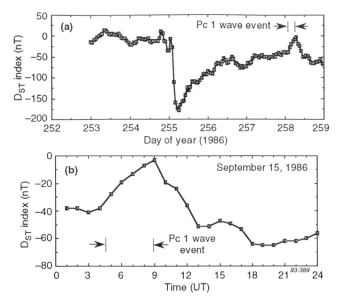

Fig. 2. D_{st} index (a) from September 9 to 16, 1986, and (b) on September 15, 1986.

September 15, 1986, an enhancement in D_{st} was observed. The D_{st} index increased for approximately 5 h until 0900 UT. The EMIC wave event discussed in this paper occurred during the positive slope in D_{st}.

The increase in D_{st} index from 0400 to 0900 UT is interpreted as resulting from an increase in solar wind dynamic pressure. The D_{st} index has been found to respond to solar wind dynamic pressure [Verzariu et al., 1972]. An empirical relationship between the D_{st} index and interplanetary conditions was investigated by Burton et al. [1975]. The observed increase in D_{st} from -38 to -2 nT would imply that the solar wind dynamic pressure increased by 5 nPa based on the empirical results of Burton et al. [1975]. Unfortunately, confirmation of a solar wind dynamic pressure increase is not possible since solar wind data are not available for this time period.

Pc 1 waves were recorded over a wide range of magnetic local times extending from the midnight sector around to the dawn and dayside sectors (Figures 3 and 4). Figure 3 contains a summary of magnetic field fluctuations in the MLT sector from 2200 to 0700 recorded at SSF, SIP, SP, and MCM from 0400 to 1000 UT. Figure 4 contains a summary of magnetic field fluctuations recorded at ROV from 0330 to 0930 UT in the 0630 to 1130 MLT sector. The waves at IVA and KIL (not shown) were lower in amplitude but had similar temporal and spectral structure. The largest wave amplitudes were recorded at SIP ($L = 4.2$) and ROV ($L = 4.8$), indicating that the source region is in this L-value range. The morphology of the waves at SIP and ROV were similar, even though the stations were separated by 7 h in MLT. The waves recorded at SIP were also very similar to those recorded at the high-latitude stations (SSF, SP, and MCM), although the waves at MCM were very low in amplitude. The similarity between Northern

and Southern Hemisphere stations is also evident. The wave event begins near 0400 UT at ROV (0630 MLT) and near 0430 UT at SIP (2330 MLT) and SP (0030 MLT). The event ends abruptly at around 0900 UT at all stations.

Magnetic and electric field fluctuations were also recorded at Viking during this event and were used to identify the source field line of the Pc 1 waves. The magnetic field data are presented in eccentric dipole coordinates, where BN is positive north, BE is positive east, and BP (not shown) is along the magnetic field direction. The electric field component, E34, is in the spin plane of the satellite. The waves were recorded from 0622 to 0637 UT when Viking crossed L-shells from $L = 3.7$ to $L = 4.5$ (Figure 5). Waves were not observed by Viking outside these L-shells. The magnetic local time of Viking at this time was 1130 MLT. Magnetic field fluctuations recorded at ROV ($L = 4.8$), IVA ($L = 5.5$), and KIL ($L = 6.0$) are shown from 0622 to 0637 UT (Figure 6). The wave amplitude was found to decrease as a function of L-value or distance from the wave source field line ($\sim L = 4$). The identification of the source field line near $L = 4$ is also consistent with the decrease in amplitude as a function of L-value observed in stations located in the post-midnight sector (SIP, SP, SSF, and MCM) (Figure 3).

The combination of Viking at 1130 MLT, the Finnish chain at 0900 MLT, and the stations in the post-midnight sector (SIP, SP, SSF, and MCM) imply that the wave source region was not localized in local time near Viking ($L = 4$ and 1130 MLT) but was extended to the dawn (ROV, IVA, and KIL) and midnight sectors (SIP, SP, SSF, and MCM). This can be seen by noting that the distance between each of the three Finnish ground stations (ROV, IVA, and KIL) and Viking were nearly the same, ruling out a single localized source at Viking (Figure 1). The maximum in amplitude and the nearly constant wave frequency observed at Viking, ROV, and SIP suggest that the source region was probably near $L = 4$ for the duration of the event. The similarity between waves recorded at SIP (post-midnight sector) and ROV (pre-noon sector) suggests that the waves were generated over a wide range in magnetic local time.

The Pc 1 waves recorded at ROV, IVA, and KIL were structured pearl pulsations with an average repetition period of 62 s (Figure 6). The waves recorded at SIP, SP, MCM, and SSF were also structured, although the structure was less defined than the ROV structure. The structure or repetition period is not as obvious in the Viking data. A repetition period of 60 s can been seen from 0622 to 0625 UT in the electric field (E34), although after 0625 UT the event appears to be unstructured. It can not be concluded, however, that pearl pulsations are features observed only on the ground, since clear examples have been observed using the Viking satellite [Erlandson et al., 1992]. The observations presented here indicate that conditions favorable for generating pearl pulsation structure may be localized in space and vary from one location to another.

The power spectra of the waves recorded at Viking were nearly identical to those recorded at ROV, IVA, and KIL (Fig-

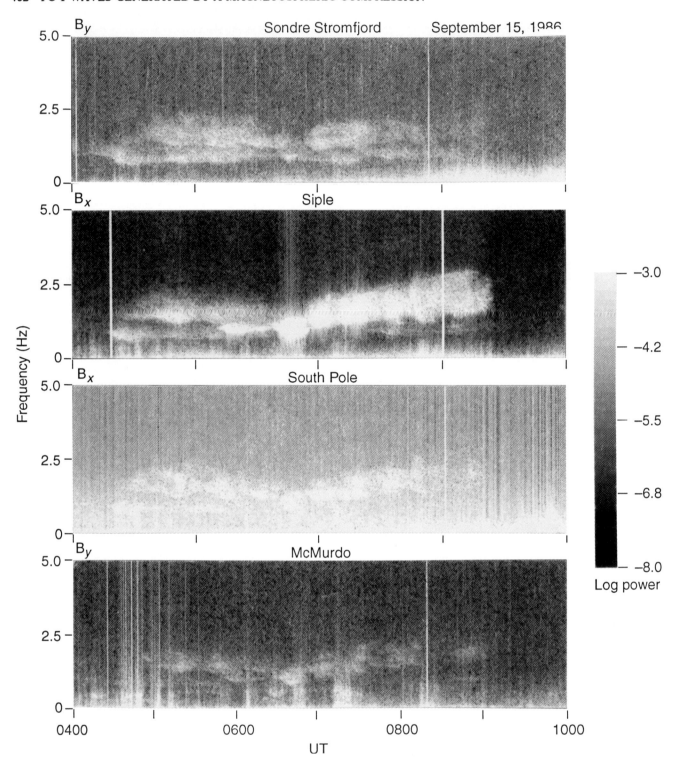

Fig. 3. Spectrograms of magnetic field fluctuations recorded at Sondre Stromfjord, Siple, South Pole, and McMurdo from 0400 to 1000 UT. The spectral power is in units of nT^{2}Hz.

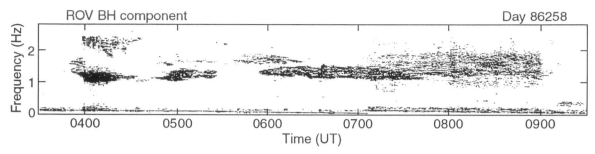

Fig. 4. Spectrogram of magnetic field fluctuations recorded at Rovaniemi (ROV) from 0330 to 0930 UT. The spectral power is in units of nT^2/Hz.

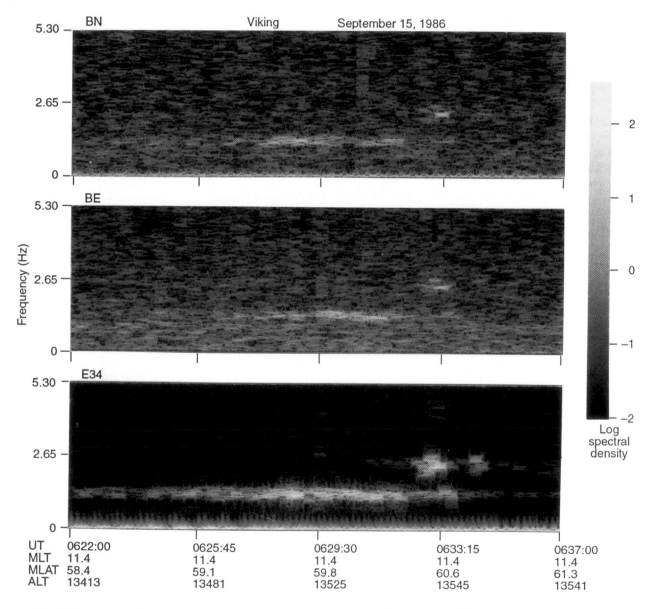

Fig. 5. Spectrograms of magnetic and electric field fluctuations in units of nT^2/Hz and $(mV/m)^2/Hz$, respectively. The data were acquired by Viking during orbit 1130 on September 15, 1986, from 0622 to 0637 UT.

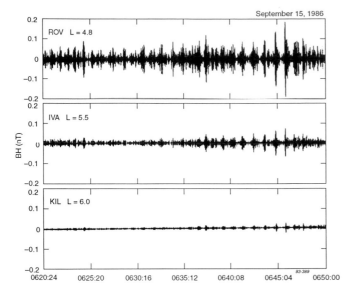

Fig. 6. Magnetic field fluctuations recorded at Rovaniemi (ROV), Ivalo (IVA), and Kilpisjarvi (KIL) on September 15, 1986, from 0620 to 0650 UT.

ure 7), even though the stations were separated from Viking by nearly 2.5 h in MLT (Figure 1). Fine structure in the power spectra was also very similar as seen in the spectral peaks at 1.3 and 1.45 Hz. The peak at 1.65 Hz observed at ROV would be expected to be below the Viking magnetometer noise level.

The observation of the wave amplitude as a function of L-value using Viking and the Finnish ground-station chain may be used to estimate the attenuation of waves in the ionosphere. The attenuation of waves in the ionosphere, as they propagate from the source region to higher latitudes based on the power spectral density at ROV, IVA, and KIL shown in Figure 7,

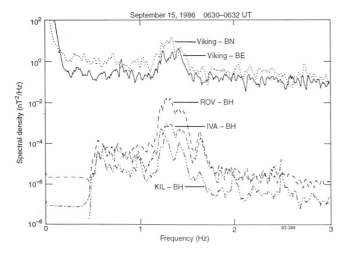

Fig. 7. Spectral comparison of Pc 1 waves recorded from 0630 to 0632 UT at Viking (BN is positive north and BE is positive east) and ROV, IVA, and KIL (BH is positive north).

was found to be approximately 0.06 dB/km. It is more complicated, however, to estimate the ionospheric attenuation between Viking (at $L = 4$) and the ground-station chain (ROV, IVA, and KIL). For example, some of the wave energy incident on the ionosphere is reflected and transmitted, representing a source of attenuation unrelated to the attenuation in the ionospheric duct. In addition, the wave amplitude at Viking's altitude (13,500 km) may not necessarily represent the amplitude of the wave at ionospheric altitudes. Ignoring these two effects, however, it is found using the data shown in Figure 7 that the average attenuation between Viking ($L = 4$) and the Finnish ground stations is 0.07 dB/km. These two estimates are very similar and are comparable to the predicted value of 0.06 dB/km from Fujita [1987], who investigated ionospheric attenuation of Pc 1 waves using the International Reference Ionosphere model. The fact that the attenuation estimates using Viking were similar to the estimates using only ground stations indicates that the increase in wave amplitude as the waves propagate to lower altitudes just happens to be compensated by the attenuation of the waves due to reflection and transmission in the ionosphere. A more detailed discussion of ionospheric attenuation, including any possible frequency dependence, will be the subject of an additional paper.

The spectrograms for SIP, SP, SSF, and MCM (Figure 3) have been expanded in Figure 8 to show the time sequence from 0610 to 0650 UT, which corresponds more closely to the time period of waves observed at Viking (see Figure 5). The time extent of the waves recorded at Viking is most likely determined by the spatial size of the source region ($\approx$59–61° invariant latitude). The ground observations, on the other hand, show the long-term temporal profile of the waves. It was shown earlier that the wave spectra recorded at Viking and on the ground in the Finnish chain were very similar, with the primary difference being the wave amplitude. It was also shown that the wave morphology at SIP, SP, MCM, and SSF was very similar (Figure 3). A more detailed comparison indicates that the wave frequency recorded at Viking was 1.1 to 1.5 Hz, whereas the wave frequency at SIP was 0.95 to 1.25 Hz at 0628 UT. Minor frequency differences might be expected, however, based on the location of the plasmapause at different local times and differences in the energetic plasma at a given local time. The waves at SIP ($L = 4.2$) and SP ($L >$ 10), which were located in the same MLT sector, were similar, although lower-frequency waves (0.5 Hz) were observed at around 0627 to 0630 UT from SP but not SIP. The lower-frequency waves observed at SP were also observed at MCM, another high-latitude station. The waves recorded at SP and SSF, stations at similar L-values but located in different hemispheres, were similar, with the primary difference being that the increase in wave amplitude after 0630 UT was observed at SP but not SSF. Differences between these two stations may be related to hemispherical differences or MLT differences between stations. This detailed comparison illustrates that although the wave morphology is similar on a long time scale (hours), significant differences may exist on shorter time scales (minutes). It may be that overall wave morphology is domi-

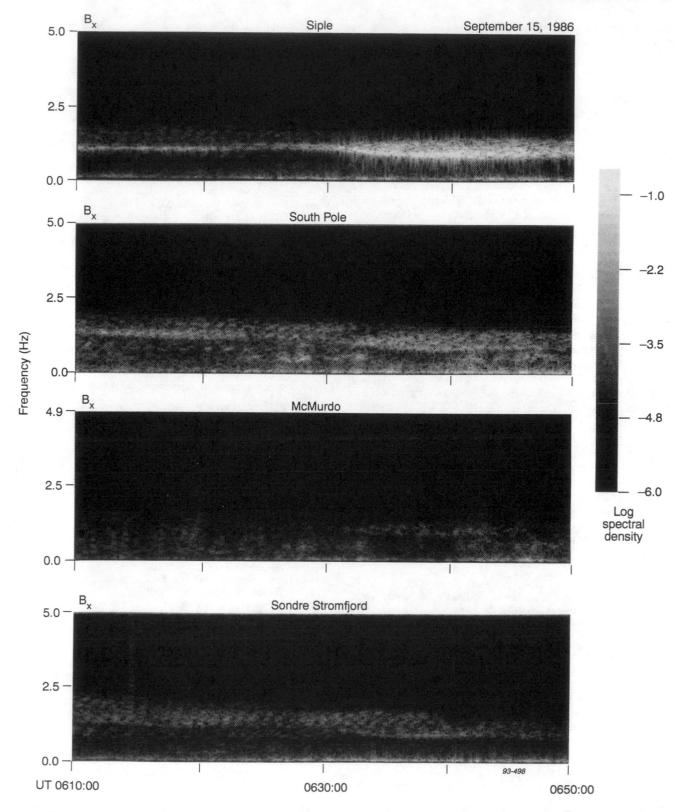

Fig. 8. Spectrograms of the magnetic field B$_x$ component at Siple, South Pole, McMurdo, and Sondre Stromfjord in units of nT²Hz from 0610 to 0650 UT on September 15, 1986.

nated by large-scale magnetospheric dynamics, whereas on shorter time scales differences result from the particular particle distributions at a given local time.

SUMMARY

The primary observations from this event are that the waves were excited during a magnetospheric compression as inferred from the D_{st} index. After the magnetosphere reached its maximum compressed state and began to relax, the waves were no longer observed. The wave event ended nearly simultaneously, both on the dayside and nightside. This observation was not necessarily expected since the magnetosphere would still be in a compressed state when the relaxation of the magnetic field began. During a compression, energetic ions that excite the waves are driven unstable through adiabatic acceleration [Olson and Lee, 1983]. The sudden end of the event, on a global scale, suggests that the energetic ions quickly become stabilized in the absence of a driving force such as a compression.

A second observation is that the event, most likely driven by a magnetospheric compression, occurred during the recovery phase of a storm. This observation suggests that the role of geomagnetic storms, in terms of Pc 1 wave generation, is to populate the magnetosphere with energetic ions. The magnetospheric compression, which occurred during the recovery phase of the storm, resulted in driving the energetic ions unstable to the EMIC instability. Therefore, the effect of a magnetospheric compression on EMIC wave generation strongly depends on the state of the 5- to 100-keV plasma in the ring current. This is consistent with the findings of Kangas et al. [1986], who suggested that the type of Pc 1 emissions during a sudden impulse reflect the state of the magnetosphere at the time of the sudden impulse. In the case study presented here, we suggest that the recovery phase of geomagnetic storms would be the most favorable condition for EMIC wave generation at low latitudes through magnetospheric compression as a result of the newly injected plasma provided during the storm.

Acknowledgments. We acknowledge the support of the National Science Foundation through support of the US–Finnish Auroral Workshops. The Viking magnetic field experiment was supported by the Office of Naval Research, and collaborative efforts were supported by the National Science Foundation. R. L. Arnoldy and M. J. Engebretson acknowledge the support of the National Science Foundation under grant DPP-89-13870.

REFERENCES

Anderson, B. J., R. E. Erlandson, and L. J. Zanetti, A statistical study of Pc 1-2 magnetic pulsations in the equatorial magnetosphere 1. Equatorial occurrence distribution, *J. Geophys. Res., 97*, 3075, 1992.

Anderson, B. J., and D. C. Hamilton, Electromagnetic ion cyclotron waves stimulated by modest magnetospheric compressions, *J. Geophys. Res.,* (in press), 1993.

Barfield, J. N., and R. L. McPherron, Investigation of interaction between Pc 1 and 2 and Pc 5 micropulsations at the synchronous orbit during magnetic substorms, *J. Geophys. Res., 77*, 4707, 1972.

Bossen, M., R. L. McPherron, and C. T. Russell, Simultaneous Pc 1 observations by the synchronous satellite ATS 1 and ground stations: Implications concerning IPDP generation mechanism, *J. Atmos. Terr. Phys., 38*, 1157, 1976.

Burton, R. K., R. L. McPherron, and C. T. Russell, An empirical relationship between interplanetary conditions and D_{st}, *J. Geophys. Res., 80*, 4204, 1975.

Cornwall, J. M., Cyclotron instabilities and electromagnetic emission in the ultra low frequency and very low frequency ranges, *J. Geophys. Res., 70*, 61, 1965.

Erlandson, R. E., L. J. Zanetti, T. A. Potemra, L. P. Block, and G. Holmgren, Viking magnetic and electric field observations of Pc 1 waves at high latitudes, *J. Geophys. Res., 95*, 5941, 1990.

Erlandson, R. E., B. J. Anderson, and L. J. Zanetti, Viking magnetic and electric field observations of periodic Pc 1 waves: Pearl pulsations, *J. Geophys. Res., 97*, 14823, 1992.

Fujita, S., Duct propagation of a short-period hydromagnetic wave based on the International Reference Ionosphere model, *Planet. Space Sci., 35*, 91, 1987.

Fukunishi, H., T. Toya, K. Koike, M. Kuwashima, and M. Kawamura, Classification of hydromagnetic emissions based on frequency-time spectra, *J. Geophys Res., 86*, 9029, 1981.

Heacock, R. R., and V. P. Hessler, Pearl-type micropulsations associated with magnetic storm sudden commencements, *J. Geophys. Res., 70*, 1103, 1965.

Heacock, R. R., Spatial and temporal relations between Pi bursts and IPDP micropulsations, *J. Geophys. Res., 76*, 4494, 1971.

Heacock, R R, and M. Kivinen, Relation of Pc 1 micropulsations to the ring current and geomagnetic storms, *J. Geophys. Res., 77*, 6746, 1972.

Hirasawa, T., Effects of magnetospheric compression and expansion on spectral structure of ULF emission, *Memoirs Nat. Inst. Polar Res. Jpn.,* Special Issue no. 18, 127, 1981.

Jacobs, J. A., Y. Kato, S. Matsushita, and V. A. Troitskaya, Classification of geomagnetic micropulsations, *J. Geophys. Res., 69*, 180, 1964.

Kangas, J., A. Aikio, and T. Pikkarainen, Multistation correlation of ULF pulsation spectra associated with sudden impulses, *Planet. Space Sci., 36*, 1103, 1986.

Kennel C. F., and H. E. Petschek, Limit on stably trapped particle fluxes, *J. Geophys. Res., 71*, 1, 1966.

Kuwashima, M., T. Toya, M. Kawamura, T. Hirasawa, H. Fukunishi, and M. Ayukawa, Comparative study of magnetic Pc 1 pulsations between low latitudes and high latitudes: Statistical study, *Memoirs Nat. Inst. Polar Res. Jpn.,* Special Issue no. 18, 101, 1981.

Liemohn, H. B., Cyclotron-resonance amplification of VLF and ULF whistlers, *J. Geophys. Res., 72*, 39, 1967.

Mauk, B. H., and R. L. McPherron, An experimental test of the electromagnetic ion cyclotron instability within the earth's magnetosphere, *Phys. Fluids, 23*, 2111, 1980.

Olson, J. V., and L. C. Lee, Pc 1 wave generation by sudden impulses, *Planet. Space Sci., 31*, 295, 1983.

Plyasova-Bakounina, T. A., and E. T. Matveyeva, Relationship between pulsations of the Pc 1 type and geomagnetic storms, *Geomagn. Aeron., 8* (Engl. trans.), 153, 1968.

Saito, T., and S. Matsushita, Geomagnetic pulsations associated with sudden commencements and sudden impulses, *Planet. Space Sci., 15*, 573, 1967.

Teply, L. R., and R. C. Wentworth, Hydromagnetic emissions, x-ray bursts, and electron bunches, 1. Experimental results, *J. Geophys. Res., 67*, 3317, 1962.

Troitskaya, V. A., Pulsations of the earth's electromagnetic field with periods of 1 to 1.5 seconds and their connection with phenomena in the high atmosphere, *J. Geophys. Res., 66*, 5, 1961.

Troitskaya, V. A., E. T. Matveyeva, K. G. Ivanov, and A. V. Gul'yelmi, Change in the frequency of Pc 1 micropulsations during a sudden deformation of the magnetosphere, *Geomagn. Aeron., 8* (Engl. trans.), 784, 1968.

Verzariu, P., M. Sugiura, and I. B. Strong, Geomagnetic field variations caused by changes in the quiet-time solar wind pressure, *Planet. Space Sci., 20*, 1909, 1972.

Wentworth, R. C., Enhancement of hydromagnetic emissions after geomagnetic storms, *J. Geophys. Res., 69*, 2291, 1964.

R. E. Erlandson and L. J. Zanetti, The Johns Hopkins University Applied Physics Laboratory, Johns Hopkins Road, Laurel, MD 20723-6099.

M. J. Engebretson, Augsburg College, Department of Physics, Minneapolis, Minnesota 55454.

R. Arnoldy, University of New Hampshire, Institute of Earth, Oceans, and Space, Durham, New Hampshire 03824.

T. Bösinger and K. Mursula, University of Oulu, Department of Physics, Oulu, Finland SF-90570.

Properties of Structured and Unstructured Pc1 Pulsations at High Latitudes: Variation Over the 21st Solar Cycle

KALEVI MURSULA, JORMA KANGAS AND TAPANI PIKKARAINEN

Department of Physics, University of Oulu, Oulu, Finland

In this paper we study the properties of the two dominant types of Pc1 micropulsations observed at high latitudes, structured and unstructured pulsations, and present quantitative results on their behaviour over the 21st solar cycle. This analysis covers the two spring equinox months (March and April) in two sunspot minimum years, 1975-76, and in two maximum years, 1979-80. A strong depletion of both Pc1 types is observed in sunspot maximum years with the amount of structured pulsations being slightly more depleted than unstructured pulsations. We present the diurnal distributions of both pulsation types for minimum and maximum years separately and note on interesting changes over the solar cycle. We have also calculated the average period of the two pulsation types during the two sunspot phases. While the average period of unstructured Pc1's remains nearly constant, the average period of structured pulsations is found to increase in sunspot maximum years. This high-latitude result is opposite to observations at low and mid-latitudes. Finally, we discuss the observed solar cycle changes in terms of the deterioration of the ionospheric wave guide in sunspot maximum years and the development of the plasmapause.

INTRODUCTION

The solar cycle variation of the occurrence of Pc1 micropulsations has been studied both at low and mid-latitudes [Benioff, 1960; Fraser-Smith, 1970 and 1981; Matveyeva et al., 1972; Strestik, 1981; Fujita and Owada, 1986; Matveyeva, 1987] and at high latitudes [Kawamura et al., 1983; Mursula et al., 1991]. The main result from these studies, roughly valid for all latitudes, is that the dominant feature in the long-term behaviour of the Pc1 activity is the sunspot cycle, and that more Pc1 pulsations occur during the sunspot minimum years than during the maximum years. Even in a more detailed comparison [Mursula et al., 1991], the long-term Pc1 activity curves at low and high latitudes were shown to have interesting similarities, following the changes in the annual sunspot activity from one solar cycle to another.

Some differences are also known to exist in the long-term Pc1 cycles at different latitudes. The inverse relation between sunspot numbers and Pc1 occurrence at low and mid-latitudes is only approximate. Fraser-Smith [1970] and Matveyeva et al. [1972] have reported that the low- and mid-latitude Pc1 events have their maximum occurrence rate during the declining phase of the sunspot cycle rather than at the minimum. This behaviour differs from the strong and direct negative correlation observed between annual sunspot numbers and the low-frequency Pc1's at high latitudes [Mursula et al., 1991] with a Pc1 maximum at sunspot minimum times. However, these seemingly contradictory observations may be reconciled when the different frequency ranges and the differences between the structured and unstructured pulsations, the two dominant pulsations types at high latitudes, are taken into account.

While structured pulsations (also called periodic or pearl emissions) are the dominant Pc1 pulsation type at low- and mid-latitude stations [Fraser-Smith, 1970; Kawamura, 1970; Kuwashima et al., 1981], the majority of Pc1 pulsations at high latitudes are unstructured pulsations, particularly of the type of hydromagnetic chorus [Nagata et al., 1980; Fukunishi et al., 1981]. Many properties, e.g. diurnal distributions and average frequencies, are known to be different for structured and unstructured pulsations [Kuwashima et al., 1981]. It has also been found that structured pulsations are intimately related to the evolution of magnetic storms [Wentworth, 1964; Plyasova-Bakunina and Matveyeva, 1968; Kuwashima et al., 1981], but the unstructured pulsations do not show such a dependence [Kuwashima et al., 1981]. These dramatic differences have led to the present view that,

Solar Wind Sources of Magnetospheric Ultra-Low-Frequency Waves
Geophysical Monograph 81

while the source region of structured pulsations is close to the plasmapause, the unstructured pulsations originate at much higher latitudes.

In the present paper we study the properties of structured and unstructured Pc1 pulsations observed at high latitudes, concentrating on the variation of these properties over the minimum and maximum phase of the 21st sunspot cycle. In the next section we will introduce the equipment used and the data collected for this analysis. Then, in section 3 we will present our results for the occurrence rates of structured and unstructured Pc1's. Section 4 deals with the diurnal distributions and their changes over the solar cycle. The results for the Pc1 period are presented in section 5. In section 6 we will discuss our observations on solar cycle changes of Pc1 properties trying to reconcile the seemingly disagreeing results on the sunspot cycle changes of Pc1 pulsations at different latitudes. Section 7 concludes the paper.

EQUIPMENT AND DATA

In this study we used data from a search-coil magnetometer situated at the high-latitude station of the Sodankylä Geophysical Observatory (geographic coordinates 67.4° Lat, 26.6° Long, corrected geomagnetic coordinates 63.9° Lat, 109° Long, L=5.2), Finland. The output signal from the magnetometer is amplitude modulated and registrated in analogue form on 6"/h and 24"/h magnetic tapes. The magnetic registrations were analyzed with an analogue/digital sonagraph (Model 5500, Kay Elemetrics Corp.), and sonagrams were made for any Pc1 range activity detected.

The main aim of the present study is to examine the variation of the properties of Pc1 pulsations during two opposite phases of the solar cycle. For that purpose, we decided to compare the two equinox months, March and April, of the two sunspot number minimum years 1975-76 with the same months of the two maximum years 1979-80 of the 21st solar cycle. Restricting the analysis to just a couple of months per year made it possible, firstly, to decrease the amount of events to be analyzed and, secondly and more importantly, to minimize the influence of seasonal variations which might mask out the solar cycle variations. (It is known since long ago [Benioff, 1960; Strestik, 1981] that e.g. the diurnal distribution and some other properties of structured Pc1's show seasonal variations). Using data from two years for both phases of the sunspot cycle, we could increase statistics and decrease random fluctuations in data. This also helps in balancing the fluctuations in magnetic activity and in the occurrence of magnetic storms that are an essential factor especially for structured pulsations. By the choice of equinox months we could maximize the number of structured Pc1 events whose annual occurrence maximizes at equinoxes [Fraser-Smith, 1970].

The observed Pc1 pulsation events were divided into five types according to the morphology of the sonagram

(dynamic spectrum) of the event. The first type consists of classical structured Pc1's which show clear repetitive intensity maxima. Unstructured pulsations, which do not show such regular intensity variation, form the second Pc1 type. They mainly consist of hydromagnetic chorus and Pc1-2 band-type events according to the classification of Fukunishi et al. [1981]. In Figure 1 sample sonagrams of these two types are presented. Structured and unstructured Pc1's form the large majority of all events observed at Sodankylä (see section 3) and the remaining three groups are just a rather small fraction of all Pc1 events observed. In the present analysis we will neglect these three small Pc1 groups and concentrate on the two dominant Pc1 types, the structured and unstructured pulsations.

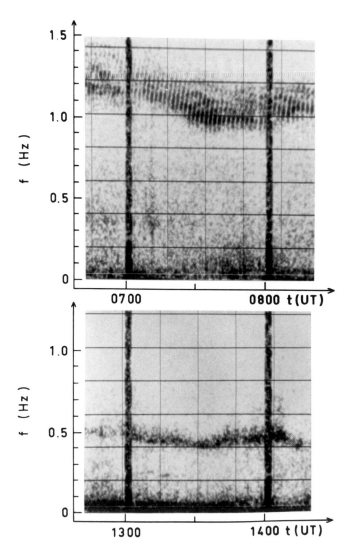

Fig. 1. Dynamic spectra for typical structured (top) and unstructured (bottom) Pc1 pulsations registered at Sodankylä on March 27, 1975, and March 6, 1976, respectively. The dark vertical lines are hour signals.

The properties of the detected Pc1 events were registered for each hour. We noted the start time and duration of each clearly separate Pc1 band, as well as its type, its maximum intensity during the hour and its possible continuation from the previous hours. Furthermore, for each hour and event, we noted the highest and lowest frequency and the mid-frequency at the hourly intensity maximum. In case of occasional multi-band Pc1 events, we treated each band as a separate event with its own properties. When studying the frequency of magnetic pulsations one should actually correct the raw data of each event with the frequency response of the magnetometer. However, because the Pc1 pulsation bands have quite a narrow frequency range of about 0.1-0.2 Hz, the average frequency of a Pc1 band, which is the main frequency property to be studied, remains practically constant in this correction. Therefore we can neglect the enormous amount of work needed for a complete correction of all Pc1 events, and use the raw data in our frequency analysis.

Pc1 OCCURRENCE IN MINIMUM AND MAXIMUM YEARS

We observed altogether 677 Pc1 events during the 8 months analyzed, with a total Pc1 active time of more than 42 days. The average length of all Pc1 events was about 90 minutes. The number of events and the total Pc1 active time in minutes are given in Table 1 separately for structured, unstructured and other pulsation types and the two intervals with minimum and maximum sunspot activity. We have also calculated the relative percentage of the two pulsation types during each interval.

First of all, one can see from Table 1 that there is a strong reduction in the overall Pc1 activity from the low sunspot years 1975-76 to high activity years 1979-80. While Pc1's appear up to about 30% of the time during minimum years, total Pc1 activity during maximum years is almost one order of magnitude lower. This is in accordance with the previous results showing a strong negative correlation between Pc1 activity at high latitudes and sunspot activity (see e.g. Mursula et al., 1991, and references therein). We would also like to note here that the Pc1 events during maximum years were found to be less intensive on an average than during minimum years.

Secondly, we can now also verify that the reduction in the occurrence of Pc1 pulsations during high sunspot number years is separately true for the two dominant Pc1 types at high latitudes, i.e. the structured and unstructured pulsations, as well as for the remaining Pc1 pulsation types in total. However, the amounts by which the different pulsation types have been reduced seem to be different. While the structured Pc1's were reduced nearly by one order of magnitude over the sunspot cycle, the corresponding factor for unstructured pulsations was about 7. The remaining Pc1 types were reduced even less, i.e. by factor of 3.

It is interesting to note that the relative fraction of

unstructured pulsations remained constant over the sunspot cycle so that 53% of all Pc1 activity both in minimum and maximum years were classified as such. On the other hand, due to the above mentioned large reduction factor, the relative fraction of structured pulsations in maximum years (30%) was smaller than in minimum years (40%). This reduction is seen as an increase in the relative amount of other types of Pc1's during maximum years. A possible explanation could be that those effects that cause the general reduction in Pc1 activity in maximum years may more strongly affect structured pulsations that are vulnerable to lose the coherence of the wave packets and thus to get misidentified.

Furthermore, using the values given in Table 1 for the Pc1 active time and the number of events, one can get an estimate for the average duration of Pc1 events of each type. In sunspot minimum years, the average durations of structured and unstructured Pc1 pulsations are approximately the same (about one hundred minutes) but the other events are shorter by roughly a factor of two. In sunspot maximum years, the average duration of structured pulsations remains approximately the same as in minimum years but the unstructured pulsations get shorter by a factor of 2, being thus approximately as short as the remaining Pc1 events whose average duration remains the same over the sunspot cycle.

DIURNAL DISTRIBUTIONS IN MINIMUM AND MAXIMUM YEARS

We have calculated the diurnal variation of the structured and unstructured Pc1's by counting the hourly duration of events of these two types separately. The results for the minimum and maximum sunspot years are presented in Figures 2 and 3, respectively. As can be seen in Figure 2, the distribution for the unstructured pulsations in minimum years attained an almost Gaussian form around the maximum at 11 UT (local time is 2 hours ahead of UT, MLT about 3 hours). Its nighttime activity was more than an order of magnitude lower than in daytime. On the

	Number of events	Total time (min/%)	Average frequency (Hz) according to	
			max. int.	band average
1975-76				
Structured	214	20975/40	0.75	0.77
Unstructured	266	28187/53	0.43	0.44
Other	72	3928/7	0.75	0.78
All	552	53090	0.58	0.60
1979-80				
Structured	21	2219/30	0.63	0.64
Unstructured	79	3936/53	0.38	0.40
Other	25	1292/17	0.70	0.69
All	125	7447	0.51	0.52

Table 1. Number of events, total times (minutes/relative percentages) and average frequencies (Hz) of structured, unstructured and other Pc1 pulsations during sunspot minimum years (1975-76) and maximum years (1979-80).

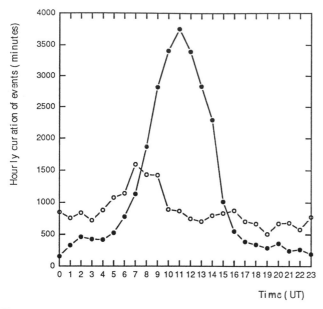

Figure 2. The diurnal distribution of the hourly duration of structured (open circles) and unstructured pulsation events (black circles) in minimum years.

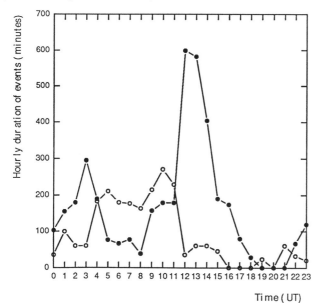

Figure 3. The diurnal distribution of the hourly duration of structured (open circles) and unstructured pulsation events (black circles) in maximum years.

other hand, the diurnal distribution of structured Pc1's in minimum years had a far less pronounced maximum in late morning hours. Its activity remained relatively high throughout the day and night, exceeding that of unstructured pulsations outside the noon-afternoon sector.

In sunspot maximum years (see Figure 3), the pronounced postnoon maximum of the unstructured pulsations also exists but seems to be shifted by 1-2 hours later.

Unstructured Pc1 activity outside this postnoon maximum was relatively larger in maximum years, in particular in the early morning sector where a secondary maximum appeared. Note that the overall activity around this secondary maximum was only reduced by a factor of 2 from minimum years, while around the postnoon maximum this reduction was considerably larger.

The diurnal distribution of structured Pc1's in maximum years was also slightly different from that in minimum years. Structured Pc1's almost exclusively appeared during morning hours. It is curious to note that the time of appearance of structured Pc1's fits very nicely with the minimum between the two maxima in the distribution of the unstructured pulsations.

AVERAGE PC1 PERIOD IN MINIMUM AND MAXIMUM YEARS

In order to study the solar cycle behaviour of the period of Pc1 pulsations we have calculated the average of the observed mid-frequency at the hourly intensity maximum for structured, unstructured and the remaining Pc1's and the two sunspot phases separately. These results are also presented in Table 1. In addition, we have repeated this procedure for the average value of the highest and lowest hourly frequencies. In Table 1 we call this the "band average". As can be seen in Table 1, the values derived for the average frequencies using the two methods are very close to each other, giving additional confidence in the results obtained.

Table 1 shows that there are notable differences in the average frequencies of the different pulsation types. For example, the results verify in a quantitative way the earlier qualitative observation [Fukunishi, 1981] that the average frequency of structured pulsations is higher than that of unstructured pulsations. In sunspot minimum years (1975-76) this difference was as large as 0.32-0.33 Hz. The average frequency of the remaining events in minimum years was nearly the same as that of structured Pc1's.

In sunspot maximum years (1979-80) the average frequencies of all pulsation were at least a little lower than the corresponding values in minimum years. This decrease was particularly large for structured pulsations while unstructured pulsations remained nearly at the same value and other pulsations experienced a smaller decrease. Accordingly, the difference between the average frequencies of structured and unstructured pulsations in maximum years was slightly smaller (about 0.24-0.25 Hz) than in minimum years.

The general tendency of decreasing Pc1 frequencies could, at least in principle, also result from an instrumental (rather than a physical) effect, e.g. due to the roaming of the frequency response of the magnetometer. However, such roaming has not been observed in the annual calibrations. Furthermore, the observed changes do not support such an interpretation. In minimum years, structured and group three (other) pulsations had identical average frequencies while in maximum years they had different average frequencies. If the frequency shift were purely an instrumental effect, the average frequencies of these two groups should change by equal amounts.

DISCUSSION

As already mentioned above, our observations verify the strong depletion of all main types of Pc1 micropulsations at high latitudes during sunspot maximum years. This is particularly interesting from the point of view of our previous analysis [Mursula et al., 1991] where a strong negative correlation between the annual low-frequency Pc1 activity and annual sunspot number was found at high latitudes over several solar cycles. However, it was not possible to make a distinction between the different pulsation types in that analysis.

The observed decrease of the average frequency of all pulsation types from 0.58-0.60 Hz in minimum years to 0.51-0.52 Hz in maximum years is also relevant for the interpretation of our earlier results [Mursula et al., 1991]. In that analysis only pulsations at the eigenfrequencies of 0.3 Hz and 0.5 Hz were detected. Thus the observed negative correlation could be explained by a possible rise of the average Pc1 frequency with increasing sunspot number, and not due to the actual depletion of Pc1 activity. Therefore the present observation of a decreasing (and not increasing) average Pc1 frequency with increasing solar activity excludes this interpretation and gives further confidence in the method used and the results obtained in that analysis.

The observed diurnal distribution of unstructured pulsations agrees well with the post-noon maximum of magnetospheric ion cyclotron waves recently detected by Anderson et al. (1992) in a large statistical analysis using the AMPTE/CCE satellite. They also observed that the probability to find these waves increases with latitude and is much higher at auroral latitudes ($L\simeq7-8$) than at mid-latitudes ($L\simeq3-4$). This also agrees with the observed increase of total ground-based Pc1 activity with latitude [Troitskaya and Gul'elmi, 1970]. Since unstructured pulsations are detected almost exclusively at high latitudes it is very likely that the waves detected by Anderson et al. (1992) are the source of unstructured pulsations.

It has been known for many years that the daily distribution of structured pulsations at low and mid-latitudes has its maximum at dawn (for a review see Saito, 1969). This has been connected with the minimum in the ionospheric ionization resulting in better propagation conditions of the ionospheric wave guide in early morning hours. However, our observation for the diurnal distribution of structured Pc1's shows that, at high latitudes, this maximum occurs later than at low or mid-latitudes and also later than the local minimum in the ionospheric ionization. This difference may be due to the fact that the high-latitude source can also produce structured pulsations to be detected on ground at high latitudes. Thus the high-latitude station of Sodankylä receives part of the structured pulsations from plasmapause which normally is located at lower latitudes, and part from the higher latitude source. This view is further supported by the fact that the average frequency decreases from morning till noon when approaching the high-latitude source. On the other hand, the mid- and low-latitude stations would, due to the longer ducting distance, only observe the lower latitude plasmapause connected events.

The differences in the diurnal distributions between sunspot maximum and minimum times may be explained by the fact that the ionospheric ducting conditions are worse in maximum years, especially at the dayside. This may be the reason e.g. to the greater depletion of unstructured pulsations around the local noon than in the early morning sector, leading to the formation of the secondary maximum in early morning hours and shifting the diurnal maximum of unstructured pulsations by 1-2 hours later. Note that the time of the secondary maximum corresponds well with the time of minimum ionospheric ionization. The fact that structured pulsations are more depleted in maximum years outside morning hours may also reflect deteriorated ducting conditions.

The observed decrease of the average frequency of high-latitude structured pulsations with solar activity is very interesting in view of the fact that, correspondingly, an increase (rather than decrease) was observed in maximum and post-maximum sunspot years at low and mid-latitudes [Strestik, 1981; Matveyeva, 1987]. Supposing that the ionospheric wave guide is deteriorated during sunspot maximum years, the stations would mainly observe pulsations whose footpoint is quite close to the station. Accordingly, in maximum years, the low- and mid-latitude stations would observe those events for which the plasmapause is located at lower latitudes than on an average while in minimum years they would register events also from further off. Since the average pulsation frequency decreases with increasing latitude for L-values up to about $L\simeq5-7$ [Erlandson et al. 1990; Anderson et al., 1992], this implies that the average frequency at low and mid-latitudes is higher in maximum than in minimum years.

On the other hand, the same mechanism implies that a high-latitude observer that is located outside the average plasmapause would, in maximum years, only detect those plasmapause connected structured events where the plasmapause is at higher latitudes than on an average. This observer would, opposite to the low-latitude observer, find that the average frequency decreases (rather than increases) in maximum years. However, due to the high-latitude source, the situation is more complicated at high latitudes than low latitudes.

There is another mechanism contributing to the difference between the average frequencies at low and high latitudes.

This mechanism is related to the change of the average plasmapause location over the solar cycle. Following the more frequent occurrence of storms, the plasmapause is, on an average, at lower latitudes in maximum years. This implies that the average frequency of structured pulsation rises in maximum years, in accordance with what is observed at low and mid-latitudes. (Note that this mechanism incorrectly predicts that more events are observed in maximum years than minimum years. Therefore this mechanism alone can not be responsible for all observations). Accordingly, the two mechanisms (ducting conditions and plasmapause location) work in the same direction as to the change of the average frequency at low and mid-latitudes.

At high latitudes, the two mechanisms work oppositely: deteriorated ducting conditions decreasing the average frequency of structured pulsations, and lower plasmapause location increasing it. The lower position of the plasmapause would also imply less of plasmapause related events at high latitudes due to longer propagation distance. Thus the relative significance of the high-latitude source with low frequency would increase leading to a lower average frequency of structured pulsations at high latitudes. The plasmapause mechanism can also explain the observation that structured pulsations are relatively more depleted at high latitudes than unstructured or other Pc1 types. (Note that this mechanism predicts correctly a depletion of structured pulsations in maximum years at high latitudes. Again, however, it is not sufficient alone since it does not explain the depletion of unstructured Pc1's).

The average frequency of unstructured pulsations was observed to remain nearly constant over the solar cycle. Since their source is at high latitudes and the average frequency of this high-latitude source remains constant with latitude beyond $L \simeq 5\text{-}7$ [Erlandson et al., 1990; Anderson et al., 1992], our observation of the constancy of the frequency of unstructured pulsations is in accordance with the above idea of deteriorating wave guide. The observed decrease of the average duration of unstructured pulsations during sunspot maximum years may result from the deteriorated ducting conditions as well, reflecting natural intensity variations that may have remained unobserved during minimum years with stronger average pulsation intensity.

Finally, we would like to note that the time interval of two months is not very long in view of storm development. Accordingly, the results on structured pulsations that depend partly on the plasmapause development may be affected by the number and nature of storms occurring during the studied interval, and need to be verified using observations from longer time intervals or other solar cycles. Such an extended analysis is underway.

Conclusions

We have studied in this paper the properties of structured and unstructured pulsations, the two dominant Pc1 types

at high latitudes, and presented new quantitative results on their behaviour over the 21st solar cycle. Strong depletion of both types of Pc1's was observed in maximum sunspot years. The main features of diurnal distributions with late morning and afternoon maxima for structured and unstructured pulsations respectively remained valid over the solar cycle. However, interesting differences between maximum and miminum years appeared in the detailed properties of these distributions. Furthermore, while the average frequency of unstructured pulsations remained nearly constant over the solar cycle, the average frequency of structured pulsations decreased slightly in maximum years, in contrast to the increase observed at low latitudes.

We discussed how these observations can be understood in terms of the deterioration of the ionospheric wave guide in sunspot maximum years. However, other effects have to be taken into account as well, in particular the development of the plasmapause because of its connection with structured Pc1 pulsations.

References

Anderson, B. J., R. E. Erlandson, and L. Zanetti, A statistical study of Pc1-2 magnetic pulsations in the equatorial magnetosphere. 1. Equatorial occurrence distributions, *J. Geophys. Res.*, *97*, 3075-3088, 1992.

Benioff, H., Observations of geomagnetic fluctuations in the period range 0.3 to 120 seconds, *J. Geophys. Res.*, *65*, 1413-1422, 1960.

Erlandson, R. E., L. Zanetti, T. A. Potemra, L. P. Block, and G. Holmgren, Viking magnetic and electric field observations of Pc1 waves at high latitudes, *J. Geophys. Res.*, *95*, 5941-5955, 1990.

Fraser-Smith, A. C., Some statistics on Pc1 geomagnetic micropulsation occurrence at middle latitudes: Inverse relation with sunspot cycle and semiannual period, *J. Geophys. Res.*, *75*, 4735-4745, 1970.

Fraser-Smith, A. C., Long-term predictions of Pc1 geomagnetic pulsations: Comparison with observations, *Planet. Space Sci.*, *29*, 715-719, 1981.

Fujita, S., and T. Owada, Occurrence characteristics of Pc1 pulsations observed at Japanese observatory network, *Mem. Natl. Inst. Polar Res., Special Issue*, *42*, 79-91, 1986.

Fukunishi, H., T. Toya, K. Koike, M. Kuwashima, and M. Kawamura, Classification of hydromagnetic emissions based on frequency-time spectra, *J. Geophys. Res.*, *86*, 9029-9039, 1981.

Kawamura, M., Short-period geomagnetic micropulsations with period of about 1 second in the middle latitudes and low latitudes, *Geophys. Mag.*, *35*, 1-53, 1970.

Kawamura, M., M. Kuwashima, T. Toya, and H. Fukunishi, Comparative study of magnetic Pc1 pulsations observed at low and high latitudes: Long-term variation of occurrence frequency of the pulsations, *Mem. Natl. Inst. Polar Res., Special Issue*, *26*, 1-12, 1983.

Kuwashima, M., T. Toya, M. Kawamura, T. Hirasawa, H. Fukunishi, and M. Ayukawa, Comparative study of magnetic Pc1 pulsations between low latitudes and high latitudes: Statistical study, *Mem. Natl. Inst. Polar Res., Special Issue*, *18*, 101-107, 1981.

Matveyeva, E. T., Cyclic variation of the activity of Pc1 geomagnetic pulsations, *Geomagn. Aeron., Engl. Transl.*, *27*, 392-395, 1987.

Matveyeva, E. T., V. A. Troitskaya, and A. V. Gul'elmi, The long term statistical forecast of geomagnetic pulsations of type Pc1 activity, *Planet. Space Sci.*, *20*, 637-638, 1972.

Mursula, K., J. Kangas, T. Pikkarainen, and M. Kivinen, Pc1 micropulsations at a high-latitude station: A study over nearly four solar cycles, *J. Geophys. Res.* *96*, 17651-17661, 1991.

Nagata, T., T. Hirasawa, H. Fukunishi, M. Ayukawa, N. Sato, R. Fujii, and M. Kawamura, Classification of Pc1 and Pi1 waves observed in high latitudes, *Mem. Natl. Inst. Polar Res., Special Issue, 16,* 56-71, 1980.

Plyasova-Bakunina, T. A., and E. T. Matveyeva, Relationship between pulsations of the Pc1 type and geomagnetic storms, *Geomagn. Aeron., Engl. Transl., 8,* 153-155, 1968.

Saito, T., Geomagnetic pulsations, *Space Sci. Rev., 10,* 322-401, 1969.

Strestik, J., Statistical properties of earth current pulsations Pc1 at the observatory of Budkov in 1965-71, *Studia geophys. et geod., 25,* 181-191, 1981.

Troitskaya, V. A. and A. V. Gul'elmi, Hydromagnetic diagnostics of plasma in the magnetosphere, *Ann. Geophys. 26,* 893-902, 1970.

Wentworth, R. C., Enhancement of hydromagnetic emissions after geomagnetic storms, *J. Geophys. Res., 69,* 2291-2298, 1964.

K. Mursula, J. Kangas and T. Pikkarainen, Department of Physics, University of Oulu, FIN-90570 Oulu, Finland.

Simultaneous Occurrence of Pc 5 and Pc 1 Pulsations in the Dawnside Magnetosphere: CRRES Observations

R. Rasinkangas[1], K. Mursula[1], G. Kremser[1,4],

H. J. Singer[2], B. J. Fraser[3], A. Korth[4], and W. J. Hughes[5]

An abrupt increase in the solar wind pressure compressed the magnetosphere on September 11, 1990, and started a long chain of intense, toroidal Pc 5 waves at 0540 UT. The waves were observed by the CRRES satellite in the dawnside magnetosphere (06 MLT, L = 7), and on the ground, e.g., by the Scandinavian magnetometer chain at about 08 MLT. Simultaneous increases in the energetic (> 20 keV) particle flux and the geomagnetic field induction were also observed by CRRES. A series of intense bursts of ion cyclotron waves (Pc 1) started soon after the onset of the compression at about 0600 UT, coinciding with an increase in the magnetospheric electron number density. The pulsations lasted until the energetic particle fluxes decreased at about 0800 UT. The observed waves have the characteristics of the newly detected dawnside Pc 1 waves on high L-shells, which get their energy from ions on open drift paths. Toward the end of the event, after a strong additional enhancement of the electron number density, the Pc 1 bursts are repeated at the Pc 5 frequency and appear at a fixed Pc 5 phase, indicating a possible modulation of the Pc 1 wave growth rate by Pc 5 waves. We discuss the particle and wave characteristics of the event and in particular the possible connection of Pc 1 and Pc 5 waves. We suggest that the Pc 1 wave modulation is produced by the electric field of the Pc 5 wave that modulates the equatorial cold plasma density by moving plasma from higher to lower L-shells and back, thus affecting the growth rate of Pc 1 waves.

INTRODUCTION

The Combined Release and Radiation Effects Satellite (CRRES) was launched into a geostationary transfer orbit in July 1990. The apogee was at 6.3 R_E and rotated from about 08 MLT at the beginning of the mission via the nightside to about 14 MLT when the satellite ceased operation in October 1991. The satellite reached L values in excess of 7 and magnetic latitudes of about ±30° due to the inclination of 18.2°. Several strong pulsation events in the Pc 5 frequency range (T ≃ 150 - 600 seconds) were observed by the magnetic field instrument during the fall of 1990, when the apogee was in the dawn sector. In this work we will investigate one such event from September

[1]Department of Physics, University of Oulu, Finland

[2]Geophysics Directorate/GPSG, Phillips Laboratory, USA

[3]Department of Physics, University of Newcastle, Australia

[4]Max-Planck-Institut für Aeronomie, Katlenburg-Lindau, Germany

[5]Center for Space Physics, Boston University, USA

Solar Wind Sources of Magnetospheric Ultra-Low-Frequency Waves
Geophysical Monograph 81
Copyright 1994 by the American Geophysical Union.

11, 1990 (orbit 115). Strong Pc 1 pulsation bursts were recorded in addition to the Pc 5 pulsations. Our main topic is the possible connection between these pulsations.

Azimuthally polarized, toroidal Pc 5 waves are often observed in the dawn sector of the Earth's magnetosphere. They are also called type A waves. At geosynchronous orbit, these waves occur most frequently between 05 and 09 LT and have typical periods of about 100-200 seconds. They have been found to be associated with high speed (>600 km/s) solar wind, and moderately disturbed conditions as judged from the AE index [Kokubun et al., 1989]. They appear to be fundamental odd mode field line resonances (FLR) with a magnetic node at the equator [Singer and Kivelson, 1979]. The horizontal spatial scale of the perturbations is large (azimuthal wave numbers are small, m < 10), and the ground-magnetosphere correlation of the waves is thus high [Kokubun et al., 1989]. However, when recorded from the ground, the pulsations show a 90° counter clockwise rotation of magnetic field variation due to modifications in the ionosphere [Hughes, 1974].

Several mechanisms have been proposed to explain the formation of waves in the Pc 5 frequency range: 1) The Kelvin-Helmholtz instability (KHI) at the magnetopause can generate boundary surface waves when the magnetosheath plasma flow velocity exceeds a threshold velocity [e.g., Cahill and Winckler, 1992]. 2) Pressure variations in the magnetosheath plasma flow can also be the source of magnetopause surface waves [see, e.g., Warnecke et al., 1990]. For example, Lysak and Lee [1992] have modelled the response of the magnetosphere to solar wind pressure pulses: compressions produce compressional (fast mode) Alfvén waves, which can couple and mode-convert into shear Alfvén waves at the point where the compressional wave frequency matches the shear mode eigenfrequency of the field line. 3) Local magnetic anomalies like flux transfer events (FTE) or other intermittent reconnection (or penetration) processes may also be associated with magnetopause boundary waves [Cahill and Winckler, 1992]. 4) Some studies show that Pc 5 pulsation events may be related to recovery phases of substorms, as the injected electrons trigger the pulsations while drifting eastward [Saka et al., 1992].

For ion cyclotron waves (or Pc 1 pulsations), the dawnside magnetosphere is a very interesting region. This is because the waves can get their energy also from other sources than the ring current ions drifting near the plasmapause. For example, it has recently been shown that Pc 1 pulsations at $L \geq 7$ in the dawnside get their energy from plasma sheet ions on open drift paths [Anderson et al., 1992a,b]. These pulsations are characterized by higher normalized frequencies ($X = f/f_{H^+} \simeq 0.4$, where f_{H^+} is the local hydrogen gyrofrequency) than elsewhere, and linear polarization in contrast to the more common left-hand polarization. At very high-latitude ground-stations, Pc 1 pulsations have been observed that may get their energy from ions injected in the cusp/cleft region and drifting westward toward dawn [Hansen et al., 1992]. Furthermore, solar wind compressions have also been shown to affect the occurrence and other properties of the Pc 1 waves [Olson and Lee, 1983] by increasing the ion anisotropy which increases the wave growth rate.

Only few studies have discussed so far the possible connection of Pc 5 and Pc 1 waves. For example, the possible amplitude-modulation of Pc 1 and Pc 2 waves by storm time compressional Pc 5 waves in the afternoon sector was studied by Barfield and McPherron [1972] with negative results. Recently Fraser et al. [1992] have reported of a possible Pc 5 modulation of Pc 2 wave generation.

INSTRUMENTATION AND DATA ANALYSIS

In this study, data are used from the Fluxgate Magnetometer and the Electron Proton Wide-Angle Spectrometer (EPAS, also known as MEB) on the CRRES satellite. The time resolution of the magnetometer is 16 samples per 1.024 seconds, and conversion to digital data is done using a 12-bit A/D converter. The least-significant bit resolution during the event studied was 0.43 nT, and the dynamic range was +850 nT. The time resolution of the data used in this study was about 0.125 seconds. For

more details of the magnetometer, see Singer et al. [1992]. The EPAS instrument measures medium energy electrons (21.5 - 285 keV, 14 energy channels) and protons (37 - 3200 keV, 12 energy channels). Electrons are measured in ten directions and ions in four directions over a total angular range of about 110° in the meridian plane of the satellite. Since the spin axis of the satellite is located "near" the orbital plane, the instrument provides an almost three dimensional particle distribution during one rotation. The sampling times for electron and proton spectra are 256 and 512 ms, respectively. The particle data were averaged over 60 seconds. The EPAS instrument is described in detail by Korth et al. [1992].

Some additional information is provided by the Plasma Wave instrument [Anderson et al., 1992] on CRRES. The instrument includes two receivers: a multichannel spectrum analyzer (5.6 Hz - 10 kHz) and a sweep frequency receiver (100 Hz - 400 kHz). The dynamic range for each of the receivers is about 100 dB beginning at the noise level. The sweep frequency receiver has four frequency bands; the lowest (50 - 400 kHz) includes the upper hybrid resonance frequency (f_{UHR}) emissions. This frequency can be used to derive the electron number density of the plasma with a time resolution of 8 seconds.

High resolution solar wind and IMF data from the IMP-8 satellite are also used with time resolution of about 1 minute and 15 seconds, respectively. On the ground, magnetometer data from IMAGE, the Scandinavian magnetometer chain [IMAGE Newsletter, 1992], and Finnish pulsation magnetometer registrations are used.

In order to investigate the characteristics of the Pc 5 pulsations, the magnetic field was converted from the Earth Centered Inertial (ECI) coordinate system into a field aligned coordinate system. Floating averages of the field measurements over 21 minutes (corresponding to about five cycles of the Pc 5 pulsations) were used to define the coordinate axes. The instantaneous measurements are then split in two components: $\mathbf{B}_\parallel$ is parallel, $\mathbf{B}_\perp$ perpendicular to the average field direction. $\mathbf{B}_\perp$ is then divided into a radial component $\mathbf{B}_r$ and an azimuthal component $\mathbf{B}_\phi$. The signs of these quantities are defined so that positive $B_\parallel$ points along the average magnetic field, positive B_r away from the Earth, and positive B_ϕ completes the right-handed coordinate system by pointing eastward. The wave polarization was studied using the method of eigenanalysis and coherency analysis of spectral matrix described, e.g., by McPherron et al. [1972], and Arthur et al. [1976, technique 1].

OBSERVATIONS

Solar Wind Pulse and Geomagnetic Activity

An abrupt increase in the solar wind density from about 20 to 40 cm^{-3} was observed by IMP-8 on September 11, 1990, at about 0528 UT (Figure 1a). The spacecraft was located roughly $X_{GSM} = 30\ R_E$, $Y_{GSM} = 5\ R_E$, and $Z_{GSM} = 15\ R_E$ during the event. The solar wind

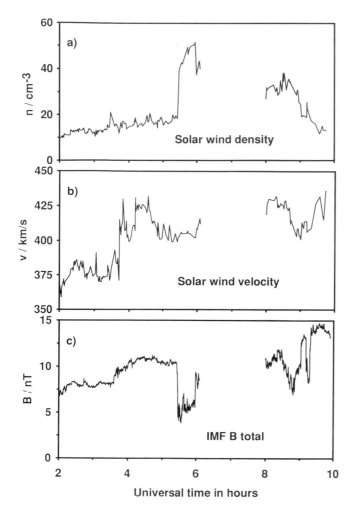

Fig. 1. Solar wind and IMF data from September 11, 1990. a) Solar wind density. b) Solar wind velocity. c) IMF B_{total}.

velocity remained almost constant at 420 km/s over the whole period studied (Figure 1b). However, a simultaneous change in the solar wind flow angle from westward to eastward was observed. Although there is a gap in the IMP-8 data between 0605 and 0800 UT, it seems likely that the density values were enhanced also during this interval. Note that even the initial value of $n_{SW} = 20$ cm^{-3} is higher than the average value, implying that the magnetosphere was already in a compressed state when the pressure pulse arrived. Dayside mid-latitude measurements on the ground (not included here) support this view. However, the main event can be attributed to the latter abrupt solar wind density enhancement. It takes a few minutes for this front to propagate from the IMP-8 location to the Earth's magnetopause. This delay agrees well with the start of the event in the magnetosphere at about 0540 UT.

The magnetic field magnitude B_{tot} of the IMF is presented in Figure 1c for the same period. B_{tot} decreases abruptly at the time of the density increase. During the preceding couple of hours B_{tot} was high indicating that the density enhancement may have compressed the IMF ahead of it. (Note, however, that the magnetic pressure is insignificant when compared with the plasma pressure during the whole event.) The IMF B_z component (not shown) was positive most of the time from about 0330 UT onwards up to the data gap; IMF had a southward component only for about 20 minutes just after the density increase (magnetic field decrease) at 0528 UT.

The D_{st} index showed positive values already prior to the event, reflecting the high solar wind plasma densities present during that time. The abrupt solar wind pulse was responsible for a strong positive maximum in the index of about 30. No clear storm period followed. The D_{st} index showed strongly fluctuating negative values of about ~ 20 during the next few days. The three hour Kp value was 4 (0600 - 0900 UT on September 11).

Onset of the Event in the Magnetosphere

During the event studied here, the CRRES satellite was located in the dawnside magnetosphere (06 MLT) at $L \simeq 7$, the magnetic latitude of the satellite being about $20°$ north. At about 0540 UT, the following changes were registered by CRRES in this region (Figures 2a-c): strong Pc 5 pulsations started with the main power in the azimuthal magnetic field component (corresponding approximately to X_{ECI} and X_{GSM} in this case), the parallel magnetic field (total field) strength increased by almost 10 nT, and energetic particle fluxes increased, particularly at pitch angles of $90°$. Strong VLF activity started also at the same time (data not shown). Somewhat later, at about 0607 UT, an increase in the electron number density was seen (as derived from the f_{UHR}, Figure 2d), and at about the same time, 0604 UT, also bursty Pc 1 emissions appeared (Figure 2e). The location of the satellite in L-value and magnetic local time during the event is seen in Figure 2f. On the ground, a strong enhancement in Pc 5 activity was observed at about 0540 UT by, e.g., the Scandinavian IMAGE magnetometer cross, which is located about 3 hours in local time ahead of CRRES. In Figure 2g the north-south magnetic field component from Kilpisjärvi is plotted. This component showed the best correlation with the pulsation data measured in space.

Pc 5 Pulsations

The Pc 5 pulsations starting at 0540 UT were observed by CRRES for several hours, but their amplitude changed considerably during the event, as can be seen from Figure 2a. Accordingly, we divide the event into five distinct intervals. Interval I (0540 - 0630 UT) shows pulsations with an amplitude of about 7 nT. Interval II (0630 - 0705 UT) shows a strong minimum in the Pc 5 activity, which is only partly recovered during interval III (0705 -

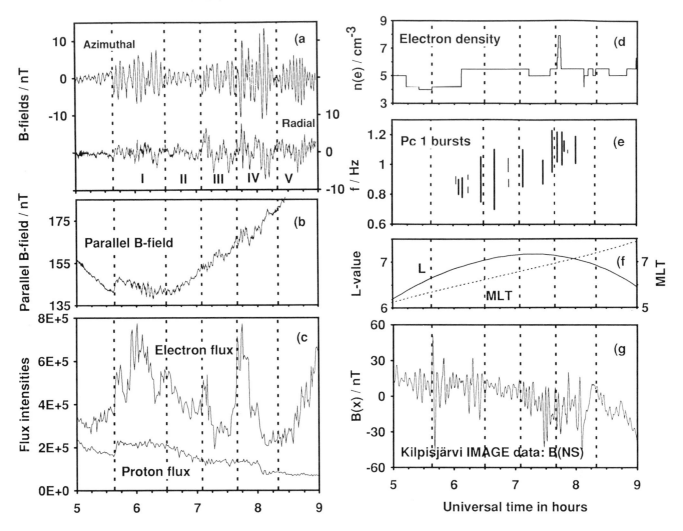

Fig. 2. The effect of the solar wind pulse in the magnetosphere (panels a to f) and on the ground (panel g). a) Azimuthal and radial magnetic field components (Pc 5 waves). b) Parallel ($\simeq$total) magnetic field strength. c) Proton (37 - 54 keV) and electron (21.5 - 31.5 keV) fluxes at 90° pitch angle in cm^{-2}s^{-1}sr^{-1} keV^{-1}. d) Electron number density. e) Location of the Pc 1 bursts during the event. f) L-value and MLT for the CRRES position. g) North-south magnetic field variations in Kilpisjärvi, Finland. The vertical dashed lines are plotted to facilitate comparison of the figures.

0740 UT), when the amplitude is about 4 nT. Interval IV (0740 - 0820 UT) shows the strongest pulsations with amplitudes up to more than 10 nT. Finally, during interval V (0820 - 0850 UT), a burst of Pc 5 pulsations with amplitudes of about 7 nT is again observed. The pulsation frequency was 4.0 mHz, and it was very stable throughout the event: only during interval V waves with 5.5 mHz frequency were measured. The predominantly linear polarization of the Pc 5 pulsations was verified by polarization calculations, according to which the absolute values of ellipticity remained smaller than 0.25 during the whole period (see also Plate 2b).

Particle Fluxes

The 90° pitch angle fluxes of protons in the energy range 37 - 54 keV and of electrons in the energy range 21.5 - 31.5 keV are plotted in Figure 2c (the fluxes at higher energies behaved similarly). The changes in these fluxes correlate partly with the changes in the Pc 5 activity (Figure 2a). The proton flux intensity shows a clear increase at the beginning of the event at 0540 UT, fairly constant values during interval I, a steady decrease during interval II, and again almost constant values during interval III. However, the beginning of interval IV is not seen in the proton flux, and the abrupt decrease of the flux intensity at about 0800 UT does not correlate with changes in the Pc 5 pulsations. The flux intensity is constant after this decrease, showing no correlation with the Pc 5 burst during interval V. A general trend in the proton flux is thus a slow decrease throughout the event after the initial enhancement at 0540 until 0805 UT. The electron fluxes show a similar general decreasing trend, but with much more variability during the event itself. First of all, they show a clear increase in intensity already at 0515 UT, well before the onset of the event at 0540 UT. After the sharp initial flux intensity increase at 0540 UT that lasts only about ten minutes there are other similar increases at about 0550, 0620, 0705 and 0740 UT. Most notably, the last two correlate with increases in the Pc 5 activity in Figure 2a, i.e., the beginnings of intervals III and IV, respectively. Also the VLF wave activity measured by CRRES (data not shown) follows these enhancements with increasing intensity. The local minimum in the electron flux between 0715 and 0735 UT may be significant, and this period can be seen as a local minimum also in electron number density (Figure 2d) and VLF emission activity.

Pc 1 Pulsations

Intense bursts of ion cyclotron waves (Pc 1 pulsations) were observed close to the CRRES apogee between about 0604 and 0802 UT, each lasting a few tens of seconds (see the schematic presentation for the appearance times of the bursts in Figure 2e). The electron number density increased to a higher level at 0607 UT (Figure 2d), which correlates rather well with the start of the Pc 1 pulsations. A strong peak in this density is seen at about 0743 UT during the period of high fluxes of energetic electron, together with a fast repetition rate of Pc 1 bursts. The abrupt end of the pulsations after 0802 UT correlates well with the decrease in the energetic particle fluxes. The average frequency of the bursts increased slowly from below 0.9 Hz to above 1.1 Hz during the event in accordance with the increase of the magnetic field intensity (Figure 2b). The local normalized frequency, $X = f/f_{H+}$, where f_{H+} is the local proton gyrofrequency, has an almost constant value of $X = 0.4$, i.e., all wave activity is above the He^+ gyrofrequency. We have estimated the normalized frequency at the equator. From the Tsyganenko's magnetic field model T89 [Tsyganenko, 1989] the ratio between the magnetic field intensities at the location of CRRES and at the equator was calculated. From this value and the measured intensity at CRRES normalized frequencies of about 0.6 - 0.7 in the equatorial region were derived.

The ellipticity of the bursts is shown in Plate 1. This data is calculated from the magnetic field measurements in the GSM coordinate system. The bursts are mainly linearly polarized, and the small amount of left-handed polarization is striking.

Weak signals of similar bursty Pc 1 activity can also be seen on the Finnish pulsation magnetometers (data not shown). However, there is no one-to-one correlation with the satellite data, and the bursts started on the ground at about 0556 UT, i.e., a little earlier than at CRRES.

Modulation of Pc 1 Pulsations by Pc 5 Waves

During the first half of interval IV the repetition rate of Pc 1 bursts matches the frequency of the Pc 5 waves. In Plate 2a a short interval between 0725 and 0805 UT is shown that contains several Pc 1 bursts. The azimuthal and radial B-field components of the Pc 5 waves are plotted in Plate 2b for comparison. It can be seen that six Pc 1 bursts out of seven occur on the falling slope of the field variation of the azimuthal component, and that between 0737 and 0754 UT the correlation is almost perfect. Note that there is a strong peak in the electron number density around this time (about 0743 UT), and that the electron flux intensities are clearly higher during the period 0737 - 0754 UT.

DISCUSSION

Processes Triggered by the Solar Wind Pressure Pulse

The observations described above suggest the presence of toroidal Pc 5 waves. The wave period of about 250 seconds is only a little longer than the typical values of 100-200 seconds observed at geosynchronous orbit, and may be attributed to a slightly larger L-value or a small component of heavy ions such as helium or oxygen. Also the ground signal is as expected for toroidal Pc 5 waves. The waves were very likely triggered by a solar wind pressure pulse in the form of an increase in the plasma density. The fact that the solar wind velocity showed no change and the IMF B_z was mainly northward rule out the Kelvin-Helmholtz instability and reconnection processes, respectively, as possible triggers for the pulsation. During the event the changes in the energetic electron fluxes occasionally correlate with the changes in the Pc 5 activity level. However, the observations here rule out the possibility that these electrons would have triggered the pulsations as discussed, for example, by Saka et al. [1992].

It is very likely that the start of the Pc 1 pulsations at about 0556 UT on the ground and at about 0604 UT at CRRES is also due to the solar wind pulse. The fact that the pulsations start earlier on the ground reflects the later local time of the Finnish pulsation magnetometer chain as compared to CRRES. The overall appearance of the Pc 1 bursts in space seems to correlate well with the time of increased proton (and electron) fluxes. The bursts start soon (but not immediately) after the particle flux increase and stop almost exactly at the time of the flux decrease.

In addition, the Pc 1 bursts were accompanied by increases in the electron number density. This is verified by the level increase of the density just after the first Pc 1 burst and also by the strong peak in density at around 0743 UT. It is well known that this increase

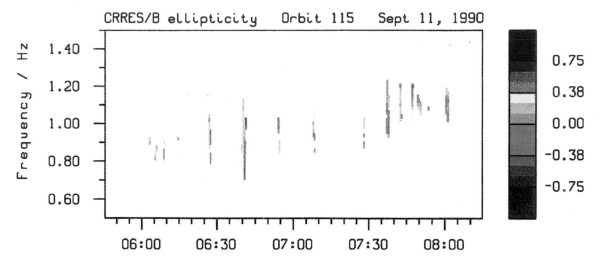

Plate 1. Ellipticity of the Pc 1 waves. Only points with sufficiently high intensity and polarization percentage are displayed. Positive (red) and negative (blue) values correspond to right-handed and left-handed polarization, respectively.

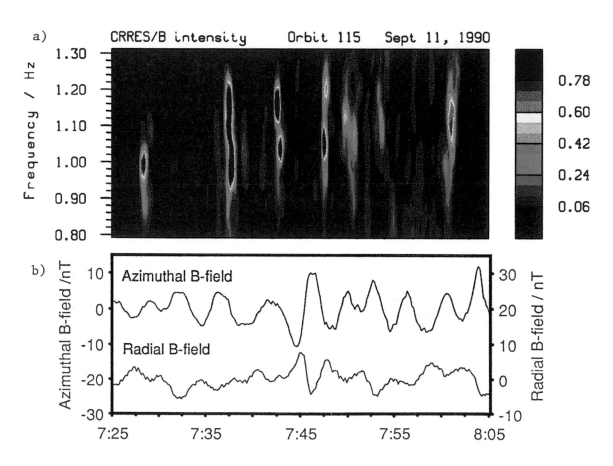

Plate 2. Modulation of the Pc 1 waves with Pc 5 pulsations. a) Intensity of the Pc 1 bursts (color surface plot) during a short interval. b) Azimuthal and radial magnetic field component showing Pc 5 pulsation cycles at the same time.

lowers the resonance energy of ions, thus allowing more ions to contribute to the ion cyclotron resonance, and enhancing the instability.

The variations in electron number density measured by CRRES may differ sizably from those occurring at the equatorial region where Pc 1 waves are generated. This is also a likely explanation for the appearance of the first Pc 1 bursts slightly before the density increase at CRRES. As drifting plasma clouds tend to be confined at the equator, the density increase is expected to occur earlier and to be larger at the equator than at 20° north of the equator where CRRES was located. Therefore, with increasing density, conditions for ion cyclotron waves were created at the equator slightly earlier than observed by CRRES. It is possible that the solar wind pressure pulse was also responsible for this plasma cloud, although we cannot verify this.

We have estimated the adiabatic effects of the magnetic field changes on the particle fluxes. The abrupt increase in the intensity of the total magnetic field at the beginning of the event at 0540 UT, undoubtedly due to the compression of the magnetosphere, is sufficient to explain the increases in energetic proton and electron fluxes by adiabatic betatron acceleration at that time. The general decreases in fluxes during the event may be attributed to the increase of the magnetic latitude during the period under study (changing from 14.5° at 0500 UT to 24.4° at 0900 UT). We will discuss the decrease in the proton flux during interval II in a separate study.

Relations Between Pc 5 Pulsations and Pc 1 Pulsations

The evidence for the modulation of Pc 1 waves by Pc 5 pulsations at 0737 - 0754 UT (Plate 2) is quite striking. The characteristics of the Pc 1 bursts are the same as found to be typical for the new Pc 1 population recently discovered by Anderson et al. [1992a, b]. These authors report on Pc 1 waves in the dawn sector at high L-values ($L \geq 7$), which are generated with linear polarization and exhibit high normalized frequencies, $X = 0.4$. All these properties are in agreement with our observations. However, due to the fairly high magnetic latitude of CRRES, we cannot argue whether the waves were already produced with linear polarization as claimed by Anderson et al. [1992a, b], or whether the polarization was changed from left-handed to linear during the propagation of the wave from the equator to the location of CRRES. In fact, a fairly small amount (of about 10 %) of He$^+$ would be enough to raise the cross-over frequency above the observed wave frequency and cause this change in wave polarization [Young et al., 1981]. Unfortunately, no data is available for the relative amount of He$^+$ for this study.

The period showing the correlation between the two wave types is somewhat different from the rest of the event. It is inside interval IV, where the azimuthal and radial amplitudes of the Pc 5 pulsation are somewhat larger than elsewhere. Also, the very strong increase in

the electron number density at about 0743 UT occur just in the middle of the period. From Plate 2b we can see how the azimuthal and radial components of the magnetic field are almost in opposite phases ($\simeq$ linear polarization), and the magnetic field vector of the Pc 5 wave moves from later local times and inner L-shells to earlier local times and outer L-shells and back. The Pc 1 bursts are only seen during the first half of this cycle (Plate 2), i.e., when field lines are brought from the inner magnetosphere toward the satellite. However, since there is a magnetic node at the equator where the Pc 1 waves are formed, this should not have much effect on them. It is likely that the radial electric field component of the Pc 5 hydromagnetic wave is strong enough to disturb the local cold plasma density in the equatorial plane in a way that favours the growth of ion-cyclotron waves during a specific phase of the Pc 5 pulsation cycle. (Unfortunately, we do not have the electric field measurements available.)

An outward electric field is expected to lead the eastward magnetic field oscillations by 90° for the fundamental toroidal resonant oscillation of the geomagnetic field [see, e.g., Potemra et al., 1989]. In the present case, this corresponds to an inward radial direction of the electric field on the falling slope of the azimuthal magnetic field, and an outward direction on the rising slope of the field. The modulation of the Pc 1 formation can be understood, if a cold plasma cloud exists at the equatorial region close to the CRRES field line, but on a higher L-shell. Then the electric field of the Pc 5 wave would transport plasma from this cloud either toward the satellite field line or away from it, depending on the phase of the cycle. When the electric field points from the plasma cloud toward the satellite field line, it transports new plasma to this field line, and lowers the resonance energy of the ions contributing to the ion cyclotron instability. The existence of such a plasma cloud is verified by the peak in the electron number density at about the time of the Pc 1 modulation. The plasma cloud could originate, e.g., from the solar wind. Note that the highest values of the density enhancement at 0743 UT are above the average solar wind density but still far below the present solar wind density. Other possibility is that the cloud is produced inside the magnetosphere because of the pressure pulse.

CONCLUSION

We have provided evidence for a modulation of Pc 1 pulsation bursts by toroidal Pc 5 waves during a period of strong magnetospheric compression. While the compression started a long chain of Pc 5 pulsations and created favourable particle distributions for Pc 1 wave growth, the modulation occurred during a period with strongest Pc 5 pulsations and an abrupt increase of the plasma density. We argue that the mainly radially directed electric field of the Pc 5 wave transported part of this high density plasma toward and away from the satellite located suitably close to the edge of the plasma cloud. The resulting strong temporal increases in plasma density lowered the resonance energy of the ions that could contribute to the ion cyclotron resonance, and enhanced the instability. In this way the Pc 1 emission bursts could be modulated with the Pc 5 frequency.

Acknowledgments. RR and KM would like to thank the Academy of Finland for its support during the course of this work. We are grateful to B. Anderson and R. Denton for very interesting discussions about the event. We would also like to thank NSSDC (J. King) and A. Lazarus and R. Lepping for IMP-8 data, R. A. Anderson for CRRES wave and density data, H. Lühr for the IMAGE data, and T. Bösinger and R. Kuula for the ground-based pulsation magnetometer data used in this study. The MEB spectrometer was designed and constructed with support from the Max-Planck Gesellschaft zur Förderung der Wissenschaften. Grants were received from the United States Air Force under AFOSR-85-0237 and from Norwegian Research Council for Science and the Humanities to the University of Bergen. The contribution of H. Michels, who provided the programs to read CRRES data tapes, is gratefully acknowledged.

References

Anderson, B. J., R. E. Erlandson, and L. J. Zanetti, A statistical study of Pc 1-2 magnetic pulsations in the equatorial magnetosphere 1. Equatorial occurrence distributions, *J. Geophys. Res., 97*, 3075-3088, 1992a.

Anderson, B. J., R. E. Erlandson, and L. J. Zanetti, A statistical study of Pc 1-2 magnetic pulsations in the equatorial magnetosphere 1. Wave properties, *J. Geophys. Res., 97*, 3089-3101, 1992b.

Anderson, R. R., D. A. Gurnett, D. L. Odem, CRRES plasma wave experiment, *J. of Spacecraft and Rockets, 29*, 5709-573, 1992.

Arthur, C. W., R. L. McPherron, and J. D. Means, A comparative study of three techniques for using the spectral matrix in wave analysis, *Radio Sci., 11*, 833-845, 1976.

Barfield, J. N., and R. L. McPherron, Investigation of Interaction between Pc 1 and Pc 2 and Pc 5 micropulsations at the synchronous orbit during magnetic storms, *J. Geophys. Res., 77*, 4707-4719, 1972.

Cahill, L. J. Jr., and J. R. Winckler, Periodic magnetopause oscillations observed with the GOES satellites on March 24, 1991, *J. Geophys. Res., 97*, 8239-8243, 1992.

Fraser, B. J., J. C. Samson, Y. D. Hu, R. L. McPherron, and C. T. Russell, Electromagnetic ion cyclotron waves observed near the oxygen cyclotron frequency by ISEE 1 and 2, *J. Geophys. Res., 97*, 3063-3074, 1992.

Hansen, H. J., B. J. Fraser, F. W. Menk, Y.-D. Hu, P. T. Newell, C.-I. Meng, and R. J. Morris, High-latitude Pc 1 burst arising in the dayside layer region, *J. Geophys. Res., 97*, 3993-4008, 1992.

Hughes, W. J., The effect of the atmosphere and ionosphere on long period magnetospheric micropulsations, *Planet. Space Sci., 22*, 1157-1172, 1974.

IMAGE Newsletter, Number 1, Finnish Meteorological Institute, November 1992.

Kokubun, S., K. N. Erickson, T. A. Fritz, and R. L. McPherron, Local time asymmetry of Pc 4-5 pulsations and associated particle modulations at synchronous orbit, *J. Geophys. Res., 94*, 6607-6625, 1989.

Korth, A., G. Kremser, B. Wilken, W. Guettler, and S. L. Ullaland, The electron and proton wide-angle spectrometer (EPAS) on the CRRES spacecraft, *J. of Spacecraft and Rockets, 29*, 609-614, 1992.

Lysak, R. L., and D. Lee, Response of the dipole magnetosphere to pressure pulses, *Geophys. Res. Lett., 19*, 937-940, 1992.

McPherron, R. L., C. T. Russel, and P. J. Coleman, Jr., Fluctuating magnetic fields in the magnetosphere, *Space Sci. Rev., 13*, 411-454, 1972.

Olson, J. V. , and L. C. Lee, Pc1 wave generation by sudden impulses, *Planet. Space Sci., 31*, 295-302, 1983.

Potemra, T. A., H. Lühr, L. J. Zanetti, K. Takahashi, R. E. Erlandson, G. T. Marklund, L. P. Block, L. G. Blomberg, and R. P. Lepping, Multisatellite and ground-based observations of transient ULF waves, *J. Geophys. Res., 94*, 2543-2554, 1989.

Saka, O., T. Iijima, H. Yamagishi, N. Sato, and D. N. Baker, Excitation of Pc 5 pulsations in the morning sector by a local injection of particles in the magnetosphere, *J. Geophys. Res., 97*, 10693-10701, 1992.

Singer, H., and M. G. Kivelson, The latitudinal structure of Pc 5 waves in space: Magnetic and electric field observations, *J. Geophys. Res., 84*, 7213-7222, 1979.

Singer, H. J., W. P. Sullivan, P. Anderson, F. Mozer, P. Harvey, J. Wygant, and W. McNeil, Fluxgate magnetometer instrument on the Combined Release and Radiation Effects Satellite (CRRES), *J. of Spacecraft and Rockets, 29*, 599-601, 1992.

Tsyganenko, N. A., A magnetospheric magnetic field model with a warped tail current sheet, *Planet. Space Sci., 37*, 5-20, 1989.

Warnecke, J., H. Lühr, and K. Takahashi, Observational features of field line resonances excited by solar wind pressure variations on 4 September 1984, *Planet. Space Sci., 38*, 1517-1531, 1990.

Young, D. T., S. Perraut, A. Roux, C. de Cilledary, R. Gendrin, A. Korth, G. Kremser, and D. Jones, Wave-particle interaction near Ω_{He+} observed on GEOS 1 and 2: 1. Propagation of ion cyclotron waves in He^+-rich plasma, *J. Geophys. Res., 86*, 6755-6772, 1981.

B. J. Fraser, Department of Physics, University of Newcastle, NSW 2308, Australia.

W. J. Hughes, Center for Space Physics, Boston University, Boston, MA 02215, USA.

A. Korth, Max-Planck-Institut für Aeronomie, D-3411 Katlenburg-Lindau, Germany.

G. Kremser, Department of Physics, University of Oulu, FIN-90570 Oulu, Finland.

K. Mursula, Department of Physics, University of Oulu, FIN-90570 Oulu, Finland.

R. Rasinkangas, Department of Physics, University of Oulu, FIN-90570 Oulu, Finland.

H. J. Singer, Geophysics Directorate/GPSG, Phillips Laboratory, Hanscom AFB, MA 01731, USA.